Antenna
Handbook

Antenna Handbook

VOLUME III APPLICATIONS

Edited by
Y. T. Lo
Electromagnetics Laboratory
Department of Electrical and Computer Engineering
University of Illinois–Urbana

S. W. Lee
Electromagnetics Laboratory
Department of Electrical and Computer Engineering
University of Illinois–Urbana

VNR VAN NOSTRAND REINHOLD
New York

Library of Congress Catalog Card Number 93-6502
ISBN 0-442-01594-1

I(T)P Van Nostrand Reinhold is an International Thomson Publishing company.
ITP logo is a trademark under license.

Printed in the United States of America.

Van Nostrand Reinhold
115 Fifth Avenue
New York, NY 10003

International Thomson Publishing GmbH
Königswinterer Str. 518
5300 Bonn 3
Germany

International Thomson Publishing
Berkshire House,168-173
High Holborn, London WC1V 7AA
England

International Thomson Publishing Asia
38 Kim Tian Rd., #0105
Kim Tian Plaza
Singapore 0316

Thomas Nelson Australia
102 Dodds Street
South Melbourne 3205
Victoria, Australia

International Thomson Publishing Japan
Kyowa Building, 3F
2-2-1 Hirakawacho
Chiyada-Ku, Tokyo 102
Japan

Nelson Canada
1120 Birchmount Road
Scarborough, Ontario
M1K 5G4, Canada

16 15 14 13 12 11 10 9 8 7 6 5 4 3 2 1

Library of Congress Cataloging-in-Publication Data

The antenna handbook/edited by Y. T. Lo and S. W. Lee.
    p. cm.
    Includes bibliographical references and indexes.
    Contents: v. 1. Fundamentals and mathematical techniques—v. 2. Antenna theory—v. 3. Applications—v. 4. Related topics.
    ISBN 0-442-01592-5 (v. 1).—ISBN0-442-01593-3 (v. 2).—ISBN 0-442-01594-1 (v. 3).—ISBN 0-442-01596-8 (v. 4)
    1. Antennas (Electronics) I. Lo, Y. T. II. Lee, S. W.
TK7871.6.A496   1993
621.382'4—dc20   93-6502
                                                       CIP

Contents

Volume III APPLICATIONS

Preface

During the past decades, new demands for sophisticated space-age communication and remote sensing systems prompted a surge of R & D activities in the antenna field. There has been an awareness, in the professional community, of the need for a systematic and critical review of the progress made in those activities. This is evidenced by the sudden appearance of many excellent books on the subject after a long dormant period in the sixties and seventies. The goal of this book is to compile a reference to *complement* those books. We believe that this has been achieved to a great degree.

A book of this magnitude cannot be completed without difficulties. We are indebted to many for their dedication and patience and, in particular, to the forty-two contributing authors. Our first thanks go to Mr. Charlie Dresser and Dr. Edward C. Jordan, who initiated the project and persuaded us to make it a reality. After smooth sailing in the first period, the original sponsoring publisher had some unexpected financial problems which delayed its publication three years. In 1988, Van Nostrand Reinhold took over the publication tasks. There were many unsung heroes who devoted their talents to the perfection of the volume. In particular, Mr. Jack Davis spent many arduous hours editing the entire manuscript. Mr. Thomas R. Emrick redrew practically all of the figures with extraordinary precision and professionalism. Ms. Linda Venator, the last publication editor, tied up all of the loose ends at the final stage, including the preparation of the Index. Without their dedication and professionalism, the publication of this book would not have been possible.

Finally, we would like to express our appreciation to our teachers, students, and colleagues for their interest and comments. We are particularly indebted to Professor Edward C. Jordan and Professor George A. Deschamps for their encouragement and teaching, which have had a profound influence on our careers and on our ways of thinking about the matured field of electromagnetics and antennas.

This Preface was originally prepared for the first printing in 1988. Unfortunately, it was omitted at that time due to a change in the publication schedule. Since many readers questioned the lack of a Preface, we are pleased to include it here, and in all future printings.

Preface to the Second Printing

Since the publication of the first printing, we have received many constructive comments from the readers. The foremost was the bulkiness of a single volume for this massive book. The issue of dividing the book into multivolumes had been debated many times. Many users are interested in specific topics and not necessarily the entire book. To meet both needs, the publisher decided to reprint the book in multivolumes. We received this news with great joy, because we now have the opportunity to correct the typos and to insert the original Preface, which includes a heartfelt acknowledgment to all who contributed to this work.

We regret to announce the death of Professor Edward C. Jordan on October 18, 1991.

PART C

Applications

Chapter 17

Millimeter-Wave Antennas

F. Schwering
US Army CECOM

A. A. Oliner
Polytechnic University

CONTENTS

Felix Schwering was born on June 4, 1930, in Cologne, Germany. He received the Diplom-Ingenieur degree in electrical engineering and the PhD degree from the Technical University of Aachen, West Germany, in 1954 and 1957, respectively.

From 1956 to 1958 he was an assistant professor at the Technical University of Aachen. In 1958 he joined the US Army Research and Development Laboratory at Fort Monmouth, New Jersey, where he performed basic research in free space and guided propagation of electromagnetic waves. From 1961 to 1964 he worked as a member of the Research Staff of the Telefunken Company, Ulm, West Germany, on radar propagation and missile electronics. In 1964 he returned to the US Army Communication Electronics Command (CECOM), Fort Monmouth, and has since been active in electromagnetic-wave propagation, diffraction and scatter theory, theoretical optics, and antenna theory. Recently he has been interested in particular in millimeter-wave antennas and propagation. At present he is also a visiting professor at Rutgers University and at New Jersey Institute of Technology.

Arthur A. Oliner was born in Shanghai, China, on March 5, 1921. He received the PhD in physics from Cornell University in 1946. He joined the Microwave Research Institute of the Polytechnic Institute of Brooklyn in 1946 and was made professor in 1957. He served as Department Head from 1966 through 1974 and Director of the Microwave Research Institute for fifteen years, from 1967 through 1982. Dr. Oliner is the author of over 150 papers and coauthor or coeditor of three books. Two of his papers have earned prizes: the IEEE Microwave Prize in 1967, and the Institution Premium, the highest award of the British IEE, in 1964. In 1982 he received the Microwave Career Award, the highest award of the IEEE Microwave Theory and Techniques Society, and he is one of only six Honorary Life Members of that society.

Dr. Oliner's research in microwaves includes network representations of microwave structures, precision measurement methods, guided-wave theory, traveling-wave antennas, plasmas, periodic-structure theory, and phased arrays. More recently he has been interested in guiding and radiating structures for the millimeter- and near–millimeter-wave ranges.

1. Introduction

The millimeter-wave region of the electromagnetic spectrum is commonly defined as the 30- to 300-GHz frequency band or the 1-cm to 1-mm wavelength range. Utilization of this frequency band for the design of data transmission and sensing systems has a number of advantages:

1. The very large bandwidth resolves the spectrum crowding problem and permits communication at very high data rates.
2. The short wavelength allows the design of antennas of high directivity but reasonable size, so that high-resolution radar and radiometric systems and very compact guidance systems become feasible.
3. Millimeter waves can travel through fog, snow, and dust much more readily than infrared or optical waves.
4. Finally, millimeter-wave transmitters and receivers lend themselves to integrated and, eventually, monolithic design approaches, resulting in rf heads which are rugged, compact, and inexpensive.

Propagation effects have a strong influence on the design and performance of millimeter-wave systems, and for this reason are briefly reviewed here. As a general rule, millimeter-wave transmission requires unobstructed line-of-sight paths, but propagation into shadow zones is possible by edge diffraction and scatter, though at a reduced signal level. Recent propagation experiments in woods and forests have shown, moreover, that under favorable conditions (trunk region with little underbrush), transmission ranges of several hundred meters can be achieved in vegetated areas.

Amplitude and angle of arrival scintillations caused by atmospheric turbulence are usually small in the millimeter region. For path lengths in the order of a few kilometers, the interesting range, they are of no consequence for most applications.* But atmospheric absorption can be pronounced. Fig. 1 shows the frequency dependence of millimeter-wave attenuation by atmospheric oxygen and water vapor, by rain, and by fog or clouds. Snow absorption is negligible at frequencies below 100 GHz but can be substantial above 140 GHz, even at moderate snowfalls.**

Absorption by rain and atmospheric gases is the dominant effect and it is evident that the choice of the operating frequency of a millimeter-wave system will depend strongly on the desired transmission range. Large transmission distances can be obtained in the low-attenuation windows at 35, 94, 140, 220, and 340 GHz.

---

*High-resolution radar systems are an exception.
**Applies to dry snow. Data on millimeter-wave attenuation by wet snow are not yet available.

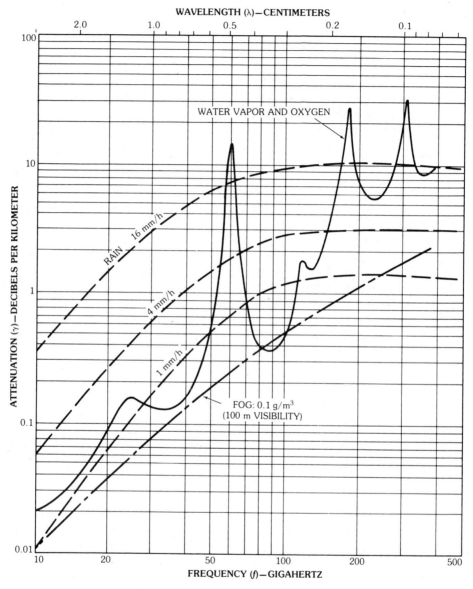

Fig. 1. Attenuation of millimeter waves by atmospheric effects.

On the other hand, operation near one of the steep absorption lines, for example the O_2 line at 60 GHz, will allow one to control the transmission distance by adjusting the frequency. In this way "overshoot" can be minimized and hence the probability of detection and interference by an unfriendly observer. Since the millimeter range, moreover, permits the design of antennas of high directivity and very small beamwidth a system of optimum transmission security can be realized.

    This security aspect, in connection with the other advantages of millimeter-wave systems, including the fact that millimeter waves travel through dust and battlefield smoke with little attenuation, makes this frequency range particularly attractive for military applications, including communication, guidance, fuzing, and radar. Also, numerous commercial and scientific applications are developing, including millimeter-wave radiometry, remote sensing and mapping, and satellite communication at 30/20 GHz, which means 30 GHz uplink, 20 GHz downlink. (The frequencies of military satellites are 44/20 GHz.) Millimeter-wave telescopes for radioastronomy have been constructed since the mid-sixties and millimeter-wave imaging seems to be an interesting possibility for the upper millimeter-wave region.

Most millimeter-wave systems require antennas of high directivity gain and these antennas have been predominantly investigated. As mentioned previously, it is one of the advantages of the millimeter-wave region that such antennas have very reasonable dimensions. Fig. 2 shows the antenna aperture area needed to achieve a given directivity. But certain applications in millimeter-wave guidance and radar also require fan-shaped radiation characteristics, and vehicular millimeter-wave communication implies the use of omnidirectional antennas providing a circular symmetric pattern in the azimuth plane and moderate directivity in the elevation plane. Such antennas are included in this chapter.

For the purpose of this discussion millimeter-wave antennas are subdivided into four classes: antennas of conventional configuration, such as horn, lens, and reflector antennas; surface-wave and leaky-wave antennas based on open waveguide technology; microstrip resonator and printed-circuit antennas; and some newer developments using integrated and monolithic design approaches. The design principles of the first class of antennas are well established at microwave frequencies, and scaling into the millimeter-wave region is straightforward in most cases. These antennas are particularly suited for frequencies below 100 GHz, although the usefulness of lens and reflector antennas extends well into the infrared and optical regions. The millimeter-wave aspects of these antennas will be summarized here. The second and third classes of antennas are of interest for both the lower and upper millimeter-wave regions. Most of these antennas have microwave counterparts, but scaling is usually not straightforward, while some are novel with no microwave heritage. These antennas will be discussed in some detail. The fourth class of antennas is still in the research stage; systematic design information is not as yet available but an attempt is made to indicate trends.

Millimeter-wave antennas are a fast-developing area, and any discussion of them must be dated. This chapter was written in 1983–84 and, except for a few later revisions, reflects the state of development at that time.

2. Antennas of Conventional Configuration

A large class of millimeter-wave antennas is obtained by wavelength scaling of well-established antenna configurations developed for the microwave and lower frequency bands [1]. This class of antennas includes reflector, lens, horn, and array antennas as radiating structures of high directivity gain, spiral antennas as broadband radiating elements of medium gain, and pillbox, geodesic, biconical, and linear antennas for fan-beam and omnidirectional applications. The design principles and performance characteristics of these antennas are discussed in pre-

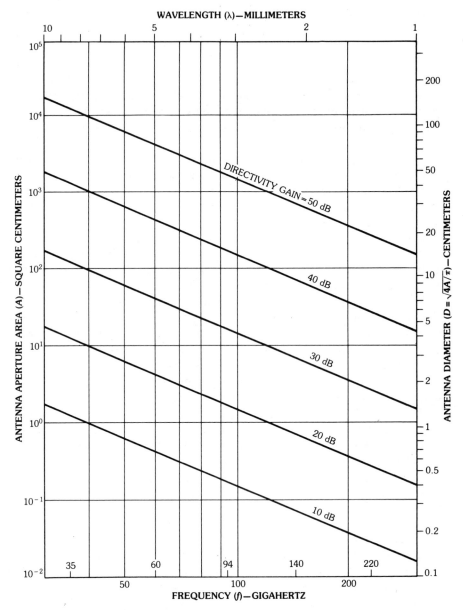

Fig. 2. Aperture area of millimeter-wave antennas as a function of directivity gain and frequency assuming an aperture efficiency of 50 percent.

ceding chapters. Scaling is, to a large degree, straightforward and the design information presented can be extended readily into the millimeter-wave region.

In practical realizations, however, conventional millimeter-wave antennas are not necessarily exact replicas of their microwave counterparts. The smaller physical size and tighter fabrication tolerances suggest modifications in their design. Structural simplicity, in particular, becomes an important design goal, while increased dimensions (as counted in wavelengths) usually do not cause problems but facilitate fabrication. The question of which type of antenna should be used for a given application may be answered differently for the microwave and millimeter-wave regions.

In the following the millimeter-wave aspects of the various conventional antenna configurations are reviewed. There is no need for a further discussion on the antenna fundamentals, which can be found in other chapters; the purpose of the review is rather to point out design options and performance limitations peculiar to the millimeter-wave region. In the process some interesting features emerge. For example, because of the much smaller size and weight of millimeter-wave antennas rapid mechanical scanning now becomes a practical option, in particular for reflector antennas. For the same reason of physical smallness, antennas made from dielectric material are much more useful in the millimeter-wave region; new types of dielectric lenses have been proposed, and dielectric horns find increasing attention.

High- and Medium-Gain Antennas

Reflector Antennas—Reflector antennas are widely used in the millimeter-wave region [2,3]. They are typically single-beam antennas of moderate or high directivity gain for communication, radar, and sensing, and monopulse antennas for tracking and guidance. Multibeam antennas are under development for millimeter-wave satellite communication [4,5].

Offset feed arrangements eliminate aperture blockage problems and, as in the microwave region, can be used to achieve low side lobes. For large antennas Cassegrain feed systems are particularly attractive from an efficiency viewpoint, since the length of waveguide runs is minimized, and from a fabrication viewpoint, since the feed horn can have a comparatively large aperture. Dual-reflector antennas with an offset feed geometry combine the advantages of both approaches and, in addition, can be designed for high aperture efficiency. Examples of such antennas are available with aperture efficiencies close to 70 percent and peak side lobes down by 30 dB from the main beam level [6]. Furthermore, depolarization effects caused by the asymmetry of the main reflector can be compensated in dual-reflector antennas, and a cross-polarization level as low as −40 dB can be achieved [6]. In most cases such antennas can be operated with conventional feed horns.

Because of their small size and weight millimeter-wave reflector antennas are suitable for *rapid mechanical scanning*, and antennas of this type may be designed for functions for which frequency-scanned arrays or phased arrays would be considered at microwave frequencies. The complexity and high cost of electronically scanned millimeter-wave arrays provide a further argument in favor of mechanically scanned antennas. Antenna configurations where the reflector is moved while the feed system remains stationary are of particular interest since no rotary joints are needed in the feeding waveguide.

Figs. 3a and 3b illustrate two different design approaches for such antennas. The design shown in Fig. 3a uses a cavity-backed spiral antenna as an offset feed to illuminate a shaped reflector [7]. The feed is stationary and provides a practically frequency-independent feed pattern of axial symmetry and circular polarization.

a

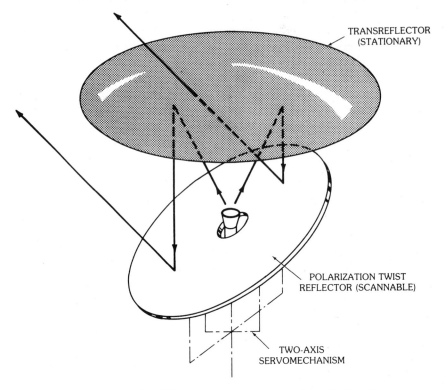

TRANSREFLECTOR
(STATIONARY)

POLARIZATION TWIST
REFLECTOR (SCANNABLE)

TWO-AXIS
SERVOMECHANISM

b

Fig. 3. Mechanically scanned millimeter-wave antennas. (*a*) Single-reflector antenna. (*Reprinted with permission of publisher from* Microwave System News, *November 1981; all rights reserved) (b)* Dual-reflector antenna. (*After Waineo and Konieczny [8],* © *1979 IEEE*)

The reflector rotates about the feed spiral axis; its shape is synthesized to yield a narrow beamwidth in the order of a few degrees in the azimuth plane and a wider beamwidth of approximately 20° in the elevation plane. Designed to operate over the entire 1- to 40-GHz band, the antenna has a height of 53 cm and a diameter of 61 cm. These rather large dimensions are determined by the lower frequency limit and could be reduced significantly for an antenna whose band of operation does not extend below the millimeter-wave region. The reflector of the present antenna rotates with a speed of up to 300 rpm. The primary application of antennas of this type is in broadband signal acquisition and direction finding [7].

The antenna depicted in Fig. 3b operates with two reflectors, a stationary parallel-wire transreflector and a scanning polarization-twist reflector [8, 9]. Pointing of the twist reflector is typically adjusted with the help of a two-axis servomechanism. The beam radiated from the feed horn is polarized to be reflected at the stationary mirror, which has a parabolic shape and serves as a highly directive antenna. The beam is reflected a second time at the polarization-twist reflector, a planar mirror, which by its orientation determines the final beam direction. At the same time, it turns the plane of polarization of the reflected beam by 90° so that the outgoing beam is now transmitted through the stationary mirror without reflection. Antennas of this type are useful as radar tracking antennas; with a quadrature feed horn they may be used as monopulse radar antennas. Dimensions depend on the desired gain of the sum channel and, to a lesser degree, on the required null depth of the difference channel and the side lobe level [8, 9].

The twist reflector, the only moving part of the antenna, can be designed as a lightweight printed-circuit plate. Since, moreover, tilting of the reflector by $\Delta\theta$ causes beam deflection by $2\Delta\theta$, these antennas permit high-speed beam scanning. A 94-GHz model [8] had an overall diameter of 18 cm and allowed for beam acceleration in excess of $20\,000°/s^2$.

Solid-state power sources have considerable size, weight, and cost advantages over electron tubes and are preferred for many applications. Their output power, however, is limited in the millimeter-wave region, and several such sources have to be combined when a comparatively large transmitter power is required as in short-range millimeter-wave radars and in satellite communication. Reflector antennas permit *space-combination*, i.e., a cluster of several closely spaced feed horns would be used, individually driven by solid-state amplifiers phase-locked to a common source [10]. This approach eliminates most of the rf losses associated with circuit techniques for power combining. Since waveguide losses increase with frequency this is an important design consideration in the millimeter-wave region. In addition, the method provides for graceful performance degradation if some of the amplifiers should fail. If, furthermore, each amplifier is coupled with a phase shifter, a very versatile feed system is obtained, permitting dynamic control of beam pointing and illumination contour. One obtains a so-called hybrid antenna, i.e., a reflector (or lens) antenna fed by a phased array. Offset dual-reflector antennas with such feed arrays are, at present, under investigation by NASA [5] for possible use as high-gain multibeam antennas for advanced future communication satellites operating in the 30/20-GHz bands. These antennas will have to provide up to twenty fixed spot beams and up to six scannable spot beams with a gain of 50 to 53 dB and a 3-dB beamwidth of 0.3°. A main objective of the NASA investigation is to demonstrate the feasibility of feed arrays where 20-GHz power amplifier and phase shifter modules are designed in monolithic microwave integrated circuit (MMIC) technology and integrated with each radiating element near its aperture. Predictable benefits derived from such feed systems include enhanced reliability inherent in the use of a large number of amplifiers, a further reduction in cost and weight, a high-speed scan capability, and improved beam isolation combined with reduced scan loss through the fine adjustment achievable with a large number of individually controlled phase shifters and amplifiers [5].

The directivity gain of reflector antennas is limited by their *size* and by their *surface errors*. Antenna size is usually not a problem in the millimeter-wave region. Fig. 2 shows, for example, that millimeter-wave antennas with gains as high as 40 or 50 dB still have diameters less than 1 m. Accuracy requirements, on the other hand, are proportional to wavelength (for constant electrical performance) and can become difficult to satisfy in the millimeter-wave region.

Theoretically the tolerance requirements of reflector antennas are well understood [11–13]. Ruze has shown that random surface irregularities reduce the directivity gain of a reflector antenna according to the relation

$$G_D = G_0 e^{-\sigma^2} \tag{1a}$$

where G_D is the directivity gain in the presence of surface errors, G_0 is the directivity of the ideal, undistorted antenna, and σ is proportional to the rms surface error ϵ_{rms}:

$$\sigma = 2\pi\Upsilon \frac{\epsilon_{rms}}{\lambda_0} \tag{1b}$$

Here, ϵ_{rms} is measured in the axial direction, λ_0 is the free-space wavelength, and Υ is an adjustment factor which is approximately 2 for a shallow reflector and 1.33 for a uniformly illuminated focal plane paraboloid. Equations 1a and 1b apply to antennas with surface errors of negligibly small correlation distances. More uniform reflector deformations with finite correlation intervals cause less scattering than uncorrelated errors of the same rms value and, as a consequence, lead to a smaller reduction in directivity gain [12]. Equations 1a and 1b can thus be regarded as a worst-case estimate.*

It is instructive to examine the frequency dependence of the directivity gain for a reflector antenna of given diameter D and surface accuracy ϵ_{rms}. The directivity of the perfect, undistorted antenna is determined by the well-known formula $G_0 = \eta(\pi D/\lambda_0)^2$, so that we have, with (1)

$$G_D = \eta\left(\frac{\pi D}{\lambda_0}\right)^2 \exp\left[-\left(\frac{4\pi\epsilon_{rms}}{\lambda_0}\right)^2\right] \propto f^2 e^{-af^2}$$

---

*The effect of surface irregularities on side lobe levels has been studied by Vu The Bao [12].

where η is the aperture efficiency and Υ is assumed to be 2. In the "low" frequency range where λ_0 remains large compared with $4\pi\epsilon_{\text{rms}}$, the gain is proportional to the aperture area measured in wavelengths and thus increases with f^2. At higher frequencies, in the region where λ_0 is in the order of $4\pi\epsilon_{\text{rms}}$, the directivity reaches a peak value

$$G_{\text{pk}} = \frac{\eta}{43}\left(\frac{D}{\epsilon_{\text{rms}}}\right)^2$$

which is proportional to the square of the precision of manufacture D/ϵ_{rms}. The peak gain occurs at $\lambda_0 = 4\pi\epsilon_{\text{rms}}$ and remains 4.3 dB below the gain G_0 of the perfect antenna. If the frequency is further increased into the range where $\lambda_0 \ll 4\pi\epsilon_{\text{rms}}$, scattering at the surface irregularities intensifies and the power of the main beam decreases exponentially with f^2. The power lost from the principal pattern reappears in the scatter pattern of the antenna. This scatter pattern is centered about the antenna axis and can be rather narrow; its beamwidth (in radians) and gain depend on the correlation distance d_c of the surface errors and for $\lambda_0 \ll 4\pi\epsilon_{\text{rms}}$ are given approximately by

$$\Delta\theta_s = \frac{7\epsilon_{\text{rms}}}{d_c} \quad \text{and} \quad G_s = \left(\frac{d_c}{2\epsilon_{\text{rms}}}\right)^2$$

respectively [11]. For reflectors with large correlation distances $d_c \gg \epsilon_{\text{rms}}$, the gain G_s can be rather high so that the main beam of the principal pattern becomes obscured by the scatter pattern.

The useful frequency band of a reflector antenna is the region below the peak gain, where the gain reduction due to scattering remains small. If this frequency band is to include the millimeter-wave region, high surface accuracy must be achieved and tolerance requirements become important for moderate-size reflectors and critical for very large reflectors. According to (1), the gain reduction will remain below 0.1 dB provided a surface accuracy of $0.012\lambda_0$ or better is maintained, i.e.,

$$\epsilon_{\text{rms}} < 0.12 \quad \text{mm at} \quad 30 \text{ GHz}$$
$$\epsilon_{\text{rms}} < 0.036 \text{ mm at} \ 100 \text{ GHz}$$

If a gain reduction of 1 dB is acceptable, the surface tolerances are relaxed to

$$\epsilon_{\text{rms}} < 0.38 \quad \text{mm at} \quad 30 \text{ GHz}$$
$$\epsilon_{\text{rms}} < 0.115 \text{ mm at} \ 100 \text{ GHz}$$

For Cassegrain antennas these tolerances must be shared between the two reflectors; a subdivision proportional to the reflector diameters is an appropriate choice [1]. On the other hand, Cassegrain antennas permit compensation of profile deformations of the primary reflector by appropriate shaping of the subreflector [14, 15].

Most millimeter-wave antennas are designed for directivity gains of less than 50 dB and diameters of less than 1 m. For reflectors of this size the tolerance requirements usually do not constitute a critical problem, but they do require precise fabrication techniques [2, 6, 16] and, in the case of high-gain antennas, a rigid support structure. Reflectors with a surface precision D/ϵ_{rms} of 2×10^4 are commercially available in diameters up to 1.2 m.

Examples of millimeter-wave systems which operate with larger antennas, having dimensions in the order of several meters, are ground stations for satellite communication and compact antenna test ranges [17]. (Compact millimeter-wave ranges are of interest not only for antenna measurements but also for scaled experiments on the microwave radar cross section of large objects.) The tolerance requirements for test ranges are particularly stringent. In order to achieve the desired plane wave distribution in the test zone with an amplitude and phase accuracy of better than 0.5 dB and 10°, respectively, a surface accuracy of $0.007\lambda_0$ at the highest frequency is required [17]. Such accuracies can be achieved by constructing the antennas from highly precise surface panels whose positions and orientations can be adjusted individually; this technique is generally applied in the design of very large reflector antennas.

The largest millimeter-wave antennas currently in operation are *radio-astronomy telescopes*. An overview with respect to the telescopes in existence in the United States and abroad in 1970 can be found in the literature [18]. The antennas typically use Cassegrain feeds and have main reflector diameters in the range from 5 m to 20 m. Receiver quality in the millimeter-wave region is by an order of magnitude poorer than in the microwave range, and large and accurate telescopes are needed to achieve high system performance. Recently, several new antennas have been designed, among them a 25-m antenna planned for the National Radio Astronomy Observatory but subsequently canceled, a 30-m antenna [19] constructed under Franco-German cooperation (IRAM) at Pico Veleta in Spain, and a 45-m antenna built by Tokyo Observatory at Nobeyama, Japan. The highest operating frequencies of these antennas are 400 GHz, 300 GHz, and 150 GHz, respectively. The configurations of the first two antennas are shown in Fig. 4; their beamwidths at their highest frequencies are 7.5 and 8.2 arc seconds. The requirements on the mechanical accuracy ($\epsilon_{rms} < 0.1$ mm) and pointing/tracking accuracy (approximately 1 arc sec) of such antennas in the presence of gravitational deformations, temperature gradients, and wind load push current technology to its limits and call for new design and fabrication methods for large reflectors [19]. One such method, termed *homology*, allows for large gravitational reflector deformations when the antenna is turned in the elevation plane but imposes the constraint that the reflector surface maintains a parabolic shape [20]. Adaptive position adjustment of the feed system ensures that the antenna remains focused at all times.

Lens Antennas—Lens antennas have found numerous applications in the millimeter-wave region [2, 3]. Together with reflector and horn antennas they are the most widely used millimeter-wave antennas.

Tolerance requirements are less stringent than for reflector antennas since lenses are usually made from materials of relatively low refractive index to minimize reflection at the lens surface. Thus the antennas are easy to machine and comparatively inexpensive. Lenses have very large bandwidth and excellent wide-angle scanning properties. Furthermore, there is no aperture blockage by the feed system, and lens antennas can be designed to have low side lobes (less than −35 dB) and a high front-to-back ratio. In the case of zoned lenses, however, the side lobe level becomes frequency sensitive.

Lens antennas are *more* attractive in the millimeter-wave region than at lower frequencies because their *dielectric weight is significantly reduced*. For given electrical performance the volume and weight of a lens decrease with λ_0^3 and are usually very reasonable at millimeter wavelengths. An example may illustrate this point: projected military satellite communication systems operating at 44/20 GHz will require multiple-beam antennas capable of directing spot beams and clusters of beams from a synchronous satellite to any point on the earth's surface. A recent design study [21] has shown that the rather stringent performance requirements on these antennas can be satisfied by an appropriately shaped zoned lens with an aperture diameter of 60 cm. This antenna would provide beams with a gain of 48 dB at 44.5 GHz, the center frequency of the uplink; the first side lobes would be more than 20 dB below the beam maximum; and pattern degradation for off-axis beams would remain small across the entire field-of-view scan range, which for a synchronous satellite has a width of ±8.6°. The weight of a polystyrene (Rexolite) antenna of this type would be less than 10 lb (4.5 kg) [21].

Because of the reduction in dielectric weight mentioned above, *spherically symmetric lenses* also become practical in the millimeter-wave region [22–30]. Luneburg lenses which focus an incident plane wave into a single point on the lens surface would provide excellent performance, but the required gradation of the refractive index profile of these lenses would probably be difficult to realize in the millimeter-wave region. Moreover, much simpler lenses, for example, homogeneous spheres, provide already good performance, and two-layer lenses consisting of a homogeneous spherical core surrounded by a concentric shell of different refractive index can be sufficiently well corrected to satisfy almost all practical requirements [28].

In addition to their simple configuration and their good electrical performance, such spherically symmetric lenses permit wide-angle beam scanning without pattern degradation. If the antennas are operated with an array of identical feed horns, beam scanning can be implemented digitally by electronic switching from feed horn to feed horn. With these advantages a further discussion of spherically symmetric lenses appears justified here, in particular since they are not widely used in the microwave region (where their weight would be excessive).

A theory of *two-layer lenses* based on ray tracing* is available in the literature and has been confirmed by experiments [28]. The theory includes homogeneous lenses, the so-called constant-K lenses, as a special case. These latter lenses [26] should be made from materials with dielectric constants in the range

$$2 \leq \epsilon_r \leq 3.5$$

---

*Wave theory approaches to the analysis of spherically symmetric lenses (based on Mie-series expansions) may be found in [29] and [30].

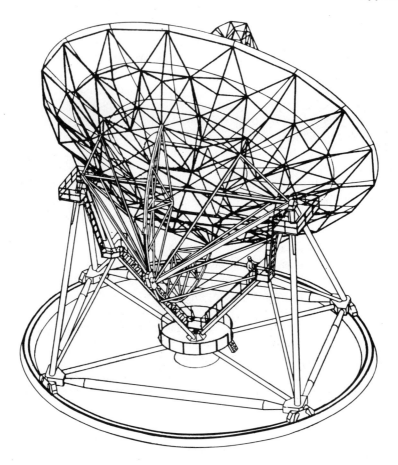

a

Fig. 4. New millimeter-wave radiotelescopes. (*a*) The 25-m antenna designed for National Radio Astronomy Observatory. (*b*) The 30-m antenna constructed at Pico Veleta, Spain. (*After Baars [19]*)

For $\epsilon_r > 3.5$ the focal point would be located in the interior of the lens, which is undesirable, and for $\epsilon_r < 2$ problems associated with large focal lengths may arise. The distance R' of the focal point from the center of a lens of radius R is approximately given by

$$R' = R \frac{\epsilon_r}{\sqrt{2}(\epsilon_r - 1)}$$

which defines the position of the feed horn. Particularly for lenses with permittivity values close to the upper limit of 3.5, spherical aberrations are small and such lenses can be designed for diameters of up to $15\lambda_0$ and a gain of up to 30 dB while maintaining good pattern quality, including a side lobe level below -20 dB.

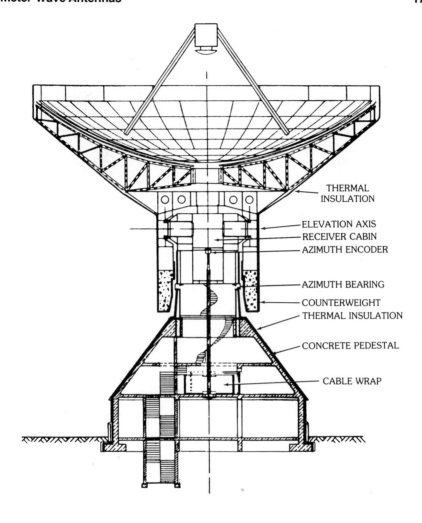

THERMAL INSULATION

ELEVATION AXIS
RECEIVER CABIN
AZIMUTH ENCODER

AZIMUTH BEARING

COUNTERWEIGHT
THERMAL INSULATION

CONCRETE PEDESTAL

CABLE WRAP

b

Fig. 4, *continued.*

As an example, Fig. 5a shows an experimental 60-GHz radio which uses two homogeneous Teflon spheres as antennas [31]. The lenses, which are fed by small waveguide horns, have a diameter $2R = 10\lambda_0 = 5$ cm. Fig. 5b shows the radiation pattern. The gain of these lenses is 26 dB, the main beam pattern is nearly circularly symmetric, and first side lobes are down by -20 dB.

The parameters of a two-layer lens can, in principle, be determined by imposing suitably defined optimum performance criteria. In the practical implementation the dielectric constants of core and shell are restricted by the limited number of available low-loss millimeter-wave materials. A lens which has shown good performance consists of a polystyrene core ($\epsilon_r = 2.46$) and a quartz shell ($\epsilon_r = 3.80$) with an outer-to-inner diameter ratio of 1.63 [28]. An experimental lens of this type with a diameter of 6.8 cm, which corresponds to approximately $16\lambda_0$ at the test frequency of 70 GHz, has yielded a gain of 32 dB and side lobes below -27 dB,

a

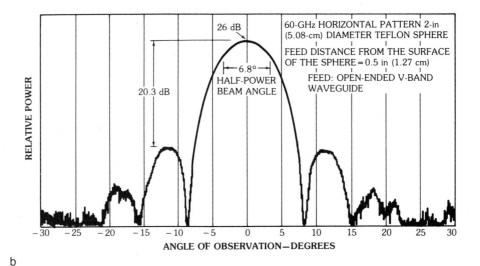

b

Fig. 5. Teflon-sphere lens antennas. (*a*) Experimental 60-GHz radios with lens antennas. (*b*) Horizontal radiation pattern at 60 GHz. (*Courtesy Chang, Paul, and Ngan [31]*)

in very good agreement with theoretical predictions.* The theory [28] shows, moreover, that lenses with ten times this diameter could be designed providing substantially increased directivity gain (51 dB) while maintaining good pattern quality. Dielectric spheres with such diameters would be large and bulky, even at

---

*It is interesting to note that for the two-layer lenses considered here, the permittivity of the shell is larger than that of the core, while in Luneburg lenses the refractive index decreases with distance from the center. In other words, these lenses should not be viewed as first-order approximations to a Luneburg lens. Their performance can rather be explained in terms of two counteracting aberrations which are caused by the core and shell, respectively [24]. If the lens parameters are chosen appropriately, these aberrations will compensate each other to a large extent, thus leading to enhanced lens performance.

millimeter wavelength, and such limiting factors as fabrication tolerances, homogeneity requirements, and dielectric losses would most likely make them impractical. But the point is that spherically symmetric lenses of simple structure are capable of providing very good electrical performance [28].

Practical requirements on spherically symmetric lenses include a very uniform permittivity throughout the lens and correctness in shape to avoid beam tilting and pattern deformations [26]. Dielectric losses in constant-K lenses can be estimated from the formula

$$L_{\text{diel}} = 36\frac{R}{\lambda_0}[\epsilon_r^{3/2} - (\epsilon_r - 1)^{3/2}]\tan\delta$$

where $\tan\delta$ is the loss factor of the lens material and L is in decibels. For a quartz lens with a diameter of $15\lambda_0$ and $\tan\delta = 0.00025$, the loss would be 0.18 dB. About the same loss would occur in a $10\lambda_0$ Teflon lens (Fig. 5) with $\tan\delta = 0.0006$. In the case of a lens consisting of core and shell, the step in refractive index at the outer surface is, in general, large enough to require an antireflection layer. A single matching layer can be effective up to large angles of incidence [28]. The step between core and shell is usually sufficiently small so that reflection is negligible, and no matching layer is needed at the inner lens surface.

Shaping techniques for millimeter-wave lens antennas have recently been developed by Lee [32, 33]. Similar to reflector shaping methods, these techniques are based on ray tracing approaches and are analytically and computationally involved. Speaking in general terms, the design procedure imposes the power conservation law to control the aperture taper, Snell's law to define the first surface of the lens (as seen from the feed point), and the path length condition to determine the second surface. The path length condition is specified by the desired phase distribution across the aperture. The radiation pattern of the lens is calculated in the usual way with scatter effects at the zoning edges neglected.

Note, however, [32] that these lens shaping methods differ from the well-established design techniques for optical lenses. Optical correction techniques are concerned primarily with spherical lenses and the compensation of their phase errors (aberrations), while the objective of millimeter-band shaping techniques is the realization of a desired phase *and* amplitude distribution across the lens aperture.

For multibeam antennas, furthermore, Lee [32, 33] has introduced a new method, termed *coma-correction zoning*, which permits one to minimize distortions of off-axis beams caused by cubic phase errors. The basic idea is that coma aberrations of a thin lens are significantly reduced provided the inner surface of the lens, i.e., the surface facing the feed horn, is spherical on the average. This condition can be satisfied by zoning. Lenses which combine contour shaping with coma-correction zoning are of particular interest for use in millimeter-wave satellite communication systems. This application requires spaceborne multibeam antennas capable of radiating a large number of closely spaced, highly directive beams of high pattern quality, which implies minimum pattern degradation for off-axis beams and low side lobes (to allow frequency reuse). In addition, a high crossover level of 4 dB is required for adjacent beams in order to ensure continuous ground coverage.*

Fig. 6 shows the cross-section profile of a shaped, coma-reduced lens designed by Lee [32]. The lens has a focal length of 24.5 cm, an aperture diameter of 20.7 cm = $30\lambda_0$, and a center thickness of 3.5 cm = $5\lambda_0$ at the design frequency of 44 GHz. The feed pattern is assumed to be the standard E-plane pattern of a square horn, i.e.,

$$g(\theta) = (1 + \cos\theta) \frac{\sin[(\pi d/\lambda_0)\sin\theta]}{(\pi d/\lambda_0)\sin\theta}$$

where θ is the angle counted from the horn axis. The horn size d is chosen to provide an edge taper of 30 dB at $\theta = 20°$, the aperture half-angle of the lens as seen from the feed location. The lens is designed to produce an aperture distribution $E(r) = [1 - (r/1.05)^2]^3$ with a uniform phase. Made from Rexolite with $\epsilon_r = 2.54$, the lens has a weight of 0.5 kg.

An experimental model of the lens has yielded a gain of 33 dB, a beamwidth of 3.3°, and a side lobe level of −30 dB for individual beams [33]. These beams could be scanned up to ±12° without significant coma degradation. For multi-beam operation the lens can be fed by a cluster of small single-beam feed horns generating closely spaced beams with the desired high crossover level of 4 dB. The aperture efficiency of the antenna is 48 percent. Total losses amount to 2 to 3 dB.

Satellite communication systems will require larger antennas of much higher directivity in many cases. But the experimental antenna demonstrates the excellent performance that can be obtained through the use of lens shaping methods.

Another type of lens antenna which should be useful for millimeter-wave applications is the *Fresnel zone-plate lens*. This device was proposed many years ago in optics as a flat version of a lens. It differs from the usual lens in that it focuses by employing diffraction instead of refraction. The zone-plate lens is based on the fact that ray bundles traced from the source point to the image point via alternate Fresnel zones of the lens plane differ in path length by a half-wavelength. The Fresnel zones form concentric rings about the optical axis; the radii of these rings are

$$r_n = \left[(n + 1/2)\lambda_0 \frac{d_1 d_2}{d_1 + d_2}\right]^{1/2}, \qquad n = 1, 2 \ldots$$

---

*If a conventional, spherical lens antenna is used, the two requirements of high crossover level and low side lobes can lead to a packaging problem of the feed horn assembly [33]. A low side lobe level implies the use of feed horns of comparatively large aperture (to achieve an appropriately tapered illumination across the lens). But horns of large aperture dimensions cannot be spaced sufficiently close to obtain a high crossover level between adjacent beams. The problem is commonly solved by using an assembly of horns of small aperture and synthesizing the desired feed pattern by exciting a subset of adjacent horns for each beam [33]. Since each beam is generated by several horns (and some horns contribute to several beams) the method requires a large, complex feed system which in the millimeter-wave region would be rather lossy. In the case of a shaped lens, on the other hand, the beam forming and side lobe control functions are shifted from the feed horns to the lens. Thus, single-beam horns of small aperture can be used and the packaging problem finds a straightforward solution.

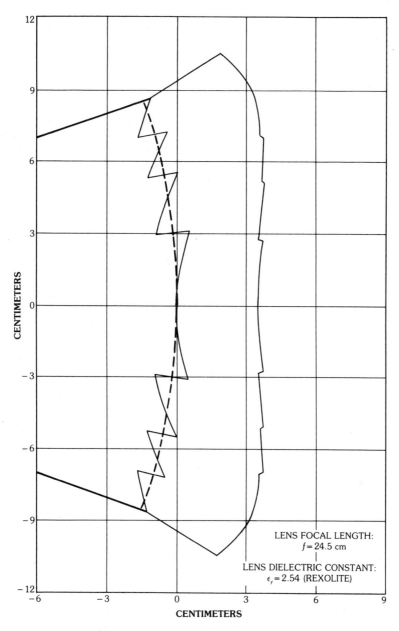

Fig. 6. Cross-section profile of coma-reduced, shaped lens for multibeam operation at 44 GHz. (*After Lee [32], © 1983 IEEE*)

where λ_0 is the free-space wavelength and d_1 and d_2 are the distances of the source point and image point from the lens plane. By blocking out each second ring the lens thus permits only those rays to pass which add constructively at the image point. Its important advantage was that the lens is flat, thereby offering an element of convenience. Its disadvantage was that roughly half of the incident light was lost because of blockage by the series of alternate rings.

In recent years, when this concept was adapted for use at microwave and then millimeter wavelength, the disadvantage mentioned above (the rejection of half the incident power) was overcome by employing a dielectric structure for the lens. The metal blocking rings were replaced by dielectric rings that were a half-wavelength thinner than the series of rings with which they alternated. In that way the field incident on the initial out-of-phase rings is no longer suppressed but is instead phase shifted, so that these rings now contribute constructively to the field at the image point, and no power is lost. The image spot is somewhat blurred, however, because of the step discontinuities at the successive rings, which introduce periodic phase errors of $\pm 90°$. These phase errors have been reduced in some advanced versions of the zone-plate lens in which the half-wavelength step is averaged in three smaller steps, each a quarter-wavelength in depth.

In these versions the complete zone-plate lens is made out of a single dielectric material. A recent alternative construction, based on the same principle, uses two different dielectric materials. In this version all of the rings have exactly the same thickness, but the phase-shifted rings are composed of the second dielectric material. The advantage of this structure is that the surface is completely flat instead of possessing small ridges. A recent article by J. C. Wiltse [34] discusses the operation of the zone-plate lens for millimeter waves and describes the various versions mentioned above.

A final remark on millimeter-wave lens antennas is concerned with an unconventional technique to effect beam steering, i.e., the use of variable permittivity media in the form of electric field–controlled liquid artificial dielectrics (CLAD) [35]. Two approaches for the realization of CLAD media have recently been pursued, i.e., forming solutions of high-molecular-weight rigid macromolecules and suspending highly asymmetric micrometer-sized metallic particles in a low-loss base liquid. Using the second approach, variable phase shifters consisting of discrete CLAD filled cells in metal waveguides* have been demonstrated at 35 and 94 GHz [35]. The experiments suggest that with careful attention to cell tolerances a 360° phase shift could be achieved with a control voltage in the order of 100 to 200 V and losses as low as 1.8 to 2.4 dB. Applied to antennas, CLAD media would permit the design of steerable lenses whose local electrical thickness is controlled with the help of a system of electrodes distributed throughout the lens volume [35]. In this way a uniform phase taper across the aperture could be effected for beam deflection, or any desired phase distribution could be achieved for adaptive beam correction. The feasibility and practicality of liquid dielectric lenses as millimeter-wave antennas remains to be established. But the number of useful CLAD media is probably very large, and only very few have been studied [35]. Some of those explored so far were both flammable and toxic.

---

*To ensure a uniform particle density throughout the CLAD medium, the liquid was kept in a continuous motion during the experiments, i.e., it was passed through the phase shifter cell at a constant flux rate. (The specific weight of the suspended metal particles is larger than that of the base liquid.)

Horn Antennas—Horn antennas for the millimeter-wave region are well developed [3, 36]. In design and performance there is little difference between these antennas and their microwave counterparts. Horns with large flare angles and lens correction provide gain levels of up to 30 dB in compact designs. Corrugated horns can be designed for very low side lobe and back lobe levels, for low cross polarization, and a practically circular symmetric radiation pattern [37–40]. For square-pyramidal and conical horns in particular, the E- and H-plane patterns are very similar, and these antennas are well suited not only for radiation at linear polarization but at circular polarization as well [41]. High mechanical precision can be achieved by using electroforming techniques. Standard gain horns are available over the complete millimeter-wave band.

As an example of a high-performance millimeter-wave horn antenna, Fig. 7a shows a *corrugated conical horn* designed for the 33-GHz band [42]. The antenna has a half-power beamwidth of 7° and a very broad bandwidth. The flare angle is 10°, and a moderate groove spacing (two grooves per wavelength) permits compact and easy construction. The choke grooves on the rim face suppress potential back lobe radiation. The H- and E-plane radiation patterns, Figs. 7b and 7c, show an extremely low side lobe level, which in the far side lobe region is more than −75 dB below the main peak. (Measurement at this level required a specially designed test range.) The antenna was originally designed for satellite-based measurements of the properties of the cosmic background radiation but it should also be very well suited as a feed horn for large parabolic reflectors.

A different type of horn antenna, i.e., *dielectric horns*, are indicated in Figs. 8a and 8b. The first figure shows a rhombic dielectric plate antenna (sectoral horn) fed by a rectangular waveguide [43], and the second figure illustrates a conical dielectric horn protruding from a circular waveguide [44]. When fabricated from materials of comparatively low dielectric constant in the range of 1.5 to 2.5, such antennas are well suited as array elements and as primary feeds for large reflectors or lenses. Dielectric horns have been found to have good bandwidth of approximately 20 percent, and a beamwidth smaller than that of metallic horns of the same flare angle and axial length. If used as a primary feed, they cause little aperture blockage. A further advantage is that dielectric horns have less critical tolerances than metal horns and are easier to machine. Furthermore, the cross-polarized component of conical dielectric horns has been shown to be very small [44]. Dielectric horns have also been utilized as Gaussian beammode launchers for quasi-optical devices operating in the upper millimeter-wave region [45].

Array Antennas—Array antennas of conventional design have not been widely used up to now in the millimeter-wave region. The technology for fixed-beam and frequency-scanned arrays is available in principle, while the utilization of phased arrays will depend on the development of suitable millimeter-wave phase shifters [2].* All of these antennas, and in particular large phased arrays, require highly

---

*Ferrite phase shifters for the 35-GHz band with characteristics suitable for systems applications have recently beome available [77].

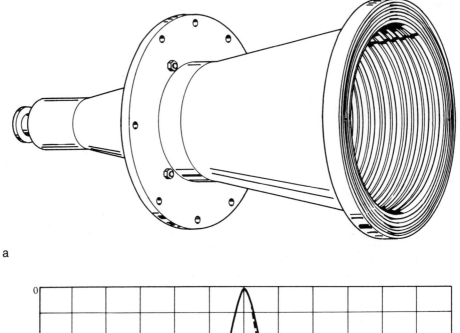

a

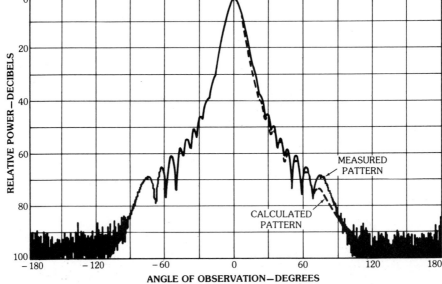

b

Fig. 7. High-performance horn antenna for the 33-GHz band. (*a*) Horn configuration. (*b*) *H*-plane pattern. (*c*) *E*-plane pattern. (*After Jansen et al. [42], © 1979 IEEE*)

precise fabrication techniques and are likely to be expensive. Now and for the immediate future, mechanically scanned fixed-beam antennas appear to be the best choice for many millimeter-wave radar, sensing, and communication uses [2]. A contributing factor is the compactness of millimeter-wave antennas which facilitates rapid mechanical scanning [7–9]. In the more distant future it is likely that the

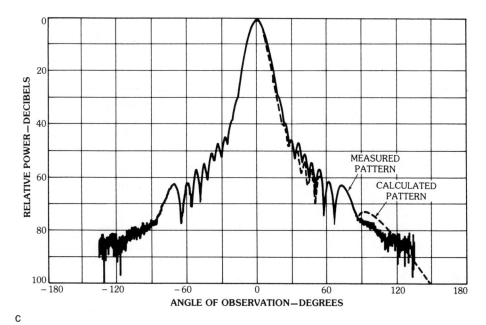

c

Fig. 7, *continued.*

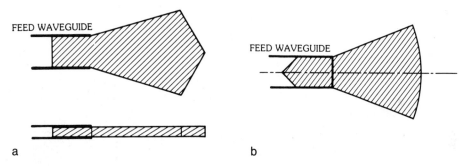

a   b

Fig. 8. Dielectric horn antennas. (*a*) Rhombic dielectric plate. (*After Ohtera and Ujiie [43]*, © *1981 IEEE*) (*b*) Conical dielectric horn. (*After Hombach [44]*)

design of millimeter-wave arrays will be dominated by the utilization of printed-circuit antenna technology and microstrip antenna techniques in particular. This holds for fixed-beam arrays as well as for frequency-scanned and phased arrays. Microstrip antennas and printed antennas in general can be conveniently and precisely fabricated by the use of etching techniques, and integrated designs are possible where feed lines, phase shifters, and eventually power amplifiers and low-noise detectors are integrated with the radiating elements on the same substrate (though possibly on different substrate layers). Advanced work in this area is in progress [8, 9, 46–72]. Recent results and future trends are discussed in Sections 4 and 5.

Returning to conventional array technology, slotted waveguide linear and planar arrays have been demonstrated at frequencies up to 94 GHz [9, 73–76]. The work on linear arrays [74], in particular, has confirmed that millimeter-wave slot antennas, including broadwall staggered-slot arrays, sidewall inclined-slot arrays, and variable offset long-slot arrays, can be designed using standard microwave procedures scaled up in frequency. Furthermore, it has been shown that high-resolution photolithographic and chemical etching techniques can be used for the fabrication of such millimeter-wave slot arrays, which will reduce their cost significantly. For example, 94-GHz linear arrays have been fabricated in this way from W-band waveguide [73]. The antennas were 21-element broadwall arrays designed for a gain of 17 dB, a 30-dB Taylor taper, and an efficiency of 85 percent. Measured results were in good agreement with the theoretical perorance except for a peak side lobe level of −24 dB (instead of −30 dB).

Planar waveguide arrays, the so-called flat plate arrays, have the advantage of smaller volume over reflector or lens antennas but they are more expensive, and in the millimeter-wave region tend to have lower efficiency. Broadwall shunt-slot arrays provide low cross polarization, and edge-slot arrays, where the radiating slots are machined into the narrow waveguide walls, have a very low side lobe capability which would make them particularly useful for military applications.

As an example, Fig. 9 shows a frequency-scanned 400-element K-band

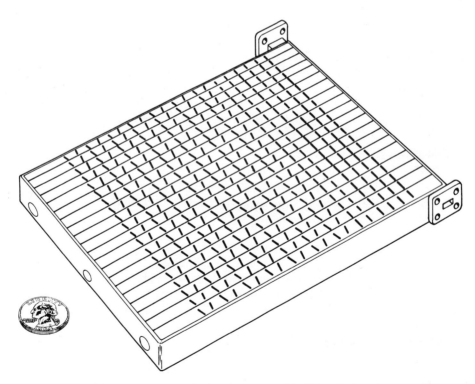

Fig. 9. A K-band frequency-scanned edge-slot array with 400 radiating elements. (*Courtesy Hilburn and Prestwood [78], © 1974 IEEE*)

edge-slot array in RG-96/U waveguide technology [78]. The array, designed for a 30-dB Taylor distribution with three equal side lobes and an overall efficiency of 65 percent, exhibits a gain of 23, 25.5, and 28.5 dB at frequencies of 26.5, 28.5, and 30 GHz, respectively. The corresponding pointing angles are 23°, 13°, and 6°, so that a scan range of 17° is obtained at a scan rate of approximately 15° per 10-percent frequency change. No side lobes or cross-polarized components were observed down to −30 dB below the main beam level. Machining tolerances for slot depth and slot-to-slot spacing (±0.025 mm and ±0.05 mm, respectively) did not present great difficulties, but precision requirements would be more difficult to achieve at higher frequencies [78]. Note that frequency scaling of such arrays is not entirely straightforward in that the wall thickness of metal waveguides, as measured in wavelengths, is larger in the millimeter-wave region than it is at microwave frequencies.

Purcell-type arrays [79, 80], though known for many years, have not found much attention up to now. They are mentioned here since their mechanical aspects, i.e., substantially increased slot dimensions and relaxed fabrication tolerances, should make them well suited for millimeter-wave applications. In these arrays a comparatively large wall thickness is used for the waveguide wall carrying the radiating slots; see Fig. 10, which shows a sectoral horn excited by a Purcell-type slotted waveguide. The slots themselves, in this case, are regarded as waveguide sections coupling the interior region of the feeding waveguide to the outside region. If the wall thickness is chosen equal to an electrical quarter-wavelength of these waveguide sections (or an odd multiple of $\lambda_g/4$), an impedance transformation will occur which in effect nearly short-circuits the slots as seen from the interior of the feeding waveguide. Hence little energy leakage will take place from each slot and the cross-sectional slot dimensions can be made comparatively large. The quarter-wavelength condition on the wall thickness limits the bandwidth of these arrays.

Spiral Antennas and Fan-Shaped Beam Antennas

Spiral Antennas—These antennas are well known as "frequency independent" antennas which provide essentially constant directivity gain, beamwidth, and impedance over broad frequency ranges. In addition, it is advantageous for many applications that the antennas operate at circular polarization.

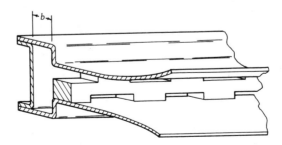

Fig. 10. A Purcell-type slot array shown feeding an *E*-plane sectoral horn. (*After Silver [80], © 1949; reprinted with permission of McGraw-Hill Book Company*)

a

Fig. 11. Millimeter-wave spiral antenna. (*a*) Antenna configuration. (*b*) Radiation pattern at 18 GHz. (*c*) Radiation pattern at 40 GHz. (*After E. M. Systems [81]; reprinted with permission of* Microwave Journal, *from the December 1975 issue,* © *1975 Horizon House–Microwave, Inc.*)

Logarithmic spiral antennas have been designed for the lower millimeter-wave region up to 60 GHz [7, 81]. Fig. 11a is a photograph of an antenna covering the 18- to 40-GHz band [81]; the spiral diameter is approximately 1.9 cm. Figs. 11b and 11c show the radiation patterns at 18 and 40 GHz. The two curves in each graph refer to horizontal and vertical polarization. The beamwidth is 70° and the axial ratio remains below 2 dB across the band.

Antennas with these characteristics are useful in particular for broadband direction-finding systems [81]. They may also be used as reflector feeds or as array elements in highly directive communication and sensing systems when a broad bandwidth is required.

Pillbox and Geodesic Lens Antennas—Pillbox antennas provide fan-shaped radiation patterns of broad beamwidth in one principal plane and narrow beamwidth in the other [82]. In their simplest configuration these antennas consist of two parallel metal plates connected by a parabolic-cylinder reflector whose focal line is directed normal to the plates; see Fig. 12a. The feed is placed in the focal region of the reflector so that it is located in the middle of the radiating aperture. The desired fan-shaped radiation pattern is obtained by choosing the aperture dimension parallel to the plates to be much larger than the plate spacing.

An improved version of these antennas, designed for the millimeter-wave region, is shown in Fig. 12b [83]; it comprises a rectangular feed horn, a doubly folded pillbox, and a corrugated flare section. In the figure the structure is bisected vertically and only one-half is shown. The doubly folded geometry permits feeding the antennas from the back. Hence aperture blockage is eliminated, resulting in

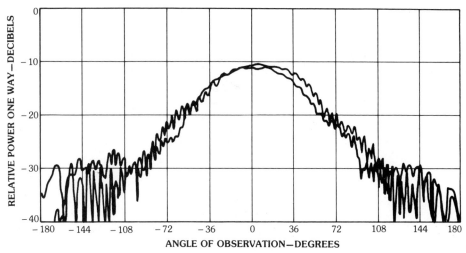

b

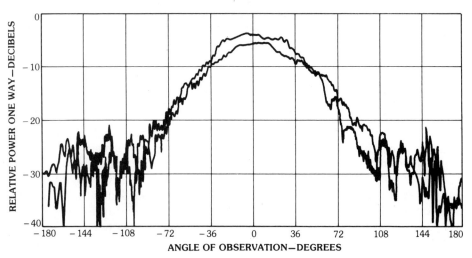

c

Fig. 11, *continued.*

improved side lobe performance, and the length of feed waveguide runs is reduced, leading to enhanced efficiency. An experimental model of the antenna designed for the 60-GHz band had aperture plane dimensions of 20 cm by 1.75 cm and provided a gain of 27 dB corresponding to a beamwidth of 2° in the H-plane and 30° in the E-plane. H-plane side lobes were −26 dB below the main beam, and it appears that by careful design a side lobe level as low as −35 dB can be achieved both in the E- and H-planes [83]. The bandwidth of the antenna is greater than 30 percent.

Geodesic lenses are a second type of antenna that can be used to produce fan-shaped radiation patterns. In addition, these antennas permit wide-angle beam

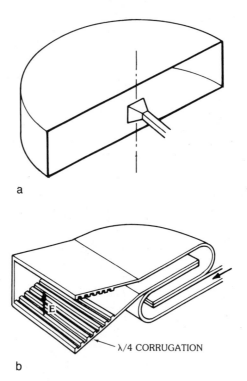

a

b

~ λ/4 CORRUGATION

Fig. 12. Pillbox antennas. (*a*) Basic configuration. (*b*) Improved version for millimeter-wave applications. (*After Chen [83], © 1981 IEEE*)

scanning without pattern degradation, and for this reason have been studied in some detail at microwave frequencies [84]. The antennas are relatively low-loss devices and, similar to pillbox antennas, have patterns that are not sensitive functions of frequency [3].

The basic antenna configuration is shown in Fig. 13a. It consists of two similar, dome-shaped metal surfaces separated by an air or dielectric layer of small but constant thickness. The antenna is fed by a small horn, usually in the form of an open waveguide, placed at the edge opening between the two surfaces. The EM waves or rays emanating from the horn are guided between the metal surfaces toward the radiating aperture on the diametrically opposite side of the edge slit. The ray paths are determined by Fermat's principle of minimum path length and follow geodesic lines. By adjusting the shape of the metal domes, the ray paths and thus the amplitude and phase distribution in the radiating aperture can be controlled. Usually these antennas are designed to transform a point source at the input into a straight, collimated line source at the output. The size of the aperture is limited in one dimension (*E*-plane) by the spacing of the metal surfaces but can be very wide in the other dimension (*H*-plane), so that the antennas are well suited to generate fan-shaped beams. Furthermore, if the metal domes are surfaces of revolution about the vertical axis, the beam direction can be scanned in the *H*-plane over a wide angular range without pattern degradation simply by moving the feed

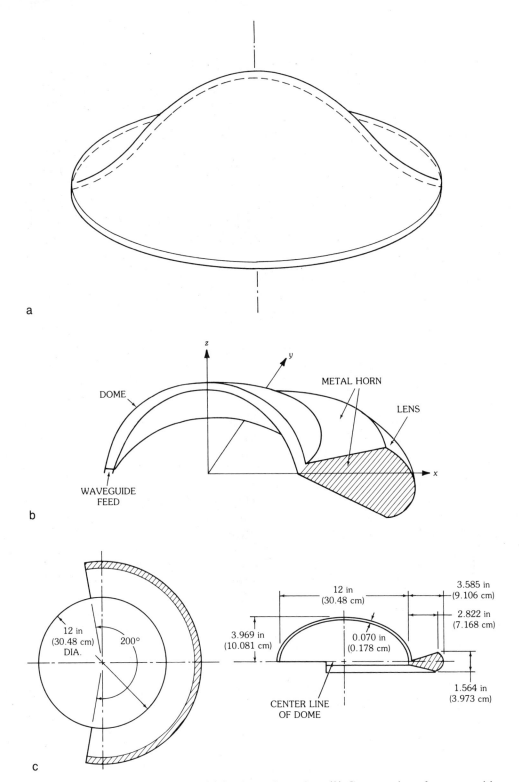

Fig. 13. Geodesic lens antennas. (*a*) Basic configuration. (*b*) Cross section of antenna with metal flares and dielectric ring lens attached to output edge. (*c*) Dimensions of experimental model for 26.5- to 40-GHz band. (*b* and *c*: *After Dufort and Uyeda [85], © 1981 IEEE*)

along the edge slot or by electronic switching among several fixed feeds. The *E*-plane pattern can be narrowed down by attaching flares to the radiating aperture to produce the effect of a sectoral horn, or by using a dielectric ring lens which follows the aperture contour.

The cross-section view of an antenna of this type [85] is shown in Fig. 13b. The feed location is indicated on the left side, the geodesic dome structure at the center, and the radiating aperture which is flared and contains a dielectric ring lens is shown on the right side. The antenna is circularly symmetric about the *z* axis over a wide-angle sector and was designed for wide-angle beam scanning. The metal domes are shaped so that the emanating beams have planar phase fronts to a good approximation. An experimental model [85], designed for the 26.5- to 40-GHz band, had the dimensions shown in Fig. 13c, and at 40 GHz provided a pattern with a beamwidth of $1.7° \times 10.7°$ and a directivity gain of 31.5 dB. The corresponding values at 26.5 GHz were $2.5° \times 15°$ and 29.5 dB. The aperture efficiency of the antenna varies between 60 and 72 percent, but side lobe levels are relatively high, i.e., above −20 dB in both principal planes. The antenna permits one to generate uniform multiple beams of similar gain and pattern behavior over a 90° sector.

Omnidirectional Antennas

Biconical and *discone antennas* are broadband antennas providing an omnidirectional pattern in the azimuth plane and moderate gain in the elevation plane. For the millimeter-wave region the same design approach can be used as in the microwave region, and antennas of this type have been developed for frequencies up to 100 GHz [86, 87]. For high efficiency, however, these antennas should be fed by a circular waveguide rather than a coaxial cable, in particular at frequencies above 60 GHz. To obtain an omnidirectional pattern the circular waveguide must be operated either in the fundamental TE_{11} mode at circular polarization or in the circularly symmetric TM_{01} mode. In the latter case, excitation of the TE_{11} mode can be inhibited by maintaining strict symmetry, or a mode filter may be used to suppress this mode. The impedance bandwidth of biconical antennas is very large; pattern bandwidth (concerning the elevation pattern) is more limited but stable patterns over a 25 percent bandwidth are easily achieved. The direction of polarization is vertical. Horizontal or circular polarization can be obtained with the help of a printed meanderline polarizer which is bent into a cylindrical shape so that it can be placed conformally into the radiating aperture of the antenna.

Dipole and *monopole antennas* are simple in shape and extremely compact in the millimeter-wave region. Monopole antennas, which operate over a metal ground plane, may be fed in the usual manner by a coaxial line or from a metal waveguide using probe or loop coupling. In the far-infrared region, linear antennas in the form of thin wires attached to detector diodes have been used as integrated antennas. Though physically short these so-called cat-whisker antennas are usually many wavelengths long; they have been shown to operate as traveling-wave antennas.

Recently, a novel integrated approach to the realization of millimeter-wave and submillimeter-wave dipole antennas has been suggested where these antennas would take the form of semiconducting strips or films grown on insulating or semiconducting substrates [88]. The primary advantage of such antennas is that their conductivity profile can be adjusted by modulating the carrier concentration along the antenna. This could be achieved, for example, through the utilization of surface field effects, biased pn junctions, or electro-optical methods [88]. In this way the current distribution of the antenna and hence its input impedance and radiation pattern can be controlled electronically. An additional advantage, which these antennas share with printed-circuit antennas, is that they could be integrated readily with other solid-state components created on the same substrate. At present, however, such semiconducting film dipoles are a concept only; an experimental model does not seem to be available as yet.

Linear arrays can be realized in the form of slotted waveguides. The slots, which are operated at resonance, are equivalent to magnetic dipoles placed tangentially on the (closed) waveguide surface. A waveguide with an axial sequence of longitudinal slots represents a collinear array, and such arrays are useful as omnidirectional antennas. For this application the waveguide axis will usually be oriented vertically, and it is desirable that maximum radiation occurs in the horizontal plane. This condition can be satisfied by operating the waveguide as a resonant array where all slots are excited in equal phase.

Because of the presence of the waveguide, however, the array configuration is not strictly symmetric about the vertical axis, and the azimuth pattern will show a certain amount of directivity. Since the waveguide has limited width this directivity will not be prohibitive, but achieving a truly omnidirectional pattern requires special care. The antenna radiates horizontal polarization.

As an example, Fig. 14 shows a broadwall, longitudinal-shunt-slot array antenna which is operated as a resonant array in the 50-GHz band [89]. In the usual manner, adjacent slots are spaced one-half guide wavelength apart and are offset by equal and opposite amounts from the center line to obtain excitation in equal phase. Eight radiating slots are provided both in the front and rear walls of the waveguide in identical arrangements. The fins attached to the narrow walls assist in achieving the desired omnidirectional pattern in the azimuth plane within a ±2-dB ripple. The beamwidth in the elevation plane is approximately 10°, corresponding to a gain of 8 dB. The bandwidth of the antenna is in the order of a few percent. Since the antenna is comparatively short the bandwidth is limited by the frequency response of the individual slot elements. For long arrays with many slots, on the other hand, the array resonance condition would be the limiting factor, restricting

Fig. 14. Slotted-waveguide linear array antenna for 50-GHz band. (*Courtesy U.S. Army CECOM, Ft. Monmouth, NJ, and Norden Systems, Norwalk, CT*)

the bandwidth to approximately $\pm 50/n$ percent, where n is the number of array elements [90].

3. Surface-Wave and Leaky-Wave Antennas Based on Open Millimeter Waveguides

Most millimeter waveguides are open guiding structures; to minimize conduction losses and simplify fabrication they do not enclose the guided fields on all sides by metal walls. Hence, if these guides are not excited in the appropriate mode or if the uniformity of the guiding structure is perturbed, energy leakage will occur and part of the guided energy will be radiated into the surrounding medium. In a waveguide, of course, radiation losses are undesirable and the design problem is to minimize energy leakage. But the leakage effect may be utilized to advantage for the design of antennas, by modifying an open guide in such a way that radiation occurs in a controlled manner. Antennas of this type will have the same advantages as the guides from which they are derived, i.e., a low profile and structural simplicity. In addition, they are directly compatible with these guides and will be well suited for integrated designs.

Considering the numerous of open waveguides which have been suggested for use in the millimeter-wave region and the various ways in which these guides may be modified to obtain radiation, it is evident that a wide variety of millimeter-wave antennas can be constructed by using this general approach. But few such antennas had been studied until recently.

Two antenna structures which have been investigated in some depth are tapered dielectric-rod antennas and periodic dielectric antennas. Although both of these antennas, which are derived from dielectric waveguides, have been known for many years, the latter antenna was not studied extensively until relatively recently and in fact in the context of millimeter waves. Several new traveling-wave antennas which are uniform rather than periodic have been proposed recently; they are based on groove guide, microstrip line, and NRD (nonradiative dielectric) guide, which is a recently described variant of H guide. All of these antennas show significant differences in their operating principles and each may be regarded as typical for a class of antennas.

In tapered-rod antennas the cross section of the dielectric waveguide is decreased monotonically in the forward direction so that a wave traveling along the guide is gradually transformed from a bounded wave into a free-space wave. The structure is a typical surface-wave antenna which radiates in the forward direction.

The remaining antennas are leaky-wave antennas. A common feature is that they radiate at an angle that depends on frequency so that the antennas may be used for frequency scanning.

Periodic dielectric antennas consist of a uniform dielectric waveguide with a periodic surface perturbation in the form of a dielectric or metal grating. Diffraction at this grating transforms a guided mode into a leaky mode and, hence, the waveguide into an antenna.

The new uniform leaky-wave antennas are based on low-loss millimeter waveguides (except for the microstrip antenna), and the leakage of radiation can be caused in two different ways. One way is to perturb the structure longitudinally (along the guiding direction) in an asymmetric but uniform fashion. The difference here from the grating dielectric antennas is that there the perturbation is periodic but here it is uniform, so that these antennas have the outward appearance of a uniform open waveguide. An advantage of the uniform perturbation is its simplicity, permitting easier fabrication at the shorter millimeter wavelengths. These uniform leaky-wave antennas are less versatile than the periodic ones, however, since they can radiate only into the forward quadrant, whereas the periodic leaky-wave antennas can radiate into the backward quadrant as well.

The second way in which radiation can be produced on uniform open waveguides is to employ a higher mode. The fundamental mode on these guides is always purely bound (if the guiding structure is unperturbed), but higher modes can leak, although sometimes only over a restricted frequency range.

Examples of uniformly perturbed millimeter waveguides that give rise to leaky-wave antennas are the groove guide and nonradiative dielectric guide. Examples of leaky higher modes which can be employed for antenna purposes occur on groove guide and on microstrip. There, the dimensions of the guide are chosen such that it supports the first higher mode, and an appropriate feed arrangement ensures that only this mode is excited.

In the following, tapered dielectric-rod antennas and periodic dielectric antennas are discussed in some detail. These antennas are already of substantial practical interest since dielectric waveguides will be employed in many applications throughout the millimeter-wave band and, in some form, are likely to be used as the basic transmission medium for integrated millimeter-wave devices operating above 100 GHz. The antennas are obtained simply by letting a dielectric waveguide terminate into a tapered section or a section with a periodic surface corrugation whose function it is to radiate out the guided energy.

The new uniform leaky-wave antennas will be reviewed with respect to their operating principles and applicational aspects in and near the millimeter-wave region. These antennas are still under study and available design information was limited when this chapter was first written. Latest available information can be found in a comprehensive report [221] that was prepared very recently. The report also contains the results of several studies concerned with linear *arrays* of such leaky-wave line-source antennas, where the individual antennas in the array are fed in parallel from one end in phased-array fashion, permitting phase-shift scanning in the cross plane.

A concluding remark is concerned with dielectric resonator antennas, a different type of millimeter-wave antenna which has been examined recently [91]. These antennas are mentioned here since, in a broad sense, they are also derived from open millimeter waveguides. Typically, they take the form of short, resonating sections of rectangular or circular dielectric waveguide placed vertically on a metal ground plane. Since these dielectric resonators are not enclosed within metal walls, all energy delivered by the feed system will ultimately be radiated out (except for dielectric and metal losses, which may be significant) and the resonators will act as antennas with relatively narrow bandwidth. If the antennas are made from a material of sufficiently high permittivity ($\epsilon_r > 5$), the dielectric cavity will support well-defined resonant modes and the radiation patterns will be determined primarily by the shape of the resonator. Hence, these antennas may be well suited for beam shaping, although the relationship between resonator configuration and mode pattern is not a simple one. The antennas should be particularly useful for the upper millimeter-wave region; at present they are studied in scaled microwave experiments.

Table 1 lists materials suitable for millimeter-wave dielectric antennas and microstrip antennas [92, 93]. The constitutive parameters shown in the table apply to the (upper) microwave range. In the millimeter-wave region the permittivity values should be approximately the same, but the loss factors may be higher. Additional values, applying to the near-millimeter-wave/far-infrared region, may be found in [94].

Tapered Dielectric-Rod Antennas

Dielectric-rod antennas have been known for many years [95–98], but they have not been widely used as microwave antennas. In the millimeter-wave region these antennas have very reasonable size and weight, and many applications can be expected because of their compatibility with dielectric waveguides.

Tapered-rod antennas can take any of the configurations shown in Fig. 15. Their cross section may be rectangular or circular, where, in the rectangular case, the taper may occur in one cross-sectional dimension or, symmetrically, in both. Millimeter-wave antennas will usually have a rectangular cross section; a linear taper, as indicated in Fig. 15a, is preferable for reasons of simplicity. Such antennas have been designed for wavelengths down to 0.12 mm, i.e., well into the submillimeter-wave region [99, 100].

In most experimental investigations reported in the literature the antennas are fed by metal waveguides, as indicated for the shaped antenna of Fig. 15c. In this case, radiation from the feed aperture can have a strong effect on pattern shape and side lobe level, and an appropriate feed arrangement must be used to minimize transition effects [101]. If the antennas are fed by dielectric waveguides made from

Table 1. Millimeter-Wave Materials

	Relative Dielectric Constant (ϵ_r)	Loss Factor ($\tan \delta$)
Teflon	2.08	0.000 6
PTFE-glass	2.17	0.000 9
	2.33	0.001 5
Polyethylene	2.26	0.000 6
Duroid 5880	2.20	0.000 6
Duroid 5870	2.33	0.000 5
Quartz Teflon	2.47	0.000 6
Polystyrene	2.54	0.001 2
Fused quartz	3.78	0.000 25
Boron nitride	4.40	0.000 3
Sapphire	9.0	0.000 1
Alumina	9.8	0.000 1
Silicon	11.8	
Gallium arsenide	13.2	0.001
Magnesium titanate	16.1	0.000 2

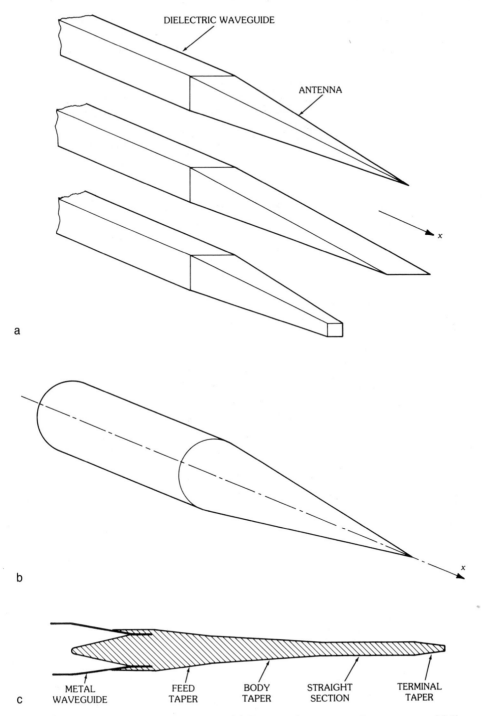

Fig. 15. Tapered dielectric-rod antennas. (*a*) Rectangular-cross-section antennas with linear taper. (*b*) Circular-cross-section antenna with linear taper. (*c*) Shaped antenna. (*After Zucker [98]*)

the same material as the antennas (Figs. 15a and 15b), the transition region will cause little radiation, and a smooth pattern can be expected, in particular for slender antennas of small apex angle, the practically interesting case.

Experimental studies on tapered-rod antennas have been conducted at microwave frequencies and, more recently, at millimeter-wave frequencies (95–106]. Theoretical methods usually involve strong simplifications and provide general design guidelines only. More accurate theories have been developed for cylindrical rod antennas (without taper) [107–113] and the two-dimensional problem of wedge type radiators [114–117], but these theories are not easily extended to the tapered antennas of finite cross section, which are of interest here. Available design information on these antennas is summarized below. The summary utilizes references [95–97, 102–106] and, in particular [98], an excellent review article on traveling-wave antennas written by Zucker more than 20 years ago.

In practical cases, the apex angle of the antennas will be small and the taper very gradual. In this case, the surface wave incident from the uniform dielectric waveguide or excited by the horn aperture of a metal waveguide will travel along the antenna with little reflection and in the process will be transformed from a strongly bounded wave whose energy is mostly confined to the dielectric region to a free-space wave propagating entirely in the air region. The phase velocity v of this wave will increase toward the free-space velocity c as the wave travels from the antenna base toward its tip. In any given cross-section plane it can be approximated by the phase velocity of a uniform dielectric waveguide having the local cross section of the antenna. For a dielectric rod of circular cross section this phase velocity is shown in Fig. 16 as a function of rod diameter d for various values of the permittivity ϵ_r. The curves of Fig. 16 apply to the fundamental HE_{11} mode of the dielectric rod. Operation in this mode will result in a radiation pattern with low side lobes and reasonable directivity [98]. The direction of maximum radiation is the end-fire direction, i.e., the x axis in Fig. 15. Operation in the HE_{11} mode is ensured when the antenna diameter at the base satisfies the condition $d_B/\lambda_0 < 0.626/\sqrt{\epsilon_r}$, where λ_0 is the free-space wavelength [98]. The next higher modes, TE_{01} and TM_{01}, are circularly symmetric about the antenna axis and would yield a pattern null in the forward direction.

To a good approximation the curves of Fig. 16 remain valid for rods of other cross sections provided the cross-section area A is the same, i.e., d in the abscissa scale may be replaced by $(4A/\pi)^{1/2}$. According to Mallach [96] and Kiely [97], for optimum performance an antenna of given electrical length L/λ_0 should have a cross section

$$A_{max} = \frac{\lambda_0^2}{4(\epsilon_r - 1)} \tag{2a}$$

at the base and

$$A_{min} = \frac{\lambda_0^2}{10(\epsilon_r - 1)} \tag{2b}$$

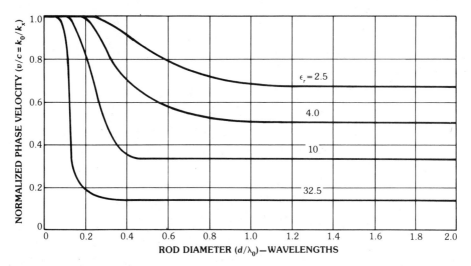

Fig. 16. Dispersion curves of fundamental HE_{11} mode of cylindrical dielectric rod. (*After Mueller and Tyrell [95]; reprinted with permission from the* AT&T Technical Journals, © *1947 AT&T*)

at the termination. For permittivity values in the range $2.5 < \epsilon_r < 20$, which covers most millimeter-wave materials, this rule corresponds roughly to letting $c/v = k_x/k_0$ be 1.1 at the feed and very close to 1.0 at the termination, where k_x is the (local) phase constant of the guided wave and k_0 is the free-space wave number.

From (2a) and (2b) it is evident that the higher the dielectric constant of the antenna, the smaller its cross section and, hence, as a general rule, its weight. On the other hand, Fig. 16 shows that the dispersion curves become steeper with increasing ϵ_r so that parameter dependences will be more sensitive and bandwidth will be reduced [98].

For tapered rods satisfying (2a) and (2b), the directivity gain is determined primarily by the antenna length L. The gain initially increases with L according to the relation [98]

$$G = p\left(\frac{L}{\lambda_0}\right) \tag{3}$$

where p is approximately 7 for long antennas having a linear taper, while p can be as large as 10 for the maximum-gain antennas discussed below.* The largest attainable gain, however, is limited to 18 to 20 dB, which corresponds to a half-power beamwidth of 17° to 20°. This gain limitation is easy to understand on the basis of the equivalence principle. The polarization current in each volume element of the antenna radiates as a Hertz dipole. The phases of these dipoles are determined by the phase velocity, of the surface wave guided by the antenna. Since this phase velocity remains below the free-space wave velocity, the contributions

---

*The value $p = 10$ applies to maximum-gain antennas of moderate length L between $3\lambda_0$ and $8\lambda_0$.

of the various volume elements of the antenna to the radiation in the forward direction will add constructively only up to a certain length of the antenna. For antennas with a gradual taper the optimum length is approximately determined by the condition

$$\int_0^L (k_x - k_0)\,dx = \pi \tag{4}$$

If this length is exceeded, destructive interference will occur and a loss of directivity must be expected.*

Experimental evidence confirms that relations (2a), (2b), and (4) may be used as working formulas in the design of linearly tapered dielectric-rod antennas [102, 106]. However, these formulas should not be taken as rigid design criteria. Increasing the cross section of the antenna at the base beyond A_{max}, such that k_x/k_0 is increased from 1.1 to 1.25–1.40, will enlarge the bandwidth of the antenna and can also be used as a method for reducing the side lobe level (at the expense of a moderate gain reduction) [98]. Furthermore, for long antennas with a gradual taper it should be of little consequence if the antenna termination at the far end is blunt as suggested by (2b) or if the antenna is physically continued to its apex. The field strength within this tip section is small since most of the energy is now traveling in the air region; and the phase velocity is practically that of free space so that the left side of (4) will not noticeably be affected. For short antennas with a length in the order of a few wavelengths, on the other hand, tapering to a sharp tip has the advantage of a reduced reflection coefficient, a feature which will be discussed later in this section.

Zucker has described a procedure based on physical reasoning and experimental evidence for the design of surface-wave antennas of maximum-gain [98]. In the case of tapered-rod antennas the basic configuration of a maximum-gain antenna takes the form shown in Fig. 15c. It includes a feed taper, a body taper, a straight section, and a terminal taper. The feed taper establishes a surface wave which is assumed to radiate continuously as it travels along the body taper and the straight section. The terminal taper reduces reflections which would be caused by an abrupt discontinuity. Details of the design procedure are given in [98]; the main results are summarized in Fig. 17.** The solid curves show the gain and beamwidth of maximum-gain antennas as a function of antenna length L/λ_0 according to experimental data reported in the literature. The half-power beamwidth, which is approximately given in degrees by

$$\Delta\theta_{HP} = 55\sqrt{\frac{\lambda_0}{L}}$$

is an average value; the beamwidth is usually slightly smaller in the E-plane and slightly larger in the H-plane [98].

---

*A formulation of condition (4) appropriate for maximum-gain antennas has been given by Zucker [98].
**Fig. 17 applies not only to tapered-rod antennas but to surface-wave antennas in general.

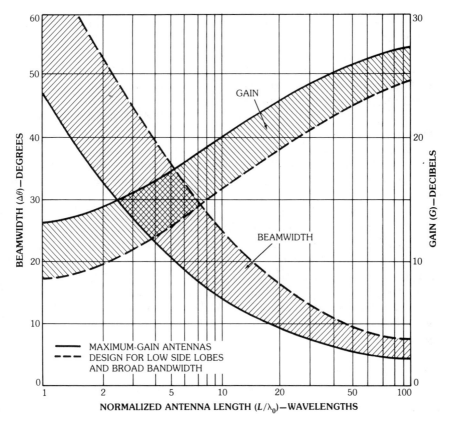

Fig. 17. Gain and beamwidth of surface-wave antennas as a function of normalized antenna length. (*After Zucker [98]*)

A problem with maximum-gain antennas is that they have comparatively small pattern bandwidth (± 10 percent) and high side lobes no more than -6 dB to -10 dB down from the main peak. Appropriate design of the body taper will solve these problems, though at the expense of a reduced gain [98]. The dashed curves in Fig. 17 characterize surface-wave antennas designed for broad bandwidth and low side lobe levels. Most tapered-rod antennas can be expected to have gain and beamwidth values in the shaded region between the dashed and solid curves. Experimental evidence [105, 106] shows, in particular, that antennas with a linear taper can be designed to have a gain not significantly lower than that of maximum-gain antennas while their pattern quality is substantially better (side lobes less than -20 dB). Since these antennas have the additional advantage of structural simplicity they should be of more interest for millimeter-wave applications than maximum-gain antennas.

With regard to the radiation patterns of linearly tapered dielectric-rod antennas the following general observations can be derived from experimental results reported in the literature [98, 102–104], in particular, by Shiau [105] and by Kobayashi, Mittra, and Lampe [106]. In the experiments of interest here, precautions were taken to minimize parasitic radiation from the waveguide-to-antenna transitions.

(*a*) Antennas of moderate length ($3\lambda_0$ to $10\lambda_0$) tend to produce a comparatively broad main beam whose sides, however, drop down rapidly toward the first zero [106]. A consequence is that the gain of these antennas is usually within the predicted range between the solid and dashed gain curves of Fig. 17, while the beamwidth may be slightly above the upper bound defined by the dashed beamwidth curve.

(*b*) By careful design the side lobe level can be kept below −25 dB. As an example, Fig. 18 shows the pattern of a pyramidal antenna of length $L/\lambda_0 = 5.5$ and permittivity $\epsilon_r = 2.1$ (Teflon). The antenna was fed by a metal waveguide with a horn launcher [101].

(*c*) The beamwidth of the *E*- and *H*-plane patterns is essentially the same [105].

(*d*) The beamwidth and side lobe level decrease monotonically with increasing axial length of the antenna. However, if an optimum length is exceeded, main beam distortions occur leading to a significantly increased side lobe level and eventually a pattern breakup [106]. Fig. 19 illustrates this trend for antennas of permittivity $\epsilon_r = 2.33$ fed by metal waveguides with horn launchers [106]. The patterns were taken at a frequency of 81.5 GHz corresponding to a wavelength $\lambda_0 =$

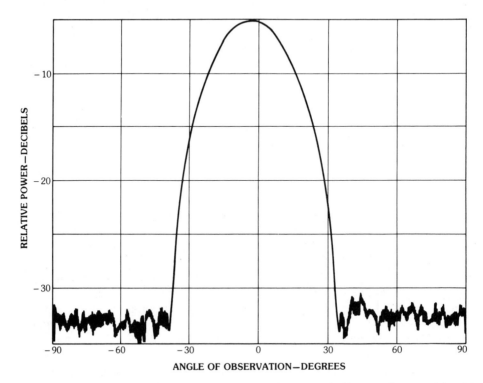

Fig. 18. *H*-plane pattern of pyramidal dielectric-rod antenna fed by metal waveguide with horn launcher. (*After Kobayashi, Mittra, and Lampe [106], © 1982 IEEE*)

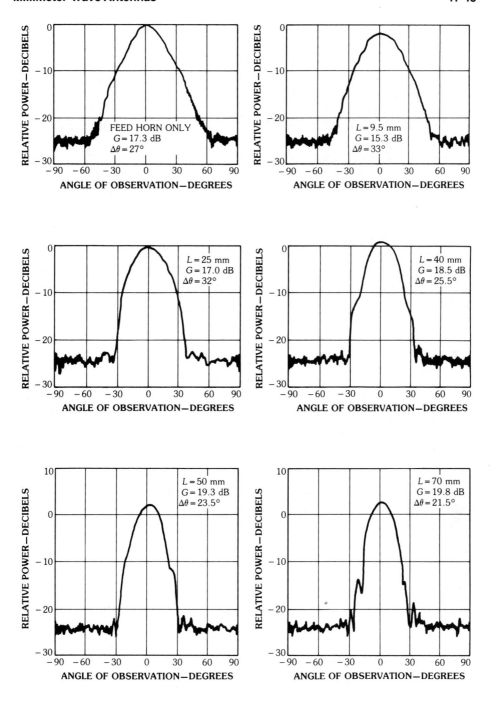

Fig. 19. Dependence of radiation pattern of tapered dielectric-rod antennas on axial length *L*. (*After Kobayashi, Mittra, and Lampe [106], © 1982 IEEE*)

3.68 mm. The first of these patterns is that of the feed horn alone; the last pattern shows the onset of main beam distortions.*

The pattern of Fig. 18 was taken with a pyramidal antenna. An antenna of the same length but tapered in only one dimension, which is mechanically stronger and easier to make than a pyramidal antenna, usually leads to comparable gain and pattern quality provided the taper occurs in the *E*-plane [106]. A taper only in the *H*-plane tends to result in degraded performance. The patterns of Fig. 19 were measured for antennas with a one-dimensional *E*-plane taper.

Comparison between a tapered-rod antenna and an aperture antenna, as for example, a parabolic dish, shows that these antennas provide approximately the same directivity gain and side lobe performance if their dimensions are related by

$$\frac{L}{\lambda_0} \cong \left(\frac{D}{\lambda_0}\right)^2$$

where L is the length of the rod and D the diameter of the dish [98]. The relation becomes obvious when one compares (3) for the gain of tapered-rod antennas with the well-known formula $G = \eta(\pi D/\lambda_0)^2$ for aperture antennas while considering that the aperture efficiency η is typically in the 50- to 70-percent range and the factor p in (3) is approximately 7.

Information on the bandwidth of tapered-rod antennas is scarce. But in the practically interesting case of a gradual taper where the reflection coefficient is small, the impedance bandwidth should be large. Furthermore, a pattern bandwidth of ±15 percent is available for most surface-wave antennas [98]. For polystyrene antennas with $\epsilon_r = 2.56$ (polyrod antennas) it has been shown that the bandwidth can be extended to ±33 percent, so that the antennas cover a 2:1 frequency band. These broadband antennas are linearly tapered, and k_x/k_0 is approximately 1.4 at the feed end; the trade-off is a decrease in gain from maximum by 2 dB. With increasing permittivity the pattern bandwidth is likely to decrease. This is concluded from Fig. 16, which shows that the dispersion curves become steeper [98] with increasing ϵ_r.

A broad impedance bandwidth requires a low reflection coefficient. The dependence of the reflection coefficient on the taper profile has been investigated for the two-dimensional case of a tapered dielectric disk antenna [118, 119]. An antenna of this type is depicted in Fig. 20a; the antenna configuration is circularly symmetric about the vertical axis and consists of a cylindrical center portion of uniform thickness $2h$ and radius a, surrounded by a tapered section of monotonically decreasing thickness, and outer radius b. When appropriately excited (at the axis) such antennas are useful as omnidirectional radiating elements providing a circularly symmetric pattern in the azimuth plane and moderate gain in the elevation plane.

---

*On the other hand, for very long tapered-rod antennas (fed by a dielectric waveguide of the same permittivity), the transformation of a surface wave traveling along the structure from a strongly bounded wave into a free-space wave will proceed very gradually and little reflection will occur. One may conclude that such antennas will radiate a smooth, highly directive pattern. But no theoretical or experimental data to confirm or refute this view seem to be available.

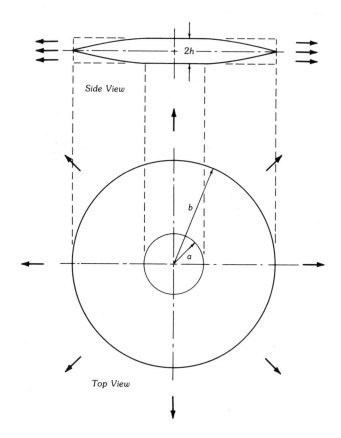

Side View

$2h$

b

a

Top View

a

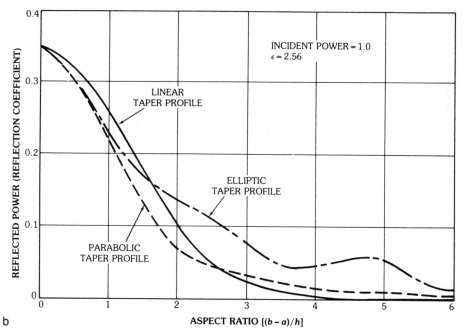

b

Fig. 20. Reflection coefficients of a tapered dielectric disk antenna. (*a*) Antenna configuration. (*b*) Reflection coefficients.

The (power) reflection coefficient is defined as the total power returning in the fundamental (guided) mode from the tapered section of the antenna, when the input power is unity. Fig. 20b shows the theoretical reflection coefficient for a linear, parabolic, and elliptic taper profile* as a function of the aspect ratio $(b - a)/h$. A polystyrene antenna ($\epsilon_r = 2.56$) is assumed having the dimensions $a = 0.2\lambda_0$ and $2h = 0.4\lambda_0$ and excited in the lowest-order TM mode. For a short taper with $(b - a)/h < 2$, the reflection coefficient can be rather large; it is largest for the linear profile, which is understandable, since this profile produces a more abrupt transition at the base of the tapered section than does a parabolic or elliptic profile. For a long, gradual taper, on the other hand, which is the practically interesting case, the linear profile results in the lowest reflection coefficient. In this case the transition from the uniform to the tapered section does not introduce a significant perturbance for any of the three profiles. But the parabolic and elliptic profiles produce a "specular" reflection point at the antenna tip while the linear taper does not. The general design principle resulting from these considerations may be applied also to the tapered dielectric-rod antennas of Fig. 15: At the antenna base the transition from a uniform dielectric-feed waveguide to the tapered antenna should be as smooth as possible, while at the forward end the antenna should terminate into a sharp tip [118, 119], which is a further argument in favor of linearly tapered antennas.

When tapered-rod antennas are fed by metal waveguides, unwanted radiation from the waveguide-to-antenna transition can be minimized by the use of a surface-wave launcher in the form of a small pyramidal horn [101] as indicated in Fig. 21. According to Malherbe and others [101] who have studied this type of launcher, the horn does not have to be long but the flare angle should be fairly large, i.e., greater than 30°. A gradual taper of the portion of the antenna extending into the waveguide will reduce the overall transmission loss to less than 0.5 dB; the length of the taper should be $1.5\lambda_0$ to $2.5\lambda_0$. None of the dimensions of the launcher are critical, so it is easy to manufacture and inexpensive [101].

To increase the directivity gain of tapered-rod antennas above the 18- to 20-dB

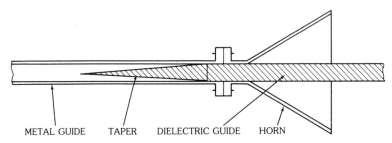

METAL GUIDE   TAPER   DIELECTRIC GUIDE   HORN

Fig. 21. Horn-launcher transition from metal waveguide to dielectric waveguide. (*After Malherbe, Trinh, and Mittra [101]; reprinted with permission of* Microwave Journal, *from the November 1980 issue,* © *1980 Horizon House–Microwave, Inc.*)

---

*For the numerical analysis the actual taper profile was replaced by a staircase approximation with ten steps.

limit, *arrays* of these antennas can be used. Since the individual array elements have already a substantial directivity, element spacing can be chosen comparatively large so that interelement coupling is minimized. According to Zucker [98], the spacing, *b*, should be in the range

$$0.5\sqrt{\frac{L}{\lambda_0}} \leqq \frac{b}{\lambda_0} \leqq \sqrt{\frac{L}{\lambda_0}}$$

where the lower bound formulates the condition that interelement coupling should be negligible while the upper bound is determined by the requirement that the first grating lobe should be substantially reduced by multiplication with the element pattern. Ideally, the first minimum of the element pattern should coincide with the first grating lobe of the array pattern. For scanned arrays the element spacing *b* must be chosen smaller, in correspondence with the desired scan range [98].

Under these circumstances the array gain and the beamwidth of the main beam of a linear array of *n* equally spaced elements are

$$G_{array} = nG_{element}$$

and

$$\Delta\theta_{array} = \frac{\lambda_0}{nb}\frac{180}{\pi}$$

respectively where $\Delta\theta_{array}$ is the 3-dB beamwidth in degrees in the plane of the array. Side lobes can be controlled by tapering the excitation of the elements. The bandwidth of the array will be smaller than that of the individual elements, probably by 5 percent [98].

The array gain, of course, can be further increased by the use of planar arrays (volume arrays) of tapered-rod antennas. It should be noted, however, that the attainable gain is determined by the overall array aperture and, therefore, is not much higher than in the case of a dipole array of the same aperture size. The difference is that highly directive end-fire elements can be spaced much farther apart than dipole elements so that the number of array elements is significantly reduced.

An interesting feed arrangement for a linear array of tapered-rod antennas is shown in Fig. 22. Suggested by Williams and others [121], the feed system is designed in insular guide technology and is shown here supporting a frequency-scanned array of eight elements operating in the 30-GHz band. Alternate array elements are fed from the main insular guide via proximity couplers. The feed circuit is repeated on the reverse side of the ground plane to feed the second half of the array. An *E*-plane power splitter is used to excite each array half with the same amplitude. The arrangement permits the use of couplers whose radii of curvature are sufficiently large to prevent excessive radiation from the curved sections but maintains the close element spacing necessary for wide-angle scanning [120]. A uniform scan performance over an angular range of more than 40° was obtained at a

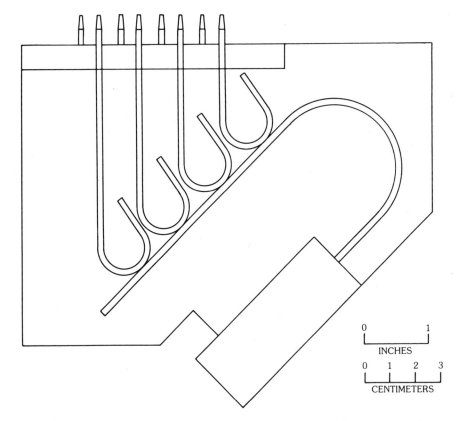

Fig. 22. Feed arrangement for eight-element frequency-scanned linear array of tapered dielectric-rod antennas. (*After Williams, Rudge, and Gibbs [121], © 1977 IEEE*)

frequency variation of 10 percent. The material from which the waveguides and antennas are made has a dielectric constant of approximately 10.

Periodic Dielectric Antennas

Periodic dielectric antennas [122–137] consist of a uniform dielectric waveguide with a periodic surface perturbation. The uniform waveguide supports a traveling wave; the surface perturbation acts as a grating that radiates out the guided energy. As is true for dielectric-rod antennas, these grating antennas are directly compatible with dielectric waveguides. Additional advantages include ruggedness and a low profile—the antennas can be installed conformally with a (planar) metal surface—and a capability for beam scanning by frequency variation.

Periodic antennas can have various configurations; recently investigated structures are shown in Fig. 23. The first of these antennas, Fig. 23a, operates with a dielectric grating in the form of a periodic surface corrugation [125–128]. The geometry of the teeth and grooves can take a variety of forms. For example, when the antenna is not backed by a metal plane, the corrugation profile can be designed to prevent radiation in the downward direction so that all radiated power is con-

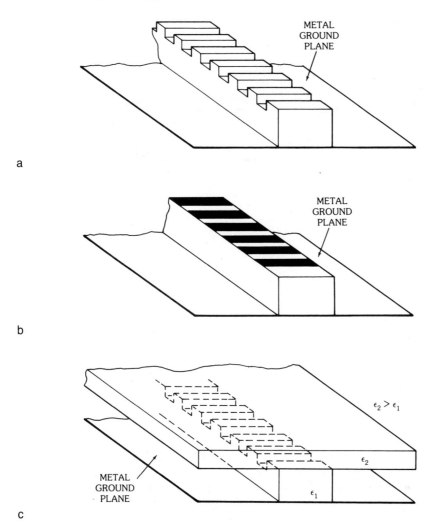

Fig. 23. Periodic dielectric antennas. (*a*) Antenna with dielectric grating. (*b*) Antenna with metal grating. (*c*) Inverted dielectric strip guide antenna. (*d*) Trapped image-guide antenna. (*e*) Image-guide–fed slot array.

centrated into a single upward-directed beam [138–141]. In the presence of a ground plane, shaping of grooves and teeth is not necessary and a rectangular profile is preferred since it is easiest to machine. A grating of metal strips, shown in Fig. 23b, is an alternative to a dielectric grating; the metal strips tend to increase the leakage constant, i.e., the radiation rate per period [122–124, 129–131]. Inverted-strip dielectric waveguide antennas, as shown in Fig. 23c, provide a very rugged design which integrates a planar radome with the antenna [132], but results in a lower leakage constant. The reason for the lower leakage constant is that the permittivity of the top plate is greater than that of the dielectric strip so that the guided energy travels primarily in the top plate in the region just above the

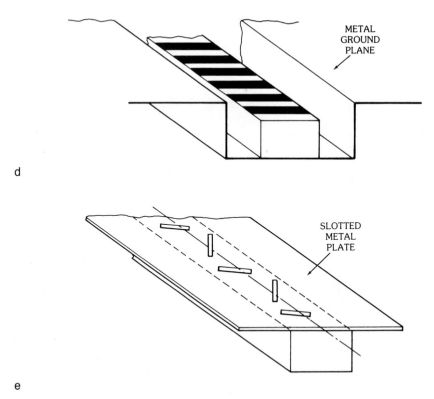

d

e

Fig. 23, *continued.*

dielectric strip. The trough guide arrangement of Fig. 23d leads to an antenna that can be flush mounted with a metal surface [133, 134]; this arrangement has the additional advantage that radiation losses from bends and corners in the feeding dielectric waveguide are lowered. The image-guide–fed slot array of Fig. 23e can be designed to provide a circularly polarized radiation pattern [142].

Review of Fundamentals—The theory of dielectric-grating antennas can be regarded as well developed due primarily to the work of Peng [125–127, 143–146]. This theory is based on a space harmonics representation of the fields guided by the antenna and should be regarded as comprising two parts. The first part applies to antennas of large width [125–127], and the second introduces modifications that allow the theory to be extended to narrower antennas [144–146], which are more likely to be employed in practice. The theory for the wide antennas is closely related to the analysis of optical grating couplers [147–152] and, to a certain extent, can be adapted from this analysis. Modifications, however, are required to accommodate the much smaller lengths of millimeter-wave devices, as expressed in wavelengths, and the effect of the much higher permittivity of millimeter-wave materials, such as silicon and alumina. Numerical evaluations have been performed primarily for antennas with dielectric gratings of rectangular profile [125–127], and design curve for these antennas are available.

For dielectric antennas with *metal* gratings, an alternative formulation in terms of an eigenvalue problem in the Fournier transform domain has been suggested by Mittra and Kastner [153], but a complete solution has not been provided. Very recently, an analysis that is complete, containing all the information required for practical design, was presented [219]. That analysis differs from the analysis below of corrugated dielectric antennas, but all of the fundamentals are similar.

For a detailed description of the theory of periodic leaky-wave antennas one may refer to the literature cited above. Here we restrict ourselves to the presentation of some of the pertinent results in order to provide a basis for the discussion on design guidelines to follow later. The present remarks apply to any of the periodic structures shown in Fig. 23, but for convenience we shall refer to the antenna of Fig. 24.

The waves supported by periodic dielectric antennas radiate as they travel along the antennas; they are leaky modes which decay exponentially in the forward direction. The lowest leaky mode is of primary interest for antenna applications. Higher leaky modes can be suppressed by choosing the antenna dimensions below the cutoff conditions of these modes or by using an appropriate feed arrangement.

A leaky mode has a phase constant β and an attenuation, or leakage, constant α, the latter being a measure of the power leaking away per unit length along the antenna. The periodicity produces an infinity of space harmonics associated with that leaky mode; the phase constants β_n of the various space harmonics are related to the phase constant β of the basic wave by

$$\beta_n = \beta + \frac{2\pi n}{d}, \qquad -\infty \leq n \leq +\infty$$

where n is the order of a given space harmonic, $n = 0$ corresponds to the basic wave, and d is the period of the modulated waveguiding structure.

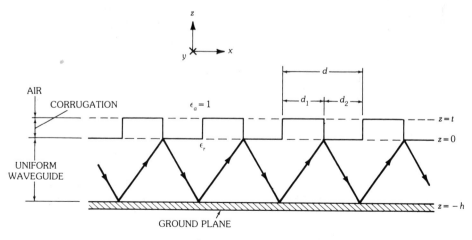

Fig. 24. Dielectric grating antenna: geometry and coordinate system. (*After Schwering and Penn [127]*, © *1983 IEEE*)

If a space harmonic is slow along the antenna surface, it is purely bound; if it is fast, it will radiate power away at an angle, given by

$$\phi_n = \sin^{-1}\left(\frac{\beta_n}{k_0}\right) = \sin^{-1}\left(\frac{\beta}{k_0} + n\frac{\lambda_0}{d}\right) \tag{5}$$

where λ_0 is the free-space wavelength, $k_0 = 2\pi/\lambda_0$ is the free-space wave number, and ϕ_n is the radiation angle measured from the positive z axis, i.e., from the normal to the antenna surface; see Fig. 24. The nonradiating space harmonics are evanescent waves that decay exponentially in the z direction. The number of these waves is infinite. The number of the radiating space harmonics, on the other hand, depends on d/λ_0 and is always finite. If d/λ_0 is large, several space harmonics are of the propagating type and the antenna will radiate in several directions simultaneously. Most antenna applications, however, require single-beam operation, which leads to the following condition* [125] on d:

$$\frac{\lambda_0}{\beta/k_0 + 1} \leq d \leq \frac{\lambda_0}{\beta/k_0 - 1}, \qquad \text{for } \beta/k_0 > 3$$

$$\frac{\lambda_0}{\beta/k_0 + 1} \leq d \leq \frac{2\lambda_0}{\beta/k_0 + 1}, \qquad \text{for } \beta/k_0 < 3 \tag{6}$$

If this condition is satisfied, only the space harmonic of order $n = -1$ will radiate. As the frequency is increased, the radiation direction

$$\phi_{-1} = \sin^{-1}\left(\frac{\beta}{k_0} - \frac{\lambda_0}{d}\right) \tag{7}$$

will scan from backfire over broadside to end-fire. If $\beta/k_0 > 3$, the full scan range is in principle available without a second beam appearing. If $\beta/k_0 < 3$, a second beam ($n = -2$) will appear near backfire as the first beam approaches the end-fire direction, and will then follow the first beam. For most millimeter-wave materials, ϵ_r is large and a reasonable scan range will be available for single-beam operation. Note, however, that these periodic antennas—as all leaky-wave antennas—do not radiate exactly in the broadside direction. As broadside conditions are approached an internal resonance develops leading to a leaky-wave "stopband." At large permittivity values, however, this "stopband" is very narrow and, since most practical antennas have a beamwidth of several degrees or more, the only effect usually noticed when scanning through broadside is an increase in vswr. For antennas with very narrow beamwidths, on the other hand, the effect can be severe.

In the case of single-beam operation, where only one space harmonic is of the radiating type, the radiation properties of the antenna are fully determined by the

---

*A more stringent condition taking into account the finite beamwidth of the antenna has been formulated by Kobayashi et al. [129].

complex propagation constant $k_x = \beta - j\alpha$. Hence β and α are the key parameters to be calculated in the theory of periodic dielectric antennas.

Phase Constant β and Leakage Constant α—The leaky modes supported by dielectric grating antennas of finite width cannot be described in terms of TE or TM waves alone. They are hybrid modes involving TE-TM coupling, and a rigorous determination of their complex propagation constants would require the solution of a vector boundary value problem [143, 144]. But approximate methods are available which provide results of good accuracy and reduce the computational burden significantly.

In the discussion below, we first present results for *wide antennas*. The phase constant β will be calculated using the so-called effective dielectric constant (EDC) method [154], which is a simple procedure that yields very accurate results for β for a large range of parameter values [125]. Furthermore, for antennas with large lateral width $w > \lambda_0/\sqrt{\epsilon_{\text{eff}} - 1}$ the leakage constant α is, to a good approximation, the same as that for an antenna of infinite width [125–127]. For this limiting case the leaky modes become TE or TM modes and the vector boundary value problem is reduced to a much simpler scalar problem. (For a precise definition of ϵ_{eff}, see below.)

The condition $w > \lambda_0/\sqrt{\epsilon_{\text{eff}} - 1}$ will not always be satisfied in practical cases. To prevent the occurrence of higher-order leaky modes, w will be chosen significantly smaller than $\lambda_0/\sqrt{\epsilon_{\text{eff}} - 1}$ in applications where a large *H*-plane beamwidth is acceptable, and it is noted that most experimental studies reported in the literature have used antennas of small w. In the second part of the discussion below, modifications of the theory will be introduced which allow the solution methods for wide antennas to be extended to *much narrower* antennas with $w < \lambda_0/\sqrt{\epsilon_{\text{eff}} - 1}$ [144–146]. In both parts it will be assumed that the antennas operate over a metal ground plane, that a dielectric grating is used with a rectangular profile of teeth and grooves, as indicated in Fig. 24, and that only the lowest-order leaky mode is present. (The principal field-strength components of this mode are E_z and H_y.) Since typical millimeter-wave materials, like silicon and gallium arsenide, have a dielectric constant of $\epsilon_r \cong 12$, this permittivity is assumed in most of the design curves presented here.

In applying the EDC method to antennas of large width the corrugation region is replaced by an equivalent layer of uniform dielectric constant. This dielectric constant is chosen to be the volume average permittivity

$$\epsilon_{\text{avg}} = \frac{\epsilon_1 d_1 + \epsilon_2 d_2}{d}$$

$$= 1 + (\epsilon_r - 1)\frac{d_1}{d}, \qquad \text{for } \epsilon_1 = \epsilon_r, \epsilon_2 = 1 \tag{8}$$

where ϵ_1, d_1 and ϵ_2, d_2 are the permittivity and width of the teeth and grooves, respectively, as shown in Fig. 24. The second expression holds if the permittivity of the teeth is that of the uniform guide and the permittivity of the grooves is that of air. This will be assumed in the following.

The EDC method [154] uses a two-step approach to derive a characteristic equation for the phase constant of the modes guided by dielectric waveguides of rectangular cross section. In the first step an effective dielectric constant, ϵ_{eff}, is defined by formulating the transverse resonance condition for the z direction, while temporarily assuming that the guide is infinitely extended in the y direction. In the second step a slab guide of permittivity ϵ_{eff} is considered which has finite width in the y direction but infinite width in the z direction. Formulation of the transverse resonance condition (in the y direction) for this guide yields the desired characteristic equation for β. The approximation consists in assuming that the geometrical discontinuities occurring at the sides of the waveguide are neglectable, in the sense that all higher transverse modes can be ignored and only the dominant vertical and horizontal transverse modes are considered. As a result one can formulate the transverse resonance conditions for the y and z directions separately. The condition for the z direction leads to the following characteristic equation for ϵ_{eff}:

$$\frac{\epsilon_{\text{avg}}}{\epsilon_r} \frac{k_z^{(\epsilon)}}{k_z^{(m)}} \tan(k_z^{(\epsilon)}h) = \frac{\epsilon_{\text{avg}}|k_z^{(o)}|/k_z^{(m)} - \tan(k_z^{(m)}t)}{1 + \epsilon_{\text{avg}}(|k_z^{(o)}|/k_z^{(m)})\tan(k_z^{(m)}t)} \qquad (9a)$$

with

$$|k_z^{(o)}| = k_0(\epsilon_{\text{eff}} - 1)^{1/2}$$
$$k_z^{(m)} = k_0(\epsilon_{\text{avg}} - \epsilon_{\text{eff}})^{1/2}$$
$$k_z^{(\epsilon)} = k_0(\epsilon_r - \epsilon_{\text{eff}})^{1/2}$$

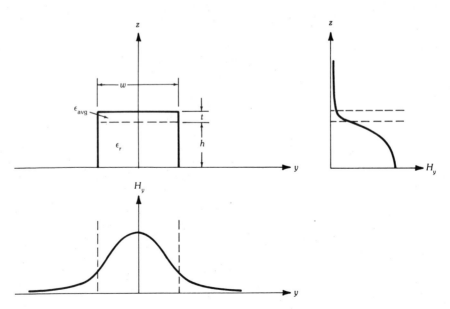

Fig. 25. Field distribution (H_y) of fundamental quasi-TM mode supported by dielectric antenna of width $w > \lambda_0\sqrt{\epsilon_{\text{eff}} - 1}$. (*After Schwering and Penn [127], © 1983 IEEE*)

and the condition for the y direction results in a second characteristic equation determining the normalized phase constant $\epsilon_{ant} = \beta^2/k_0^2$:

$$\frac{k_y^{(\epsilon)}}{|k_y^{(o)}|} \tan(k_y^{(\epsilon)}w/2) = 1 \qquad (9b)$$

with

$$|k_y^{(o)}| = k_0(\epsilon_{ant} - 1)^{1/2}$$
$$k_y^{(\epsilon)} = k_0(\epsilon_{eff} - \epsilon_{ant})^{1/2}$$

The square roots in these equations are all positive real, with the exception of $k_z^{(m)}$, which is either positive real or negative imaginary (depending on whether the field in the corrugation region is of the propagating or evanescent type). The superscripts (o), (m), and (ϵ) indicate the air region, modulation region, and dielectric region, respectively. The fundamental mode of the antenna* is associated with the lowest-order solutions of (9a) and (9b). With these solutions, ϵ_{eff} and ϵ_{ant}, determined, the phase constant of the fundamental mode is obtained as

$$\beta = k_0\sqrt{\epsilon_{ant}} \qquad (9c)$$

The fundamental mode does not have a cutoff frequency; its cross-sectional field distribution is indicated in Fig. 25, with the corrugation region of the antenna replaced by a layer of uniform permittivity ϵ_{avg}. The *effective dielectric constant* for this mode is shown in Fig. 26 as a function of the *corrugation depth t* for various values of the height h of the uniform portion of the antenna [126, 127]. The teeth and grooves of the corrugation are assumed to have equal width $d_1 = d_2 = d/2$, resulting in an average permittivity of 6.5 for the corrugation layer. Fig. 27** shows the normalized phase constant $\epsilon_{ant} = (\beta/k_0)^2$ as a function of w. Evidently, when w is large, ϵ_{ant} does not deviate significantly from ϵ_{eff} and the width w of the antenna has little effect on the phase constant β, which can be approximated by $k_0\sqrt{\epsilon_{eff}}$.

Returning to Fig. 26 it is seen now that the phase constant β of a wide antenna does not significantly vary with the corrugation depth t, provided t and h are

---

*In the form in which the characteristic equations (9a) and (9b) are presented here, they apply to the fundamental mode of the antenna and to those higher modes whose polarization and symmetry properties are similar to those of the fundamental mode. The principal field-strength components of this mode are E_z and H_y, and these components are of even symmetry with regard to the planes $y = 0$ and $z = -h$ in Fig. 24. (We assume here that the plane $y = 0$ bisects the antenna which, thus, extends laterally by $w/2$ to either side of this plane.)

**As explained above, the curves of Figs. 26 and 27 are the lowest-order (fundamental mode) solutions of the characteristic equations (9a) and (9b) which formulate the transverse resonance conditions for the z and y directions, respectively. In formulating the former condition, the antenna is temporarily extended to infinity in the y direction (see insert in Fig. 26) and the fundamental mode, whose principal field components are E_z and H_y, becomes a pure TM wave. In formulating the latter condition, the antenna is infinitely extended in the z direction (see insert in Fig. 27) and the fundamental mode becomes a pure TE wave. The purpose of the legends "TM mode" in Fig. 26 and "TE mode" in Fig. 27 is to express this situation.

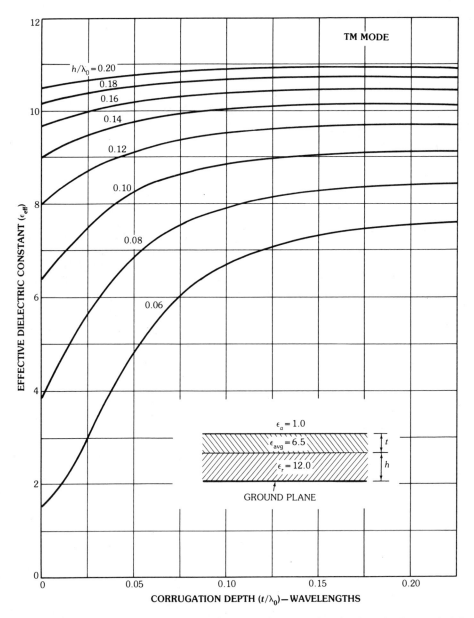

Fig. 26. Effective dielectric constant as a function of corrugation depth and substrate height for $d_1/d = 0.5$. (*After Schwering and Penn [127], © 1983 IEEE*)

sufficiently large, i.e., $t > 0.05\lambda_0$, $h > 0.1\lambda_0$. If the surface corrugation is sufficiently deep, the actual value of t is not critical.

The *grating period d* has little influence on β, but it has a determining influence on the *radiation angle* ϕ_{-1}. The dependence of ϕ_{-1} on both d and h is shown in Fig. 28 for $t = 0.05\lambda_0$ and $d_1 = d_2$. The requirement of single-beam operation is satisfied in the range below the dashed curve $\phi_{-2} = -90°$. Above this curve the $n = -2$

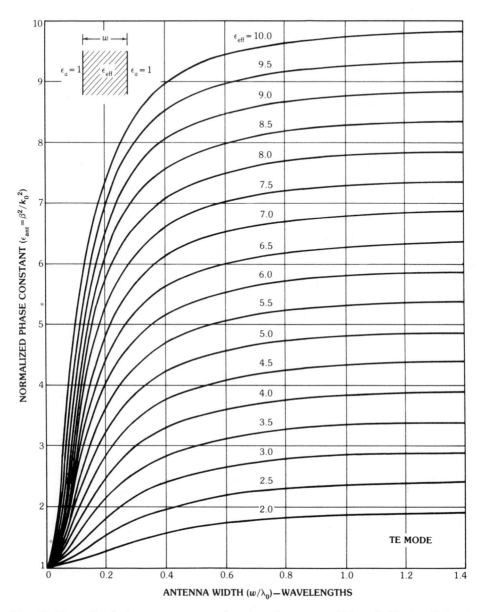

Fig. 27. Normalized phase constant as a function of antenna width and effective dielectric constant. (*After Schwering and Penn [127], © 1983 IEEE*)

space harmonic is transformed from an evanescent wave into a propagating wave, and the antenna will radiate two and eventually three and more beams simultaneously. In the range below the (solid) curve $\phi_{-1} = -90°$, where the $n = -1$ space harmonic becomes evanescent, the antenna ceases to radiate. An antenna of given dimensions h and d is characterized in Fig. 28 by a straight line through the origin. The short-dashed line, for example, represents an antenna with $d = 2.5h$,

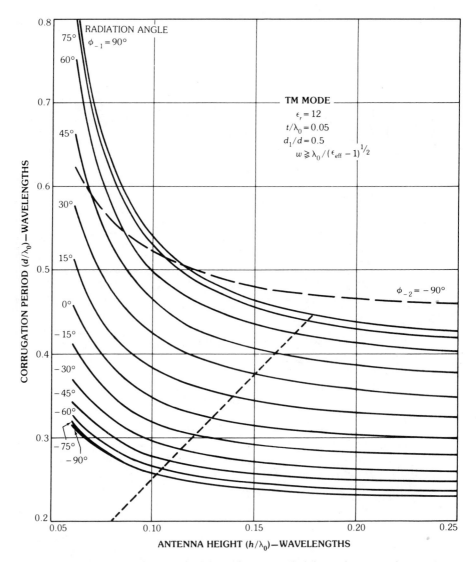

Fig. 28. Radiation angle as a function of antenna height and corrugation period.

and the frequencies associated with given scan angles ϕ_{-1} can be read conveniently from either the abscissa or the ordinate scale so that the scan characteristic of the antenna is readily established.

The *leakage constant* α of wide antennas is found by solution of the (scalar) boundary value problem for infinite width, mentioned above. The leakage constant of the fundamental leaky mode and its dependence on the antenna parameters $h, t, d,$ and d_1/d is shown in Figs. 29–32 [126–127]. For completeness, the radiation angle ϕ_{-1} is indicated in these figures by a curve or else numerically at a few discrete points. The graphs are discussed below.

Approximate closed-form expressions for $\alpha\lambda_0$ have been derived by Tamir and

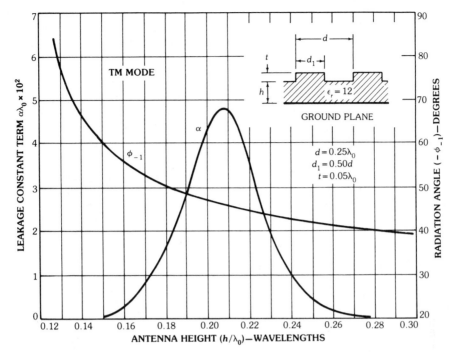

Fig. 29. Dependence of leakage constant α on antenna height. (*After Schwering and Penn [127], © 1983 IEEE*)

Peng [152]. These expressions apply to antennas with a rectangular grating profile whose teeth and grooves have equal width $d_1 = d_2 = d/2$. It is assumed, furthermore, that $\epsilon_{\text{eff}} > \epsilon_{\text{avg}}$, so that the leaky mode guided by the antenna is evanescent in the corrugation region; for a grating profile with the assumed aspect ratio of $d_1/d = 0.5$ this is usually the case. The expression for α takes a different form for small corrugation depths, where α increases proportional to t^2, and for large corrugation depths, where α reaches a saturation level and (apart from a superimposed oscillation of modest amplitude) is practically independent of t.

For $2\pi(\epsilon_{\text{eff}} - \epsilon_{\text{avg}})^{1/2}\left(\dfrac{t}{\lambda_0}\right) \lesssim 1$ in the region of small t, we have

$$\alpha\lambda_0 = \frac{2}{\sqrt{\epsilon_{\text{eff}}}}(\epsilon_r - \epsilon_{\text{eff}})(\epsilon_r - 1)\left(\frac{Q(1)}{\Gamma\tau(1)}\right)\left(\frac{t^2}{h\lambda_0}\right)$$

and for $2\pi(\epsilon_{\text{eff}} - \epsilon_{\text{avg}})^{1/2}\left(\dfrac{t}{\lambda_0}\right) \gtrsim 1$ in the region of large t, we have

$$\alpha\lambda_0 = \frac{1}{2\pi^2\sqrt{\epsilon_{\text{eff}}}}\frac{(\epsilon_r - \epsilon_{\text{eff}})(\epsilon_r - 1)^2}{(\epsilon_r - \epsilon_{\text{avg}})(2\sqrt{\epsilon_{\text{eff}}} - \lambda_0/d)}\left[\frac{Q(\epsilon_{\text{avg}})}{\Gamma\tau(\epsilon_{\text{avg}})}\right]\left(\frac{d}{h}\right)$$

where

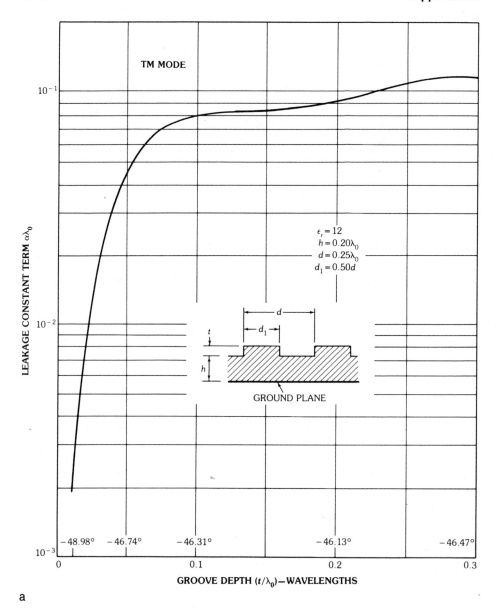

a

Fig. 30. Dependence of leakage constant α on groove depth. (*a*) For h constant. (*b*) For $h + t$ constant. (*After Schwering and Penn [127],* © *1983 IEEE*)

$$\Gamma = [\epsilon_{avg} - (\sqrt{\epsilon_{eff}} - \lambda_0/d)^2]^{1/2}$$

$$\tau(\epsilon_{avg}) = \begin{cases} 1 + \dfrac{1}{k_0 h(\epsilon_{eff} - \epsilon_{avg})^{1/2}} & \text{for TE modes} \\[3ex] 1 + \dfrac{1}{k_0 h(\epsilon_{eff} - \epsilon_{avg})^{1/2}} \cdot \dfrac{1}{(\epsilon_{eff}/\epsilon_r \epsilon_{avg})(\epsilon_r - \epsilon_{avg}) - 1} & \text{for TM modes} \end{cases}$$

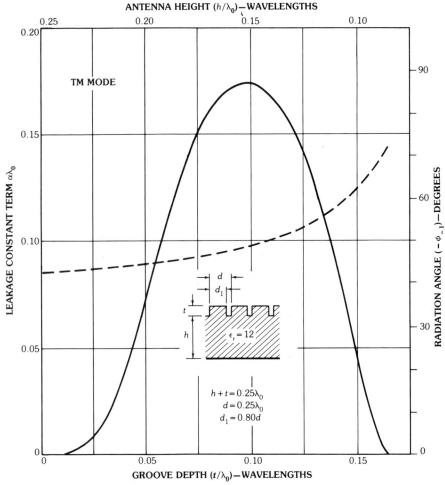

Fig. 30, *continued.*

$$Q(\epsilon_{avg}) = \begin{cases} 1 & \text{for TE modes} \\ \dfrac{\epsilon_r - \epsilon_{avg}}{\epsilon_r \epsilon_{avg}} \cdot \dfrac{\Gamma^2(\epsilon_r - \epsilon_{avg}) + \epsilon_{eff}(\sqrt{\epsilon_{eff}} - \lambda_0/d)^2}{(\epsilon_r - \epsilon_{avg}) + (\epsilon_{avg}/\epsilon_r)^2(\epsilon_r - \epsilon_{eff})} & \text{for TM modes} \end{cases}$$

The expression for Γ holds in both regions. The expressions for τ and Q apply to the formula for $\alpha\lambda_0$ for large t. In the formula for small t, ϵ_{avg} in τ and Q is replaced by unity (except where it occurs in Γ).

Comparison with values for $\alpha\lambda_0$ obtained by numerical solution of the scalar boundary value problem mentioned earlier has shown that the approximate formulas yield results of good accuracy for antennas of low permittivity in the order of 2 to 4. For high-permittivity antennas with $\epsilon_r = 10$ to 12, the formulas may still

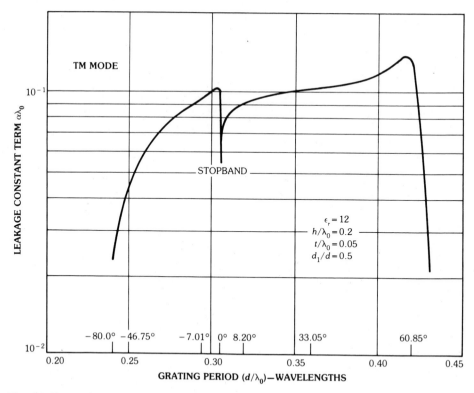

Fig. 31. Dependence of leakage constant α on grating period. (*After Schwering and Penn [127], © 1983 IEEE*)

be used in the range of small t, but they become inaccurate for large t, leading to errors by a factor of 2 or greater [125].

For low-permittivity antennas the dependence of α on the aspect ratio d_1/d approaches a sine-squared relationship with the maximum occurring at $d_1/d = 0.5$, the value assumed in the above formulas. We have in good approximation [152]

$$\alpha\left(\frac{d_1}{d}\right) = \alpha\left(\frac{1}{2}\right)\sin^2\left(\pi\frac{d_1}{d}\right)$$

For antennas of large permittivity and large corrugation depths the sine-squared relationship is no longer valid; the design curves [126, 127] shown below indicate that for $\epsilon_r = 10$ to 12 the maximum of α is shifted to $d_1/d = 0.7$ to 0.8.

Design Guidelines for Wide Dielectric Grating Antennas—An efficient grating antenna must have an axial length L sufficiently large so that a major portion of its input power is radiated before it reaches the antenna termination at the far end, where the remaining power must be absorbed to avoid pattern distortions. To obtain high efficiency at reasonable antenna length the leakage constant must be sufficiently large. This, in turn, implies the use of high-permittivity materials, such

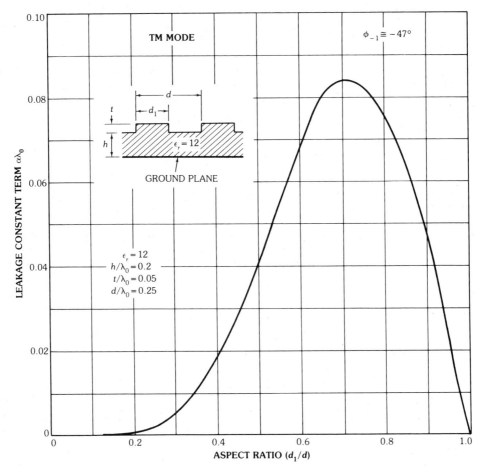

Fig. 32. Dependence of leakage constant α on aspect ratio of surface corrugation. (*After Schwering and Penn [127], © 1983 IEEE*)

as silicon or gallium arsenide [125]. Under optimized conditions, $\alpha\lambda_0$ is typically in the order of 0.1 in this case, while for lower-permittivity materials, such as boron nitride and polystyrene, $\alpha\lambda_0$ would be an order of magnitude smaller.

Neglecting dielectric losses, the antenna efficiency is given by

$$\eta = 1 - e^{-2\alpha L} \tag{10}$$

If $\alpha\lambda_0 = 0.1$, then

$$\frac{L}{\lambda_0} = \begin{cases} 11.5 & \text{for } \eta = 90 \text{ percent} \\ 23 & \text{for } \eta = 99 \text{ percent} \end{cases}$$

which leads to reasonable antenna dimensions even in the lower millimeter-wave region near $\lambda_0 = 1$ cm. The corresponding *E*-plane beamwidth (*xz* plane) is in the order of a few degrees, which is appropriate for most millimeter-wave applications.

In order to maximize the leakage constant α, the antenna dimensions should be chosen according to the following guidelines [126, 127]:

(a) The *height* h/λ_0 of the uniform portion of the antenna structure should be in the range 0.15 to 0.21 for $\epsilon_r = 12$. The upper value corresponds to an aspect ratio of the corrugation profile $d_1/d = 0.5$, the lower value to $d_1/d = 0.8$; see Figs. 29 and 30b.

The height h is a critical design parameter in maximizing α, as is easy to understand. If h is small, most of the guided energy travels in the air region outside the antenna and the surface corrugation will cause little radiation. In the opposite case that h is large the energy will be confined primarily to the interior of the antenna and again the surface corrugation has little effect. In the intermediate range, however, there must be a height h for which the energy density in the corrugation region reaches a peak value and, in this situation, α will be at a maximum.

(b) The *depth* t of the surface corrugation should be chosen sufficiently large, but it is not a critical design parameter. Fig. 30a shows that in the range of small groove depths the leakage constant increases significantly with t, but that it reaches a saturation value at large t. For an antenna with $\epsilon_r = 12$ the normalized groove depth t/λ_0 should be equal to or greater than 0.05. Under this condition the phase constant β also shows little variation with t; see Fig. 26.

In Fig. 30a it is assumed that the height h of the uniform waveguide portion of the antenna remains constant while t varies. For later reference, Fig. 30b shows α for the alternate case that the *total* height of the antenna, $h + t$, is fixed. Hence h decreases as t is increased, and α reaches a peak at the value of h for which the field intensity in the corrugation section is at a maximum. This optimum h is approximately $0.15\lambda_0$ for $d_1/d = 0.8$, the aspect ratio assumed in the figure. In other words, the strong variation of the α curve reflects its dependence on h rather than t.

(c) The *period* of the surface corrugation, d, is determined primarily by the desired radiation angle and the condition (6) for single-beam operation, but it is not a parameter available for maximizing α. Fig. 31 shows, moreover, that over much of the angular range, α does not vary strongly with d/λ_0, and is fairly constant over a wide range of d values. The stopband which occurs when broadside conditions are approached is very narrow.

(d) The *aspect ratio* d_1/d of the surface corrugation should be chosen close to 0.7, i.e., the width of the grooves should be significantly smaller than that of the teeth; see Fig. 32. This is a consequence of the high permittivity considered here. For low-permittivity antennas, maximum energy leakage would occur near $d_1/d = 0.5$.

(e) As pointed out before, the effect of the *lateral width* w of the antenna on α is small for $w > \lambda_0/\sqrt{\epsilon_{eff} - 1}$, the case considered here.

Design Procedure for Narrow Dielectric Grating Antennas—In the above discussion on periodic dielectric antennas of *large width* ($w > \lambda_0/\sqrt{\epsilon_{eff} - 1}$) it was pointed out that the propagation constant of these antennas can be obtained simply and accurately by the use of the EDC method, and that their leakage constant is practically the same as that of an antenna of infinite width (with the leaky modes traveling in the direction normal to the surface grooves). This equivalence reduces the computational burden substantially. Except for $w \to \infty$, the infinite antenna has the same geometrical parameters and the same permittivity as the finite antenna. The discussion has shown, furthermore, that α depends strongly on the fraction of the guided energy traveling within the corrugation region of the antenna.

For antennas of *narrow width* ($w < \lambda_0/\sqrt{\epsilon_{eff} - 1}$), the propagation constant can still be obtained in good accuracy by the EDC procedure, but the leakage constant can no longer be approximated by that of an infinitely wide antenna with the same permittivity as the finite antenna. With decreasing w, the phase velocity of the leaky modes guided by the actual antenna increases; furthermore, an increasing portion of the guided energy will now travel in the air region on both sides of the antenna, thus reducing the energy density within the corrugation region. Both effects combine to decrease the leakage constant of the antenna of narrow width.

However, an equivalent antenna of infinite width yielding approximately the same α as the narrow antenna can still be defined with the help of the EDC procedure [146] which allows one to replace an antenna of finite width by an equivalent antenna of infinite width having an effective dielectric constant which is smaller than the permittivity of the original antenna.* The decreased permittivity reduces the perturbation of the guided wave caused by the surface modulation and hence the leakage constant of the infinite antenna. The geometrical parameters of this equivalent antenna are the same as those of the original antenna (apart, of course, from w). There is a great advantage in the availability of an equivalent antenna of infinite. width. As pointed out before, for $w \to \infty$, the leaky modes become pure TE or TM modes, depending on their polarization, so that the boundary value problem, from which α is determined, is reduced from a vector problem to a much simpler scalar problem. Since, in addition, the equivalent antenna has the same height and corrugation profile as the original antenna, the formulas and graphs for α and β, already available and discussed before, can be readily utilized.

In order to determine the equivalent structure of infinite width we use the EDC procedure to carry out a transverse resonance in the lateral (y) direction [146]. In doing so, we temporarily extend the structure in the vertical direction to positive and negative infinity, producing a *uniform* dielectric layer, as shown in the inset of Fig. 33a. The transverse resonance then results in the dispersion relation (9b), which yields dispersion curves of the form shown in Fig. 33a as a function of the width w of the fictitious uniform layer and therefore of the original antenna.** The

---

*Note, however, that the effective dielectric constant is determined from the condition that both antennas have the same *phase* constant.

**Note the difference in the way in which the effective dielectric constant is determined here and in the section on the antennas of large width. In that section the effective permittivity is calculated from the transverse resonance condition in the z direction, while here it is determined from the resonance condition for the y direction. To signify this difference the symbol $\bar{\epsilon}_{eff}$ is used here instead of ϵ_{eff}. This also means that in calculating $\bar{\epsilon}_{eff}$ from (9b), the quantities $k_y^{(o)}$ and $k_y^{(\epsilon)}$, which appear on the left side of this equation, must be rewritten as

$$|k_y^{(o)}| = k_0(\bar{\epsilon}_{eff} - 1)^{1/2} \quad \text{and} \quad k_y^{(\epsilon)} = k_0(\epsilon_r - \bar{\epsilon}_{eff})^{1/2}$$

The equation then applies to TE polarization, i.e., the polarization associated with the fundamental mode. The dispersion relation for TM polarization is obtained if, in addition, the left side of (9b) is multiplied by $1/\epsilon_r$.

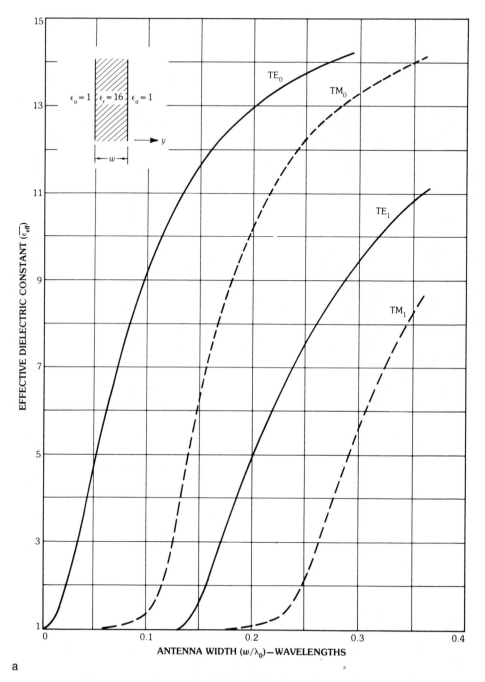

a

Fig. 33. Approximate procedure for calculation of the leakage constant of dielectric grating antennas of narrow width. (*a*) Dispersion curves for uniform dielectric slab, to be used in obtaining the equivalent antenna of infinite width. (*b*) Comparison between theoretical values and experimental results for the leakage constant of a 30-GHz antenna of $w = 1.3$ mm and $\epsilon_r = 16$. (*After Peng et al. [146],* © *1984 IEEE*)

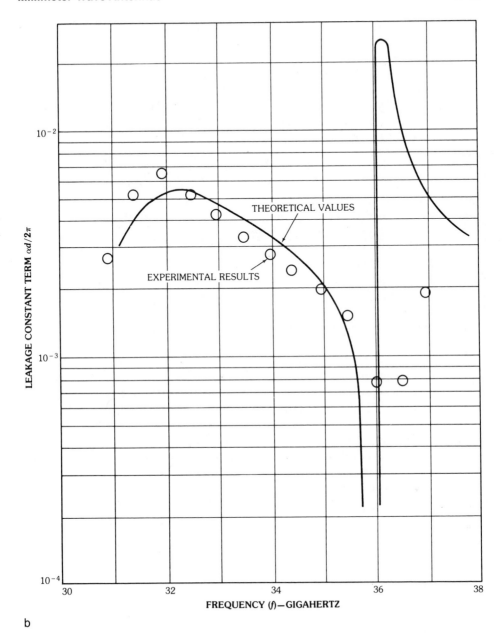

LEAKAGE CONSTANT TERM $\alpha d/2\pi$

THEORETICAL VALUES

EXPERIMENTAL RESULTS

FREQUENCY (f)—GIGAHERTZ

b

Fig. 33, *continued.*

actual curves appearing in Fig. 33a correspond to a dielectric constant of 16 for the layer, for both TE polarization (solid lines) and TM polarization (dashed lines). The ordinate is $\bar\epsilon_{eff}$, the effective dielectric constant of the equivalent antenna of infinite width.

To show how the curves are to be used in the present context, consider as an example the fundamental leaky mode for which the main electric field component is oriented in the vertical (z) direction. We have, therefore, a TE mode transverse resonance in the lateral (y) direction. For an antenna width $w = 0.11\lambda_0$, the value of $\bar{\epsilon}_{\text{eff}}$ for the lowest TE mode is obtained from Fig. 33a to be 9.25. This (reduced) permittivity implicitly accounts for the effect of the antenna width, and the original antenna of actual finite width w and dielectric constant $\epsilon_r = 16$ may be replaced by an antenna of *infinite* width which has the *effective* dielectric constant of 9.25.

In determining the phase constant of antennas of finite width the EDC method can be used with reasonable confidence. But it is not immediately obvious that the α values obtained by the present approach will show a similar degree of accuracy, and the usefulness of the method requires experimental confirmation. Fig. 33b presents a comparison between theoretical values for α and some experimental data obtained with a 30-GHz antenna of width $w = 1.3$ mm and dielectric constant $\epsilon_r = 16$. The leakage constant is normalized to the period d of the corrugation and is plotted as a function of frequency. It is evident that the agreement is rather good, except in the vicinity of the sharp dip and rise in the theoretical curve. That region corresponds to broadside radiation, at which a leaky-wave "stopband" occurs, as mentioned above in connection with Fig. 31. The agreement is not expected to be good there since the sharp behavior predicted by the theoretical structure of infinite length would be greatly softened by the finite length of the practical structure on which the measurements were made.

Design Information for Antennas With Metal Gratings—The design guidelines presented above apply to antennas with dielectric gratings. Experimental studies which have been reported so far have been concerned mostly with antennas with metal gratings [122–124, 129–131]. A systematic comparison of these two versions of periodic antennas is not as yet available. It appears, however, that metal gratings can be designed to produce substantially stronger energy leakage. Thus, as a general rule: While dielectric grating antennas should be configured for maximum α, antennas with metal gratings should be designed for "reasonable" radiation. Important design parameters are the height h of the antenna and the width s of the metal strips forming the grating; see Fig. 34. The height h should be chosen to

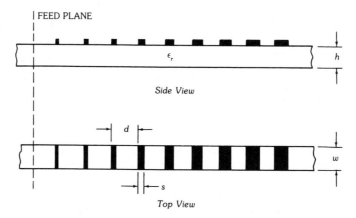

Fig. 34. Antenna with metal grating of variable strip width s.

ensure that a large field intensity exists on the grating surface, and the strip width s can then be adjusted to obtain a desired radiation rate. The optimum height is, thus, roughly the same as in the case of an antenna with a dielectric grating, though it should be taken into account that a metal grating tends to have a stronger influence on the field distribution in the antenna than a surface corrugation, in particular if the metal strip width s is large.

Trinh and others have shown in a recent experimental study [130, 131] that the optimum strip width s is approximately $0.4\lambda_g$, where $\lambda_g = \lambda_0(k_0/\beta)$ is the guide wavelength. If s is much smaller, i.e., less than $0.2\lambda_g$, a noticeable amount of residual radiation in the end-fire direction has always been observed, even for long antennas (50 strips). If the metal strips are too wide, $s > 0.5\lambda_g$, a major portion of the guided power is radiated by the first few strips and the effective antenna aperture is very small, leading not only to a large beamwidth but to high side lobes, which are probably due to a large mismatch at the waveguide-to-antenna transition [130].

Furthermore, it was found that linear tapering of s from period to period, i.e., starting with a small strip width at the feed point and gradually increasing s until it reaches the final value of $0.4\lambda_g$, substantially reduces the side lobes to a level less than -25 dB below the main beam [129]. An optimization study [130, 131] has resulted in the following empirical relation for the width of the nth strip:

$$s_n = \begin{cases} [0.145 + 0.015(n - 1)]\lambda_g & \text{for } n \leq 18 \\ 0.4\lambda_g & \text{for } n > 18 \end{cases}$$

The first few strips in this case produce little radiation since their function is apparently limited to providing an impedance match in the transition region.

Similar improvements in pattern quality can be expected for antennas with dielectric gratings, when the groove width or depth are appropriately varied from period to period. Design procedures for such gratings are discussed under the heading "Pattern Shaping."

Beam Scanning—The scan performance of antennas with dielectric gratings may be seen from Fig. 28. It is discussed here in more detail for an antenna with the specific dimensions indicated in Fig. 35 [125]. These dimensions apply to beam scanning in the 3-mm band (94 GHz) but can be scaled, of course, to any other frequency region. The figure shows the radiation angle ϕ_{-1} and the leakage constant α for two antennas, one designed for radiation in the forward direction ($d = 1.05$ mm), the other for radiation in the backward direction ($d = 0.81$ mm). For a 10-percent variation in wavelength, say from $\lambda_0 = 2.83$ mm to 3.15 mm, the main beam direction sweeps almost linearly over an angular range of about 25° for backward radiation and 20° for forward radiation. In both cases the leakage constant reaches a maximum near $\lambda_0 = 3$ mm and decreases almost symmetrically when λ_0 is varied in either direction. This is a very desirable characteristic. If one designates $\lambda_0 = 3$ mm as the center wavelength of a frequency scan, the radiation constant and hence the beamwidth will change comparatively little as the antenna is scanned and a stable radiation pattern with little beam degradation can be expected [125].

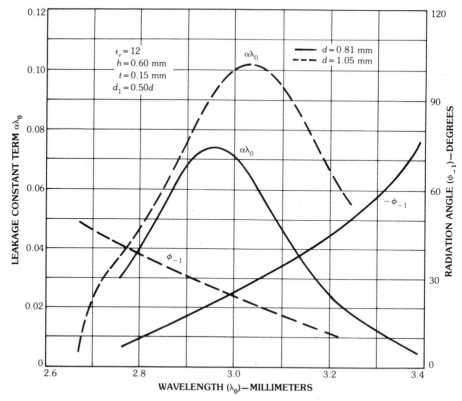

Fig. 35. Scan performance of dielectric grating antennas: variation of radiation angle ϕ_{-1} and leakage constant α with wavelength.

Recent experiments indicate that the theoretical scan rate of 20° to 25° per 10-percent change in frequency is conservative. For grating antennas made from silicon ($\epsilon_r = 12$) and magnesium titanate ($\epsilon_r = 16$), a scan rate of approximately 40° per 10-percent frequency variation was measured [122, 128]. The silicon antenna used a metal grating while the magnesium titanate antenna operated with a dielectric grating with a rectangular profile of large aspect ratio d_1/d.

Under the heading "Review of Fundamentals" it was pointed out that periodic dielectric antennas do not radiate in the broadside direction; as broadside conditions are approached, an internal resonance develops which prohibits radiation. According to a method proposed by James and Hall [155] this problem can be alleviated, however, by the use of a surface modulation whose elements are *pairs* of grooves or metal strips rather than *single* scatter elements. The spacing between the elements of each pair is equal to a quarter guide wavelength at the design frequency. In this case the wave reflected by the first element of each pair will nearly be canceled by the wave reflected from the second element, so that the total reflected power is greatly reduced. The pair-to-pair spacing (i.e., the overall period of the surface modulation), of course, will be chosen at one guide wavelength at the design frequency in order to obtain broadside radiation.

The validity of the method has been confirmed experimentally by Solbach and

Adelseck [156], who used an image-guide antenna made from Duroid, which was designed to operate as a frequency-scanned antenna over the 60- to 75-GHz band. A periodic surface perturbation was provided by a sequence of 27 pairs of small, circular metal disks placed on the upper surface of the guide. The measured reflection coefficient remained below −20 dB over most of the band, but showed an increase to about −12 dB in the range of broadside radiation near 70 GHz. While this increase still indicates the existence of a "stopband," the absolute value of the reflection coefficient near 70 GHz remained by an order of magnitude below that measured for a similar antenna using a sequence of single disks. Furthermore, the gain of the new antenna showed a smooth frequency dependence and did not exhibit a noticeable dip near broadside.

For many applications an antenna would be desirable which permits electronic beam scanning at constant frequency, i.e., by other means than frequency variation. A promising approach, suggested by Jacobs, is indicated in Fig. 36. The basic idea is to vary the phase constant β of the antenna, and thus its radiation angle, by electronically controlling the antenna cross section. This is accomplished with the help of semiconducting fins attached to the antenna on one or both sides. In a recent experimental study [123, 124], these fins were made from silicon (same ϵ_r as antenna) which was processed so that the fins could be operated as distributed pin diodes. Through biasing currents applied with the help of electrodes placed on the top and bottom of the diodes one can control the conductivity of the fins and hence the depth to which the fields guided by the antenna extend into the fin region. In this way one can adjust the effective cross section of the antennas and, in turn, its phase constant and radiation angle.

Conduction losses suffered by the fields extending into the fin region reduce antenna efficiency. But these losses are small in the two limiting cases of very high and very low conductivity (i.e., full bias and off bias) so that the method appears particularly suited for digital beam scanning.

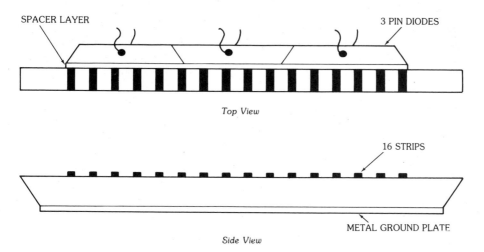

Fig. 36. Antenna with metal grating and distributed pin diodes for electronic beam scanning at constant frequency. (*After Horn et al. [123], © 1980 IEEE*)

The experimental antenna [123, 124] was designed for the 60-GHz band and consisted of a 1-mm × 1-mm silicon rod with 16 metal strips. With three pin diodes attached to one side, a scan range of 8° to 10° was obtained at low power consumption (1 V × 100 mA per diode). In some cases the beam shift was as large as 15°. To minimize losses the diodes were separated from the antenna by a thin layer of a low-permittivity insulating material.

An alternate approach to beam steering at constant frequency has been suggested by Bahl and Bhartia [157]. This approach would use a periodic antenna structure containing a liquid artificial dielectric medium whose permittivity can be controlled by a biasing electric field. The practicality of this approach remains to be established, but a theoretical study using data for existing artificial dielectrics has predicted good performance for the structure under investigation, i.e., a large scan range of 40° to 50° for a moderate change in permittivity, and a practically constant beamwidth.

Radiation Pattern—The main-beam direction of a dielectric-grating antenna is found from (7) once the phase constant β is known. Calculation of the radiation pattern is a more difficult task. An approximate method is described in the following that should yield fairly accurate results for the main beam and near-in side lobe regions. Neglected effects, however, may strongly influence the radiation pattern in the far side lobe region.

Fig. 37 shows a grating antenna of length L and width w, where it is assumed that $L \gg \lambda_0$ and $w > \lambda_0/\sqrt{\epsilon_{\text{eff}} - 1}$. The terms θ and ϕ are the angular coordinates of a spherical coordinate system whose axis coincides with the y axis. The tangential magnetic field distribution of the fundamental leaky mode in the antenna aperture, i.e., in the portion of the plane $z = t$ just above the antenna, can be expressed in terms of the space-harmonics representation of this mode. We have

$$H_y(x, y, t) = e^{-\alpha x} \sum_{n=-\infty}^{+\infty} I_n e^{-j\beta_n x} \cos(\pi y/w) \tag{11}$$

for $0 \leq x \leq L$, $-w/2 \leq y \leq +w/2$, with $\beta_n = \beta + 2\pi n/d$. The y dependence is approximate but should be of good accuracy since it is assumed that $w > \lambda_0/\sqrt{\epsilon_{\text{eff}} - 1}$. For the same reason the x component of the magnetic field strength is expected to be small compared with the y component and has been neglected.* In the air half-space the space harmonics are decoupled and travel as independent waves. For single-beam operation, the practically interesting case, the space harmonic of order $n = -1$ is the only wave of the propagating type, while all other space harmonics are evanescent, i.e., they decrease exponentially in the z direction. Hence, for the purpose of determining the radiation pattern of the antenna, the field distribution in the plane $z = t$ may be approximated by that of the $n = -1$ space harmonic alone:

---

*The x component of the aperture illumination, $H_x(x, y, t)$, determines the cross-polarized part of the radiation pattern. It is antisymmetric in y so that the cross-polarized radiation vanishes in the E-plane, but it is different from zero elsewhere. For antennas of narrow width the cross-polarized radiation would have to be taken into account since H_x can be significant when $w < \lambda_0/\sqrt{\epsilon_{\text{eff}} - 1}$.

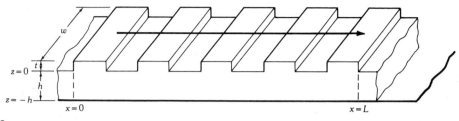

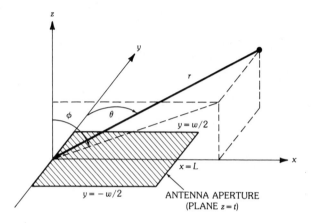

Fig. 37. Radiation pattern calculation. (*a*) Antenna geometry. (*b*) Antenna aperture and spherical coordinate system. (*After Schwering and Penn [127], © 1983 IEEE*)

$$H_y(x, y, t) = \begin{cases} I_{-1}e^{-\alpha x}\, e^{-jk_0 x \sin\phi_{-1}} \cos(\pi y/w) & \text{for } 0 \leq x \leq L, \quad |y| \leq w/2 \\ 0 & \text{elsewhere} \end{cases} \tag{12}$$

Equation 7 was used here to replace $\beta_{-1} = \beta - 2\pi/d$ by $k_0 \sin\phi_{-1}$. For a discussion of the approximations involved in (12) we refer to the literature [127].

The power radiation pattern associated with aperture distribution (12) is given by

$$G(\theta, \phi) = G_D S(\cos\theta)\, T(\sin\phi \sin\theta)$$

where

$$G_D = \frac{16}{\pi^3} k_0^2 \frac{w}{\alpha} \tanh\left(\frac{\alpha L}{2}\right) \cos\phi_{-1} \tag{13a}$$

is the directivity gain of the antenna,

$$S(\cos\theta) = \left(\frac{\pi}{2}\right)^4 \frac{\cos^2[(k_0 w/2)\cos\theta]}{\{(\pi/2)^2 - [(k_0 w/2)\cos\theta]^2\}^2} \sin^2\theta \tag{13b}$$

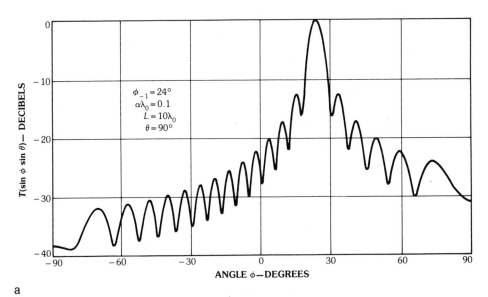

a

Fig. 38. Theoretical *E*-plane radiation patterns of dielectric grating antennas. (*a*) Antenna length $L = 10\lambda_0$. (*b*) Antenna length $L = 20\lambda_0$. (*c*) Antenna length $L = 150\lambda_0$. (*After Schwering and Penn [127], © 1983 IEEE*)

is the *H*-plane pattern, and

$$T(\sin\phi\sin\theta) = \left(\frac{\alpha L}{1 - e^{-\alpha L}}\right)^2 \frac{1 - 2e^{-\alpha L}\cos[k_0 L(\sin\phi\sin\theta - \sin\phi_{-1})] + e^{-2\alpha L}}{(\alpha L)^2 + (k_0 L)^2(\sin\phi\sin\theta - \sin\phi_{-1})^2}$$

(13c)

determines the *E*-plane pattern. Both *S* and *T* are normalized to unity in the main beam direction $\theta = 90°$, $\phi = \phi_{-1}$.

The directivity gain is proportional to the antenna aperture area *wL* in the case that $\alpha L \ll 1$, where the aperture illumination is practically uniform (and the antenna efficiency is low). In the opposite case that $\alpha L \gg 1$, the effective aperture area is determined by the decay constant α and independent of *L*; we have $G_D \propto w/\alpha$.

The H-plane pattern is the well-known radiation characteristic of a cosine-tapered aperture distribution. The 3-dB beamwidth is $\Delta\theta = 1.2\lambda_0/w$ (in radians) and first side lobes are down by −23 dB. The *E*-plane pattern is that of an exponentially tapered source distribution. Figs. 38a through 38c show this pattern for $\alpha\lambda_0 = 0.1$ and $L/\lambda_0 = 10$, 20, and 150. The corresponding antenna efficiencies are $\eta = 86.5$, 98.2, and approximately 100 percent; see (10). The beamwidths of these patterns are in the order of a few degrees, which is appropriate for most millimeter-wave applications. A smaller beamwidth can be realized by designing the antenna to have a smaller leakage constant α and by correspondingly increasing the antenna length *L*. Designing for a broader beamwidth would be more of a problem. Either α would have to be raised, which is not easily done since the assumed value

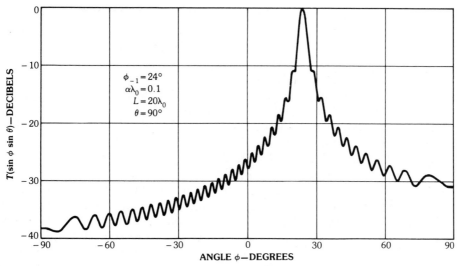

b

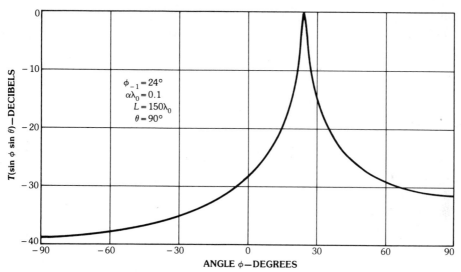

c

Fig. 38, *continued.*

of $\alpha\lambda_0 = 0.1$ is already close to maximum for antennas with $\epsilon_r = 12$, or the antenna length would have to be reduced at the expense of a decreased efficiency. The problem can be solved by the use of a metal grating instead of a dielectric grating. It was mentioned earlier that a metal grating will produce a strong leakage constant when the width of the metal strips is chosen sufficiently large.

The first side lobes of the E-plane pattern are down by only -13 dB. The high side lobe level is caused by the discontinuity in the aperture distribution (12) at the termination $x = L$. Note, however, that as the antenna length is increased and the discontinuity at $x = L$ is reduced, the side lobes degenerate into small shoulders only. For $\alpha L \gg 1$, Fig. 38c, the side lobes completely disappear and, in the approximation used here, a very smooth pattern is obtained.

Pattern Shaping—Pattern shaping implies control of the field distribution along the radiating aperture of the antenna. Usually a linear phase progression is required to obtain a narrow main beam, and a tapered amplitude distribution to achieve low side lobes. Periodic antennas provide the desired linear phase progression, and their amplitude distribution can be controlled by varying the groove depth or the aspect ratio of the surface corrugation from period to period. While this variation will have a strong effect on the leakage constant, which will become a function of x, the phase constant will not be significantly affected as long as the corrugation period remains a constant, the antenna height is sufficiently large ($h + t > 0.2\lambda_0$), and the maximum excursions of t and d_1 remain within reasonable limits. Under these conditions the linear phase variation across the antenna aperture will be preserved while tapering of the leakage constant will permit beam shaping through modification of the amplitude distribution.

Thus, assuming that α varies with x while β remains constant, equation (12) for the field distribution of the $n = -1$ space harmonic in the aperture plane of the antenna is modified by replacing the exponential amplitude distribution $e^{-\alpha x}$ by the more general function

$$F(x) = F(0)\left\{\frac{a(x)}{a(0)}\right\}^{1/2} \exp\left(-\int_0^x a(x)\,dx\right) \tag{14}$$

which applies when $\alpha = \alpha(x)$. The phase factor $\exp(-jk_0 x \sin\phi_{-1})$ in (12) remains unchanged since the radiation angle ϕ_{-1} depends on only β and d, which are not varied. Hence

$$H_y(x,y,t) = \begin{cases} I_{-1}F(x)\,e^{-jk_0 x \sin\phi_{-1}}\cos(\pi y/w) & \text{for } 0 \leq x \leq L, \quad |y| \leq w/2 \\ 0 & \text{elsewhere} \end{cases} \tag{15}$$

Assuming that the antenna is sufficiently long and the variation of α from period to period is sufficiently gradual, the leakage constant is treated here as a continuous function rather than a step function of x. The directivity gain and the E-plane radiation pattern associated with aperture illumination (15) are obtained from conventional antenna theory as

$$G_D = \left(\frac{2}{\pi}\right)^3 k_0^2 w \, \frac{\left(\int_0^L F(x)\,dx\right)^2}{\int_0^L F(x)^2\,dx} \cos\phi_{-1} \tag{16a}$$

$$T(\sin\phi\sin\theta) = \left|\frac{\displaystyle\int_0^L F(x)\exp\{jk_0x(\sin\phi\sin\theta - \sin\phi_{-1})\}\,dx}{\displaystyle\int_0^L F(x)\,dx}\right|^2 \tag{16b}$$

The H-plane pattern is independent of $F(x)$ and, as before, determined by (13b).

When the E-plane pattern $T(\sin\phi\sin\theta)$ is prescribed, the aperture illumination $F(x)$ is, in principle, a known function of x, which can be determined from (16b), to good accuracy, by standard methods. The distribution of the leakage constant $\alpha = \alpha(x)$, which will produce a given $F(x)$, is found by inversion of (14). It is not difficult to show that

$$a(x) = \frac{1}{2}\frac{F^2(x)}{F(0)^2/2a(0) - \displaystyle\int_0^x F^2(x)\,dx} = \frac{1}{2}\frac{F^2(x)}{(1/\eta)\displaystyle\int_0^L F^2(x)\,dx - \displaystyle\int_0^x F^2(x)\,dx} \tag{17}$$

where η is the antenna efficiency, i.e., the ratio of the radiated power to the input power of the antenna. The variation of groove depth or aspect ratio from period to period, which will yield a given α profile, can be read from the curves $\alpha = \alpha(t/\lambda_0)$ and $\alpha = \alpha(d_1/d)$ as they are shown in Figs. 30 and 32 for the example of an antenna with $\epsilon_r = 12$ and $d = 0.25\lambda_0$. Together with (17) these curves provide the design information necessary for pattern shaping of dielectric grating antennas.

Fig. 39 shows an example. An aperture illumination with a cosine-squared amplitude taper

$$F(x) = \cos^2\left[\frac{\pi}{2}\left(\frac{2x}{L} - 1\right)\right], \qquad 0 \le x \le L \tag{18}$$

is a desirable aperture distribution; it produces a radiation pattern with side lobes below -30 dB. The function $F(x)$ is indicated in Fig. 39a. Corresponding α profiles for antennas with efficiencies of 75 and 95 percent are shown by the solid curves in Figs. 39b and 39c, respectively. Note that these curves do not show the same symmetry as the aperture illumination $F(x)$. The energy density of the field traveling along the antenna decreases with x due to radiation, and the maximum of the α profile must be shifted to the right to compensate for the decreasing field amplitude and, thus, to produce the desired symmetric aperture illumination.

Assuming that the α profile of Fig. 39b is realized by varying the aspect ratio of the surface corrugation, and the α profile of Fig. 39c is obtained by tapering the groove depth, the dashed curves in these figures show the corresponding d_1 and t variations along the antenna aperture. The curves were obtained from Figs. 30b and 32, and they apply to a silicon antenna having a total height $h + t = 0.25\lambda_0$ and a length $L = 25\lambda_0 = 100d$. The main beam direction ϕ_{-1} is $-47°$. As in the case of α, the d_1 and t profiles are shown as continuous rather than as step functions of x. Fig. 39c suggests that a simple triangular groove depth distribution has the capability of providing very low side lobes.

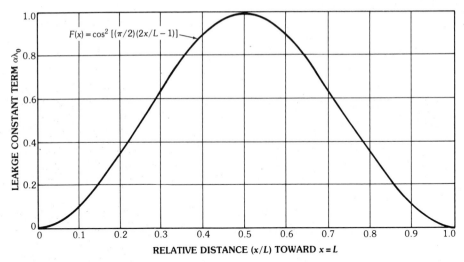

a

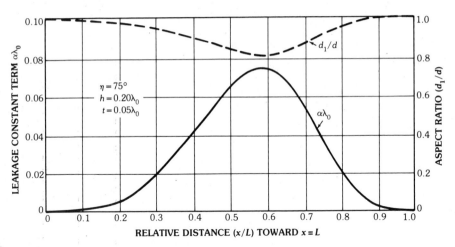

b

Fig. 39. Control of amplitude distribution in aperture plane of periodic dielectric antenna. (a) Cosine-squared amplitude distribution. (b) Leakage constant α and aspect ratio d_1/d versus x for antenna of efficiency $\eta = 75$ percent. (c) Leakage constant α and groove depth t/λ_0 versus x for antenna of efficiency $\eta = 95$ percent.

   Close inspection shows that the radiation angle ϕ_{-1} changes slightly with d_1 and t. These deviations could, in principle, be compensated by introducing an appropriate moderate variation in the period of the surface corrugation as well. But in view of the other approximations involved in the pattern calculations, it is not clear how much improvement in pattern quality could be gained by such refinements.

   While grating antennas produce E-plane patterns of narrow beamwidth, their H-plane patterns are usually not very directive since the width w of the antennas is

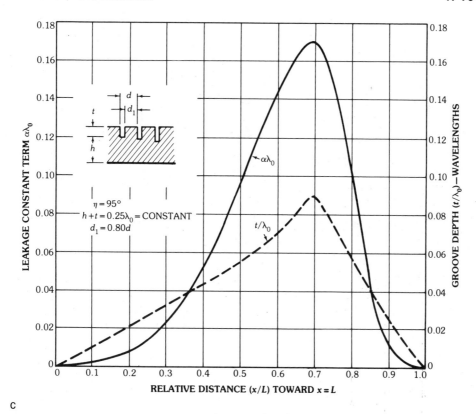

c

Fig. 39. *Continued*

limited to prevent the occurrence of higher-order modes. A small beamwidth in the *H*-plane can be realized by using an array of several antennas in parallel as indicated in Fig. 40. Separating the antennas by metal fins will minimize mutual coupling. The use of arrays has the additional advantage that the excitation of the individual antennas can be controlled independently, thus permitting scanning of the array pattern both in the *E*-plane (by frequency variation) and in the *H*-plane (by phasing).

An alternate method for enhancing *H*-plane directivity has been suggested by Trinh and others [130, 131]. A single grating antenna is embedded in a rectangular metal trough with flares attached to each side. The resulting *H*-plane sectoral horn configuration is shown in Fig. 41; the dimensions apply to *E*-band (81.5 GHz) where this concept was tested. Directivity can be maximized by appropriate choice of the flare angle γ, which should be approximately 15° in the case of near broadside radiation. The experimental antenna used a dielectric guide of permittivity $\epsilon_r = 2.47$ with a grating of 32 metal strips and a guide wavelength of 2.7 mm. The gain was measured to be 29 dB and side lobes were at least −25 dB below the main beam. The *E*- and *H*-plane half-power beamwidths were 4° and 13°, respectively.

A closely related approach which permits one to reduce the *H*-plane beamwidth even further would use a periodic antenna of small width as a feed for a

Fig. 40. Array of dielectric grating antennas.

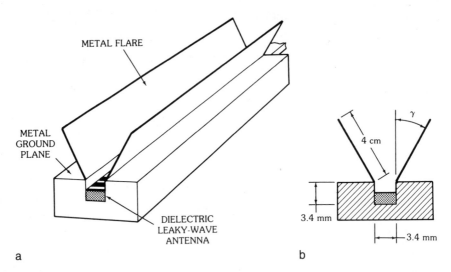

Fig. 41. Dielectric grating antenna of enhanced *H*-plane directivity. (*After Trinh, Mittra, and Paleta [130], © 1981 IEEE*)

reflector antenna. The reflector in this case would be a parabolic cylinder with the dielectric antenna placed along its focal line. Antennas of this general type have been investigated by Ore [158, 159].

Omnidirectional Dielectric Grating Antennas—Omnidirectional versions of periodic dielectric antennas are shown in Fig. 42 [119, 160, 161]. The antennas possess rotational symmetry about the vertical axis and are assumed to be fed by uniform dielectric waveguides of circular cross section. Fig. 42a shows an antenna with a dielectric grating in the form of a periodic surface corrugation, and Fig. 42b an antenna with a metal grating consisting of a sequence of equally spaced metal rings. Design versatility can be enhanced by combining both options, i.e., by the

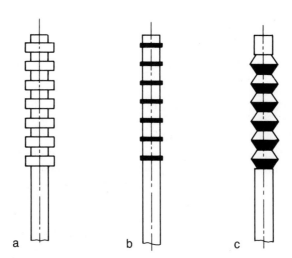

Fig. 42. Omnidirectional dielectric grating antennas. (*a*) Dielectric grating. (*b*) Metal grating. (*c*) Partially metallized dielectric grating. (*After Ore [159], © 1972 IEEE*)

use of an antenna with a partially metallized dielectric grating as indicated in Fig. 42c. Such antennas have been studied by Ore [158, 159].

An omnidirectional radiation pattern is obtained when the antennas are excited in the circular symmetric TM_{01} or TE_{01} modes of the feeding waveguide. Mode TM_{01} operation leads to vertical polarization, TE_{01} operation to horizontal polarization of the radiated field. As an alternative the antennas may be excited in the fundamental HE_{11} mode, provided this mode is circularly polarized. The radiation field will then show elliptical polarization.

The main beam of these antennas has a conical shape. The axis of the radiation cone coincides with the antenna axis, and the apex angle, being determined by (7), varies with frequency; scanning from backfire over broadside to end-fire is, in principle, possible. As with all grating antennas these omnidirectional antennas develop a "stopband" when broadside conditions are approached, i.e., in theory they do not radiate in the azimuth plane and in practice an increased vswr will result.

When a pattern maximum in the horizontal plane is desired, which is a common requirement for omnidirectional antennas, this deficiency can be corrected by the technique discussed under the heading "Beam Scanning," which utilizes a surface modulation consisting of *pairs* of grooves or metal rings, spaced $\lambda_g/4$ apart, to reduce the reflection coefficient of the antenna. An alternate solution would employ a conically shaped radome in which the antenna is embedded such that its axis coincides with the cone axis. It is assumed here that the radome consists of a solid cone of refractive material and that the apex of the cone is at the upper end, its base at the lower end (ground plane) of the antenna. The antenna may then be designed for radiation under an oblique angle (within the radome); but refraction at the radome-to-air interface will redirect the emanating beam into the horizontal direction, provided the aperture half-angle γ_R of the radome is appropriately related to the radiation angle θ of the antenna field within the radome. The condition is that

$$\tan \gamma_R = \frac{n_R \sin \theta}{n_R \cos \theta - 1}$$

where n_R is the refractive index of the radome and θ is counted from the horizontal plane. The use of a radome has the additional advantage of enhancing the mechanical strength of the antenna.

The beamwidth of the elevation pattern of omnidirectional grating antennas is determined primarily by the leakage constant and the axial length of the antennas, and it can be made very narrow. Side lobes can be suppressed by pattern shaping techniques as discussed under the heading "Pattern Shaping."

The theory of omnidirectional grating antennas is closely related to the theory of planar periodic antennas and leads to similar parameter dependencies and the same general design principles. In the case of TM_{01} and TE_{01} excitation the field distribution of the antennas can be described in terms of TM or TE waves alone, which simplifies the analysis considerably.

Uniform-Waveguide Leaky-Wave Antennas

As is the case for tapered-rod and periodically perturbed dielectric antennas, the radiating structures discussed here are derived from open millimeter waveguides. They differ from the above-mentioned antennas in that these guiding structures are maintained *uniform* in the longitudinal direction. It is true that for many applications the aperture dimensions of these antennas will be tapered slightly in order to control the radiation side lobes, but we must understand that it is not necessary to taper these structures or to modulate them periodically in order to produce the radiation.

The radiation is the result of leakage from these open millimeter waveguides; depending on the guiding structure this leakage can be produced in two basic ways. One way is to *perturb* the structure *longitudinally* in a uniform fashion, thereby transforming an initially purely bound mode into a leaky mode. The second way is to employ a *higher mode* on the open waveguide. The dominant mode is generally purely bound, but for certain guides, higher modes, including the lowest mode of the opposite polarization, may leak [162, 163, 172].

An important advantage associated with these uniform leaky-wave antennas is their *simplicity*, permitting easier fabrication at the higher millimeter-wave frequencies where the antenna dimensions become very small. In fact, these antennas have the outward appearance of a uniform open waveguide (except for the slight taper which controls the side lobe level and distribution). On the other hand, these uniform leaky-wave antennas are less versatile than the periodic leaky-wave antennas, since they can radiate into the forward quadrant only, whereas the periodic structures can radiate into the backward quadrant as well and even radiate near the broadside direction.

In principle, uniform leaky-wave antennas may be based on any one of a variety of open waveguides. So far, however, only three waveguides have been considered for this purpose: groove guide, NRD guide, and microstrip line. The first two waveguides are *low-loss* guides specially designed for millimeter wavelengths; they are therefore particularly suitable for the shorter millimeter wavelengths, where the waveguide material losses may otherwise be so great as to compete with the leakage loss, thereby distorting the antenna performance and reducing antenna efficiency. We should recall that moderately lossy waveguides are still acceptable for waveguide circuit components which are only a wavelength or so long, but they may not be suitable for leaky-wave antennas, which may typically be $20\lambda_0$ to $100\lambda_0$ long, where λ_0 is the free-space wavelength. The third waveguide, microstrip line, is not a low-loss guide, but the leaky-wave antenna based on it is so simple that its possible use is worth further consideration.

Each of these leaky-wave antennas may be used as a line-source element in a linear array of such elements arranged in parallel, providing a two-dimensional scanning antenna where the scanning in one direction (the leaky-wave direction) can be controlled electronically or by changing the frequency, and in the orthogonal direction by using phased array methods. This type of phase/frequency two-dimensional scanning antenna was also discussed briefly in connection with the periodically grooved dielectric waveguide treated in the preceding subsection. Such two-dimensional arrays for millimeter waves have not yet been built or tested, and are mentioned here only as a concept likely to be useful.

In view of the small size of waveguiding structures at millimeter wavelengths, and the fabrication difficulties at these wavelengths associated with more conventional line source antennas, such as slot arrays, we believe that leaky-wave antennas, of both the uniform and periodic types, form a *natural* class of antennas for millimeter waves.

Our discussion of uniform leaky-wave antennas follows the waveguides on which the antennas are based. First, we treat the *groove guide*, where leakage can be produced in both of the ways mentioned above, namely, by a uniform longitudinal perturbation and by the use of a higher mode. Second, we consider the *NRD (nonradiative dielectric) guide*, where the leakage is obtained by two basically different perturbation methods. Third, we describe an antenna based on *microstrip line*, where the leakage is produced when the *first higher mode* operates in a narrow frequency range near cutoff.

Leaky-Wave Antennas Based on the Groove Guide—Groove guide is one of several waveguiding structures proposed for millimeter wavelengths about 20 years ago in order to overcome the higher attenuation occurring at these higher frequencies.

The cross section of the groove waveguide is shown in Fig. 43, together with an indication of the electric field lines present; the field is in effect vertically polarized. One should note that the structure resembles that of rectangular waveguide with most of its top and bottom walls removed. Since the currents in these walls contribute significantly to the waveguide losses, the overall attenuation of groove waveguide is less than that for rectangular waveguide. The reduced attenuation loss will therefore interfere negligibly with the leakage loss of any leaky-wave antenna based on groove guide.

The greater width in the middle, or central, region was shown by T. Nakahara [164, 165] to serve as the mechanism that confines the field in the vertical direction, much as the dielectric central region does in H guide. The field thus decays

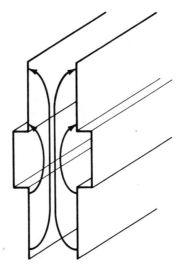

Fig. 43. Cross section of symmetrical, nonradiating groove waveguide. (*After Lamporiello and Oliner [170],* © *1985 IEEE*)

exponentially away from the central region in the narrower regions above and below. If the narrower regions are sufficiently long, it does not matter if they remain open or are closed off at the ends.

Work on the groove guide progressed in Japan and in the United States until the middle 1960s, but then stopped until it was revived and developed further by D. J. Harris and his colleagues [166, 167] in Wales. The first contribution to *antennas* based on the groove waveguide was made by Oliner and Lampariello [168, 169], who proposed and analyzed the first of the leaky-wave antennas to be described. A pair of papers [170, 171] that present the theory and design considerations in detail have also appeared.

(*a*) *Antenna employing an asymmetric strip*—The first of these leaky-wave antennas is shown in Fig. 44. The basic difference between the structures in Figs. 43 and 44 is that in Fig. 44 a *continuous metal strip* of narrow width has been added to the guide in *asymmetrical* fashion. Without that strip the field of the basic mode of the symmetrical groove waveguide is evanescent vertically, so that the field has decayed to negligible values as it reaches the open upper end. The function of the asymmetrically placed metal strip is to produce some amount of net *horizontal* electric field, which in turn sets up a mode akin to a TEM mode between parallel plates. The field of that mode propagates all the way to the top of the waveguide, where it leaks away. It is now necessary to close up the bottom of the waveguide, as seen in Fig. 44, to prevent radiation from the bottom, and (nonelectrically) to hold the structure together. Of course, the upper walls could end as shown in Fig. 44 or they could attach to metal flares or a horizontal ground plane.

We now have available a leaky-wave line-source antenna of simple construction. The value of the phase constant β of the leaky wave is governed primarily by the properties of the original unperturbed groove guide, and the value of the leakage constant α is controlled by the width and location of the perturbing strip.

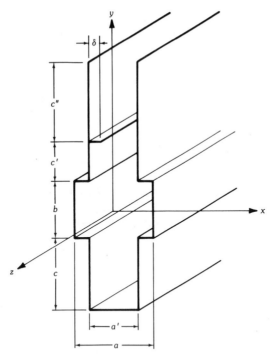

Fig. 44. Leaky-wave antenna derived from groove guide by the introduction of an asymmetric, continuous metal strip. (*After Lamporiello and Oliner [170], © 1985 IEEE*)

As with any leaky-wave antenna one can, by suitably changing the dimensions and the frequency, obtain a variety of scan angles and leakage constants. Let us choose a *typical case* where $a'/a = 0.70$, $b/a = 0.80$, $c/a = 1.215$, $c'/a = 0.145$, $c''/a = 1.50$, $\delta/a = 0.21$, $\lambda_0/a = 1.20$. For this particular set of dimensions, calculations [169] yield $\beta/k_0 = 0.749$ and $\alpha/k_0 = 6.24 \times 10^{-3}$. This value of leakage constant α yields a leakage rate of about 0.34 dB per wavelength, resulting in an antenna about $30\lambda_0$ long if, as is customary, 90 percent of the power is to be radiated, with the remaining 10 percent dumped into a load. The resulting beamwidth of the radiation is approximately 2.9°, and the beam radiates at an angle of about 49° to the normal. At a frequency of 50 GHz, for example, dimension a would be 0.50 cm and the antenna would be 18 cm long.

The leaky-wave antenna has been analyzed [170] by deducing the proper transverse equivalent network, deriving simple closed-form expressions for the various parameters of this network, and then obtaining the dispersion relation for the complex propagation constant of the leaky mode from the lowest resonance of the transverse equivalent network. The resulting expression for the complex propagation constant is also in closed form. From this expression, numerical calculations were made [171] of the antenna's performance characteristics, and their dependence on the various dimensional parameters of the antenna.

As indicated above, it is the added continuous strip of width δ that introduces *asymmetry* into the basic groove guide and creates the leakage. The strip gives rise to an additional transverse mode and couples that mode to the original transverse mode, which by itself would be purely bound. The transverse equivalent network for the cross section of the structure shown in Fig. 44 must therefore be based on these two transverse modes, which are coupled together by the narrow asymmetrical strip. These coupled transverse modes then combine to produce a net TE longitudinal mode (in the z direction) with a complex propagation constant, $\beta - j\alpha$. To assist in the discussion below regarding design considerations, let us call the *original* bound transverse mode the $i = 1$ transverse mode and the *additional* transverse mode the $i = 0$ transverse mode, in accordance with their field variations in the x direction (see Fig. 44). In the transverse equivalent network, each of these transverse modes is associated with a corresponding transmission line.

In order to systematically design radiation patterns one must be able to taper the antenna aperture amplitude distribution while maintaining the phase linear along the aperture length, i.e., one must be able to vary α while keeping β the same. Fortunately, several parameters can be varied that will change α while affecting β hardly at all; the best ones are δ and c, if c is long enough.

Since the $i = 0$ transverse mode is *above* cutoff in both the central and outer regions of the guide cross section, however, a *standing-wave* effect is present in the $i = 0$ transmission line. As a result a short circuit can occur in that transmission line at the position of the coupling strip of width δ, and the value of α then becomes zero. Hence we must choose the dimensions to avoid that condition, and in fact to optimize the value of α.

In the design one first chooses the width a and adjusts a' and b to achieve the desired value of β/k_0, which is determined essentially by the $i = 1$ transverse mode. That value of β/k_0 immediately specifies the angle of the radiated beam. It is then recognized that the value of α can be increased if the coupling strip width δ is increased, or if the distance c' between the step junction and the coupling strip is decreased, since the coupling strip is excited by the $i = 1$ transverse mode, which is evanescent away from the step junction in the outer regions. After those dimensions are chosen, the length c must be determined such that the standing-wave effect mentioned above optimizes the value of α. If c is sufficiently long, it will affect only the $i = 0$ transmission line and influence β negligibly. The length c'' also affects α strongly and β weakly, and it also must be optimized because another, although milder, standing wave exists between the coupling strip and the radiating open end.

We present here, in Figs. 45a and 45b respectively, a curve of α/k_0 as a function of c', and the value of $c + c'$ that must be selected so as to achieve the value of α determined from Fig. 45a. In effect, therefore, Fig. 45b indicates the value of c required once α (via c') is specified. It is interesting to note that $c + c'$ is almost constant for optimization. The curves in Fig. 45 apply to an antenna with $b/a = 0.8$, $a'/a = 0.7$, $c''/a = 1.5$, $\delta/a = 0.21$, and $\lambda_0/a = 1.2$.

It is important to realize that the dimensions for optimization are *independent of frequency*, since the transverse wave numbers are all frequency independent. Of course, when the frequency is altered the values of β and α will change, but the dimensional optimization is undisturbed. In fact, for the dimensions discussed above, the radiated beam can be scanned with frequency from about 15° to nearly 60° from the normal before the next mode begins to propagate.

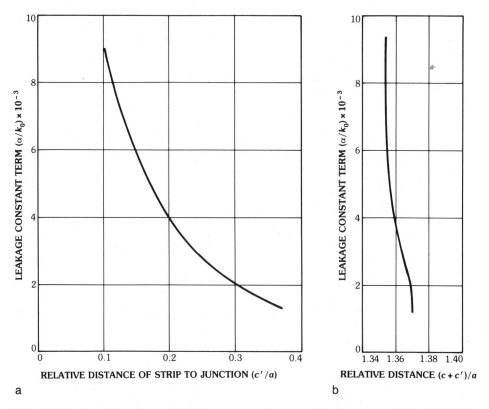

Fig. 45. Design of groove guide leaky-wave antenna employing asymmetric metal strip. (a) Leakage constant α/k_0 (optimum value) as function of distance c'/a of perturbing strip from the step junction. (b) Value of $(c' + c)/a$ required to achieve the optimum value of α/k_0. (After Lamporiello and Oliner [170], © 1985 IEEE)

The leaky-wave antenna discussed above is, therefore, straightforward to understand and amenable to systematic design. It is also sufficiently flexible; a reasonably wide range of pointing angles and beamwidths can be achieved by appropriate adjustment of its dimensional parameters.

(b) *Antenna based on the use of leaky higher modes*—Recent studies by Lampariello and Oliner [172, 173] have shown that only the dominant mode of groove guide is purely bound, and that all of the higher modes of that guide are *leaky*. These higher modes thus provide the basis for new leaky-wave antennas.

In particular, interesting results were obtained [172, 173] for the first higher even mode and the first higher odd mode. These modes are distinguished from the dominant mode by their transverse variations (in the x direction in Fig. 46). The dominant ($n = 1$), first higher even ($n = 2$), and first higher odd ($n = 3$) modes possess, respectively, a half sine wave, a full sine wave, and a three-halves sine wave variation across the width a.

An earlier paper by Nakahara and Kurauchi [165] examined the simple relations among the wave numbers for the odd higher modes and demonstrated that these modes must be leaky. These authors did not indicate the magnitude of

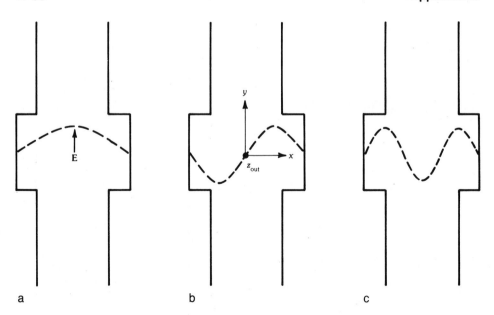

Fig. 46. Transverse field distribution of groove guide modes. (*a*) Dominant mode ($n = 1$). (*b*) First higher even mode ($n = 2$). (*c*) First higher odd mode ($n = 3$).

the leakage constant, however. The recent studies mentioned above [172, 173] first proved that a corresponding statement is valid for the even higher modes. In addition, these studies derived transverse equivalent networks for the first higher odd mode ($n = 3$) and the first higher even mode ($n = 2$). From these transverse equivalent networks dispersion relations were derived in the form of transcendental relations all of whose constituents are in a simple closed form. The leakage constants for each higher mode can be found readily from these dispersion relations.

Let us first consider the propagation behavior of the first higher *odd* mode. To understand that behavior we must recognize that the cross-section dimensions must be large enough to permit both the dominant and the first higher odd longitudinal modes to propagate. Looking in the y direction in Fig. 46, we then see that the guide can be initially excited such that the incident power is basically either in the $i = 1$ transverse mode, for which the x dependence is a half sine wave, or in the $i = 3$ transverse mode, for which the x dependence contains three half sine waves. These excitations result in the $n = 1$ and $n = 3$ longitudinal modes, respectively. In either case, we observe that at the step junction an $i = 1$ or $i = 3$ transverse mode will couple to all other transverse modes of the same symmetry; for example, the $i = 3$ mode will couple to all of the $i = 1, 5, 7, \ldots$ transverse modes. In the transverse equivalent network, the $i = 1$ and $i = 3$ transverse modes are separated out, and separate transmission lines are furnished for each of them. The coupling susceptance between the transmission lines was derived using small obstacle theory in a multimode context.

When groove guide is excited in *dominant mode* fashion, the $i = 1$ transmission line is propagating transversely in the central region of width a, but evanescent in the outer narrower regions of width a'. Also, it can be shown that the $i = 3$ transmission line is below cutoff everywhere, so that the dominant longitudinal mode is purely bound. On the other hand, *when the groove guide is excited in the first higher odd longitudinal mode*, corresponding to the $i = 3$ transverse mode, the $i = 3$ transmission line is propagating in the central region but evanescent in the outer regions. But, the $i = 1$ transmission line can now be shown to be propagating in *both* the central and the outer regions. The result is that the first higher odd mode is *leaky*, but with the interesting feature that the energy that leaks has the variation in x of the *dominant* mode, not of the first higher mode. These coupled transverse modes combine to produce a net TE longitudinal mode (in the z direction) with a complex propagation constant, $\beta - j\alpha$.

Some numerical results are presented next for the behavior of the phase constant β and the attenuation constant α of the leaky mode that results when the groove guide is excited in the first higher odd longitudinal mode (the $i = 3$ transverse mode). In Figs. 47a and 47b we note the variation of β and α as a function of frequency. We see from Fig. 47a that β is almost linear with frequency at the higher frequencies, but shows substantial curvature near cutoff. The variation of α with frequency, in Fig. 47b, is seen to be almost hyperbolic at the higher frequencies; nearer to cutoff, α is seen to rise substantially.

These variations follow directly from the wave-number relationship

$$k^2 = k_x^2 + k_y^2 + k_z^2$$

where k_x and k_y are determined by the waveguide dimensions and $k_z = \beta - j\alpha$. By taking the real part, and noting that the transverse wave numbers are independent of frequency, we find that β is approximately linearly proportional to the frequency

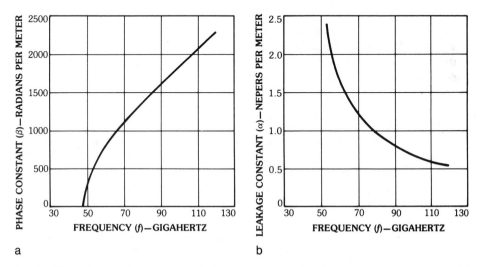

a   b

Fig. 47. Variations with frequency of phase constant and leakage constant of groove guide excited in first higher odd longitudinal mode. (*a*) Phase constant β. (*b*) Leakage constant α. (*After Lamporiello and Oliner [173], © 1983 IEEE*)

when α is small, which occurs for the higher frequencies. Such behavior is in agreement with that in Fig. 47a. When we take the imaginary part, we find that the product $\alpha\beta$ should remain independent of frequency. In the frequency range for which β is proportional to frequency, we thus find that α must vary as the reciprocal of the frequency, in agreement with Fig. 47b. For Fig. 47 the guide dimensions are $a = 1.0$ cm, $a'/a = 0.7$, and $b/a = 0.4$.

The behavior described above applies to the spectrum of *odd* higher modes. Qualitatively similar leakage behavior is found for the *even* higher modes, although certain important differences occur. The step junction then couples only the even transverse modes, of course; in particular, for the first higher even ($n = 2$) mode, the $i = 2$ transverse mode will couple at the step junction to the $i = 0$ transverse mode, which is a TEM-like mode. Thus, when we excite the $n = 2$ mode in the groove guide it will leak, but the transverse form of the leakage is *TEM-like*, and not of the form of the exciting second mode. In a sense this leakage behavior is similar to that found for the first higher odd ($n = 3$) mode, but the polarization of the electric field of the leakage energy is *horizontal* here whereas it was *vertical* there.

This difference in the polarization of the leakage energy is very interesting because it can form the basis for two new types of leaky-wave antenna, with vertical and horizontal polarization, respectively.

An additional interesting observation can be made with respect to the even higher mode case. As shown in Fig. 48a, the vertical midplane is an electric wall, or short-circuit wall, by virtue of the symmetry of the excitation. The waveguide can therefore be bisected, and a metal wall can be placed along the electric-wall midplane without affecting the electric field distribution in either bisected half. The resulting structure, of half the width, as shown in Fig. 48b, may now be viewed as an *asymmetric* groove guide of normal width. That asymmetry serves the same function as the continuous asymmetric strip in the antenna shown in Fig. 44, namely, to couple the exciting $i = 1$ transverse mode to the TEM-line $i = 0$ transverse mode, which then leaks power away.

The bisected structure in Fig. 48b may alternatively be viewed as an *E*-plane *tee junction* on its side with its stub guide, of length $a - a'$, terminated by a short circuit. This viewpoint leads to completely new analysis of the behavior of this antenna, in which the transverse equivalent network is based on the above-mentioned tee junction. An accurate network representation for this tee network was derived, in which the parameters are in closed form and yet are more accurate than the expressions given in the *Waveguide Handbook* [220]. The mode in the transmission lines representative of the main guide and the stub guide is the TEM mode propagating at an angle. The transverse equivalent network is thus simpler in form than the one discussed above that requires the coupling between two transverse modes. Using this new network, a dispersion relation in closed form was derived, and numerical results for the phase constant β and the leakage constant α were obtained for the structure shown in Fig. 48b for various dimensional ratios [174].

It was found that the values of α for this structure are somewhat larger than one would like, with the result that narrow beams are not readily obtained. To overcome this limitation, two additional modifications were introduced, as

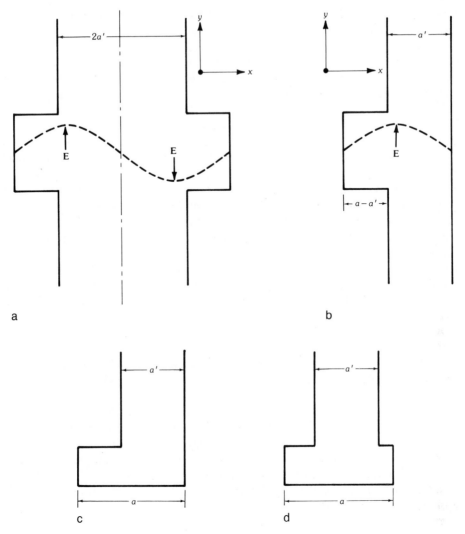

Fig. 48. Groove guide excited in the first higher even ($n = 2$) mode. (*a*) Full guide, showing electric wall symmetry for the vertical midplane. (*b*) Bisected guide, corresponding to an asymmetric groove guide excited in the dominant mode. (*c*) Further bisection of the structure in (*b*), using the fact that the horizontal bisection plane is an electric wall. (*d*) Modification of the structure in (*c*) in which the open stub guide is moved nearer to the central position.

shown in Figs. 48c and 48d. The first of these modifications, in Fig. 48c, produces a structural change but no change in the electrical performance except for some small additional metal loss; it is based on the recognition that, because of symmetry, the horizontal line bisecting the stub guide in Fig. 48b is an electric wall. This structure now resembles an L where the vertical portion of the L is an open-ended guide that permits the leakage of radiation. The second change, shown in Fig. 48d, moves the vertical open-ended guide from the end location to a position somewhere between that end and the centered position.

When the open-ended guide is in the exact central position, the structure resembles groove guide bisected horizontally, and no radiation is produced. At the other extreme, with this open-ended guide at one end, as seen in Fig. 48c, the leakage is quite strong. The leakage rate can therefore be controlled by locating the open-ended guide at intermediate positions.

The modified structure in Fig. 48d can also be analyzed accurately by employing the transverse equivalent network based on the tee network discussed above, but with different parameters since the structure has been rotated through 90° so that stub and main guides are interchanged. Such an analysis has been performed [175], and the dispersion relation was obtained in closed form. Very good numerical agreement was obtained with earlier results for the two extreme positions. That is, the values of β for various dimensional ratios for groove guide agreed almost exactly with those obtained when the open-ended guide in Fig. 48d is centered, and the values of β and α for the end position agreed very well with those derived in a different way for the structure in Fig. 48c (or b). Numerical results for intermediate positions [175] demonstrate that the values of leakage constant α can indeed be controlled over a wide range while the values of phase constant β vary very little. This new structure, with its added degree of freedom, thus yields great *flexibility* in the antenna beamwidth.

It is interesting that this new antenna evolved from a leaky higher mode in groove guide by employing two bisections and then an additional modification. When one looks at the final structure, however, it may be viewed as a rectangular waveguide, fed in its dominant mode, with a vertical open-ended stub guide present all along its length. The analysis performed for the antenna [175] did in fact utilize a tee junction approach consistent with this viewpoint. Because of its very simple cross section, and the fact that it can be fed from a rectangular waveguide carrying the dominant mode, the antenna shown in Fig. 48d has the potential of being a practical one for the millimeter-wave range.

Measurements were taken on this antenna at millimeter wavelengths, over the frequency range 40 GHz to 60 GHz, and the results agreed very well with the above-mentioned analysis. The antenna was fed from a rectangular waveguide, and measurements were made first of the phase and leakage constants and then of the radiation patterns, with the results verifying the theory well in all cases [221].

Leaky-Wave Antennas Based on the Nonradiative Dielectric (NRD) Guide—Two papers appeared recently [176, 177] which proposed a new type of waveguide for millimeter waves, and showed that various components based on it can be readily designed and fabricated. By a seemingly trivial modification, the authors, T. Yoneyama and S. Nishida, transformed the old well-known H guide, which had languished for the past decade and appeared to have no future, into a practical waveguide with attractive features. The old H guide stressed its potential for low-loss long runs of waveguide by making the spacing between the metal plates large, certainly greater than half a wavelength; as a result, the waveguide had lower loss, but discontinuities or bends would produce leakage of power away from the guide. Yoneyama and Nishida simply observed that when the spacing is reduced to less than half a wavelength all the bends and discontinuities become purely reactive; they therefore call their guide "nonradiative dielectric waveguide," or NRD guide. As a result of this modification many components can be constructed easily, and in an integrated circuit fashion, and these authors proceeded to demonstrate how to fabricate some of them, such as feeds, terminations, ring resonators, and filters.

These papers [176, 177] treat only reactive circuit components, and no mention is made of how this type of low-loss waveguide can be used in conjunction with antennas. In this section, two new types of leaky-wave antenna are described which can be readily fabricated with NRD guide and, in fact, can be directly connected to NRD guide circuits in integrated circuit fashion, if desired. The first of these two antenna types was discussed recently by Sanchez and Oliner [178, 179], and the second in two different talks by Oliner, Peng, and Sheng [180] and by Shigesawa, Tsuji, and Oliner [181]. The leakage mechanisms in these two antenna types are different; in the first antenna type the leakage is produced by foreshortening the metal plates on one side of the guide, whereas in the second the leakage is caused by introducing asymmetry in the dielectric strip cross section. In addition, the polarizations of the radiation are also different, being vertical in the first and horizontal in the second. In both types, however, the basic guided wave must be operated in the fast-wave range.

(a) *Antenna produced by foreshortening one side of the guide*—The new waveguide, shown in Fig. 49a, looks like the old H guide except that the spacing between plates is less than half a wavelength to ensure the nonradiative feature. In the vertical (y) direction, the field is of standing-wave form in the dielectric region and is exponentially decaying in the air regions above and below. The guided wave propagates in the z direction. The leaky-wave antenna based on this waveguide is shown in Fig. 49b; the antenna is created simply by decreasing the distance d between the dielectric strip and the top of the metal plates. When the distance d is small, the fields have not yet decayed to negligible values at the upper open end, and therefore some power *leaks away*. The upper open end forms the antenna

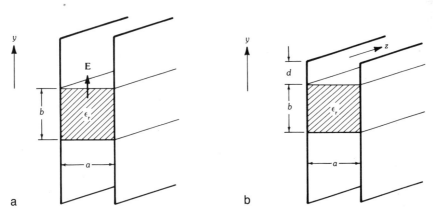

Fig. 49. Leaky-wave antenna derived from nonradiative dielectric (NRD) guide by foreshortening one side of the guide. (*a*) NRD guide. (*b*) Antenna. (*After Sanchez and Oliner [178], © 1984 American Geophysical Union*)

aperture. The amplitude distribution in this aperture can be controlled by varying the distance d as a function of the longitudinal coordinate z. This may be accomplished either by appropriately shaping the upper edge of the metal plates, or by slightly curving the dielectric strip waveguide so that its distance from the edge of the plates varies in a prescribed fashion with z; the edge in this case would be straight. This leads to a very simple and easy-to-build antenna configuration which is indicated in Fig. 50. Furthermore, the antenna is directly compatible with transmit and receive circuits designed in NRD guide technology. The antenna radiates with vertical polarization.

A leaky-wave antenna *for H guide* of the type shown in Fig. 49b was proposed some years ago by Shigesawa and Takiyama [182, 183]. The principle of operation is identical with that described here; however, the structure presented by them was symmetrical, so that it radiated from both sides, although that was not a necessary feature. On the other hand, their structure would not permit a bending of the dielectric strip to taper the amplitude distribution, as in Fig. 50, because additional radiation would be produced due to the bend. Amplitude tapering could be achieved by keeping the dielectric strip straight and cutting back the open end, however. Additional leakage would occur at the feed and load junctions, in any case, because of the large spacing between the plates. The utilization of the NRD guide instead of the H guide makes this antenna type much more practical.

The NRD guide antenna, in the form shown in Fig. 49b, was analyzed [178, 179] as a leaky waveguide that possesses a complex propagation constant $\beta - j\alpha$, where β is the phase constant and α is the attenuation or leakage constant. An accurate transverse equivalent network for the cross section of the antenna was established, and the dispersion relation for the β and α values was obtained from the resonance of this network. This dispersion relation contains elements all of which are in closed form, thus permitting easy calculation.

The various parametric dependences of α and β on the dimensions and on the dielectric constant have been obtained [184] in order to clarify design information. Here we discuss only a single typical case, for which the parameter values are indicated in Fig. 51. The figure shows the behavior of β and α as a function of the

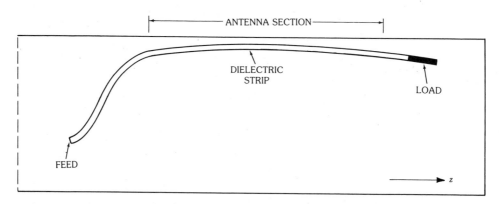

Fig. 50. Side view of NRD waveguide antenna. (*After Sanchez and Oliner [178]*, © *1984 American Geophysical Union*)

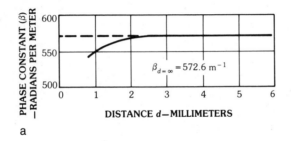

a

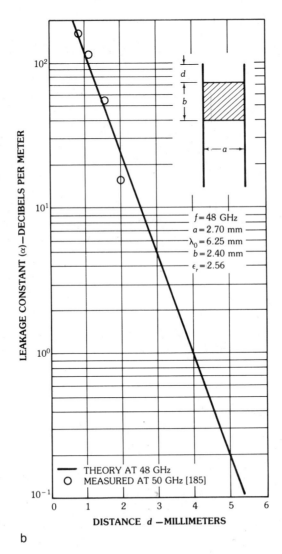

b

Fig. 51. Phase constant and leakage constant of NRD guide antenna as functions of distance of dielectric strip from antenna aperture. (*a*) Phase constant. (*b*) Leakage constant. (*After Sanchez and Oliner [178], © 1984 American Geophysical Union*) (*c*) Comparison with measurements by Han et al. [186]. (*After Han et al. [186], © 1987 IEEE*)

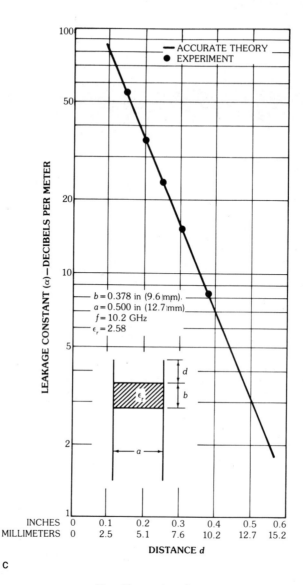

Fig. 51, *continued.*

distance d of the dielectric strip from the antenna aperture; see the inset in Fig. 51b. For distances $d > 2$ mm, one sees from Fig. 51a that the value of β remains essentially unchanged, as desired. Fig. 51b shows that α increases as d is shortened, as expected since the field decays exponentially away from the dielectric region. Thus the values of α that one can achieve span a very large range.

Leaky-wave antennas are usually designed so that 90 percent of the incident power is radiated, and the remaining 10 percent is dumped into a load. Following this criterion, if one selects $d = 2.0$ mm for this geometry, the length of the antenna will be about 40 cm, and the beam will radiate at an angle of about 35° from the normal, with vertical electric field polarization, and with a beamwidth of approximately 1°. A larger (or smaller) value of d will result in a narrower (or wider) beam whose width can be calculated from the curve of α in Fig. 51b.

Measurements of the leakage constant α as a function of distance d have been taken by Yoneyama [185] at a frequency of 50 GHz and by Han, Sanchez, and Oliner [186] at frequencies in the vicinity of 10 GHz on a scaled structure. All of the measurements agree very well with theoretical values. Some of the measurements taken by Yoneyama are superimposed as points on the theoretical values which are represented by the solid line in Fig. 51b. Although the theoretical curve corresponds to a frequency of 48 GHz and the measured points to a frequency of 50 GHz, these frequencies are sufficiently close to permit a comparison between measurement and theory. The scaled measurements by Han and colleagues [186], on the other hand, permitted greater precision and demonstrated excellent agreement with the theory, as seen in Fig. 51c, where the measured points fall almost directly on the theoretical solid curve over a rather large range of values of leakage constant α. Such agreement provides confidence that the theoretical results are reliable for design purposes.

(*b*) *Antenna produced by asymmetry*—Two antenna structures, based on NRD guide, that leak because of the introduction of asymmetry are shown in Figs. 52a and 52b. The physical mechanism that produces the leakage is the same as that present in the antenna of Fig. 44, where an asymmetric strip was introduced to perturb the symmetry of the open waveguide. In all of these antennas, the asymmetry causes mode conversion to an additional transverse mode of the TEM type that propagates at an angle in the parallel plate region of the cross section, thereby creating radiation polarized with the electric field parallel to the aperture. For the orientation of the structures in Fig. 52 the polarization is horizontal; in contrast, the radiation from the antenna configuration in Fig. 49b is vertically polarized.

The structure in Fig. 52a has been analyzed [180, 181] by means of mode

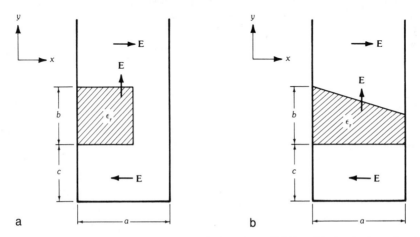

Fig. 52. Two uniform leaky-wave antennas derived from NRD guide by the introduction of asymmetry. (*a*) Antenna with dielectric strip of rectangular cross section and air gap. (*b*) Antenna with dielectric strip of trapezoidal cross section.

matching at the air-dielectric interfaces, and numerical results were obtained as a function of the geometrical parameters. When the NRD guide is operated as a nonradiating waveguide, as in Fig. 49a, the dielectric strip fills the waveguide cross section uniformly. The introduction of a small air gap produces the asymmetry that results in mode conversion to a TEM mode that propagates away at an angle in the outside air-filled parallel plate regions. The air gap does not have to be very large to produce significant leakage; the gap shown in Fig. 52a would yield a large value of α, with a consequent wide beam in the radiation pattern. The geometry can thus be controlled easily to achieve a large range of values for the leakage constant α, and therefore a large range of desired beamwidths.

For certain values of length c in Fig. 52a, additional complications can arise in the radiation behavior. These problems can be avoided by taking care in the selection of c, but most easily by making $c = 0$ and halving the value of b, because the metal plane at $c = 0$ then acts as an imaging mirror. This last modification has the added virtue of a simpler (and perhaps more rugged) structure. In addition, the finite length of open-ended parallel-plate guide on the upper side can cause the introduction of another leaky wave, a modification of the well-known channel guide mode. As a result, interesting coupling effects have been found [181, 221] to occur between the NRD guide leaky mode and this channel guide mode, but the antenna can be designed to avoid such coupling effects so that they need not present a practical difficulty.

In summary, care must be taken in the practical design of this antenna type, and it appears that the simplest structural form is the easiest and safest to design. Accurate theoretical results [180, 181] have been obtained for the phase and leakage behavior, so that such antennas can be designed reliably. By modifying the width of the air gap, a large range of beamwidths can be achieved. No measurements have been taken as yet on this antenna type.

The structure in Fig. 52b, on the other hand, has not been analyzed, but some preliminary measurements [187] have been made on it. The measurements show that radiation indeed occurs, and that good patterns result, but parameter optimization has not yet been accomplished.

Leaky-Wave Antenna Based on the First Higher Mode on Microstrip Line— Although microstrip line is not a low-loss waveguide, its simplicity makes it attractive as a guide on which to base antennas. The dominant mode on a uniform microstrip line is a slow wave relative to free space, so that the dominant mode cannot furnish a way to achieve a uniform leaky-wave antenna. Many antennas have been conceived and built, however, based on short lengths of dominant-mode microstrip line which are operated as open resonators; these antennas, as they relate to millimeter-wavelength applications, are discussed in the next section.

It is possible, on the other hand, to create a *uniform* leaky-wave antenna based on microstrip line if one employs a *higher mode* in an appropriate range of operation. The most convenient higher mode is the *first* higher mode. The electric field distribution of that mode in the microstrip line cross section is shown in Fig. 53. In contrast to the dominant mode the first higher mode has a nonzero cutoff frequency, which depends on the guide width.

Ermert [188, 189] conducted a careful numerical study of the propagation

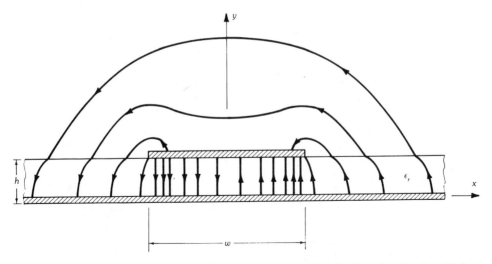

Fig. 53. Cross section of microstrip line, showing electric-field lines for the first higher mode.

characteristics of the dominant and first two higher modes of microstrip line employing an accurate mode-matching procedure, but the microstrip had a top cover which permitted him to employ discrete higher modes in the transverse representation. He found that in a range close to cutoff for the higher modes it was not possible to obtain real values for the propagation wave number; he therefore termed this range the "radiation" range, but he did not interpret it any further except to indicate that this "radiation region" was characterized by a continuous spectrum. One of the figures presented by Ermert [188, 189] is reproduced here, with modifications, as Fig. 54. (For this figure the microstrip line dimensions are: strip width, 3.00 mm; dielectric layer thickness, 0.635 mm; $\epsilon_r = 9.80$; and the height of the top cover is five times the dielectric layer thickness.) Ermert's curves are the solid ones, shown for the normalized propagation wave number β/k_0 for the lowest mode and the first two higher modes of covered microstrip line. All of his wave-number values are real, meaning that the modes are purely bound in those ranges. In the region shown lined, which he called the "radiation region," no real solutions exist. The dashed lines in Fig. 54 do not appear in his figure.

Oliner and Lee [190, 191] added these dashed lines and pointed out that they correspond to complex solutions (of course, only the real part is plotted) which signify physically that the modes have become *leaky* in this "radiation region." Ermert selects a spectral distribution for the modes of microstrip, and in his second paper [189] he rejects any inclusion of leaky-modes since they are nonspectral. Although that statement is a mathematically correct one, his rejection of leaky modes was unfortunate because it prevents one from understanding certain practical consequences. An alternative representation in terms of leaky modes can be obtained from his continuous spectrum by employing the steepest descent representation, which is not a spectral one. Such an investigation [190, 191] shows that the continuous spectrum in Ermert's radiation region is indeed characterized by essentially a *single leaky mode*. The dashed lines in Fig. 54 show the results for

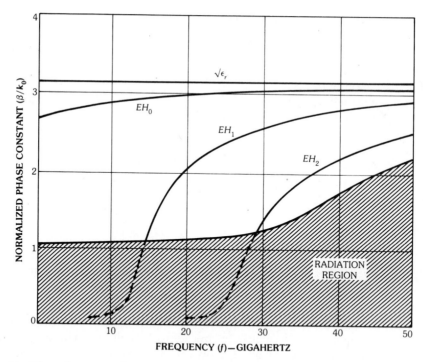

Fig. 54. Dispersion curves for the lowest mode and the first two higher modes in microstrip line with a top cover. (*After Oliner and Lee [190], © 1986 IEEE*)

the appropriate leaky-mode solution corresponding to each of the two higher modes.

Oliner and Lee [190] have also shown that leakage can occur in two forms: a surface wave and a space wave. Furthermore, the onset of leakage for each form is given by simple conditions.

Let us envision a top view of the strip and the dielectric region around it. With this picture, we examine the case of leakage away from the strip in the form of a *surface wave* on the dielectric layer outside of the strip region. When there is leakage into the surface wave, the modal field propagates axially (in the z direction) with phase constant β, and the surface wave propagates away (on both sides) at some angle with phase constant k_s. The surface-wave wave number k_s has components k_z and k_x in the z and x directions, respectively, where k_z must be equal to β, since all field constituents are part of the same leaky modal field. We may therefore write

$$k_x^{\,2} = k_s^{\,2} - \beta^2$$

For actual leakage, k_x must be real, so that the condition for leakage is $k_x^{\,2} > 0$. Applying this condition, we find that for leakage,

$$\beta < k_s$$

This condition defines the lined region in Fig. 54; the upper boundary of that region is actually the dispersion curve for the surface wave, of wave number k_s, that can be supported by the dielectric layer on a ground plane, if the microstrip line is open above, or by the dielectric layer between parallel plates, if there is a metal top cover. At the onset of the surface wave, it emerges essentially parallel to the strip axis, consistent with the condition $\beta = k_s$.

As the frequency is lowered so that β is decreased below the value of k_s, power leaks away in the form of a surface wave, as discussed above. As β is decreased further, power is then *also* leaked away in another form, the *space wave*. If the microstrip line is *open above*, this space wave actually corresponds to radiation at some angle, the value of this angle changing with the frequency. At the onset of this space wave, the wave emerges essentially parallel to the strip axis, so that $\beta = k_0$ then, where $k_0 = (2\pi/\lambda_0)$ is the free-space wave number. This boundary corresponds to the horizontal line $\beta/k_0 = 1$ in Fig. 54. For values of

$$\beta < k_0$$

power will leak into a space wave in addition to the surface wave.

What happens when the microstrip line has a *top cover*, of height H? If $H < \lambda_0/2$, approximately, such that only the surface wave can propagate in the dielectric-loaded parallel-plate region, then all the other modes are below cutoff, and power can leak away *only* in surface-wave form. If the plate spacing is increased, then some of the non–surface-wave modes are above cutoff, and these modes can also carry away power. The "space wave" then corresponds to the sum of those modes.

A separate study has shown that as the height H of the top cover increases, the proportion of power in the "space wave" also increases. When the top cover is removed far away and the dielectric substrate is thin, the percentage of power going into the surface wave becomes very small indeed. When there is no top cover present, therefore, we should expect that, for frequencies for which $\beta < k_0$, almost all of the power is radiated away in the form of a space wave. A leaky-wave antenna based on a microstrip higher mode would therefore be very efficient, and very little power would be lost to surface waves.

We next recognize that a traveling-wave antenna based on the first higher mode on microstrip line has actually been built and tested by Menzel [192]. It could have been operated as a leaky-wave antenna, but it was instead made very short (a typical length being $2.23\lambda_0$) as a competitor to microstrip resonator antennas. This antenna is therefore not characterizable as a true leaky-wave antenna, but it is novel and interesting, nevertheless, and represents a pioneering structure in this category.

In order to support the first higher mode above cutoff, the strip upper conductor is made somewhat wider than a half-wavelength in the substrate material. The dominant mode, which can be present simultaneously, of course, is suppressed, so that the traveling wave field corresponds only to that of the first higher mode. Based on this traveling wave field, but not taking any attenuation due to leakage into account, Menzel [192] also performed some elementary calculations with respect to the radiation pattern and beamwidth. He also conducted some measurements on typical structures.

In his approach, Menzel assumed that the propagation wave number of the first higher mode was real in the very region where Ermert said no such solutions exist; since his guided wave, with a real wave number, was fast in that frequency range, Menzel presumed that it should radiate. His approximate analysis and his physical reasoning were therefore incomplete, but his proposed antenna was valid and his measurements demonstrated reasonably successful performance. Once we recognize the relevance of leaky modes to the "radiation" region of microstrip line higher modes, it becomes clear that Menzel's antenna is a leaky-wave antenna in principle, even though he did not recognize this fact and did not discuss the antenna's design or behavior in those terms.

Oliner and Lee [190, 191] derived an accurate transverse resonance formulation for the propagation characteristics of the higher modes, both in the purely bound range (real wave numbers) and the "radiation" range (complex wave numbers). In this derivation, they employed a rigorous (Wiener-Hopf) solution derived by D. C. Chang and E. F. Kuester [193] for the reflection from one side of that microstrip line. They made a parametric dependence study of the leakage and phase constants of the first three higher modes in the "radiation" range; they found that the leakage rate α grows rapidly as the mode approaches "cutoff," as expected, but that the phase constant β, after approaching zero, slowly increased again and continued to increase as the frequency was lowered further. Such behavior modifies the usual understandings of the nature of "cutoff" for microstrip line higher modes in open regions.

What is puzzling at first with regard to Menzel's antenna, however, is why the antenna radiated so well in traveling-wave fashion even though it was so short $(2.23\lambda_0)$; leaky-wave antennas are usually much longer. To answer this question, Oliner and Lee [191] employed the accurate expression mentioned above for the propagation characteristics of microstrip line higher modes, and obtained the variations of phase constant β and leakage constant α shown in Figs. 55a and 55b for Menzel's structure as a function of frequency.

On use of the relation $\sin \theta_m \cong \beta/k_0$, where θ_m is the angle of the beam maximum measured from the broadside direction, we see from Fig. 55a that after the onset of leakage the beam moves from end-fire toward the broadside direction. From Fig. 55b we note that as the frequency is lowered (and the beam swings up from end-fire) the value of α increases rather substantially, so that the beam width will increase strongly as the beam swings up.

The large value of α/k_0 ($= 0.0378$) explains why Menzel's antenna radiated so well despite its short length $(2.23\lambda_0)$; a quick calculation shows that over this short length about 65 percent of the power was actually radiated. The remaining 35 percent would be largely reflected from the end, producing a large back lobe at the same angle from broadside. A look at Menzel's experimental pattern (his Fig. 11 in [192]) indeed verifies that such a back lobe is present with an amplitude about 0.4 of that of the main beam.

Since leaky-wave antennas are designed to radiate 90 percent or so of the incident power, one simply increases the length of the strip appropriately to accomplish that end. For the same cross section as that of Menzel's antenna, the strip length L would be increased to 21.7 cm from 10.0 cm for 90-percent power radiation. In wavelengths, L changes from $2.23\lambda_0$ to $4.85\lambda_0$; the half-power beamwidth then reduces from 26° to about 14°. It is interesting that in this case one needs only to slightly more than double the strip length to substantially improve the efficiency, essentially eliminate the back lobe, and reduce the beamwidth to a more practical value.

The discussion above applies to the radiation pattern in the H-plane. The E-plane pattern is very wide, of course, being dependent on the microstrip line strip width, but it can be narrowed down substantially, if desired, by using an array of parallel microstrip lines etched on the same substrate and excited with equal phase. Varying the phase would permit beam scanning in the E-plane.

A practical antenna would require the suppression of the dominant mode. That suppression can be accomplished by exciting the upper conductor of the microstrip line directly in antisymmetric fashion, using a pair of stubs fed from below in ± fashion, for example. Another way to produce suppression is to use an asymmetric feed arrangement, as shown in Fig. 56, together with a sequence of transverse slits on the center line of the antenna to inhibit the propagation of the dominant mode [192], which possesses a strong longitudinal current in the center portion of the upper conductor (whereas the longitudinal current there would be zero for the first higher mode). A quarter-wavelength transformer can be used to match the impedance of the antenna to that of the much thinner feed line. In this way, an impedance bandwidth of 10 percent can be obtained within a vswr of 2:1 [192].

Menzel states in his paper [192] that he proposed this antenna because it could yield improved bandwidth relative to microstrip resonator antennas. That is why it is made so short; as a result, as mentioned above, the antenna is not efficient, and a substantial reflected wave from the end is found experimentally, manifesting itself in a large back lobe. If the same structure were made longer and designed as a leaky-wave antenna, one should expect to obtain a uniform leaky-wave antenna of *simple configuration* that would yield a narrower beamwidth and be efficient.

4. Microstrip Resonator Antennas and Other Printed-Circuit Antennas

This section deals with printed-circuit antennas and their use in the millimeter-wave region. The best-known antennas of this type are microstrip antennas, which consist of metal patches or dipoles printed on a single, thin dielectric substrate backed by a ground plane. These antennas are operated *at resonance*, which distinguishes them from the traveling-wave microstrip antennas reviewed in the preceding section. The basic microstrip geometry is the same for both the resonant and the traveling-wave microstrip antennas, of course, but they differ in their design principles and modes of operation, so that treatment in different sections is appropriate.

While microstrip traveling-wave antennas are a comparatively new development, resonant microstrip antennas have been investigated extensively during the past decade at microwave frequencies. In particular, microstrip patch resonator antennas can be regarded as well understood by now although papers on this subject are still appearing quite frequently in the literature. This technique has

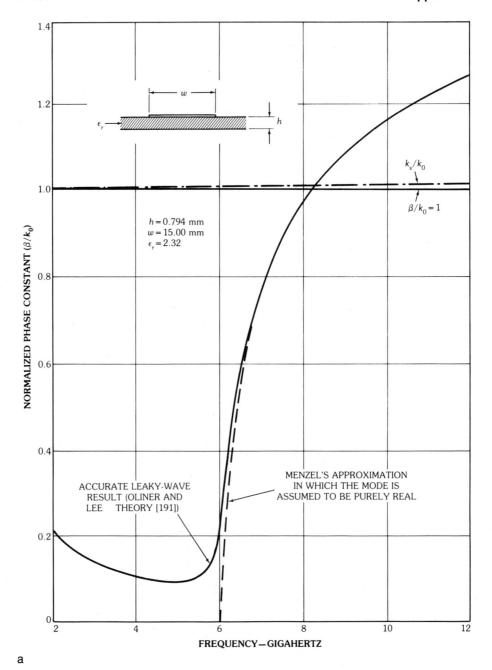

a

Fig. 55. Normalized phase constant β/k_0 and leakage constant α/k_0 as functions of frequency for Menzel's antenna structure [192]. (*a*) Phase constant β/k_0. (*b*) Leakage constant α/k_0. (*After Oliner and Lee [191], © 1986 IEEE*)

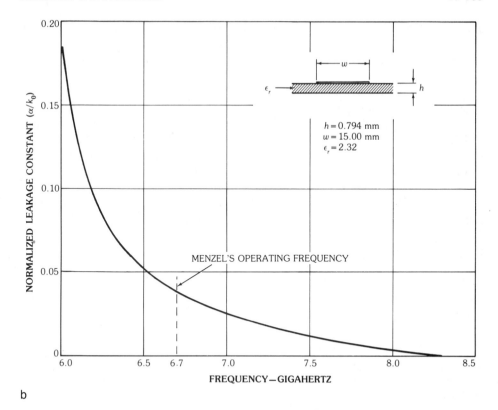

Fig. 55, *continued.*

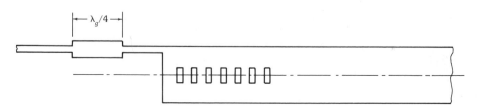

Fig. 56. Asymmetrically excited microstrip traveling-wave antenna. (*After Menzel [192]*)

been used very successfully for the design of single antennas and for a wide variety of antenna arrays for the uhf through microwave regions. The technique is described in detail in Chapter 12; it is this technique which is commonly associated with the term microstrip antennas.

Microstrip patch and dipole antennas possess numerous advantages, including a very low profile, suitability for conformal installation, convenient fabrication by etching techniques, a high degree of reproducibility, ruggedness, low cost, and light weight. Moreover, the microstrip technique is very well suited for the design of integrated antennas. Transmission lines, phase shifter circuits, hybrid couplers, mixer circuits, and filters may be printed with the antennas on the same substrate, and solid-state devices can be created monolithically in this substrate if materials such as silicon and gallium arsenide are used. The only problem with these antennas is their narrow bandwidth (less than 7 percent), which would not permit their use in broadband systems.

In recent years several efforts have been made to extend the microstrip antenna technique into the millimeter-wave region [9, 46–56, 61–65]. The planar geometry amenable to integrated and monolithic design approaches makes this technique particularly attractive for millimeter-wave applications. Extension into the millimeter-wave band, however, is not simply a matter of straightforward wavelength scaling; new problems, as well as new opportunities, appear. Major problems include attaining good electrical efficiency and high dimensional accuracy for large arrays. But recent experimental studies have shown that these problems can be solved in a satisfactory manner, and microstrip patch arrays which combine fairly high gain with good pattern quality and reasonable efficiency have been demonstrated for frequencies up to 100 GHz [49–56, 61–65]. Theoretical work has shown, moreover, that the use of electrically thick substrates, which becomes practical in the millimeter-wave region, permits one to extend the bandwidth of microstrip antennas substantially [57–60, 71, 72, 194]. In general, it appears that microstrip patch resonator arrays will be useful antennas for the lower millimeter-wave range up to a frequency of 100 GHz and, possibly, 140 GHz [62]. The usefulness of microstrip dipole antennas, on the other hand, may extend well into the upper millimeter-wave region, particularly if these antennas are printed on electrically thick substrates. The upper frequency limit will depend on the design of efficient, easy-to-fabricate feed systems for arrays of these antennas.

Stripline antennas, a second class of printed-circuit antennas, consist of three, rather than two, metal layers separated by thin dielectric substrates. The center conductor usually has small width, and its function is similar to that of the center conductor of a coaxial line. If it couples to a resonant slot cut into the upper conductor, the device operates as an antenna [52]. Such stripline antennas have been designed in various configurations as microwave antennas and arrays. But they have found little interest up to now in the millimeter-wave region.* The probable reason is the greater structural complexity of these tri-plate devices. The stripline technique has been used, however, in the design of feed systems for microstrip millimeter-wave arrays [52, 53]. The tri-plate approach permits one to minimize energy leakage, which can be a problem particularly near the array feed point where power levels on the transmission lines are high.

A third type of printed-circuit antenna, i.e., holographic antennas [195], which are rather different from the others, is reviewed briefly. Similar to stripline antennas, these antennas have been studied, up to now, at microwave frequencies. But they appear well suited for millimeter-wave applications and are, therefore, included in the discussion.

Permittivity and loss tangent values for typical substrate materials are listed in Table 1.

---

*There are exceptions, however. For example, Sedivec and Rubin have designed a 4 × 4 element stripline array for the 44-GHz band [196]. The array has the advantage of a large bandwidth.

Microstrip Antennas with Electrically Thin Substrates

Small Arrays of Microstrip Patch Antennas: Feed Arrangements and Tolerance Problem—Fig. 57 shows the metallization pattern of a typical 4 × 4 element patch resonator array including a microstrip feed system. The wide portions of the feed lines are quarter-wavelength impedance transformers matching the element impedances of approximately 120 Ω (for square patches at resonance) to the 50-Ω input port of the overall antenna.

The length dimension b of the patches is determined by the resonance condition which requires that b (with an appropriate edge effect correction) is equal to a half-wavelength in the substrate material. The width w determines the input admittance of the patch resonators and can be utilized to control the excitation amplitudes of the array elements on an individual basis. The substrate thickness h is typically in the order of 5 to 10 mils for antennas printed on low-loss, low-permittivity substrates with $\epsilon_r = 2$ to 4 and designed for operation in the 30- to 100-GHz band. The choice of h involves a trade-off. Greater substrate height will increase bandwidth and reduce feed line losses; smaller height will help to minimize feed system radiation, improve pattern quality, and reduce cross-polarized

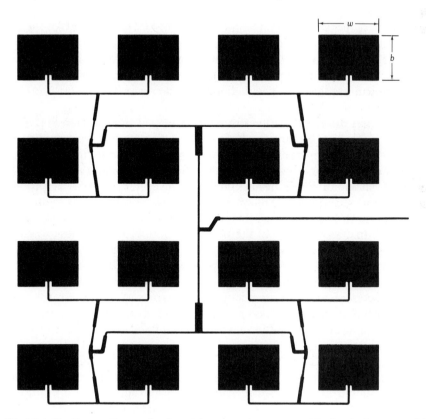

Fig. 57. Metallization pattern of microstrip resonator array with 4 × 4 elements. (*After Weiss and Cassell [49]*)

radiation. Note, however, that the radiation resistance of the antenna patches is practically independent of h as long as the substrate remains electrically thin.

The patch dimensions b and w are very small in the millimeter-wave region, and the width of the microstrip transmission lines feeding the patch elements is even smaller; it is typically in the order of 0.1 mm in the 30- to 100-GHz band and very accurate etching techniques are required for the fabrication of such arrays. Weiss has shown that close tolerances can be achieved through the use of glass-plate negatives and by choosing the thinnest possible copper cladding for the substrate [48–50]. Glass plates provide greater stability and higher resolution than the film negatives typically used in etching microwave antennas; and for typical substrate materials such as Duroid, Polyguide, and quartz the use of ¼ oz (7.09 g, 0.4 mil) copper cladding—instead of the more conventional 1 oz (28.35 g, 1.4 mils) cladding—resolves the "undercutting" problem at the edges of the microstrip patches and feed lines and thus permits maintaining close tolerances. As a general rule, devising precise fabrication techniques should be regarded as an integral part of the design of microstrip millimeter-wave antennas. It is advisable, also, to fabricate and optimize these antennas first at a lower scale frequency, which allows a much greater sensitivity in refining antenna performance [49]. Reduction of the final production mask to the appropriate millimeter-wave dimensions is easily and accurately done by photography.

Using these techniques, Weiss has designed 4×4 element arrays for the 35- and 60-GHz bands [48, 50]. The arrays are etched on Duroid 5880 substrates, 10 mils and 5 mils thick, respectively, and have the general layout shown in Fig. 57. Measured gain (17 dB) and radiation patterns were found to be in good agreement with theoretical predictions. The radiation efficiency is in the 75- to 80-percent region.

As a second example of a microstrip millimeter-wave antenna, Fig. 58 shows a 4 × 8 element 35-GHz array which can be operated independently in two orthogonal polarizations; the array was designed by Lalezari [61]. A dual corporate feed system is used whose input ports can be seen near the top and bottom of the array. The layout of the transmission lines solves the topological problem of accommodating two printed feed systems on the same substrate without "crossed wires." Note furthermore that in either feed system the antenna patches are excited in pairs such that the currents in the transmission line sections close to each pair have opposite directions. Hence feed-line radiation is compensated to a large degree and a low cross-polarization level can be achieved. For the array of Fig. 58 cross-polarized radiation remains 28 dB below the main beam level, which results in a port-to-port isolation greater than 30 dB. Arrays of this type can be used as building blocks for large arrays and permit full-duplex operation, i.e., simultaneous transmission and reception in the same frequency band. The array is printed on a Duroid substrate.

Large Arrays of Microstrip Patch Antennas: Feed Arrangements and Efficiency Problem—Small arrays of microstrip millimeter-wave antennas, for example, 4×4 or 8×8 element arrays, can be designed to have good efficiency and pattern quality. But in large arrays, which are needed to obtain high directivity gain, losses can be substantial. Most of these losses occur in the feed system of the array. Microstrip lines are not low-loss lines and power dissipation can be large, in particular in a

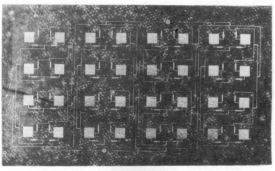

Fig. 58. Dual-polarization microstrip array for 35-GHz band. (*Courtesy Ball Aerospace Systems Division*)

printed corporate feed system consisting of long runs of microstrip lines. On the other hand, losses in the radiating elements of a microstrip array usually are negligible in comparison to feed system losses. In the lower millimeter-wave region, the radiation efficiency of microstrip patch resonators typically exceeds 80 percent. A second problem associated with microstrip arrays with long feed lines is that inhomogenieties in the substrate material can lead to phase errors in the excitation of the array elements ("detuning" of feed system) and, thus, to pattern distortions.

Accurate formulas for the calculation of the propagation and attenuation constants of microstrip lines are available in the literature. These formulas are lengthy, but they are shown here since in large microstrip arrays, as in all large arrays, the design of the feed system is a critical problem. The equations given below were taken from the book by Gupta, Garg, and Chadha [197], where also references to the earlier literature on this subject can be found. The characteristic impedance Z_0 of a microstrip line is determined by the permittivity ϵ_r and height h of the substrate, and the width w and thickness t of the metal strip.* According to [197],

$$
Z_0 = \begin{cases}
\dfrac{1}{2\pi}\dfrac{\eta}{\sqrt{\epsilon_{\text{eff}}}}\ln\left(\dfrac{8h}{w_e} + 0.25\dfrac{w_e}{h}\right) & \text{for } w/h \leq 1 \\[2ex]
\dfrac{\eta}{\sqrt{\epsilon_{\text{eff}}}}\left[\dfrac{w_e}{h} + 1.393 + 0.667\ln\left(\dfrac{w_e}{h} + 1.444\right)\right]^{-1} & \text{for } w/h \geq 1
\end{cases} \qquad (19a)
$$

---

*In the microwave region the thickness of the metal strip is usually very small compared with the substrate height, and the dependence of the line parameters on t amounts to a small correction only. In the millimeter-wave region, t/h may be as large as 0.1 to 0.2 and strip thickness effects cannot be neglected. Equation 19 for Z_0 and the subsequent equations for α_{cond} account for these effects approximately; expressions which would yield the t dependence precisely do not seem to be available as yet in the literature.

where

$$\epsilon_{\text{eff}} = \frac{\epsilon_r + 1}{2} + \frac{\epsilon_r - 1}{2}\left[\left(1 + 10\frac{h}{w}\right)^{-1/2} - \frac{1}{2.3}\frac{t}{\sqrt{wh}}\right] \tag{19b}$$

$$\frac{w_e}{h} = \frac{w}{h}\left[1 + \frac{1.25}{\pi}\frac{t}{w}\left(1 + \ln\left(\frac{2B}{t}\right)\right)\right] \tag{19c}$$

$$B = \begin{cases} 2\pi w & \text{for } w/h \leq 1/2\pi \\ h & \text{for } w/h \geq 1/2\pi \end{cases} \tag{19d}$$

$$\eta = 120\pi \text{ ohms}$$

Usually, the substrate parameters ϵ_r and h and the thickness t of the metal cladding are given, and Z_0 is prescribed. Equation 19 can then be regarded as implicit relations for the determination of w. With w determined, the propagation constant β is found from

$$\beta = k_0\sqrt{\epsilon_{\text{eff}}}$$

where ϵ_{eff}, the effective dielectric constant of the feed lines, is given by (19b). The attenuation constant α consists of two terms

$$\alpha = \alpha_{\text{cond}} + \alpha_{\text{diel}}$$

where α_{cond} accounts for the conduction loss and α_{diel} for the dielectric loss. We have

$$\alpha_{\text{cond}} = \begin{cases} 1.38A\dfrac{R_s}{Z_0}\dfrac{\lambda_0}{h}\dfrac{32 - (w_e/h)^2}{32 + (w_e/h)^2}\ \text{dB}/\lambda_0 & \text{for } w/h \leq 1 \\[2ex] 6.1 \times 10^{-5}A\,R_s Z_0 \epsilon_{\text{eff}}\dfrac{\lambda_0}{h}\left[\dfrac{w_e}{h} + \dfrac{0.667w_e/h}{w_e/h + 1.444}\right]\text{dB}/\lambda_0 & \text{for } w/h \geq 1 \end{cases}$$

$$\alpha_{\text{diel}} = 27.3\frac{\epsilon_r}{\epsilon_r - 1}\frac{\epsilon_{\text{eff}} - 1}{\sqrt{\epsilon_{\text{eff}}}}\tan\delta \ \text{dB}/\lambda_0$$

In these expressions,

$$A = 1 + \frac{h}{w_e}\left[1 + \frac{1}{\pi}\ln\left(\frac{2B}{t}\right)\right]$$

$$R_s = \left(\frac{\omega\mu_0}{2\sigma}\right)^{1/2}$$

Characteristic impedance Z_0 and surface resistance R_s are in ohms, σ is the conductivity of the strip, $\tan\delta$ is the loss factor of the substrate, and ϵ_{eff}, w_e/h, and

B are given by (19b), (19c), and (19d), respectively. It is convenient to express α in decibels per free-space wavelength.

Typically, a well-designed 50-Ω feed line on a 5-mil substrate will have an attenuation coefficient [63]

$$\alpha = \begin{cases} 0.12 \text{ dB}/\lambda_0 & \text{for Duroid} \\ 0.14 \text{ dB}/\lambda_0 & \text{for quartz} \\ 0.28 \text{ dB}/\lambda_0 & \text{for alumina} \end{cases}$$

These values apply to the 30- to 100-GHz band and include both conduction and dielectric losses, with the former losses yielding the dominant contribution. Microstrip losses tend to increase with line impedance, and for a given impedance the conduction loss increases with decreasing substrate height, while the dielectric loss remains approximately constant. It is interesting to note, however, that in monolithic circuits where microstrip lines are created on silicon or gallium arsenide substrates, the dielectric loss is much higher than the conduction loss [198].

James and Hall [62] have pointed out that surface roughness of the microstrip metal surfaces can have a substantial effect on the conduction loss. In the presence of surface roughness, α_{cond} in the above expression for α must be replaced by

$$\alpha'_{\text{cond}} = \alpha_{\text{cond}} \left\{ 1 + \frac{2}{\pi} \tan^{-1} \left[1.4 \left(\frac{\Delta}{\delta_s} \right)^2 \right] \right\}$$

where Δ is the rms surface roughness and δ_s the skin depth. Apparently, when Δ significantly exceeds the skin depth the conduction loss will be increased by a factor of 2. The expression for α'_{cond} was derived for a triangular roughness profile but should yield good results also for other profiles. Estimates for Δ were obtained by James and Hall for commercially available Duroid substrates with 1- and ½-oz (28.35-g and 14.17-g) copper claddings [62]. These estimates were made from electron microscope photographs of the surfaces and, for electrodeposited copper, yielded Δ values of 2.5 µm and 1.25 µm for the 1- and ½-oz claddings, respectively. Corresponding values for rolled copper were close to 0.42 µm for both copper weights. Comparison of these values with the skin depth of copper, which is 0.35 µm at 30 GHz and 0.21 µm at 100 GHz, indicates that α'_{cond} will not be far away from the upper limit of $2\alpha_{\text{cond}}$. The numerical values for α given above, therefore, may be somewhat optimistic. For polished copper, on the other hand, Δ will be much smaller than δ_s and surface roughness effects should be insignificant [62].

Fig. 59 shows a 32 × 32 element patch resonator array for the 38-GHz band which was designed for broadside radiation with a directivity of 35 dB and a beamwidth of approximately 3° [49]. The array is etched on a 5-mil-thick Duroid substrate and uses a corporate feed system. The problems associated with long runs of microstrip feed lines were minimized, in this case, by subdividing the array into four equal subarrays of 16×16 elements each, and by using a four-way power divider in metal waveguide to feed the subarrays in parallel. This technique also eliminates parasitic radiation from bends and corners of the feed system near the input terminal where in large arrays the power level is high.

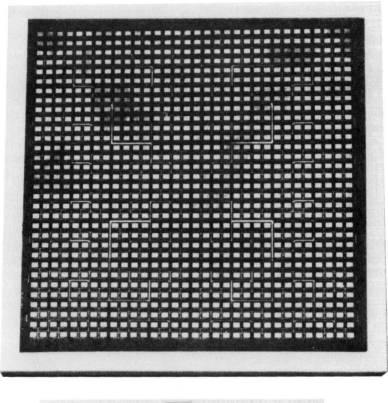

a

Fig. 59. A 32 × 32 element microstrip antenna array for 38-GHz band. (*a*) Front view. (*b*) Back view. (*Courtesy Weiss and Cassell [49]*)

The power divider is milled into the aluminum base plate whose thickness of 6.5 mm is appropriate to accommodate the waveguide. At 38 GHz, losses in the waveguide legs of the power divider (the small screws in Fig. 59b outline its contour) are smaller than 0.008 dB/cm as compared with a loss of approximately 0.24 dB/cm in the microstrip feed lines.* An additional advantage of this solution is that the increased thickness of the base plate enhances the mechanical strength of the antenna.

Experiments with a first model of the array indicated efficiency and pattern quality problems, but subsequent investigations have shown that arrays of this type, when carefully designed, are capable of providing good pattern quality, low vswr over a bandwidth in the order of a few percent, and acceptable efficiency. The

*The array shown in Fig. 59 was a first model. By careful design, losses in the microstrip lines could probably be reduced by a factor of 2.

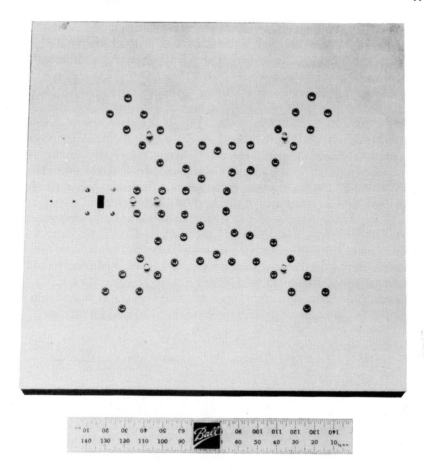

b

Fig. 59, *continued.*

efficiency could be further improved by the use of a higher-order eight- or sixteen-way power divider, i.e., by further extending the metal waveguide portion of the feed system at the expense of the microstrip portion.

An alternative approach to enhancing antenna efficiency is the use of *series-fed* arrays where the length of feed-line runs is substantially reduced in comparison to a corporate feed system [51, 63–65]. Fig. 60 shows the metallization pattern of a linear array of this type. The excitation amplitudes of the various array elements

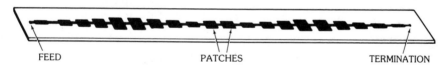

Fig. 60. Series-fed microstrip patch array. (*After Jones, Chow, and Seeto [65], © 1982 IEEE*)

are controlled by their widths and the excitation phases by the element spacings. Parameter dependencies are well established and such arrays are well suited for accurate pattern shaping. The arrays may be designed as resonant arrays or as traveling-wave arrays. In a *resonant* array the radiating patches are spaced at intervals of one guide wavelength and the array radiates in the broadside direction. The bandwidth, however, is narrow, which is not surprising since both the individual array elements and the overall array are operating at resonance. In *traveling-wave* arrays, on the other hand, the spacing of the radiating patches is not critical and the bandwidth is larger. In principle, by varying the resonance frequencies of the patches from element to element, an impedance bandwidth of 10 to 20 percent can be achieved which substantially exceeds that of the individual elements [64]. But in such arrays the portion of the antenna which effectively radiates shifts with frequency and it is difficult to maintain good pattern quality over the full frequency band so that the pattern bandwidth rather than the impedance bandwidth will become the limiting factor. Traveling-wave arrays where all elements resonate at the same frequency have smaller impedance bandwidth but the advantage of superior pattern quality.

A certain disadvantage of traveling-wave arrays is that the main beam will scan with frequency.* The scan rate in degrees is given by the formula

$$\Delta\theta = \frac{180°}{\pi} \frac{d/df(f \cdot n_{\text{eff}}) - \sin\theta}{\cos\theta} \frac{\Delta f}{f} \cong \frac{180°}{\pi} \frac{\sqrt{\epsilon_r} - \sin\theta}{\cos\theta} \frac{\Delta f}{f} \qquad (20)$$

where $\Delta f/f$ is the relative frequency change, θ is the main beam direction counted from broadside,** n_{eff} is the effective propagation wave number of the traveling-wave array relative to the free-space wave number k_0, and ϵ_r is the permittivity of the substrate; n_{eff} does not vary strongly with frequency and for the present purpose may be approximated by $\sqrt{\epsilon_r}$. A typical microstrip array will be printed on a low-loss substrate of permittivity ϵ_r between 2 and 3 and will be designed to produce a main beam not too far away from broadside. Under these conditions, (20) predicts a scan rate of approximately 1° per 1 percent of frequency change, which is in good agreement with experimental evidence [66]. For high-permittivity substrates, such as alumina and silicon, the scan rate would be about twice as large. The beam-scanning effect may appreciably narrow down the usable frequency band of a high-gain traveling-wave array with a beamwidth in the order of a few degrees, but it is not likely to noticeably affect the performance of a broad-beam, low-gain antenna.

A planar array of series-fed microstrip antennas can be obtained by printing several linear arrays in parallel on the same substrate and by connecting their feed points to a corporate feed or, simply, a common microstrip feed line. A center-fed array of this type is shown in Fig. 61. The array is of the traveling-wave type, with

---

*Applications where beam scanning is desirable usually require a larger scan range than can be accommodated within the limited bandwidth of microstrip antennas. These antennas appear better suited for fixed-beam operation.
**The sign of θ is positive (negative) in the range of forward (backward) directions.

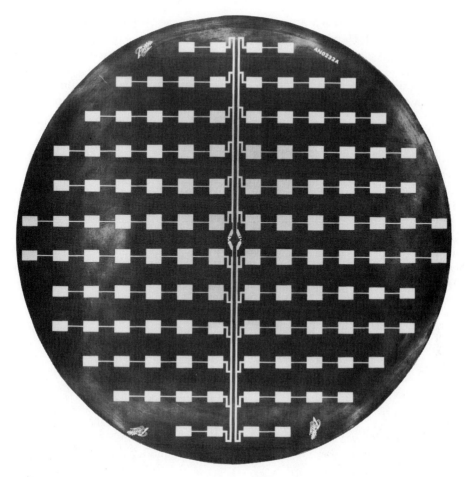

Fig. 61. Series-fed microstrip array for broadside radiation. (*Courtesy Ball Aerospace Systems Division*)

all elements resonating at the same frequency. The antenna is designed for radiation in the broadside direction and is interesting for several reasons:

(*a*)   The short transmission line runs of the series feed-system result in low losses and good antenna efficiency.

(*b*)   The widths of the radiating elements and hence their excitation amplitudes decrease away from the array center in a roughly circular symmetric fashion. The effect is reinforced by the power reduction from element to element due to radiation. Thus a tapered aperture illumination is obtained, resulting in low side lobes.

(*c*)   The array elements are spaced to be excited in equal phase at the design frequency so that the array radiates in the broadside direction. Away from the design frequency the excitation phase will vary from element to element in a linear fashion. But it is easily seen from the symmetry of the array configuration that the two partial beams produced by the right and the left half of the array will scan in opposite directions (while their pattern shape is the same). Hence the main beam of the composite pattern will continue to be pointed into the broadside direction and no beam scanning will occur. The beamwidth, however, will increase with increasing deviation from the design frequency and side lobe levels will rise, eventually resulting in a split main beam. The pattern bandwidth of the array is therefore limited.

The particular array shown in Fig. 61 was designed for use in an X-band radar system. The gain of the array is 27 dB, the efficiency is 75 percent, the side lobes are more than 20 dB below the main beam level, and the bandwidth is 1 percent. Maximum power is 1 kW and the weight is 100 g. The same design principle, of course, can be used in the millimeter-wave band to obtain fixed-beam antennas of high gain and good efficiency, although the efficiency and power-handling capability will be somewhat lower, and the weight will be smaller. Fig. 62 shows a second example of a series-fed traveling-wave type array. Laid out in a four-quadrant configuration, the array operates as a monopulse antenna in the 35-GHz band. The monopulse comparator visible at the center is integrated with the array elements on the same substrate. The arrays shown in Figs. 61 and 62 were designed by Lalezari. Both antennas are etched on Duroid substrates.

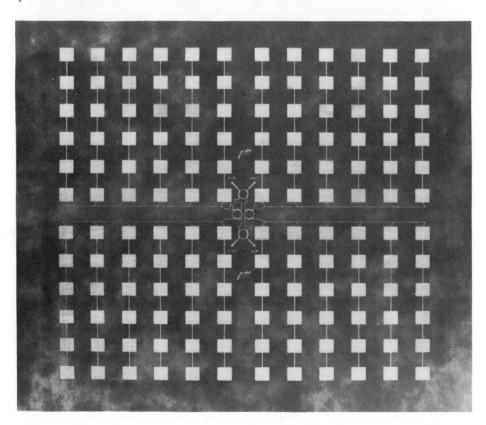

Fig. 62. A 35-GHz microstrip monopulse array with integrated comparator network. (*Courtesy Ball Aerospace Systems Division*)

A completely different technique for reducing feed system losses has recently been suggested by James and others [62, 67, 68]. As illustrated by Fig. 63a, microstrip feed lines are eliminated; they are replaced by a low-loss open dielectric waveguide (in the form of an insular guide) whose fringing fields excite the microstrip antenna patches. By varying the distances of the microstrip patches from the feed guide, as indicated in Fig. 63b, the excitation amplitudes of the array elements can be adjusted and a degree of pattern control exercised. Improvements in feed system efficiency that can be achieved by this approach may be seen from Table 2 [69], which applies to the specific example of an alumina feed guide placed on a quartz substrate, where it replaces a microstrip feed line with a characteristic impedance of 50 Ω. The table compares the theoretical attenuation coefficients of the two guides for three frequencies in the lower millimeter-wave band. The attenuation coefficient of silver-plated metal waveguide is included in the table for comparison. While metal waveguide has the lowest attenuation (and the largest size and production costs), insular guide has a substantial efficiency advantage over microstrip line and these improvements tend to increase with frequency. A second advantage of this feed method is its simplicity; the dielectric guide has a reasonable cross section and is easy to fabricate, and no contact has to be established between radiating elements and feed line [67].

Experiments with a 2×40-element array of this type have confirmed the usefulness of the concept [62]. Etched on a Duroid substrate, the array was dimensioned for broadside radiation at 90 GHz. An alumina feed guide was used with a cross section of 0.35×1.10 mm^2. The guide was operated in the fundamental HE_{11} mode, and the spacing of the radiating elements from the feed guide was tapered for uniform aperture illumination. Feed line losses and cross-polarization levels were low. But launching losses and side lobe levels remained a problem requiring further improvements.

The next logical step would be the use of insular guide not only for the feed line of the array but for the radiating elements as well, i.e., the microstrip patches in

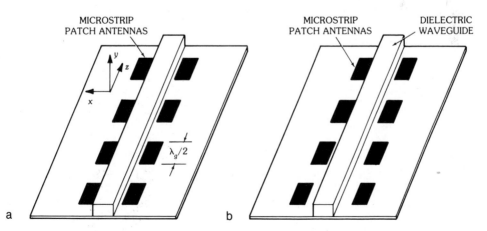

Fig. 63. Microstrip patch resonator arrays fed by dielectric image line. (*a*) Basic configuration. (*b*) Configuration permitting pattern control. (*After James and Henderson [68]*)

Table 2. Theoretical Attenuation Coefficient of Microstrip Line, Insular Guide, and Metal Waveguide in the Lower Millimeter-Wave Region (*After Knox* [69], © *1976 IEEE*)

Feed System		Frequency (GHz)	h (cm)	a (cm)	α (dB/cm)
Microstrip line (gold on fused quartz: 50 Ω)		30	0.027	0.054	0.056
		60	0.014	0.027	0.154
		90	0.009	0.018	0.280
Insular waveguide (alumina on fused quartz)		30	0.027	0.268	0.022
		60	0.013	0.134	0.055
		90	0.009	0.090	0.096
Metal waveguide (silver plated)		30		1.067	0.007
		60		0.376	0.016
		90		0.254	0.030

Figs. 63a and 63b would be replaced by dielectric resonators. These resonators may take the form of small rectangular or cylindrical dielectric blocks placed on the substrate. Arrays of this type have been studied by Birand and Gelsthorpe [70].

Microstrip Dipole Antennas—Microstrip array antennas which use printed half-wave dipoles rather than patch resonators as radiating elements have been suggested by James and others [54, 55] and by Williams [56]. The arrays consist of a linear sequence or a planar matrix of (unbroken) horizontal dipoles end-fed by high-impedance microstrip lines. The metallization patterns of two such arrays, a resonant array and a traveling-wave array, are shown in Fig. 64. The resonant array radiates in the broadside direction and pattern control can be exercised by varying the width of the dipoles across the array. The dipole spacing is fixed at one guide wavelength at the design frequency. The traveling-wave array radiates under an angle against the broadside direction and the main beam will scan with frequency. The radiation pattern can be controlled by adjusting the spacing and width of the dipoles and hence their excitation phase and amplitude. For both resonant and traveling-wave arrays the bandwidth is limited by the resonance behavior of the dipoles; since these dipoles are operating very close to the ground plate their impedance bandwidth is small. Measurements at 36 and 70 GHz have shown that such arrays are capable of providing good performance in the lower milli-meter-wave region, possibly up to 94 GHz. But further study is needed to improve

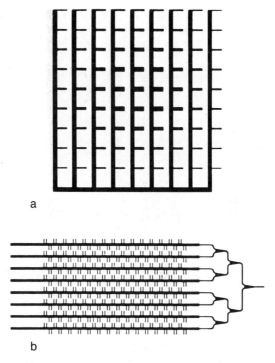

a

b

Fig. 64. Metallization patterns of printed horizontal half-wave dipole arrays fed by microstrip lines. (*a*) Resonant array. (*b*) Traveling-wave array. (*After James et al. [55]*)

pattern quality at 70 GHz and above [53]. In particular, radiation from the corporate feed of these arrays seems to be a problem necessitating an enclosed, tri-plate layout of the feed system near the array feed point, where power levels are high.

Fig. 65 shows a center-fed array with a diagonal layout of the feed system [56]. The characteristics of this array are similar to those of the patch array shown in Fig. 61: The use of a series feed-system results in short transmission line runs and good efficiency; power division at the feed-line branching points and radiation losses at the dipoles lead to a tapered amplitude distribution across the array aperture and low side lobes; and a symmetric distribution of the excitation phase about the array center ensures that the main beam remains pointed in the broadside direction when the frequency is shifted away from the design frequency. But the beamwidth increases with Δf and the pattern bandwidth is limited. An experimental 16×16 element array for the 36-GHz band printed on a polythene substrate of 0.8-mm thickness has shown a gain of 25 dB and side lobe and cross-polarization levels below -20 dB [56]. The array size was 6.8×6.8 cm^2. A rather good radiation efficiency of 60 percent was measured at the center frequency, but the bandwidth (corresponding to a 3-dB reduction in gain) was limited to 700 MHz.

Antenna-to-Waveguide Coupling—An effective technique for coupling a microstrip millimeter-wave antenna to a metal waveguide (mounted on the back of the base

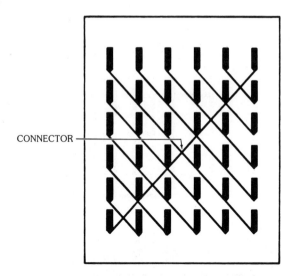

CONNECTOR

Fig. 65. Center-fed printed dipole array. (*After Williams [56], previously published in* Electronics Letters, *March 1978, by The Institution of Electrical Engineers*)

plate) is the use of a vertical probe which is in electrical contact with the microstrip feed line of the antenna and extends into the waveguide at a quarter-wavelength distance from the short-circuited end of this guide [48, 50]; see Fig. 66. With appropriate design, a vswr better than 2:1 can be achieved easily. The bandwidth of this coupling technique exceeds that of the antenna. Coupling to a coaxial line is straightforward and also leads to good results when an arrangement is used where the line feeds the antenna from the back [48]. Both methods can be combined by feeding a microstrip antenna from a metal waveguide via a microstrip-to-coax-to-waveguide transition. The transition would include a short section of coaxial line only so that the high losses normally associated with these lines at millimeter-wave frequencies are avoided. A transition of this type is often used by necessity since a probe which extends from the microstrip feed line into a metal waveguide (Fig. 66) has to pass through a small aperture in the base plate of the antenna. The thickness of the base plate is easily in the order of a quarter-wavelength or more in the millimeter-wave region so that the probe section in the aperture will act as a short piece of coaxial line [63]. Efficient coupling of microstrip antennas to other millimeter-wave guides remains to be studied. Work in this area is in progress and includes the examination of alternative arrangements for feeding these antennas from metal waveguides and coaxial lines [62, 199, 200].

Computer-Aided Design (CAD)—In many cases the design of microstrip array antennas to specifications can be handled efficiently by computer. In particular, the pattern synthesis problem can be solved in this way. The parameter dependencies of microstrip antennas are well understood; their radiation and impedance characteristics depend sensitively on their dimensions; and the overall array configuration (including substrate parameters) is usually specified beforehand so that the pattern synthesis problem reduces to the constrained adjustment of a well-

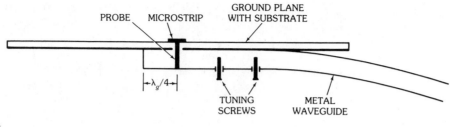

a

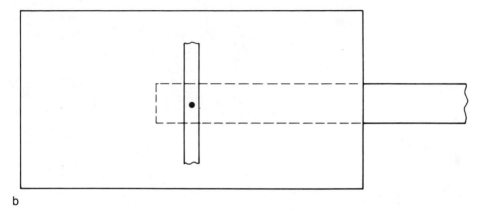

b

Fig. 66. Microstrip to metal waveguide (probe) transition. (*a*) Side view. (*b*) Top view. (*After Weiss [48]*)

defined set of geometrical parameters. Computer design may be supplemented by computer-controlled fabrication of the array where the fabrication control data are generated as part of the CAD procedure. In this case there is no need for drawings or layout artwork and the engineering process is considerably simplified [64].

As an example a computer code devised by Campi is mentioned here which permits the design and the generation of fabrication control data for series-fed linear arrays of rectangular microstrip patch antennas [64]. The code generates an equivalent network representation of the array, from which the radiation pattern can be calculated. Using an optimization subroutine the code then continuously varies the conductance values of the array elements until the computed radiation pattern approximates in the least-square sense the desired pattern specified by its directivity gain, main beam direction, and side lobe structure, which may be characterized by a Tschebychev, Taylor, binominal, or other distribution function. The code includes provisions for parametric and sensitivity studies of the pattern, a feature useful for engineering and optimization analyses. Measurements on antennas produced with the help of this code have demonstrated the usefulness of the CAD procedure for the 1- to 100-GHz region [64]. Fig. 67 shows a 28-element array designed for the 94-GHz band.

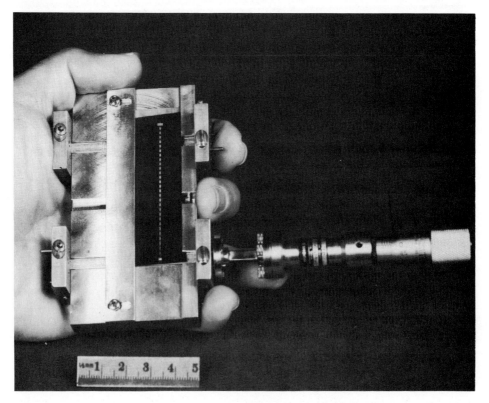

Fig. 67. Computer-designed 94-GHz microstrip antenna array. (*Courtesy Campi [64]*)

Microstrip Antennas with Electrically Thick Substrates

In the microwave region the thickness of typical substrates used in the design of printed antennas is a small fraction of a wavelength only. Thin substrates imply the use of resonator antennas of high Q in order to raise the radiation resistance to reasonable values. This, in turn, leads to a narrow bandwidth. In the millimeter-wave region, on the other hand, a substrate of low geometrical height may have an electrical thickness in the order of a quarter-wavelength to several wavelengths. Hence there is no need for using high-Q antennas, and other antenna configurations can be employed which provide substantially larger bandwidth. An example would be a printed dipole or loop antenna of large width w; Fig. 68a shows a dipole antenna of this type. The use of thick substrates has the additional advantage that fabrication tolerances become less critical.

A problem associated with substrates of large electrical thickness is the generation of *surface waves*. These surface waves are guided away by the substrate even though they do not radiate;* their power must therefore be counted as antenna loss. The design problem is to obtain a large bandwidth while maintaining good efficiency.

---

*Except for producing uncontrolled radiation at substrate edges or at other geometrical discontinuities.

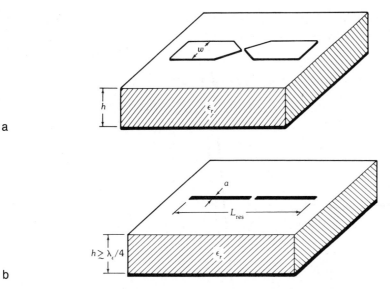

Fig. 68. Printed dipole antennas on electrically thick substrates. (*a*) Dipole of large width *w*. (*b*) Thin-wire dipole.

The theory of microstrip antennas on thick substrates has recently been studied by Alexopoulos, Katehi, and Rutledge [58, 59], who have established the basic parameter dependencies and provided systematic design information for these antennas. The results obtained so far apply to the case of a single printed dipole of small width *w* (wire dipole).* For most millimeter-wave applications high-gain antennas of reasonably broad bandwidth are required, and an array of dipoles of large width would be of interest. A theory of such arrays, however, is not as yet available and the results of the single-dipole study are discussed here in some detail since they are not only interesting in themselves, but provide conditions for substrate optimization which should apply to arrays as well. In addition, the study yields at least a worst-case estimate (thin dipole case) on attainable bandwidth. Alexopoulos [58] has pointed out that the analytical expressions for the field radiated by a microstrip dipole are obtained as products of a substrate factor which is independent of the antenna dimension and an antenna factor which is independent of the substrate parameters. This suggests that substrate optimization arguments will hold for all printed antennas regardless of their shape [58]. An independent study by Pozar [194] has yielded theoretical results in substantial agreement with [58, 59].

The following discussion draws freely from the cited papers by Alexopoulos and colleagues. The antenna configuration considered is shown in Fig. 68b. A center-fed wire dipole is placed on a substrate of permittivity ϵ_r and height h; the substrate is backed by a metal ground plane. The dipole is assumed to have the resonance length corresponding to this situation. Its resonance impedance R_{res}

---

*Pairs of dipoles and their mutual interaction effects have been studied in a separate investigation by Alexopoulos and Rana [201].

consists of two parts, $R_{res} = R_r + R_s$, where R_r is the radiation resistance and R_s the surface-wave excitation resistance. The radiation *efficiency* η of the microstrip dipole is

$$\eta = \frac{R_r}{R_r + R_s}$$

where substrate losses have been neglected. The dependence of R_{res}, R_r, and R_s on the substrate height h is shown in Fig. 69 for the example of a quartz substrate with $\epsilon_r = 4$. The cutoff h/λ_0 values of the various surface-wave modes supported by the substrate are indicated on the upper edge of the figure. These cutoff values are given by

$$\frac{h}{\lambda_0} = \frac{m}{4} (\epsilon_r - 1)^{-1/2} \tag{21}$$

where $m = 2n$ for the nth TM mode and $m = 2n + 1$ for the nth TE mode. As before, λ_0 is the free-space wavelength.

The important result illustrated by Fig. 69 is that the radiation efficiency reaches a maximum just below the cutoff thickness of the lowest-order TE surface-wave mode;* careful numerical studies [58] have shown that this result holds not only for quartz but also for materials of any (reasonable) permittivity. From the physical viewpoint this is easy to understand. The lowest-order surface-wave mode, TM_0, does not have a cutoff frequency and is always present, but for substrates of low height it is only weakly excited by a horizontal dipole antenna, while the radiated power increases with h/λ_0. The TE_0 surface-wave mode, on the other hand, would be strongly excited by the antenna. Hence maximum η occurs just before this mode can exist.

With (21), the optimum substrate thickness h_{opt} satisfies

$$h_{opt} \lesssim \frac{\lambda_0}{4\sqrt{\epsilon_r - 1}} \tag{22}$$

The optimum thickness and the corresponding maximum radiation efficiency η_{max} are shown in Fig. 70 as functions of substrate permittivity ϵ_r. For completeness the resonance length L_{res} of the dipole is included in the figure. The graph shows that the requirement of high radiation efficiency implies the use of a substrate of relatively low permittivity. For η_{max} to exceed 50 percent, ϵ_r must be smaller than 4.5. Hence quartz would be a useful substrate material while such high-permittivity materials as silicon and gallium arsenide would not be suitable for this application. The optimum substrate thickness h_{opt} is moderately larger than a quarter-wavelength (in the substrate material) and remains within reasonable limits even in the lower millimeter-wave region. For example, at 38 GHz, h_{opt} would be approximately 1.2 mm for quartz and approximately 1.8 mm for Duroid ($\epsilon_r = 2.20$).

---

*The radiation efficiency approaches 100 percent for $h \to 0$. But the radiation resistance vanishes at this point so that this first maximum of η is meaningless.

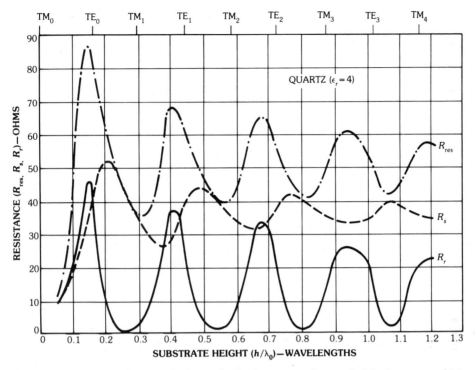

Fig. 69. Resonance resistance of microstrip dipole versus substrate height for quartz. (*After Alexopoulos, Katehi, and Rutledge [58], © 1983 IEEE*)

The impedance *bandwidth* (bw) of a dipole antenna near resonance can be expressed as

$$\text{bw} = \frac{\Delta f}{f_0} = \frac{2R_{\text{res}}}{f_0(dX/df)_{\text{res}}} \tag{23}$$

where $(dX/df)_{\text{res}}$ is the derivative of the input reactance of the antenna at resonance.* The dependence of $\Delta f/f_0$ on substrate thickness h/λ_0 is shown in Fig. 71 for $\epsilon_r = 4$. A dipole of radius $a/\lambda_0 = 10^{-4}$ is assumed. Maximum bandwidth, similar to maximum efficiency, occurs just below the cutoff thickness of the TE$_0$ surface-wave mode [58]. This is typical for low-permittivity substrates, with $\epsilon_r < 5$. For large ϵ_r, on the other hand, maximum bandwidth would be obtained at a substrate thickness noticeably greater than h_{opt}. This is illustrated by Fig. 72. The solid curve shows the maximum attainable bandwidth as a function of substrate permittivity; the dashed curve shows the bandwidth at $h = h_{\text{opt}}$, i.e., under the

---

*In computing the curves of Figs. 71 and 72, Alexopoulos and colleagues [58] have used a modified version of (23), which expresses the bandwidth in terms of the parameters discussed above, i.e., $\Delta f/f_0 = 2R_{\text{res}}/[L_\lambda(dX/dL_\lambda)]_{\text{res}}$, where $L_\lambda = L/\lambda_\epsilon$ is the antenna length normalized to wavelength in the dielectric, and the term in square brackets is taken at resonance. This modified definition of bandwidth utilizes the fact that the antenna reactance near resonance changes more rapidly with antenna length than with substrate height.

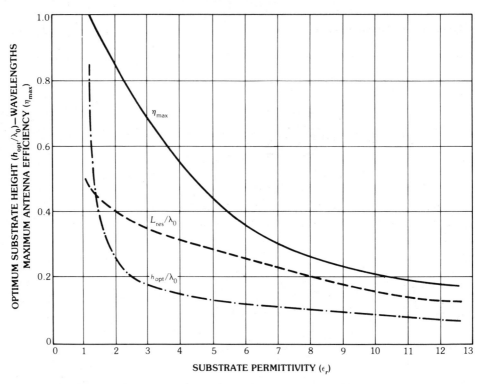

Fig. 70. Optimum substrate height and maximum antenna efficiency versus substrate permittivity. (*After Alexopoulos, Katehi, and Rutledge [58], © 1983 IEEE*)

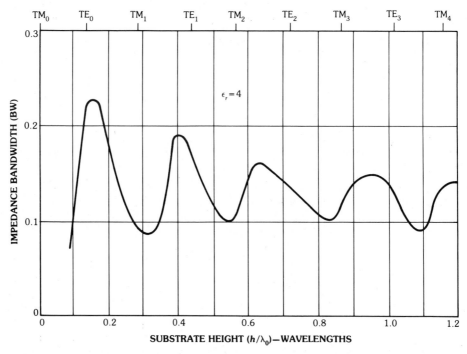

Fig. 71. Bandwidth of thin printed dipole versus substrate height for $\epsilon_r = 4$. (*After Alexopoulos, Katehi, and Rutledge [58], © 1983 IEEE*)

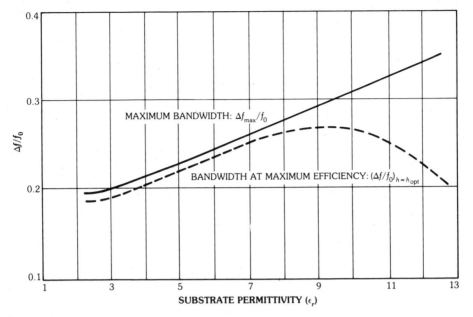

Fig. 72. Bandwidth as a function of substrate permittivity. (*After Alexopoulos, Katehi, and Rutledge [58], © 1983 IEEE*)

maximum efficiency constraint. In this case, $\Delta f/f_0$ reaches a maximum of 26 percent at $\epsilon_r = 9.4$. In the range of low permittivities (less than 4.5), where high radiation efficiencies can be obtained, the bandwidth is limited to approximately 20 percent. This, however, is a substantial improvement over the bandwidth of microstrip patch antennas on thin substrates, which is on the order of a few percent only. Note, furthermore, that this bandwidth applies to wire dipoles of very small radii. Printed dipoles of large width can be expected to have substantially greater bandwidth.

The theoretical radiation pattern of a dipole of length $L = 0.3\lambda_0$ placed on a quartz substrate of thickness $h = 0.2\lambda_0$ is shown in Fig. 73. The pattern is essentially hemispherical, but both the E- and H-plane patterns have a null in the horizontal plane $\theta = \pm 90°$. Evidently any radiation in this plane occurs in the form of surface waves.

Latest theoretical results by Alexopoulos and Jackson [71, 72] show that the performance of microstrip dipole antennas can be significantly improved by the use of a low-loss superstrate whose permittivity (or permeability) exceeds that of the substrate.* It appears that by careful selection of the substrate and superstrate parameters substantial improvements in gain and radiation resistance can be realized over a relatively large bandwidth. Furthermore, these parameters can be adjusted such that a nearly omnidirectional E- or H-plane pattern is achieved. Alternatively, a very high radiation efficiency—close to 100 percent with no surface

---

*Both infinitesimal dipoles [71] and dipoles of finite length and width [72] have been considered in this study.

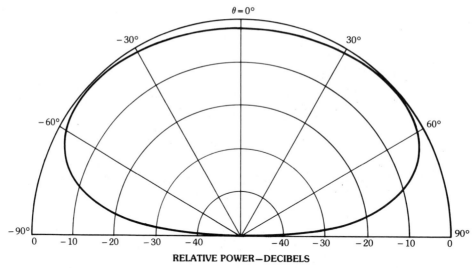

a

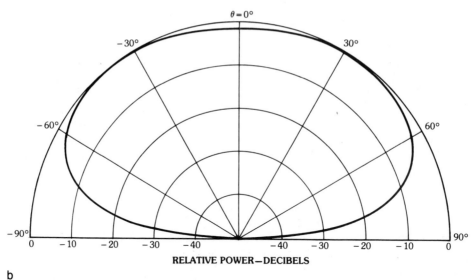

b

Fig. 73. Power radiation pattern of microstrip dipole on quartz substrate, with $h/\lambda_0 = 0.2$, $L/\lambda_0 = 0.3$, and $\epsilon_r = 4$. (*a*) *E*-plane pattern. (*b*) *H*-plane pattern. (*After Alexopoulos, Katehi, and Rutledge [58],* © *1983 IEEE*)

wave excited—can be obtained over a limited bandwidth. An additional very practical advantage is that a superstrate will act as a protective cover for the antenna.

Millimeter-wave antennas are usually designed for high directivity gain, which implies the use of arrays of printed dipoles. This leads to the additional question of mutual coupling between closely spaced printed dipoles—surface waves supported by the substrate are likely to increase this coupling—and to the problem of devising efficient feed systems for such arrays. Ideally, i.e., for simplicity and ease of fabrication, the feed system should be printed with the dipoles on the same substrate. But a balanced, two-wire feed system would be complicated and not advantageous for integrated designs, while an unbalanced single-wire system would lead to feed-line leakage, multimoding, and impedance matching problems because of the large substrate height. The use of a dual substrate appears more promising. In this case a single-line microstrip feed system would be printed on the lower portion of the substrate while the (unbroken) dipoles would be etched on the upper portion, and coupling would be effected electromagnetically, i.e., without conductive contact between feed lines and antennas. In the microwave region, where the substrate thickness is only a small fraction of a wavelength, this method has been used successfully [202–204]. Its feasibility for feeding antennas on fairly thick substrates has been shown recently by Katehi and Alexopoulos for the single-dipole case [205]. But for large microstrip millimeter-wave arrays on thick substrates, the design of efficient, easy-to-fabricate feed systems remains up to now an open question requiring further study. (The Alexopoulos group is investigating this problem at present.) A similar question remains for the problem of mutual coupling in such arrays. A theoretical study on the interaction between two printed dipoles has confirmed that mutual coupling is a significant effect which cannot be neglected [201]. Suppression of surface waves through the use of a superstrate of high ϵ_r or μ_r [71, 72] seems to be a promising approach to this problem deserving further attention.

Holographic Antennas

When a hologram made in accordance with the interference pattern of two propagating waves is illuminated by one of these waves, the amplitude and phase distribution of the other wave is reconstituted on the surface of the hologram. A holographic plate may thus be used as a beam-shaping antenna; it permits one to derive a desired aperture distribution (across the hologram) from a given illumination incident from a primary feed. In the microwave and millimeter-wave regions a hologram can be approximated by an appropriately shaped metallization pattern etched on a printed-circuit board. The use of a metal pattern will, in general, permit one to closely approximate the desired phase distribution across the holographic aperture, whereas the amplitude distribution (which is less critical) may not be entirely independent of that of the illuminating field.

Iizuka and coworkers have described an antenna of this type; the following discussion is based on a paper by these authors [195]. The antenna, depicted in Fig. 74, consists of a holographic plate with a metallization pattern made by photoetching a printed-circuit board in accordance with the interference pattern of a spherical wave originating from the horn aperture C and a plane wave incident normal to the surface of the printed-circuit board. The holographic pattern, in this case, consists of a set of concentric circular rings spaced one wavelength apart. When the plate is illuminated by a spherical wave radiated from the horn aperture the waves scattered from the metal rings have equal phase, and two beams emanate from the plate in opposite directions normal to the plate surface. In holographic terminology the true image and the conjugate image are reconstructed [195].

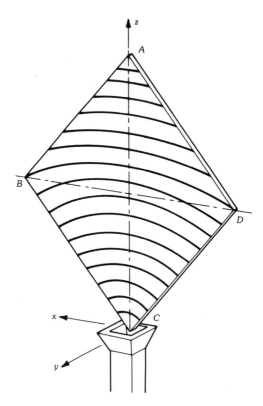

Fig. 74. Holographic antenna. (*After Iizuka et al. [195], © 1975 IEEE*)

If two holographic plates are used in a parallel arrangement, the conjugate beam traveling in the backward direction can be suppressed while the main beam in the forward direction is reinforced [195]. The condition is that the two plates are spaced a quarter-wavelength apart and that the metal rings on the second plate have radii a quarter-wavelength larger than those on the first plate. Figs. 75a and 75b illustrate these conditions. Fig. 75c shows, moreover, that in the presence of the second plate the wave reflected toward the feed horn (downward direction) is canceled. Hence the impedance characteristic of the antenna is improved, resulting in increased bandwidth. Radiation in the upward direction can be reduced to a tolerable level by adjusting the position of the plates relative to the feed horn [195]. Beam shaping is possible by modifying the metallization pattern of the plates.

Fig. 76 compares the measured E- and H-plane patterns of a one-plate and a two-plate antenna. The plates had a size of $12\lambda_0 \times 12\lambda_0$ with sixteen metal rings printed on each board by photoetching. A pyramidal horn with aperture size $0.56\lambda_0 \times 0.64\lambda_0$ was placed such that its apex was located near the common center point of the metal rings of each plate. The dual-plate antenna has a unidirectional pattern with a front-to-back ratio of 20 dB and a main lobe level 7 dB higher than that of the single-plate antenna. Back lobe radiation can be reduced further by increasing the number of holographic plates [195].

By replacing the feed horn with a printed antenna etched with the holographic

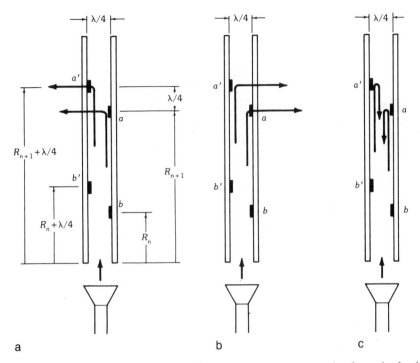

Fig. 75. Dual-plate holographic antenna. The spherical wave emanating from the feed horn is scattered at the metal rings of the two plates. (*a*) In the forward direction the scattered fields of the two plates have the same path lengths and add in phase. (*b*) In the backward direction the path lengths differ by $\lambda_0/2$; the scattered fields are in antiphase and cancel each other. (*c*) In the downward direction the path lengths also differ by $\lambda_0/2$ and the wave reflected toward the feed horn is eliminated. (*After Iizuka et al. [195], © 1975 IEEE*)

pattern on the same circuit board, a fully integrated antenna would be realized [195]. This possibility appears particularly attractive for millimeter-wave applications. The experiments were conducted at 19 GHz.

5. Integrated Antennas

Fundamental contributions to the state of the art in millimeter-wave antennas are currently being made by the development of integrated antennas where printed circuits and active solid-state components, usually in monolithic form, become an integral part of the radiating structure. Such antennas serve not only as electromagnetic radiators, but include additional functions (e.g., mixing, amplification, or phase shifting) in a highly structured but very compact device. Integrated antennas are of interest not only for the lower millimeter-wave region but are particularly well suited for the so-called near–millimeter-wave range above 100 GHz [88, 99, 100, 206–213]; and it can be expected that eventually they will be extended into the sub-millimeter-wave region as well. The small size of radiating elements, components, and circuits in these frequency bands strongly suggests the use of integrated planar design approaches.

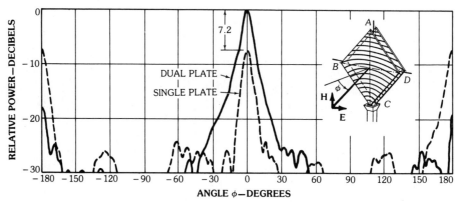

a

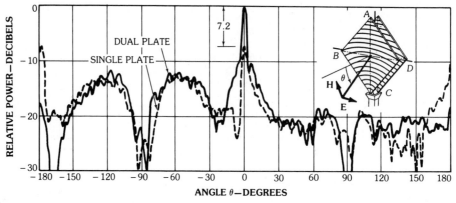

b

Fig. 76. Radiation patterns of single-plate and dual-plate holographic antennas. (*a*) *E*-plane pattern. (*b*) *H*-plane pattern. (*After Iizuka et al. [195], © 1975 IEEE*)

The microstrip antenna technique lends itself to the design of array antennas where transmission lines, phase shifter circuits, and receiver or transmitter circuits are integrated with the radiating elements on the same substrate (which may consist of several layers when circuits for more complex functions have to be accommodated). Both dielectric antennas and printed-circuit antennas are suited for monolithic designs using silicon or gallium arsenide (GaAs) technology. While silicon has been used extensively in the past, it appears that future monolithic designs will prefer gallium arsenide, in particular at millimeter-wave frequencies. Silicon is mechanically stronger and easier to work with. The primary advantage of gallium arsenide is its high electron mobility, which is critical for the performance of monolithic active components at higher frequencies.

Devising precise fabrication techniques is an essential step in the design of integrated antennas and must be regarded as an integral part of the design procedure. It is evident that the feasibility of these small but highly structured devices depends on the availability of appropriate fabrication methods and that the quality and performance of the finished device are ultimately determined by the accuracy of the manufacturing process.

The crucial problem is the design and fabrication of the integrated active components. In comparison to a monolithic mixing or switching diode, for example, the antenna elements themselves are large in size and simple in structure. Practical problems include heat sinking for silicon components, and mechanical stress when gallium arsenide, a mechanically fragile material, is used. Also, available space for integrated components may be limited in the case of printed-circuit antennas since the antenna patches, though physically small, are large from an IC viewpoint and together with their feed lines take up a substantial portion of the substrate surface. In addition, a certain clearance is required between the radiating elements and feed lines on the one hand and any integrated circuits and components on the other in order to minimize undesired near-field coupling into these circuits and avoid pattern distortions. Because of these and similar constraints the electrical performance of integrated antennas may remain somewhat below the state of the art. But this will be traded against substantial advantages in compactness, ruggedness, and cost [100].

Current studies are usually directed at developing a proof-of-concept model to demonstrate a specific point with potentially important future implications, but an overall research pattern is not easily identified. Instead of trying to present a general overview over the field, an attempt is made here to indicate trends by discussing three typical examples: a dielectric antenna with a monolithically integrated silicon mixer diode useful for short-distance, line-of-sight communication at high data rates [209]; a steerable-beam monolithic microstrip antenna array with integrated phase shifters for radar and guidance applications [210]; and an integrated microstrip dipole array for millimeter-wave imaging [212]. The discussion draws freely from the cited publications.

Tapered Dielectric-Rod Antenna with Integrated Mixer Diode

The antenna was designed and investigated by Yao, Schwarz, and Blumenstock [209]. An earlier version, designed for the far-infrared region and using microbolometers as detectors, was developed by Rutledge and others [99, 100]. The configuration of the present antenna is shown in Fig. 77. It consists of a monolithic assembly of a dielectric antenna, a dielectric waveguide, a coupler, and a

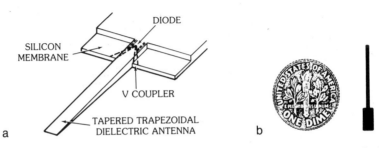

Fig. 77. Dielectric antenna with monolithically integrated Schottky mixer diode. (*a*) Configuration. (*b*) Size comparison with US dime. (*After Yao, Schwarz, and Blumenstock [209], © 1982 IEEE*)

mixer diode. The antenna is a tapered dielectric rod, etched together with the dielectric waveguide from a silicon wafer. A V-shaped metallic coupler printed on the backside of the waveguide serves to concentrate the received electromagnetic energy toward the mixer diode located at the apex. No external metal structure is needed to achieve this coupling; the V-shaped coupler is the only metal part of the device. The mixer diode, which is the critical component of the integrated antenna, is a planar Schottky diode. The diode configuration is indicated in Fig. 78; details on its fabrication process may be found in [209]. The finished antenna, designed for the 100-GHz band, is 17 mm long, 0.21 mm high, and 1.1 mm wide at the position of the coupler. The membrane thickness is 0.03 mm. The polarization of the incident field is assumed to be parallel to the plane of the coupler.

The device was tested at 85 GHz. A 3-dB beamwidth of 49° in the E-plane and 56° in the H-plane was measured, which corresponds to a directivity gain of 10 dB. The bandwidth of the device is very broad. The lower frequency limit is primarily determined by the antenna and waveguide dimensions, the upper limit by the rolloff of the responsivity of the Schottky diode above a cutoff frequency that depends on the details of its fabrication. Designed for operation in the 100-GHz band, the diode in the experimental antenna leads to a measured maximum responsivity of 35 V/W at the test frequency of 85 GHz.

In its present form, the device has poor conversion efficiency; an overall conversion loss* of 35 dB was measured. But the investigators expect that substantial improvements can be achieved by the use of epitaxial silicon or gallium arsenide. From the available data for these materials they predict a conversion loss of 16 dB for epitaxial silicon and a loss as low as 10 dB for epitaxial gallium arsenide [209].

The device should be regarded as a first step toward the development of fully integrated receiver front ends where, in addition to a mixer, a hybrid coupler, a filter, an if amplifier, and possibly a local oscillator would be monolithically integrated with a dielectric antenna and waveguide. The most likely application of an integrated antenna of this type would be in short-haul terrestrial communication and data transmission in the 100- to 300-GHz band [209]. The device would be used in this case as a feed system for a highly directive lens or mirror.

The diode used in the present antenna has a relatively low cutoff frequency. But Yao and colleagues [209] point out that planar surface oriented diodes have been demonstrated with a zero bias cutoff frequency several hundred times higher, i.e., in the 3200- to 4500-GHz band. Integrated antennas of this kind should, therefore, be feasible for the infrared region as well.

Monolithic Microstrip Antenna Phased Array

The feasibility of building phased arrays with monolithically integrated phase shifters in microstrip technology has recently been demonstrated by Stockton through the design of a 4×4 element array operating at 18 GHz [210]. This frequency is still below the millimeter-wave region, but it is sufficiently close to prove the concept.

---

*Overall conversion loss equals the power delivered to the if amplifier divided by the received radiative power.

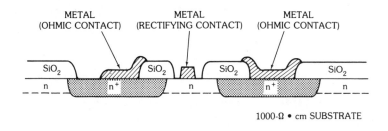

a   1000-Ω • cm SUBSTRATE

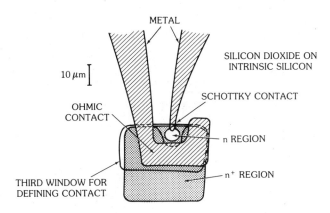

b

Fig. 78. Surface-oriented Schottky diode. (*a*) Diagram. (*b*) Top view. (*After Yao, Schwarz, and Blumenstock [209],* © *1982 IEEE*)

Fig. 79 shows the metallization pattern of the array. The antenna is printed on a 5-cm–diameter silicon wafer of high resistivity (equaling or greater than 5000 Ω-cm) and includes 96 planar beam lead pin diodes which are fabricated in situ in this wafer. The diodes serve as rf switching devices in the integrated phase shifters of the array. The phase shifters are 3-bit devices with two diodes required per bit; there is one phase shifter for each array element. (In the usual manner the first, second, and third stages of each phase shifter provide phase changes of 180°, 90°, and 45°, respectively, thus permitting phasing in increments of 45°.) After fabrication of the monolithic diodes a thin metal layer is deposited by thin film sputtering on the wafer, and the radiating elements, feed networks, and phase shifter circuits of the array are created by photoetching of this layer. All metal parts are fabricated in a single metallization and etching step which also produces the diode interconnections and dc bias lines. Fig. 79 shows the photo mask for the etching process; the dark areas and gray lines define the metal pattern of the array remaining on the wafer after etching.

Array efficiency is estimated at 25 percent, determined by a loss of 3 dB in the complete, 3-bit phase shifters and a loss of 0.35 dB/cm in the 50-Ω microstrip feed lines (based on a postprocessing substrate resistivity of 2500 Ω-cm). The directivity gain of the array is 17.5 dB. Experiments have shown good pattern quality in the *H*-plane. In the *E*-plane certain pattern distortions occurred when the antenna was scanned. These distortions are under study at present.

This investigation leads the way toward the development of large monolithic

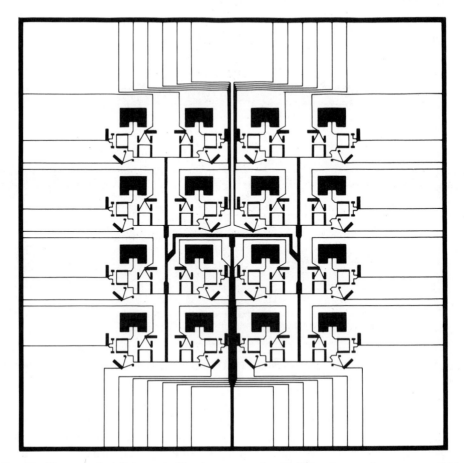

Fig. 79. Photo mask defining metallization pattern for 18-GHz, sixteen-element monolithic phased array in microstrip technology. (*Courtesy Ball Aerospace Systems Division*)

phased arrays of high directivity gain and an operating frequency in the lower millimeter-wave region. The size of the present array is limited by the diameter (approximately 5 cm) of currently available silicon wafers. At millimeter-wave frequencies gallium arsenide provides advantages over silicon in the monolithic fabrication of active devices and is likely to be used as the substrate material.

Eventually, in the design of future high-gain, multielement phased arrays, it will become desirable to integrate not only phase shifters but power amplifiers and low-noise detectors with the radiating elements or else at a suitably chosen subarray level, in order to ensure high efficiency in the transmit mode and high signal-to-noise ratio in the receive mode. Monolithic integration of these devices as well will be the ultimate goal. The use of distributed, individually controlled amplifiers and detectors in phased arrays would have the additional advantages that beam shaping becomes possible, that low side lobes can be achieved over a broad scan range and a sizable frequency band, and that graceful performance degradation will occur if some of the active components should fail. Work on monolithic millimeter-wave transmit and receive modules for array applications is in progress [214,215]. A main objective is to demonstrate that gallium-arsenide FET technology is capable of meeting the requirements for the 30- to 100-GHz band.

Integrated Near–Millimeter-Wave Imaging Array

The Rutledge group [212] has recently demonstrated an integrated millimeter-wave imaging antenna in the form of a linear array of eight parallel dipoles, each with its own detector, monolithically fabricated on a planar substrate of fused quartz ($\epsilon_r = 3.78$). The array is designed for the 1.2-mm band and, in its present form, provides one-dimensional imaging. The multidetector approach overcomes the limited sensitivity and speed of more conventional imaging systems using a single, scanned detector.

The operating principle is illustrated in Fig. 80. The objective lens projects the object plane onto the substrate surface (image plane), where the receiving array is located. Plotting the output of the antenna detectors produces the received image. In addition to serving as base plate for the array, the planar quartz substrate has the beneficial effect of directing the radiation patterns of the array elements into the direction of the incoming waves* [58, 208, 216]. But the surface waves supported by the (planar) substrate would counteract this effect by reducing the effective receiving cross section of the array elements and by increasing mutual coupling. The primary purpose of the substrate lens is to minimize these ill effects by eliminating the conditions under which surface waves can be trapped in the substrate [216]. Absorption and reflection losses of the substrate lens are minimized by choosing a low-loss dielectric (the lens is made from the same material as the substrate) and by using a quarter-wavelength antireflection coating.

In the practical realization the antennas are bow-tie dipoles spaced 310 µm apart (a half-wavelength in quartz) so that the eight-element array has a width of 2.5 mm; see Fig. 81. The dipoles are fabricated from evaporated silver by photography and lift-off. The detectors are bismuth bolometers located at the antenna feed points, where they connect to the two arms of the bow-ties. The

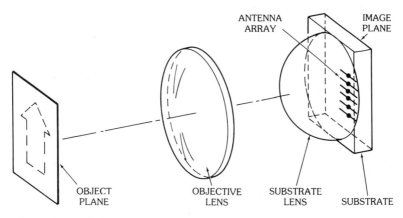

Fig. 80. Operating principle of near–millimeter-wave imagining array. (*After Neikirk et al. [212]*)

---

*The array is located on the *far* side of the substrate.

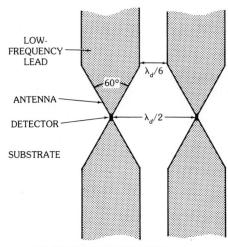

λ_d = WAVELENGTH IN SUBSTRATE

Fig. 81. Bow-tie antennas for imaging array. (*After Rutledge and Muha [208],* © *1982 IEEE*)

dimensions of these microbolometers are 2 μm by 5 μm. The bow-tie angle of 60° results in an antenna impedance of 150 Ω, matching that of the detectors. A special feature of the antenna is that the extension arms of the bow-tie dipoles serve as low-frequency leads with no additional rf isolation required. The substrate lens had a radius of 6.6 mm and the lens-substrate combination a thickness of 10 mm. In the presence of the lens about 50 percent of the available power is coupled into the array.

The element responsivity was measured at 1 to 2 V/W, but with recent improvements a system responsivity of 10 V/W can be expected as a reasonable estimate [217]. The experiments have confirmed, moreover, that the array resolution approaches the diffraction limit.

The device is a first step toward the realization of fully integrated two-dimensional radiometers and imaging radars. However, according to Rutledge and others [212], for such applications the microbolometers should probably be replaced by more sensitive diode detectors.

Work in this area is currently in progress. Rutledge, Schwarz, and their co-workers have recently demonstrated imaging arrays which use Schottky diodes as integrated detectors. The arrays [213] were designed for operation at 69 and 94 GHz. Their configuration is similar to the one shown in Fig. 80 and also uses eight parallel bow-tie dipoles spaced a half-wavelength (in quartz) apart. The quartz substrate supports a thin gallium arsenide wafer to permit monolithic fabrication of the diodes. The thinness of this wafer is crucial so as to avoid the excitation of surface waves;* the dipoles are printed on this wafer. A substantially improved

---

*Due to the asymmetry that the gallium-arsenide wafer is bordered by quartz on one side and by air on the other, *all* surface waves supported by this wafer have cutoff frequencies and can be suppressed [218]. The substrate lens, however, is still neede to eliminate surface waves in the (much thicker) quartz substrate.

system responsivity was obtained; the highest measured value was 330 V/W.

Fig. 82 shows the diode configuration and a micrograph of a diode integrated with one of the dipoles of the array. The fabrication process is described in [213]. The cutoff frequency of the diodes is estimated at 500 to 700 GHz. Fig. 83 is a photograph of a recent version of the 94-GHz array. The device shown here operates with nine dipoles and incorporates some further improvements in design which have resulted in a still higher system responsivity of 600 V/W [218]. The

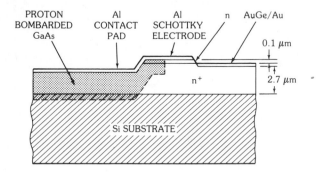

a

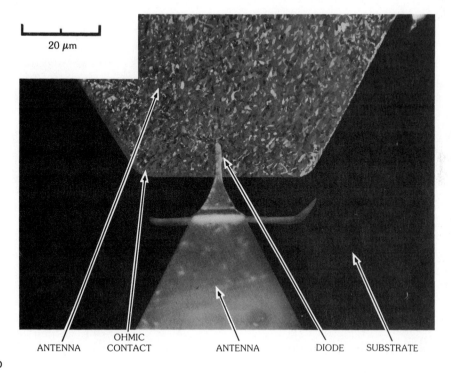

b

Fig. 82. Monolithic Schottky diode for imaging array. (*a*) Diode configuration. (*b*) Diode integrated with bow-tie dipole. (*Courtesy Rutledge, after Rav-Noy et al. [213],* © *1983 IEEE*)

a

b

Fig. 83. Imaging array for 94-GHz band. (*a*) Monolithic 9-element array on gallium arsenide wafer. (*b*) Finished device with substrate lens. (*Courtesy, Rutledge, after Zah et al. [218],* © *1984 IEEE*)

width of the nine-element array is about 2 wavelengths in quartz, i.e., approximately 7 mm.

While the use of diode detectors leads to superior system responsivity, the investigators still expect that microbolometers will provide a better noise performance. The peculiar advantage of diode detectors is that they will permit heterodyning [217].

6. References

[1] A. F. Kay, "Millimeter-wave antennas," *Proc. IEEE*, vol. 54, pp. 641–647, April 1966.

[2] R. B. Dybdal, "Millimeter-wave antenna development," *Proc. 1982 Antenna Appl. Symp.*, University of Illinois, September 22–24, 1982.

[3] E. K. Reedy and G. W. Ewell, "Millimeter-wave radar, in infrared and millimeter waves," in *Infrared and Millimeter Waves, Vol. 4,* ed. by K. J. Button and J. C. Wiltse, New York: Academic Press, 1981, pp. 23–24.

[4] R. W. Myhre, "Advanced 30/20-GHz multiple-beam antennas for communications satellites," *Proc. 1982 Antenna Appl. Symp.*, University of Illinois, September 22–24, 1982.

[5] J. Smetana, "Application of MMIC modules in future multibeam satellite antenna systems," *Proc. 1982 Antenna Appl. Symp.*, University of Illinois, September 22–24, 1982.

[6] N. Williams and N. A. Adatia, "Millimeter-wave antennas," *Proc. Mil. Microwave Conf.*, London, UK, October 22–24, 1980.

[7] R. L. Powers, K. D. Arkind, and D. G. LaRochelle, "Extended design yields compact 18–40 GHz antenna," *Microwave Systems News*, pp. 89–96, November 1981.

[8] D. K. Waineo and J. F. Konieczny, "Millimeter-wave monopulse antennas with rapid scan capability," *IEEE/AP-S 1979 Intl. Symp. Dig.*, pp. 477–480, Seattle, June 18–22, 1979.

[9] O. B. Kessler and J. George, "94-GHz antenna techniques," *Tech. Rep. AFWAL-TR-80-1222*, Texas Instruments, February 1981.

[10] L. M. Schwab, A. R. Dion, and D. L. Washington, "Space-fed, offset, plane-wave Cassegrainian system for ehf applications," *Proc. 1982 Antenna Appl. Symp.*, University of Illinois, September 22–24, 1982.

[11] J. Ruze, "Antenna tolerance theory—a review," *Proc. IEEE*, vol. 54, pp. 633–640, April 1966.

[12] B. E. Vu The Bao, "Influence of correlation interval and illumination taper in antenna tolerance theory," *Proc. Inst. Electr. Eng.* (London), vol. 166, pp. 195–202, 1969.

[13] H. Zucker, "Gain of antennas with random surface deviations," *Bell Syst. Tech. J.*, vol. 47, pp. 1637–1651, 1968.

[14] P. R. Cowles and E. A. Parker, "Reflector surface error compensation in Cassegrainian antennas," *IEEE Trans. Antennas Propag.*, vol. AP-23, pp. 323–328, May 1975.

[15] E. N. Davies, "Proposals for electronic compensation of surface profile errors in large reflectors, design and construction of large steerable aerials," *IEE Conf. Pub. 21*, pp. 80–83, 1966.

[16] R. A. Semplak and R. H. Turrin, "Pressure formed parabolic reflectors for millimeter wavelengths," *IEEE Trans. Antennas Propag.*, vol. AP-16, pp. 762–764, November 1968.

[17] A. G. Repjar and D. P. Kremer, "Accurate evaluation of a millimeter-wave compact range using planar near-field scanning," *IEEE Trans. Antennas Propag.*, vol. AP-30, pp. 419–425, May 1982.

[18] J. R. Cogdell et al., "High-resolution millimeter reflector antennas," *IEEE Trans. Antennas Propag.*, vol. AP-18, pp. 515–529, July 1970.

[19] J. W. M. Baars, "Design of large millimeter-wave radio telescopes," *Proc. 1980 Intl. URSI Symp.*, Munich, Germany, pp. 143A/1–4, August 26–29, 1980.

[20] S. V. Hoerner and W.-Y. Wong, "Gravitational deformation and astigmatism of tiltable radio telescopes," *IEEE Trans. Antennas Propag.*, vol. AP-23, pp. 689–695, September 1975.

[21] W. Rotman, "Ehf dielectric lens antenna for multiple beam MILSATCOM applications," *1982 Intl. IEEE/AP-S Symp. Dig.*, Albuquerque, pp. 132–135, June 1982.

[22] J. T. Mayhan and A. J. Simmons, "A low side lobe K_a-band antenna-radome study," *IEEE Trans. Antennas Propag.*, vol. AP-23, pp. 569–572, July 1975.

[23] G. Bekefi and G. W. Farnell, "A homogeneous dielectric sphere as a microwave lens," *Can. J. Phys.*, vol. 34, 1956.

[24] G. Toraldo diFrancia, "New stigmatic system of the concentric type," *J. Opt. Soc. Am.*, vol. 47, p. 566, June 1957.

[25] G. Toraldo diFrancia, "Spherical lenses for infrared and microwaves," *J. Appl. Phys.*, vol. 32, p. 2051, 1961.

[26] T. C. Cheston and E. J. Luoma, "Constant-K lenses," *APL Tech. Dig.*, April 1963.

[27] S. Cornbleet, "A simple spherical lens with external foci," *Microwave J.*, vol. 8, p. 65, May 1965.

[28] T. L. ApRhys, "The design of radially symmetric lenses," *IEEE Trans. Antennas Propag.*, vol. AP-18, pp. 497–506, July 1970.

[29] H. Mieras, "Radiation pattern computation of a spherical lens using Mie series," *IEEE Trans. Antennas Propag.*, vol. AP-30, pp. 1221–1224, November 1982.

[30] M. S. Narasimhan and S. Ravishankar, "Radiation from aperture antennas radiating in the presence of a dielectric sphere," *IEEE Trans. Antennas Propag.*, vol. AP-30, pp. 1237–1240, November 1982.

[31] Y. W. Chang, J. A. Paul, and Y. C. Ngan, "Millimeter-wave integrated circuit modules for communication interconnect systems," *US Army R&D Tech. Rep. ECOM76-1353-2*, November 1977.

[32] J. J. Lee, "Dielectric lens shaping and coma-correction zoning, Part I: Analysis," *IEEE Trans. Antennas Propag.*, vol. AP-31, pp. 211–216, January 1983.

[33] J. J. Lee and R. L. Carlise, "A coma-corrected multibeam shaped lens antenna, Part II: Experiments," *IEEE Trans. Antennas Propag.*, vol. AP-31, pp. 216–220, January 1983.

[34] J. C. Wiltse, "Fresnel zone-plate lenses," *SPIE Proc.*, vol. 544, Millimeter-Wave Technology III, July 1985.

[35] H. T. Buscher, "Electrically controllable liquid artificial dielectric media," *IEEE Trans. Microwave Theory Tech.*, vol. MTT-27, pp. 540–545, May 1979.

[36] R. Blundell and M. C. Carter, "Millimeter-wave aerials for full illumination radars," *Proc. Mil. Microwaves*, London, UK, October 22–24, 1980.

[37] R. Baldwin and P. A. McInnes, "A rectangular corrugated feedhorn," *IEEE Trans. Antennas Propag.*, vol. AP-23, pp. 814–817, November 1975.

[38] H. P. Coleman, R. M. Brown, and B. D. Wright, "Parabolic reflector offset fed with a corrugated conical horn," *IEEE Trans. Antennas Propag.*, vol. AP-23, pp. 817–819, November 1975.

[39] T.-S. Chu and W. E. Legg, "Gain of corrugated conical horns," *IEEE Trans. Antennas Propag.*, vol. AP-30, pp. 698–703, July 1982.

[40] B. M. Thomas and K. J. Greene, "A curved aperture corrugated horn having very low cross-polar performance," *IEEE Trans. Antennas Propag.*, vol. AP-30, pp. 1068–1072, November 1982.

[41] C. A. Mentzer and L. Peters, "Pattern analysis of corrugated horn antennas," *IEEE Trans. Antennas Propag.*, vol. AP-24, pp. 304–309, 1976.

[42] M. A. Jansen, S. M. Bednarczyk, S. Gulkis, H. W. Marlin, and G. F. Smoot, "Pattern measurement of a low-sidelobe horn antenna," *IEEE Trans. Antennas*

Propag., vol. AP-27, pp. 551–555, July 1979.

[43] T. Ohtera and H. Ujiie, "Radiation performance of a modified rhombic dielectric plate antenna," *IEEE Trans. Antennas Propag.*, vol. AP-29, pp. 660–662, July 1981.

[44] A. Hombach, "Dielectric feeds with low cross polarization," *Proc. 1980 Intl. URSI Symp.*, Munich, Germany, August 26–29, 1980.

[45] N. Nakajima and R. Wantanabe, "A quasioptical circuit technology for short millimeter-wavelength multiplexers," *IEEE Trans. Microwave Theory Tech.*, vol. MTT-29, pp. 897–905, September 1981.

[46] R. J. Eckstein et al., "35-GHz active aperture," *Tech. Rep. AFWAL-TR-81-1079*, Motorola, June 1981.

[47] M. F. Durkin, R. J. Eckstein, M. D. Mills, M. S. Stringfellow, and R. A. Neidhard, "35-GHz active aperture," *1981 IEEE MTT-S Intl. Microwave Symp. Dig.*, Los Angeles, June 1981.

[48] M. A. Weiss, "Microstrip antennas for millimeter waves," *R&D Tech. Rep. ECOM-76-0110-F*, October 1977.

[49] M. A. Weiss and R. B. Cassell, "Microstrip millimeter-wave antenna study," *R&D Tech. Rep. CORADCOM-77-0158-F*, April 1979.

[50] M. A. Weiss, "Microstrip antennas for millimeter waves," *IEEE Trans. Antennas Propag.*, vol. AP-29, pp. 171–174, January 1981.

[51] T. Metzler, "Microstrip series arrays," *Proc. Workshop on Printed-Circuit Antenna Technology*, New Mexico State University, Las Cruces, NM, pp. 21-1–16, October 17–19, 1979.

[52] K. R. Carver and J. W. Mink, "Microstrip antenna technology," *IEEE Trans. Antennas Propag.*, vol. AP-29, pp. 2–24, January 1981.

[53] R. J. Mailloux, J. F. McIlvenna, and N. P. Kernweis, "Microstrip array technology," *IEEE Trans. Antennas Propag.*, vol. AP-29, pp. 25–37, January 1981.

[54] J. R. James, P. S. Hall, C. Wood, and A. Henderson, "Some recent developments in microstrip antenna design," *IEEE Trans. Antennas Propag.*, vol. AP-29, pp. 124–128, January 1981.

[55] J. R. James, P. S. Hall, C. Wood, and A. Henderson, "Microstrip antenna research at the Royal Military College of Sciences," *Proc. Workshop on Printed Circuit Antenna Technology*, New Mexico State University, Las Cruces, pp. 1-1–10, October 17–19, 1979.

[56] J. C. Williams, "A 36-GHz printed planar array," *Electron. Lett.*, vol. 14, pp. 136–137, March 1978.

[57] I. E. Rana and N. G. Alexopoulos, "Current distribution and input impedance of printed dipoles," *IEEE Trans. Antennas Propag.*, vol. AP-29, pp. 99–105, January 1981.

[58] N. G. Alexopoulos, P. B. Katehi, and D. B. Rutledge, "Substrate optimization for integrated-circuit antennas," *IEEE Trans. Microwave Theory and Tech.*, vol. MTT-31, pp. 550–557, July 1983.

[59] P. B. Katehi and N. G. Alexopoulos, "On the effect of substrate thickness and permittivity on printed-circuit dipole properties," *IEEE Trans. Antennas Propag.*, vol. AP-31, pp. 34–39, January 1983.

[60] W.-C. Chew and J.-A. Kong, "Analysis of circular microstrip disk antenna with a thick dielectric substrate," *IEEE Trans. Antennas Propag.*, vol. AP-29, pp. 68–76, January 1981.

[61] F. Lalezari, "Dual-polarized high-efficiency microstrip antenna," U.S. patent no. 322930, November 1981.

[62] J. R. James and C. M. Hall, "Investigation of new concepts for designing millimeter-wave antennas," final technical report on Contract DAJA37-80-C-0183, US Army European Research Office, September 1983.

[63] F. Lalezari and T. Pett, "Millimeter microstrip antennas for use in mil-spec environment," final report to Battelle Columbus Labs/US Army CECOM, November 21, 1983.

[64] M. Campi, "Design of microstrip linear array antennas by computer," 1982 Army

Science Conference, West Point, New York, June 1982.

[65] B. B. Jones, F. Y. M. Chow, and A. W. Seeto, "The synthesis of shaped patterns with series-fed microstrip patch arrays," *IEEE Trans. Antennas Propag.*, vol. AP-30, pp. 1206–1212, November 1982.

[66] F. Lalezari, private communication.

[67] A. Henderson, A. E. England, and J. R. James, "New low-loss millimeter-wave hybrid microstrip antenna array," Eleventh European Microwave Conference, Amsterdam, The Netherlands, September 1981.

[68] J. R. James and A. Henderson, "A critical review of millimeter planar arrays for military applications," Military Microwave Conference, London, England, October 20–22, 1982.

[69] R. M. Knox, "Dielectric waveguide microwave integrated circuits—an overview," *IEEE Trans. Microwave Theory Tech.*, vol. MTT-24, pp. 806–814, November 1976.

[70] M. T. Birand and R. V. Gelsthorpe, "Experimental millimetric array using dielectric radiators fed by means of dielectric waveguide," *Electron. Lett.*, vol. 17, no. 18, pp. 633–635, September 1981.

[71] N. G. Alexopoulos and D. R. Jackson, "Fundamental superstrate effects (cover) on printed-circuit antennas," *Integrated Electromagnetics Lab Rep.*, No. 10, UCLA Rep. no. ENG-83-50, October 14, 1983.

[72] D. R. Jackson and N. G. Alexopoulos, "Superstrate (cover) effects on printed-circuit antennas," *Dig. 1984 Intl. IEEE-APS/URSI Symp.*, pp. 563–565, Boston, June 1984.

[73] B. R. Rao, "94-gigahertz slotted waveguide array fabricated by photolithographic techniques," *Dig. 1983 Intl. IEEE-APS/URSI Meeting*, pp. 688–689, Houston, May 23–26, 1983.

[74] F. G. Farrar, "Millimeter-wave *W*-band slotted waveguide antennas," *1981 IEEE/ AP-S Intl. Symp. Dig.*, Los Angeles, pp. 436–439, June 16–19, 1981.

[75] M. C. Carter and E. R. Cashen, "Linear arrays for centimetric and millimetric wavelengths," *Proc. Mil. Microwave Conf.*, London, UK, October 22–24, 1980.

[76] C. A. Strider, "Millimeter-wave planar arrays," *1974 Millimeter-Wave Techniques Conf.*, NELC, San Diego, pp. B6/1-11, March 26–28, 1974.

[77] C. A. Boyd, Jr., "Practical millimeter-wave ferrite phase shifters," *Microwave J.*, vol. 25, pp. 105–108, December 1982.

[78] J. L. Hilburn and F. H. Prestwood, "K-band frequency scanned waveguide array," *IEEE Trans. Antennas Propag.*, vol. AP-22, pp. 340–342, March 1974.

[79] R. C. Honey, "Line source and linear arrays for millimeter wavelengths," *Proc. Symp. Millimeter Waves*, Polytechnic Institute of Brooklyn, NY, March 31–April 2, 1959.

[80] S. Silver, ed., *Microwave Antenna Theory and Design*, New York: McGraw-Hill Book Co., 1949.

[81] E. M. Systems, Inc., "Millimeter-wave spiral antenna," *Microwave J.*, vol. 18, p. 28, December 1975.

[82] K. S. Kelleher, "High-gain reflector-type antennas," chapter 12 in *Antenna Engineering Handbook*, ed. by H. J. Jasik, New York: McGraw-Hill Book Co., 1961.

[83] C. C. Chen, "High-efficiency *V*-band fan beam antenna," *1981 Intl. Symp. Dig.*, Los Angeles: IEEE Antennas and Propagation Society, pp. 124–126, June 16–19, 1981.

[84] K. S. Kelleher, "Scanning antennas," chapter 15 in *Antenna Engineering Handbook*, ed. by H. J. Jasik, New York: McGraw-Hill Book Co., 1961.

[85] E. C. Dufort and H. Uyeda, "A wide-angle scanning optical antenna," *IEEE Trans. Antennas Propag.*, vol. AP-31, pp. 60–67, January 1983.

[86] F. E. Ore, "The modified biconical horn millimeter-wave antenna," *Tech. Rep. AFAL-TR-65-156*, Univ. of Illinois, Urbana, July 1965.

[87] A. D. Munger and J. H. Greg, "Design and development of biconical horn antennas for 18–100 GHz," *NOSC Tech. Note 580*, San Diego, December 1978.

[88] F. C. Jain, R. Bansal, and C. V. Valerio, Jr., "Semiconductor antenna: a new device in millimeter- and submillimeter-wave integrated circuits," *IEEE Trans. Microwave*

Theory Tech., vol. MTT-32, pp. 204–207, February 1984.

[89] "Final technical report for 54.5-GHz omni radio system," *Rep. No. 1301 R 0001*, US Army CORADCOM Contract No. DAAK80-79-C-0765, Norden Systems, October 1980.

[90] M. J. Ehrlich, "Slot antenna arrays," chapter 9 in *Antenna Engineering Handbook*, ed. by H. J. Jasik, New York: McGraw-Hill Book Co., 1961.

[91] S. A. Long, M. V. McAllister, and L. C. Shen, "The resonant cylindrical dielectric cavity antenna," *IEEE Trans. Antennas Propag.*, vol. AP-31, pp. 406–412, May 1983.

[92] T. E. Nowicki, "Microwave substrates, present and future," *Proc. Workshop on Printed Circuit Antenna Technology*, New Mexico State University, Los Cruces, NM, October 17–19, 1979.

[93] H. Howe, Jr., "Dielectric material development," *Microwave J.*, vol. 21, pp. 39–40, November 1978.

[94] P. F. Goldsmith, "Quasi-optical techniques," Chapter 5 in *Infrared and Millimeter Waves, Vol. 6*, ed. by K. J. Button, p. 335, New York: Academic Press, 1982.

[95] G. E. Mueller and W. A. Tyrell, "Polyrod antennas," *Bell Syst. Tech. J.*, vol. 26, pp. 837–851, October 1947.

[96] P. Mallach, "Notes from unpublished German documents," Central Radio Bureau Library, London.

[97] D. G. Kiely, *Dielectric Aerials*, London: Methuen & Co., 1952.

[98] F. J. Zucker, "Surface- and leaky-wave antennas," chapter 16 in *Antenna Engineering Handbook*, ed. by H. J. Jasik, New York: McGraw-Hill Book Co., 1961.

[99] D. B. Rutledge et al., "Antennas and waveguides for far-infrared integrated circuits," *IEEE J. Quantum Electron.*, vol. QE-16, pp. 508–516, May 1980.

[100] S. E. Schwarz and D. B. Rutledge, "Moving toward near mm-wave integrated circuits," *Microwave J.*, vol. 23, pp. 47–67, June 1980.

[101] J. A. G. Malherbe, T. N. Trinh, and R. Mittra, "Transition from metal to dielectric waveguide," *Microwave J.*, vol. 23, pp. 71–74, November 1980.

[102] B. J. Levine and J. E. Kietzer, "Hybrid millimeter-wave integrated circuits," *US Army R&D Tech. Rep.* ECOM-74-0577-F, October 1975.

[103] H. S. Jones, Jr., "Conformal and small antenna designs," *Tech. Rep. HDL-TR-1952*, US Army ERADCOM, Adelphi, Maryland, April 1981.

[104] D. C. Chang and R. Mittra, "Workshop report on modern millimeter-wave systems," *Scientific Rep. 63*, University of Colorado, Boulder, May 1981.

[105] Y. Shiau, "Dielectric-rod antennas for millimeter-wave integrated circuits," *IEEE Trans. Microwave Theory Tech.*, vol. MTT-24, pp. 869–872, November 1976.

[106] S. Kobayashi, R. Mittra, and R. Lampe, "Dielectric tapered-rod antennas for millimeter-wave applications," *IEEE Trans. Antennas Propag.*, vol. AP-30, pp. 54–58, January 1982.

[107] S. P. Schlesinger and A. Vigants, "HE_{11} excited dielectric surface-wave radiators," *Tech. Rep. AFCRC-TN-59-573*, Columbia University, June 1959.

[108] C. M. Angulo and W. S. C. Chang, "A variational expression for the terminal admittance of a semi-infinite dielectric rod," *IEEE Trans. Antennas Propag.*, vol. AP-7, p. 207, July 1959.

[109] F. J. Zucker, "Electromagnetic boundary waves," *Tech. Rep. AFCRL-63-165*, AF Cambridge Research Laboratories, June 1963.

[110] J. R. James, "Theoretical investigation of cylindrical dielectric-rod antennas," *Proc. IEE* (London), vol. 114, pp. 309–319, March 1967.

[111] A. D. Yaghjian and E. D. Kornhauser, "A modal analysis of dielectric-rod antennas excited in the HE_{11} mode," *IEEE Trans. Antennas Propag.*, vol. AP-20, pp. 122–128, January 1972.

[112] J. R. Blakey, "A scattering theory approach to the prediction of dielectric-rod antenna radiation patterns: the TM_{01} mode," *IEEE Trans. Antennas Propag.*, vol. AP-23, pp. 577–579, July 1975.

[113] F. J. Zucker, "Surface-wave antennas," chapter 21 in *Antenna Theory, Part II*, ed. by R. E. Collin and F. J. Zucker, New York: McGraw-Hill Book Co., 1969.

[114] L. B. Felsen, "Radiation from a tapered surface-wave antenna," *IRE Trans. Antennas Propag.*, vol. AP-8, pp. 577–586, November 1960.

[115] P. Balling, "Radiation from the dielectric wedge," Lic. Tech. Dissertation, Technical University of Denmark, December 1971.

[116] P. Balling, "Surface fields on the source-excited dielectric wedge," *IEEE Trans. Antennas Propag.*, vol. AP-21, pp. 113–115, January 1973.

[117] P. Balling, "On the role of lateral waves in the radiation from the dielectric wedge," *IEEE Trans. Antennas Propag.*, vol. AP-21, pp. 247–248, March 1973.

[118] S. T. Peng and F. Schwering, "Effect of taper profile on performance of dielectric taper antennas," *Dig. 1979 Natl. Radio Sci. Mtg. and Intl. IEEE-APS Symp.*, p. 96, Seattle, June 18–22, 1979.

[119] S. T. Peng and F. Schwering, "Omni-directional dielectric antennas," 9th DARPA/Tri-Service Millimeter-Wave Conference, Huntsville, Alabama, October 20–22, 1981.

[120] T. Itoh, "Dielectric waveguide type millimeter-wave integrated circuits," in *Infrared and Millimeter Waves, Vol. 4*, ed. by K. J. Button, pp. 199–273, New York: Academic Press, 1981.

[121] N. Williams, A. W. Rudge, and S. E. Gibbs, *Proc. IEEE MTT-S Intl. Microwave Symp.*, pp. 542–544, San Diego, 1977.

[122] K. L. Klohn, R. E. Horn, H. Jacobs, and E. Freibergs, "Silicon waveguide frequency scanning linear array antenna," *IEEE Trans. Microwave Theory Tech.*, vol. MTT-26, pp. 764–773, October 1978.

[123] R. E. Horn, H. Jacobs, E. Freibergs, and K. L. Klohn, "Electronic modulated beam-steering silicon waveguide array antenna," *IEEE Trans. Microwave Theory Tech.*, vol. MTT-28, pp. 647–653, June 1980.

[124] R. E. Horn, H. Jacobs, K. L. Klohn, and E. Freibergs, "Single-frequency electronic-modulated analog line scanning using a dielectric antenna," *IEEE Trans. Microwave Theory Tech.*, vol. MTT-30, pp. 816–820, May 1982.

[125] S. T. Peng and F. Schwering, "Dielectric grating antennas," R&D technical report, CORADCOM-78-3, Fort Monmouth, July 1978.

[126] F. Schwering and S. T. Peng, "Design of periodically corrugated dielectric antennas for millimeter-wave applications," *Proc. 1982 Antenna Appl. Symp.*, Univ. of Illinois, Urbana, September 22–24, 1982.

[127] F. Schwering and S. T. Peng, "Design of dielectric grating antennas for millimeter-wave applications," *IEEE Trans. Microwave Theory Tech.*, vol. MTT-31, pp. 199–209, February 1983.

[128] J. Borowick, W. Bayha, R. A. Stern, and R. W. Babbitt, "Dielectric waveguide antennas," 1982 Army Science Conference, West Point, New York, June 1982.

[129] S. Kobayashi, R. Lampe, R. Mittra, and S. Ray, "Dielectric-rod leaky-wave antennas for millimeter-wave applications," *IEEE Trans. Antennas Propag.*, vol. AP-29, pp. 822–824, September 1981.

[130] T. N. Trinh, R. Mittra, and R. J. Paleta, "Horn image-guide leaky-wave antenna," *1981 IEEE-MTT-S Intl. Microwave Symp. Dig.*, Los Angeles, June 1981.

[131] T. N. Trinh, R. Mittra, and R. J. Paleta, "Horn image-guide leaky-wave antenna," *IEEE Trans. Microwave Theory Tech.*, vol. MTT-29, pp. 1310–1314, December 1981.

[132] T. Itoh, "Application of gratings in dielectric waveguides for leaky-wave antennas and band-reject filters," *IEEE Trans. Microwave Theory Tech.*, vol. MTT-25, pp. 1134–1138, December 1977.

[133] T. Itoh and B. Adelseck, "Trapped image guide for millimeter-wave circuits," *IEEE Trans. Microwave Theory Tech.*, vol. MTT-28, pp. 1433–1436, December 1980.

[134] T. Itoh and B. Adelseck, "Trapped image-guide leaky-wave antennas for millimeter-wave applications," *IEEE Trans. Antennas Propag.*, vol. AP-30, pp. 505–509, May 1982.

[135] K. Solbach, "*E*-band leaky-wave antenna using dielectric image line with etched radiating elements," *1979 MTT-S Intl. Microwave Symp. Dig.*, pp. 214–216, April 1979.

[136] K. Solbach, "Slots in dielectric image line as mode launchers and circuit elements," *IEEE Trans. Microwave Theory Tech.*, vol. MTT-29, pp. 10–16, January 1981.

[137] W. V. McLevige, T. Itoh, and R. Mittra, "New waveguide structures for millimeter-wave and optical integrated circuits," *IEEE Trans. Microwave Theory Tech.*, vol. MTT-23, pp. 788–794, October 1975.

[138] S. T. Peng and T. Tamir, "Effects of groove profile on the performance of dielectric grating couplers," *Proc. Symp. Opt. Acoust. Micro-Electron.*, Polytechnic Press, Brooklyn, 1974.

[139] D. Marcuse, "Exact theory of TE-wave scattering from blazed dielectric gratings," *Bell Syst. Tech. J.*, vol. 55, pp. 1295–1317, 1976.

[140] K. C. Chang and T. Tamir, "Simplified approach to surface-wave scattering by blazed dielectric gratings," *Appl. Opt.*, vol. 19, pp. 282–288, 1980.

[141] A. Gruss, K. T. Tam, and T. Tamir, "Blazed dielectric gratings with high beam-coupling efficiencies," *Appl. Phys. Lett.*, vol. 36, pp. 523–526, 1980.

[142] T. Hori and T. Itanami, "Circularly polarized linear array antenna using a dielectric image guide," *IEEE Trans. Microwave Theory Tech.*, vol. MTT-29, pp. 967–970, September 1981.

[143] S. T. Peng, "Oblique guidance of surface waves on corrugated dielectric layers," *Proc. URSI Symp. Electromag. Waves*, Paper No. 341B, Munich, Germany, August 1980.

[144] S. T. Peng, A. A. Oliner, and F. Schwering, "Theory of dielectric grating antennas of finite width," *IEEE AP-S Intl. Symp. Dig.*, pp. 529–532, Los Angeles, June 1981.

[145] M. J. Shiau, S. T. Peng, and A. A. Oliner, "Simple and accurate perturbation procedure for millimeter-wave dielectric grating antennas of finite width," *1982 IEEE/AP-S Symp. Dig.*, pp. 648–651, Albuquerque, May 24–28, 1982.

[146] S. T. Peng, M. J. Shiau, A. A. Oliner, J. Borowick, W. Bayha, and F. Schwering, "A simple analysis procedure for dielectric grating antennas of finite width," 1984 IEEE-AP-S Symposium, Boston, June 25–28, 1984.

[147] T. Tamir, *Integrated Optics*, New York: Springer-Verlag, 1975.

[148] S. T. Peng, T. Tamir, and H. L. Bertoni, "Theory of periodic dielectric waveguides," *IEEE Trans. Microwave Theory Tech.*, vol. MTT-23, p. 123, 1975.

[149] M. Neviere, R. Petit, and M. Cadilhac, "About the theory of optical grating coupler-waveguide systems," *Opt. Commun.*, vol. 8, pp. 113–117, 1973.

[150] K. Honda, S. T. Peng, and T. Tamir, "Improved perturbation analysis of dielectric gratings," *Appl. Phys.*, vol. 5, p. 325, 1975.

[151] S. T. Peng and T. Tamir, "TM mode perturbation analysis of dielectric gratings," *Appl. Phys.*, vol. 7, p. 35, 1975.

[152] T. Tamir and S. T. Peng, "Analysis and design of grating couplers," *Appl. Phys.*, vol. 14, pp. 235–254, 1977.

[153] R. Mittra and R. Kastner, "A spectral domain approach for computing the radiation characteristics of a leaky-wave antenna for millimeter waves," *IEEE Trans. Antennas Propag.*, vol. AP-29, pp. 654–656, July 1981.

[154] R. M. Knox and P. P. Toulios, "Integrated circuits for the millimeter through optical frequency range," *Proc. Symp. Millimeter Waves*, Polytechnic Institute of Brooklyn, March 31–April 2, 1970.

[155] J. R. James and P. S. Hall, "Microstrip antennas and arrays, part 2: new array-design technique," *IEE J. Microwaves, Optics and Antennas*, no. 1, pp. 175–181, 1977.

[156] K. Solbach and B. Adelseck, "Dielectric image line leaky-wave antennas for broadside radiation," *Electron. Lett.*, vol. 19, pp. 640–644, August 1983.

[157] I. J. Bahl and P. Bhartia, "Leaky-wave antennas using artificial dielectrics at millimeter-wave frequencies," *IEEE Trans. Microwave Theory Tech.*, vol. MTT-28, pp. 1205–1212, November 1980.

[158] F. R. Ore, "A millimeter-wave receiving antenna with an omnidirectional or

directional scannable azimuthal pattern and a directional vertical pattern," *Tech. Rep. AFAL-TR-72-282*, Univ. of Illinois, September 1971.

[159] F. R. Ore, "A millimeter-wave receiving antenna with an omnidirectional or directional scannable azimuthal pattern and a directional vertical pattern," *IEEE Trans. Antennas Propag.*, vol. AP-20, pp. 481–482, July 1972.

[160] G. E. Mueller, "A broadside dielectric antenna," *Proc. IRE*, vol. 40, pp. 71–75, July 1952.

[161] S. T. Peng, "Omnidirectional dielectric antennas," CECOM R&D report in preparation.

[162] S. T. Peng and A. A. Oliner, "Guidance and leakage properties of a class of open dielectric waveguides, part I: mathematical formulations," *IEEE Trans. Microwave Theory Tech.*, vol. MTT-29, pp. 843–855, September 1981.

[163] A. A. Oliner, S. T. Peng, T. I. Hsu, and A. Sanchez, "Guidance and leakage properties of a class of open dielectric waveguides, part II: new physical effects," *IEEE Trans. Microwave Theory Tech.*, vol. MTT-29, pp. 855–869, September 1981.

[164] Polytechnic Institute of Brooklyn, Microwave Research Institute, *Monthly Performance Summary, Rep. PIBMRI-875*, pp. 17–61, 1961.

[165] T. Nakahara and N. Kurauchi, "Transmission modes in the grooved guide," *J. Inst. Electron. Commun. Eng. Japan*, vol. 47, no. 7, pp. 43–51, July 1964. Also in *Sumitomo Electr. Tech. Rev.*, no. 5, pp. 65–71, January 1965.

[166] D. J. Harris and K. W. Lee, "Groove guide as a low-loss transmission system for short millimetric waves," *Electron. Lett.*, vol. 13, no. 25, pp. 775–776, December 8, 1977. Professor Harris and his colleagues have published many papers on this topic, of which this is one of the first.

[167] D. J. Harris and S. Mak, "Groove-guide microwave detector for 100-GHz operation," *Electron. Lett.*, vol. 17, no. 15, pp. 516–517, July 23, 1981.

[168] A. A. Oliner and P. Lampariello, "Novel leaky-wave antenna for millimeter waves based on groove guide," *Electron. Lett.*, vol. 18, pp. 1105–1106, December 1982.

[169] P. Lampariello and A. A. Oliner, "Theory and design considerations for a new millimeter-wave leaky groove-guide antenna," *Electron. Lett.*, vol. 19, pp. 18–20, January 1983.

[170] P. Lampariello and A. A. Oliner, "A new leaky wave antenna for millimeter waves using an asymmetric strip in groove guide, part I: theory," *IEEE Trans. Antennas Propag.*, vol. AP-33, pp. 1285–1294, December 1985.

[171] P. Lampariello and A. A. Oliner, "A new leaky wave antenna for millimeter waves using an asymmetric strip in groove guide, part II: design considerations," *IEEE Trans. Antennas Propag.*, vol. AP-33, pp. 1295–1303, December 1985.

[172] P. Lampariello and A. A. Oliner, "Bound and leaky modes in symmetrical open groove guide," *Alta Frequenza*, vol. LII, no. 3, pp. 164–166, 1983.

[173] P. Lampariello and A. A. Oliner, "Leaky modes of symmetrical groove guide," *Dig. IEEE/MTT-S Intl. Microwave Symp.*, pp. 390–392, Boston, May 30–June 3, 1983.

[174] A. A. Oliner and P. Lampariello, "A new simple leaky wave antenna for millimeter waves," *Dig. 1985 North American Radio Sci. Meeting*, p. 57, Vancouver, Canada, June 17–21, 1985.

[175] A. A. Oliner and P. Lampariello, "A simple leaky wave antenna that permits flexibility in beam width," *Dig. Natl. Radio Sci. Meeting*, p. 26, Philadelphia, June 9–13, 1986.

[176] T. Yoneyama and S. Nishida, "Nonradiative dielectric waveguide for millimeter-wave integrated circuits," *IEEE Trans. Microwave Theory Tech.*, vol. MTT-29, no. 11, pp. 1188–1192, November 1981.

[177] T. Yoneyama and S. Nishida, "Nonradiative dielectric waveguide circuit components," International Conference on Infrared and Millimeter Waves, Miami, December 1981.

[178] A. Sanchez and A. A. Oliner, "Accurate theory for a new leaky-wave antenna for millimeter waves using nonradiative dielectric waveguide," *Radio Sci.*, vol. 19, no. 5, pp. 1225–1228, September–October 1984.

[179] A. Sanchez and A. A. Oliner, "Microwave network analysis of a leaky-wave

structure in nonradiative dielectric waveguide," *Dig. IEEE MTT-S Intl. Microwave Symp.*, pp. 118–120, San Francisco, May 30–June 1, 1984.

[180] A. A. Oliner, S. T. Peng, and K. M. Sheng, "Leakage from a gap in NRD guide," *Dig. 1985 IEEE Intl. Microwave Symp.*, pp. 619–622, St. Louis, June 3–7, 1985.

[181] H. Shigesawa, M. Tsuji, and A. A. Oliner, "Coupling effects in an NRD guide leaky wave antenna," *Dig. Natl. Radio Sci. Meeting*, p. 27, Philadelphia, June 9–13, 1986.

[182] H. Shigesawa and K. Takiyama, "Study of leaky H-guide," *Paper No. M1-7*, International Conference on Microwaves, Circuit Theory and Information, Tokyo, 1964. More complete version in K. Takiyama and H. Shigesawa, "On the study of a leaky H-guide," *Sci. Eng. Rev. Doshisha University*, vol. 7, no. 4, pp. 203–225, March 1967 (in English).

[183] K. Takiyama and H. Shigesawa, "The radiation characteristics of a leaky H-guide," *J. Inst. Electr. Commun. Eng. Japan* (J.I.E.C.E.), vol. 50, no. 2, pp. 181–188, February 1967 (in Japanese).

[184] A. Sanchez and A. A. Oliner, "A new leaky waveguide for millimeter waves using nonradioactive dielectric (NRD) waveguide, part I: accurate theory," *IEEE Trans. Microwave Theory Tech*, vol. MTT-35, pp. 737–747, August 1987.

[185] Y. Yoneyama, letter to A. A. Oliner, July 4, 1983.

[186] Q. Han, A. A. Oliner, and A. Sanchez, "A new leaky waveguide for millimeter waves using nonradioactive dielectric (NRD) waveguide, part II: comparison with experiments," *IEEE Trans. Microwave Theory Tech.*, vol. MTT-35, pp. 748–752, August 1987.

[187] T. Yoneyama, T. Kuwahara, and S. Nishida, "Experimental study of nonradiative dielectric waveguide leaky wave antenna," *Proc. 1985 Intl. Symp. Antennas Propag. (ISAP)*, Kyoto, August 1985.

[188] H. Ermert, "Guided modes and radiation characteristics of covered microstrip lines," *Archiv für Electronik und Übertragungstechnik*, vol. 30, pp. 65–70, February 1976.

[189] H. Ermert, "Guiding and radiation characteristics of planar waveguides," *Microwaves, Optics and Acoustics*, vol. 3, pp. 59–62, March 1979.

[190] A. A. Oliner and K. S. Lee, "The nature of the leakage from higher modes on microstrip line," *Dig. 1986 IEEE Intl. Microwave Symp.*, pp. 57–60, Baltimore, June 2–4, 1986.

[191] A. A. Oliner and K. S. Lee, "Microstrip leaky wave strip antennas," *Dig. 1986 IEEE Intl. Antennas Propag. Symp.*, pp. 443–446, Philadelphia, June 8–13, 1986.

[192] W. Menzel, "A new traveling-wave antenna in microstrip," *Archiv für Electronik und Übertragungstechnik*, vol. 33, pp. 137–140, April 1979.

[193] D. C. Chang and E. F. Kuester, "Total and partial reflection from the end of a parallel-plate waveguide with an extended dielectric loading," *Radio Sci.*, vol. 16, pp. 1–13, January–February 1981.

[194] D. M. Pozar, "Considerations for millimeter-wave printed antennas," *IEEE Trans. Antennas Propag.*, vol. AP-31, pp. 740–747, September 1983.

[195] K. Iizuka, M. Mizusawa, S. Urasaki, and H. Ushigome, "Volume-type holographic antenna," *IEEE Trans. Antennas Propag.*, vol. AP-23, pp. 807–810, November 1975.

[196] D. F. Sedivec and B. H. Rubin, "A wideband 44-GHz printed-circuit array antenna," *Tech. Rep. RADC-TR-83-198*, Rome Air Development Center, Rome, N.Y., October 1983.

[197] K. C. Gupta, R. Garg, and R. Chadha, *Computer-Aided Design of Microwave Circuits*, Dedham: Artech House, 1981.

[198] T. Itoh and J. Rivera, "A comparative study of millimeter-wave transmission lines," chapter 2, vol. 9, of *Infrared and Millimeter Waves*, ed. by K. J. Button, New York: Academic Press, 1983.

[199] J.-F. Miao and T. Itoh, "Coupling between microstrip line and image guide through small apertures in the common ground plane," *IEEE Trans. Microwave Theory Tech.*, vol. MTT-31, pp. 361–363, April 1983.

[200] R. E. Neidert, "Waveguide-to-coax-to-microstrip transitions for millimeter-wave

monolithic circuits," *Microwave J.*, vol. 27, pp. 93–101, June 1983.

[201] N. G. Alexopoulos and I. E. Rana, "Mutual impedance computation between printed dipoles," *IEEE Trans. Antennas Propag.*, vol. AP-29, pp. 106–111, January 1981.

[202] H. G. Oltman and D. A. Huebner, "Electromagnetically coupled microstrip dipole arrays," *IEEE Trans. Antennas Propag.*, vol. AP-29, pp. 151–157, January 1981.

[203] R. S. Elliott and G. J. Stern, "The design of microstrip dipole arrays including mutual coupling," part I: theory; part II: experiments," *IEEE Trans. Antennas Propag.*, vol. AP-29, pp. 757–765, September 1981.

[204] A. Sabban, "A new broadband stacked two-layer microstrip antenna," *Dig. 1983 Intl. IEEE Symp. Antennas Propag.*, pp. 63–66, University of Houston, May 1983.

[205] P. B. Katehi and N. G. Alexopoulos, "A generalized solution to a class of printed-circuit antennas," *Dig. 1984 Intl. IEEE-APS/URSI Symp.*, pp. 566–568, Boston, June 1984.

[206] K. Mizuno, Y. Daiku, and S. Ono, "Design of printed resonant antennas for monolithic diode detectors," *IEEE Trans. Microwave Theory Tech.*, vol. MTT-25, pp. 470–472, June 1977.

[207] D. B. Rutledge and S. E. Schwarz, "Planar multimode detector arrays for infrared and millimeter-wave applications," *IEEE J. Quantum Electron.*, vol. QE-17, pp. 407–414, March 1981.

[208] D. B. Rutledge and M. S. Muha, "Imaging antenna arrays," *IEEE Trans. Antennas Propag.*, vol. AP-30, pp. 535–540, July 1982.

[209] C. Yao, S. E. Schwarz, and B. J. Blumenstock, "Monolithic integration of a dielectric millimeter-wave antenna and a mixer diode: an embryonic millimeter-wave IC," *IEEE Trans. Microwave Theory Tech.*, vol. MTT-30, pp. 1241–1247, August 1982.

[210] R. J. Stockton, "A monolithic phased array at *K*-band," EHF SATCOM Technology Workshop, San Diego, August 1981.

[211] R. J. Stockton, "Monolithic integrated antenna system—a new trend," Microwave and Millimeter-Wave Monolithic Circuit Symposium, Dallas, June 1982.

[212] D. P. Neikirk, D. B. Rutledge, M. S. Muha, H. Park, and C.-X. Yu, "Far-infrared imaging antenna arrays," *Appl. Phys. Lett.*, vol. 40, pp. 203–205, 1982.

[213] Z. Rav-Noy, C. Zah, U. Schreter, D. B. Rutledge, T. C. Wand, S. E. Schwarz, and T. F. Kuech, "Monolithic Schottky diode imaging arrays at 94 GHz," *Dig. Infrared and Millimeter-Wave Conf.*, Miami Beach, December 1983.

[214] T. A. Midford, M. Feng, R. Hackett, J. M. Schellenberg, E. Watkins, and H. Yamasaki, "Advanced GaAs FET technology for ehf monolithic arrays," *NOSC Contractor Rep. 225*, February 1984.

[215] C. R. Seashore and D. R. Singh, "Millimeter-wave ICs for precision guided weapons," *Microwave J.*, vol. 26, pp. 51–65, June 1983.

[216] C. Zah, R. C. Compton, and D. B. Rutledge, "Efficiencies of elementary integrated-circuit feed antennas," *Electromagnetics*, special issue on printed-circuit antennas and devices, pp. 239–254, March 1983.

[217] D. B. Rutledge, private communication.

[218] C. Zah, W. Lam, J. S. Smith, Z. Rav-Noy, and D. B. Rutledge, "Progress in monolithic Schottky-diode imaging arrays," Ninth International Conference on Infrared and Millimeter Waves, Osaka, November 1984.

[219] M. Guglielmi and A. A. Oliner, "A practical theory for image guide leaky-wave antennas loaded by periodic metal strips," *Proc. 17th European Microwave Conference*, pp. 549–554, Rome, Italy, September 7–11, 1987.

[220] N. Marcuvitz, *Waveguide Handbook, Vol. 10*, MIT Radiation Laboratory Series, Sec. 6.1, McGraw-Hill Book Co., New York, 1951.

[221] A. A. Oliner, "Scannable Millimeter Wave Arrays," Final Report on Contract No. F19628-84-K-0025, Rome Air Development Center, Hanscom Field, MA, December 1, 1987.

Chapter 18

Practical Aspects of Phased Array Design

Raymond Tang
Hughes Aircraft Company

CONTENTS

Raymond Tang is the manager of the Antenna Array Laboratory, Electromagnetic Laboratories, Surveillance and Sensor Systems Division, of Hughes Aircraft Company. He received the BSEE degree from the Polytechnic Institute of Brooklyn, and the MSEE degree from the University of Southern California.

During his 30 years at Hughes Aircraft Company Mr. Tang has been concerned with microwave antennas and components. He has participated in the design and development of various types of electronic scanning antennas, such as frequency-scan, phase-scan, and optically scanning lenses. In recent years he has been actively involved in the development of wideband phased arrays and limited-scan phased arrays. He has also been engaged in the development of diode and ferrite phase shifters, switches, and solid-state transmit/receive modules.

Mr. Tang holds 12 patents and has authored or coauthored 15 technical publications. He is a member of the IEEE, PGAP, PGMTT, PGED, and Eta Kappa Nu fraternity.

1. Introduction

The intent of this chapter is to provide the reader with a basic understanding of the practical aspects of phased array antenna design. The theory of phased arrays has been covered in the preceding chapters. In this chapter a treatment of the various design considerations and trade-offs is given, so that the antenna designer can arrive at an optimum antenna configuration in order to meet a given set of radar system requirements. The topics that will be covered are

(a) design specifications and procedure for phased array antennas
(b) selection criteria for array components
(c) effects of component errors on array performance

To establish a common basis of understanding, let us first define the basic components in a phased array antenna. As shown in Fig. 1, the phased array antenna consists of an array of radiating elements with each radiating element connected to a phase shifter. The phase shifters control the phase of the radiated signals at each element to form a beam at the desired direction θ_0. A beam-forming network, commonly called a *feed network*, is used to distribute the output signal from the transmitter to the radiating elements and to provide the required aperture distribution for beam shape and side lobe control. Phase shifter drivers provide the required control/bias currents and voltages for each phase shifter for steering the beam to the desired scan angle θ_0. The control signals (or phase words) for the drivers are calculated by the beam-steering computer and stored in the serial shift registers. When the beam is ready to be scanned, the beam-forming trigger signal causes the stored phase words in the serial shift registers to dump into the parallel latching registers, which in turn set the drivers and phase shifter for the desired scan angle. Using this type of phased array the radar is capable of performing the following functions:

1. Rapid and accurate beam scanning; typically the beam-switching time is 10 to 40 µs
2. Search and automatic target tracking over a hemispherical scan coverage by using four planar array faces
3. Perform multiple functions such as surveillance, multiple target tracking, target illumination and missile guidance, terrain following and avoidance, ground mapping, etc.
4. Pulse-to-pulse frequency and/or beam agility
5. Beam shaping and/or polarization flexibility
6. Low peak and average side lobe levels, typically −40 dB peak and −55 dB average

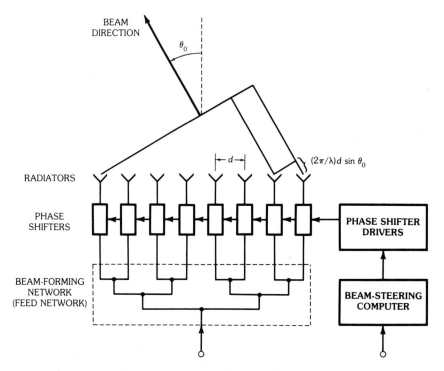

Fig. 1. Basic components of a phased array antenna.

7. High power transmission with multiple distributed transmitters
8. Electronic beam stabilization on moving platforms

With this simplified description of the basic functions of a phased array antenna, let us now proceed to the discussion of the design specifications of phased arrays.

2. Design Specification and Procedure of Phased Array Antennas

The design specifications of a phased array antenna are usually determined by the overall radar system requirements. These requirements are given in terms of radar performance requirements, physical requirements, operating environmental requirements, producibility, maintainability, and reliability requirements. A detailed listing of all the requirements, showing the breakdown of each category, is given in Chart 1.

The antenna engineers, working together with the system and mechanical design engineers, must perform a trade-off study of these requirements (as shown in Chart 1) in order to establish a set of antenna specifications that would satisfy all the radar system requirements at minimum cost. In other words, the antenna requirements must not be overspecified so that the cost would not be affordable. For example, the number of simultaneous beams, beam-switching speed, beamwidth, and total transmitted power can be traded against each other to simplify the

Chart 1. Design Requirements of Phased Array Antennas

1. Performance Requirements
   spatial scan coverage
   tunable and instantaneous bandwidth
   beamwidth
   peak and average side lobe level
   antenna gain
   polarization
   peak and average power
   beam-switching speed
   prime power
   number of simultaneous beams
   beam shape

2. Physical Requirements
   size
   weight
   transportability
   mobility—setup time and march time

3. Environmental Requirements
   operating temperature range
   shock and vibration loads
   humidity, salt, fog, and fungus
   overpressure

4. Producibility, Maintainability, and Reliability

5. Cost (Affordability)

antenna requirements without sacrificing the required power aperture product for a given radar search mode. Once the antenna requirements are specified, the next step is to formulate and draw an overall antenna schematic diagram showing all the functional subassemblies of the antenna system. The subsequent step is to select the array components for the various functional subassemblies that would best meet the system requirements in terms of performance and cost. Following the component selection, the next step is to perform an error analysis to determine the allowable tolerances of these components that would meet the antenna performance requirements. These tolerance requirements are then used as the performance and physical specifications for the design, development, fabrication, and acceptance testing of these components. Detailed discussions on the selection criteria of array components and error analysis follow in Sections 3 and 4.

3. Selection Criteria of Array Components

The three major components of a phased array antenna are the radiators, phase shifters, and beam-forming feed network. Some of the commonly used criteria in selecting these components for a given phased array application are shown in Chart 2.

The requirements in Chart 2 influence the selection of array components in different ways. For example, the selection of the type of radiator is mainly

Chart 2. Selection Criteria of Array Components

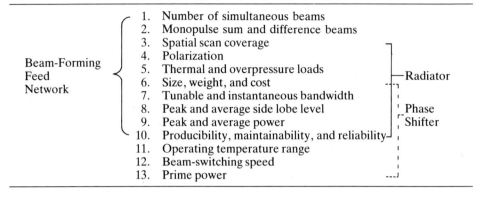

determined by the requirements of items 3 through 10, whereas the selection of the type of phase shifter is mainly determined by the requirements of items 6 through 13, and the selection of feed network is determined by items 1 through 10. As shown in Chart 2, some of the requirements are unique to only one of the components while the others affect two or all three components. For example, the requirements of beam-switching speed, operating temperature range, and prime power affect only the phase shifter selection, whereas the frequency bandwidth, side lobe, and power level requirements affect all three components. In general, there are many combinations of antenna component specifications that can result in the same overall system performance. These nonunique component requirements allow the antenna designer to perform trade-offs in selecting the component to arrive at an optimum design in terms of meeting performance requirements at minimum cost. A detailed discussion of the various trade-offs available in component selection is given in the following sections.

Radiator Selection

Types of Radiators—Before we begin the discussion on radiator selection, let us first review the various types of radiators that are commonly used in phased arrays. The basic types of phased array radiators are listed below:

1. Open-ended waveguide radiators
2. Dipole radiators
3. Waveguide slot radiators
4. Disk and patch radiators

The open-ended waveguide radiator comes in two basic forms, namely, the rectangular and the circular waveguide radiators. The rectangular waveguide radiator is usually used for linear polarization applications, whereas the circular waveguide radiator is frequently used for dual linear or circular polarization applications. The rectangular waveguide radiator can be packaged into very close spacing in the E-plane of the waveguide (less than 0.5λ) by using reduced height waveguide, thus allowing a very large scan coverage in the E-plane. The H-plane scan is restricted by the width of the guide as determined by the desired ratio of the operating frequency to the cutoff frequency. The rectangular waveguide radiator, however, can be packaged into an equilateral triangular element lattice arrangement in order to provide conical scan coverage as in the case of the circular waveguide radiators, thus allowing an increase in *H*-plane scan coverage. The various possible lattice arrangements for both the rectangular and circular waveguide radiators are shown in Fig. 2. The dimension *d* in Fig. 2 is usually in the order of one-half wavelength at the high end of the operating frequency band. For all cases of lattice arrangements the spacing between the elements can be reduced by dielectrically loading and/or ridge loading the waveguides.

      The two common types of dipole radiators are the microstrip dipole [1, 2] and the coaxial dipole as shown in Fig. 3. In the case of the microstrip dipole the dipole wings are either etched or printed on a dielectric substrate, such as a copper-clad Teflon-fiberglass board, or on an alumina (Al_2O_3) substrate. These dipole wings

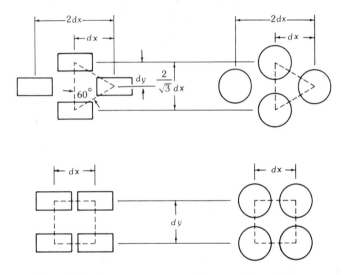

Fig. 2. Lattice arrangements of waveguide radiators.

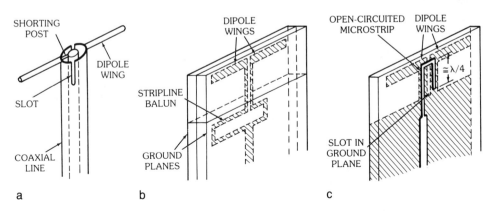

Fig. 3. Common types of dipole radiators. (*a*) Coaxial dipole. (*b*) Stripline dipole. (*c*) Microstrip dipole.

are excited by means of a microstrip balun. In the case of the coaxial dipole the dipole wings are excited by means of a slotted coaxial transmission line as shown in Fig. 3. These dipole radiators can be packaged in a large two-dimensional array just as in the case of the rectangular waveguide radiators.

There are many different types of slot radiators. In the case of slots excited by waveguide transmission lines, there are shunt slots [3] which are cut along and parallel to the centerline of the broadwall of the waveguide and series inclined [4] as well as noninclined slots [5] which are cut along the sidewall of the waveguide (see Fig. 4). The noninclined slot, which is a magnetically coupled transverse slot [5], does not have cross polarization as in the case of the inclined slot. There are also slot radiators which are etched on the ground plane side of a microstrip transmission line. The operating bandwidth of the slot radiators, as in the case of the dipole radiators, is less than that of the open-ended waveguide radiators.

The disk and patch radiators are not commonly used in ground-based or shipboard applications. These radiators are more commonly used in conformal arrays for airborne or missile applications where the depth of the array is of cardinal importance. The two common types are the microstrip excited-patch radiator [6, 7] and the coaxial excited-disk radiator [8] as shown in Fig. 5. The coaxial excited-disk radiator has much larger bandwidth than that of the microstrip excited-patch radiator.

Selection Considerations of Radiators—With the knowledge of the various types of available radiators, let us now proceed to a discussion on how to select a radiator for a given application. In general, the radiator must be selected on the basis of meeting all the antenna performance, physical packaging, and environmental requirements at minimum cost. Some of these requirements and the corresponding selection considerations are given in Chart 3.

The allowable area per element requirement is determined by choosing the proper element spacing and lattice to avoid the formation of grating lobes over the entire volumetric scan coverage. The element lattice is usually chosen to maximize

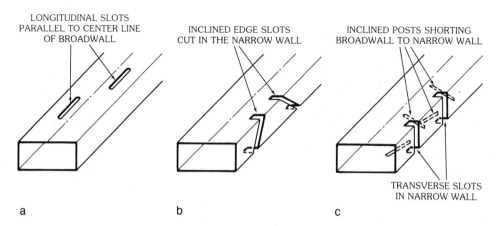

a   b   c

Fig. 4. Various types of slot radiators. (*a*) Broadwall shunt slots. (*b*) Inclined edge slots. (*c*) Magnetically coupled transverse slots.

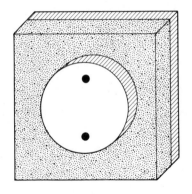

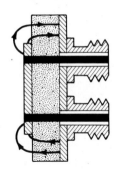

a

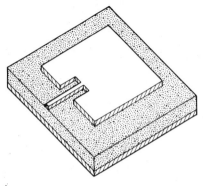

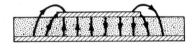

b

Fig. 5. Microstrip radiators. (*a*) Coaxial excited-disk radiator. (*After Byron [8]; reprinted by permission, © 1972 Artech House, Inc.*) (*b*) Microstrip excited-patch radiator. (*After Carver and Mink [1], © 1960 IEEE*)

the allowable area per element corresponding to the required scan coverage. For example, a triangular lattice should be used for a conical scan coverage and a rectangular lattice for a rectangular scan coverage. The grating lobe locations for a rectangular lattice are given by

$$\sin\theta\,\cos\phi - \sin\theta_0\,\cos\phi_0 = \pm\frac{\lambda}{d_x}p \tag{1a}$$

$$\sin\theta\,\sin\phi - \sin\theta_0\,\sin\phi_0 = \pm\frac{\lambda}{d_y}q \tag{1b}$$

where

θ_0, ϕ_0 = the scan direction of the main beam

d_x, d_y = element spacing along the x and y axis

$p, q = 0, 1, 2, \ldots$

Chart 3. Selection Considerations of Radiators

Design Requirements	Selection Considerations
Allowable area per element element spacing lattice operating frequency	Dipoles, disk, and patch radiators for X-band or lower. Waveguide and slot radiators for S-band or higher
Aperture impedance matching scan coverage frequency bandwidth	Dipoles and waveguides for 10- to 25-percent bandwidth. Waveguides for 10-percent to octave bandwidth
Polarization linear or circular single or dual	Dipole and rectangular waveguide for linear. Cross dipole and circular waveguide for dual linear or dual circular
Power-handling capacity peak and average	Dipole and waveguide for 1 kW or less. Waveguide for 1 kW or more
Environmental thermal, shock, vibration, etc.	Dipole and waveguide for non-nuclear hardened. Waveguide for nuclear hardened
Cost, reliability, and producibility	Dipoles for integrated subarray modular construction

For a triangular lattice the grating lobes are located at

$$\sin\theta \cos\phi - \sin\theta_0 \cos\phi_0 = \pm \frac{\lambda}{2d_x} p \qquad (2a)$$

$$\sin\theta \sin\phi - \sin\theta_0 \sin\phi_0 = \pm \frac{\lambda}{2d_y} q \qquad (2b)$$

where $p + q$ is even.

In order to prevent the grating lobe formation the maximum projected element spacing d along a given scan plane must satisfy the following formula:

$$\frac{d}{\lambda} \leqq \frac{1}{1 + \sin\theta_{max}} \qquad (3)$$

For a conical scan coverage of a 60° half-angle cone the maximum allowable area per element is approximately $0.3\lambda^2$ as shown in Fig. 2. Once the area per element is established, a radiator must be selected to fit within that area. Since the element spacing is directly proportional to the wavelength of the operating frequency, the dipole radiators are usually used for X-band or lower frequencies and waveguide/slot radiators are used for S-band or higher frequencies. Another requirement for radiator selection is aperture impedance matching over the required scan coverage and operating frequency bandwidth. For a 60° half-angle cone coverage the dipole and waveguide radiators can be reasonably well matched over a 10-percent bandwidth, whereas waveguide radiators [9] can be matched over almost an octave bandwidth. In general, some of the other considerations in radiator selection for meeting the requirements of polarization, power-handling capacity, environment, cost, etc., are given in Chart 3.

Development Procedure of Phased Array Radiators—Once the radiator selection is made in accordance with the requirements of Chart 3, the usual step-by-step procedure to develop this radiator in a two-dimensional array environment is given below:

Step 1
Select an element spacing and lattice that does not formulate grating lobes or surface-wave resonances over the required scan coverage and the operating frequency band.

Step 2
Formulate an analytical model of the radiating aperture and optimize the performance by varying the design parameters, such as element dimensions.

Step 3
Perform aperture matching, such as by using an inductive iris or a dielectric plug in the opening of the waveguide radiator, metallic fences around dipole radiators, or using dielectric sheets in front of the radiating aperture.

Step 4
Using the analytical results from steps 2 and 3, fabricate waveguide simulators to measure and verify the radiation impedance at discrete scan angles.

Step 5
Fabricate a small test array (usually 9×9 elements) and measure the active element pattern of the central element over the required frequency band. This active element pattern is defined as the pattern of an element with all the neighboring elements terminated into matched loads. It describes the variation in array gain G (including aperture mismatch) as a function of beam scan angle:

$$G = \frac{4\pi A}{\lambda^2} \eta \cos \theta \, (1 - |\Gamma(\theta)|^2) \tag{4}$$

where

A = area of element multiplied by the total number of elements in the array antenna

η = aperture efficiency corresponding to the amplitude distribution across the array aperture, with $\eta = 1$ for uniform amplitude distribution

θ = beam scan angle

$|\Gamma(\theta)|$ = magnitude of reflection coefficient at scan angle θ

The net gain of the antenna is given by the above gain minus all the other ohmic losses and mismatch losses, such as the phase shifter loss, beam-forming feed network loss, etc.

Step 6
Establish the performance characteristics of the final radiator design by combining the measured impedances of step 4 with the measured active element patterns of step 5.

Phase Shifter Selection

The selection criteria of a phase shifter for a particular phased array application are listed below:

1. Operating frequency and bandwidth (tunable and instantaneous)
2. Peak and average rf power
3. Insertion loss
4. Switching time (reciprocal and nonreciprocal)
5. Drive power
6. Size and weight
7. Phase quantization error
8. Cost and producibility

Some of these criteria are established by radar requirements, while the others are used to compare the relative merits of various phase shifter types. Among all the above criteria the six most often used criteria for selecting a particular phase shifter are the operating frequency, peak and average rf power, switching time, size, weight, and cost. The cost is an important consideration since the phase shifters contribute to one-third of the cost of most phased arrays. The other two-thirds of the total cost are contributed almost equally by the phase shifter drivers and the beam-forming feed network plus the radiators. A more detailed discussion on the selection criteria of the phase shifters will be given after the various available phase shifter types are described.

Phase Shifter Types—The two types of commonly used phase shifters are the semiconductor diode phase shifters [10, 11, 12] and the ferrite phase shifters [10, 11, 12]. It is not the intention here to describe the theory of operation of these phase shifters since this subject is well covered elsewhere. This section will, however, describe in detail the performance and physical characteristics of these phase shifters so that a comparison can be made to select the proper phase shifter for a particular radar application. The diode phase shifters are generally digital phase shifters, i.e., the phase states of the phase shifter are in discrete binary phase increments. Fig. 6 shows the binary phase states of a 4-bit diode phase shifter. The

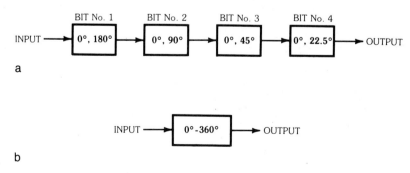

Fig. 6. Basic types of electronic phase shifters. (*a*) Semiconductor diode phase shifter (digital). (*b*) Ferrite phase shifter (digital or analog).

diode phase shifter uses pin diodes to provide phase shifting either by switching in different line length across a transmission line or by changing from inductive to capacitive loading across the transmission line. Typical phase shifter circuits using pin diodes are shown in Fig. 7. There are analog diode phase shifters [13, 14] containing either varactor diodes or pin diodes. However, these analog diode phase

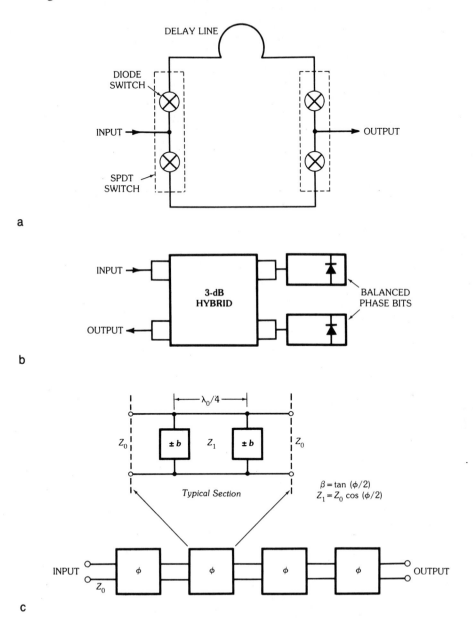

Fig. 7. Circuit designs for diode phase shifter. (*a*) Switched-line phase bit. (*b*) Hybrid-coupled phase bit. (*c*) Loaded-line phase bit. (*After Stark [10], © 1984; reprinted with permission of McGraw-Hill Book Company*)

shifters are extremely low power devices, typically in the milliwatts range. On the other hand, the ferrite phase shifters can be either digital or analog devices as illustrated in Fig. 8. In general, the two types of commonly used ferrite phase shifters are the nonreciprocal, toroidal ferrite phase shifter [15, 16] and the reciprocal, dual-mode ferrite phase shifter [17, 18]. Fig. 8 shows the basic configurations of these two types of ferrite phase shifters. The toroidal ferrite phase shifter employs the use of a toroidal shaped ferrite bar placed in a rectangular waveguide. A drive wire is inserted longitudinally through the center core of the toroid to provide transverse magnetization in the toroid. Phase shifting is achieved by varying the current in the drive wire, hence varying the biasing magnetic field in the toroid. This phase shifter can operate as an analog device by using a long toroid and varying the phase shift by means of changing the holding current in the drive wire. It can also operate as a digital device by magnetically latching (no

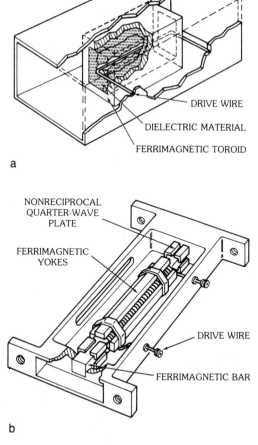

Fig. 8. Basic ferrite phase shifters. (*a*) Nonreciprocal twin-slab toroidal type. (*b*) Reciprocal dual-mode type. (*After Tang and Burns [11]*)

holding current) the toroid to the various minor hysteresis loops. The amount of phase shift is determined by the product of the magnitude and the time duration of the voltage pulse. In applications where fast switching speed is required, the long toroid can be split into smaller sections with the length of each section corresponding to a binary phase bit. In this case each section is magnetized into saturation and is quickly switchable since the volume of ferrite material for each section is significantly smaller. This type of phase shifter is nonreciprocal; hence the phase shift must be reset between the transmit and receive modes of the radar. The dual-mode ferrite phase shifter, however, is reciprocal, and it does not require resetting between transmit and receive. This phase shifter consists of a long metallized ferrite bar in between two nonreciprocal quarter-wave plates. The quarter-wave plate at each end of the ferrite bar converts the incident linearly polarized electric field into a circularly polarized field. This circularly polarized field interacts with the longitudinally magnetized biasing field in the ferrite bar to produce Faraday rotation, resulting in a net phase shift of the wave propagating through the ferrite bar. The amount of phase shift is controlled by the magnitude of the biasing field in the ferrite bar. This device can be used as a digital or analog phase shifter as in the case of the nonreciprocal toroidal phase shifter. However, the switching time of this phase shifter is much longer (50 µs compared with 10 µs for the toroidal phase shifter) due to the eddy current effects of the metal wall around the ferrite bar. There are other types of reciprocal ferrite phase shifters such as the Reggia-Spencer phase shifter [19] and Fox phase shifter [20]. The details of these phase shifters can be found in the references.

Performance Characteristics of Phase Shifters—A comparison of the general characteristics of diode and ferrite phase shifters is shown in Table 1. As shown in this table, most of the diode phase shifters that have been built operate over the frequency range of uhf to X-band, whereas the ferrite phase shifters have been built to operate over the frequency range of S-band to W-band. Ferrite phase shifters of the reciprocal, dual-mode type have been built at 95 GHz with insertion loss of approximately 2.5 dB for 360° of phase shift. The insertion loss of the diode phase shifter varies typically from 0.5 dB at L-band to 1.4 dB at X-band. On the other hand, the insertion loss of the ferrite phase shifter varies typically from 0.6 dB at S-band to approximately 1.0 dB at X-band. The insertion loss of the

Table 1. General Characteristics of Diode and Ferrite Phase Shifters

Parameter	Diode (Digital)	Remanent Ferrite (Analog and Digital)
Frequency	Uhf to X-band	S- to W-band
Insertion loss/2π	0.5 to 1.4 dB	0.6 to 2.5 dB
Temperature sensitivity	Negligible	<0.03 to 0.3%/°C
Peak power	≦8 kW	≦100 kW
Average power	≦300 W	≦800 W
Bandwidth	10% to 25%	10% to octave
Switching speed	10 ns to 30 µs	1 to 50 µs
Control power	0.1 W to 0.5 W	20–800 µJ

ferrite phase shifter goes up to approximately 1.4 dB at 30 GHz. A comparison of the insertion loss of the diode and ferrite phase shifters as a function of frequency is shown in Fig. 9. At *L*-band frequencies or below, the diode phase shifters are used because of their low insertion loss and lower cost. At *S*-band frequencies the insertion loss of the diode phase shifter is quite comparable to that of the ferrite phase shifter. For example, the insertion loss of the ferrite phase shifter is approximately 0.2 to 0.3 dB lower. Above *S*-band, however, the disparity in insertion loss becomes significantly more in favor of the ferrite phase shifter. The temperature sensitivity of the diode phase shifter is nil compared to 0.03%/°C to 0.3%/°C for the ferrite phase shifter, depending on the type of ferrite material used. The peak and average power handling capacity of most of the diode phase shifters is approximately 8 kW and 300 W, respectively, compared with 100 kW and 800 W, respectively, for the ferrite phase shifter. The bandwidth of most of the diode phase shifters is typically 10 percent. However, an octave bandwidth can be achieved by using Schiffman coupled circuits [21]. In the case of the ferrite phase shifter an octave bandwidth can also be achieved by using multiple stages of ridge-loaded waveguide transformers. The switching speed of the diode phase shifter varies from 0.5 ns to approximately 30 µs depending on the capacitance of the diode and the type of driver used to switch the diode from forward bias to reverse bias. Typically, the switching speed is approximately 30 µs, using a simple resistive pull-up type of driver circuit. However, the switching speed can be reduced to 10 µs or less when an active pull-up circuit is used. The switching speed for the ferrite phase shifter varies from 1 µs for the individual toroidal bit type to about 50 µs for the long Faraday rotator type. The required control power for the diode phase shifter varies from 0.1 to 0.5 W depending on the amount of forward bias current needed to set the quiescent state of each bit to the constant resistance region of diode *I-V* characteristics. Forward bias currents can be reduced from the above settings at the expense of a slight increase in insertion loss. The required control energy for the ferrite phase shifter varies from 20 to 800 µJ, depending on the speed with which the phase shifter has to be switched.

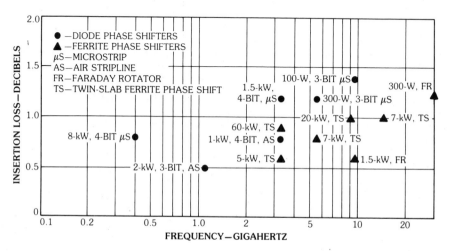

Fig. 9. Insertion loss of diode and ferrite phase shifters. (*After Tang and Burns [11]*)

General Observations in Selecting Phase Shifter Type—Above *S*-band and at peak power level of 2 kW or more per element, ferrite phase shifters are preferred due to their lower insertion loss. However, in situations where the required rf power is low (less than 2 kW) and size and weight constraints dictate a compact and lightweight package, diode phase shifters are preferred due to their simplicity in construction and lower production cost. For example, cost can be minimized by combining several dipole radiators, diode phase shifters, feed networks, drivers, and logic circuits on a common alumina substrate as an integral subarray module.

Beam-Forming Feed Network Selection

Among all the selection criteria, as stated in Chart 2, the most influential criteria in the selection of a beam-forming feed network for a particular radar application are the following:

1. Number of simultaneous beams
2. Monopulse sum and difference beams
3. Peak and average side lobe level
4. Tunable and instantaneous bandwidth
5. Peak and average power
6. Size, weight, and cost

For most applications the above criteria can be narrowed down from the multitude of possible design choices to a few practical design selections. For example, if a large number of multiple simultaneous beams are required, the possible design choices for the beam-forming feed network would be either a multiple-beam optical feed (such as the circular pillbox feed, Rotman lens feed, etc.) or a constrained matrix feed (such as the Butler or Blass matrix feed). Once the choices are narrowed down to only a few possible design approaches, a comparison of these approaches can then be made in terms of side lobe performance, bandwidth, size, weight, and cost in order to determine the optimum design approach. In order to select the optimum feed for a particular application it is necessary for the antenna designer to have a broad and comprehensive knowledge of the various types of available beam-forming feed networks. This subsection summarizes some of the basic types of beam-forming feed networks and their performance capabilities and limitations. A detailed discussion of the various feed network designs is given in Chapter 19, on beam-forming feed networks.

In general, all the beam-forming feed networks can be classified into the following three basic categories:

Category 1—Space feeds:
      transmission type
      reflection type
Category 2—Constrained feeds:
     series feed
     parallel feed
Category 3—Hybrid feeds (a combination of space and constrained feeds)

Space Feeds—In the category of space feeds there are two basic types, namely, the transmission type [22] and the reflection type [23]. In the case of the transmission type the array elements and phase shifters are connected to an array of pickup elements, which are in turn illuminated by a feed horn located at a given focal distance away from the aperture of the pickup array (see Fig. 10). The ratio of the focal distance to the diameter of the radiating aperture varies nominally from 1/2 to 1. The array of pickup elements in conjunction with the phase shifters and the radiating elements forms an electronic scanning feedthrough lens (transmission lens). The phase shifters are set to provide the required phase increments between the radiating elements for beam scanning and for correcting the phase error introduced by the spherical phase front from the feed horn. When digital phase shifters (constant phase with frequency type) are used in the lens this spherical phase front correction approach can suppress the peak error side lobe introduced by the phase quantization error of the digital phase shifter. A detailed discussion of the effect of phase quantization error on array performance is given by Miller [24]. The tunable and instantaneous bandwidths of this space-fed antenna are typically 10 percent and 40 MHz, respectively. When the time-delay type of phase shifters are used, this antenna system has extremely wide instantaneous bandwidth limited only by the performance of the components. An instantaneous bandwidth of 1000 MHz is achievable. Monopulse sum and difference beams can be formed by using a cluster of feed horns at the focal point. For example, a cluster of 2×2 feed horns combined with magic tees can provide a sum beam, an elevation difference beam, and an azimuth difference beam as shown in Fig. 10. Multiple simultaneous beams can also be formed by using feed horns displaced from the focal point. The number of multiple beams that can be formed, however, is limited by the spherical aberration effects. Various methods of correcting the spherical aberration are treated by Rotman and Turner [25]. Since the signal distribution from the feed horns to the elements is through free space, this space feed is probably the simplest

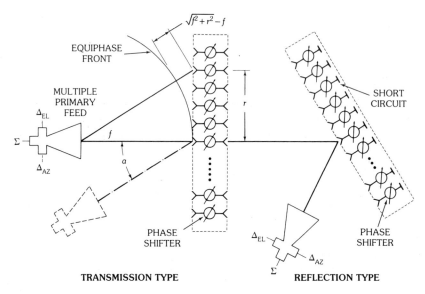

Fig. 10. Space feed systems. (*After Cheston and Frank*, Array Antennas *[22] TG-956, JHU/APL, Laurel, MD, March 1968*)

way to form multiple simultaneous beams. In the case of the reflection type of space feed (see Fig. 10), the concept is the same as that of the transmission type except a short is placed behind each phase shifter so that the signals are reflected and reradiated from the pickup elements. Since the signal travels back and forth through the phase shifters twice, the amount of required phase shift at each element is half of that of the transmission case for the same scan angle. However, the peak power requirement for the phase shifter is quadrupled because of the standing wave. In order to minimize the blockage effect of the feed horn, the feed horn has to be offset with respect to the axis of the reflective lens.

Space feeds have several disadvantages. One of them is the large physical volume required by the space feed. Another is the multiple reflection effects between the mismatches of the two lens apertures as a function of beam scan angle. This effect can be minimized by performing the best impedance matching possible for the two apertures over the required scan coverage and frequency band. A less obvious problem with space feeds is the difficulty of connecting control wires of the phase shifters in the lens because of the lack of access from the front or the back of the lens. The only access is through the peripheral edge of the lens.

Constrained Feeds—In the category of the constrained feeds there are two basic types, namely, the series feeds and the parallel feeds. By definition the radiating elements are fed serially in a series feed, while they are fed in parallel in a parallel feed. Typical examples of the series feed are shown in Fig. 11. The basic form of a series feed is shown in Fig. 11a. The input signal is fed from one end of the feed and the other end is terminated into a matched load. The input signal is then coupled serially through directional couplers to the phase shifters and radiating elements. Since the transmission line length increases from the input to the following radiating elements, there is a progressive phase change between the radiating elements with frequency variations; thus the beam scans with frequency change. The amount of beam squint with frequency is given by

$$\text{beam squint} = \frac{\Delta f}{f} \frac{1}{\cos \theta_0} \tag{5}$$

where Δf is the amount of frequency change and θ_0 is the nominal scan angle. A detailed discussion of the bandwidth limitations of series and parallel feeds is given by Frank [26]. The beam squint, however, can be brought back to the original position by resetting the phase shifters at each radiating element. One of the problems with this type of series feed is that the mismatches from the radiating elements, phase shifters, and couplers can all add up in phase at the frequencies when the path length between the elements is a multiple of a half-wavelength. In order to provide sum and difference beams for monopulse tracking the feed input is moved from the end to the center of the series feed as shown in Fig. 11b. The two halves of the feed are fed by a magic tee so that the sum port of the magic tee provides in-phase excitation of the two halves and the difference port provides out-of-phase excitation to generate a difference beam. Since the same coupling coefficients of the couplers are used for both the in-phase and the out-of-phase excitations, the side lobes for the sum and difference beams cannot be op-

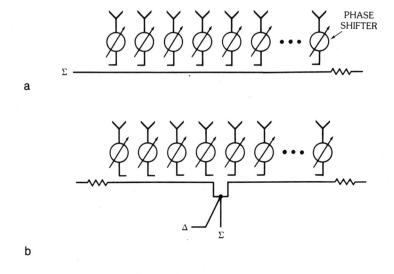

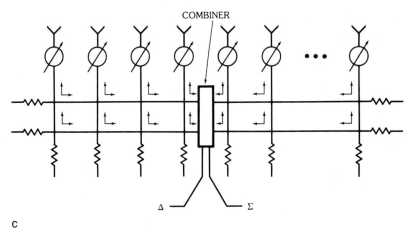

Fig. 11. Constrained feeds (series feed networks). (*a*) End feed. (*b*) Center feed. (*c*) Center fed with separately optimized sum and difference channels. (*d*) Equal path length feed. (*e*) Series phase shifters. (*After Cheston and Frank,* Array Antennas *[22], TG-956, JHU/APL, Laurel, MD, March 1968*)

timized simultaneously. For low side lobes the sum beam requires an even distribution, such as the Taylor distribution [27], and the difference beam requires an odd distribution, such as the Bayliss distribution [28]. In order to achieve low side lobes for both sum and difference beams a center-fed dual series feed (see Fig. 11c) is used. The excitations of the two parallel series feeds are adjusted to optimize the side lobes for both the sum and the difference beams. A detailed description of the dual series feed is given by Lopez [29]. At broadside beam position the two halves of the center-fed series feed scan in opposite directions with frequency. This results in a beam broadening with no change in direction. If the two halves scan too

d

e

Fig. 11, *continued.*

far apart, it could even result in a splitting of the beam. Therefore the bandwidth of the center-fed series feed is significantly worse than that of the parallel feed. However, at large scan angles such as 60° from broadside the bandwidth of the center-fed series feed is quite comparable to that of a parallel feed (see Frank [26]). In order to broaden the bandwidth at broadside for a center-fed array the path length from the feed input to the radiating elements can be made equal as shown in Fig. 11d. In this case the bandwidth is only limited by the in-phase addition of the mismatches of the couplers. For a series-fed array the phase shifters can be also inserted serially between the couplers along the series feed line as shown in Fig. 11e. In this design the amount of required phase shift for each phase shifter is greatly reduced compared with that of the phase shifters at the radiating elements. However, the insertion losses of these serial phase shifters are additive, resulting in a reduction in array gain and an increase in side lobe level from the asymmetrical amplitude distribution.

   Typical examples of the parallel feeds are shown in Fig. 12. The basic form of a parallel feed is shown in Fig. 12a. In this basic form the input signal is divided in a corporate tree fashion to all the radiating elements. The path lengths from the input to each output are made equal. The bandwidth at broadside is ideally infinite, except for the practical limitations of such components as the couplers, phase shifters, and radiators. When the beam is scanned away from broadside the beam scans with frequency. The amount of beam squint with frequency is given by

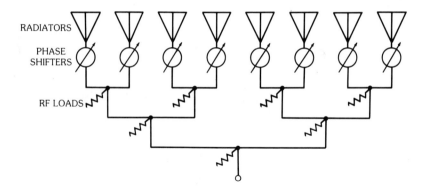

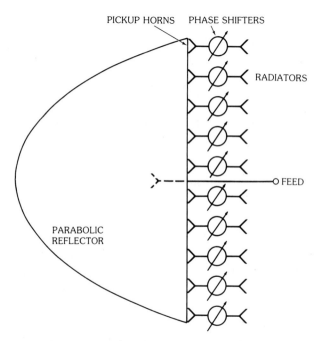

Fig. 12. Constrained feeds (parallel feed networks). (*a*) Corporate feed. (*b*) Pillbox feed. (*After Rotman [34],* © *1958 IEEE*) (*c*) Butler matrix feed (multiple beams). (*After Butler and Lowe [30]*) (*d*) Time-delay matrix feed (multiple beams). (*After Blass [31],* © *1960 IEEE*)

$$\text{beam squint} = \frac{\Delta f}{f} \tan \theta_0 \tag{6}$$

The bandwidth at 60° scan (see Frank [26]) is given by

$$\text{percent bandwidth} = \text{beamwidth in degrees}$$

When magic tees or hybrid couplers are used at each level of the corporate feed the mismatches from the radiating elements are reasonably well isolated from each other. Hence the parallel corporate feed does not have the additive effect as in the

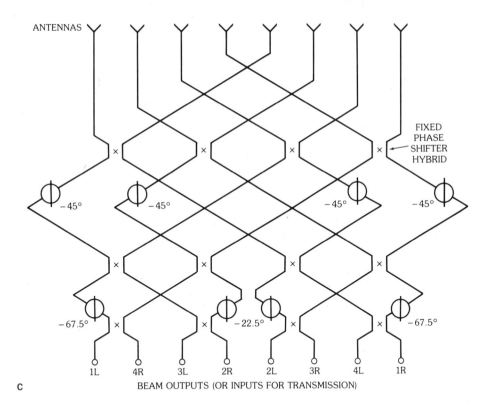

ANTENNAS

FIXED
PHASE
SHIFTER
HYBRID

−45° −45° −45° −45°

−67.5° −22.5° −67.5°

1L 4R 3L 2R 2L 3R 4L 1R

c

BEAM OUTPUTS (OR INPUTS FOR TRANSMISSION)

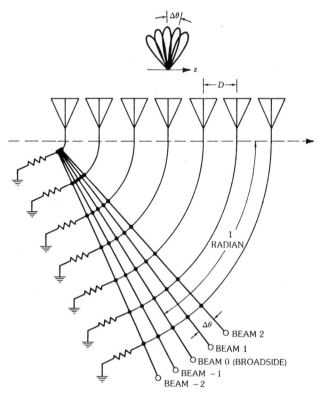

$\Delta\theta$

z

D

1
RADIAN

$\Delta\theta$

BEAM 2

BEAM 1

BEAM 0 (BROADSIDE)

BEAM −1

BEAM −2

d

Fig. 12, *continued.*

case of the series feed. An optical form of the corporate feed is shown in Fig. 12b. In this optical configuration the input feed horn illuminates a parabolic reflector, which in turn produces a reflected plane wave illuminating the pickup horns located at the aperture of the reflector. The signals received by the pickup horns are fed into the phase shifters and radiating elements the same way as the corporate feed. The fact that the power distribution from the feed input to the radiating elements is accomplished optically with the parabolic reflector, instead of with the hybrid couplers as in the case of the corporate feed, enhances the bandwidth of the feed system. However, the freedom in controlling the amplitude taper of the signals at the radiating elements is restricted to those provided by the primary illumination pattern of the feed horn. Typically the amplitude distribution is of the form of a truncated $\cos^n x$ distribution, where n is an integer greater than 1. This form of amplitude taper is achieved by using multiple feed horns at the focal point of the reflector. For example, a cosine distribution is achieved by using two feed horns displaced symmetrically on both sides of the focal point. Higher-order cosine distributions can be achieved by employing three or more feed horns. When a large number of feed horns are used to generate the desired amplitude taper, the aperture blockage effect caused by the feed horns can be circumvented by using a folded pillbox feed. The feed horns are placed at one level of the folded pillbox feed, while the pickup horns are placed at the next level. Multiple beams can also be formed by using multiple-feed horns. The number of multiple beams, however, is limited by the defocusing effect of the parabolic reflector. Typically, scanned beams can be formed to approximately two beamwidths from either side of the focal beam (broadside beam). In order to form more simultaneous beams the shape of the reflector must be changed from a parabolic to a circular configuration, and the feed horns must be arranged along a circular arc, concentric to the reflector surface. For this case the f/D ratio is nominally 1/2. Due to the circular symmetry the radiation patterns of the feed horns are essentially identical except for the edge or truncation effect of the reflector. The radiation patterns, however, are deteriorated slightly by the spherical aberration effect of the circular reflector.

Multiple simultaneous beams can also be formed by using constrained parallel feeds such as the Butler matrix feed [30] and the Blass matrix feed [31]. The Butler matrix feed, as shown in Fig. 12c, consists of layers of 90° hybrids (3-dB couplers) interconnected with transmission lines and fixed phase shifters. The maximum number of simultaneous beams that are formed by the Butler matrix feed is equal to the total number of radiating elements in the array. For example, an eight-element array has eight simultaneous beams. All the beams are orthogonal to each other, and they cover the entire radiation space from one end-fire direction to the opposite end-fire direction. Due to the drop-off in the active element pattern of the radiating elements the beams close to end-fire directions are generally not used. The amplitude distribution across the radiating elements corresponding to any beam position is uniform; hence the crossover point of the adjacent beams is approximately 4 dB for orthogonal beams. The beam-pointing direction and the beamwidth change with frequency so that the orthogonality is maintained. Due to the large number of components required in the matrix feed for a large array and the complexity of packaging these components, the practical usage of the Butler matrix feed is generally limited to a sixteen-element array.

In order to prevent the beam from scanning with frequency a time-delay matrix feed of the form shown in Fig. 12d can be used. This time-delay matrix feed is a special form of the Blass matrix feed [31]. For the broadside beam (beam 0 in Fig. 12d) all the path lengths from the input to the radiating elements are equal. For all the scanned beams the difference in path lengths between two adjacent feed lines from the beam input terminal to the radiating elements is exactly equal to the required time delay between these elements for that particular scan angle. Hence this time-delay matrix feed has extremely wide bandwidth, and it is only limited by the bandwidth of the couplers in the feed. One of the practical design problems of this time-delay matrix feed is the coupling between the feed lines. The amount of coupling is determined by the crossover level of the corresponding beams and the directivity of the couplers. High crossover level between beams (higher than the level corresponding to orthogonal beams) appears as a cross-coupling loss which in turn degrades the antenna gain. Poor directivity in the couplers causes circulating power between feed lines, which produces amplitude and phase errors at the radiating elements. In order to minimize amplitude and phase errors it is imperative that the beams are spaced as closely to the orthogonal condition as possible, and the directivity of the couplers is as high as possible.

Independently controlled sum and difference beams for monopulse tracking can be formed in a parallel constrained feed as shown in Fig. 13. The signals from symmetrical pairs of radiating elements located diametrically opposite from the centerline are combined in 180° hybrids (magic tees) to form in-phase and out-of-phase signals. The in-phase signals are combined in a feed network with the proper amplitude weighting to form a sum beam. The out-of-phase signals from the magic

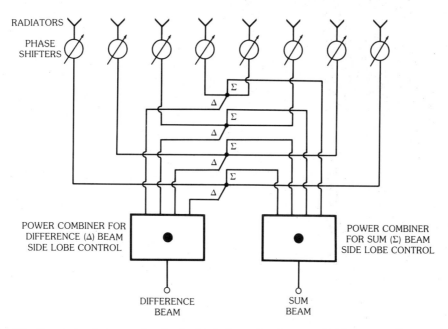

Fig. 13. Constrained monopulse feed with independently controllable sum and difference beam side lobes. (*After Tang and Burns [11]*)

tees are combined in a separate feed network to form a difference beam. Different amplitude distributions, such as the Taylor distribution for the sum beam feed network and the Bayliss distribution for the difference beam feed network, can be used to independently control the side lobes of both beams. When the lengths of the transmission lines from the input to the radiating elements are made equal, the bandwidth of this monopulse feed network is the same as that of a corporate feed.

General Observations in Selecting Beam-Forming Feeds—In general, the series feeds are more frequency sensitive than the parallel feeds; hence the series feeds are more bandwidth limited. Furthermore, the mismatches from the couplers in a series feed are additive when the transmission lines between the couplers are integer multiples of a half-wavelength. However, the lengths of transmission lines between the couplers in a parallel feed are nonuniform so that the mismatches do not add in phase. Also, the mismatches from the couplers can be isolated by the magic tees in order to prevent the reflected signal from circulating in the feed network and, as a result, cause high-power breakdown problems. In order to avoid the resonance problem in the series feed the length of transmission line in the series feed is chosen to avoid resonance over the operating band. One advantage of the series feed is that it is physically more compact than the parallel feed; hence it is more useful for applications where antenna volume is of prime importance.

4. Effect of Component Errors on Array Performance

One of the most important considerations in the design of a phased array antenna is the effect of errors in the array components on the antenna performance. These errors are mainly caused by the manufacturing tolerances of the components and the batch-to-batch variations in material consistency. In general, all the errors can be identified and grouped into two basic types: namely, the systematic type of error and the random type of error. The systematic errors are deterministic errors resulting from some inherent characteristics of the components. For example, the use of quadrature couplers in a beam-forming network results in fixed 90° errors across the outputs of the feed network. These phase errors can be trimmed out by storing the required phase correction terms in the programmable read-only memories (PROMs) of the beam-steering computer, and then by correcting these errors with the phase shifters at the radiating elements. The net required setting of the phase shifter is the sum of the phase correction term and the incremental beam-steering phase term. Any systematic type of amplitude error can also be trimmed out by fine-tuning the power split ratios of the couplers. The random errors, on the other hand, are not deterministic, and they are not correlated from element to element. Therefore, it would not be practical to correct the error at each element with the phase shifter since they are all different. These random errors can be minimized by controlling the manufacturing toierances of the components. Therefore, in the remainder of this section, we will address the effects of random errors on antenna performance.

Before we can analyze the effect of random errors on array performance the following assumptions relating to the characteristics of these errors are made:

1. Errors in each radiating element are statistically independent from those in all the other elements.
2. Errors in all elements possess the same statistic.

Based on the above assumptions the array performance can be calculated in terms of the net rms amplitude and phase errors at each element of a large two-dimensional array. The net rms error at each element is determined by statistically summing all the contributions from the various components to that element using the central limit theorem. Detailed treatments of the error effects on array performance were done by Ruze [32], Allen [33], Miller [24], etc. They have expressed the array performance in terms of the rms error as follows:

1. The rms side lobe power, $\bar{p}$, is

$$\bar{p} = \frac{\sigma^2}{\eta N} \tag{7}$$

where

$\sigma^2 = \sigma_a^2 + \sigma_p^2$

σ_a = rms amplitude error

σ_p = rms phase error in radians

η = aperture efficiency, which is less than or equal to 1

N = total number of elements

2. The peak side lobe level,* p_{pk}, is

$$p_{pk} = |S_0| + 2\sigma/\sqrt{\eta N} \tag{8}$$

where S_0 is the amplitude of the design side lobe without errors.

3. The reduction in antenna gain is

$$\frac{G}{G_0} = \frac{1}{1 + (3\pi/4)(d/\lambda)^2\sigma^2} \tag{9}$$

where

d/λ = element spacing in wavelengths

G_0 = antenna gain with the absence of errors

4. The beam-pointing error is

$$\frac{\delta_\psi}{\Delta\theta_{rad}} = \sqrt{\frac{3}{N}} \frac{\sigma}{0.88\pi} \tag{10}$$

---

*Estimate is based on power not to be exceeded with 98-percent probability.

where

δ_ψ = rms beam-pointing error in radians

$\Delta\theta_{rad}$ = beamwidth in radians

The above formulas can be used in conjunction with actual radiation pattern calculations to estimate the allowable error budgets for the array components in meeting a given set of array performance specifications. Conversely, if the amplitude and phase errors of the components are known, then the array performance can be estimated by the above formulas. An example which illustrates the effects of errors on the array performance is given below. This example is for the case of an array of 2000 elements with element spacing d of 0.5λ. The error-free side lobe level $(p_{pk})_0$ is asumed to be -40 dB; correspondingly, the aperture efficiency η is equal to 0.64. Assuming the composite rms amplitude error σ_a and the rms phase error σ_p are 0.5 dB and 4°, respectively, then the net rms error σ is given by

$$\sigma^2 = \sigma_a^2 + \sigma_p^2 = 0.0035 + 0.0049 = 0.0084$$

The average side lobe level of the array with errors is given by

$$\bar{p} = \frac{\sigma^2}{\eta N} = 6.6 \times 10^{-6} \quad \text{or} \quad -51.8 \text{ dB}$$

The peak side lobe level p_{pk} is obtained from $p_{pk} = |S_{pk}|^2$, where

$$S_{pk} = S_0 + \frac{2\sigma}{\sqrt{\eta N}} = 0.01 + 0.005 = 0.015$$

so that

$$p_{pk} = -36.5 \text{ dB}$$

The gain reduction is given by

$$\frac{G}{G_0} = \frac{1}{1 + (3\pi/4)(d/\lambda)^2\sigma^2} = 0.995 \quad \text{or} \quad -0.02 \text{ dB}$$

The beam-pointing error is given by

$$\frac{\delta_\psi}{\Delta\theta_{rad}} = \sqrt{\frac{3}{N}} \frac{\sigma}{0.88\pi} = 0.001\,28 \quad \text{or} \quad \delta_\psi = \Delta\theta_{rad}/800$$

5. References

[1] K. R. Carver and J. W. Mink, "Microstrip antenna technology," *IEEE Trans. Antennas Propag.*, vol. AP-29, no. 1, pp. 2–24, January 1981.

[2] R. J. Mailloux, J. F. McIlvenna, and N. P. Kernweis, "Microstrip array technology," *IEEE Trans. Antennas Propag.*, vol. AP-29, no. 1, pp. 25–37, January 1981.

[3] R. J. Stegen, "Longitudinal shunt slot characteristics," *Tech. Memo. No. 261*, Hughes Aircraft Company, 1951.

[4] A. F. Stevenson, "Theory of slots in rectangular waveguide," *J. Appl. Phys.*, no. 19, pp. 24–38, 1948.

[5] J. S. Ajioka, "Frequency-scan antennas," *Antenna Engineering Handbook*, 2nd ed., ed. by R. C. Johnson, New York: McGraw-Hill Book Co., 1984, pp. 19-1–19-30.

[6] R. S. Munson, "Microstrip antennas," *Antenna Engineering Handbook*, 2nd ed., ed. by R. C. Johnson, New York: McGraw-Hill Book Co., 1984, pp. 7-1–7-28.

[7] J. Q. Howell, "Microstrip antennas," in *Dig. Intl. Symp. Antennas Propag. Soc.*, pp. 177–180, Williamsburg, Virginia, December 1972.

[8] E. V. Byron, "A new flush-mounted antenna element for phased array application," in *Proc. 1970 Phased Array Antenna Symp.*, pp. 187–192, Polytechnic Institute of Brooklyn, June 1970. Reprinted in *Phased Array Antennas*, ed. by A. A. Oliner and G. H. Knittel, Dedham: Artech House, 1972.

[9] C. C. Chen, "Broadband impedance matching of rectangular waveguide phased arrays," *IEEE Trans. Antennas Propag.*, vol. AP-21, pp. 298–302, May 1973.

[10] L. Stark, R. W. Burns, and W. P. Clark, "Phase shifters for arrays," *Radar Handbook*, ed. by M. I. Skolnick, New York: McGraw-Hill Book Co., 1970, pp. 12-1–12-65.

[11] R. Tang and R. W. Burns, "Phased arrays," *Antenna Engineering Handbook*, 2nd ed., ed. by R. C. Johnson, New York: McGraw-Hill Book Co., 1984, pp. 20-1–20-67.

[12] D. H. Temme, "Diode and ferrite phaser technology," in *Proc. Phased Array Antenna Symp.*, 1970, pp. 212–218, reprinted in *Phased Array Antennas*, ed. by A. A. Oliner and G. H. Knittel, Dedham: Artech House, 1972.

[13] C. A. Liecht and G. W. Epprechi, "Controlled wideband differential phase shifters using varactor diodes," *IEEE Trans. Microwave Theory Tech.*, vol. MTT-15, pp. 586–589, October 1967.

[14] J. F. White, "Figure of merit for varactor reflection type phase shifters," *NEREM Rec.*, pp. 206–207, November 1965.

[15] W. J. Ince et al., "The use of manganese-doped iron garnets and high dielectric constant loading for microwave latching ferrite phasers," *G-MTT Dig.*, pp. 327–331, 1970.

[16] D. H. Temme et al., "A low cost latching ferrite phaser fabrication technique," *G-MTT Dig.*, pp. 88–96, 1969.

[17] R. G. Roberts, "An *X*-band reciprocal latching Faraday rotator phase shifter," *G-MTT Dig.*, pp. 341–345, 1970.

[18] C. R. Boyd, Jr., "A dual-mode latching reciprocal ferrite phase shifter," *G-MTT Dig.*, pp. 337–340, 1970.

[19] F. Reggia and E. G. Spencer, "A new technique in ferrite phase shifting for beam scanning of microwave antennas," *Proc. IRE*, vol. 45, pp. 1510–1517, November 1957.

[20] A. G. Fox, "An adjustable waveguide phase changer," *Proc. IRE*, vol. 35, pp. 1489–1498, December 1947.

[21] B. M. Schiffman, "A new class of broadband microwave 90-degree phase shifters," *PGMTT-MTT-6*, pp. 232–237, 1958.

[22] T. C. Cheston and J. Frank, "Array antennas," *Tech. Memo. TG-956*, The Johns Hopkins University Applied Physics Laboratory, March 1968.

[23] R. Tang, R. W. Burns, and N. S. Wong, "Phased array antenna for airborne application," *Microwave J.*, vol. 14, no. 1, pp. 31–38, January 1971.

[24] C. J. Miller, "Minimizing the effects of phase quantization errors in an electronically

scanned array," *Proc. 1964 Symp. Electronically Scanned Array Techniques and Applications*, RADC-TDR-64-225, vol. 1, pp. 17–38.

[25]   W. Rotman and R. F. Turner, "Wide angle microwave lens for line source," *IEEE Trans. Antennas Propag.*, pp. 623–632, November 1963.

[26]   J. Frank, "Bandwidth criteria for phased array antennas," *Proc. Phased Array Antenna Symp.*, pp. 243–253, Polytechnic Institute of Brooklyn, June 1970. Reprinted in *Phased Array Antennas*, ed. by A. A. Oliner and G. H. Knittel, Dedham: Artech House, 1972.

[27]   T. T. Taylor, "Design of line-source antennas for narrow beamwidth and low sidelobes," *IRE Trans. Antennas Propag.*, vol. AP-3, pp. 16–28, 1955.

[28]   E. T. Bayliss, "Design of monopulse antenna difference patterns with low sidelobes," *Bell Syst. Tech. J.*, pp. 623–650, May–June 1968.

[29]   A. R. Lopez, "Monopulse networks for series feeding an antenna," *IEEE Trans. Antennas Propag.*, vol. AP-16, pp. 436–440, June 1968.

[30]   J. Butler and R. Lowe, "Beamforming matrix simplifies design of electronically scanned antennas," *Electron. Des.*, vol. 9, no. 7, pp. 170–173, April 1961.

[31]   J. Blass, "The multi-directional antennas: a new approach to stacked beams," *IRE Conv. Proc.*, vol. 8, pt. 1, pp. 48–51, 1960.

[32]   J. Ruze, "Physical limitations on antennas," *Tech. Rep. No. 248*, Massachusetts Institute of Technology, October 30, 1952.

[33]   J. L. Allen, "The theory of array antennas," *Tech. Rep. No. 323*, Lincoln Lab, Massachusetts Institute of Technology, July 25, 1963.

[34]   W. Rotman, "Wide-angle scanning with microwave double-layer pillboxes," *IRE Trans. Antennas Propag.*, vol. AP-6, pp. 96–105, January 1958.

Chapter 19

Beam-Forming Feeds

J. S. Ajioka
Hughes Aircraft Company

J. L. McFarland
(Late) Hughes Aircraft Company

CONTENTS

James S. Ajioka was born in Thornton, Idaho. He received the BS and MS degrees in electrical engineering from the University of Utah in 1949 and 1951, respectively.

From 1949 to 1955 he was a group leader and project engineer in the Antenna Design Section of the Navy Electronics Laboratory. In 1955 he joined Hughes Aircraft Company, where he is manager of the Electromagnetics Laboratories, Hughes Ground Systems Group, in Fullerton, California. He has been active in the design and development of electronically scanning arrays, waveguide slot arrays, geodesic antennas, and wide-angle multiple-beam optical antenna systems, and also low-noise high-efficiency feeds, broadband multioctave multiplexing feeds for reflectors and lenses, and millimeter-wave phased arrays.

Mr. Ajioka has more than 20 patents awarded and several are pending. He has presented and published numerous papers on antenna and microwave systems and has contributed to several books. He has also won awards for outstanding achievements in antenna engineering. Mr. Ajioka is a Fellow of the IEEE.

Jerry L. McFarland was born in Iraan, Texas. He received the BS and MS degrees in electrical engineering, with the latter being from the University of Southern California, Los Angeles.

From 1958 to 1964 he was a staff engineer with Hughes Aircraft Company involved with the development of frequency scanning antennas and research on multiple-beam geodesic lenses. From 1964 to 1972 he was engaged in similar work with Autonetics, a division of Rockwell International. In 1972 he became the technical director of EMP, Chatsworth, California, developing antennas and single-axis trackers. In 1977 he joined Lockheed Missiles and Space Company, where he was concerned with adaptive arrays and multiple-beam antennas. He rejoined Hughes Aircraft Company, Fullerton, California, in 1982 as a senior scientist and was most recently involved in antenna development and investigation of nonconventional slot radiators.

Mr. McFarland's untimely death is a great loss to his profession and to all his friends and associates.

1. Introduction

When phased arrays were relatively simple, antenna subassemblies were easy to identify as feed networks, phasors, and radiating aperture. Modern phased arrays, however, have become quite complex, with a wide variety of designs and physical implementations depending on the particular application. With simple phased arrays the feed network was a passive network of branching transmission lines to distribute the power from a single transmitter to each of the radiating elements in the array via the phasors, and, conversely on receive, it combined the power received by each of the radiating elements to the input to a single receiver. Modern phased arrays may have multiple distributed transmitters, multiple distributed preamplifiers, multiple duplexing switches, and multiple simultaneous beam ports, each with its own final receiver. In addition, adaptive arrays may have adaptive control loops distributed throughout the feeding network with a significant amount of signal processing done within the antenna. For these reasons, general categories and general definitions become somewhat ambiguous. However, since generality is necessary to discuss phased arrays in general, an attempt is made to organize feed systems into general categories, and the reader should be aware of the shortcomings.

2. Constrained Feeds (Transmission-Line Networks)

The simplest method of feeding an array is to use simple passive transmission-line networks which take the transmitter power from a single source, or possibly multiple sources, and to distribute the power to each radiating element via transmission lines and associated passive microwave devices, such as hybrids, magic-Ts, or directional couplers, etc. The network itself is usually a combination of directional couplers, hybrids, and/or Ts in waveguide, coax, stripline, or microstrip. Printed-circuit techniques are popular for some of the approaches to be discussed. Fig. 1 is a simplified diagram of a constrained feed showing the basic subassemblies. In the case of multiple beams, there would be more than one input and more than one beam. In the following sections the series-fed, parallel-fed, time-delay-fed, and multiple-beam matrix-fed approaches are discussed.

Series Feed Networks

The simplest power distribution feed is an end-fed transmission line in which power is coupled off at (usually) periodic intervals to the radiating elements as shown schematically in Fig. 2. Although the series feed is simple, low cost, and packages easily, it has a drawback in that it "frequency scans" because there is an interelement progressive phase delay equal to $k_g d$, which is proportional to frequency, where $k_g = 2\pi/\lambda_g = \omega/v_g$ is the wave number of the feeding transmission line and d is the line length between branch line couplers. Although the phase

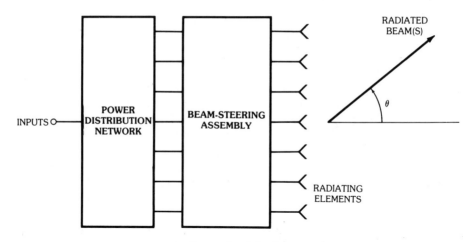

Fig. 1. Constrained feed approach.

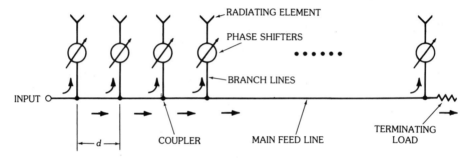

Fig. 2. Schematic diagram of a series feed.

shifters are usually reset for each frequency change, the instantaneous or signal bandwidth is degraded because the antenna beams corresponding to each of the frequency components in a narrow pulse will be pointing in slightly different directions in space. Thus the composite antenna beam is "smeared out" in angular space, thereby broadening the effective beamwidth to reduce the antenna gain and angular resolution.

There are many coupler designs that can couple the desired amount of power from the main feed line to the branch lines. The couplers can be nondirective three-port reactive couplers, such as pure series or pure shunt couplers, nondirective matched couplers, or directive four-port couplers. Nondirective couplers are simple and inexpensive to fabricate. Pure series or shunt couplers are the simplest but are inherently mismatched. For arrays with a large number of elements (on the order of 50 or more elements) the individual reflections due to coupler mismatch are quite small because the coupling is very loose. For large arrays the maximum voltage coupling coefficient is on the order of 0.1. These coupler mismatch reflections, for the most part, add randomly in phase at the input to give a low feed input vswr except at or very near the resonant frequency, that is, the frequency corresponding to a broadside beam. At the resonant frequency the couplers are an integral number of wavelengths apart (or half-wavelengths if the phase of the couplers is alternately 0° and 180°) and all the reflections add in phase to result in a very high input vswr. For purely series couplers the impedances add, and for purely shunt couplers the admittances add. The input vswr versus frequency curve shows high resonance for large arrays, and the width of the vswr versus frequency curve is inversely proportional to the number of elements in the array. In short, the input vswr versus frequency curve resembles an antenna pattern in all respects except the independent variable is frequency instead of spatial angle. That is, tapered aperture distributions result in lower side lobes and broader beamwidth in antenna patterns, and tapered coupling coefficients result in lower side lobes and broader resonance bandwidths in the vswr curve.

Because of the high vswr at resonance the resonant frequency is usually designed to be out of the operating frequency range when simple nondirective couplers are used. Matched nondirective couplers have been designed in waveguide that allow operation at the resonant frequency [1]. Although they are more complex, more costly, and, in general, more difficult to package, directional couplers have superior performance because they are inherently matched at all four ports.

In designs using nondirective couplers the reflections from the couplers can couple into the preceding branch arms and cause spurious side lobes in the antenna pattern; hence they are called *reflection lobes*. In particular, a reflection from a poorly matched terminating load will cause a reflected beam in the conjugate direction from broadside as the desired beam. Reflections from mismatches in the branch lines, such as reflections from the input end of the phase shifters or connectors on the feed side of the phase shifter, will usually reflect specularly as a whole and the reflected wave reflects back into the feed main line and will be phased such that it will be dissipated in the terminating load of the series feed.

Parallel Feed Networks

The simplest parallel feed network consisting of branching transmission lines is commonly known as a corporate feed (see Fig. 3). Since the path lengths from the feed point to the radiating elements are equal, there is no progressive phase delay between radiating elements, and hence no frequency scan, so that the instantaneous bandwidth is much greater. If the corporate feed were made up nondispersive TEM transmission line, the feed itself would be true time delay and would, in principle, have unlimited bandwidth.

The power dividing branch points of the feed are usually matched four-port hybrid junctions instead of reactive three-port T junctions. Since a matched four-port junction is impedance matched in all four ports, spurious reflections from connectors, phase shifters, radiating apertures, etc., will not be scattered from the various junctions back into the radiating aperture to cause undesired side lobes and other antenna pattern degradation. Instead, these spurious reflections are absorbed in the terminating loads of the hybrids. Also, multiple reflections among the junctions can cause high resonance effects that can lead to high-power breakdown in the feed network. The use of directional couplers helps alleviate this potential problem.

As with the use of directional couplers in the series feed, the unused terminal

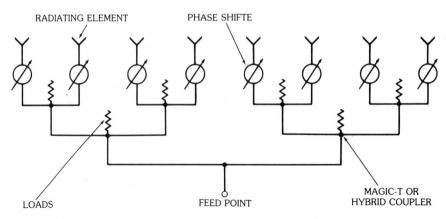

RADIATING ELEMENT   PHASE SHIFTE

LOADS   FEED POINT   MAGIC-T OR
                                             HYBRID COUPLER

Fig. 3. Diagram of a corporate feed, phase shifters, and radiating elements.

(the port that is normally terminated in a matched load) can be used to form a multitude of auxiliary antennas such as side lobe blanking and coherent side lobe cancellation antennas [2].

Again, as in the series feed, when a plane wave is incident on the antenna from the direction of the main beam, most of the receive power arrives at the input port of the feed and very little power ends up in the terminating loads. In an ideally lossless feed network, that portion of the power in the incident plane wave which ends up in the terminating loads accounts for the fact that the aperture efficiency η is less than unity for aperture distributions other than uniform. For uniform aperture distributions none of the received power is dissipated in the terminating loads. For tapered aperture distributions most of the received power that does not arrive at the feed input is dissipated in the terminating loads of the junctions that have the highest power split ratios which are near the edges of the aperture. With reactive three-port power splitters the receive power that is not received at the input port of the feed is reflected back into free space and can result in a high radar cross section. In any case in which matched four-port junctions are used in the feed, the radar cross section of the ideal impedance-matched aperture antenna is zero. Of course it is assumed that the incident radar wave has the same polarization and frequency as the antenna. That is, all the energy incident in the direction of the main beam of the antenna is absorbed at the antenna input (receiver or duplexer load)—none of it is reflected back into space. For plane waves incident on the antenna from any other direction than in the main beam, the incident power nearly all (except for the small amount of energy associated with side lobes of the antenna pattern) ends up in the terminating or "unused" ports of the hybrid junctions. That small amount of incident power (if there is any) associated with the side lobe in that direction ends up at the input port of the antenna feed. Hence the unused ports of the hybrid junctions can be combined in various ways to create a large variety of antenna pattern shapes outside the region of the main beam of the main antenna. Such patterns can be used as auxiliary antennas such as for coherent side lobe cancellation or for communications outside the region of the main beam of the main antenna. All these auxiliary antenna patterns would have essentially a null in the direction of the main beam of the main antenna [2].

True Time-Delay Feeds

A true time-delay antenna is in which that the time delay from an incoming wavefront to a feed point is the same for every path from the wavefront via each of the radiating elements, phase shifters, etc., to the feed point. This is illustrated in Fig. 4. The equal time delay ensures that signals via all paths add in phase at the feed point for every frequency component in the pulse. In a non-true-time-delay antenna the time delays for the various paths may differ by an integral number of rf cycles of the center frequency in the pulse. In general, other frequencies in the pulse do not add exactly in phase at the feed point. Hence there is a reduction in the peak of the received pulse and the pulse is smeared out or broadened in time. From the transmit point of view the interelement phase shift ψ is correct only at the center frequency to have its main beam point in the desired direction in space. Antenna patterns corresponding to other frequencies in the pulse will be scanned off from the desired direction according to the expression $\sin\theta = c\psi/\omega d$, which shows that the beam-pointing angle θ depends on the frequency ω since ψ is constant. The antenna is a bandpass filter in that the frequency spectrum of the

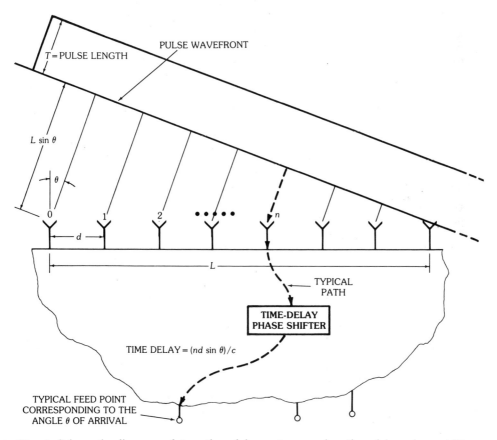

Fig. 4. Schematic diagram of true-time-delay antenna using time-delay phase shifters. (*Courtesy Hughes Aircraft Co., Fullerton, Calif.*)

pulse radiating toward a target is modified. The frequency component corresponding to the center frequency of the pulse is weighted most heavily because its corresponding antenna pattern is, by design, pointing in the direction of the target, whereas the other frequency components in the short pulse have their antenna patterns scanned off from the direction of the target; hence their weighting is reduced by the amount that their antenna patterns are scanned off. By the principle of reciprocity this spectral transformation on receive is the same as that previously discussed for transmit.

In a true time-delay array the interelement phase shift ψ must not be constant but must be proportional to frequency so that all frequency components in the pulse will have their corresponding antenna patterns all pointing in the same direction.

Examples of true time-delay feeding techniques include the use of the circular folded pillbox covered in Section 3, under "Pillbox," and the Meyer geodesic lens covered in Section 3, under "Meyer Lens," and in Section 6, under "Hughes Matrix-Fed Meyer Geodesic Lens," among others.

Multiple-Beam Matrix Feeds

Butler Matrix—Probably the most widely known multiple-beam matrix feed is the Butler matrix. It is well documented in the literature [3–8] and will be discussed only briefly here. Fig. 5 is a schematic representation of an eight-element Butler matrix.

The Butler matrix has N outputs (aperture elements) and N inputs (beam ports). Unit excitation at the beam ports results in N orthogonally spaced $\sin u/u$ type patterns.* The aperture distribution $B(n)$ is related to the beam inputs by

$$B_{nm} = \frac{\exp j\{[n - (N + 1)/2][m - (N + 1)/2]2\pi/N\}}{\sqrt{N}}$$

where B_{nm} is the field amplitude of the nth aperture element when beam port m is excited with unit amplitude and N is the number of aperture elements, which is also the number of beam ports. In the above form of B_{nm}, the phase distribution for each beam is symmetrical about the center of the aperture array. Any Butler matrix can be put into this form because an arbitrary aperture phase gradient can be applied to the aperture. This merely causes the whole beam cluster to scan as a whole, retaining the same beam spacing in $\sin\theta$ space. The quantity that is invariant is the difference in phase gradients between adjacent beams, which is always $2\pi/N$. The aperture amplitude distribution is uniform for all beams with $(\sin u)/u$ type patterns and each beam derives full 100-percent directivity from the projected aperture of the common array. The beams cross over at $E = 2/\pi$ or 3.92 dB down from the beam peaks and the peak of any beam peak falls on the nulls of all other beams. The beams are orthogonal and there is no beam coupling. The crossover level is independent of frequency; hence the beams must frequency scan by an amount

---

*More accurately, for discrete arrays it is $(\sin n\psi)/n \sin\psi$, where $\psi = (\pi d/\lambda)\sin\theta$, which is the counterpart for $(\sin u)/u$ for continuous distributions.

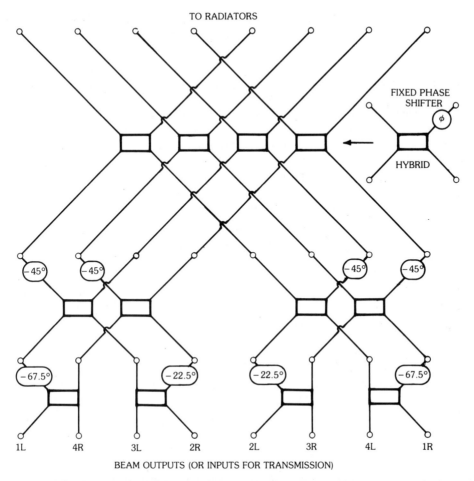

Fig. 5. Eight-port Butler matrix. (*Courtesy Hughes Aircraft Co., Fullerton, Calif.*)

proportional to their separation from broadside in order to retain orthogonality and a fixed crossover level. For large arrays the beams span an angular coverage of $2\sin^{-1}(\lambda/2d)$, where d is the element spacing and λ is the free-space wavelength. For $d > \lambda/2$, grating lobes will fill in the region near end-fire, and for $d < \lambda/2$, some beams will be in imaginary space. In all cases all of visible space is covered.

As exemplified by Fig. 5, numerous hybrid junctions and fixed phase shifters must be cascaded to generate a Butler matrix. Specifically [5], the number of hybrids is equal to $(N/2)\log_2 N$ and the number of fixed phase shifters is equal to $(N/2)\log_2(N-1)$. Moody [8] gives a systematic design procedure for generating such matrices.

It should be mentioned that the multiple-beam matrices of the type described in [5] (and most others in the literature) use all 90° 3-dB or all 180° 3-dB couplers. In these matrices the number of elements/beams is a power of 2 and requires fixed phase shifts. However, since any fixed phase shift can be generated by a combination of 90° and 180° hybrids, matrices without separate fixed phase shifters can be designed. Jones and Van Blaricum [7] give a design procedure for these matrices. In general, the hybrids are not necessarily a 3-dB power split and the number of elements/beams is not necessarily a power of 2. An example of a multiple-beam matrix that uses both 90° and 180° hybrids with no fixed phase shifters is the one used to feed the radial transmission line described in Section 3. The preceding expression for B_{nm} is general and is applicable to all lossless multiple-beam matrices.

There are numerous applications of the Butler matrix, not only as an antenna itself, or as an antenna feed, but as a multimode generator. When the outputs are placed on a circle, modes of the form $e^{\pm jm\phi}$ are generated (m = integer, ϕ = circumferential angle). Due to the unitary nature of this matrix, a complete set of orthogonal modes is created. Most of the time the number of beams required is much less than the number of elements; consequently the full capability (and associated complexity) of a Butler matrix is not needed. A later section on parallel-plate optics using a multimode radial transmission line describes a scheme in which the number of elements and the number of beams are totally independent. The next section describes a multiple-beam transmission-line matrix that also has independence between the number of elements and the number of beams. This scheme also allows arbitrary beam spacing. Also see Hansen [4], vol. 3, ch. 3.

Blass Matrix—Another multiple-beam matrix invented by J. Blass [9] that uses serially coupled transmission lines instead of the parallel coupled lines of the Butler matrix is briefly described here (see Fig. 6). This configuration uses one feeder line for each beam with branch guides coupled serially by directional couplers. The transmission lines may be waveguide, coax, or stripline. For a specified inter-element spacing d, the beam position in space is determined by two things: the tilt angle of the feed line and the propagation constants of the branch lines and the feed line in question (may be all the same or all different). Each feed line produces a linear (ideal) phase gradient at the array aperture and forms a frequency scanning beam in space. By proper choice of the feed line tilt angles, propagation constants, and coupling distributions, multiple beams can be generated in space having whatever crossover value the designer chooses. These multiple beams, in general, may not be precisely orthogonal over a large frequency band but can be designed to be nearly so from a loss point of view. All the beams frequency scan together. For a small number of beams (two or so) this array works quite well. For a large number of beams, or for very low side lobes, care must be taken to account for the phase and amplitude changes that occur in the transmission coefficient as the wave passes by a multiplicity of nonideal directional couplers. The larger the number of beams, the more difficult this becomes. As can be seen from Fig. 6, feed line 1 produces a branch guide field that does not have to pass by any other couplers, while beam M must pass by $M - 1$ other couplers before entering free space; hence perturbations are present in feed line M that are not there for feed line 1. These perturbations will manifest themselves to some degree in the form of radiation pattern degradation and/or excessive power dissipated in the terminating loads. These properties are manifestations of the fact that ideal beam orthogonality is not guaranteed as in the Butler matrix or radial line approaches. It is conceivable that orthogonality could be forced by a design procedure similar to the Gram-Schmidt procedure [10], but in

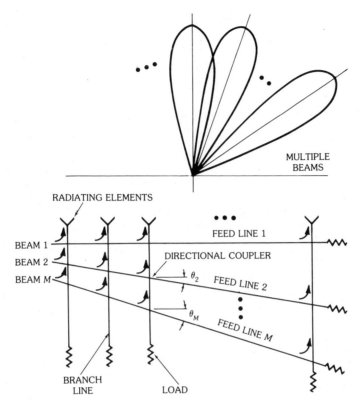

Fig. 6. Multiple-beam Blass matrix. (*After Butler and Lowe [3], © 1961; reprinted with permission of Hayden Publishing Co.*)

doing so, some of the advantages of the Blass matrix, such as simplicity and arbitrary beam spacing, could be restricted. However, this allows arbitrary beam spacing.

Various derivatives of the Blass matrix are also given in Hansen [4], including a true time-delay version of this matrix. The reader is referred to this reference for more details.

Two-Dimensional, Isosceles, Triangularly Spaced, Multiple-Beam Matrix—This class of two-dimensional multiple-beam matrices introduced by Chadwick and Glass has been extensively investigated by Chadwick, McFarland, Charitat, Gee, and Hung of the Lockheed Missile and Space Company [11, 12]. The analysis of this relatively new class of two-dimensional multiple-beam matrices is quite involved and, compared with other topics of this handbook, the required length of a self-contained explanation would not be justified. For this reason, only an acknowledgment that such an antenna technique does exist and a discussion of its possible advantage over the more established techniques are given. For further information the reader is referred to the references cited.

The Butler matrix multiple-beam antenna is one-dimensional and applies to multiple-beam linear arrays. A two-dimensional array is made by an array of these linear arrays. The Butler matrix is a special case of the triangularly spaced array in which the triangle collapses into a line. This two-dimensional matrix forms a family of lossless, isosceles, triangularly spaced, orthogonal beams.

This two-dimensional matrix offers more freedom for the designer to choose the number of elements (beams) with a variety of array shapes. These shapes are derived from the basic triangle and include such shapes as irregular hexagons, V shapes, Z shapes, hourglass shapes, and parallelograms. Since there is a one-to-one correspondence between array space and beam space, a variety of beam coverage geometries are available. This may be advantageous where irregular beam coverage is desired, such as the coverage of certain geographical areas on earth from a spacecraft antenna. A desired beam coverage might be approximated by one of the allowed beam coverage shapes. Butler matrices achieve an irregular coverage by terminating the beam ports corresponding to beams that fall outside the desired coverage area. This is wasteful in that the total number of components in the matrix is greater than for a matrix that utilizes all the available beams. Due to the variety of beam coverage shapes available, the relative number of "wasted" beams should be smaller for this two-dimensional matrix array.

Multimode Element Array Technique

For electronically scanning antennas with requirements of high directivity and high gain but with limited field of view (typically on the order of $10°$), it is very wasteful to have radiating elements with phase shifters at half-wavelength intervals over the entire radiating aperture. The limited scan requirement allows a reduction in these devices (which may include power amplifiers and low-noise amplifiers in an active array) in proportion to the limited solid angle of scan. Limited-scan antennas are designed to minimize the number of these costly devices with minimal degradation of antenna performance. Constrained feeding techniques using large directive elements spaced much greater than at $\lambda/2$ intervals are discussed in this section. Limited-scan techniques using optical type devices are described in Section 4 on unconstrained optical feeds and in Section 5 on optical transform feeds.

The use of large directive elements (e.g., waveguide horns) in a phased array usually results in high grating lobe levels as scanning is performed; however, for limited scan, Mailloux and Forbes [13, 14] have found that by properly exciting waveguide flared horns with not only the dominant LSE_{10} mode but with controlled higher-order odd modes as well, the grating lobe that would ordinarily be the worst can be suppressed. For example, consider first that array scanning is to be performed in the E-plane only, with elements as depicted by Fig. 7. The array factor will be scanned by controlling the phase shifter labeled η, while the odd-mode amplitude control will utilize the phase shifter labeled $\eta + \Delta$. An example of far-field radiation patterns for the LSE_{10} and LSE_{11} modes is shown in Fig. 8. This particular example is predicated on forming a null at $\eta = -0.75$. This null location would correspond to a grating lobe position for a positive main beam scan angle. The null location can be controlled by an appropriate linear combination of these two modes. By linearly combining the two modes, a scanned element pattern that suppresses the worst grating lobe can be achieved. An example of this is given by Fig. 9 for waveguide horns that are 2.9λ by 2.9λ on a side. Also shown, for reference purposes, is the element pattern produced by the dominant LSE_{10} mode only.

Fig. 10 is a photograph of an experimental eight-element array taken from

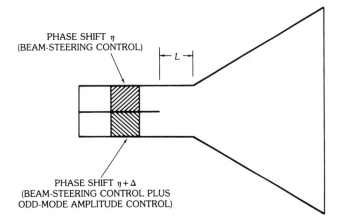

Fig. 7. Even/odd-mode power divider circuit for *E*-plane scanning. (*After Mailloux [13]*)

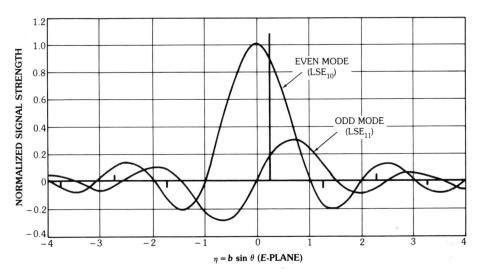

Fig. 8. *E*-plane field patterns for LSE_{10} and LSE_{11} waveguide modes. (*After Mailloux [13]*)

Mailloux [13]. It is designed for *E*-plane scan of $\pm 12°$. The amplitude distribution is such that the center four are uniform, the second element in from each end is -3 dB, and the outer elements are -6 dB. Examples of the measured far-field patterns are given in Figs. 11 and 12, where for comparison the latter also shows the calculated pattern without odd-mode control. Without odd mode control, the highest grating lobe for $12°$ scan of the main beam would be about 4 dB higher than the main beam. With odd-mode control, however, this grating lobe is suppressed by about 20 dB for $\pm 12°$ scan.

For odd-mode control in both planes the configuration shown by Fig. 13 can be used [14]. This configuration would allow large directive elements to be used in both dimensions, while suppressing the offending grating lobe ordinarily encountered. For more details the reader is referred to Mailloux and Forbes [13, 14].

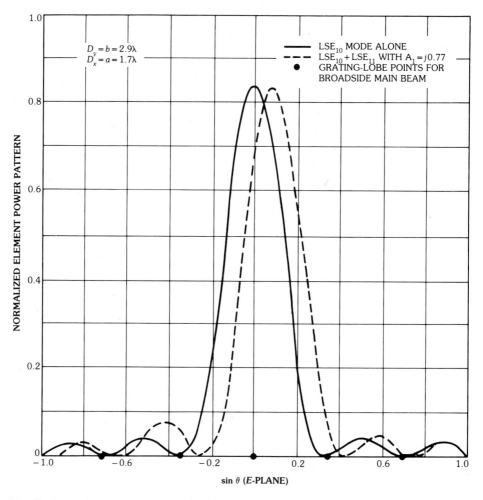

Fig. 9. Array element patterns of fundamental mode ($A_1 = 0$) and of fundamental mode with odd mode ($A_1 = j0.77$), with 2.9λ element separation in the *E*-plane.

3. Semiconstrained Feeds (Parallel-Plate Optics)

Feed systems that are constrained in one dimension but are not constrained in the other dimension are used because of simplicity, low cost, low loss, and high power handling capability. The volume occupied, however, is generally greater than that for fully constrained feeds. These feeds utilize propagation in the quasi-TEM mode between closely spaced (less than $\lambda/2$) parallel metallic surfaces. The surfaces need not be planar but can also be singly or doubly curved as long as the spacing between the surfaces is small in terms of wavelength and the radii of curvature are large with respect to the wavelength. Under these conditions the wave is considered to be propagating on the mean surface. In the dimension of the mean surface the wave is essentially unbounded and is not constrained by guiding structures. The rf power "radiates" from the primary feed along the mean surface to a pickup array that, in turn, feeds the radiating elements of the antenna. Hence

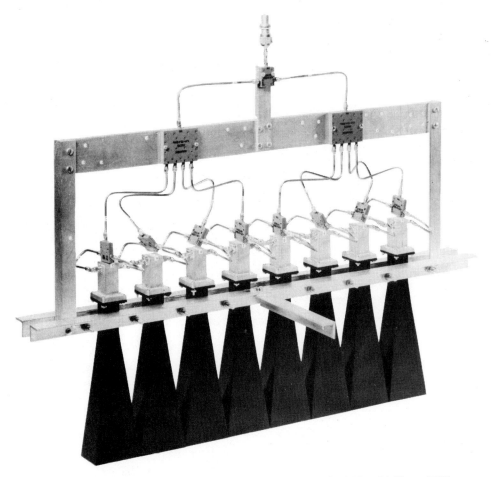

Fig. 10. Prototype array for *E*-plane scan ($W = 3.01$). (*After Mailloux [13]*)

the aperture distribution of the antenna array is determined by the radiation pattern of the primary feed and the geometric properties of the mean surface. If the mean surface is planar, the aperture illumination is determined by the feed pattern in the usual sense. However, if the mean surface is doubly curved or if the index of refraction is a function of position, the feed pattern is modified, as will be discussed later.

Under the restrictions of closely spaced surfaces and other dimensions large compared with the wavelength, the laws of geometric optics are quite valid on the mean surface. That is, Fermat's principle stating that the path of a ray in geometric optics is stationary applies. Stated mathematically,

$$\delta \int_{\text{ray path}} n \, ds = 0$$

where n is the index of refraction of the medium between the surfaces. The index of refraction may vary as a function of position on the mean surface. This is the same

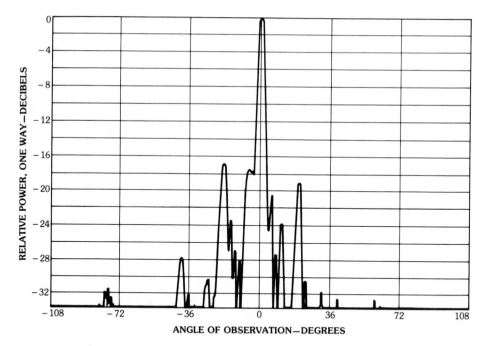

Fig. 11. Array radiation patterns (broadside). (*After Mailloux [13]*)

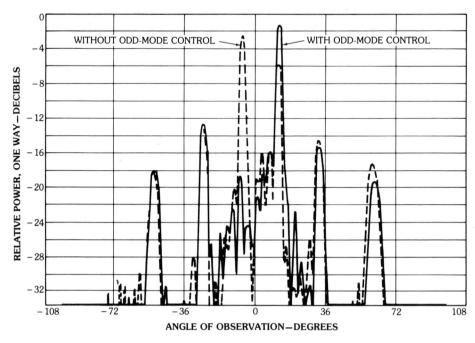

Fig. 12. Array radiation patterns (12° scan). (*After Mailloux [13]*)

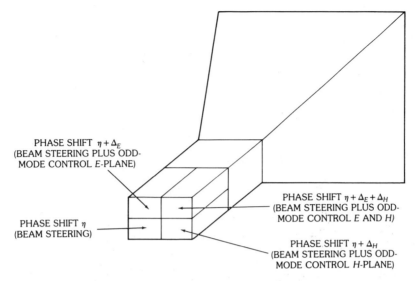

PHASE SHIFT $\eta + \Delta_E$
(BEAM STEERING PLUS ODD-
MODE CONTROL E-PLANE)

PHASE SHIFT $\eta + \Delta_E + \Delta_H$
(BEAM STEERING PLUS ODD-
MODE CONTROL E AND H)

PHASE SHIFT η
(BEAM STEERING)

PHASE SHIFT $\eta + \Delta_H$
(BEAM STEERING PLUS ODD-
MODE CONTROL H-PLANE)

Fig. 13. Four-mode horn and power divider circuit. (*After Mailloux [13]*)

as saying that the time of transit of a ray between two points on the mean surface is an extremum (usually a minimum) and the rays are geodesics [15]. For this reason such feeds are generally called *geodesic feeds* or *lenses*.

The simplest and probably the earliest geodesic feed for phased arrays is the folded or two-layered pillbox [16–19]. It consists of two parallel-plate regions connected by a 180° bend that is parabolic in the plane of the plates. A point source feed is located in one layer. A cylindrical wave radiated from the feed goes around the 180° bend to enter the second layer, where it is collimated into a plane wave by the parabolic shape of the bend. The bend may be abrupt [16, 17], circular, or mitered [19].

The mean surface of geodesic lenses may be planar, singly curved, doubly curved, or composite. For planar surfaces the geodesics are straight lines. For singly curved (cylindrical) surfaces the geodesics are straight lines on the developed (flattened into a plane without stretching or tearing) surface. Doubly curved surfaces cannot be flattened into a plane without stretching or tearing the surface. The geodesics on doubly curved surfaces are the curves that a tightly stretched string would take with the string constrained to stay on the surface everywhere. Mathematically the geodesic is an extremum, which is the shortest or the longest path length on the surface between two points. As a simple example the shortest arc of a great circle through two points on a sphere is a geodesic, as is the longest arc of the same great circle through the points. Usually the geodesics of interest are the shortest paths between two points.

If two surfaces intersect* along a line, a geodesic (ray) across the intersection obeys Snell's laws of reflection and transmission, namely,

---

*Of course, any arbitrary "cut" in a geodesic surface can be considered as an intersection but most often the intersection is between two portions that are each developable.

(1)   The angle of reflection is equal to the angle of incidence, $\theta_r = \theta_i$.
(2)   The angle of refraction (transmission) is related to the angle of incidence by

$$\frac{\sin \theta_t}{\sin \theta_i} = \frac{\sqrt{\epsilon_i}}{\sqrt{\epsilon_t}} = \frac{n_i}{n_t}$$

where ϵ_i is the relative dielectric constant in the parallel-plate region of the incident ray and ϵ_t is that in the region of the transmitted (refracted) ray. The terms n_i and n_t are the corresponding indexes of refraction. All angles are measured from the common normal.

As will be discussed later, there are many practical geodesic lens surfaces that are composites of developable surfaces.

The parallel-plate Luneburg lens is not applicable for efficiently feeding a linear array. This is clearly evident from Fig. 14. Although the Luneburg lens is perfectly focused for all scan angles, it is very inefficient for feeding a fixed linear array because the illumination scans or translates off the linear array aperture as

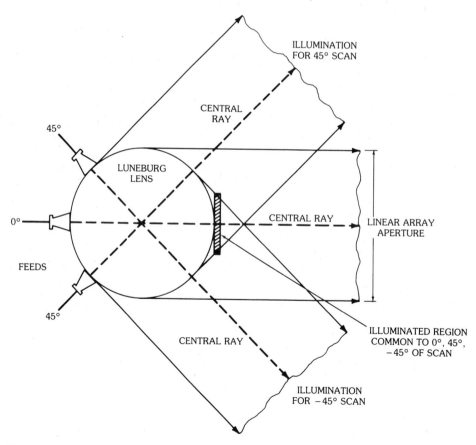

Fig. 14. Luneburg lens illuminating a linear array. (*Courtesy Hughes Aircraft Co., Fullerton, Calif.*)

the beam is scanned. The darkened portion of the linear array is all that is common for angles of scan out to ±45°. The amount of feed spillover and illumination asymmetry that results with beam scan is intolerable. Note that the central ray does not pass through the center of the linear array for scanned off beams. This results in asymmetric array illumination for tapered distributions. The circular pillbox and Meyer lens (these lenses are discussed later) do not suffer this shortcoming, as can be seen in Fig. 15, which shows that for the Meyer lens (the same is true for the circular folded pillbox) the central ray for all beams passes through the center of the linear array, i.e., the illumination does not translate across the linear array aperture. It should be mentioned that since the projected aperture of the linear array decreases as the cosine of the angle of scan, to keep the illumination efficiency from degrading, the feed illumination pattern should be made more directive according to the secant of the scan angles. This requires larger feeds as the scan increases. Fortunately, there is just enough space to do this (in terms of beamwidths of scan) because the radiated beamwidth also increases as the secant of the scan angle. This effect is true for *all* optical type feeding techniques for linear or planar arrays.

Pillbox

Fig. 16 shows a sketch of a parabolic folded pillbox [16–19]. This device has been used since the early 1950s for IFF and air traffic control antennas and as a feed for phased arrays.

The input primary feed is usually an open-ended waveguide, or two of them side by side for monopulse operation. Sometimes electric probes fed by coaxial lines backed by a quarter-wave short or cavity are used for the primary feed. The primary feed radiates in the region between a pair of parallel plates, as depicted by Fig. 16. The phase center of the primary feed is located at the focus of the parabola. The field then propagates in the TEM parallel-plate mode to the 180° bend. The bend is designed such that the "reflected" (or more accurately, transmitted around

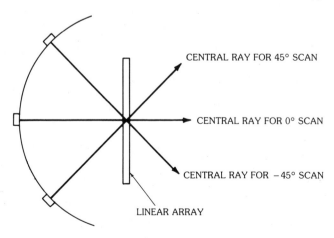

Fig. 15. Meyer lens illuminating a linear array. (*Courtesy Hughes Aircraft Co., Fullerton, Calif.*)

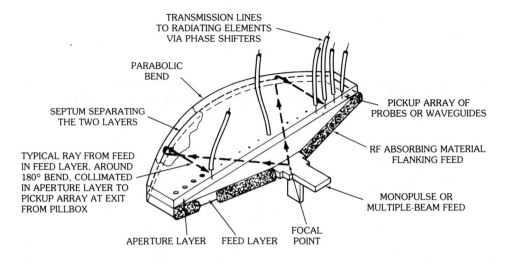

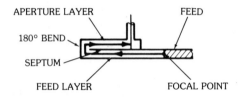

Fig. 16. Folded pillbox feed for phased array. (*Courtesy Hughes Aircraft Co., Fullerton, Calif.*)

the bend) field emerges between the upper set (ideally) of TEM parallel plates, and is collimated because of the parabolic shape. The output is located in the upper set of parallel plates at any convenient terminal plane. The output is transitioned to coax waveguide or stripline, thus forming the pickup array. Radio-frequency absorbing material should be used in the lower set of parallel plates to absorb any reflected power that did not go around the 180° bend [18]. This is especially important if low side lobes are required. Using the parabolic folded pillbox, low side lobe radiation patterns (e.g., 30 dB or better) are fairly straightforward to obtain [18]. Moreover, the loss through the structure is very low, being less than an equal path length run of waveguide of the same height as the parallel-plate separation.

The reason this device has been used so extensively throughout industry is because of its high performance, simplicity, and form factor. Its flat, thin profile usually makes it easy to package (e.g., on the back side of a planar array). In any application where a parallel-fed corporate feed is required, the parabolic folded pillbox can be used instead. Moreover, monopulse operation is simple to implement in the pillbox by feeding it with a dual feed and a magic-T or with a multimode feed with independent sum and difference modes. It is much less complicated (mechanically) than the corporate feed and does not require the multiplicity of components (hybrid junctions or directional couplers, bends, etc.) inherent in the corporate feed.

If wide-angle-coverage multiple beams are desired, the circular folded pillbox [19] can be used instead of the parabolic folded pillbox. Fig. 17 is a photograph of a 29-beam circular folded pillbox at X-band. Its principle of operation is similar to that of the parabolic folded pillbox, except that a circular arc is used at the 180° bend rather than a parabolic arc, and the feed locus is a circle whose radius is slightly greater than half the radius of the circular reflector (at the 180° bend). Without dielectric loading, the usable portion of the output aperture is restricted to about one-half the diameter, or slightly greater. The reason for this is that spherical aberration produces phase errors that limit the usable portion of the aperture. Most of the phase errors at the aperture can be corrected by trimming the line lengths at the output aperture, similar to a Schmidt correction in optics. The spherical aberration for the scanned beams resembles the spherical aberration for the on-axis beam; hence the spherical aberration that is common to all beams can be removed at the output aperture by line length adjustment.

For example, consider a design that forms multiple beams over a 90° sector ($\pm 45°$). Fig. 18 shows the calculated spherical aberration for scanned beams over $\pm 45°$. The curves all resemble each other, grossly speaking; consequently the mean value of the spherical aberration can be removed. The long and short dashed curve is the mean value, or compensation curve. The negative of this curve is inserted at the aperture by line length adjustments in the pickup array. The maximum phase error after compensation is shown dashed. The spherical aberration has been reduced by a factor of about 5 or 6. This factor depends on the total field of view over which multiple beams are formed, and the allowable degree of spherical aberration (after compensation) that can be tolerated.

With partial dielectric loading of the entrance layer the optics can be modified such that nearly 80 percent of the output diameter is usable aperture. For example, the pillbox in Fig. 17 uses Rexolite 1422 loading ($\epsilon_r = 2.56$) between the feed arc and $0.8R$ in the entrance layer and gives respectable phase error over 80 percent of D. The extreme scanned beams suffered from what appeared to be residual spherical aberration phase error, indicating that the use of 80 percent of D is

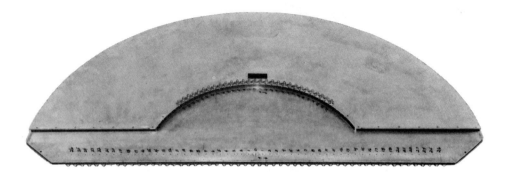

Fig. 17. Circular, folded, multiple-beam pillbox, shown with 12-in (30.48-cm) ruler. (*Courtesy North American Rockwell*)

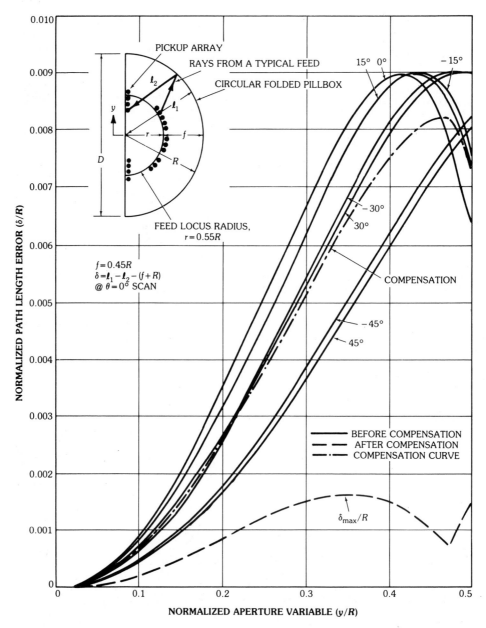

Fig. 18. Path length error versus aperture variable before compensation, and maximum path length error after compensation. (*Courtesy Hughes Aircraft Co., Fullerton, Calif.*)

probably a little too much. There was no mechanism for focusing (i.e., feed probe positions were not adjustable). In this design electric probes are used as feeds, which yields a very nearly uniform amplitude distribution at the aperture, with beam crossover levels of about 4 dB and side lobe levels consistent with a uniform amplitude distribution (13 dB) for most of the beams. The crossover level varies with frequency consistent with the beamwidths' dependence on frequency. The 29-beam peak positions remain fixed in space independent of frequency. If waveguides were used as feeds with focusing (radial feed position adjustability) capability, improved performance would undoubtedly be realized.

Radial Transmission Line

This section describes a simple, inexpensive, and ideally lossless multiple-beam-forming device whose cost and complexity do not increase rapidly with the number of radiating elements [20–22].

In physical appearance it resembles a constrained lens using parallel-plate optics. From the microwave circuit point of view it is similar to the Butler hybrid matrix but it differs in that the number of beams does not have a definite mathematical relationship with the number of radiating elements. The number of elements is arbitrary, in contrast with the Butler matrix in which the number of beams equals the number of elements. In practice the number of beams is quite limited in the radial transmission-line scheme but the number of radiating elements can be increased arbitrarily with little extra complexity and at a cost that varies only linearly with the number of elements.

Consider a parallel-plate radial transmission line terminated on its periphery by an array of "pickup" probes connected to the radiating elements of a linear array with equal lengths of transmission line as shown in Fig. 19. Suppose that besides the TEM mode, higher-order cylindrical modes with circumferential phase variation could be excited in the radial line. The circumferential variation of these modes can be characterized by $A_m \exp(jm\phi)$, where A_m is the amplitude of the mth mode and m is an integer (positive or negative). Because of the orthogonality properties of the modes they do not couple. By virtue of equal line length connection between the circumferentially dispersed pickup probes and the elements of the linear array,

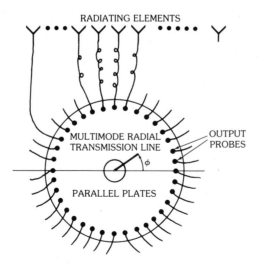

Fig. 19. Radial-transmission-line multiple-beam-forming network. (*Courtesy Hughes Aircraft Co., Fullerton, Calif.*)

the circumferential phase variation $\exp(jm\phi)$ is transformed to a linear progressive phase $\exp(j2m\pi x/L)$, where L is the length of the array and x is the aperture variable ($\phi = 0$ corresponds to $x = 0$ and $\phi = 2\pi$ corresponds to $x = L$). Thus there is a beam for each value of m. Positive and negative values of m correspond to left and right beams, respectively. For $m = 0$, there is a beam at broadside. The amplitude distribution for each of the beams is uniform and the far-field pattern is given by

$$E(\psi) = \frac{\sin(N\psi/2)}{N\sin(\psi/2)}$$

where

N = number of elements in the array

$\psi = 2\pi d/\lambda \sin(\theta - \alpha)$

θ = angle of beam from broadside

$d = L/(N - 1)$, interelement spacing

$\alpha = 2m\pi/N$, interelement phase shift for the mth beam

This is exactly the same as a Butler matrix array of N elements with m beams being used.

The $m = 0$ mode can be excited at the center of the radial transmission line with a circular waveguide operating in the TM_{01} or with a coaxial TEM mode as shown in Fig. 20. The method of Fig. 20 also generates the $m = \pm 1$ modes. The peak power-handling capability is limited by the E-plane magic-T. Fig. 21 shows a very high power transition using a multimode turnstyle junction to excite these modes. The TE_{11} excitation results in a $\cos\phi$ circumferential variation. An orthogonal TE_{11} mode phased 90° will give a $j\sin\phi$ circumferential variation. Hence, by adding or subtracting the two orthogonal TE_{11} modes, the $\exp(\pm j\phi)$ circumferential variation in the radial transmission line is achieved. In other words, an excitation of right and left circular polarizations in the circular waveguide feed will result in a right and a left antenna beam. Orthogonal TE_{12} modes as shown in Fig. 22 will excite $\exp(\pm 2j\phi)$ modes in the radial transmission line. In this figure the solid lines represent the field configuration of the TE_{21} mode ($\cos 2\phi$ circumferential variation). The dashed lines represent the field configuration of the orthogonal TE_{21} mode ($\sin 2\phi$ variation). By combining them in a 90° phase relationship, $\exp(j2\phi)$ and $\exp(-j2\phi)$ modes are created. The radial line can be fed with a multimode turnstyle type junction fed by hybrid circuitry as in Fig. 21a. A photograph of an S-band radial line fed by this method is shown in Fig. 21b. This hybrid-fed waveguide method of excitation of higher-order modes in the radial transmission line has the advantage of simplicity, low loss, and high power handling capability. For more beams, feeding at essentially one point becomes impractical. For a greater number of beams a circular array of probes with less than half-wave spacing fed with a Butler matrix would probably be the most efficient. The obvious question may be raised as to why not dispose of the radial line and just use the

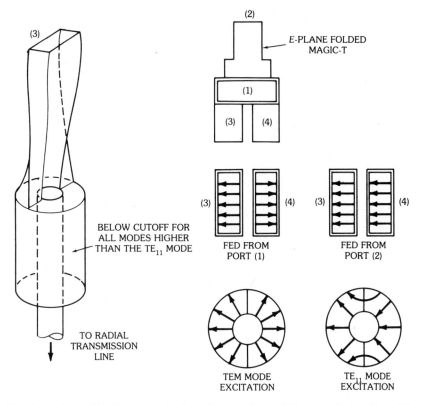

Fig. 20. Operation of high-power, dual-mode feed for radial transmission line. (*Courtesy Hughes Aircraft Co., Fullerton, Calif.*)

Butler matrix [3, 5]. The advantage of using the radial line is that regardless of the number of beams the number of outputs from the radial line can be increased to any arbitrary number by merely enlarging the diameter of the radial line to accommodate a larger number of output elements. In many systems applications requiring large arrays the number of simultaneous beams required is often much less than the number of radiating elements. For example, many phased-array systems may require only two beams for monopulse capability with beam steering achieved with ferrite or diode phase shifters. Obviously in this extreme case a radial line feed would be much simpler and much less expensive than a Butler matrix. In other applications a relatively small multiple-beam cluster that can be scanned in synchronism may satisfy the system requirements.

As an example of a possible application of the multimode radial line beam-forming network, suppose that on transmit a uniform distribution is desired and on receive a 30-dB side lobe tapered aperture distribution is desired with monopulse capability. A feeding arrangement that is still different from those of Figs. 20 and 21 is shown in Fig. 23 and is used for this example. The radial transmission line has a coaxial input (excites $m = 0$ mode) and a dual orthogonal mode (TE_{11} mode and orthogonal TE_{11} mode) waveguide input which can excite the $m = \pm 1$ mode. On transmit the coaxial line input only is used, resulting in a uniform aperture

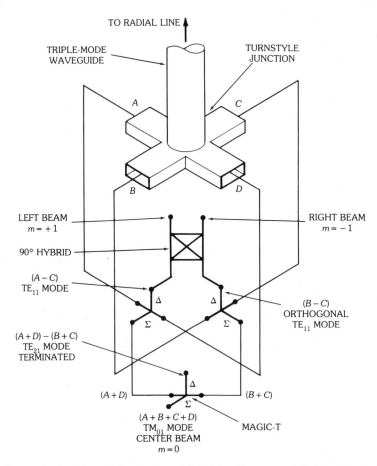

Fig. 21. Turnstyle junction and antenna. (*a*) High-power turnstyle junction to generate three beams. (*b*) Multiple-beam antenna using a multimode radial transmission line fed by a hybrid turnstyle junction. (*Courtesy Hughes Aircraft Co., Fullerton, Calif.*)

distribution on transmit. On receive, the $m = 0$ (uniform) mode and the $m = 1$ ($\cos\phi$) mode are combined in the power ratio of 2 to 1 to form a voltage distribution of $1 + \cos\phi$, which is equal to $2\cos^2(\phi/2)$, which is theoretically a -32-dB side lobe level distribution. This creates the low side lobe "sum" monopulse pattern. By changing the power division ratio a more efficient and lower side lobe cosine squared on a pedestal distribution is easily achieved. The "difference" monopulse pattern is taken from the orthogonal TE_{11} mode port, which gives a $\sin\phi$ aperture distribution which results in a difference pattern with good side lobe characteristics because there is no abrupt discontinuity in the aperture distribution as in the case of split aperture phase monopulse schemes.

A technique for a larger number of beams uses a small circular array of probes concentric with the center of the radial line, which are properly phased to generate the different modes. A seven-beam system was designed and tested [22]. This feed system utilized a hybrid network of 90° and 180° 3-dB couplers. A Butler matrix of

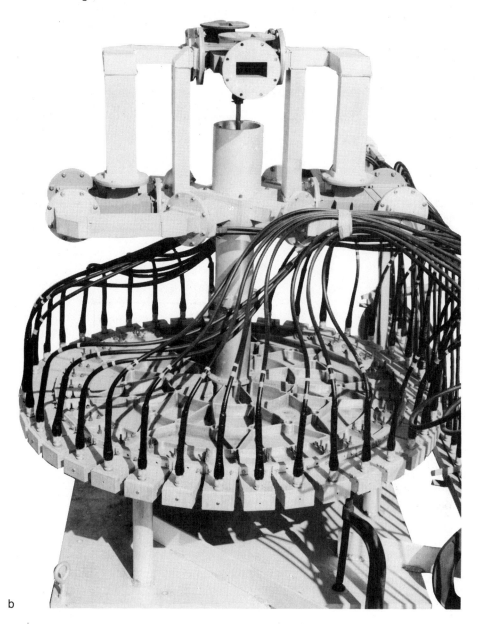

b

Fig. 21, *continued.*

the type shown in Fig. 5 could have been used just as well. It is interesting to note that, as mentioned in Section 2, the matrix in Fig. 24 does not require fixed phase shifters, in contrast to that of Fig. 5. Fig. 24 is a schematic diagram of the network; the 180° (magic-T) is designated by the letter T and the 90° coupler by the letter H. The phase progression around the circular array for each antenna

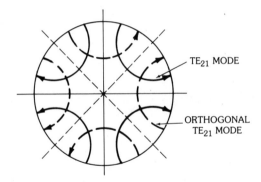

Fig. 22. Waveguide excitation of $m = 2$ modes. (*Courtesy Hughes Aircraft Co., Fullerton, Calif.*)

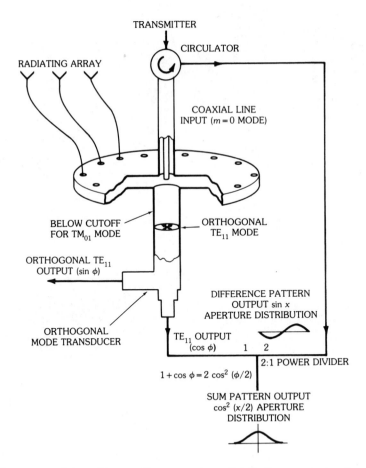

Fig. 23. Arrangement to achieve uniform aperture distribution on transmit and cosine-squared distribution on receive with monopulse capability. (*Courtesy Hughes Aircraft Co., Fullerton, Calif.*)

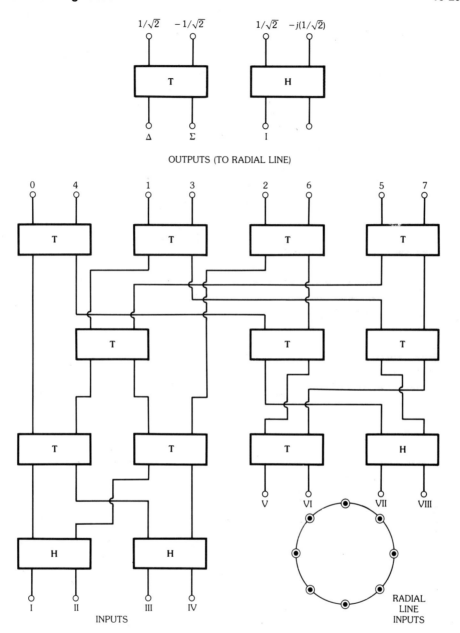

Fig. 24. Block diagram of feed circuitry. (*Courtesy Hughes Aircraft Co., Fullerton, Calif.*)

beam (mode) is given in Table 1. Note that input V, which corresponds to output phases that are alternately 0° and 180°, gives rise to a split end-fire beam that results in two actual opposing end-fire beams. This input is not used and is terminated in a matched load. Fig. 25 is a photograph of the stripline network. Because of the finite number of probes the $\exp(\pm jm\phi)$ variation is only discretely approximated, and if the probe separation is too great, undesirable higher-order modes are also

Table 1. Phase Progression Around Circular Array for Each Antenna Beam

Input	\multicolumn{8}{c}{Outputs}							
	0	1	2	3	4	5	6	7
I	0°	135°	270°	45°	180°	315°	90°	225°
II	0°	225°	90°	315°	180°	45°	270°	135°
III	0°	45°	90°	135°	180°	225°	270°	315°
IV	0°	315°	270°	225°	180°	135°	90°	45°
V	0°	180°	0°	180°	0°	180°	0°	180°
VI	0°	0°	0°	0°	0°	0°	0°	0°
VII	0°	90°	180°	270°	0°	90°	180°	270°
VIII	0°	270°	180°	90°	0°	270°	180°	90°

Radial Transmission-Line Modes Corresponding to Each Input

Input I	$m = +3$	Input V	(terminated)
Input II	$m = -3$	Input VI	$m = 0$
Input III	$m = +1$	Input VII	$m = +2$
Input IV	$m = -1$	Input VIII	$m = -2$

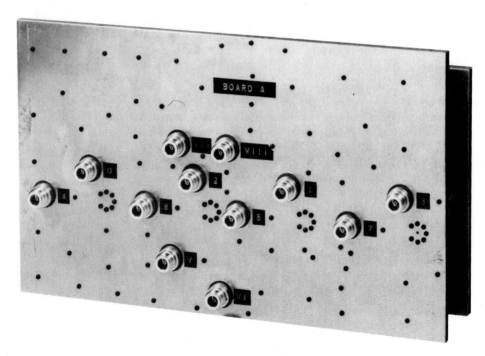

Fig. 25. Stripline hybrid-feed network. (*Courtesy Hughes Aircraft Co., Fullerton, Calif.*)

generated. However, if the feed probe spacing is somewhat closer than $\lambda/2$, the undesirable higher-order modes are "below cutoff." This corresponds to $n > kR_0$ in the cylindrical modal expansion, where n is any undesired higher-order mode number ($n > m$), R_0 is the radius of the circular array of probes, and k is $2\pi/\lambda$. In this case the higher-order modes are "below" cutoff and do not radiate from the feed circle. The undesired higher-order modes, if they did radiate, would result in extraneous beams in directions corresponding to these modes which can also be identified as grating lobes from the linear array due to the periodic amplitude ripple in the aperture distribution. With a $\lambda/2$ probe spacing, the undesired higher-order modes were not sufficiently suppressed; hence a $3\lambda/8$ spacing was used in the experimental model. It should be emphasized that by higher-order modes we mean modes that are higher than the desired modes intentionally generated by the feed. A photograph of the experimental model including the hybrid feed network, the radial line, and the linear array is shown in Fig. 26. The radial line outputs are waveguides and the linear array consists of open-ended waveguide elements with a common horn.

Fig. 27 gives the measured patterns of the seven beams superposed. The patterns were taken without individual tuning or gain adjustment of the different beams. Hence the patterns as recorded indicate the relative gains of the separate beams including impedance-mismatch, circuit, and scan losses. As with the waveguide method of excitation the patterns agree quite well with the theoretical patterns. The vswr was less than 1.3 for all beams. To verify the beam-combining technique to produce tapered aperture distributions, three beams were combined to give a cosine-squared function on a pedestal aperture distribution. Fig. 28 is a typical antenna pattern which shows that low side lobe distributions are achievable. Also, as with other multiple-beam antennas, sector beams and other shaped

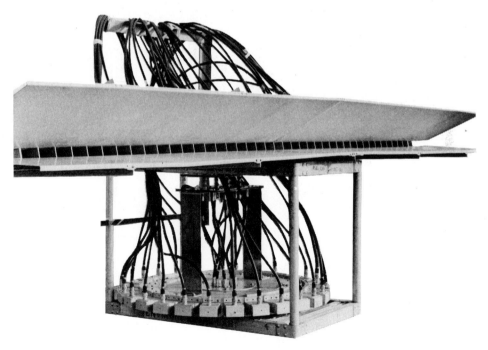

Fig. 26. Experimental model of seven-beam antenna. (*Courtesy Hughes Aircraft Co., Fullerton, Calif.*)

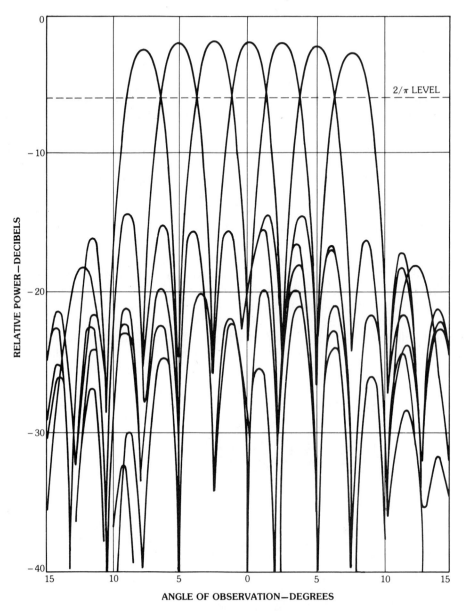

Fig. 27. Measured patterns from the seven-beam antenna system. (*Courtesy Hughes Aircraft Co., Fullerton, Calif.*)

patterns are achievable by combination of beams in the proper relative amplitudes and phases.

Meyer Lens

The principle of the Meyer lens is easily seen by considering a point-source feed radiating in the region between parallel plates as shown in Fig. 29a. The phase

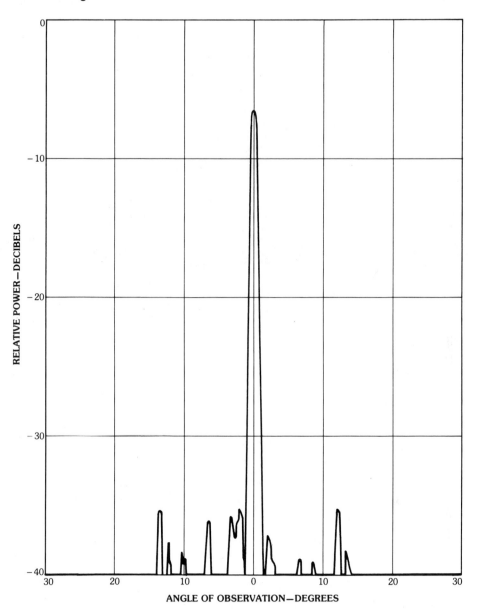

Fig. 28. Measured pattern of cosine-squared function on a pedestal distribution. (*Courtesy Hughes Aircraft Co., Fullerton, Calif.*)

front in the parallel plates is circular and the rays are radial from the feed. The circular phase front can be made linear to collimate the rays by simply curving the parallel plates in the form of a cylinder and putting a 90° bend at the base of the cylinder to direct the collimated rays normal to the axis of the cylinder as shown in Fig. 29b. The Meyer concept is more general in that the bend angle need not be 90° but is arbitrary. The shape of the curve of the cylinder for perfect focus from a

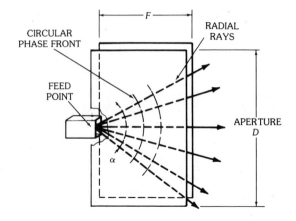

a

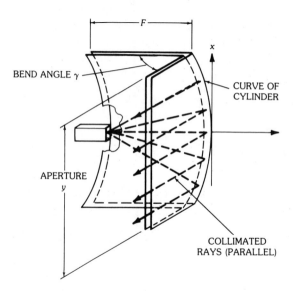

b

Fig. 29. Principle of the Meyer lens. (*a*) For circular phase front. (*b*) For collimated rays. (*Courtesy Hughes Aircraft Co., Fullerton, Calif.*)

given point-source feed is a catenary [23] given by

$$\frac{y}{\delta} = \cosh\left(\frac{x}{\delta}\right) - 1, \quad \delta = f(\sec\phi + \tan\phi), \quad \gamma = \frac{\pi}{2} - \phi$$

where γ is the angle of the bend. If the feed is moved laterally and rotated about its phase center to maintain the proper illumination of the common aperture, the beam will scan from that of the original beam by an amount proportional to lateral displacement, just as with a parabolic antenna. In general, as with all optical devices, the wide-angle scan capability improves with larger F/D, or more precisely, the smaller the feed subtending angle α. By making the bend angle γ larger, keeping the aperture size D the same, the F/D ratio is made larger, the feed angle α is smaller, and the wide-angle scannability is improved. As with all optical devices, however, the feed must be larger and the whole structure less compact by virtue of larger F.

It is interesting to note that if the bend angle γ is made 0°, the catenary degenerates into a parabola and the Meyer lens becomes the familiar folded pillbox described in Section 3, under the heading "Pillbox." For bend angles on the order of 90° or greater the catenary curve is a much better fit to a circle than is a parabola; hence the Meyer lens designed with a nonperfectly focusing circular cylinder has better aperture phase characteristics than a circular pillbox with the same F/D. Due to their circular geometry both have unlimited scannability. A modified 360° Meyer lens with dielectric loading can be used to feed a cylindrical array. This is discussed in the last part of Section 6.

The circular Meyer lens [24] is depicted by Figs. 30 and 31 in its undeveloped and developed states, respectively.

A feed may be placed and properly oriented at any point along the feed circle, and the beam radiated from the lens will point in the direction of a principal ray from the feed point through the center of the lens. The analysis may be more easily understood by developing the lens onto a flat surface, as in Fig. 31. In the developed lens all the geodesic paths become straight lines, making analysis very simple.

Two developed* surfaces can be connected together by joining corresponding points of the two surfaces with equal lengths of TEM cable. In this case Snell's laws become

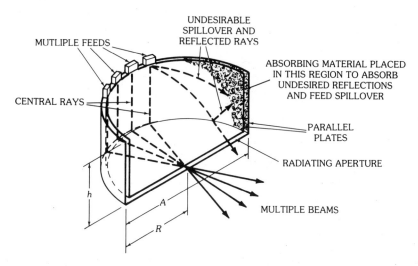

Fig. 30. Schematic diagram of undeveloped multiple-beam geodesic line source. (*Courtesy Hughes Aircraft Co., Fullerton, Calif.*)

---

*The joining surfaces need not, in general, be developable or even be physically joinable but in most practical cases the parts are separated at an intersection in the undeveloped configuration.

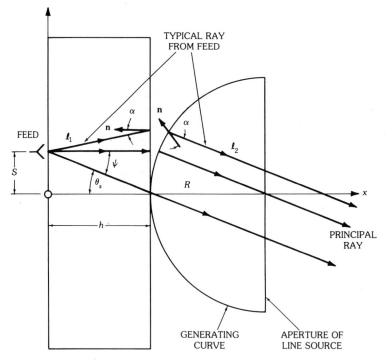

Fig. 31. Developed geodesic lens. (*Courtesy Hughes Aircraft Co., Fullerton, Calif.*)

$$\alpha_i = \alpha_r$$

$$\frac{\sin \alpha_t}{\sin \alpha_i} = \frac{\sqrt{\epsilon_1}\, ds_1}{\sqrt{\epsilon_2}\, ds_2} = \frac{n_1 ds_1}{n_2 ds_2}$$

where

α_i = the angle of incidence

α_r = the angle of reflection

α_t = the angle of transmission (refraction)

ϵ_1, ϵ_2 = the relative dielectric constants of regions 1 and 2, respectively

n_1, n_2 = corresponding indexes of refraction

ds_1, ds_2 = the local spacing of connecting points of regions 1 and 2, respectively

It can be seen from these relationships that the relative spacings ds_1 and ds_2 of the connecting points have the same effects as the relative index of refraction, n_1 and n_2.

An obvious further generalization can be made to make the cable lengths not necessarily equal. Then the differential line length $d\ell$ between adjacent lines times the index of refraction $\sqrt{\epsilon_3}$ in the cables must be added to give the relation

$$\sqrt{\epsilon_1}\,ds_1\sin\alpha_1 = \sqrt{\epsilon_2}\,ds_2\sin\alpha_2 + \sqrt{\epsilon_2}\,d\ell$$

The law of reflection remains essentially the same with the assumption that the major portion of the reflected energy is due to the parallel plate–to-cable transition mismatch.

Rotman and Turner [25] used these concepts to design wide-angle multiple-beam lenses using planar parallel plates interconnected by cables. Consequently, these types of lenses are known as *Rotman lenses*. Excerpts from a paper by Rotman and Turner are given in the next section.

In Fig. 31 the beam direction is determined by the principal ray which passes through the center of the aperture. The phase across the aperture may be found by comparing the geodesic path length $\ell = \ell_1 + \ell_2$ for any ray to the principal ray of path length $h + R$.

This type of lens is not perfectly focusing, that is, it will not provide a perfectly plane phase front over the entire aperture. It does, however, provide a reasonably flat phase front over approximately 70 to 80 percent of the aperture. The spherical aberration of this lens is similar to that of the circular folded pillbox covered earlier, but smaller as explained before, allowing more usable aperture.

For phasing a linear array, a pickup array is located at the output aperture which could be waveguide, coax, or stripline. The line lengths at the pickup array would be adjusted to remove the portion of the spherical aberration that is common to all beam positions in exactly the same fashion as discussed previously for the circular folded pillbox.

A multiplicity of feeds located on the circular feed arc is used to create multiple simultaneous beams. The pointing angles of the multiple beams are frequency independent, which implies that the crossover level is frequency dependent, since the beamwidths depend on frequency.

The main advantage of this lens over the circular folded pillbox is that it has more usable aperture. Its main disadvantage is its form factor (in its undeveloped form), which could represent a packaging problem for some applications. However, the circular portion of this lens could be folded, thereby halving its height. It could also be folded more than once to decrease its height even more, but there is certainly a point of diminishing returns.

Radio-frequency absorbing material should be placed in the curved plates at places that do not interfere with the principal optical paths of any feed. This is to absorb any reflected power that does not go around the 90° bend. Materials such as synthane (linen base phenolic) can be used as structural members to hold the plate separation fixed while also acting as rf absorbers. The loss through the Meyer lens is very low, similar to that of the folded pillbox.

Rotman and Turner Line Source Microwave Lens

The Rotman and Turner lens [25] is a parallel-plate constrained lens consisting of a focal arc on which multiple feeds are placed (see Figs. 32 and 33), a set of parallel plates whose plate separation is less than $\lambda/2$ into which the feeds radiate, a pickup array along a surface designated Σ_1 in Fig. 32, a set of interconnecting cables of variable line lengths, and a radiating array designated Σ_2 along a straight line. Fig. 33 illustrates the physical configuration. Four independent conditions are

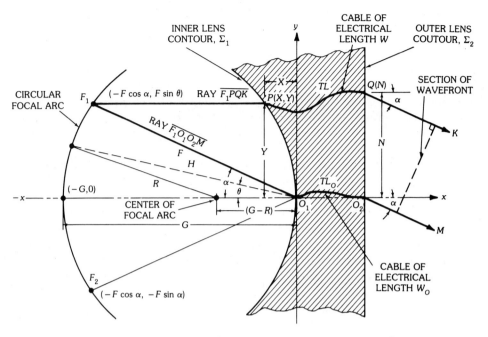

Fig. 32. Microwave lens parameters. (*After Rotman and Turner [25], © 1963 IEEE*)

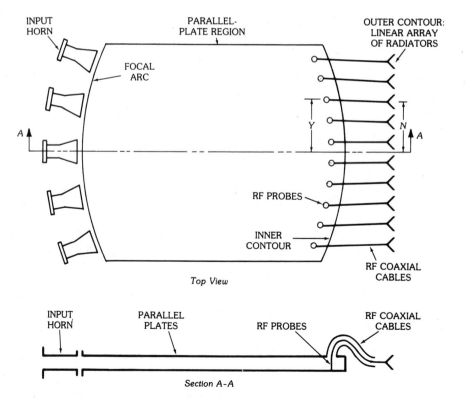

Fig. 33. Parallel-plate microwave lens. (*After Rotman and Turner [25], © 1963 IEEE*)

imposed on the lens system to uniquely determine its configuration: the four conditions imposed are a straight front face, two symmetrical off-axis focal points, and an on-axis focal point. This lens is similar to Ruze's [26] lens except that an additional degree of freedom is available (that is, $y \neq N$, see Fig. 33) which will manifest itself in improved performance for very large apertures. In Ruze's design the lens has two perfect off-axis symmetrical focal points and an on-axis focal point for which the second, but not higher-order, phase deviation is zero. One hundred beamwidths of scan are possible. For these no-second-order lenses, both the front and back lens contours are curved. Ruze [26] also discusses a straight front-face design that has two perfect symmetrical off-axis foci and one highly corrected on-axis focal point. For very large apertures there is some degree of residual higher-order coma aberration in the design.

Rotman and Turner use the generalized lens design principles developed by Gent [27] and others and impose the four previously mentioned conditions to arrive at their configuration.

The following three equations condition perfect focusing at the three* foci using a straight radiating face (see Fig. 32):

$$\overline{(F_1P)} + W + N \sin \alpha = F + W_0 \qquad (1)$$

$$\overline{(F_2P)} + W - N \sin \alpha = F + W_0 \qquad (2)$$

and

$$\overline{(GP)} + W = G + W_0 \qquad (3)$$

where

$$\overline{(F_1P)}^2 = F^2 + X^2 + Y^2 + 2FX \cos \alpha - 2FY \sin \alpha, \qquad (4)$$

$$\overline{(F_2P)}^2 = F^2 + X^2 + Y^2 + 2FX \cos \alpha + 2FY \sin \alpha \qquad (5)$$

and

$$\overline{(GP)}^2 = (G + X)^2 + Y^2 \qquad (6)$$

A set of parameters is normalized relative to the focal length F:

$$\eta = N/F, \quad x = X/F, \quad y = Y/F \qquad (7)$$

$$\omega = \frac{W - W_0}{F}, \quad g = G/F \qquad (8)$$

Also,

---

*Very recently a quadrufocal bootlace lens has been reported by Rappaport and Zaghloul [28].

$$a_0 = \cos \alpha \qquad b_0 = \sin \alpha \tag{9}$$

After normalizing and using definition (9), Equations 4, 5, and 6 can be combined with (7):

$$\frac{\overline{(F_1P)^2}}{F^2} = 1 + x^2 + y^2 + 2a_0x - 2b_0y \tag{10}$$

$$\frac{\overline{(F_2P)^2}}{F^2} = 1 + x^2 + y^2 + 2a_0x + 2b_0y \tag{11}$$

$$\frac{\overline{(GP)^2}}{F^2} = (g + x)^2 + y^2 \tag{12}$$

The normalized forms of (1) and (10) are combined to yield

$$\begin{aligned}
\frac{\overline{(F_1P)^2}}{F^2} &= (1 - \omega - b_0\eta)^2 \\
&= 1 + \omega^2 + b_0^2\eta^2 + 2b_0\omega\eta - 2\omega - 2b_0\eta \\
&= 1 + x^2 + y^2 + 2a_0x - 2b_0y
\end{aligned} \tag{13}$$

Since the off-axis focal points are located symmetrically about the center axis, the lens contours must also be symmetrical. Therefore, (13) remains unchanged and can be separated into two independent equations if η is replaced by $-\eta$ and y by $-y$. One equation contains only odd powers of y and η while the other contains the even powers. Thus,

$$-2b_0\eta + 2b_0\omega\eta = -2b_0y \tag{14}$$

or

$$y = \eta(1 - \omega) \tag{15}$$

Also,

$$x^2 + y^2 + 2a_0x = \omega^2 + b_0^2\eta^2 - 2\omega \tag{16}$$

Equations 3 and 6 relating to on-axis focus together with definitions (8) and (9) are similarly combined to give

$$\frac{\overline{(GP)^2}}{F^2} = (g - \omega)^2 = (g + x)^2 + y^2 \tag{17}$$

or

$$x^2 + y^2 + 2gx = \omega^2 - 2g\omega \tag{18}$$

After algebraic manipulation, (15), (16), and (18) give the following relations between ω and η:

$$a\omega^2 + b\omega + c = 0 \tag{19}$$

where

$$a = \left[1 - \eta^2 - \left(\frac{g-1}{g-a_0} \right)^2 \right] \tag{20}$$

$$b = \left[2g \left(\frac{g-1}{g-a_0} \right) - \frac{(g-1)}{(g-a_0)^2} b_0^2 \eta^2 + 2\eta^2 \right] \tag{21}$$

and

$$c = \left[\frac{g b_0^2 \eta^2}{g-a_0} - \frac{b_0^4 \eta^4}{4(g-a_0)^2} - \eta^2 \right] \tag{22}$$

For a fixed set of values of the design parameters α and g, ω can be computed as a function of η from (19). These values of ω and η are then substituted into (14) and (18) to give x and y, which completes the solution of the lens design.

Choosing $\alpha = 30°$, Figs. 34a, 34b, and 34c show the computed lens shape and mapping function $\eta(y)$, as well as the feed locus, for values of $g = 1.10, 1, 1.137$, respectively. The corresponding path length errors $\Delta\ell$ for scan angles of $5°$, $10°$, $20°$, $30°$ are given in Fig. 35, illustrating the extremely small degree of error over these scan angles.

Fig. 36 illustrates the hardware implementation of this technique, while Figs. 37a, 37b, and 37c illustrate the measured scanned radiation patterns.

According to Rotman and Turner's calculations this type of antenna design is capable of scanning 800 beamwidths of scan without appreciable degradation. For further details, the reader is referred to Rotman and Turner [25].*

Rinehart-Luneburg Lens

Up to this point, this section has dealt with the feeding of linear or planar phased arrays. Although the Rinehart-Luneburg lens is not particularly applicable to the feeding of linear or planar phased arrays, which is the subject of Section 2, under "Multiple-Beam Matrix Feeds," it is introduced here because it is a parallel-plate device. It is applicable for feeding circular or cylindrical arrays and is discussed more fully for that application in Section 6.

The Rinehart [29] geodesic lens is an analog of the nonuniform index of

---

*The reader should be warned that there are some typographical errors in some of the equations in [25]. The most important error is in the expressions for b (Equation 21 of this chapter). The term $-2g$ is missing in the corresponding equation (12) in [25].

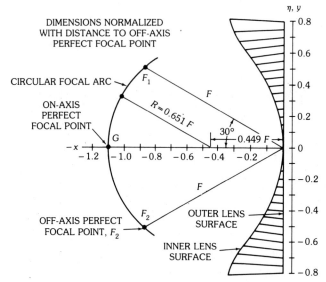

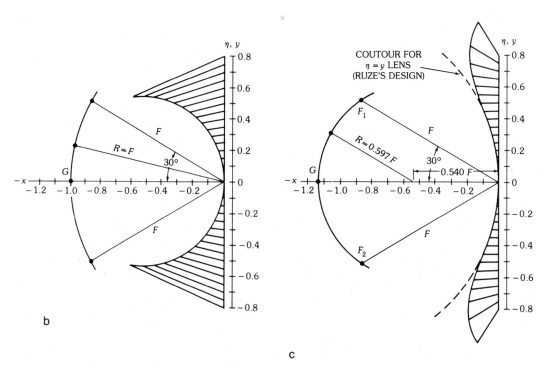

Fig. 34. Microwave lens contours. (*a*) With *g* = 1.10. (*b*) With *g* = 1. (*c*) With *g* = 1.137. (*After Rotman and Turner [25], © 1963 IEEE*)

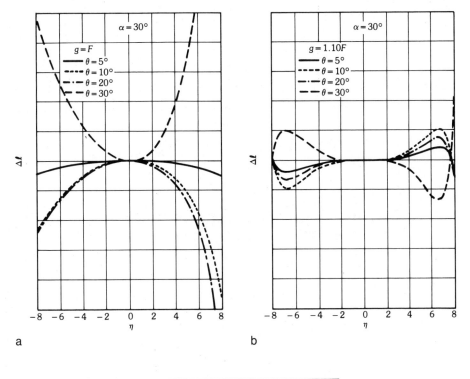

a

b

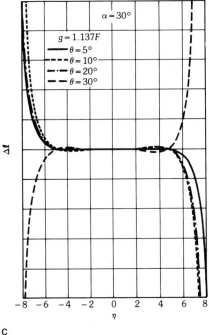

c

Fig. 35. Path length errors in a microwave lens. (*a*) With $g = 1.00$. (*b*) With $g = 1.10$. (*c*) With $g = 1.137$. (*After Rotman and Turner [25], © 1963 IEEE*)

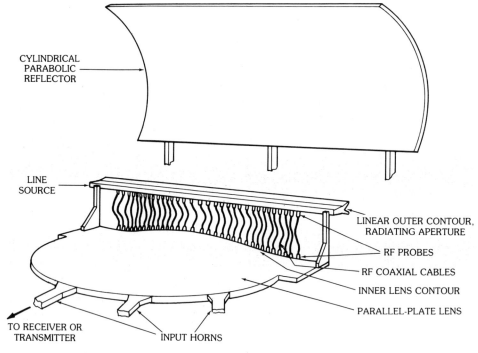

CYLINDRICAL
PARABOLIC
REFLECTOR

LINE
SOURCE

LINEAR OUTER CONTOUR,
RADIATING APERTURE

RF PROBES

RF COAXIAL CABLES

INNER LENS CONTOUR

PARALLEL-PLATE LENS

TO RECEIVER OR
TRANSMITTER   INPUT HORNS

Fig. 36. Parallel-plate lens and reflector. (*After Rotman and Turner [25], © 1963 IEEE*)

refraction spherical lens due to R. K. Luneburg [30]. Because of its shape it is commonly referred to as the "tin-hat" Rinehart lens.

Figs. 38 and 39 illustrate two versions of the geodesic analog of the Luneburg lens, commonly referred to as the tin-hat or Rinehart and flat-plate Luneburg lens, respectively. Both of these operate in the TEM mode. The electrical parallel-plate separation is less than $\lambda/2$ for both lenses.

The tin-hat derives its focusing properties from the physical length that the rays must traverse in propagating from the feed point to the linear aperture as depicted. The feed is usually an open-ended waveguide or waveguide horn, although other types may be used.

The main advantages of the tin-hat are that it contains no dielectric materials, is very simple in construction, and is perfectly focusing over 360°. Moreover, it is very low loss. Since it operates in the TEM mode it is fairly forgiving of small perturbations in shape because slight feed defocusing can partially correct for slowly varying errors. The amplitude transformation that occurs through this lens is such that an inverse taper occurs, e.g., a primary feed with a $\cos\theta$ power pattern transforms into a uniform distribution on emerging from the lens, as illustrated by the bunching of rays near the edges of the output aperture in Fig. 38. Consequently a high gain factor is realizable; however, because of this phenomenon, ultralow side lobes may be difficult to achieve with a simple feed, and a highly directive feed is required to produce very low side lobes.

The bandwidth capability of the tin-hat and flat-plate Luneburg lens is

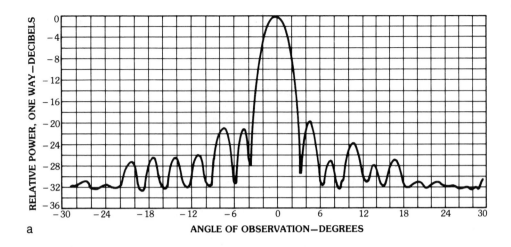

a

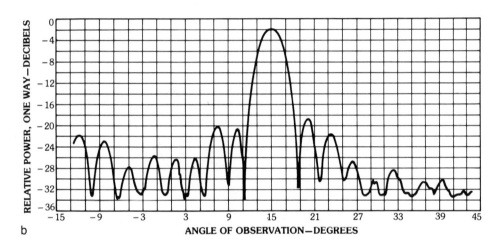

b

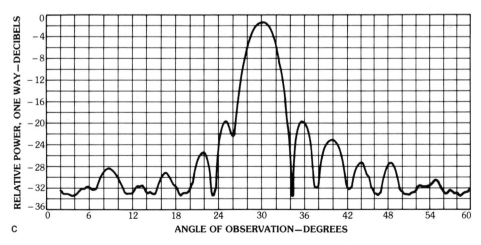

c

Fig. 37. Radiation patterns of microwave lens antenna. (*a*) For $\theta = 0°$ (on axis). (*b*) For $\theta = 15°$. (*c*) For $\theta = 30°$. (*After Rotman and Turner [25], © 1963 IEEE*)

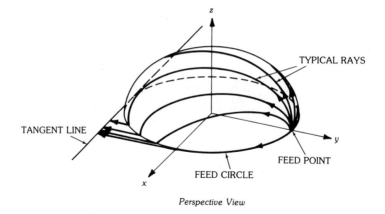

Perspective View

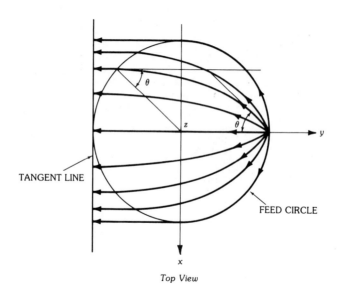

Top View

Fig. 38. Paths taken by rays on mean surface of geodesic Rinehart lens.

essentially limited only by the bandwidth of the components used in conjunction with them, since the lens is a true time-delay device.

The main advantage of the flat-plate Luneburg lens lies in its form factor, i.e., being flat, it is relatively easy to package.

If either of these lenses is to form multiple simultaneous beams over 360°, then switchable circulators would have to be used, one for each element. The reason for this is that each element in the pickup array around its periphery has to act as a feed as well as a transfer element. Fig. 40 illustrates this situation. The radiating array would have to be on the same radius as the pickup array to retain the correct phase distribution. The two arrays are depicted as lying on different radii for clarity only.

If 360° coverage is needed, but not simultaneously, then diode switches could be used rather than circulators. The reason for this is because, at any given

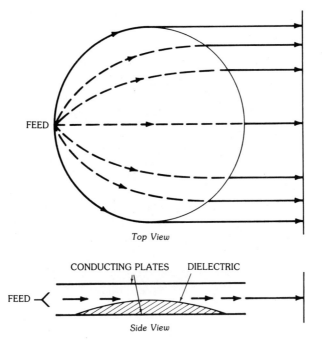

Fig. 39. Flat-plate Luneburg TEM-mode lens.

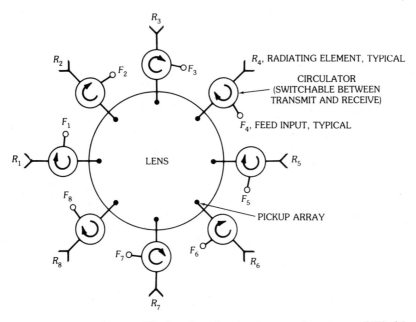

Fig. 40. Flat-plate Luneburg or Rinehart lens for simultaneous beams over 360°. (*Courtesy Hughes Aircraft Co., Fullerton, Calif.*)

moment, each element either acts as a feed or as a transfer element but not simultaneously. Fig. 41 is a photograph of a cylindrical array of line sources fed by a Rinehart geodesic lens.

There are some limited-scan applications whereby these lenses are used to phase a line source. In that case the planar output parallel plates must be extended to form a line source. (See Fig. 19.) The output amplitude distribution moves with scan, however, and the physical length of the line source is longer than its active length.

There are numerous derivatives of these lenses, among them the TE_{10} mode Luneburg that uses the TE_{10} mode between nearly parallel plates. The central plate spacing is greater than at the periphery. Dielectric is usually used to fill the separation between plates to structurally act as plate separators. The bandwidth is much less than its TEM mode counterpart because the ideal plate spacing for the

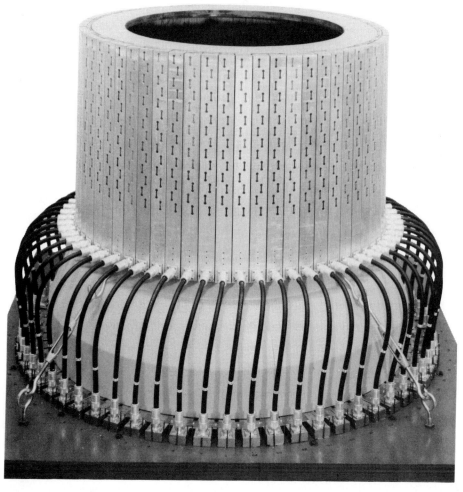

Fig. 41. Cylindrical array fed by a Rinehart geodesic lens. (*Courtesy Hughes Aircraft Co., Fullerton, Calif.*)

TE$_{10}$ mode lens depends on wavelength. Since the correct spacing can be specified only at center frequency, it results in a design that is correct only at the center frequency. The actual bandwidth is a function of the size.

The folded Luneburg lens is, in principle, geodesically the same as the nonfolded one. It resembles a convex dome with a concave dent in the top. From a practical point of view the problem that arises is that of reflections occurring at the points where the slope changes. A mitered bend could probably be used to impedance-match the rays for certain angles of incidence; however, it is doubtful that such a design would work well for all the angles of incidence that the lens requires (up to grazing angles of incidence). Using a smooth bend in the region where the slope changes could be attempted but this changes the geodesics. Little or no information is available to draw definite conclusions regarding the folded Luneburg.

DuFort-Uyeda Lens

DuFort and Uyeda [31] reported a modified Rinehart metal dome antenna with a dielectric ring lens at the output for decreased beamwidth in the elevation plane. The ring lens provides beam collimation in the plane perpendicular to the azimuth scan plane and results in a much smaller antenna than a conventional Rinehart lens fitted with an *E*-plane flared horn.

Even for moderate directivity in the plane perpendicular to the plane of the multiple beams, the length of a low-phase-error, low-flare-angle horn is quite an appreciable fraction of the radius of the dome. This added radius does not contribute to the effective aperture in the azimuth plane which, of course, is equal to the diameter of the dome only. If a shorter horn with a larger flare angle were used, a dielectric lens would be required to collimate the circular phase front of the horn. With dielectric in the toroidal horn it becomes clear by tracing a few rays that the Rinehart lens will no longer focus in the azimuthal plane. The DuFort-Uyeda lens focuses perfectly in the azimuthal plane with an effective aperture equal to the total diameter of the dome and dielectric ring-horn while achieving directivity in the perpendicular plane.

An experimental model was built and tested at K_a-band (26.5 to 40 GHz). It provided a 1.7° by 10.7° beam at 40 GHz. Pattern and gain data show excellent performance.

Fig. 42 shows the top view and cross section of the experimental model. Fig. 43 is a photograph of the experimental model. Fig. 44 shows a typical beam measured at 32 GHz using reduced-height WR28 waveguides for feeds. Fig. 45 depicts an *H*-plane multiple-beam overlay of five contiguous beams measured at 32 GHz. Fig. 46 shows a lower side lobe pattern measured at 32 GHz using a 3/4-inch (1.905-cm) *H*-plane flared horn for the feed.

Aperture efficiencies of between 60 and 72 percent were obtained over the 26.5- to 40-GHz band. Uniform multiple beams of similar gain and pattern behavior can be generated over a 90° sector [31].

4. Unconstrained (Optical) Feeds

An *unconstrained feed* is one in which free space exists between the feed(s) and the radiating aperture. The rf power distribution from feed to aperture is achieved

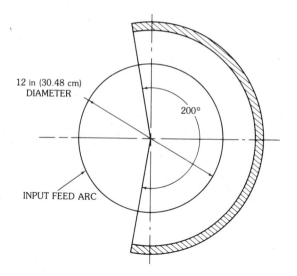

a

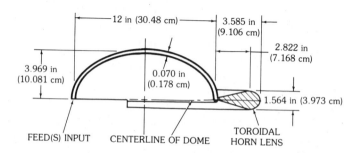

b

Fig. 42. Design for K_a-band geodesic dome and lens aperture. (*a*) Top view. (*b*) Cross section. (*Courtesy Hughes Aircraft Co., Fullerton, Calif.*)

Fig. 43. Experimental model DuFort-Uyeda lens. (*Courtesy Hughes Aircraft Co., Fullerton, Calif.*)

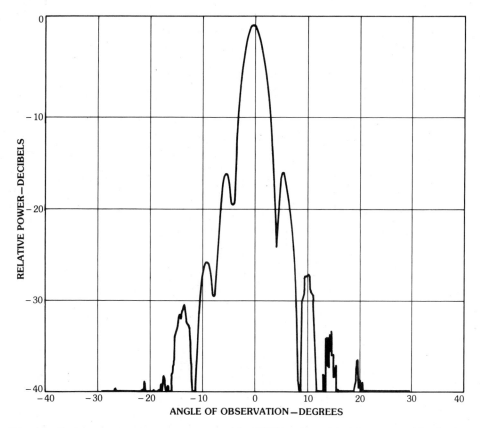

Fig. 44. *E*-plane pattern measurement with WR28 reduced-height waveguide feed at 32 GHz. (*Courtesy Hughes Aircraft Co., Fullerton, Calif.*)

by unconstrained radiation from the feed to the aperture. Hence the aperture distribution is essentially determined by the radiation pattern of the feed. Fig. 47 illustrates such a feed in its most general form. For example, the feed could be a one- or two-dimensional array of radiating elements while the beamformer which collimates the beams could be a reflector(s) or lens which could limit the FOV (field of view) to a relatively narrow sector of space because of off-axis aberrations. As another example the feed could be a simple monopulse feed and the beamformer could be a two-dimensional, nominally half-wavelength-spaced pickup array and radiating array with a phase shifter between a pickup element and a radiating element. In the latter case the FOV is limited only by the wide angle of aperture matching and grating lobe formation (see Chapter 13). A FOV of, say, 90° to 120° cone or greater can be achieved with proper aperture design. For a specified gain (or beamwidth) one would probably use entirely different techniques for achieving a large FOV as opposed to a narrow FOV, or limited scan. In general, the larger the FOV, the greater the complexity and cost of the antenna system. Also, the greater the instantaneous bandwidth, the greater the complexity and cost. Trade-offs can be made between FOV and instantaneous bandwidths.

   Any attempt to categorize the various unconstrained-feed approaches is

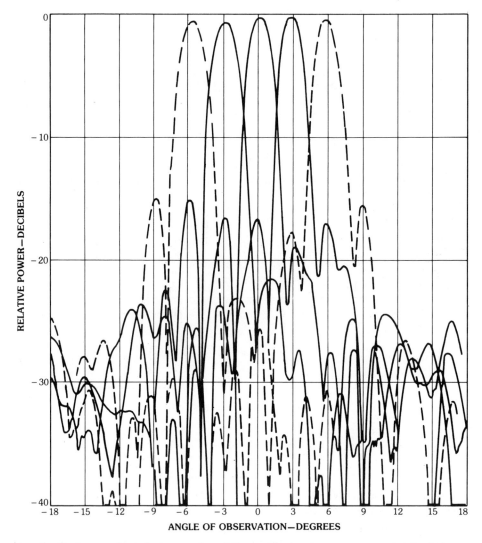

Fig. 45. *H*-plane multiple-beam overlay of five contiguous beams measured at 32 GHz with WR28 reduced-height feeds. (*Courtesy Hughes Aircraft Co., Fullerton, Calif.*)

difficult because there are overlaps in the categories. Nevertheless, a review of the various techniques suggests four categories for the purpose of discussion. These are the following:

1. Wide FOV (0° to 360° phase shifters, nominally half-wavelength-spaced elements
2. Limited scan (0° to 360° phase shifters and optics with aperture magnification)
3. Subarray or partial time delay using overlapping or non-overlapping subarrays, 0° to 360° phase shifters, and time delay at the subarray level. *Note*: Totally constrained limited-scan techniques are discussed in Section 3 of Chapter 13,

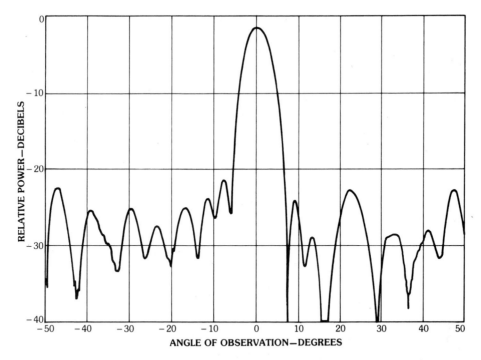

Fig. 46. Low side lobe *H*-plane pattern measured at 32 GHz with 0.75-in (1.905-cm) *H*-plane flared horn. (*Courtesy Hughes Aircraft Co., Fullerton, Calif.*)

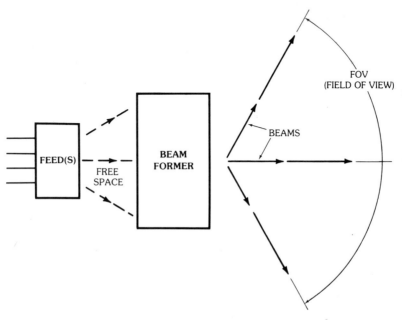

Fig. 47. Unconstrained feed. (*Courtesy Hughes Aircraft Co., Fullerton, Calif.*)

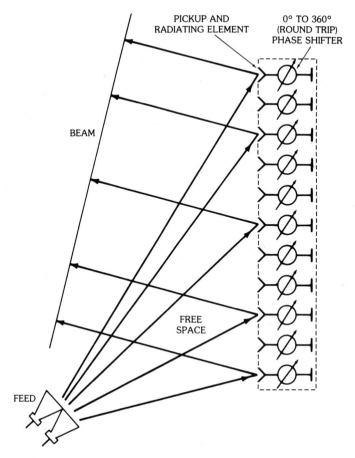

Fig. 48. Space-fed reflectarray. (*a*) Offset-fed. (*b*) Center-fed. (*Courtesy Hughes Aircraft Co., Fullerton, Calif.*)

under the heading "Array Organization: Subarrays and Broadband Feeds," and also in Section 2 of this chapter.
4.   True time delay (true time-delay phasing of all elements)

Wide Field of View (Nontrue Time Delay)

The wide field of view technique implies a space-fed two-dimensional array of nominally half-wavelength-spaced elements. The offset-fed and center-fed reflec-tarray of Fig. 48 depicts examples of this approach [32]. In these cases the feed would typically be a four-horn monopulse feed (only two shown). Offset feeding offers less aperture blockage but causes an asymmetry in the illumination. The spherical wavefront incident on the array is converted to a scanned beam by 0° to 360° type phase shifters. Center feeding eliminates the illumination asymmetry but presents aperture blockage which would cause some side lobe level degradation. Many systems do not require ultralow side lobe levels, and this feed approach offers a straightforward solution. The Raytheon MTR Radar and Radome Antenna

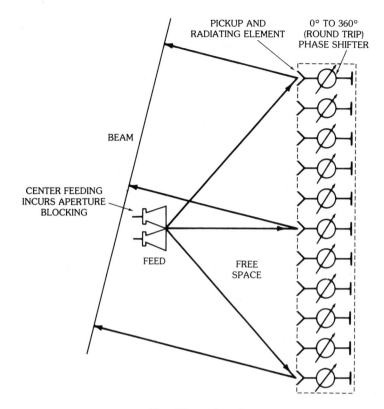

b

Fig. 48, *continued.*

and RF Circuitry (RARF) array are examples of the center-fed reflectarray [32]. In the case of the RARF array six waveguides (monopulse sum and two difference channels for two orthogonal polarizations) are routed to the feed in order to achieve polarization agility. Other examples are the NRL *S*-band Reflect Array Radar (RAR) developed by Blass Antenna Electronics Company [32] and the Hughes Aircraft Company Electronically Scanning Airborne Intercept Radar Antenna (ESAIRA).

Aperture blockage is eliminated entirely using a space-fed feed through lens array as depicted in Fig. 49. Since no aperture blockage is presented, the feed could be as large and complex as desired, without degraded side lobe level effects. There are numerous advantages to this feeding scheme. The Raytheon Sam-D Radar uses this approach [32]. It enables a space-duplexing feature to be employed; that is, on transmit the phase shifters are set to focus on the transmit feed, and on receive the phase shifters are set to focus on a separate receive feed. Moreover, two transmitter feeds could be employed using two transmitters for redundancy, where different phase shifter settings are used, depending on which transmitter is being used [32]. (The offset-fed reflectarray could also have this feature without increased aperture blockage.) Since there is room available, optimized sum and difference aperture illuminations could be realized by a sophisticated monopulse feed design. A multiple stack of vertical feed horns could be combined for elevation beam

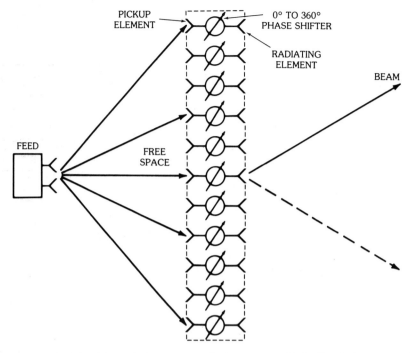

Fig. 49. Space-fed feedthrough lens array. (*After Patton [34], © 1972; reprinted with permission of Artech House*)

shaping, and azimuth patterns could be generated by using multimode sum and difference extraction of each horn in the azimuth plane [32].

Although Fig. 49 depicts the space-fed feed through array with both faces flat, they need not be.

The sperry HAPDAR radar uses a curved pickup face and flat radiating face, with phase shifters in the transmission lines between faces. This structure is called TACOL, for thinned aperture computed lens [32, 33]. The feed is a five-horn cluster that provides independent control over the sum and difference illuminations, using an *F/D* ratio of 1. All five horns are used for the sum pattern but only the outer four are used for the difference patterns. In this way, low side lobe sum and acceptable difference patterns are realized.

Limited Scan

High-gain antennas with wide field of view and/or wide instantaneous bandwidths are very costly due to the large number of active devices. For phased arrays a phase shifter and driver are required for each radiating element, which may total to many thousands for a high-gain array. For multiple-beam antennas with beam switching, a similar number of switching elements and drivers are required. In many radar applications wide field of view is not required although high angular resolution and high gain are. Examples of such limited-scan applications are the airport precision-approach radar, weapon-locating radars, and earth-coverage antennas for synchronous satellites.

It is the objective of limited-scan antenna design to take advantage of this limited field of view to design an antenna with the fewest number of active devices but retain the narrow beamwidth and high gain. To achieve high gain and narrow beamwidth requires a large antenna aperture, and a large field of view requires a phase shifter and drivers for every nominally half-wavelength-spaced radiating element. Hence high gain and large FOV result in an extremely costly antenna system. By restricting the FOV the cost can be greatly reduced.

Limited-scan antennas can employ optical (unconstrained) feeds, totally constrained feeds, or combinations of these. In this section on unconstrained feeds, only those antennas employing optical techniques are discussed. Limited-scan techniques using totally constrained feeds in which elements of an array are grouped together to form identical subarrays with a phase shifter per subarray are discussed in Chapter 13, Section 3, on array organization.

Optically Fed Aperiodic Array—An array of large equal-size directive elements (e.g., a nonscanning subarray) whose element pattern is tailored to the required FOV could be considered, using a phase shifter for each large element or subarray. For regular element spacing, however, this approach produces excessively high grating lobes as the beam is scanned, which would not meet most system requirements. The grating lobes result from the fact that radiation from periodic widely spaced elements add in phase at angles in space other than the desired direction of the main beam. If the periodicity is broken up, the energy concentrated in the grating lobes is smeared out "randomly" in all directions, resulting in lower peak values. The overall antenna gain, however, is not improved since it is only a redistribution of the grating lobe energy. In most cases there is a slight though usually negligible reduction in gain.

One method of minimizing the grating lobe problem, while still using large directive elements, is given by Patton [13, 34]. In this approach he uses concentric rings of equal-area subarray elements, as shown in Fig. 50 (depicted for only a few subarrays). Fig. 51 depicts a large number of subarrays of equal area. The subarrays are dual-polarized dipoles above a ground plane. Each polarization is summed (in phase) within each subarray. Thus each subarray has two outputs— one for each polarization. Fig. 52 illustrates how the concentric ring array is fed. For each subarray pair of cables there is a corresponding pair of terminals (one for each polarization) located on the surface of a partial sphere. The cables must track in phase over the frequency band of the system. The feed is located at the center of the spherical cap. The output final aperture distribution is thus determined by the primary Σ and Δ patterns produced by the primary feed. Since dual-polarized elements are used, the final radiated polarization is determined solely by the polarization of the primary feed. There are phase shifters in every semirigid coaxial line between the concentric ring array and the spherical cap.

By using this technique the periodicity of the usual array factor (for regular spacing) is broken up; consequently, grating lobes are suppressed. The degree of suppression is determined by the total number of subarrays used. Patton [34] shows calculated data for 10 ft and 30 ft diameter (3.0 and 9.1 m) arrays at C-band using a $\pm 5°$ FOV. For the 30 ft diameter case, using 1000 subarrays, the vestigial grating lobe is suppressed to about -21 dB for maximum scan of $\pm 5°$. The 10 ft diameter case (100 subarrays) corresponds to -15-dB suppression at $\pm 5°$ scan.

Predicted and measured loss, including illumination taper (10-dB tapered

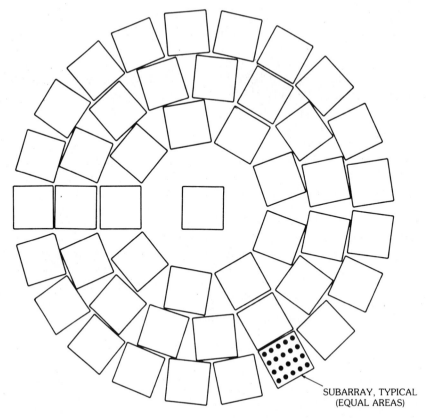

SUBARRAY, TYPICAL
(EQUAL AREAS)

Fig. 50. Concentric ring array of equal-area subarrays. (*After Patton [34], © 1972; reprinted with permission of Artech House*)

Gaussian) for the 10 ft diameter model are 5.94 dB and 5.58 dB, respectively, at midband. Measured first side lobes were about −15 dB, due to effective aperture blockage. The gain decreased with scan by about 2.5 dB at ±5° from broadside.

The primary feed is a multimode monopulse feed, dual polarized for the sum channel, and horizontally polarized only for the difference. Conventional reflector feed techniques for achieving dual monopulse, dual polarization, for both Σ and Δ channels are applicable here; moreover, techniques for achieving independent control over the Σ and Δ patterns can also be used. Since there is room available, a more complex feeding structure could be considered for any additional advantages that may be realized.

Phased Array with Paraboloid—Fig. 53 depicts a paraboloidal reflector fed by a phased reflectarray, after Winter [35]. The reflectarray radiating face is not located at the focus, but is shifted toward the vertex of the paraboloid. Conceptually viewed on receive, the idea is to place the reflectarray aperture in the region forward of the focus, pick up the converging field, and phase-shift it to refocus on the primary feed(s). Geometrical optics was used to determine the phase shifter values, as a function of scan angle θ_s. As θ_s varies, the converging field

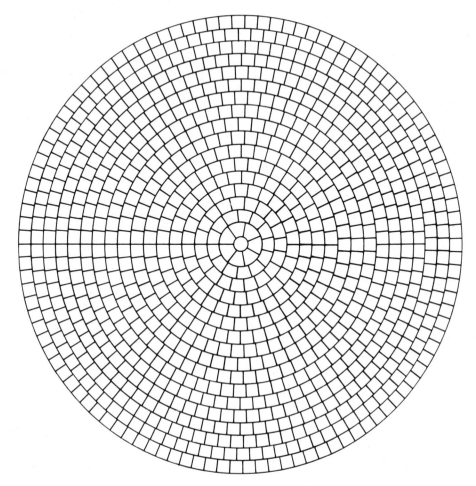

Fig. 51. Aperiodic array for limited scan. (*After Patton [34], © 1972; reprinted with permission of Artech House*)

moves, but for small enough θ_s, most of the reflected power will still be incident on the reflectarray face. Thus one can still change the phase of the converging field to refocus on the primary feed. Note that only the phase is controllable, not amplitude. As θ_s gets larger, only part of the reflected incident field is intercepted by the radiating aperture of the reflectarray antenna; hence, spillover occurs (on receive), causing irretrievable loss. Furthermore, for large θ_s, the effects of aberrations are to cause the reflectarray amplitude distribution to become skewed, introducing both an even and an odd part. The simple primary feed horn couples only to the even part, and the energy in the odd part is lost. Ideally, a horn containing an odd-mode component as well as the even mode, or a small array with the proper amounts of even and odd parts, could be employed to regain that lost energy. For maximum gain, one would use a simple feed designed to produce an illumination across the reflectarray face that best fits the even part of the received converging illumination across the reflectarray face over some specified range of

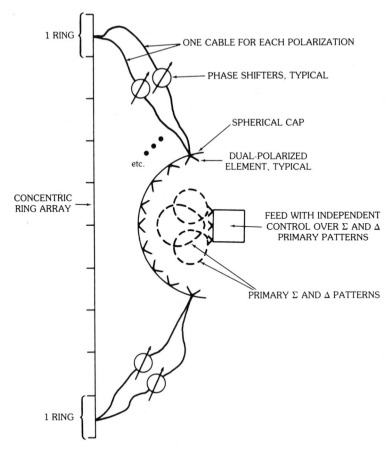

Fig. 52. Concentric ring array showing feed mechanization and power transfer for each polarization. (*After Patton [34], © 1972; reprinted with permission of Artech House*)

scan angles $\pm\theta_s$. This would produce the best amplitude fit that is achievable using a simple primary feed.

Viewing the process on transmit, the primary feed and reflectarray could be designed to illuminate the entire reflector* for a broadside beam; however, as scanning is performed, because of the optics of the system, only a portion of the reflector can be illuminated correctly to form a beam in an arbitrary given direction. Thus the illuminated portion of the reflector moves with scan, which diminishes the efficiency of the antenna system. The aperture blockage presented by the relatively large reflectarray causes a further reduction in efficiency, as well as increased side lobe levels. One of Winter's [35] experimental models used a 988-element reflectarray and simple feed horn and achieved eight beamwidths of scan with 15-dB side lobes in the *H*-plane and 10 dB in the *E*-plane. The reflector was

---

*In Winter's experiment 80 inches (203 cm) of the 96-inch (244-cm) diameter was used for a broadside beam.

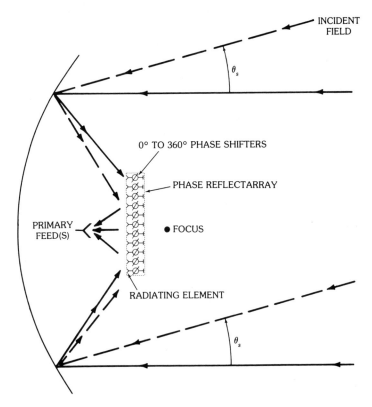

Fig. 53. Winter's approach, using phased array with paraboloid. (*After Winter [35], © 1968 IEEE*)

8 ft (244 cm) in diameter with a 51-inch (130-cm) focal length, offset fed as shown in Fig. 54. His results show that it is feasible to utilize a phased array in conjunction with a reflector to achieve a limited degree of scan. The high blockage presented by the phased array could be circumvented by using the phased array in conjunction with a transmission type feedthrough lens, or by offset-feeding to a greater degree. By using a more complex primary feeding arrangement, such as a small active array that is commanded to change with scan, more control over the amplitude would be possible. Thus the radiation characteristics would be expected to improve, provided the design is not blockage limited, as appears to be the case in Winter's experiment.

Offset Phased Array with Hyperbolic Reflector—By moving the phased array out of the FOV of the reflector the aperture blockage would be eliminated and improved radiation characteristics obtained. The GCA System TPN-19 uses such an approach [36], utilizing 824 phase shifters in conjunction with a 9- × 11½-ft (2.74- × 3.5-m) hyperbolic main reflector. Improved side lobe levels of 22 to 24 dB were obtained. The aperture efficiency was 30 percent at broadside and dropped 3½ dB over ±10 beamwidths (elevation) by ±7.15 (azimuth) beamwidths of scan. Azimuth and elevation beamwidths are 1.4° by 0.75°, respectively. The illumination

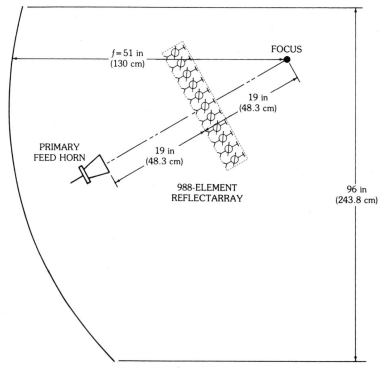

Fig. 54. Winter's experimental model. (*After Winter [35], © 1968 IEEE*)

moves over the surface of the reflector as scanning is performed; hence the aperture efficiency is low at the extreme scanned beam positions. The system operates at X-band over a 2-percent bandwidth, with an on-axis gain of 42.5 dB.

Near-Field Cassegrain System—Fig. 55 depicts the near-field Cassegrain antenna system. A typical application of this system would have a limited field of view (LFOV) of 5° to 10°, and a very large electrical aperture, e.g., $D/\lambda > 250$. It is composed of two confocal paraboloids with $f/d = F/D$, and is fed by a planar two-dimensional phased array of 0° to 360° type phase shifters. Since the subreflector is in the near field of the phased array, the rays remain collimated from the phased array to the subreflector such that there is no space attenuation. The optics of this system are such that the amplitude distribution emerging from the main reflector is the same as the amplitude distribution across the phased-array feed. In particular, a uniform amplitude distribution over the phased-array feed transforms into a uniform amplitude distribution emerging from the main dish. There is, however, a relatively large blockage by the subreflector or by the effective blockage (d_e) presented by the phased-array feed. Minimum blockage occurs when these two blockages are equal. A typical design [36] would have d/D ratios between 0.25 for $D/\lambda = 400$ and 0.35 for $D/\lambda = 250$. (Also see reference 37.)

The near-field Cassegrain system can be considered as the limiting case of the standard Cassegrain system as the feed point focus of the hyperboloid recedes to

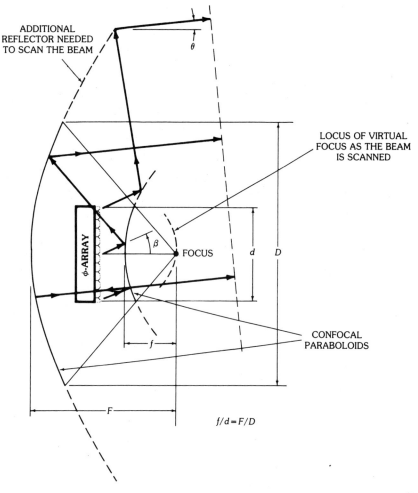

Fig. 55. Near-field Cassegrain system. (*Courtesy Hughes Aircraft Co., Fullerton, Calif.*)

infinity, when the hyperboloid subreflector becomes paraboloidal. The main-beam scan angle θ is related to the angle β by $\tan\beta = (D/d)\tan\theta$. The remotely located primary feed (at infinity) is replaced by the phased-array feed near the vertex of the main reflector with a linear phase gradient to produce rays making an angle β with respect to the z axis, which scans the main beam by the angle θ. Only linear phase gradients are required to scan the main beam; thus one has the advantage of row and column beam steering. A major disadvantage of this scheme is that the illuminated portion of the large reflector moves with beam scan, which causes inefficient usage of the large reflector aperture. This is illustrated in Fig. 55.

Lincoln Lab's Offset-Fed Gregorian System—Fig. 56 depicts a cross-sectional view of the Lincoln Lab's offset-fed Gregorian antenna system. The main reflector is a paraboloid and the subreflector is also a paraboloid with a common focus. Fig. 57

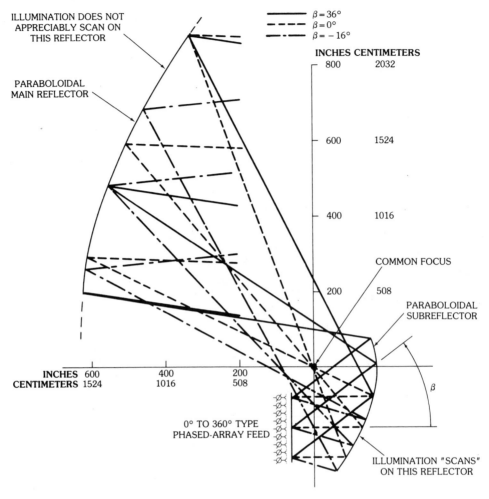

Fig. 56. Preliminary design of an offset-fed Gregorian system and ray tracing for three scan angles. (*After Interim Report [38]*)

shows an artist's concept of such a system. The primary feed for this system is a planar two-dimensional phased array using 0° to 360° type phase shifters. The basic idea here is to use two offset-fed paraboloids in such a way that the off-axis aberrations tend to cancel each other. If this could be achieved, then a small, truncated plane wave leaving the relatively small phased-array feed would be converted to a large, truncated plane wave emerging from the large main reflector. Fig. 56 illustrates ray tracings [38] for three angles β of scan for the phased array. (Also see references 39 and 40.) Although not perfectly parallel, the corresponding rays leaving the main reflector remain nearly collimated, where the scan in terms of beamwidths for the system as a whole would be about the same as the scan in terms of beamwidths for the phased-array feed alone. As can be seen from the ray tracings most of the main reflector is utilized as aperture for all β scan angles; consequently, the efficiency of this system is good for all scan angles. Mailloux and

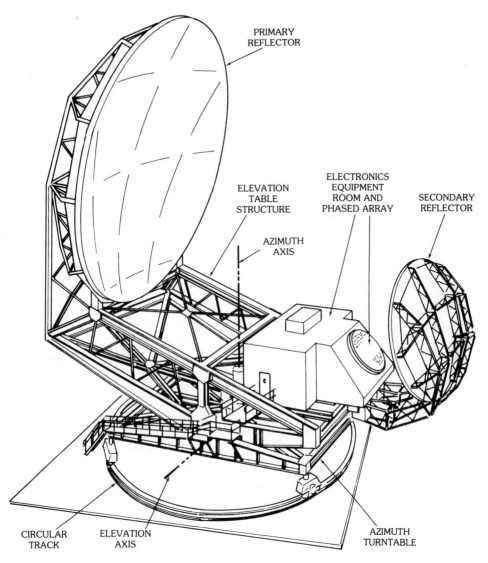

Fig. 57. Physical configuration of the antenna. (*Courtesy Hughes Aircraft Co., Fullerton, Calif.*)

Blacksmith [36] cite an example of a ½°-beamwidth system being scanned 14 beamwidths using a $45\lambda \times 45\lambda$ phased-array feed. Side lobe levels were -15 to -17 dB for all scan angles. The generalized [36] F/D ratio was 1.5.

A study [41, 42] to develop a computer program that determines the optimum main reflector and subreflector contours for maximum gain for a given scan range showed that appreciable improvement over the confocal paraboloidal system is possible. For details the reader is referred to references 41 and 42.

Improved performance can be obtained by matching the caustics of the main reflector and the subreflector, arriving at a configuration in which the smaller parabola of Fig. 56 is rotated in a clockwise direction by 45°. For more details, see Sletten [43].

Raytheon Dual-Lens Limited-Scan Concept—Fig. 58 depicts a cross-sectional view of the Raytheon dual-lens, limited-scan concept [44]. It is composed of a relatively large, constrained-aperture lens fed by a relatively small, constrained matching lens which in turn is fed by a two-dimensional planar phased-array feed using 0° to 360° type phase shifters. The contour of the output face of the matching lens corresponds to the focal surface of the aperture lens. The main aperture lens is an equal path length lens with a two-point correction (0°, ±10°). The planar feed array uses 437 radiating elements with a 23-dB tapered Gaussian amplitude distribution. The phase shifters are set to focus the array output to a small spot on the inside surface of the matching lens. The function of the matching lens is to transfer the focused field to the output face of the matching lens which is, by design, the focal surface of the main aperture lens. The transferred focused field then acts as a primary feed on the focal surface of the aperture lens, that is, the field diverges from the focused spot, illuminating the main aperture lens to form a scanned beam in space.

The inner surface of the matching lens is elliptical in shape and contains 2617 elements. The main aperture lens is 65 wavelengths in diameter. This system scans an FOV of ±10° with a 1.2° beamwidth for all angles of scan. The generalized F/D ratio [36] is about 1.7. Using this technique, the element-use factor (defined by

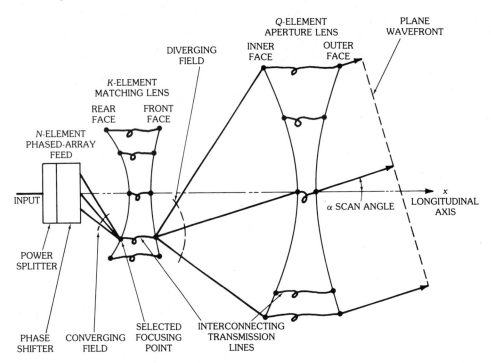

Fig. 58. Dual-lens, limited-scan concept. (*After C. H. Tang, Raytheon Co.*)

Patton [34]) is very good (about 1.5) because of the highly efficient use of the main aperture lens.

Hughes Reflector-Lens Limited-Scan Concept—Fig. 59 shows a multiple-beam constrained lens feeding an offset parabolic reflector. Ideally, points from the feed array are mapped to points on the aperture of the reflector. These points are conjugate focii of the overall optical system as shown by ray tracing, say, from feed point A to aperture point A'. Other points, such as B to B', are likewise mapped. This mapping is not perfect due to the optical aberrations of the system but is adequate for ±10° field of view. It is well known [45] that the focal spot of con-

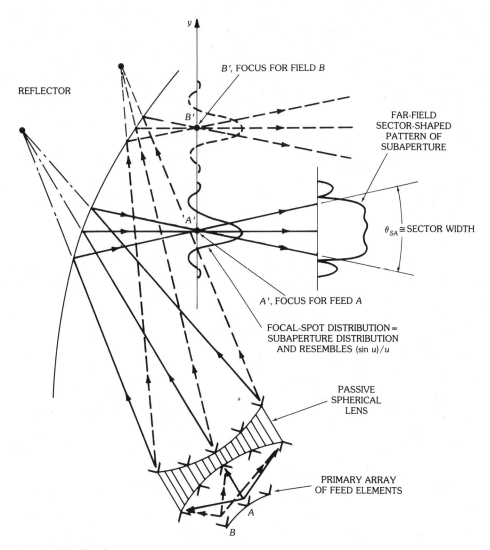

Fig. 59. Idealized geometrical mapping of rays from feed to final aperture. (*Courtesy Hughes Aircraft Co., Fullerton, Calif.*)

verging or diverging waves uniform over a sector has a focal plane distribution like $J_1(u)/u$ for spherical waves and like $(\sin u)/u$ for cylindrical waves. These distributions form overlapping $(\sin u)/u$-like subapertures that can be made orthogonal, or nearly so, by the proper amount of overlap, which is controlled by primary feed spacing along the feed arc. To be truly orthogonal, $(\sin x)/x$ distributions must be all of equal width and crossover such that the peak of any one distribution occurs at the nulls of all others. Proper radial positioning of the feed is made so that the conjugate focal point falls on the aperture plane. The $(\sin u)/u$-like subaperture distributions give rise to sector-shaped far-field patterns whose width is approximately that of the geometrical angle subtended by the rays focused at the aperture, as shown in Fig. 59.

An experimental two-dimensional parallel-plate model was built and tested [46]. An artist's concept is shown in Fig. 60 and a photograph of the experimental model is shown in Fig. 61. Calculated and measured subaperture patterns are given in Fig. 62.

Now consider the primary feeds fed by a Butler matrix (or equivalent) with intervening phase shifters. This allows multiple simultaneous beams that can be scanned in unison by the phase shifters. This is possible because a linear progressive phase across the array of lens feed elements will cause a corresponding progressive phase across the subapertures by virtue of the one-to-one correspondence explained previously. The beam can be scanned within the limits of the subaperture pattern (approximately $\pm 10°$). Fig. 63 shows representative patterns both measured and calculated for beam scans of $0°$, $-5°$, and $-9°$—all within the subaperture pattern of $10°$.

Partial-Time-Delay Systems—One of the earlier partial-time-delay systems is the MUBIS (multiple-beam interval scanner) proposed by Rotman and Franchi.* Widely spaced (several beamwidths) true-time-delay beams from a multiple-beam lens or from a true-time-delay network are synchronously scanned in the interval between the true-time-delay beams by means of $0°$–$360°$ phase shifters in the lens aperture. The system has true time delay at the true-time-delay (phase shifters set to zero) beam positions and almost true time delay if a small amount about the true-time-delay angles is scanned. Thus the field of view is increased by the number of time-delay beams for a given required instantaneous bandwidth. This technique is fully described in Chapter 13.

Completely Overlapped Space-Fed Subarray Antenna System—The completely overlapped space-fed subarray antenna system developed at Hughes Aircraft Company [47, 48] is illustrated in Fig. 64. It is composed of a planar feedthrough lens with $0°$ to $360°$ type phase shifters in every element. It is fed by a planar-feed array, which in turn is fed by a Butler or Blass multiple-beam matrix. Time-delay type phase shifters are placed at the inputs to the Butler or Blass matrix with a corporate summing feed network. The subarray inputs to the matrix form uniform

---

*See Chapter 13, Section 3, under the heading "Broadband Array Feeds With Time-Delayed Offset Beams."

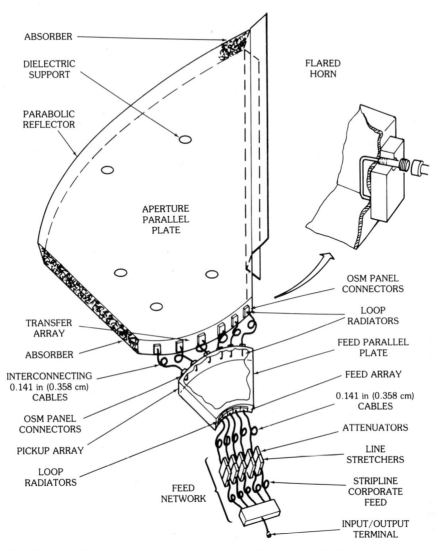

Fig. 60. Artist's concept of the experimental reflector-lens, limited-scan antenna.

distributions at the output of the feed array, which in turn form an orthogonal set of $(\sin u)/u$ type primary radiation patterns that illuminate the pickup array. Except for the slight amount of inverse distance-squared power decay, the emerging distributions from the radiating array form a near-orthogonal set of amplitude subarray distributions which completely overlap each other over the entire radiating aperture face as shown in Fig. 64. The subarray far-field radiation patterns form near-rectangular patterns superimposed in space but whose phase centers are displaced across the face of the radiating aperture. In order to minimize gain loss as a function of scan and to apply grating lobes over a wide range of scan angles and wide bandwidth, the ideal subarray pattern would be rectangular in shape but narrower than the angular separation between the first grating lobes [47].

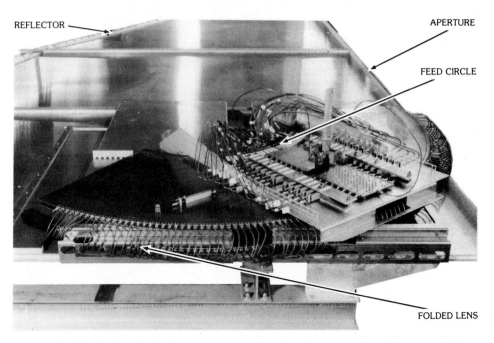

Fig. 61. Reflector-lens, limited-scan antenna, showing lens assembly. (*Courtesy Hughes Aircraft Co., Fullerton, Calif.*)

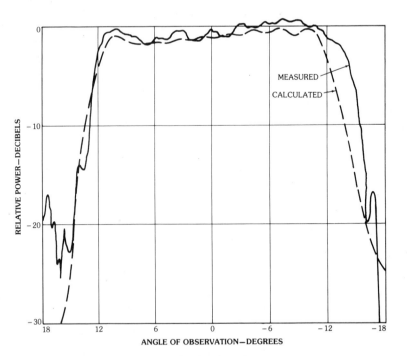

a

Fig. 62. Subaperture far-field patterns. (*a*) Subaperture centered at $y = 0$. (*b*) Subaperture centered at $y = -11.5\lambda$. (*c*) Subaperture centered at $y = -23\lambda$. (*Courtesy Hughes Aircraft Co., Fullerton, Calif.*)

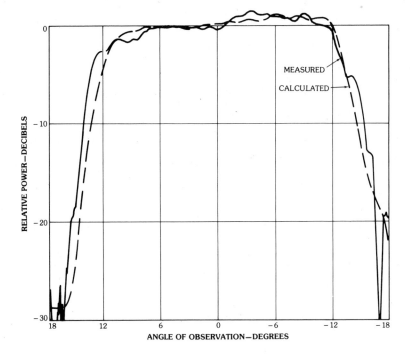

b

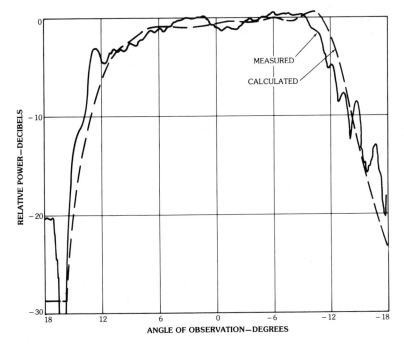

c

Fig. 62, *continued.*

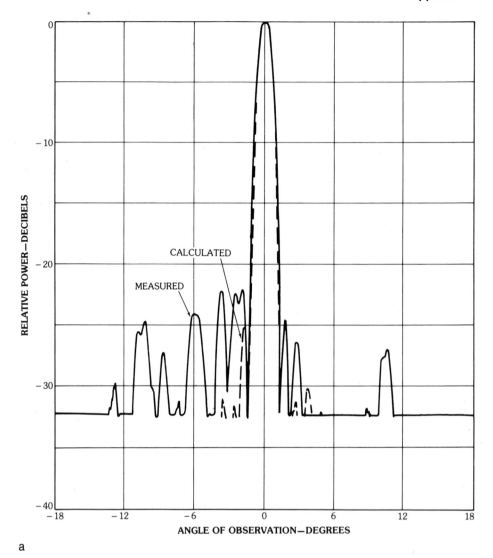

a

Fig. 63. Representative patterns for various beam scans at $f = 9.5$ GHz. (*a*) For a beam scan of $0°$. (*b*) For a beam scan of $-5°$. (*c*) For a beam scan of $-9°$. (*Courtesy Hughes Aircraft Co., Fullerton, Calif.*)

Fig. 65 depicts the effect of scanning the ideal subarray pattern with respect to the array pattern. This ideal subarray pattern results in no gain reduction as the main beam is scanned within the subarray pattern and has complete grating lobe suppression outside the subarray pattern. The completely overlapped space-fed subarray technique achieves a close approximation to the ideal subarray pattern. The amplitude distribution over the radiating aperture for each subarray input terminal is a discrete truncated $(\sin x)/x$ function which gives an approximate rectangular subarray pattern.

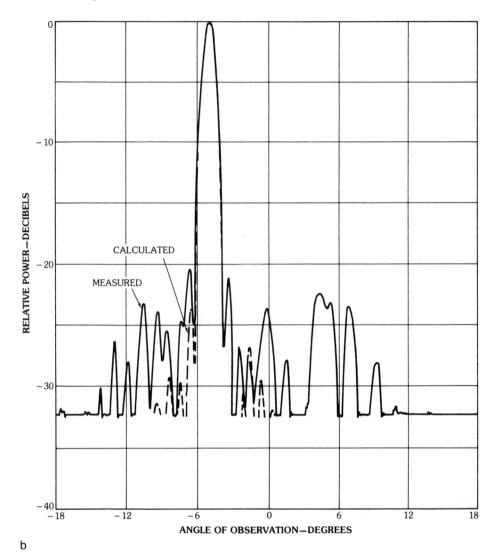

b

Fig. 63, *continued.*

Completely overlapped subarrays using transform networks are described in Chapter 13.

DuFort's Optical Technique—DuFort's optical technique [49] for broadbanding a phased array is illustrated in Fig. 66. It is composed of a relatively large feed-through aperture lens with 0° to 360° type phase shifters, an intermediate passive lens with fixed phase shifters and fixed time delayers, a small feed array with variable phase shifters and variable time delayers, and a summing network.

The aperture lens and the feed lens have a common focus, with a magnification $m = F/f$.

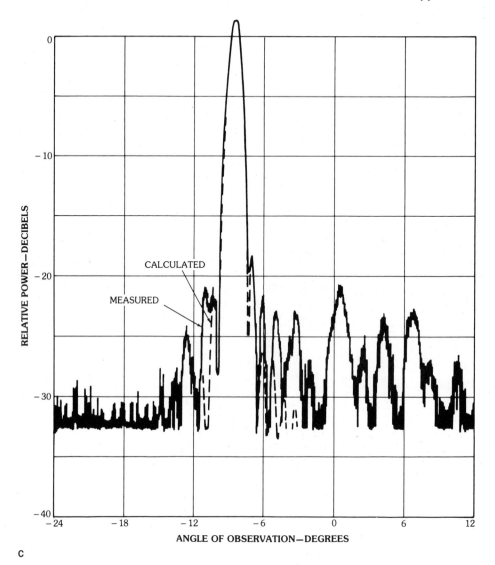

c

Fig. 63, *continued.*

The feed array is located at a distance h behind the confocal feed lens as shown.

The operation of this lens technique is as follows: A plane wave is incident on the lens aperture* x_1, making some general angle θ_0 with respect to the aperture lens normal. The 0° to 360° type phase shifters in the aperture lens are set such that at center frequency the received rays are focused to the center-frequency focal point as shown. These focused rays then diverge and are incident as a spherical

---

*Note that x_1 is an axis or aperture, not a specific point.

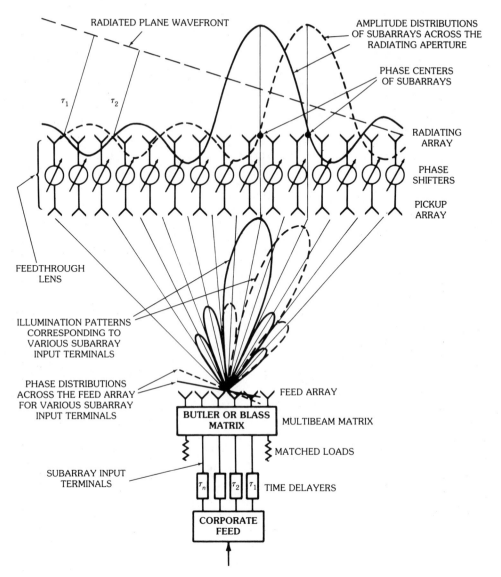

Fig. 64. Completely overlapped subarray pattern antenna system. (*After Tang [47]*)

phase front on the confocal feed lens. The fixed phase shifters and fixed time delayers in the confocal feed lens are such that the divergent incident rays are collimated and normal to the feed lens aperture or the x_2 axis. The feed array then receives these normally incident rays and coherently sums them through the appropriate settings of variable phase shifters and variable time delayers. Off the center frequency the aperture-lens phase shifter settings cause the received plane wave to be defocused from the center-frequency focal point and shifted laterally to an off-frequency focal region.

The divergent rays incident on the confocal feed lens are displaced along the x_2

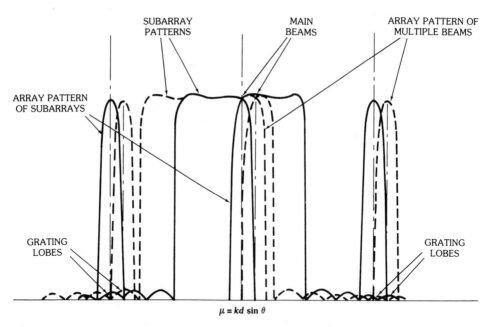

SUBARRAY
PATTERNS

MAIN
BEAMS

ARRAY PATTERN OF
MULTIPLE BEAMS

ARRAY PATTERN
OF SUBARRAYS

GRATING
LOBES

GRATING
LOBES

$\mu = kd \sin \theta$

Fig. 65. Effect of scanning the ideal "rectangular" subarray pattern with respect to the array pattern. (*Courtesy Hughes Aircraft Co., Fullerton, Calif.*)

axis; however, by a judicious choice of the fixed phase shifters and fixed time delayers in the confocal feed lens, the displaced rays emerge from the x_2 axis still focused to the feed array aperture, to the first order in fractional frequency bandwidth. This design thus preserves the even-amplitude aperture lens distribution and minimizes the spillover loss. DuFort analyzes several cases and compares the results with the results one would obtain using a discrete subarray approach. He concludes that this optical approach provides less gain than the corresponding discrete subarray system, but that the side lobe performance of the optical approach is substantially better.

For example, a sample case has been calculated for cosine amplitude distribution, $F/D = 1/\sqrt{2}$, using eleven time delayers, at the extreme scan angle of 50° and at a frequency 5 percent above center frequency. The radiation pattern for the optical case is given by Fig. 67. The normalized gain is down about 1.7 dB from the maximum value of 0.816 due to cosine tapering. The far-out side lobe level is −18.5 dB, with other side lobes mostly below −25 dB. Spillover loss is only 1.1 percent. The gain loss is due to off-axis phase error that causes beam broadening. The corresponding case for the discrete subarray is shown in Fig. 68. The main beam is narrower than the optical case and the gain is down only 0.7 dB, but the grating lobe level is up to −9.9 dB, with several above −20 dB.

For more details the reader is referred to DuFort [49].

Wide Field of View True-Time-Delay Antenna Systems

Most unconstrained feeds that offer true-time-delay performance are optical in nature. The transit time between the incident wavefront, from any angle within the

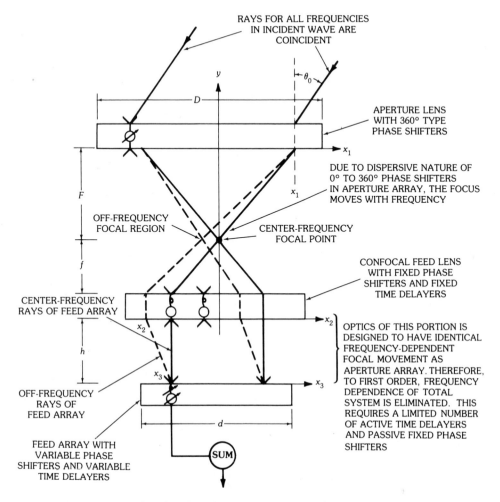

RAYS FOR ALL FREQUENCIES
IN INCIDENT WAVE ARE
COINCIDENT

θ_0

D

y

APERTURE LENS
WITH 360° TYPE
PHASE SHIFTERS

x_1

DUE TO DISPERSIVE NATURE OF
0° TO 360° PHASE SHIFTERS
IN APERTURE ARRAY, THE FOCUS
MOVES WITH FREQUENCY

F

x_1

OFF-FREQUENCY
FOCAL REGION

CENTER-FREQUENCY
FOCAL POINT

f

CONFOCAL FEED LENS
WITH FIXED PHASE
SHIFTERS AND FIXED
TIME DELAYERS

CENTER-FREQUENCY
RAYS OF FEED ARRAY

x_2

x_2

h

OPTICS OF THIS PORTION IS
DESIGNED TO HAVE IDENTICAL
FREQUENCY-DEPENDENT
FOCAL MOVEMENT AS
APERTURE ARRAY. THEREFORE,
TO FIRST ORDER, FREQUENCY
DEPENDENCE OF TOTAL
SYSTEM IS ELIMINATED. THIS
REQUIRES A LIMITED NUMBER
OF ACTIVE TIME DELAYERS
AND PASSIVE FIXED PHASE
SHIFTERS

x_3

x_3

OFF-FREQUENCY
RAYS OF
FEED ARRAY

d

FEED ARRAY WITH
VARIABLE PHASE
SHIFTERS AND VARIABLE
TIME DELAYERS

SUM

Fig. 66. Components of optical broadband phased array. (*Courtesy Hughes Aircraft Co., Fullerton, Calif.*)

FOV, to the corresponding collecting feed is a constant, regardless of the actual path that is taken for any given ray. Consequently, an extremely short time pulse, approaching a delta function, could, in principle, be transmitted or received by such an antenna system without distortion or spreading in the time domain. In practice, however, the instantaneous bandwidth is determined by the components that are used in devising such a system. If waveguides are used, or other types of elements, they will not possess infinite bandwidth when impedance matching or grating lobe phenomena are addressed. If a switching matrix is used to switch between beams, it will function only over some finite bandwidth. Thus the instantaneous bandwidth limitation of a true-time-delay system will not be established by the optics of the system but rather by the bandwidth of the components used in implementing it. For example, all parabolic reflector systems and dielectric lenses that are not zoned are true-time-delay antenna systems. Their

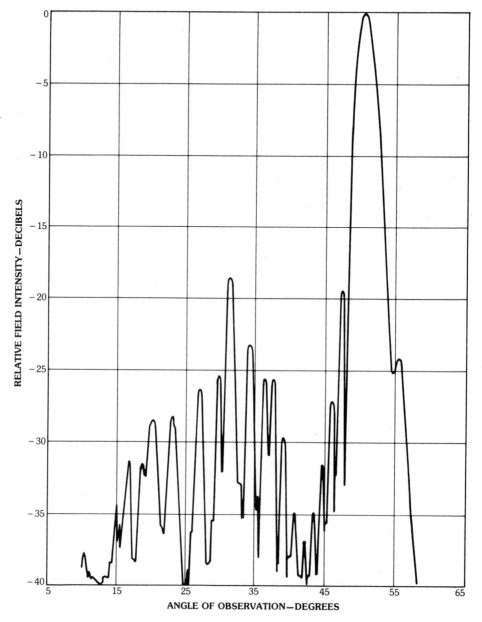

Fig. 67. Optical system radiation pattern. (*Courtesy Hughes Aircraft Co., Fullerton, Calif.*)

bandwidth, both instantaneous and tunable, is determined by the bandwidth of the feed system, or, perhaps to some extent in the case of dielectric lenses, the bandwidth may be determined by the impedance-matching transformers on the lens surfaces, as well as the primary feed(s).

Conventional monopulse feeds for conventional reflectors and lenses are well documented in the literature and will not be covered here, since most of these

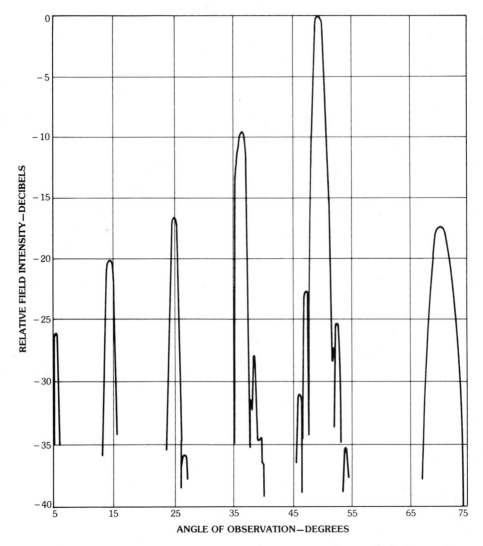

Fig. 68. Discrete subarray radiation pattern. (*Courtesy Hughes Aircraft Co., Fullerton, Calif.*)

antenna systems do not offer wide-angle performance. The purpose of this sub-section is to consider and illustrate some true-time-delay approaches that can be used to feed and phase a planar array over a wide angular FOV, e.g., a 90° to 120° conical FOV. Since this section is on unconstrained feeds it may seem out of place to discuss systems that are partially constrained; however, if free space exists between the feeds and the final aperture, then they are considered unconstrained in this context.

True-time-delay antennas can be further subdivided into those that have simultaneous multiple-beam capability and those that do not. Simultaneous multiple-beam capability is important for high-data-rate systems. An example of a true-time-delay system without simultaneous multiple-beam capability is a phased array with true-time-delay phase shifters for each of the radiating elements. An example of a true-time-delay phase shifter is a digital phase shifter in which the phase shift is achieved by switching in the proper length of TEM transmission line. In this example there is only one beam for a given set of phase shifter settings. Examples of antennas with simultaneous multiple beams which are not true-time-delay system are the Maxson (Blass)* tilt transmission-line-fed traveling-wave array [9] and the Butler hybrid-matrix multiple-beam array [3]. Both of these produce multiple beams in only one plane.

True-time-delay, high-gain, simultaneous-multiple-beam antennas with polarization diversity are necessarily quite complex because they are in reality many antennas sharing a common radiating aperture. The advantages of this over using many separate antennas is obvious because so many of the costly components, e.g., transmit/receive modules of an active array, are shared by each of the antennas that make up the multiple-beam antenna system.

Wide-Angle Multiple-Beam Constrained Lens—The design of a true-time-delay, simultaneous-multiple-beam, constrained-lens system with polarization diversity is described here. Other antenna systems that achieve the same goals are the Luneburg lens and spherical reflector systems. Each system has its advantages and disadvantages, depending on the particular application in which cost is usually the determining factor. Studies have shown that the optimum choice depends on the desired beamwidth, the frequency band of operation, and the desired coverage. A general statement as to the superiority of one method over another cannot be made. The system described here, however, has one advantage over most other systems in that the beam-positioning (feed) elements are separate from the pickup or transfer elements. This facilitates component design and packageability as well as offering electrical advantages.

The true-time-delay, multiple-beam, spherically symmetrical, constrained-lens antenna described here can be used as a radiating antenna by itself or can be used as a phasing device for a multiple-beam planar array. For extremely high-range resolution a true-time-delay system is required to accommodate the wide instantaneous frequency spectrum of extremely short pulses. If the device is to be used as a phasing device for a planar array as shown in Fig. 69, it should illuminate the array aperture with the same amplitude distribution for all beam positions. That is, the illuminated portion of the aperture should not scan as the beam is scanned. Concentric (spherical) lenses or reflector systems, such as the hemispherical reflector system shown in Fig. 69, automatically satisfy the nonscanning-aperture requirement because the central ray for every beam passes through the center of the aperture.

Concentric systems have inherent wide-scan capability because they are spherically symmetrical, and hence the beam-forming device, by itself, has phase errors independent of scan angle. However, when the latter is used as a phasing

---

*The Maxson or Blass array can be made true time delay but in doing so, much of its appealing simplicity, low cost, and compact packaging is sacrificed. See Hansen [4], vol. 3, p. 254.

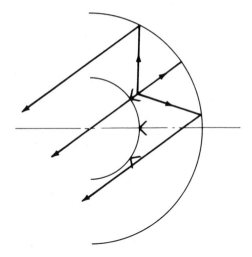

a

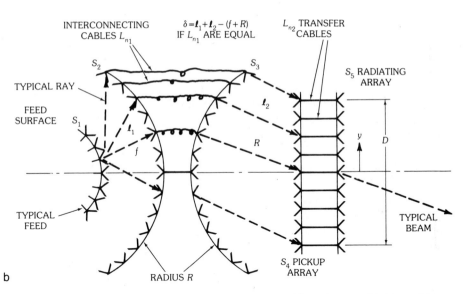

b

Fig. 69. Hemispherical reflector and lens counterpart. (*a*) Reflector. (*b*) Lens. (*Courtesy Hughes Aircraft Co., Fullerton, Calif.*)

device for a planar array the phase errors may actually decrease as the beam is scanned off broadside. The reason for this is that only the central portion of the phase front is intercepted by the planar array. This is true to the extent that phase errors at one plane can be projected to phase errors at another plane.

A hemsipherical lens [50] developed at Huges Aircraft Company, Fullerton, California, is the transmission (lens) counterpart to a hemispherical reflector system, as shown in Fig. 69a. If the current distribution on the reflector of the hemisphere were transferred via pickup elements with equal line lengths to an identical transfer array, the radiation of the transfer array would be identical with that from the reflector as indicated in the schematic diagram, Fig. 69b. The diagram shows such a lens used as a phasing device for a planar array. Because of spherical symmetry the basic lens has virtually only even (e.g., quadratic, quadric, etc.) phase errors. These even phase errors can be corrected exactly for any single beam position by adjustment of line lengths L_{n_1} or L_{n_2}. By doing this, however, one causes the system to deviate from being spherically symmetrical and the phase error now depends on scan position. In fact, odd phase errors may now arise for scanned beams. By line length adjustment, however, the total phase errors, both odd and even, can be made considerably less than that for the uncompensated lens.

The optimum parameter values depend on the size of the antenna in wavelengths and the desired amount of maximum scan. For narrow beamwidths and large maximum scan angle, S_1, S_2, and S_3 should be spherical.

If an angular correspondence between feed position angle and beam direction is desired, S_4 and S_5 should have the same size. That is, the ratio of the beam scan angle to feed position angle is the ratio of the linear dimensions of S_4 and S_5. With no line length phase compensation the path length error curves in the plane of scan for various scan angles are given in Fig. 70 (for a typical f/R). As mentioned before, because of spherical symmetry the aperture phase errors are essentially even (e.g., quadratic quadric) for all scan angles. A similar set of phase error curves can be calculated in the plane perpendicular to the plane of scan.

The common portion of the even phase errors can be removed by adjustment of the output line lengths L_{n_2}, e.g., the line lengths L_{n_2} are adjusted to remove the path length labeled "compensation curve" shown in Fig. 70; it also shows a plot of maximum path length error after line length compensation for a typical set of parameters. To illustrate the relative sizes involved, for a 70λ aperture (1° beam for typically tapered distribution), the diameter of the radiating array is equal to the radius of the lens sphere and the maximum phase error would be about $\lambda/16$ for beam scan to ±45°.

An experimental model of such a beam-forming lens was built and tested. Figs. 71a and 71b are photographs of this model. Antenna patterns were taken for scan in vertical, horizontal, and 45° planes. Fig. 72a shows typical patterns for various scan angles in the plane of scan. Fig. 72b shows typical patterns taken in planes normal to the plane of scan. Fig. 72c shows patterns for 45° diagonal scan over a range of frequencies covering approximately a 40-percent bandwidth. These test results establish the feasibility of this type of three-dimensional, multiple-beam-forming device.

Below is a summary of the properties of the lens:

1. Wide angular coverage
2. Simultaneous multiple true-time-delay beams
3. Nonscanning aperture, hence efficient as phasing device for planar array
4. Polarization diversity capability. With dual-polarized lens elements the polarization is determined solely by the polarization of the feed
5. The number of beams and beam positions are independent of the number and locations of the transfer or pickup elements of the lens since these elements are not used for a dual purpose of feed and transfer or pickup elements

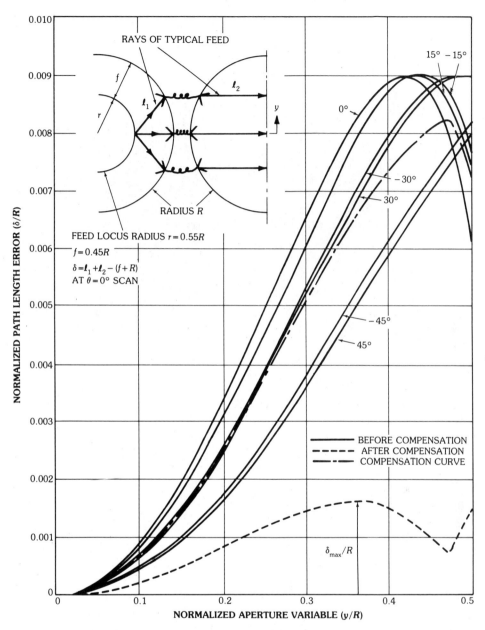

Fig. 70. Path length error versus aperture variable before compensation, and maximum path length error after compensation. (*Courtesy Hughes Aircraft Co., Fullerton, Calif.*)

A consequence of the property stated above is that the beams can be placed in any desired manner in space by proper placement of feeds on surface S_1 of Fig. 69b. For example, if the device is used to phase a planar array, the feeds can and should have variable spacing to keep a constant "crossover level" between adjacent beams since the radiated beam broadens as the beam is scanned off broadside to the

TYPICAL ARRAY ELEMENT
(DIPOLE IN CAVITY)

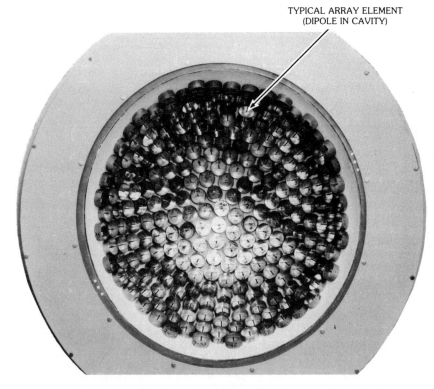

a

Fig. 71. Beam-forming lens. (*a*) Front view of lens. (*b*) Side view of partially interconnected lens. (*Courtesy Hughes Aircraft Co., Fullerton, Calif.*)

planar array. This capability of arbitrary beam positioning facilitates the implementation and computation of monopulse operation between adjacent beams for increased angular accuracy. Since the projected aperture decreases as the cosine of the scan angle, the feed directivity must increase in like manner to keep the feed spillover loss a constant. Thus the feed directivity (hence the feed aperture) must increase as the secant of the scan angle, and for a constant beam crossover level this is just the amount of space available for the feed. Thus the independence of feed and lens element allows the antenna efficiency to be the same (equal feed spillover) for all beam positions.

For the detailed design procedure see McFarland and Ajioka [50].

Modifications of the Basic Lens—Two modifications of the basic lens have been studied in an attempt to improve the aperture efficiency or to increase the usable portion of the approximate plane wave. The first scheme is similar to a Cassegrain technique, as illustrated in Fig. 73a.

This scheme slightly increases the usable portion D of the S_3 surface diameter $2R$, as is shown by Fig. 73b, where it is compared with the basic lens. However, it has the disadvantage of using the radiators covering surface S_2 for the dual purpose of acting as feeds and transfer radiators. To accomplish this function would require the appropriate circuitry in the interconnecting cables L_{n_1}.

PICKUP   INTERCONNECTING LINES L_{n_1} TRANSFER
HEMISPHERE S_2 HEMISPHERE S_3

b

Fig. 71, *continued.*

The second scheme is similar to a Mangin Mirror, as shown in Fig. 74. It uses a constant concentric dielectric correcting lens at the feed side of the basic lens.

This scheme appreciably increases the usable portion D of the S_3 surface diameter $2R$, as is shown by Fig. 73b, where it is compared with the basic lens.

This modification retains all of the merits of the basic lens while considerably improving the output phase.

A similar design procedure can be followed in designing the modified versions of the lens as has been used for the basic lens.

High-Resolution Hemispherical Reflector Antenna (HIHAT)—The HIHAT multiple-beam antenna system was invented [51] at Hughes Aircraft Company in the early 1960s by Louis Stark of the Ground Systems Group. Fig. 75 illustrates the

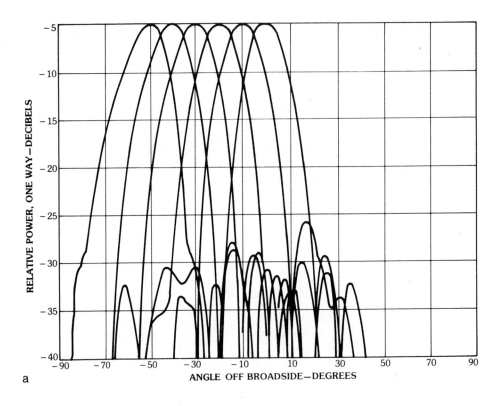

a

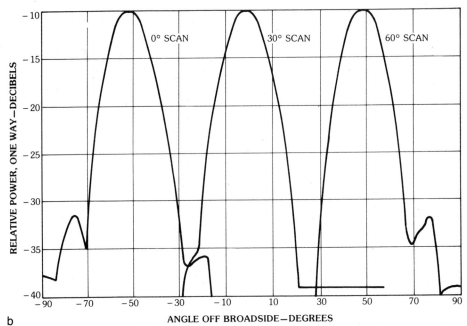

b

Fig. 72. Typical beam-forming lens antenna patterns. (*a*) In plane of scan, showing wide-scan capability. (*b*) Normal to plane of scan for various scan angles. (*c*) Over frequency range (beam scanned 45° in diagonal plane). (*Courtesy Hughes Aircraft Co., Fullerton, Calif.*)

19-86

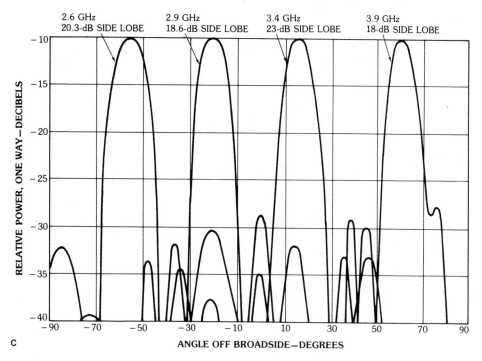

Fig. 72, *continued.*

basic concept. A spherical reflector is fed by a concentric array of feeds located on a sphere of slightly greater than half that of the spherical reflector. This concentric array of duplexing feeds is dual circularly polarized. This novel feature enables the aperture blockage that would ordinarily be presented by the feed array to become essentially "invisible"; thus a given feed or feed cluster is selected by a beam-switching matrix to be excited as a circularly polarized primary feed, e.g., right circular. The primary feed(s) illuminates a portion of the spherical reflector and becomes nearly collimated with the opposite sense circular polarization (e.g., left circular) after reflection off the spherical reflector. The dualpolarized duplexing feed array picks up the nearly collimated rays in the left circularly polarized channel (for this example) and transfers the nearly collimated wavefront to a spherical transfer array. The spherical transfer array is designed to have its radiators' phase centers lie on the same radius sphere as for the duplexing array. Equal line lengths are used for this transfer. From the spherical transfer array, radiation into free space could take place, or it could be used to phase a planar array, as shown in Fig. 76. In the latter case the common portion of the spherical aberration could be removed prior to radiation into free space, thereby improving the flatness of the radiated phase front. This technique was discussed in the previous section and is applicable here. Excellent radiation patterns are obtainable since the central portion of the reflected phase front is highly correctable by using variable line lengths between the pickup and radiating planar array. Fig. 77 shows typical measured Σ and Δ patterns using two adjacent feeds for primary illumination.

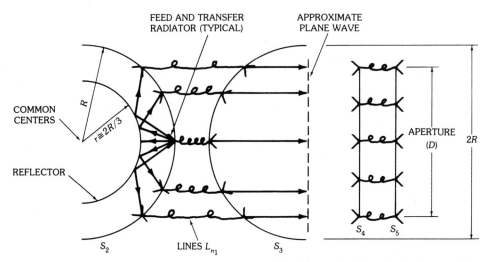

Note: S_4, in reality, is located at the diameter of S_3

a

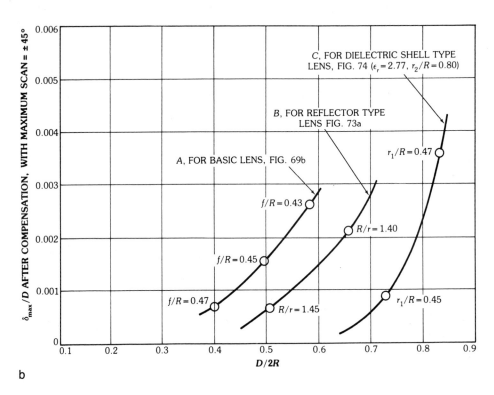

b

Fig. 73. Cassegrain modification of basic lens. (*a*) Cassegrain reflector type variation of basic lens. (*b*) Normalized maximum path length error, after compensation, versus usable portion of the output hemisphere diameter as radiating aperture. (*Courtesy Hughes Aircraft Co., Fullerton, Calif.*)

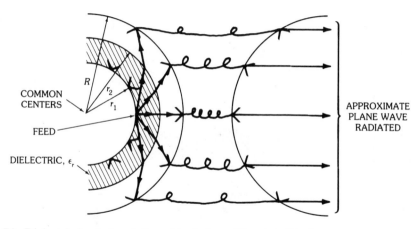

Fig. 74. Dielectric-loaded variation of basic lens. (*Courtesy Hughes Aircraft Co., Fullerton, Calif.*)

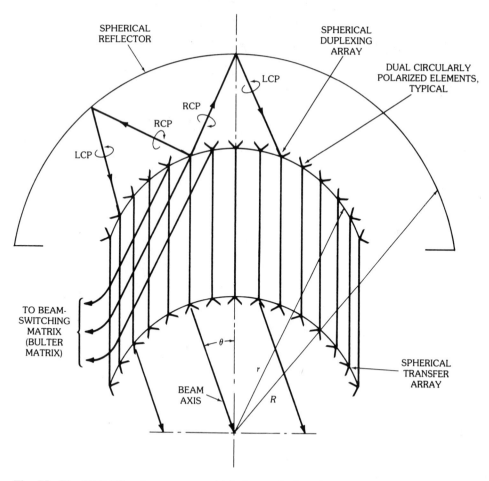

Fig. 75. The HIHAT antenna as a multiple-beam device. (*Courtesy Hughes Aircraft Co., Fullerton, Calif.*)

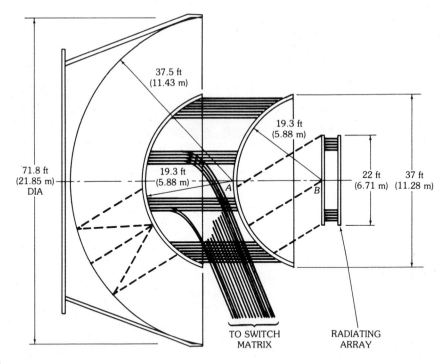

a

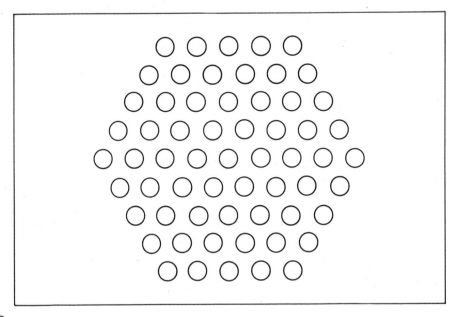

b

Fig. 76. The HIHAT used to phase a planar array. (*a*) System setup. (*b*) Radiating array. (*Courtesy Hughes Aircraft Co., Fullerton, Calif.*)

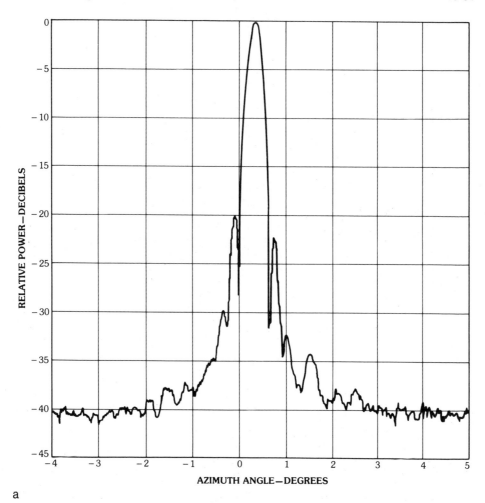

a

Fig. 77. Typical measured Σ and Δ patterns using two adjacent feeds (in the plane of scan). (*a*) Σ pattern. (*b*) Δ pattern. (*Courtesy Hughes Aircraft Co., Fullerton, Calif.*)

5. Optical Transform Feeds

In certain feed systems the input to the feed and the resulting aperture distribution of the array are related by one or more Fourier transforms. This has been evident in many of the limited-scan techniques that were described in Section 4. These feeds are called *transform feeds* [47, 49, 52]. As stated earlier it is prohibitively expensive to have active elements, such as phase shifters,* power amplifiers, low-noise amplifiers, or adaptive control loops, for every radiating element in a large phased array. But if we back down to having active elements only

---

*For broad instantaneous bandwidths, true-time-delay phase shifters may be required.

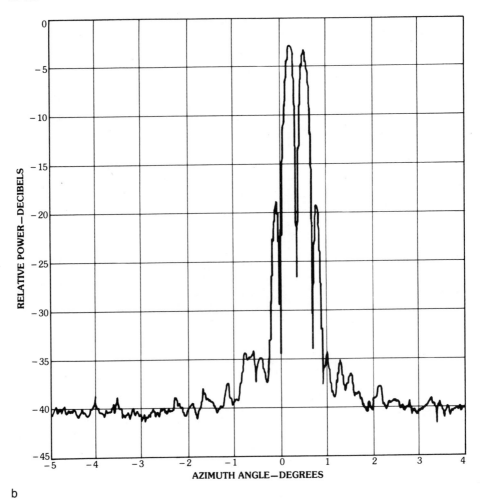

b

Fig. 77, *continued.*

at one feed point of the antenna, we do not have an electronically steerable antenna at all. It is therefore necessary for us to back down to an intermediate level where the number of devices becomes affordable. In doing so, a certain amount of performance must be sacrificed, i.e., the amount of scan coverage is reduced by the ratio of the number of active elements used to the number of active elements required to scan the larger field of view.

Two basic concepts of transform feeds that involve several Fourier transforms are depicted in Figs. 78 and 79. The first concept is a fully constrained feed that uses a small Butler matrix to feed a large Butler matrix whose outputs feed the radiating elements of the antenna. The second concept is an example of optical Fourier transformers in which a large lens is fed by a small lens. The active elements (e.g., phase shifters) are placed in a small phased array that feeds or illuminates the small

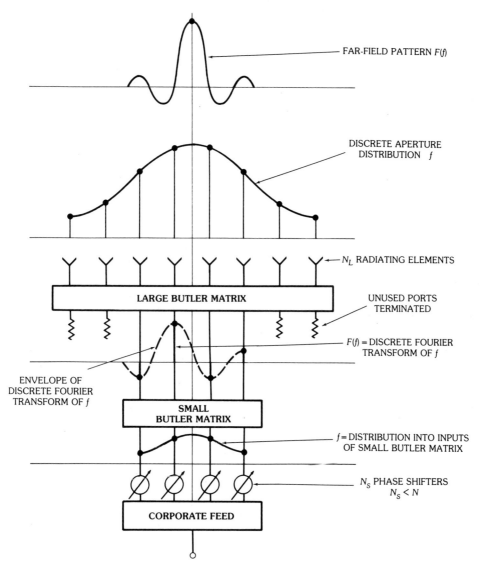

FAR-FIELD PATTERN $F(f)$

DISCRETE APERTURE
DISTRIBUTION f

N_L RADIATING ELEMENTS

LARGE BUTLER MATRIX

UNUSED PORTS
TERMINATED

$F(f) =$ DISCRETE FOURIER
TRANSFORM OF f

ENVELOPE OF
DISCRETE FOURIER
TRANSFORM OF f

**SMALL
BUTLER MATRIX**

$f =$ DISTRIBUTION INTO INPUTS
OF SMALL BUTLER MATRIX

N_S PHASE SHIFTERS
$N_S < N$

CORPORATE FEED

Fig. 78. Example of constrained transform feed. (*Courtesy Hughes Aircraft Co., Fullerton, Calif.*)

matrix or optical device. Figs. 78 and 79 are conceptual schemes only and may not represent practical antenna systems. For practical transform antennas refer to Section 4 and the quoted literature.

The ideal Butler matrix is a perfect discrete Fourier transformer. A reflector or lens that focuses a plane wavefront to a point is also a Fourier transformer, but not perfect. The Butler matrix will be briefly discussed, while the optical transformers will be discussed in greater length because of their imperfections.

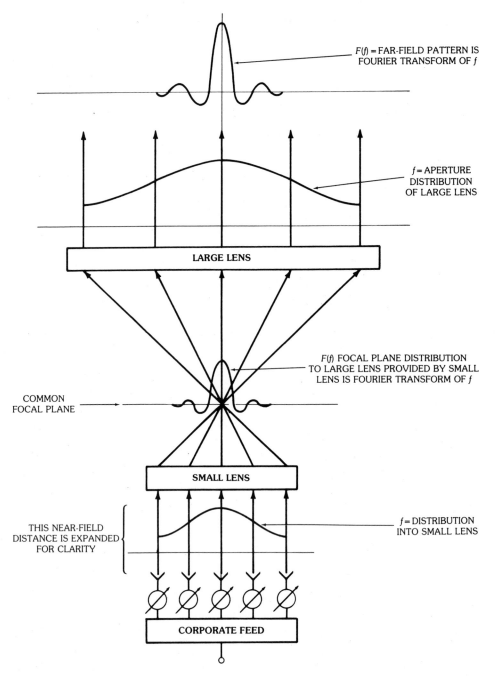

Fig. 79. Example of optical transform feed. (*Courtesy Hughes Aircraft Co., Fullerton, Calif.*)

Butler Matrix as a Fourier Transformer

The inputs and outputs of an N-element Butler matrix are related by (Section 2)

$$g_{nm} = \frac{\exp j\{(2\pi/N)[n - (N + 1)/2][m - (N + 1)/2]\}}{\sqrt{N}}$$

referenced to the center of the array, $(N + 1)/2$ spaces, where g_{nm} is the field amplitude at the nth output element due to unit excitation at the mth input or beam port. That is, a delta-function input (each port is a discrete delta function) results in a plane wave (linear progressive phase) over all the outputs. By the principle of superposition an arbitrary input distribution $f(m)$ will result in a superposition of discrete plane waves weighted by $f(m)$, resulting in the Fourier transform

$$g(n) = \frac{1}{\sqrt{N}} \sum_{m}^{N} f(m) \exp j\{(2\pi/N)[n - (N + 1)/2][m - (N + 1)/2]\}$$

Fig. 78 illustrates the use of Butler matrices in a transform feed application. The figure is self-explanatory.

Optical Devices as Fourier Transformers

Fig. 79 is an idealized optical analog to the matrix method of Fig. 78. With idealized assumptions to be discussed shortly, Fig. 79 is self-explanatory. Refer to Fig. 80 for the following discussion.

For purposes of discussion consider a wide-angle lens or reflector that will focus over a relatively large region of the focal plane (see Section 4, under "Wide-Angle Multiple-Beam Constrained Lens"). A point-source feed at some point on the focal plane will radiate a spherical wave that will be collimated into a continuous plane wave over the aperture of the lens. Then, just as in the Butler matrix, a point-feed (delta function) input will result in a plane wave output. Again, by the principle of superposition, any focal plane distribution can be generated by a superposition of point feeds over the usable portion of the focal plane.* Again, we have an expansion into plane waves radiating in different directions in space. In general, any device that can focus electromagnetic energy from one point to another is a Fourier transformer. Usually one of the points is infinity.

It should be mentioned that the focal region field is planar on the focal plane [45]. A beam waveguide is an example where both focii are finite. A Gaussian distribution where f and $F(f)$ are identical is used in this application.

Roughly speaking, the primary feed pattern that illuminates a lens or reflector is the Fourier transform of the feed aperture distribution. After collimation this primary feed pattern becomes the aperture distribution of the lens or reflector. Now the Fourier transform of this aperture distribution is the far-field pattern of the lens or reflector. Thus, for real, even feed distributions, the far-field pattern should be identical in shape with the aperture distribution of the primary feed. It is

---

*The term "usable portion" means that off-focal point aberrations are tolerable over that region.

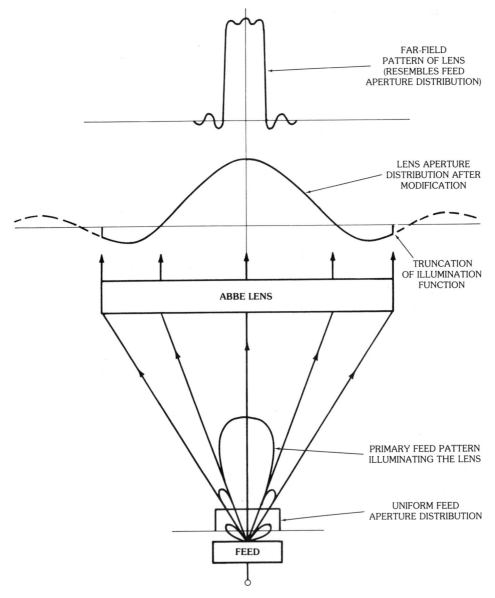

FAR-FIELD
PATTERN OF LENS
(RESEMBLES FEED
APERTURE DISTRIBUTION)

LENS APERTURE
DISTRIBUTION AFTER
MODIFICATION

TRUNCATION
OF ILLUMINATION
FUNCTION

ABBE LENS

PRIMARY FEED PATTERN
ILLUMINATING THE LENS

UNIFORM FEED
APERTURE DISTRIBUTION

FEED

Fig. 80. Lens configuration used for discussion. (*Courtesy Hughes Aircraft Co., Fullerton, Calif.*)

assumed that the lens or reflector is in the far field of the feed, the feed spillover effects are negligible, and the amplitude modification due to the $1/r^2$ spreading of the rays and the modification of the amplitude due to the lens configuration are also negligible. The latter two assumptions are the most significant. Since all the radiation from the primary feed is not intercepted by the finite lens, the primary illumination function is truncated. The effect of this truncation is discussed later.

The other major effect is the $1/r^2$ effect and the modification of amplitude of the illumination function by the lens design. Depending on the lens configuration the net result may be an additional taper to the incident illumination, or it may add an additional inverse taper. For example, a parabolic reflector or a hyperbolic dielectric lens with one flat face will give an additional tapering effect, while an elliptical dielectric lens with a spherical inner surface results in an inverse tapering effect. Lenses obeying the Abbe sine condition have an inverse tapering effect in power that varies as the secant of the angle between the axis of the lens and a general ray from the focal point. The Abbe sine condition states that the distance of a collimated exit ray from the axis of the lens is proportional to the sine of the angle θ that the corresponding ray from the focal point makes with the lens axis. This results in a modifying factor $k\sqrt{\sec\theta}$ to the illumination function.

To help one get a "feel" for the aforementioned effects a special case of a uniform feed distribution is calculated. This is a relatively stringent test because highly tapered distributions would be less affected by feed pattern truncation due to the low amplitude at the point of truncation. Aside from these effects the uniform feed distribution should give a uniform (sector-shaped) far-field pattern from the lens. Fig. 81 shows a calculated far-field pattern from a lens that satisfies the Abbe

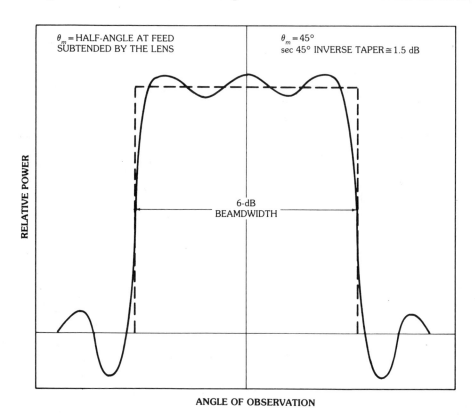

Fig. 81. Far-field pattern of a lens obeying the Abbe sine condition when the primary feed pattern is truncated to the peaks of the first side lobes of the primary pattern. (*Courtesy Hughes Aircraft Co., Fullerton, Calif.*)

sine condition. The lens truncated the $(\sin u)/u$ type primary feed pattern at the peaks of the first side lobes. This is a rather severe truncation. One would expect that at least several side lobes of the illuminating function would have to be included. However, it is seen that, even so, the resulting far-field pattern does resemble a sector-shape pattern with small ripples. In this case the inverse tapering effect of the Abbe lens was 1.5 dB. It has also been shown that the sector shape and width are nearly frequency independent. This is also a consequence of the Fourier transform. At a higher frequency the primary pattern is more directive, which effectively illuminates less of the lens aperture, thereby causing the final far-field pattern to broaden to compensate for the narrowing of the pattern due to the increase in frequency.

A three-dimensional case was both measured and calculated. A circular aperture feed with uniform aperture distribution was used to feed a lens. The far-field pattern of the lens was measured and calculated by the Fourier transform methods. A comparison of the measured and calculated far-field patterns is shown in Fig. 82. It can be seen that the expected sector shape was obtained and the measured and calculated patterns show excellent agreement.

This example, both measured and calculated parts, lends justification to the concept of an optical type antenna as a Fourier transformer.

6. Cylindrical Array Feeds

Numerous techniques have been conceived for the scanning of a cylindrical array [53–56]. The application of such an array may be for air traffic control, radar, and other uses requiring 360° of azimuthal coverage. Electronic scanning may be required in azimuth only, with perhaps some particular beam shape in elevation, or electronic scanning may be required in both planes with or without special elevation (and perhaps azimuth) beam shaping. Each particular system has its own set of requirements.

For the purpose of discussion, consider that the antenna is composed of a cylindrical array of identical line sources parallel to the z axis, as depicted by Fig. 83. The line sources may have phase shifters behind every element, or they may be passively fed with an azimuthal beam-forming network or lens. For good radiation patterns it is apparent that only a portion (180° sector or less) of the total number of line sources should be fed (and phased) for any given azimuthal beam direction. Consequently, some means must be devised to position a given amplitude distribution around the circumference of the array as a function of the desired azimuth direction of the main beam. The main differences between the various concepts are the techniques for generating and positioning this excitation.

Before proceeding further some terms should be defined, and assumptions stated. The term $P(R, \theta, \phi)$ represents the spherical coordinates of the point of observation in the far field (Fig. 83b). The term $P'(a, \phi'_n, z'_p)$ represents the cylindrical coordinates on a cylinder of radius a of a typical radiating element. The terms n and p are integers associated with (ϕ'_n, z'_p), using the notation of [53], except for the angle θ. It is assumed that the element is impedance matched for wide-angle scanning and the interelement spacing in both directions is close enough

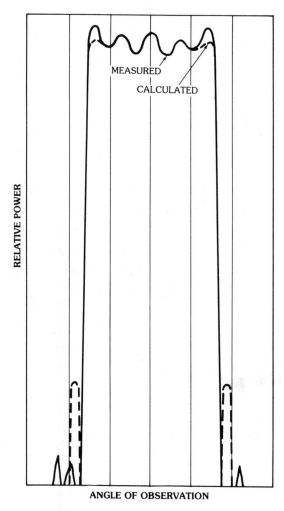

Fig. 82. Measured and calculated far-field patterns from a lens with a circular aperture feed with a uniform distribution. (*Courtesy Hughes Aircraft Co., Fullerton, Calif.*)

(near $\lambda/2$) to produce an ideal element factor $\sqrt{\cos\gamma}$, where γ is the angle between the normal to the array and the point of observation.

The far-field pattern is related to the surface illumination by

$$E(\theta,\phi) = \iint \varepsilon(\phi',z') \frac{e^{-jkR}}{R} \sqrt{\cos\gamma}\, d\phi'\, dz' \tag{23}$$

Let the complex array illumination function be

$$\varepsilon(\phi_n, z_p') = \ell(\phi_n')f(z_p') \tag{24}$$

Then the far field for the array is given [54] by

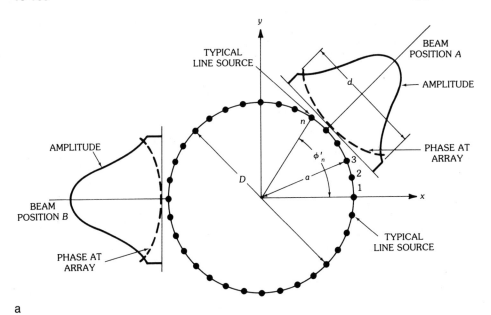

Fig. 83. Antenna as a cylindrical array of line sources. (*a*) Cylindrical array plan view. (*b*) Cylindrical array oblique view. (*Courtesy Hughes Aircraft Co., Fullerton, Calif.*)

$$E(\theta) = \sqrt{\cos\theta} \sum_p f(z_p') e^{jkz_p'\cos\theta} \sum_n \ell(\phi_n') \sqrt{\cos(\phi - \phi_n')} e^{jka\sin\theta\sqrt{\cos(\phi-\phi_n')}} \quad (25)$$

or

$$E(\theta, \phi) = L(\theta) R(\theta, \phi) \quad (26)$$

where

$$L(\theta) = \sqrt{\cos\theta} \sum_p f(z_p') e^{jkz_p'\cos\theta} \quad (27)$$

and

$$R(\theta, \phi) = \sum_n \ell(\phi_n') \sqrt{\cos(\phi - \phi_n')} e^{jka\sin\theta\cos(\phi-\phi_n')} \quad (28)$$

Thus the required array excitation to generate a main beam at (θ_m, ϕ_m) is given by

$$|f(z_p')||\ell(\phi_n')| \, e^{-jk[z_p'\cos\theta_m + a\sin\theta_m\cos(\phi_m-\phi_n')]} \quad (29)$$

Various techniques for generating this cylindrical array excitation will be discussed briefly. For purposes of clarity some of the diagrams to follow show the use of only eight-array elements.

Matrix-Fed Cylindrical Arrays

The various matrix-fed arrays can be readily understood by use of the following tutorial model.

Consider two back-to-back Butler matrices with controllable phase shifters in between them as shown in Fig. 84a. If an input port a_n of Butler matrix A is excited with any amplitude and phase, a corresponding port b_n of Butler matrix B will be excited with equal amplitude and phase when the phase shifters are all set equal. If the a_ns are weighted in amplitude and phase to any desired distribution, a corresponding distribution is achieved at the outputs b_n. Now if a linear phase distribution with a total phase of $m2\pi$ (m an integer) is applied across the array of phase shifters, the entire distribution over b_n will translate across the output in increments of one element spacing for each successive value of m. The translation of the output distribution by an increment less than one element spacing requires a total phase that is a nonintegral multiple K of 2π. That is, $|m| < |K| < |m| + 1$. In these cases there is no longer a one-to-one correspondence between an input port excitation and an output port excitation. Instead, if one input port is excited, not one but all output ports are excited to some extent. However, as would be expected the two adjacent ports corresponding to phases of $m2\pi$ and $(m + 1)2\pi$ are excited most strongly since they straddle the "phantom" element corresponding to the phase of $K2\pi$. In fact, the output distribution is a discrete sampling of a sharp $(\sin x)/x$ distribution as shown in Fig. 84b. The width of the "main lobe" of the

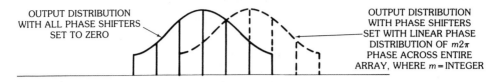

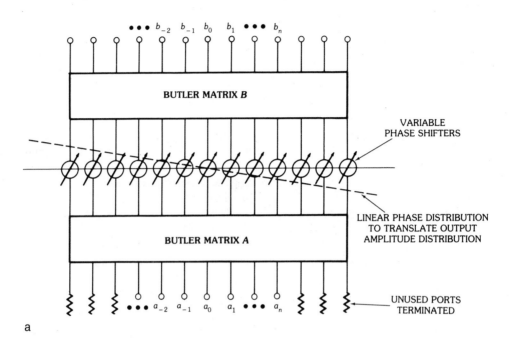

a

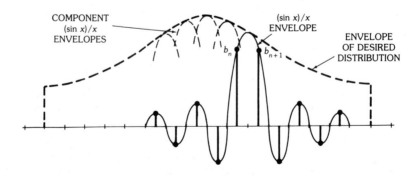

b

Fig. 84. Back-to-back Butler matrix feeding technique. (*a*) Tutorial model of matrix feeding. (*b*) Output excitation when only one input is excited and linear phase across bank of phase shifters is not an integral multiple of 2π. (*Courtesy Hughes Aircraft Co., Fullerton, Calif.*)

envelope $(\sin x)/x$ is two elements and the side lobe width is one element. The output distribution corresponding to phases of integral multiples of 2π consists of discrete values identical with those at the input to the first Butler matrix A because of the one-to-one correspondence between a_ns and b_ns. This is consistent with the sampled $(\sin x)/x$ output in these special cases; the sampling points are all at the nulls of the $(\sin x)/x$ function except at the peak of the main lobe, hence only that element is excited. For intermediate cases corresponding to nonintegral multiples of 2π, the output distribution is a superposition of sampled $(\sin x)/x$ functions of the kind just described. The net result is that the envelope of this output distribution is almost identical with that of the envelope of the designed output distribution, which is identical with the input distribution of matrix A.

Now, if the output ports b_n are connected by equal line lengths to a similar number of elements arranged uniformly along a circle of radius R, linear phase shifter adjustments will cause the distribution over the b_ns to move along the periphery of the circular array. If the phase distribution is such as to radiate a plane wavefront from the circular array, a directive beam is formed from the circular array that can be steered by means of changing the linear progressive phase of the phase shifters. It is obvious that this peripherally scanned distribution should not extend more than a semicircle because elements to the rear should not be excited. The input ports corresponding to these rearward-looking elements are shown terminated in Fig. 84a. In practice, less than 180° of the circular (or cylindrical) array is excited for one beam direction because of active element patterns, local grating lobe formation, and aperture matching problems associated with the edge elements that are locally phased for near end-fire. Also, for intermediate phasing, the elements near the edge of the active portion of the cylinder may have their corresponding sampled $(\sin x)/x$ lobes to the rear that are greater than desired for low side lobe and rear lobe patterns.

Collimation of the beam from the circular array requires a phase distribution of $\phi_n = kR(1 - \cos n\Delta\theta)$, where $\Delta\theta$ is the angle subtended by one-element spacing, R is the radius of the cylinder, and $k = 2\pi/\lambda$. This phase is easily incorporated in the weighting of the a_ns.

It is interesting to note that the amount of beam scan in terms of antenna beamwidths for each 2π of total phase change is equal to one beamwidth if the element spacing Δs is $\lambda/2$:

$$\Delta s = R\,\Delta\theta = \frac{\lambda}{2}, \qquad \Delta\theta = \frac{\lambda}{2R}$$

This is also the beamwidth for a uniform array of aperture $2R$.

The foregoing serves as a tutorial basis for describing some of the feeding techniques for cylindrical arrays to be discussed.

Sheleg Method—In the preceding discussion it is readily seen that the output of the first Butler matrix A (input to the bank of phase shifters) has an amplitude and phase distribution that is fixed (independent of beam steering); therefore it can be replaced by a passive feed network that produces that same distribution. This reduces to the Sheleg approach. A simplified schematic diagram of the approach for an eight-element array is shown in Fig. 85. Sheleg arrived at this solution from a different point of view. He expressed the far-field pattern in cylindrical modes in the form

$$E(\phi) = \sum_{-N}^{N} C_m e^{jm\phi}$$

where the number of elements is $2N$, and

$$C_m = 2\pi K j^m I_m J_m(ka)$$
$$K = \text{constant}$$

Sheleg made the observation that it is not necessary or even desirable to use all the m modes of the Butler matrix. This concurs with the foregoing tutorial discussion in that modes that excite elements to the rearward direction of the cylinder should not be used. As stated earlier, in practice less than half the number of available modes should be used to avoid local end-fire radiation at the extremes of the active semicircle. The Sheleg scheme is depicted for an eight-element array in Fig. 85.

This technique can be used to produce uniform or tapered illuminations for the cylindrical array. Sheleg gives computed and measured results for uniform, cosine, and cosine-squared array distributions with quite close correlation between

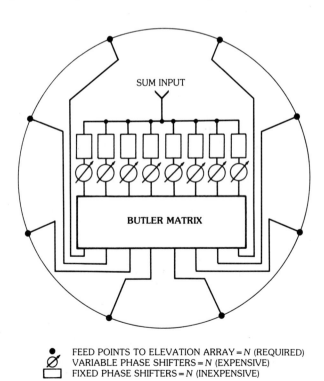

SUM INPUT

BUTLER MATRIX

● FEED POINTS TO ELEVATION ARRAY = N (REQUIRED)
Ø VARIABLE PHASE SHIFTERS = N (EXPENSIVE)
▭ FIXED PHASE SHIFTERS = N (INEXPENSIVE)

Fig. 85. Sheleg's approach. (*Courtesy Hughes Aircraft Co., Fullerton, Calif.*)

measurement and theory except in the remote side lobe region. He shows the patterns as a function of the number of modes used. Up to a point the patterns improve as more modes are used; then, beyond this point, the radiation pattern degrades somewhat. This has been explained by use of the tutorial model previously discussed.

For uniform distribution the azimuth pattern shape as a function of azimuth scanning of the main beam is shown to be reasonably invariant, i.e., it makes little difference whether the peak of the array distribution falls on an element, midway between two elements, or any fractional part thereof. This is also explained by use of the tutorial model. For more details the reader is referred to Sheleg's paper [56].

Wheeler Lab Approach—Fig. 86 depicts the Wheeler Lab approach [53, 54] to feeding a cylindrical array. It consists of a power divider followed by fixed phase shifts, variable phase shifters, a sequencing switch, an array of diode switches, and the radiating array.

This technique utilizes a *switching network* to step the amplitude distribution around the cylindrical array in coarse steps equal to the interelement angular spacing. The phase and amplitude distribution are established by the power divider and a set of fixed and variable phase shifters. *The variable phase shifters provide fine steering.*

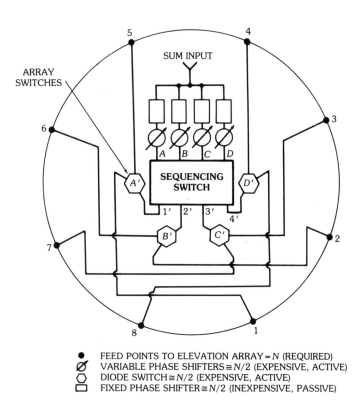

FEED POINTS TO ELEVATION ARRAY = N (REQUIRED)
VARIABLE PHASE SHIFTERS ≅ N/2 (EXPENSIVE, ACTIVE)
DIODE SWITCH ≅ N/2 (EXPENSIVE, ACTIVE)
FIXED PHASE SHIFTER ≅ N/2 (INEXPENSIVE, PASSIVE)

Fig. 86. Wheeler Lab approach. (*Courtesy Hughes Aircraft Co., Fullerton, Calif.*)

In this technique the ratio of the total number of azimuth elements to the number of active elements used at any beam position must be an integer.

In Fig. 86 the number of elements used at a given time is four since the corporate feed is shown as having four outputs. The desired output amplitude and phase distribution is designated A, B, C, D. For a given state of the sequencing switch, for example, elements $1, 2, 3, 4$ are excited with the complex amplitude A, B, C, D, respectively. Now suppose it is desired to move the distribution by one element, say, elements $2, 3, 4, 5$. With no sequencing switch the output amplitude distribution would be B, C, D, A, which is not the desired distribution. The sequencing switch reswitches the distribution back to A, B, C, D, the desired output. The same technique will allow the desired distribution, A, B, C, D, to be positioned all the way around the array in steps of one element at a time.

For the example shown, the sequencing switch would consist of four dpdt pin diode switches, as depicted by Fig. 87a. The complex inputs to the sequencing switch are labeled A, B, C, D. The outputs of the sequencing switch are labeled $1', 2', 3', 4'$. Each of the four pin diode switches has two possible states, labeled a or

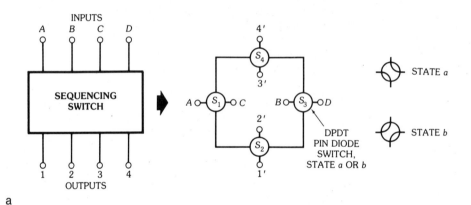

a

Arrangement	Sequencing Switch Outputs				Pin Diode Switching States			
	(Normal)				S_1	S_2	S_3	S_4
	1'	2'	3'	4'				
1	A	B	C	D	a	a	a	a
2	B	C	D	A	b	b	a	b
3	C	D	A	B	b	a	b	a
4	D	A	B	C	a	b	b	b
	(Inverted)							
5	D	C	B	A	b	b	b	b
6	A	D	C	B	a	a	b	a
7	B	A	D	C	a	b	a	b
8	C	B	A	D	b	a	a	a

b

Fig. 87. Sequencing switch circuit and logic. (*a*) Sequencing switch circuit. (*Courtesy Hughes Aircraft Co., Fullerton, Calif.*) (*b*) Logic of switch circuit.

b. The table of Fig. 87b shows four normal combinations and four inverted combinations of complex distributions emerging from the output of the sequencing switch, along with their corresponding switching states for the four dpdt switches labeled S_1, S_2, S_3, S_4. By using the inverted as well as the normal arrangements, beam positions mirror-imaged about the center line of the active cylindrical array normal are generated; consequently the number of phase states required by the phase shifters is halved. The switches labeled array switches in Fig. 86 have a single input with one of two possible outputs, depending on which radiating elements are to be excited.

For fine beam steering, for beam positions corresponding to that between elements for example, the variable phase shifters are used while the elements excited remain fixed at the same position as for a beam position corresponding to an angle through an element.

The power divider as depicted in Fig. 86 shows only a sum input. In actuality, both a sum and a difference port would typically be available by appropriate design of the power divider network.

Hughes Phased-Lens–Fed Approach—This technique utilizes a rotationally symmetric lens such as a Luneburg or Rinehart lens to perform the azimuth beam forming, a switching system to select lens feed points for coarse steering, and phase shifters at each lens output to the cylindrical array to do the fine steering. Fig. 88 schematically depicts this approach.

The sum (and difference, not shown) input is power-divided by the number of lens probes to be fed simultaneously, forming the lens primary feed illumination, two for the example of Fig. 88. Suppose lens probes 1 and 2 are chosen to form a beam diametrically opposite to the position lying between probes 1 and 2. Then switch *A* would select probe position 1 while switch *B* would select position 2. All the probes dispersed over the arc labeled 4, 5, 6, 7 would then pick up the focused field and switches labeled D, C, J, I would switch the fields to the elements labeled 4, 5, 6, 7 via phase shifters behind each element. The coarse beam position is determined by choosing which probes to excite. The fine beam steering is accomplished by the phase shifters behind each element.

The number of actively excited probes for a given beam position must divide integrally into the total number of elements; that is, the ratio of the total number of azimuth elements to the number of simultaneously excited lens probes must be an integer.

For example, if three probes needed to be excited simultaneously to achieve the desired primary lens feed distribution, then the total number of elements must be a multiple of three. If nine elements were chosen to satisfy this requirement, then the sum (and difference) input would have to be divided into three outputs followed by three switches each with three positions, rather than four as depicted in Fig. 88.

The lens probes are seen to have to perform the dual function of not only acting as primary feed elements but as transfer elements as well.

Matrix-Fed Conventional Lens Approach—For coarse beam stepping this technique is identical with the phased lens–fed approach described previously. For

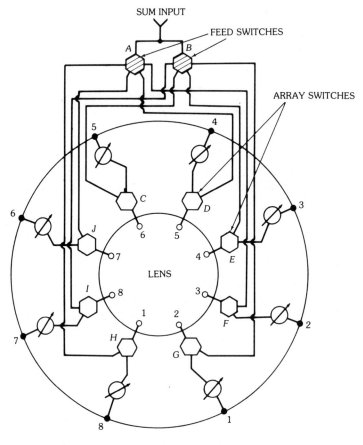

● FEED POINT TO ELEVATION ARRAY = N (REQUIRED)
Ø VARIABLE PHASE SHIFTERS = N (ACTIVE, EXPENSIVE)
◌ DIODE SWITCH = N (2:1) (ACTIVE, EXPENSIVE)
⬡ DIODE NETWORK = F (A:1) (ACTIVE, EXPENSIVE)
○ LENS FEED PROBE = N (INEXPENSIVE, PASSIVE)
N = TOTAL NUMBER OF RADIATING ELEMENTS
F = NUMBER OF LENS ELEMENTS USED TO FORM FEED PATTERNS OF LENS
A = NUMBER OF ACTIVE RADIATING ELEMENTS A < N/2

Fig. 88. Phased-lens–fed approach. (*Courtesy Hughes Aircraft Co., Fullerton, Calif.*)

fine steering a "Sheleg" technique is used on a relatively small feed array using an F-element Butler matrix and F variable phase shifters, where F is the small number of lens elements used to form the desired feed pattern. The feed distribution is translated as described before in Fig. 84a, which translates the phase center of the feed for fine steering.

Hughes Matrix-Fed Meyer Geodesic Lens—This technique [54, 57] is a variation of the matrix-fed conventional lens approach discussed in the previous section but utilizes a modified Meyer geodesic lens (figure of revolution) for the beam forming to eliminate active switches and uses passive diplexers or circulators instead. The

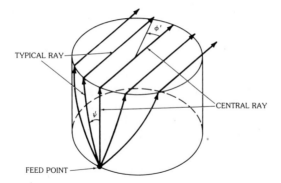

a

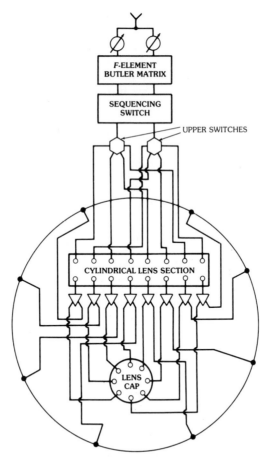

b

- ● FEED POINT TO ELEVATION ARRAY = N (REQUIRED)
- ∅ VARIABLE PHASE SHIFTER = F (ACTIVE, EXPENSIVE)
- ⊖ DIODE SWITCH = F (ACTIVE, EXPENSIVE)
- ○ LENS FEED PROBE = N (PASSIVE, INEXPENSIVE)
- ▽ DIPLEXER = N (PASSIVE, INEXPENSIVE)

Fig. 89. Meyer geodesic lens. (*a*) Basic Meyer lens. (*b*) Matrix feed. (*Courtesy Hughes Aircraft Co., Fullerton, Calif.*)

basic Meyer lens for 360° coverage showing typical ray paths from a feed point is given pictorially in Fig. 89a.

The matrix feed is depicted in Fig. 89b. The sum-and-difference corporate power divider, input phase shifters to the Butler matrix, the amplitude sequencing switch, and feed switches are identical to those of the conventional lens above. The remainder of this approach includes passive diplexers, a cylindrical portion of the Meyer lens, a lens cap, and the radiating elements. A somewhat more physical depiction of this approach is illustrated in Fig. 90.

The cylindrical portion of the lens may be as shown in Fig. 90 or it may be folded, since folding does not change the geodesic paths. The rf paths on transmit and receive are depicted in Fig. 91, showing the folding of the cylindrical portion of the lens. A schematic of the diplexer/circulator is given in Fig. 92. A functional diagram showing transmit and receive signal flow in the diplexer/circulator is shown in Fig. 93.

The Meyer lens, as usually used (discussed in Section 3) phases a line source; however, by completing the cylindrical portion to 360°, it can be used to phase a cylindrical array. It is then possible to feed the lens at any point on the 360° feed circle, and the output of the lens may be taken along the circumference of the circular cap. Furthermore, if the lens cap is dielectrically loaded, it can be smaller

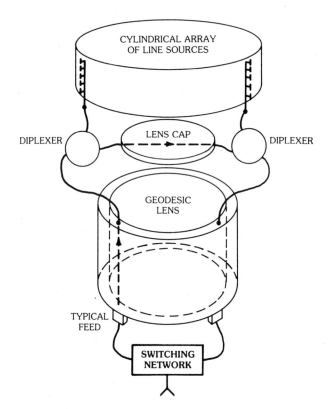

Fig. 90. Meyer system configuration. (*Courtesy Hughes Aircraft Co., Fullerton, Calif.*)

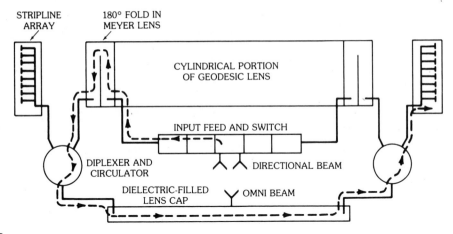

a

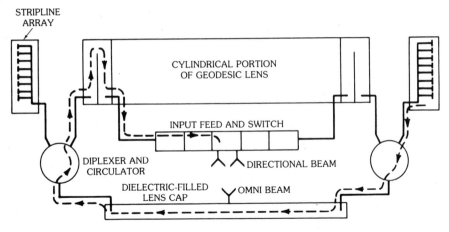

b

Fig. 91. Functional block diagram showing rf paths of Meyer lens. (*a*) Directional-beam signal flow. (*b*) Receive signal flow. (*Courtesy Hughes Aircraft Co., Fullerton, Calif.*)

by the square root of the relative dielectric constant. This allows for easier packaging of the associated components.

A brief analysis of the dielectrically loaded lens is given and experimental results are presented. Since the cylinder is a singly curved surface, the cylinder and flat circular portion can be developed, or broken apart and flattened for the sake of easy analysis. Refer to Fig. 94. The geodesics all become straight lines in the developed case. As in Fig. 94 equal lengths of transmission line are used for the rf connection between the developed cylinder and the cap, where the arc length s is preserved, or is linearly mapped onto the lens cap through some proportionality constant α. The choice of α determines the physical size of the lens cap and the necessary dielectric loading ϵ_2 to correctly phase the radiating elements. The same is true between the cap and the radiating elements, but the mapping here preserves

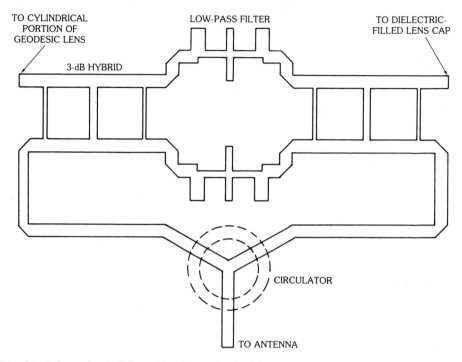

TO CYLINDRICAL
PORTION OF
GEODESIC LENS

LOW-PASS FILTER

TO DIELECTRIC-
FILLED LENS CAP

3-dB HYBRID

CIRCULATOR

TO ANTENNA

Fig. 92. Schematic of diplexer/circulator circuit. (*Courtesy Hughes Aircraft Co., Fullerton, Calif.*)

the angle designated ϕ'. The electrical path length $L(\phi')$ from the feed point to the plane wavefront in space is given by geometrical optics:

$$L(\phi') = \ell_1\sqrt{\epsilon_1} + \ell_2\sqrt{\epsilon_2} + \ell_3 \tag{30}$$

The reference path length through the center is ℓ_0, given by

$$\ell_0 = h\sqrt{\epsilon_1} + 2b\sqrt{\epsilon_2} \tag{31}$$

Thus the electrical path length error δ is given by

$$\delta = L(\phi') - \ell_0 \tag{32}$$

The output phase error $\Phi(\phi')$ is then

$$\Phi(\phi') = k\delta(\phi') \tag{33}$$

Snell's law is obeyed because of equal line lengths; thus

$$\frac{\sin\gamma}{\sin\psi} = \frac{n_1}{an_2} \tag{34}$$

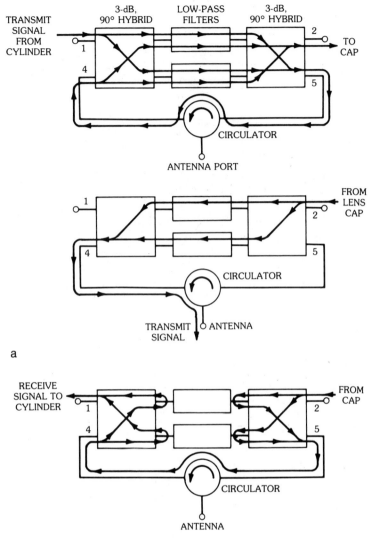

a

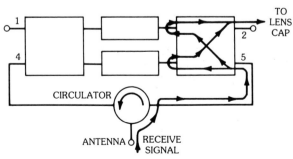

b

Fig. 93. Functional diagram of diplexer/circulator, showing signal flow. (*a*) Transmit signal paths. (*b*) Receive signal paths. (*Courtesy Hughes Aircraft Co., Fullerton, Calif.*)

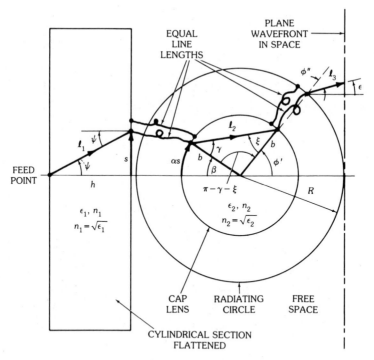

Fig. 94. Developed full cylindrical lens. (*Courtesy Hughes Aircraft Co., Fullerton, Calif.*)

The following relations can be seen from Fig. 94:

$$\ell_1 = h \sec \psi \qquad (35)$$

$$s = h \tan \psi \qquad (36)$$

$$\alpha s = b\beta \qquad (37)$$

Thus

$$\beta = \frac{\alpha}{b} h \tan \psi \qquad (38)$$

From the law of sines,

$$\frac{b}{\sin \gamma} = \frac{b}{\sin \xi} = \frac{\ell_2}{\sin(\gamma + \xi)} \qquad (39)$$

Thus

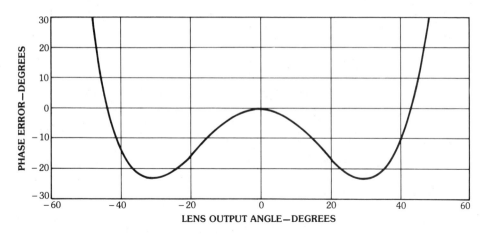

Fig. 95. Phase error at output of geodesic lens. (*Courtesy Hughes Aircraft Co., Fullerton, Calif.*)

a

Fig. 96. Geodesic lens and patterns. (*a*) Modified Meyer lens. (*b*) Pattern (eight adjacent beams). (*c*) Sum pattern. (*d*) Expanded sum and difference pattern. (*Courtesy Hughes Aircraft Co., Fullerton, Calif.*)

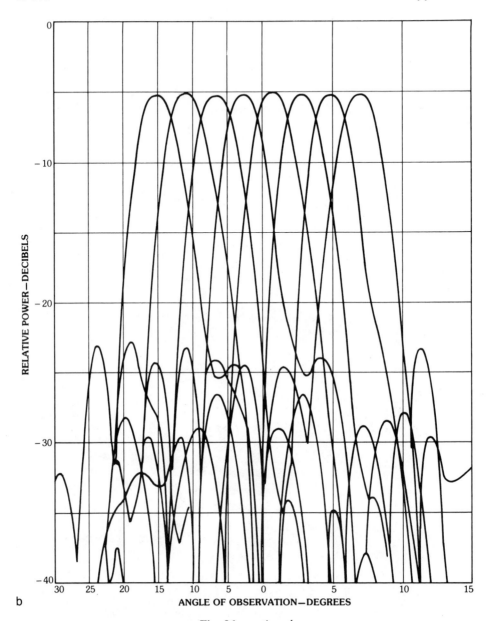

Fig. 96, *continued.*

$$\xi = \gamma \tag{40}$$

$$\ell_2 = \frac{b \sin 2\gamma}{\sin \gamma} = 2b \cos \gamma \tag{41}$$

From geometry it follows that

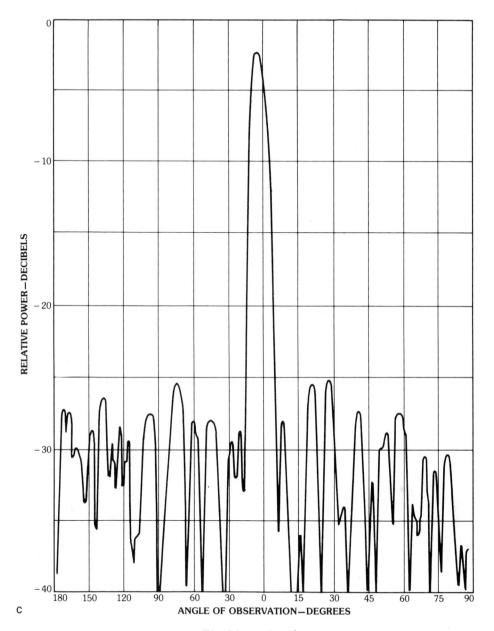

Fig. 96, *continued.*

$$\phi' = 2\gamma - \beta \qquad (42)$$

Again using Snell's law, it follows that

$$\frac{\sin \phi''}{\sin \xi} = \frac{bn_2}{R} \qquad (43)$$

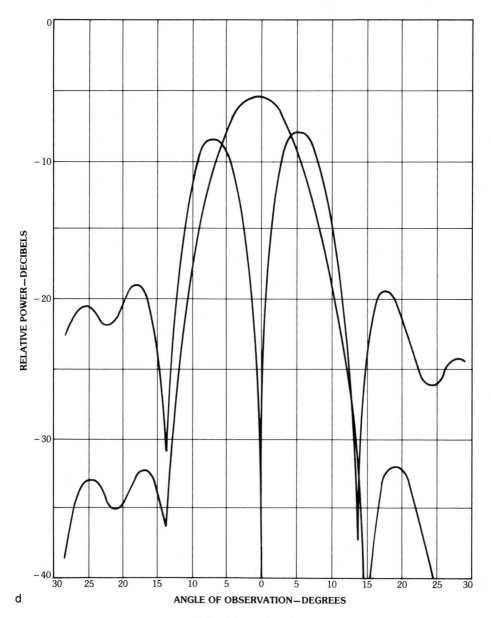

Fig. 96, *continued.*

From geometry,

$$\epsilon = \phi'' - \phi' \tag{44}$$

$$\ell_3 = R(1 - \cos\phi')\sec\epsilon \tag{45}$$

$$y = R\sin\phi' + R(1 - \cos\phi')\tan\epsilon \tag{46}$$

Relations 34 through 46 then enable one to compute the path length error for all the rays, evaluated on the hypothetical plane wavefront in space. The parameters h/b, α, b/R, n_1, and n_2 are selected to minimize the overall path length error, as well as to establish physical sizes.

For example, Fig. 95 is the calculated phase error for a 256-element array, illustrating that the spherical aberration phase error is less than $\pm 20°$ over a $\pm 45°$ sector. Thus over 70 percent of the cylindrical array diameter is nearly perfectly phased. Moreover, the associated amplitude distribution falls off rapidly beyond $\pm 45°$. Consequently, a low-azimuth side lobe distribution is realizable using this technique. A high degree of amplitude distribution control is inherent, for both the sum and difference distributions independently, since an eight-element array of probes was used, each of which is amplitude controllable.

To demonstrate the feasibility of this approach, a 128-element array was constructed and tested at X-band. The beamwidth of the model was $5°$. Fig. 96a shows the actual model. In the interest of time and funding, the lens was fabricated in one piece rather than connecting the cap and cylinder with cables; however, the array was coupled to the lens by coaxial probes.

The measured patterns of this model are shown in Figs. 96b through 96d. Fig. 96b shows several adjacent beams superposed. Fig. 96c shows a typical sum pattern, and Fig. 96d shows expanded sum and difference patterns. For economy of cost and time the feed used was a simple dual open-ended waveguide. Time did not allow an array-type feed which would have given more precise illumination with independent control over the sum and difference illuminations. Had an array-type feed been used, better patterns would have resulted, especially for the difference pattern.

Acknowledgments

The authors of this chapter gratefully acknowledge their colleagues in the Electromagnetics Laboratories of Hughes Aircraft Company, Fullerton, California, for their consultation and constructive criticism and especially Mrs. Lynda Schenet for typing, revising, and editing many cycles of the manuscript.

7. References

[1] L. A. Gustafson, "S-band two-dimensional slot array," *Tech. Memo 462*, Hughes Aircraft Company, Culver City, California, March 1957.

[2] R. M. Brown, "Performance of an antenna sharing the aperture of a frequency scanned array," *Rep. 8226*, Naval Research Lab, Washington, DC, May 1978.

[3] J. Butler and R. Lowe, "Beam forming matrix simplifies design of electronically scanned antenna," *Electron. Design*, vol. 9, no. 8, pp. 170–173, April 12, 1961.

[4] R. C. Hansen, *Microwave Scanning Antennas*, vols. 1, 2, 3, Array Systems, New York: Academic Press, 1966.

[5] W. P. Delaney, "An rf multiple beam forming technique," *IRE Trans. Mil. Electron.*, pp. 179–186, April 1962.

[6] J. P. Shelton and K. S. Kelleher, "Multiple beams from linear arrays," *IRE Trans.*, vol. AP-9, pp. 154–161, March 1961.

[7] W. R. Jones and G. F. Van Blaricum, "Multiple-beam forming hybrid networks,"

Hughes Aircraft Co., Ground Systems Group, *Tech. Memo. No. TP71-14-2*, April 7, 1970.

[8] H. J. Moody, "The systematic design of the Butler matrix," *IEEE Trans. Antennas Propag.*, vol. AP-12, pp. 786–788, November 1964.

[9] J. Blass, "The multidirectional antenna: a new approach to stacked beams," *1960 IRE Conv. Rec.*, pt. 1, pp. 48–50, 1960.

[10] J. Matthews and R. L. Walker, *Mathematical Methods of Physics*, 2nd ed., Menlo Park: Benjamin-Cummings, pp. 152–153, 1970.

[11] G. G. Chadwick, W. Gee, P. T. Lam, and J. L. McFarland, "An algebraic synthesis method for RN^2 multi-beam matrix network," internal publication of Lockheed Missile and Space Co., Sunnyvale, Calif., 94086. Also in *Proc. 1981 Antenna Appl. Symp.*, Allerton Park, Univ. of Illinois, September 23, 1981.

[12] J. L. McFarland, "The RN^2 multiple beam array family and beam forming matrix," *Proc. 1981 Antenna Appl. Symp.*, Allerton Park, Univ. of Illinois, September 23–25, 1981.

[13] R. J. Mailloux, "Array techniques for limited-scan applications," *Phys. Sci. Res. Papers, No. 503*, AFCRL-72-0421, Air Force Cambridge Research Laboratories, July 19, 1972.

[14] R. J. Mailloux and G. R. Forbes, "Experimental studies of a multiple mode array technique for limited-scan applications," *Phys. Sci. Res. Papers, No. 575*, AFCRL-TR-73-0686, Air Force Cambridge Research Laboratories, November 6, 1973.

[15] G. T. DiFrancia, "A family of perfect configuration lenses of revolution," *Optica Acta*, vol. 1, no. 4, pp. 157–163, February 1955.

[16] L. J. Chu and M. A. Taggart, "Pillbox antenna," US Patent No. 2,688,546, May 12, 1945.

[17] M. A. Taggart and E. C. Fine, "Parallel-plate bends," *MIT Radiation Lab Rep. 760*, September 5, 1945.

[18] J. S. Ajioka, "The development of an integral mkx iff antenna for low-frequency radar," *Rep. No. 534*, US Navy Electronic Lab, San Diego, January 1955.

[19] W. Rotman, "A study of microwave double-layer pillboxes," *AFCRC-TR-54-102*, Air Force Cambridge Research Center, Cambridge, Massachusetts, July 1954.

[20] J. S. Ajioka, "A multiple beam forming network using a multimode radial transmission line," *1963 NEREM Conv. Rec.*

[21] J. S. Ajioka, "A multiple beam forming antenna apparatus," US Patent No. 3,290,682, December 6, 1966.

[22] J. S. Ajioka and H. A. Uyeda, "Experimental performance of a multimode radial transmission line beam-forming network," *Microwave J.*, pp. 53–56, December 1968.

[23] J. L. McFarland, "Catenary geodesic lens antenna," US Patent No. 3,383,691, assigned to Hughes Aircraft Co., May 14, 1968.

[24] S. B. Meyer, "Parallel-plate optics for rapid scanning," *J. Appl. Phys.*, vol. 18, pp. 221–229, 1947.

[25] W. Rotman and R. F. Turner, "Wide-angle microwave lens for line source applications," *IEEE Trans. Antennas Propag.*, pp. 623–632, November 1963.

[26] J. Ruze, "Wide-angle metal plate optics," *Proc. IRE*, vol. 38, pp. 53–58, January 1950.

[27] H. Gent, "The bootlace aerial," *Royal Radar Establishment J.*, pp. 47–57, October 1957.

[28] C. M. Rappaport and A. I. Zaghoul, "Optimized three-dimensional lenses for two-dimensional scanning," *IEEE AP-S 1984 Symp. Dig.*, vol. II, June 25–29, 1984.

[29] R. F. Rinehart, "A solution of the problem of rapid scanning for radar antennae," *J. Appl. Phys.*, vol. 19, September 1948.

[30] R. K. Luneburg, *Mathematical Theory of Optics*, Providence: Brown University, pp. 189–213, 1944.

[31] E. C. DuFort and H. A. Uyeda, "A wide-angle scanning optical antenna," *IEEE Trans. Antennas Propag.*, vol. AP-31, no. 1, p. 60, January 1983.

[32] R. T. Hill, "Phased array feed systems," in *Phased Array Antennas*, ed. by A. A.

Oliner and G. H. Knittel, Dedham: Artech House, pp. 197–211, April 1972.

[33] J. R. Kahrilas, "HAPDAR—an operational phased array radar," *Proc. IEEE*, November 1968.

[34] W. T. Patton, "Limited-scan arrays," *Proc. 1970 Phased Array Antenna Symp.*, ed. by A. A. Oliner and G. H. Knittel, Dedham: Artech House, pp. 332–343, 1970.

[35] C. Winter, "Phase-scanning experiments with two reflector antenna systems," *Proc. IEEE*, vol. 56, no. 11, pp. 1984–1999, 1968.

[36] R. J. Mailloux and P. Blacksmith, "Array and reflector techniques for precision approach radars," *AGARD Conf. Proc. No. 139 on Antennas for Avionics*, NATO 26–30, November 1973.

[37] W. D. Fitzgerald, "Limited electronic scanning with a near-field Cassegrainian system," *Tech. Rep. 484*, ESD-TR-71-271, Lincoln Lab, 1971.

[38] Hughes Aircraft Company, Ground Systems Group, "Tradex S-band phased array design study," *Interim Rep. FP71-14-3*, Fullerton, California, May 27, 1971.

[39] W. D. Fitzgerald, "Limited electronic scanning with an offset-feed near-field Gregorian system," *Tech. Rep. 486*, ESD-TR-272, Lincoln Lab, MIT, 1971.

[40] Lincoln Lab, "A KREMS phased array radar," report dated April 26, 1971.

[41] C. J. Miller and D. Davis, "LFOV optimization study, final report," prepared by Westinghouse Defense and Electronics Systems Center, Systems Development Division, for MIT Lincoln Lab, Contract No. F19628-70-C-0230, May 1, 1972.

[42] Hughes Aircraft Company Ground Systems Group, "Study program for reflector surface optimization LFOV system (TRADEX/KREMS)," *FP No. 71-14-138*, Fullerton, California, August 19, 1971.

[43] C. J. Sletten, "Caustic matching: a new technique for improving limited-scan antennas," US Air Force Cambridge Research Laboratories, Hanscom Air Force Base, Massachusetts, December 17, 1974.

[44] C. H. Tang and C. F. Winter, final report on Contract No. AF19628-72-C-0213.

[45] M. Born and E. Wolf, *Principles of Optics*, New York: Pergamon Press, 1959.

[46] Hughes Aircraft Company Ground Systems Group, "Limited-scan antenna techniques study," final report on Contract No. F1962B-73-C-0129, Fullerton, California 92634, August 14, 1975.

[47] R. Tang, "Survey of time delay beam steering techniques," *Proc. 1970 Phased Array Antenna Symp.*, ed. by A. A. Oliner and G. H. Knittel, Dedham: Artech House, pp. 254–260, 1972.

[48] N. S. Wong, R. Tang, and E. E. Barber, "A multielement high-power monopulse feed for low side lobes and high aperture efficiency," *IEEE Trans. Antennas Propag.*, vol. AP-22, pp. 402–407, May 1974.

[49] E. C. DuFort, "Optical technique for broadbanding phased arrays," *IEEE Trans. Antennas Propag.*, vol. AP-23, pp. 516–523, July 1975.

[50] J. L. McFarland and J. S. Ajioka, "Multiple beam constrained lens design," *NEREM Rec.*, November 1962.

[51] L. Stark, "High-resolution hemispherical reflector antenna," US Patent No. 3,852,748, assigned to Hughes Aircraft Company, December 3, 1974.

[52] G. V. Borgiotti, "An antenna for limited scan in one plane: design criteria and numerical simulation," *IEEE Trans. Antennas Propag.*, vol. AP-25, no. 2, March 1977.

[53] R. J. Giannini, "An electronically scanned cylindrical array based on a switching and phasing technique," *1969 G-AP Intl. Symp. Program Dig.*, pp. 199–204, 1969.

[54] A. E. Holley, E. C. DuFort, and R. A. Dell-Imagine, "An electronically scanned beacon antenna," *IEEE Trans. Antennas Propag.*, vol. AP-22, no. 1, pp. 3–12, January 1974.

[55] J. E. Boyns, C. W. Gorham, A. D. Munger, J. H. Provencher, J. Reindel, and B. I. Small, "Step-scanned circular array antenna," *IEEE Trans. Antennas Propag.*, vol. AP-18, no. 5, pp. 590–595, September 1970.

[56] P. Sheleg, "A matrix-fed circular array for continuous scanning," *Proc. IEEE*, vol. 56, no. 11, November 1968.

[57] "Proposal for air traffic control radar beacon system (ATCRBS) electronic scan antenna," *RFP No. WA5R-1-0059*, submitted by Hughes Aircraft Company, Ground Systems Group, Fullerton, California, March 19, 1971.

Chapter 20

Antennas on Aircraft, Ships, or Any Large, Complex Environment

W. D. Burnside
The Ohio State University ElectroScience Laboratory

R. J. Marhefka
The Ohio State University ElectroScience Laboratory

CONTENTS

Walter D. Burnside is a professor in the Electrical Engineering Department at The Ohio State University (OSU), from which he received his BSEE, MS (1968), and PhD (1972) in electrical engineering. He graduated valedictorian of his undergraduate class and received from the IEEE (Institute of Electrical and Electronic Engineers) Antennas and Propagation Society the R. W. P. King Award (1975), the Best Application Paper Award (1975), and the Best Paper Award (1980). From OSU, he received the Teacher of the Year Award in Electrical Engineering (1980) and the College of Engineering Harrison Award for Outstanding Young Faculty Member (1983). Dr. Burnside is a Fellow of the IEEE (1985); is a member of Sigma Xi, Eta Kappa Nu, Phi Eta Sigma, and Tau Beta Pi; and was Associate Editor of the *IEEE Transactions on Antennas and Propagation* for five years. He has engaged in a wide variety of electromagnetic studies, which include both theoretical and experimental efforts. His main contributions have been in the application of electromagnetic analysis or measurement techniques to solve complex real-world problems; and, more recently, he has been very active in the analysis, measurement, and control of scattering from complex targets.

Ronald Joseph Marhefka was born in Cleveland, Ohio, on June 2, 1947. He received the BSEE degree from Ohio University, Athens, in 1969, and the MS and PhD degrees in electrical engineering from The Ohio State University, Columbus, in 1971 and 1976, respectively.

Since 1969, he has been with The Ohio State University Electro-Science Laboratory, where he currently is a research scientist. His research interests are in the areas of developing and applying high-frequency asymptotic solutions, such as the uniform geometrical theory of diffraction, hybrid solutions, and other scattering techniques. He has applied these methods to numerous practical antenna and scattering problems, including airborne, spacecraft, and shipboard antenna analysis and radar cross-section prediction.

He is the author of the user-oriented computer code, the NEC-Basic Scattering Code. He has written over 70 technical reports and papers. In 1975, he coauthored a paper that won the IEEE Antennas and Propagation Society's Best Application Paper and one that won the R. W. P. King Award.

Dr. Marhefka is a member of IEEE, Tau Beta Pi, Eta Kappa Nu, Phi Kappa Phi, and Sigma Xi. He has served as Secretary-Treasurer, Vice Chairman, and Chairman of the Columbus joint chapter of the IEEE Antennas and Propagation and Microwave Theory and Techniques societies during 1977–1980, respectively.

1. Introduction

The overall capability of an electromagnetic radiating system is dependent on its ability to operate effectively in a complex environment, in that its pattern performance can be adversely limited by pattern distortion effects, such as blockage and structural scattering. In many cases these detrimental effects can be minimized by judiciously locating the antennas. This task is complicated by the large number of systems that are competing for prime locations on, for example, a modern military ship. Without an efficient means to position such systems one normally attempts to use locations similar to previous designs, which may be inexpensive but are certainly not optimum. As a result there is a great need for electromagnetic tools that can efficiently evaluate the pattern performance of radiating systems in their proposed environment.

As with most engineering performance questions the antenna engineer can resort to experimental as well as theoretical solutions. Before applying either approach the engineer needs to examine its properties so that he can most effectively determine and evaluate his design. Due to the size and complexity of most structures it is rather obvious that experimental results will be expensive; yet they are very pleasing in that they potentially provide real-world results. For simple structures, such as a horn antenna radiating in free space, one can gain some insight into modifications necessary to improve its performance. For a horn antenna radiating on a military aircraft, however, the radiation patterns are normally far too complex to allow one to relate structural effects to corresponding measured results. Therefore experimental results are not very useful as a diagnostic tool but are most appropriate for evaluating the performance of various competing systems. Theoretical solutions, on the contrary, are used to numerically simulate the real-world situation and provide insight into the scattering mechanisms creating the resulting pattern performance. From the previous comments it is apparent that, whenever possible, it is most appropriate to use both a theoretical and an experimental approach to design a radiating system for application in a complex environment. The theoretical solution would be used to efficiently evaluate various alternatives and suggest a few candidate systems. These systems could then be evaluated in detail using experimental measurements. Of course, this combined approach sounds very attractive; however, it is predicated on the availability of a theoretical solution. The development of such numerical solutions is the main theme of this chapter.

Before the advent of digital computers, numerical electromagnetic solutions were limited to more classical geometries such as cylinders, cones, spheres, etc. With the ready access to powerful computer systems a designer is now able to construct numerical electromagnetic solutions for much more complex structures. These computer solutions have also required the development of new techniques which are more accurate and efficient for numerical calculations. Two major numerical solutions which have found great success in such applications are the method of moments and the uniform geometrical theory of diffraction (UTD). These two theories complement each other nicely in that the former is used to solve structures which are small in terms of the wavelength, whereas the latter is applicable for electrically large geometries. This chapter is devoted to electrically large structures which can be solved using the UTD; however, many of the general statements also apply to the method of moments solutions.

The uniform geometrical theory of diffraction (UTD) is used here based on its ability to simulate many complex structures [1–8] as well as being numerically amenable, accurate, and efficient. It is based on a ray optical format which is used to determine components of the field incident on and diffracted by the various structures. Components of the diffracted field are found using the UTD solutions in terms of individual rays which are summed with the geometrical optics terms in the far field of each scattering center. Note that one can be in the far field of all the scattering centers yet in the near field of the structure. The rays from a given scatterer tend to interact with other structures, causing various higher-order terms. These various possible combinations of rays that interact between scatterers can be traced out with only the dominant ones included in the final solution. Thus one need not be concerned with all other higher-order terms. This method normally leads to accurate and efficient computer codes that can be systematically written and tested, in that complex problems can be built up from simpler pieces.

The limitations associated with the UTD numerical solutions result mainly from the basic nature of the theoretical analysis. This point is discussed further in Chapter 4; in addition, the limitations associated with the application of the UTD for practical results are presented throughout this chapter.

The UTD simulation models are based on representing the actual structure in terms of a collection of much simpler parts, such as flat plates and elliptic cylinders. The use and validity of these simulation models are presented and verified through numerous applications and comparisons with experimental results. The theory only limits the structures to be locally defined by simple canonical shapes; however, the simple structures mentioned lead to less complex computer codes. The Airborne Code and the NEC-Basic Scattering Code (BSC) are used to demonstrate the usefulness and accuracy of UTD user-oriented numerical solutions.

In order to represent a large variety of antennas it is most appropriate to develop a Green's function type solution for infinitesimal current radiators. If the current distribution of the antenna is known, such as from a moment method analysis, an iteration scheme, or other means, the field from the various weighted current segments can be summed to give the total field. If the scattering centers are in the far zone of the antenna as a whole, the antenna can be represented by its pattern factor or, in the case of an array of antennas, by its array factor. This can save a large amount of computation time by eliminating multiple calculations for each infinitesimal current element.

The antenna can be mounted either on or off the various structures; however, if the antenna is mounted on a curved surface, then one must use the UTD radiation solutions. On the other hand, if the antenna is not located on or near a curved surface, then the UTD scattering solutions are applicable. Consequently these two problems require two different analyses and result in distinct computer codes as will be more apparent later.

The UTD theoretical solutions simply require the receiver to be in the far field of the individual scattering centers. This is extremely significant in that one can develop a UTD numerical solution in terms of a general receiver location, or a solution which does not change as the receiver moves from the near to the far field. This allows one to put a patch type antenna, for example, on an actual aircraft and measure any convenient pattern in the near field of the structure. Then one could construct a simulation structure, verify the model, and compute the desired far-field pattern. This could also greatly reduce the cost associated with obtaining experimental results.

Once the UTD numerical solutions are constructed for general structures one could envision an antenna designer sitting at a graphics display terminal, trying various configurations and determining some candidate antennas and sites for the desired application. These various candidates could then be evaluated using appropriate near-field experimental results. This approach should allow an antenna designer to integrate his antenna efficiently into a complex environment.

2. Numerical Simulation of the Antenna

The general-purpose solutions discussed here can be used with a wide range of antennas. They are basically a Green's function solution; that is, the solution has been developed for infinitesimally small current elements. Any antenna can be modeled, therefore, as a array of fundamental radiators if its current distribution is known. This can be quite a challenge in practice, since many of the antennas used are heuristically designed and are of a very complex physical nature. The current distribution of an antenna can be obtained in many different ways, such as a method of moments analysis [9–11], a synthesis scheme [12, 13], known aperture distributions, measurements, or other means.

The degree to which the current and/or radiation pattern of the antenna under investigation needs to be known depends on the problem being solved. The impedance of an antenna is highly dependent on the exact current distribution on the antenna, since the impedance is proportional to the voltage divided by the current at the terminals. If the impedance of an antenna in the presence of the complex environment is desired, it is necessary to use a method of moments technique or a hybrid method of moments and UTD approach [14] from which the currents can be calculated. The radiation pattern of an antenna, however, is not as sensitive to the current distribution on the antenna. This is because the radiation pattern is an end result of an integration of the currents, which averages out small variations in the distribution. The question is then: given a complex antenna, what is the best method to reproduce a good facsimile of the radiation pattern in the most efficient way possible? One assumption that can be made here is that the radial component of the elemental radiators is not needed. If the near field of an antenna is desired, it is necessary to use infinitesimal elements and integrate them numerically, reproducing the near field for the antenna as a whole.

If the scattering centers are in the far zone of the antenna as a whole, the antenna can most efficiently be represented by its pattern factor or, in the case of an array of antennas, by its array factor [15, 16]. This can save a large amount of computation time by eliminating multiple calculations for each infinitesimal current element. The general form of the transmitted electric field **E** from an antenna at a distance r using pattern factors is given by

$$\mathbf{E} = j Z_0 I_m k \mathbf{h} \frac{e^{-jkr}}{4\pi r} \qquad (1)$$

for an electric source type and

$$\mathbf{E} = -j K_m k \mathbf{h} \frac{e^{-jkr}}{4\pi r} \qquad (2)$$

for a magnetic source type, where k is the wave number and Z_0 is the wave impedance of free space. The function **h** is the vector effective height of the antenna, which can be given as

$$I_m \mathbf{h} = \hat{\boldsymbol{\theta}} L w J_m \sin \theta \; F_x(\theta, \phi) \; F_z(\theta) \; F_a(\theta, \phi) \qquad (3)$$

for an electric source type. The magnetic source type is the same except $\hat{\boldsymbol{\phi}}$ is substituted for $\hat{\boldsymbol{\theta}}$, K_m for I_m, and M_m for J_m. The length of the source is given as L and the width as w. The $\sin \theta$ term comes from the pattern factor of a elemental radiator. The function F_z is due to the length, F_x is due to the width, and F_a is an array pattern factor. For a line source $w \to 0$; therefore $w J_m \to I_m$ and $F_x = 1$. The form of the pattern factors depends on the current distribution [15, 16] and will not be given here.

The array pattern factor is a useful concept if the scattering centers are relatively far away from the elements of an array or the infinitesimal elements of a discretized current distribution. A number of the elements can be clustered together with a common phase center, hence a single ray origin. Weaker elements in an array can be clustered more easily since their overall effect on the pattern is less critical.

Many times the pattern of the antenna in free space or on a ground plane is the only information that is known about an antenna. It is possible to use a synthesis scheme to find an equivalent distribution across the aperture of an antenna that will reproduce the antenna pattern reasonably well. The synthesis scheme could be found from the far-zone power pattern of the antenna [12] or more accurately from near-zone measurements, which include phase information [13]. It should be noted that it is not necessary that the calculated currents be the exact currents on the antenna if all that is desired is a facsimile of the pattern.

Interpolation of the measured pattern data is another possible scheme that can be used if that is all that is available. This method is not as desirable as some of the other methods because usually only the principal-plane patterns have been measured. These patterns cannot always be used in a manner to reproduce the patterns well in the off principal planes.

The spatial coupling between antennas can be obtained approximately by using the vector antenna height concept. This can be derived from the generalized reaction principle for antennas in a scattering environment. The total open-circuit voltage at the terminals of the receiver can be obtained by adding the dot product of the **E** fields from the various rays incident on the receiving antenna by the vector effective height of the receiving antenna in that direction, that is,

$$V_{oc} = \mathbf{h}(\theta, \phi) \cdot \mathbf{E}^i \qquad (4)$$

This quantity can be normalized to a more useful term, such as power out over power in, by the following equation [7]:

$$\frac{P_{out}}{P_{in}} = \frac{|1/2\, I_m\, V_{oc}|^2}{P_{rt} P_{rr}} \left[\frac{R_r R_L}{|Z_r + Z_L|^2} \right] \qquad (5)$$

where P_{rt} is the power radiated by the transmitter and P_{rr} is the power radiated by the receiver if it were a transmitter. The current weight at the receiving terminals is I_m, the radiation impedance is Z_r, and the load impedance is Z_L. The Rs are the resistive parts of the impedances. Note that the power radiated terms can be obtained by a method of moments code or they can be approximated by integrating the volumetric radiation pattern of the antenna in free space or on a ground plane if necessary. This coupling formula is a more general statement of Friis' transmission formula that takes into account the scattering of the environment around the antennas. It is a useful approximation if the antennas are sufficiently far apart from each other or are not near large scattering structures, that is, so that the currents on the antennas are not significantly modified by each other or the environment. This is reasonably accurate as long as the antennas are more than a half-wavelength apart and if they are about a quarter-wavelength from an edge [14]. Of course, if an antenna is over a large ground plane or some other large reflecting structure, the method applied to find the currents can use image theory to include the ground effect.

3. Numerical Simulation of the Environment

There are many questions that must be addressed before one embarks on the development of a numerical electromagnetic solution for a complex structure. The first problem considered here is usually the last question addressed by the electromagnetic analyst; however, it is of great concern to the antenna designer. Using a numerical solution the designer has to specify the geometry to the computer, which implies by necessity that the numerical model simply simulates the actual antenna environment. How well, then, does the computer model actually represent the real world? The most obvious simulation model would include all features associated with the actual structure; in addition it would represent each feature to a high degree of accuracy.

If one attempts to define the geometry associated with a shipboard antenna, it would quickly become very clear that the complete structure can not be simulated to a high degree without a great deal of effort. Even if one could describe such a geometry to the computer the electromagnetic analysis needed to solve such a configuration would be overwhelming to develop, very inefficient to run, and most likely would be limited to specific configurations. Since measurements can ultimately provide the real-world results, in most cases the efficiency of a numerical solution is its most outstanding attribute. With an efficient solution an antenna designer can very quickly evaluate various configurations and narrow design choices down to a few practical solutions.

Since the efficiency of the numerical solution plays such a dominant role in its usefulness, let us examine the features of the solution which dictate its execution time. It has already been mentioned that the geometry used in the analysis must be of such a form that it can be easily defined by the user; however, it must be general enough to describe a complex antenna environment, such as an aircraft. In addition it is most appropriate if the computer model represents the basic shapes being analyzed by the code. This is very useful in terms of evaluating the accuracy of the resulting patterns. With these concepts in mind it is advantageous to divide a complex structure into simpler parts or substructures. These substructures should then be the basic theoretical geometries analyzed by the code so that they can be very efficiently calculated as well as being easily input to the computer.

The evaluation of the substructures used in the simulation model revolves around the electromagnetic analysis needed to solve the class of problems being studied. Since this chapter is devoted to electrically large structures the UTD as described in Chapter 4 will be exclusively applied here. For the sake of brevity the actual equations used to analyze the various structures will not, in general, be given, except to introduce the basic concepts of the UTD approach.

In order to demonstrate the implementation of a UTD solution let us consider the radiation pattern of a short dipole in the presence of a finite flat plate as shown in Fig. 1. Note that the flat-plate geometry is considered here because it can be easily input to the computer and can be efficiently analyzed using the UTD. The total UTD solution consists to the first order of the superposition of the incident, reflected, edge-diffracted, and corner-diffracted fields as described in the following.

The incident field is basically the line-of-sight signal going directly from the dipole to the receiver. Using the UTD methodology the incident field is set to zero if the ray from the dipole intercepts the plate before reaching the receiver, as shown in Fig. 2. Thus the incident field is discontinuous at the incident shadow boundaries, where the field abruptly goes to zero.

The reflected field from the finite flat plate is determined using image theory. To begin the solution the receiver image position is found as illustrated in Fig. 3. The source solution is then used to compute the electric field $\mathbf{E}^S$ at the image position. This field would exist at that point if the plate were not present. The reflected field $\mathbf{E}^r$ is then given by

$$\begin{bmatrix} E_x^r \\ E_y^r \\ E_z^r \end{bmatrix} = \begin{bmatrix} T_{xx} & T_{xy} & T_{xz} \\ T_{yx} & T_{yy} & T_{yz} \\ T_{zx} & T_{zy} & T_{zz} \end{bmatrix} \begin{bmatrix} E_x^S \\ E_y^S \\ E_z^S \end{bmatrix} \qquad (6)$$

where $[T]$ represents the reflected field polarization transformation matrix which satisfies the boundary conditions on the plate. This matrix is determined using the normal to the plate such that

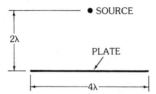

Fig. 1. Geometry for a source in the presence of a plate.

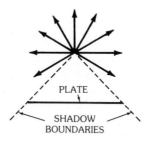

Fig. 2. Illustration of source-ray paths.

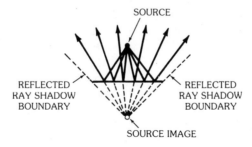

Fig. 3. Illustration of plate-reflected ray paths.

$$\mathbf{E}^r = (\hat{\mathbf{n}} \cdot \mathbf{E}^S)\hat{\mathbf{n}} - (\hat{\mathbf{t}} \cdot \mathbf{E}^S)\hat{\mathbf{t}}$$
$$(\hat{\mathbf{t}} \cdot \mathbf{E}^S)\hat{\mathbf{t}} = \mathbf{E}^S - (\mathbf{E}^S \cdot \hat{\mathbf{n}})\hat{\mathbf{n}} \qquad (7)$$
$$\mathbf{E}^r = 2(\hat{\mathbf{n}} \cdot \mathbf{E}^S)\hat{\mathbf{n}} - \mathbf{E}^S$$

or

$$T_{xx}\hat{\mathbf{x}} + T_{yx}\hat{\mathbf{y}} + T_{zx}\hat{\mathbf{z}} = 2(\hat{\mathbf{n}} \cdot \hat{\mathbf{x}})\hat{\mathbf{n}} - \hat{\mathbf{x}}$$
$$T_{xy}\hat{\mathbf{x}} + T_{yy}\hat{\mathbf{y}} + T_{zy}\hat{\mathbf{z}} = 2(\hat{\mathbf{n}} \cdot \hat{\mathbf{y}})\hat{\mathbf{n}} - \hat{\mathbf{y}} \qquad (8)$$
$$T_{xz}\hat{\mathbf{x}} + T_{yz}\hat{\mathbf{y}} + T_{zz}\hat{\mathbf{z}} = 2(\hat{\mathbf{n}} \cdot \hat{\mathbf{z}})\hat{\mathbf{n}} - \hat{\mathbf{z}}$$

where the wedge (^) denotes a unit vector. Since the **T** matrix is independent of the receiver location it is stored in order to optimize the computational eficiency of the numerical solution. The total reflected field is then computed using (6), unless the reflection point does not occur on the flat plate, in which case it is set to zero. Thus the reflected field is discontinuous at the reflection shadow boundaries. The incident plus reflected fields form the geometrical optics (GO) solution, which is inherently composed of discontinuities that must be corrected before the solution is complete.

The first step in the development of the edge-diffraction solution is to find the edge-diffraction point. It is known that for a given receiver location there is only one point along an infinitely long straight edge at which the diffracted field can emanate for a near-zone source and/or receiver. Thus this point must be found for each of the edges that describe the flat plate. There are many ways of finding this diffraction point, one of which is described here. Since it is known that $\beta_0 = \beta_0'$ (see Fig. 4), it is obvious that

$$\hat{\mathbf{e}} \cdot \hat{\mathbf{i}} = \hat{\mathbf{e}} \cdot \hat{\mathbf{d}} \tag{9}$$

where ($\hat{\mathbf{e}}$, $\hat{\mathbf{i}}$, and $\hat{\mathbf{d}}$) are, respectively, the edge unit vector, incident direction unit vector, and diffraction direction unit vector. Note that the edge-diffracted rays emanating from a point on an edge form a conical surface making the half-cone angle β_0. In order to solve for the unique edge-diffraction point one merely projects the source and receiver onto the edge in a normal to the edge vector sense. Then by using similar triangles one can determine the isolated diffraction point. If the edge-diffraction point does not fall within the limits of the actual edge, then the diffracted field from that edge is set to zero. If the edge-diffraction point is on the edge, the diffracted field using the UTD formulation developed by Kouyoumjian and Pathak [17] is given by

$$\mathbf{E}^d(s) \cong \mathbf{E}^i(Q_E) \cdot \bar{\bar{\mathbf{D}}}_E A(s) e^{-jks} \tag{10}$$

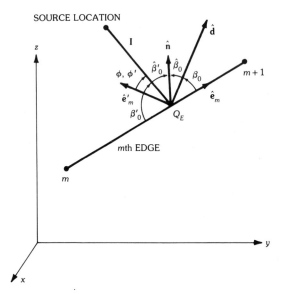

Fig. 4. Edge coordinate system at point of diffraction (Q_E).

where $A(s)$ is the spread factor given by $A(s) = \sqrt{s'/s(s + s')}$ with s and s' the distance from the edge to the receiver and source, respectively. Note that the diffraction coefficient is given by

$$\bar{\bar{D}}_E = -\hat{\beta}_0'\hat{\beta}_0 D_s - \hat{\phi}'\hat{\phi} D_h \tag{11}$$

and

$$D_{s,h}(\phi, \phi', \beta_0) = \frac{-e^{-j\pi/4}}{2n\sqrt{2\pi k}\,\sin\beta_0}\left[\cot\left(\frac{\pi + \beta^-}{2n}\right)F[kLa^+(\beta^-)]\right.$$
$$+ \cot\left(\frac{\pi - \beta^-}{2n}\right)F[kLa^-(\beta^-)] \mp \left\{\cot\left(\frac{\pi + \beta^+}{2n}\right)\right.$$
$$\left.\times F[kLa^+(\beta^+)] + \cot\left(\frac{\pi - \beta^+}{2n}\right)F[kLa^-(\beta^+)]\right\}\right] \tag{12}$$

and

$$F(X) = 2j|\sqrt{X}|e^{jX}\int_{|\sqrt{X}|}^{\infty} e^{-j\tau^2}\,d\tau$$
$$\beta^\mp = \phi \mp \phi' \tag{13}$$

where $F(X)$ is called the *transition function* and the parameters $a^\pm(\beta^\pm)$ are defined in Reference 17 and Chapter 4. The geometry associated with this solution is illustrated in Fig. 5. Note that $\mathbf{E}^i(Q_E)$ is the field incident on the edge at the diffraction point. The complexity of the edge-diffraction solution is lumped into the

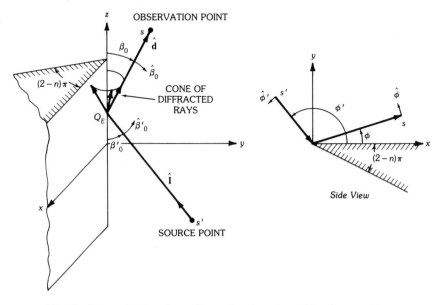

Fig. 5. Geometry for three-dimensional wedge-diffraction problem.

transition function which is illustrated in Fig. 6. Due to the simplicity of this term the edge-diffracted field is rapidly computed, as will be discussed later.

The edge-diffracted field is also abruptly discontinuous at the incident and reflection shadow boundaries in just the precise way to make the total field continuous. On the lit side of the shadow boundary the diffracted field approaches minus one-half the shadowed field, whereas on the shadow side the diffracted field simply changes sign. Thus at the shadow boundary the shadowed plus edge-diffracted field is approximately one-half the shadowed field magnitude.

The first-order UTD solution is not complete until the corner-diffracted field is added. Since the edge-diffracted field is also abruptly discontinuous when the diffraction point falls off the actual edge, the corner term is needed to compensate for this behavior.

The corner-diffraction problem is illustrated in Fig. 7. The corner-diffracted fields associated with one corner and one edge in the near field with spherical-wave incidence are given by [3, 18]

$$
\begin{Bmatrix} E_\parallel^c \\ E_\perp^c \end{Bmatrix} = \begin{Bmatrix} IZ_0 \\ MY_0 \end{Bmatrix} \frac{\sqrt{\sin\beta_c \, \sin\beta_{0c}}}{(\cos\beta_{0c} - \cos\beta_c)} F[kL_c a(\pi + \beta_{0c} - \beta_c)] \frac{e^{-jks}}{4\pi s} \tag{14}
$$

where

$$
\begin{Bmatrix} I \\ M \end{Bmatrix} = -\begin{Bmatrix} E_\parallel^i(Q_C) \\ E_\perp^i(Q_C) \end{Bmatrix} \begin{Bmatrix} C_s(Q_E)Y_0 \\ C_h(Q_E)Z_0 \end{Bmatrix} \sqrt{\frac{8\pi}{k}} e^{-j(\pi/4)} \tag{15}
$$

and

$$
C_{s,h}(Q_E) = \frac{-e^{-j(\pi/4)}}{2\sqrt{2\pi k}\,\sin\beta_0} \left\{ \frac{F[kLa(\beta^-)]}{\cos(\beta^-/2)} \left| F\left[\frac{La(\beta^-)/\lambda}{kL_c a(\pi + \beta_{0c} - \beta_c)} \right] \right| \right.
$$
$$
\left. \mp \frac{F[kLa(\beta^+)]}{\cos(\beta^+/2)} \left| F\left[\frac{La(\beta^+)/\lambda}{kL_c a(\pi + \beta_{0c} - \beta_c)} \right] \right| \right\} \tag{16}
$$

The function $F(X)$ was defined earlier, $a(\beta^\pm) = 2\cos^2(\beta^\pm/2)$, where $\beta^\pm = \phi \pm \phi'$ and $L = s's''\sin^2\beta_0/(s' + s'')$ and $L_c = s_c s/(s_c + s)$ for spherical-wave incidence. The function $C_{s,h}$ is a modified version of the diffraction coefficient for the half-plane case $(n = 2)$. The modification factor

$$
\left| F\left[\frac{La(\beta^\pm)/\lambda}{kL_c a(\pi + \beta_{0c} - \beta_c)} \right] \right| \tag{17}
$$

is an empirically derived function that ensures that the diffraction coefficient will not change its sign abruptly when it passes through the shadow boundaries of the edge. There is also a corner-diffraction term associated with the other edge forming the corner, and it is found in a similar manner. Even though further study is needed to improve the description of this diffraction mechanism, the writers believe that the benefits obtained from its inclusion warrant its use here.

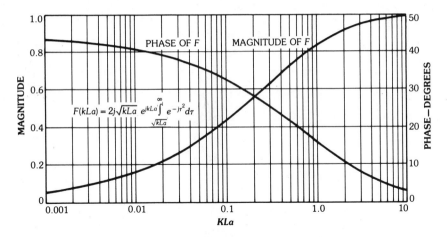

Fig. 6. Transition function.

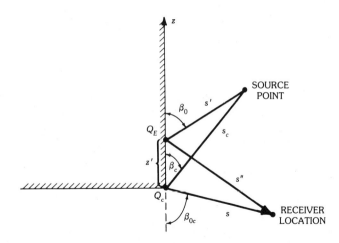

Fig. 7. Geometry for corner-diffraction problem.

This example illustrates very nicely the various attributes associated with UTD numerical solutions. First, the complexity of the solution is lumped into various diffraction coefficients which can be written in subroutine form. Since these coefficients consist of trigonometric and transition function terms, they are very easily and rapidly computed.

This UTD solution for a short dipole radiating in the presence of a finite flat plate can be used to compute the radiation pattern in either the near or far field of the structure. The only limitation is that the receiver not get any closer than a wavelength to an edge or the dipole. The total far-zone *E*- and *H*-plane patterns are shown in Fig. 8. Note the very good agreement obtained between the calculated and measured results in the *H*-plane pattern. In order to illustrate the significance of the various UTD terms they are individually plotted in Fig. 9. The individual terms are shown to the same radiation level so one can acquire a feel for the

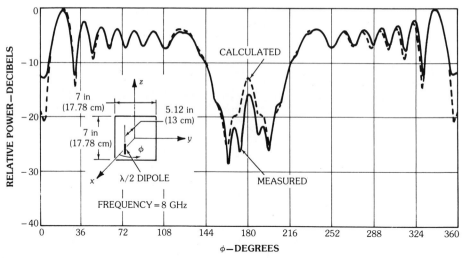

a

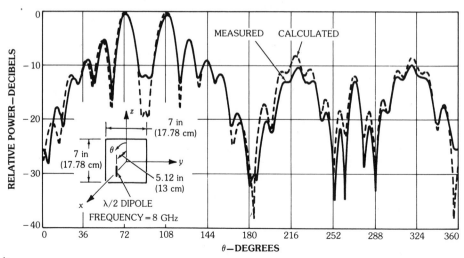

b

Fig. 8. Measured and calculated *H*- and *E*-plane patterns for a half-wave dipole located above a square plate. (*a*) *H*-plane patterns: $E_{\theta p}$. (*b*) *E*-plane patterns: $E_{\phi p}$.

significance of each mechanism. In addition the various discontinuities associated with each term are clearly illustrated.

The total field solution consists of the superposition of a few terms, each of which is associated with an isolated scattering point on the structure. This leads to two useful aspects of the solution: (1) the pattern indicates when the solution fails in that each term introduces a discontinuity which must be corrected in a precise way, and (2) pattern problem sectors can be associated with particular mechanisms which in turn can pinpoint regions on the structure from which the problem term or

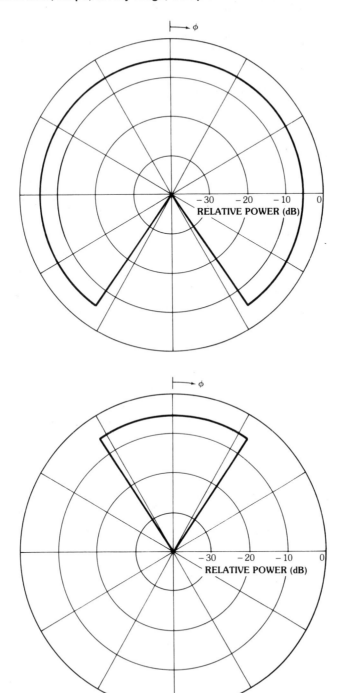

Fig. 9. H-plane patterns due to various ray mechanisms for a half-wave dipole located above a square plate. (*a*) Source field: S. (*b*) Reflected field: R. (*c*) Diffracted field: D. (*d*) $S + R + D$.

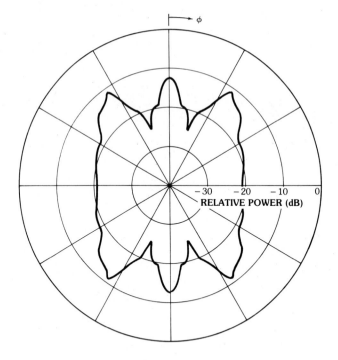

c

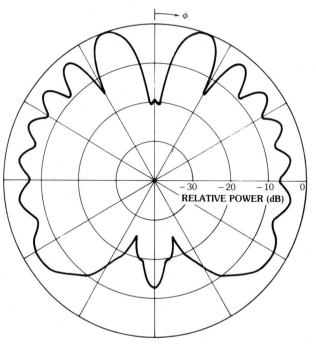

d

Fig. 9, *continued.*

terms emanate. By using the former, one can tell by his pattern calculation if the solution is valid theoretically. By using the latter, one could consider using corrective measures to eliminate various unwanted terms.

The UTD wedge-diffraction solutions have been extended in [19] to treat finite flat plates covered with a thin dielectric or absorber layer. The absorber could be used here to reduce the reflected-field magnitude, which in turn would decrease the strong interference pattern shown in Fig. 10a. A thin ferrite absorber was placed on the flat plate, and the calculated and measured results are shown in Fig. 10b. Again the comparison between results is very encouraging. One should note that with the introduction of the absorber model into the numerical solution he or she can begin

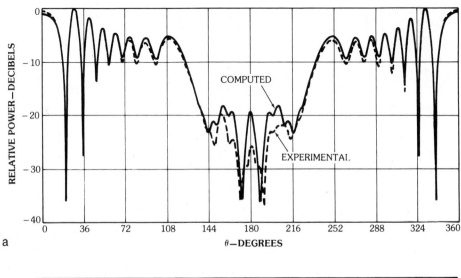

a

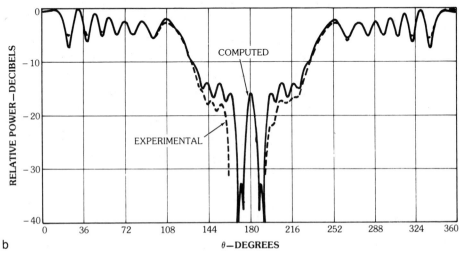

b

Fig. 10. Measured and calculated $E_{\theta p}$ patterns. (*a*) For a perfectly conducting metallic plate. (*b*) For a thin ferrite absorber on a metallic plate.

to take realistic measures to improve pattern performance by using properly located absorber panels. In addition the absorber model does not add significantly to the computation time, so an antenna designer can quickly examine possible alternatives.

On a standard high-speed digital computer a complete pattern for the previous geometry can be run in less than 10 s. As a result one could envision an antenna designer sitting at a graphics display terminal, which is used to draw the input geometry and resulting patterns, trying various configurations and determining some realistic geometries which provide the desired performance. Since the computation time is so short the antenna designer can observe the cause and effect associated with various changes in the structure. In such a case he or she will be able to quickly eliminate poor designs of the same general class in that they will all have the same characteristic problem, i.e., plate reflections too strong, for example. The motive here is to avoid inherently poor designs and converge rapidly to a few practical configurations which have acceptable pattern performance. At this point it would be appropriate, if possible, to take a few measurements in order to verify the numerically simulated results.

So far it has been ascertained that the UTD can be used to efficiently and accurately compute the patterns for a fundamental antenna radiating in the presence of a finite flat plate. Even though the flat plate represents a challenging electromagnetic problem, it is too great of a simplification to be used in simulating various complex structures. On the other hand the UTD solutions for the flat plate can be extended to treat multiple flat plates which could be used to box out a given structure. Such a representation is satisfactory if the structure is not doubly curved, in which case a very large number of plates would be needed to simulate the configuration. Even though the computer running time is very short for one plate it grows exponentially for multiple plates. In addition the multiple-plate simulation of a curved surface leads to analysis problems which have not been uniformly resolved. As a result it is not in general profitable to simulate doubly curved structures by multiple plates. Thus one is forced to use the UTD electromagnetic solutions for curved surfaces.

Let us examine the properties of the UTD curved surface solutions and evaluate their potential use in a numerical solution. First, there are three distinct UTD solutions for curved surfaces, i.e., scattering, radiation, and coupling. Since the geometry adjacent to the antenna and receiver affects the electromagnetic performance most significantly, different simulation models can be used to analyze various types of problems. For example, multiple flat plates can be used to simulate the scattering by an aircraft wing; however, a curved surface is better for modeling a fuselage mounted antenna radiation problem. That is to say, the radiation problem requires that the simulation surface very accurately represent the actual structure nearest to the antenna since the radiated field strength is largest near the antenna. As a result curved-surface numerical solutions normally fall into one of the three categories (scattering, radiation, and coupling) in that the simulation models for each type of category can be drastically different. Needless to say, the theoretical solutions which apply for each category are also different, which implies that one must associate the appropriate type of problem with an efficient representation of the curved surface.

For the plate geometries examined earlier the ray paths are basically straight lines either going directly from the source to the receiver or going from the source, scattering off the plates, and then arriving at the receiver. As the energy expands outward from the source on a curved surface the rays traverse the surface along *geodesic* paths, a geodesic path being the shortest path between the initial and final points of the trajectory, i.e., it is a great circle on a sphere and a helix on a cylinder. The geodesic paths are defined, for example, by a pair of nonlinear coupled differential equations, and as a result they are very inefficient to track. Because of this inefficiency one is tempted to use as simple a curved surface as possible so that the geodesic paths can be found in some rapid sense. For the coupling problem where both the source and receiver are mounted on the curved surface, one is forced to very accurately represent the complete surface between the two antennas. This requires that one solve the geodesic equations for a very complete representation of the geometry, which implies that coupling solutions are rather inefficient. This is not necessarily the case in that for a coupling problem the source and receiver locations are fixed, so there is basically a small set of geodesic paths needed for the total calculation. Referring now to the radiation problem, one has to find new geodesic paths for each different radiation direction; however, only the source is mounted on the curved surface, so only that portion of the surface needs to be accurately simulated. With this in mind it has been shown in [20, 21] that a perturbation solution can be used to very efficiently determine the geodesics on elongated prolate spheroids and ellipsoids. These solutions are very rapid since they relate the pattern direction to a specific geodesic path; in addition they base the next geodesic path on the last path in that they change very little from one pattern angle to the next. For example, the complete radiation pattern for a short monopole mounted on a prolate spheroid can be run in less than 10 s. The basic curved surface solutions have been treated in Chapter 4, so they will not be reiterated here.

In order to use the UTD to represent complex structures it is necessary to combine the curved-surface- and wedge-diffraction solutions so that both curved surfaces and flat plates can be used in the simulation model. Using these two basic geometries in this analysis let us consider development of the radiation and scattering numerical solutions.

Basic Model Simulations—Antennas Mounted on a Curved Surface

This section describes the development of UTD simulations for antenna problems in which the antenna is mounted on a curved surface. Most of this discussion will involve aircraft shapes since there are many such problems for which this theory can be applied in the development of airborne antenna systems. Although the examples tend to aerodynamic shapes one can use this solution approach to solve a wide variety of problems, as will be noted as the simulation structures evolve. For example, this approach could be used to determine the pattern performance for a mast-mounted antenna radiating in the presence of other ship structures. In another case it could be used to examine the pattern distortion associated with a satellite-mounted antenna.

In order to begin our simulation studies with a simple two-dimensional problem let us consider the antena problem illustrated in Fig. 11. In this example an electric line monopole is mounted on the surface of a perfectly conducting circu-

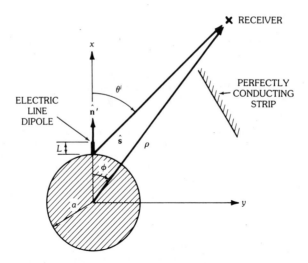

Fig. 11. Electric line dipole mounted on a circular cylinder and radiating in the presence of a perfectly conducting strip.

lar cylinder. In addition a strip is located in close proximity to the cylinder. Before developing the various scattering terms let us review the basic radiation mechanisms associated with antennas mounted on a curved surface. First, the analysis is divided into two parts, these being the lit and shadow region solutions. In the lit region the radiation appears to come directly from the source and has a field value given by

$$H_{\text{lit}}^c(\varrho,\phi) \cong \left\{ \frac{\cos(kL\hat{\mathbf{n}}\cdot\hat{\mathbf{s}}) - \cos kL}{[1 - (\hat{\mathbf{n}}\cdot\hat{\mathbf{s}})^2]\sin kL} \right\}(\sin\theta^i)G(\xi^\ell)\frac{e^{-jks}}{\sqrt{s}} \qquad (18)$$

where $G(\xi)$ is known as a Fock function, which is defined in Reference 18 and in Chapter 4. In the shadow region the radiation appears to come from the two effective sources shown in Fig. 12. These two effective sources result from the two creeping waves excited by the source which propagate outward from the source in a clockwise and counterclockwise sense. The shadow region field value for either of the two creeping waves is given by

$$H_{\text{shadow}}^c(\varrho,\phi) \cong \left[\frac{1 - \cos kL}{\sin kL} \right]G(\xi)e^{-jk\ell}\frac{e^{-jks}}{\sqrt{s}} \qquad (19)$$

where ℓ is the propagation length of the creeping-wave path. Since the radiated wavefront appears to come from the source or effective sources, these apparent radiation centers can be used to compute the reflected and diffracted fields from the strip. Furthermore, (18) and (19) can be used as the field incident on the strip and its associated edges. Note that the incident field, that is, the field going from the antenna (possibly around the cylinder) to the receiver as given by (18) and (19), is computed provided the ray going from the effective source to the receiver does not

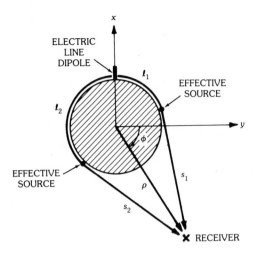

Fig. 12. Radiation geometry with receiver in shadow region.

intersect the strip. From this point forward the effective source terminology is used to indicate the apparent radiation center for a given ray, which implies it is the actual source if the receiver is in the lit region.

The field reflected from the strip is computed using the image position of the receiver in the strip as shown in Fig. 13. The field incident on the strip is computed using (18) and (19) with the observation point located at the image position. For an electric line dipole the field at the image position is identical with the field at the receiver, provided the ray going to the image from the effective source as shown in Fig. 13 intersects the plate. In fact the intersection point is actually the point of reflection. Note that the reflected field will only exist for a small sector of space; however, the strip edge-diffracted fields radiate in all directions, as shown in Fig. 14, and are discussed next.

The edge-diffracted field for the top edge of the strip is considered first because it is illuminated by a lit region field shown in Fig. 14a. In this case the source of the edge-diffracted field is the source itself, and its field value is given by

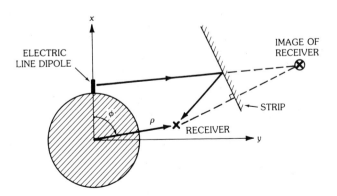

Fig. 13. Strip-reflected field geometry.

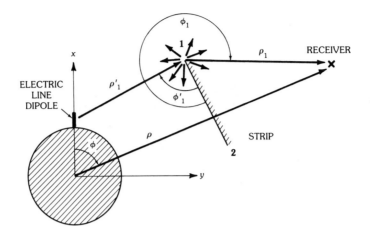

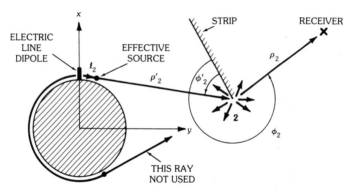

Fig. 14. Diffracted field geometry. (*a*) For edge 1. (*b*) For edge 2.

$$H_1^d = D_h [H_{\text{lit}}^c]_{\text{at edge 1}} \frac{e^{-jk\varrho_1}}{\sqrt{\varrho_1}} \tag{20}$$

where

$$
\begin{aligned}
D_h = &D(\varrho_1'\varrho_1/(\varrho_1' + \varrho_1), \phi_1 - \phi_1', n = 2) \\
&+ D(\varrho_1'\varrho_1/(\varrho_1' + \varrho_1), \phi_1 + \phi_1', n = 2)
\end{aligned} \tag{21}
$$

Since the incident field is computed in the lit region, H_{lit}^c at edge 1 is given by (18). The diffracted field from the second edge is computed for an incident field in the shadow region. This implies that two possible contributions which result from the two creeping-wave terms need to be treated. This is not the case in that one ray normally travels a much shorter distance on the circular cylinder; consequently one need only treat the creeping wave with the shorter geodesic path on the curved surface such as shown in Fig. 14b. In this case the diffracted field from edge 2 is given by

$$H_2^d = D_h[H_{\text{shadow}}^c]_{\text{at edge 2}} \frac{e^{-jk\varrho_2}}{\sqrt{\varrho_2}} \tag{22}$$

where

$$
\begin{aligned}
D_h = & D(\varrho_2'\varrho_2/(\varrho_2' + \varrho_2), \phi_2 - \phi_2', n = 2) \\
& + D(\varrho_2'\varrho_2/(\varrho_2' + \varrho_2), \phi_2 + \phi_2', n = 2)
\end{aligned} \tag{23}
$$

using the geometry illustrated in Fig. 14b. Even though the plate-scattered field can be subsequently scattered by the circular cylinder, this type of interaction is not considered here. It is useful in a practical sense, however, to shadow the plate-scattered field if the rays appear to intersect the cylinder. This is done so that a discontinuity in the resulting pattern can be observed, which is indicative of the strength of the higher-order scattering terms which are missing in the solution. The total solution is then the sum of the incident, strip-reflected, and two edge-diffraction terms which are obtained using (18) through (23).

Before considering the next example, recall that there is the possibility of multiple diffractions between the two edges of the strip. Normally one need only include up to double diffraction, which is significant when the source or receiver and the two edges align since the edge diffraction from the first edge can be nearly as significant as the source field in terms of the second edge illumination. Since double diffraction is normally a secondary contributor it is quite often neglected in the total solution. Again the effect of neglecting double diffraction will be apparent in the computed patterns in that a discontinuity will exist in the directions of the strip orientation, and the magnitude of this jump will indicate its importance.

If the model illustrated in Fig. 11 is to be used to simulate an aircraft structure, it is apparent that the strip has to be attached to the cylinder as shown in Fig. 15. Using this configuration the previous procedure is followed to find the reflected field from the strip and the diffracted field from edge 1. A complication arises in determining the diffracted field from the junction edge formed by the cylinder and strip. Recall that the wedge-diffraction coefficient is not valid for this type of structure in that the field incident on the junction edge results from a creeping wave propagating around the cylinder. In order to develop an approximate solution to this problem the solution can be modified based on the properties associated with this type of structure as shown in Reference 18.

An example is presented which illustrates the different UTD mechanisms included in the solution to this type of problem. For this purpose a single flat plate is attached to a circular cylinder as shown in Fig. 16. The various terms treated using the previous solution are shown in Fig. 17. Note that each pattern is normalized to the same level such that the relative significance of each term can be seen. In order to better illustrate how the total solution evolves, various terms are superimposed in Fig. 18. Note that the double-diffraction term is introduced to the solution to illustrate its significance. An interesting result is shown in Fig. 18a, where the source and reflected fields are superimposed. These two terms form a GO solution which is far from complete, as can be observed from the discon-

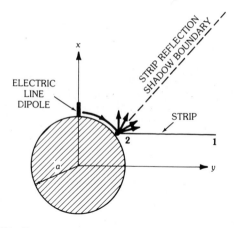

Fig. 15. Geometry of a strip attached to the cylinder.

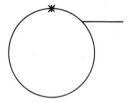

Fig. 16. Geometry of a flat plate attached to a circular cylinder.

tinuities in the pattern. The total UTD solution shown in Fig. 18d is continuous and should agree with experimental results, be presented later.

In order to simulate the wide variety of structures found in practice, more components need to be added into the total UTD solution. Since introducing more perfectly conducting strips into the solution is rather straightforward and efficient to analyze, consider a more general solution in which a cylinder-mounted antenna is radiating in the presence of multiple strips. Using this general approach, more, complex situations can be treated by boxing out the various structures making up the environment surrounding the antenna. The various UTD mechanisms used in this solution are depicted by the ray paths illustrated in Fig. 19. The field solutions used for each of these terms follow the same format applied in the previous cylinder/strip problem.

For completeness consider the analysis used to develop the diffracted-reflected field term illustrated in Fig. 20a. Since a reflection from strip 2 is being considered, the first step in this solution is to find the image of the receiver in strip 2 as shown in Fig. 20b. The diffracted field from the junction edge is then computed to find the field at the image point. That is the same field that would be incident on the receiver provided that the diffracted ray from the junction edge to the image intersects the virtual limits of strip 2. Note that the diffracted field incident on the receiver image position is found with strip 2 removed. The actual diffracted-reflected field is then found using the method described earlier (6). Finally, the

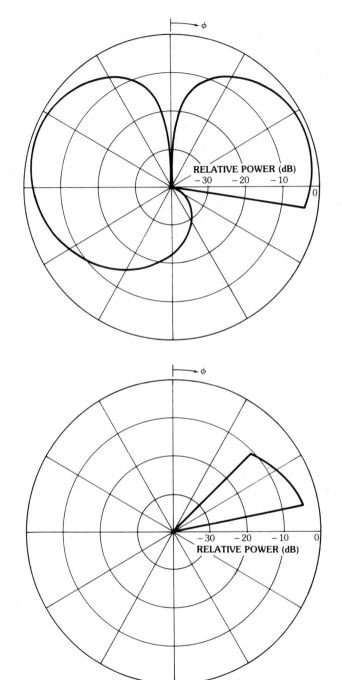

a

b

Fig. 17. Radiation patterns due to various ray mechanisms. (*a*) Source field: *S*. (*b*) Reflected field: *R*. (*c*) Diffracted field: *D*. (*d*) *S* + *R* + *D*.

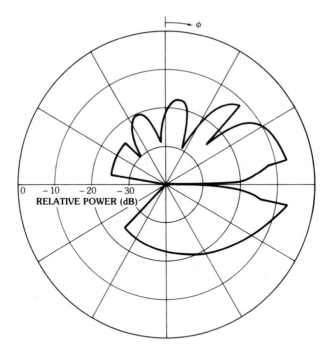

c

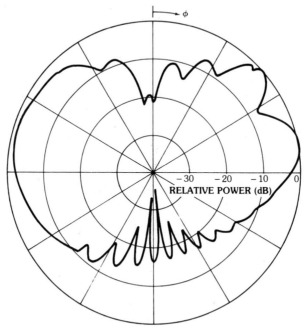

d

Fig. 17, *continued.*

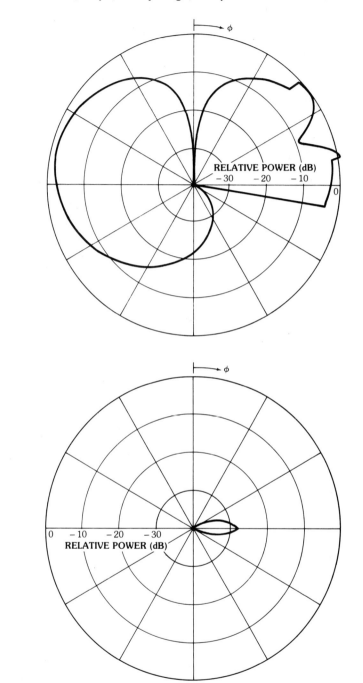

Fig. 18. Radiation patterns due to various combinations of the UTD terms. (*a*) *S* + *R*. (*b*) Double diffracted field: *DD*. (*c*) *S* + *R* + *D*. (*d*) *S* + *R* + *D* + *DD*.

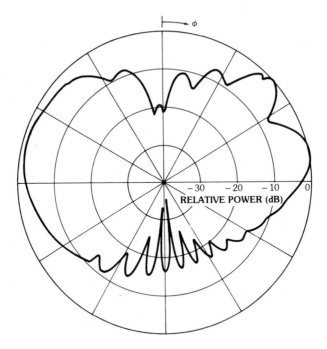

c

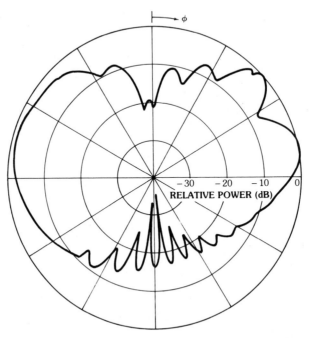

d

Fig. 18, *continued.*

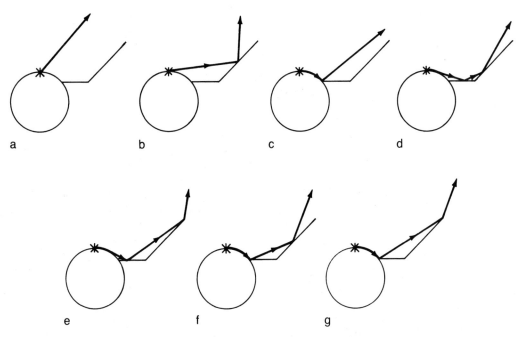

Fig. 19. Various UTD terms. (*a*) Source field. (*b*) Reflected field. (*c*) Diffracted field. (*d*) Reflected-reflected field. (*e*) Reflected-diffracted field. (*f*) Diffracted-reflected field. (*g*) Diffracted-diffracted field. (*After Burnside, Rudduck, and Marhefka [4], © 1980 IEEE*)

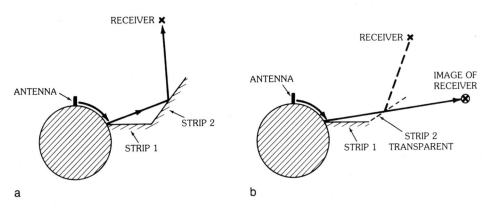

Fig. 20. Analysis for a diffracted-reflected field. (*a*) Diffracted-reflected field component. (*b*) Image receiver in strip 2 to find the diffracted-reflected field.

complete actual ray path must be traced out in order to ensure that some other strip does not block the ray; if so, that field component is set to zero. Note that other field solutions basically follow the same format.

The next example is used to illustrate the additional UTD mechanisms added to treat multiple strips. This geometry is shown in Fig. 21, in which a second plate is added to the model treated in the previous discussion. The source, reflected,

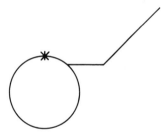

Fig. 21. Bent plate attached to a circular cylinder.

and diffracted terms used in our previous solution are shown in Fig. 22. The superimposed pattern using these three terms is illustrated in Fig. 22d. Note the discontinuities in this pattern at approximately 14° and 45°. These discontinuities are compensated for by the higher-order interaction terms illustrated in Fig. 23. As before, these terms are plotted relative to the same signal level so that one can observe their relative importance. It is clear that these higher-order terms can be significant in certain sectors of the pattern. In order to demonstrate how the complete solution creates the desired total pattern, various combinations of terms are illustrated in Fig. 24. The GO solution is shown in Fig. 24b and, as before, several discontinuities exist in this pattern. The total superposition of the recently introduced higher-order terms is shown in Fig. 24c. It is very interesting to recall that the patterns illustrated in Figs. 22d and 24c superimpose to give the total pattern shown in Fig. 24d. As with any UTD solution for a complex structure, one can compute higher- and higher-order terms. There are, however, two major problems with adding such terms beyond those included here: (1) the accuracy of the basic diffraction solutions becomes questionable, and (2) the numerical solution becomes very inefficient. Furthermore, based on cases examined to date it appears that the majority of structures found in practice can be adequately solved using the interaction terms considered here.

The previous example was used to illustrate the various higher-order terms applied in the present analysis. This next problem is used to show the various UTD terms for a more realistic aircraft geometry as shown in Fig. 25. In this case additional plates are added which could be used to simulate the engine housing for a private aircraft. The first-order terms (source, reflected, and diffracted) are shown in Fig. 26, and they represent the modification to the analysis just discussed. Finally, various combinations of UTD terms are illustrated in Fig. 27. Again it is very interesting how all the different discontinuous terms combine to give the nice, smooth total pattern shown in Fig. 27d.

In order to investigate the accuracy of the present solution a long structure was constructed which included a cylinder, flat-plate wing, and movable engine housing shape. The patterns are then taken in the plane perpendicular to the cylinder axis to simulate the two-dimensional solutions. The first example is used to verify the numerical results generated for the geometry shown in the insert of Fig. 28 along with the calculated and measured roll-plane patterns. Note that a roll-plane pattern is taken in the plane perpendicular to the axis of the aircraft. In the next case the engine plate model is moved to the wing tip, and the measured and calculated roll-

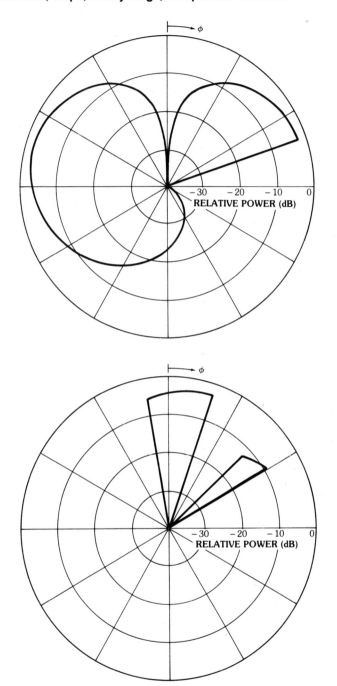

Fig. 22. Radiation patterns due to various UTD terms for Fig. 21. (*a*) Source field: *S*. (*b*) Reflected field: *R*. (*c*) Diffracted field: *D*. (*d*) *S* + *R* + *D*. (*After Burnside, Rudduck, and Marhefka [4], © 1980 IEEE*)

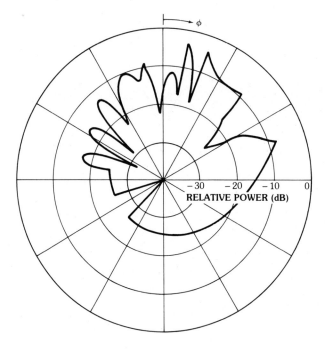

c

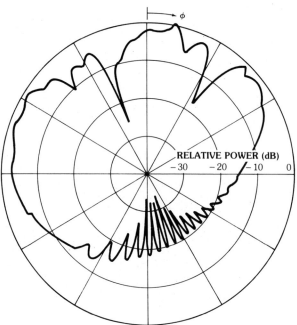

d

Fig. 22, *continued.*

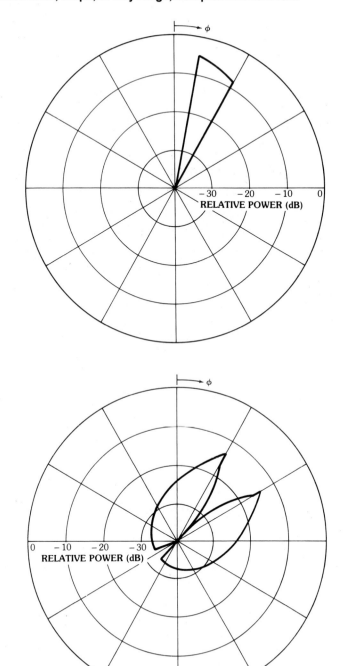

a

b

Fig. 23. Radiation patterns due to second-order-interaction UTD terms for Fig. 21. (*a*) Reflected-reflected field: *RR*. (*b*) Reflected-diffracted field: *RD*. (*c*) Diffracted-reflected field: *DR*. (*d*) Diffracted-diffracted field: *DD*. (*After Burnside, Rudduck, and Marhefka [4], © 1980 IEEE*)

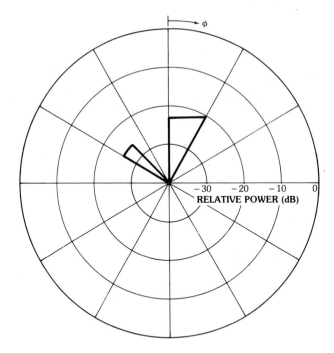

c

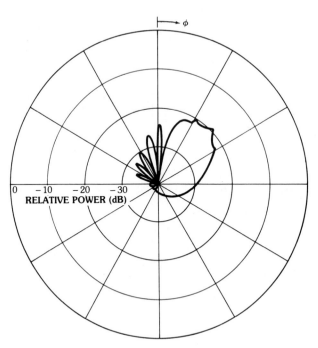

d

Fig. 23, *continued.*

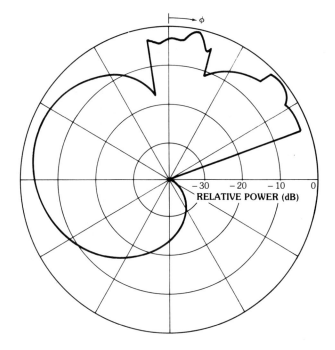

a

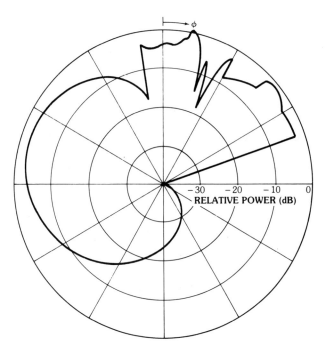

b

Fig. 24. Radiation patterns due to various combinations of UTD terms for Fig. 21. (a) $S + R$. (b) GO solution: $S + R + RR$. (c) Second-order-interaction GTD terms: $RR + RD + DR + DD$. (d) Total solution. (*After Burnside, Ruddeck, and Marhefka [4], 1980 IEEE*)

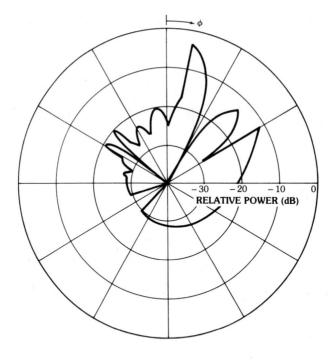

c

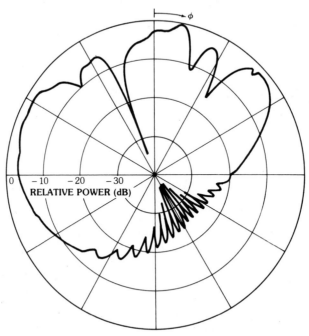

d

Fig. 24, *continued.*

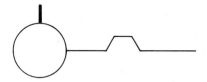

Fig. 25. Flat plate model for the wing and engine.

plane patterns for this geometry are shown in Fig. 29. In each case excellent agreement between measured and calculated results is obtained.

The UTD solution for the three-dimensional solutions follows the same format as for the two-dimensional ones except that the ray paths are not as easily computed. As a result the efficiency of a three-dimensional solution is dictated by the ray trajectory calculations, in which case one is referred to [5, 18, 20, 21, 22].

Basic Model Simulations—Antennas Not Mounted on a Curved Surface

When simulating a scattering structure, such as an aircraft or ship mast with a yardarm, it has been shown that it is very convenient to use both a cylinder and a plate in the model. If the source is not mounted on the curved surface, the field terms differ slightly from those. In the following analysis the various field terms that are needed to find the UTD scattered fields by a two-dimensional cylinder and plate model with the source off the cylinder will be discussed. This is basically an extension of the discussions given in the section on plates. In this two-dimensional analysis the interpretation of the fields and their parameters will be emphasized. The discussion on the means of finding the ray paths in terms of the three-dimensional model can be found in [5]. As before, however, one must keep in mind that any ray must be shadowed that intersects a plate or cylinder that is not a part of the specific mechanism being analyzed. This means that the shadowed ray will be set to zero and will not be included in the sum total of the pattern. As has been demonstrated earlier, the UTD compensates for these discontinuities and produces a continuous field up to the order of the solution.

As a review of the fields that have been discussed previously, the ones needed here will be briefly outlined and others will be introduced. Each of these field terms is illustrated by the example which follows this discussion. The source field is given by

$$u^i = u_0 \, \frac{e^{-jks^i}}{\sqrt{s^i}} \tag{24}$$

where it can represent either an electric or magnetic line source, and u_0 represents a complex excitation constant. It is used as the direct field from the source to the receiver or as the field incident on a scattering structure producing the fields to follow. Based on various comparisons with experimental results it has been determined that there are five dominant scattered field terms which must be included in a multiple-plate analysis. These are the reflected, edge-diffracted, doubly reflected, reflected-diffracted, and diffracted-reflected fields. The plate-reflected field is given by

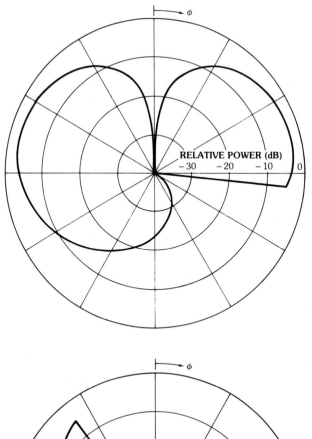

Fig. 26. Radiation patterns due to various UTD terms for Fig. 25. (*a*) Source field: *S*. (*b*) Reflected field: *R*. (*c*) Diffracted field: *D*. (*d*) *S* + *R* + *D*.

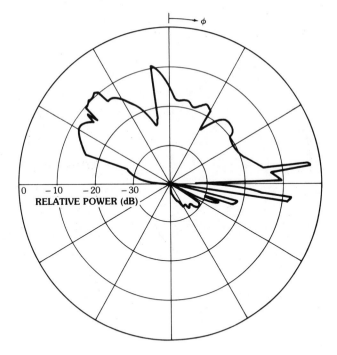

c

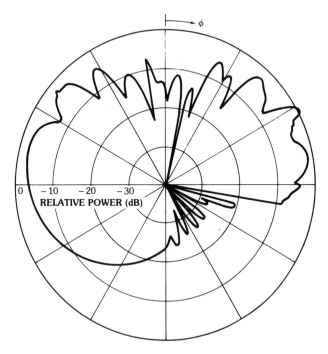

d

Fig. 26, *continued.*

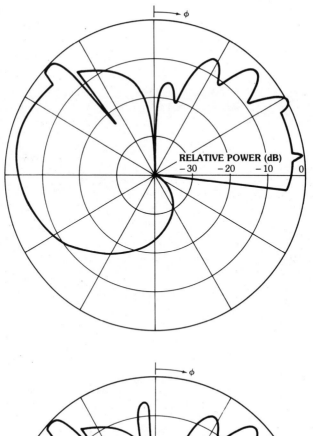

RELATIVE POWER (dB)
−30   −20   −10   0

a

RELATIVE POWER (dB)
−30   −20   −10   0

b

Fig. 27. Radiation patterns due to various combinations of the UTD terms for Fig. 25. (*a*) *S* + *R*. (*b*) GO solution: *S* + *R* + *RR*. (*c*) Second-order-interaction GTD terms: *RR* + *RD* + *DR* + *DD*. (*d*) Total solution.

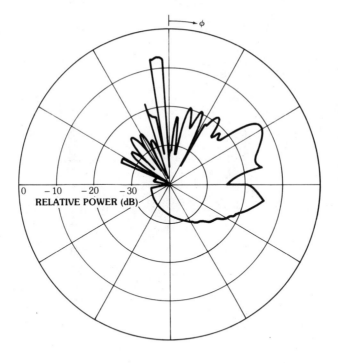

c

d

Fig. 27, *continued.*

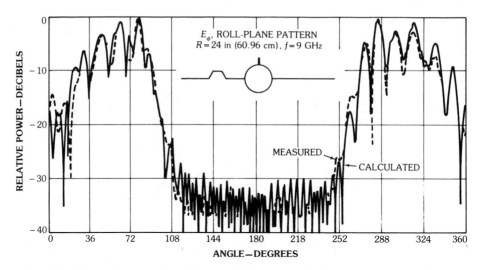

Fig. 28. Measured and calculated near-field patterns for the test geometry shown.

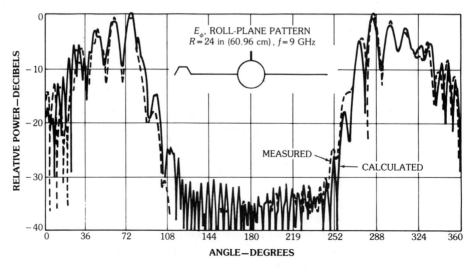

Fig. 29. Measured and calculated near-field patterns for test geometry with engine plate model moved to the wing tip as shown.

$$u_p^r = u_0 R \, \frac{e^{-jks^r}}{\sqrt{s^r}} \tag{25}$$

The plate-diffracted field is given by

$$u_p^d = u^i(Q_d) D \, \frac{e^{-jks^d}}{\sqrt{s^d}} \tag{26a}$$

or

$$u_p^d = u_0 D \; \frac{e^{-jk(s^i + s^d)}}{\sqrt{s^i s^d}} \tag{26b}$$

The plate doubly reflected field is given by

$$u_{pp}^{rr} = u_0 \; \frac{e^{-jks^{rr}}}{\sqrt{s^{rr}}} \tag{27}$$

The plate reflected-diffracted field is given by

$$u_{pp}^{rd} = u_p^r(Q_d)D \; \frac{e^{-jks^d}}{\sqrt{s^d}} \tag{28a}$$

or

$$u_{pp}^{rd} = u_0 R D \; \frac{e^{-jk(s^r + s^d)}}{\sqrt{s^r s^d}} \tag{28b}$$

The plate diffracted-reflected field is given by

$$u_{pp}^{dr} = u^i(Q_d)D \; \frac{e^{-jks^d}}{\sqrt{s^d}} \; R \sqrt{\frac{s^d}{s^d + s^r}} \; e^{-jks^r} \tag{29a}$$

or

$$u_{pp}^{dr} = u_0 D R \; \frac{e^{-jk(s^i + s^d + s^r)}}{\sqrt{s^i(s^d + s^r)}} \tag{29b}$$

This is like an imaged wedge, that is

$$u_{pp}^{dr} = u_0 D R \; \frac{e^{-jk(s^i + s^{dr})}}{\sqrt{s^i s^{dr}}} \tag{29c}$$

where $s^{dr} = s^d + s^r$. Note that R is the soft or hard reflection coefficient, and D is the soft or hard diffraction coefficient. Recall that "soft" or "hard" indicates whether an electric or a magnetic line source is being considered.

The uniform reflected field from the cylinder is given by

$$u_c^r = u^i(Q_R)\mathcal{R} \sqrt{\frac{\varrho^r}{\varrho^r + s_c^r}} \; e^{-jks_c^r} \tag{30a}$$

or

$$u_c^r = u_0 \mathscr{R} \sqrt{\frac{\varrho^r}{s^i(\varrho^r + s_c^r)}}\, e^{-jk(s^i+s_c^r)} \tag{30b}$$

In addition, the cylinder has a second field component, known as the *uniform diffracted field*, which is given by

$$u_c^d = u^i(Q_1)\mathscr{T}\, \frac{e^{-jks_c^d}}{\sqrt{s_c^d}} \tag{31a}$$

or

$$u_c^d = u_0 \mathscr{T}\, \frac{e^{-jk(s^i+s_c^d)}}{\sqrt{s^i s_c^d}} \tag{31b}$$

Note that $\mathscr{R}$ is the modified soft and hard reflection coefficient, and $\mathscr{T}$ is the modified curved surface diffraction coefficient found in Chapter 4. The definition of ϱ^r is given in (37a). It is also possible to have cylinder-to-cylinder interactions.

The combination of plates and cylinders gives rise to fields that interact between them. The fields that are reflected from a plate as one of the interactions are fairly straightforward since image theory can be applied. These terms include the field reflected by the plate and then reflected or diffracted by the cylinder, and the fields reflected or diffracted by the cylinder and then reflected by a plate. The terms that deal with a diffraction off a plate are more complicated, as will be seen below.

The field reflected by a plate and then reflected by a cylinder uses image theory for the reflection off the plate and the uniform reflection coefficient for the reflection off of the cylinder. This field component is given by

$$u_{pc}^{rr} = u_p^r(Q_R)\mathscr{R} \sqrt{\frac{\varrho^r}{\varrho^r + s_c^r}}\, e^{-jks_c^r} \tag{32a}$$

or

$$u_{pc}^{rr} = u_0 R\mathscr{R} \sqrt{\frac{\varrho^r}{s^r(\varrho^r + s_c^r)}}\, e^{-jk(s^r+s_c^r)} \tag{32b}$$

The various parameters are defined in Chapter 4, with the source distance being defined from the source image point. The field reflected by a plate and then diffracted by a cylinder follows in a similar manner as above, that is, the field is given by

$$u_{pc}^{rd} = u_p^r(Q_1)\mathscr{T}\, \frac{e^{-jks_c^d}}{\sqrt{s_c^d}} \tag{33a}$$

or

$$u_{pc}^{rd} = u_0 R \mathscr{T} \frac{e^{-jk(s^r + s_c^d)}}{\sqrt{s^r s_c^d}} \tag{33b}$$

The clockwise and counterclockwise terms are included for the creeping-wave terms. The field that uniformly reflects from the cylinder and then reflects from a plate also follows in a straightforward manner. This field is given by

$$u_{cp}^{rr} = u_c^r(Q) R \sqrt{\frac{\varrho^r + s_c^r}{\varrho^r + s_c^r + s^r}} \, e^{-jks^r} \tag{34a}$$

where the square-root term follows from the fact that the source of the plate reflection is the image of the cylinder reflected-field caustic. This can also be written as

$$u_{cp}^{rr} = u_0 \frac{e^{-jks^i}}{\sqrt{s^i}} \mathscr{R} \sqrt{\frac{\varrho^r}{\varrho^r + s_c^r}} \, e^{-jks_c^r} R \sqrt{\frac{\varrho^r + s_c^r}{\varrho^r + s_c^r + s^r}} \, e^{-jks^r} \tag{34b}$$

or

$$u_{cp}^{rr} = u_0 \mathscr{R} R \sqrt{\frac{\varrho^r}{s^i(\varrho^r + s_c^r + s^r)}} \, e^{-jk(s^i + s_c^r + s^r)} \tag{34c}$$

which makes this image relationship quite apparent. Similarly, the field diffracted by the cylinder and then reflected by the plate is given by

$$u_{cp}^{dr} = u_c^d(Q) R \sqrt{\frac{s_c^d}{s_c^d + s^r}} \, e^{-jks^r} \tag{35a}$$

or

$$u_{cp}^{dr} = u_0 \mathscr{T} R \frac{e^{-jk(s^i + s_c^d + s^r)}}{\sqrt{s^i(s_c^d + s^r)}} \tag{35b}$$

Again, the parameters needed for the above fields can be found in Chapter 4, with the scattered distance being replaced with the distance from the scatter point to the image of the observation point.

The field terms dealing with the interaction between a plate edge and a cylinder have an added complication as compared with the terms discussed above. It is assumed in the UTD coefficients that the incident wavefronts are in the form of a ray optical field. This is true for the reflected field off a cylinder and the diffracted field off a plate edge when the fields do not fall within the transition region of the structures. This is one reason why the fields reflected from the cylinder are assumed to be of the geometrical optics type here. The field reflected by a cylinder and diffracted by a plate, therefore, is composed of a geometrical optics field incident on an edge and is given by

$$u_{cp}^{rd} = u_c^r(Q_d)D \, \frac{e^{jks^d}}{\sqrt{s^d}} \tag{36a}$$

or

$$u_{cp}^{rd} = u_0 \mathscr{R} D \sqrt{\frac{\varrho_r}{s^i s^d (\varrho^r + s_c^r)}} \, e^{-jk(s^i + s^d + s_c^r)} \tag{36b}$$

where the reflection coefficient for the cylinder $\mathscr{R}$ approaches ± 1. The reflected-field caustic distance is given by

$$\varrho^r = \frac{s^i R_1 \cos \theta^i}{R_1 \cos \theta^i + 2s^i} \tag{37a}$$

However, the distance parameter for the wedge-diffraction coefficient must take into account that the incident wavefront looks like it is coming from the caustic of the reflected field off the cylinder. This means that the distance parameter is given by

$$L = \frac{\varrho^i s^d}{\varrho^i + s^d} \tag{37b}$$

where

$$\varrho^i = \varrho^r + s_c^r \tag{37c}$$

This will produce a continuous field at the shadow boundary of the wedge. This solution will not be valid, however, when the reflection point is in the transition region of the cylinder, that is, near the shadow boundary of the cylinder.

The field diffracted by a plate and reflected by the cylinder is given by

$$u_{pc}^{dr} = u_p^d(Q_R)\mathscr{R} \sqrt{\frac{\varrho^r}{\varrho^r + s_c^r}} \, e^{-jks_c^r} \tag{38a}$$

or

$$u_{pc}^{dr} = u_0 \mathscr{R} D \sqrt{\frac{\varrho^r}{s^i s^d (\varrho^r + s_c^r)}} \, e^{-jk(s^i + s^d + s_c^r)} \tag{38b}$$

The distance parameter for this case is simply given by

$$L = \frac{s^i s^d}{s^i + s^d} \tag{39a}$$

as before. However, the reflection caustic distance is now given by

$$\varrho^r = \frac{s^d R_1 \cos \theta^i}{R_1 \cos \theta^i + 2s^d} \qquad (39b)$$

This assumes that the incident field looks like it is emanating from the plate edge. This will only be true when the diffracted ray is not in the transition region of the edge. In the transition region the true representation is much more complicated and beyond the scope of this discussion. It is assumed that the fields in the transition region using the chosen parameters will give a reasonable engineering representation for the field in any event. In practice these approximations appear to give usable answers to this complicated problem, since the regions in which they are not valid are rather small in extent.

The total field for a problem consisting of plates and cylinders is then composed of all the fields described above from the various scattering parts of the structure. The fields must be shadowed properly even though this may result in a discon-

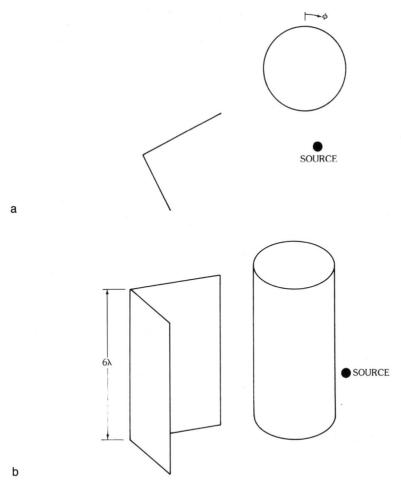

Fig. 30. Illustration of source and scattering elements. (*a*) Top view. (*b*) Side view.

tinuous pattern because all the higher-order interactions have not been included. These discontinuities give a gauge to the accuracy of the solution, which will be seen in the examples that follow.

The following example illustrates how the various field terms previously discussed individually contribute to produce a usable continuous pattern for a complex shape. The example consists of two plates forming a corner reflector and a circular cylinder as shown in Fig. 30. The pattern is taken in the plane normal to the cylinder so it is assumed that a two-dimensional analysis applies here. This is reasonable to assume since in many cases the top and bottom plate edges will not

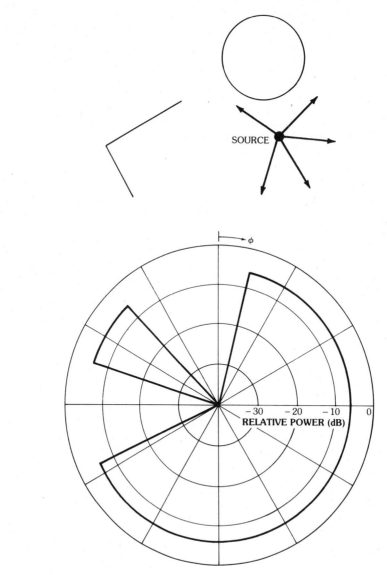

Fig. 31. Source fields. (*a*) Ray paths. (*b*) General pattern.

contribute significantly to the pattern cut illustrated by the top view. The patterns are plotted with a scale from 0 to −40 dB, and they are normalized to the maximum value of the total pattern. The first-order terms associated with the plate (source, reflected, and diffracted fields) are shown in Figs. 31 and 32.

In addition to first-order plate terms there are other first-order terms, such as: (1) the scattered (reflected and diffracted) fields from the cylinders, (2) the field reflected from the end caps, and (3) the fields diffracted by the end cap rims. These are shown in Figs. 33 to 35. Note that in the geometry presented in Fig. 30, end cap reflections will not occur. Therefore a different geometry is shown in Fig. 35 to demonstrate end cap reflections. Note that with more than one body present, individual terms are often shadowed by other bodies in the structure, creating discontinuities as shown in many of the figures (as in Fig. 33 for the cylinder scattered field).

In addition to first-order mechanisms second-order scattering occurs where a ray is scattered by one body and then scattered by a second one. Several different double-scattering (or second-order) terms are computed. Double reflection, where a ray is reflected by one plate and then by another, is shown in Fig. 36.

Another second-order scattering mechanism involving plates is reflection-diffraction, where a ray is reflected from one plate and is then diffracted as shown in Fig. 37. The inverse mechanism, diffraction-reflection, illustrated in Fig. 38, involves fields diffracted from a plate edge and then reflected by another plate.

A number of the scattering mechanisms involve interactions between the cylinder and one of the plates. Two such terms result from scattering of the fields by the cylinder and then reflection by a plate, and vice versa. Fig. 39 illustrates the ray paths and fields of rays which are reflected from a plate and then scattered by the cylinder. Fig. 40 illustrates ray paths and fields resulting from rays scattered by the cylinder and then reflected from a plate.

Another second-order scattering mechanism involves fields reflected by the cylinder and then diffracted by a plate edge. The ray paths and fields for this term

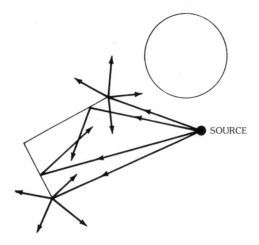

a

Fig. 32. First-order terms associated with the plate. (*a*) Ray paths. (*b*) Field due to plate reflection. (*c*) Field due to plate diffraction.

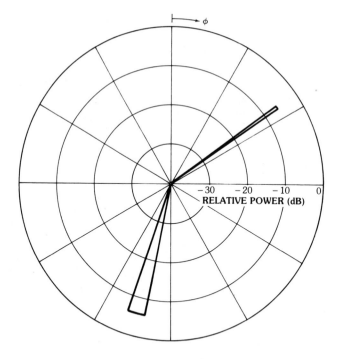

b

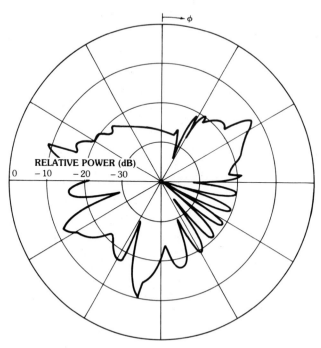

c

Fig. 32, *continued.*

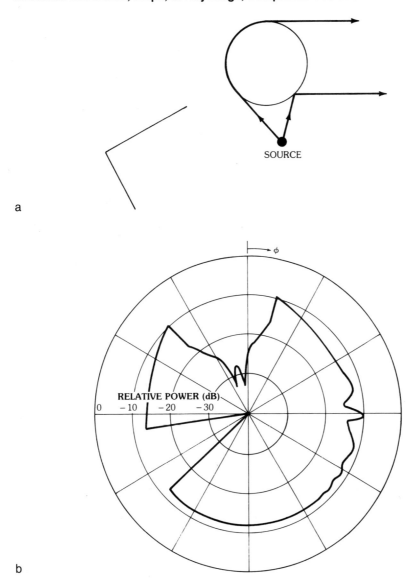

Fig. 33. First-order terms associated with the cylinder. (*a*) Ray paths. (*b*) Scattered field from cylinder's curved surface.

are illustrated in Fig. 41. The reverse of this term is the fields of rays diffracted by a plate edge and then reflected from the cylinder, as shown in Fig. 42.

The total pattern is obtained by summing the field components for the mechanisms mentioned previously. The total field pattern is illustrated in Fig. 43. One should especially note the continuity of the total pattern, which indicates that that is calculated properly and that higher-order terms are not significant for this configuration.

Higher-order scattering terms can also be computed, which will in some cases

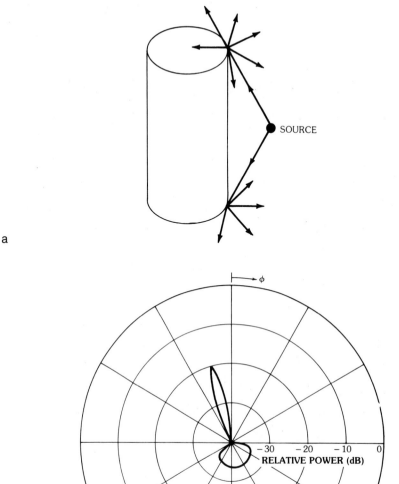

Fig. 34. First-order terms associated with the end cap diffracted fields. (*a*) Ray paths. (*b*) Fields due to end cap diffraction.

improve the accuracy of the field computations. Generally it is found that such terms are negligible in magnitude as well as being difficult to compute and therefore need not be included. The presence of discontinuities in a final field pattern, however, indicates the presence of regions where higher-order terms are needed. An approximation to the true pattern can often be obtained by visually averaging

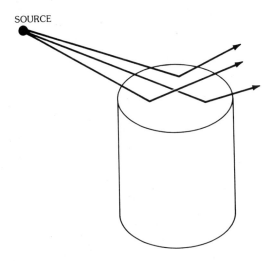

SOURCE

Fig. 35. Illustration of ray paths for end cap reflected fields.

the discontinuities. In any event a pattern generated using UTD gives good visual clues as to its accuracy.

The analysis of three-dimensional models consisting of multiple flat plates and finite elliptic cylinders is a natural extension of these solutions. The analysis has progressed in a systematic fashion by defining the basic elements of the problem and building them into the necessary tools to model real-world scattering structures. The two-dimensional analysis given above is intended to give a feeling for the way the fields of the various terms fit together. More information on the three-dimensional analysis, such as the methods used to find the ray paths, can be found in References 5 and 18.

4. Far Field of Antenna Versus Far Field of Structure

To be in the true far field the receiver has to be an infinite distance away from the transmitting antenna. This implies that the various scattered rays emanating from the antenna and its environment are all parallel. From a practical point of view, a measurement of the far-field pattern of an antenna in its environment must be made at a reasonable distance. The question then arises as to what range is necessary to measure a very good estimate of the true far-field pattern.

Based on a reasonable phase error for nearly parallel rays it is clear that an antenna in free space can be measured at a range given by

$$R = 2D^2/\lambda$$

where D is the maximum extent of the antenna and λ is the wavelength. If the same antenna is placed on or near a significant structure, then one must distinguish the antenna far field from the antenna-and-structure far field. To illustrate this distinction consider a parabolic reflector antenna. The far field of the feed may be a matter of a few feet (or meters), whereas the far field of the feed-and-reflector

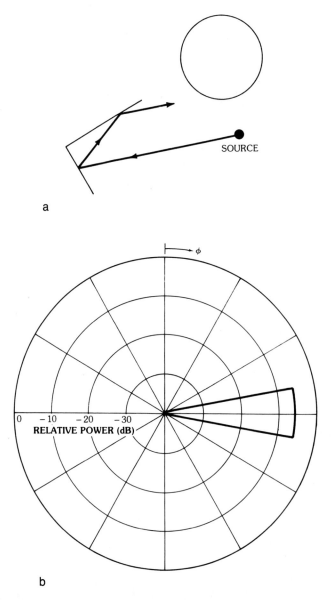

Fig. 36. Illustration of plate double reflection. (*a*) Ray path. (*b*) Fields due to doubly reflected rays.

system may be a mile (or kilometer) or more. In this case the feed is designed to illuminate the reflector in a precise way to generate a narrow-beam far-field pattern. So the reflector is actually an integral part of the antenna system, and the true far-field range is $2D_r^2/\lambda$, where D_r is the diameter of the reflector.

    When an antenna radiates in the presence of a large structure that significantly influences the antenna radiation pattern the structure dimensions must be included in the distance (D) used to define the far-field range. In fact the antenna and its

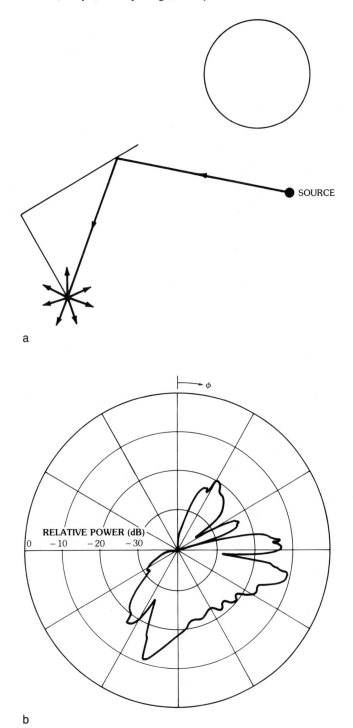

a

b

Fig. 37. Illustration of plate reflection-diffraction. (*a*) Ray path. (*b*) Fields resulting from plate reflection-diffraction.

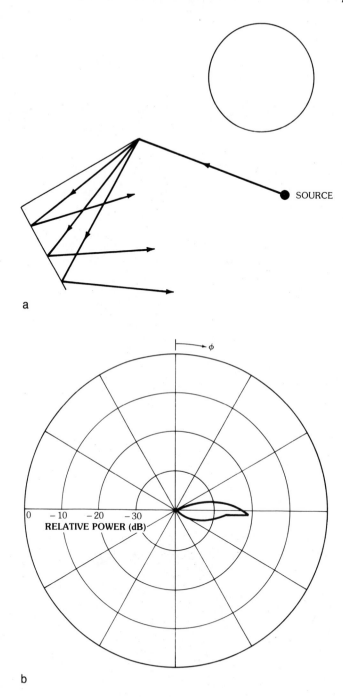

Fig. 38. Illustration of plate diffraction-reflection. (*a*) Plate diffracted-reflected ray paths. (*b*) Fields due to plate diffraction-reflection.

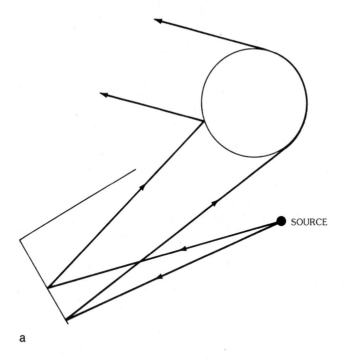

a

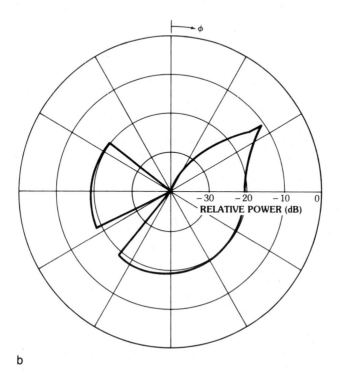

b

Fig. 39. Illustration of reflection from a plate and scattering by a cylinder. (*a*) Ray paths. (*b*) Pattern of fields resulting from (*a*).

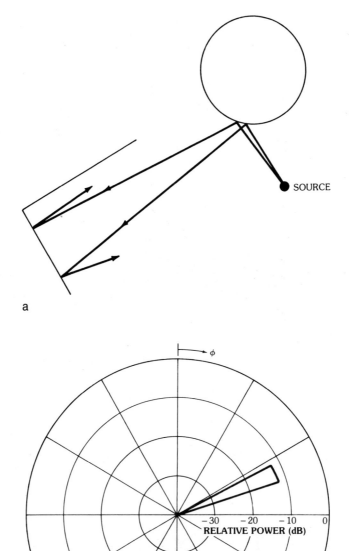

Fig. 40. Illustration of scattering by a cylinder and reflection from a plate. (*a*) Ray paths. (*b*) Fields resulting from (*a*).

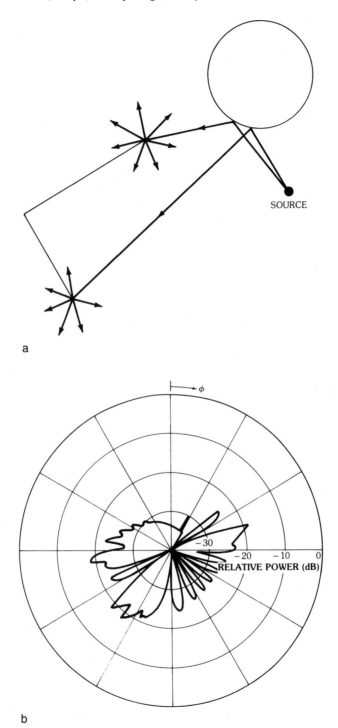

Fig. 41. Illustration of reflection from a cylinder and diffraction by plate edges. (*a*) Ray path. (*b*) Fields due to reflection from cylinder and diffraction by plate edges.

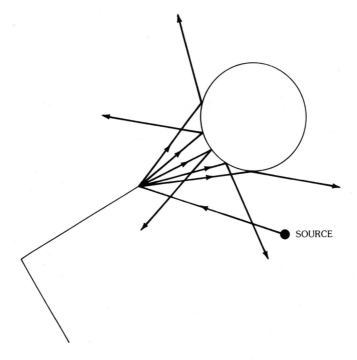

a

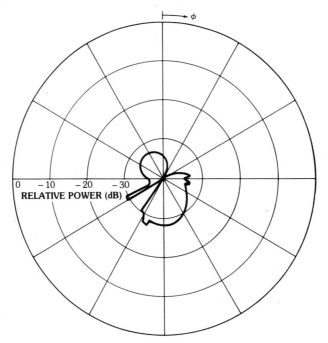

b

Fig. 42. Illustration of diffraction by a plate edge and reflection from a cylinder. (*a*) Ray paths. (*b*) Fields resulting from (*a*).

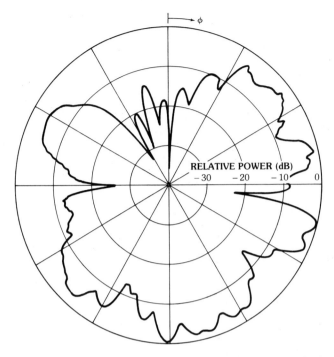

Fig. 43. Total fields of the source in the presence of scattering bodies.

significant environment should be encompassed within a sphere of diameter D; then the far-field range of the antenna-and-structure is given by

$$R = 2D^2/\lambda$$

For a low-gain communications antenna mounted on an aircraft the complete structure is strongly illuminated, and as a result it significantly affects the antenna radiation patterns. Thus the entire aircraft has to be considered as an integral part of the antenna, and the receiver must be moved to the far-field range of the aircraft if the true far field is to be examined. This point will be discussed in more detail later.

As a result of the excessive ranges needed to measure large antenna-and-structure systems, there has been a great deal of interest in determining far-field patterns based on near-field pattern measurements since near-field measurements can be taken efficiently in small anechoic chambers. In order to determine the desired far-field patterns based on these near-field measurements, most of the attention has focused on plane-, cylindrical-, and spherical-wave spectrum approaches. However, each of these spectrum approaches must transform the near-field measured data to the desired far-field result. This transform is basically an integral relationship which in itself can be tedious and expensive. The real solution to this problem lies in a direct approach that simply converts near-field data to the desired far-field results. Thus the following dilemma prevails: far-field patterns are desired but cannot be easily measured directly; near-field patterns are much easier to measure but cannot be transformed simply to the far field.

This dilemma has plagued airborne-antenna designers for many years. The majority of antenna systems operate at microwave frequencies which require far-field ranges in excess of 1000 ft (about 300 m). Such outdoor ranges usually exist in a hilly or mountainous terrain with a transmitter on one peak and a receiver on another and with a deep valley in between to reduce the ground effect. This approach, one would think, is much more expensive than a near-field measurement taken at a range comparable to the size of the aircraft.

The concept applied here is to use a theoretically derived solution to this problem which is valid in both the near and far fields and is such that the near-field solution can be easily verified by a near-field measurement. Once this near-field verification is accomplished, this solution can be directly applied to the far field without the need of a transformation. This concept is based on UTD, which is a ray optical approach such that as the receiver is moved from the far field to the near field, the various ray paths going to the receiver are no longer parallel. This does not violate the UTD postulates in that the receiver is essentially in the far field of each isolated specular point. For instance the receiver might not be in the far field of a flat plate, yet it is sufficiently removed from each of the edge-diffraction points so that the solution is valid (i.e., the receiver is at least a wavelength away from the isolated diffraction point). Consequently a UTD solution can be developed which efficiently solves for near-field patterns. This solution can then be used directly to compute a far-field pattern simply by using the receiver range dictated by the far-field criterion. If one is selective in taking near-field measurements, the ray paths and associated specular points do not deviate greatly from the near field to the far field. This point will be stressed later.

The calculated and measured data presented in Fig. 44 serve as a verification of this near-field analysis, showing results for a test geometry composed of a monopole on a circular cylinder with one attached plate. As indicated, the plane of the plate is tilted 16° with respect to the cylinder axis such that a curved edge exists at the plate-cylinder junction. A 124° conical pattern is presented at a range of 24 in (60.96 cm) from the origin. Note that the receiver moves within about a wavelength of the wing tip for this case. With this in mind the agreement between measured and calculated results is very encouraging.

A practical example demonstrating the improved results obtainable with the near-field analysis is depicted in Fig. 45, which compares measured and calculated elevation plane patterns for a monopole antenna mounted over the cockpit of a Boeing 737 aircraft. The measured data shown were taken in an anechoic chamber using a highly detailed 1:11 scale model aircraft. The receiver range used for the measured data was 26 ft (7.9 m). (This is equivalent to a range of 286 ft or 87.2 m, on a full-scale 737 aircraft.) Two computed patterns are also shown in Fig. 45 for comparison. One of the computed patterns was obtained using the far-field range, while the other was obtained using the near-field analysis. Both of the computed solutions employ the same cylinder and plates model (depicted in Fig. 45a) to simulate the aircraft structure, and comparable scattering terms are included in both solutions. The near-field solution, of course, employs the same receiver range as the measured data. The improved result obtained with the near-field analysis

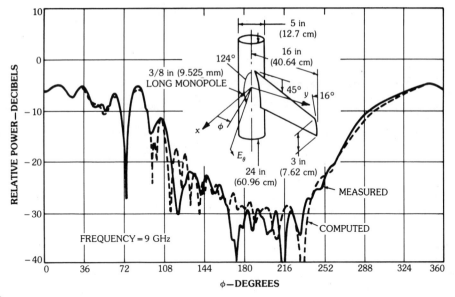

a

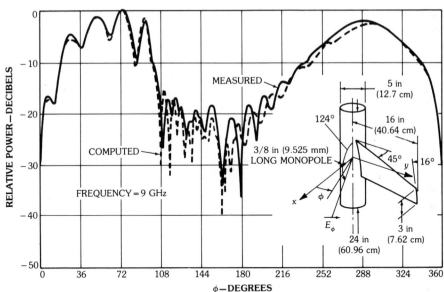

b

Fig. 44. Measured and calculated near-field patterns for the test geometry shown. (*a*) Component E_θ. (*b*) Component E_ϕ.

is especially apparent in the lower left quadrant of the pattern. From the previous results it is clear that the UTD solution can be used to accurately compute the near fields of airborne antennas.

There are many ways to simulate a given aircraft structure using this approach

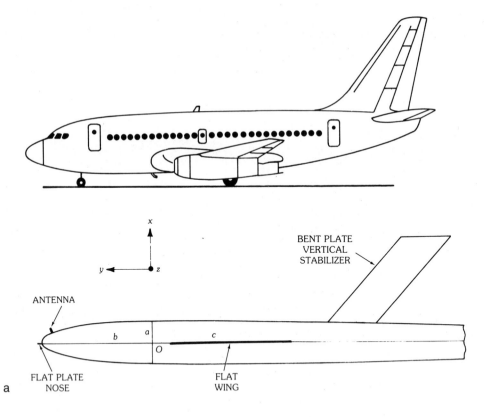

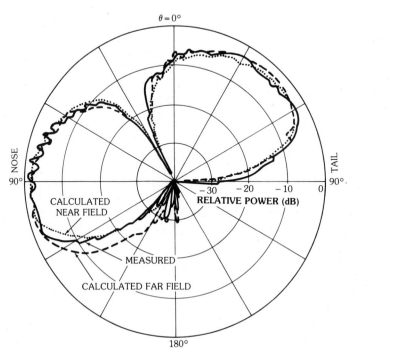

Fig. 45. Boeing 737 aircraft and elevation-plane patterns of a monopole antenna mounted on it. (*a*) Profile views of actual aircraft and model used for analysis. (*b*) Measured and calculated patterns. (*After Burnside, Wang, and Pelton [3], © 1980 IEEE*)

even though specific guidelines are suggested—for example, how much detail of a specific configuration must be included in a simulation model. With this in mind let us assume that an antenna designer has used the code to specify an appropriate antenna and location. At this point it would be satisfying to verify that he or she did achieve the desired far-field pattern performance. Taking the true far-field pattern is out of the question because such a measurement range is not available; however, the designer can simply take some near-field patterns by placing the antenna on the actual aircraft skin. If he or she is going to use this approach, it would be most satisfying to take a near-field pattern whose radiation mechanisms most nearly resemble those of the far field. In order to examine this situation consider the series of elevation-plane patterns taken at a constant range about the three pattern origins shown in Fig. 46. The near-field pattern effects are very noticeable, as shown in Fig. 47. Since the line-of-sight signal from the source to the receiver dominates the significant radiation portion of the pattern, it is not surprising that the near-field pattern should be taken about an origin located at the antenna position. This conclusion is very clearly emphasized by the near-field patterns, where the

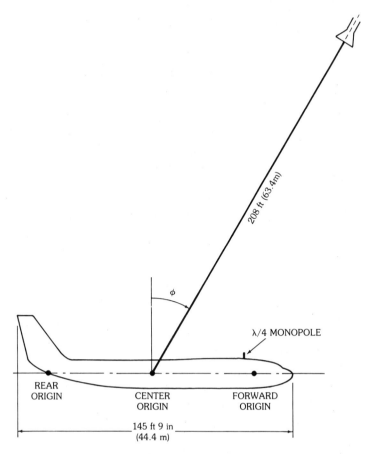

Fig. 46. Geometry associated with near-field pattern computations for the KC-135 aircraft. (*After Burnside, Wang, and Pelton [3],* © *1980 IEEE*)

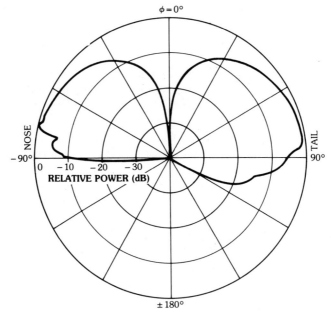

a

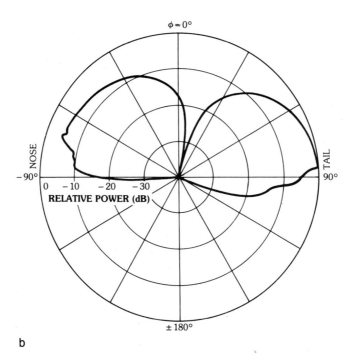

b

Fig. 47. Various elevation-plane patterns taken on a KC-135 aircraft using the geometry in Fig. 46. (*a*) Forward origin. (*b*) Center origin. (*c*) Rear origin. (*d*) Far-field result. (*After Burnside, Wang, and Pelton [3], © 1980 IEEE*)

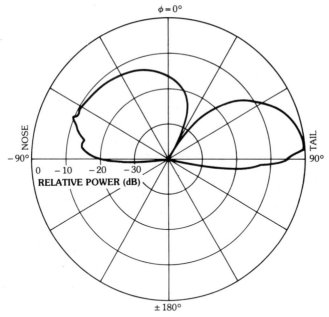

c

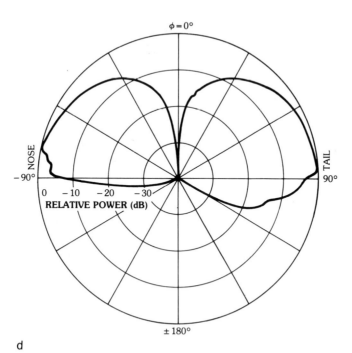

d

Fig. 47, *continued.*

forward-location pattern (closest to the antenna) most nearly resembles the true far-field pattern.

The examples treated so far have concentrated on airborne antenna applications; this procedure, however, can be applied to a very wide variety of problems. Using the NEC BSC, antennas radiating in the presence of a complex environment can be analyzed if the antenna is not mounted on a curved surface. With this code the effects of using various receiving antennas in the near field can be studied.

A simple example of the coupling between a transmitting and a receiving antenna is illustrated in Fig. 48. The coupling between two dipole antennas is shown as they move apart with a square plate placed midway between them. This figure also shows the free-space coupling between the two antennas for comparison. The method of moments solution [10] is also shown on this graph. The UTD solution is obtained by taking the currents of the dipoles in free space and using the diffraction from the four edges and corners of the plate to compute the coupling. The modification of the currents due to the presence of the plate is not taken into account. Since the moment method does take this into account, the ripple that is present in the method of moments result must be due to this effect.

Notice that the coupling between the antennas actually decreases after a certain point as they get closer together. This is due to the distance that the rays must travel to get around the plate even though the dipoles are closer together. In addition the pattern factor of the dipole affects the amplitude of the fields incident on the top and bottom edges as the dipoles get closer.

This next example shows the effect of a receiving antenna's pattern in determining the fields around a shiplike environment. The geometry consists of a plate and cylinder with a dipole between them, as shown in the insert of Fig. 49a.

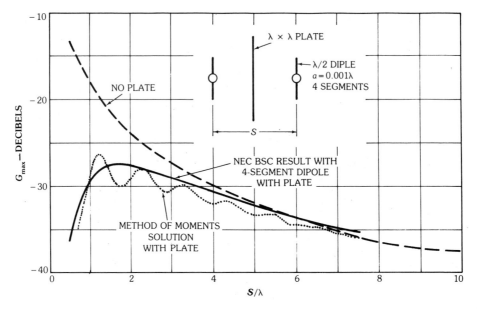

Fig. 48. The coupling between two dipoles in the presence of a plate compared with method of moments results [10].

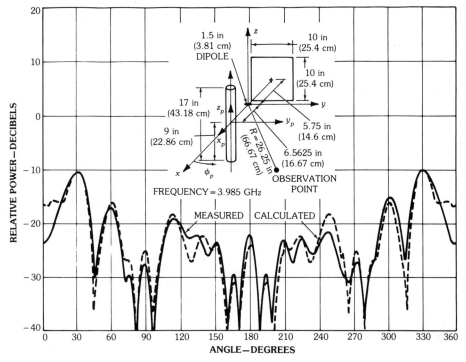

a

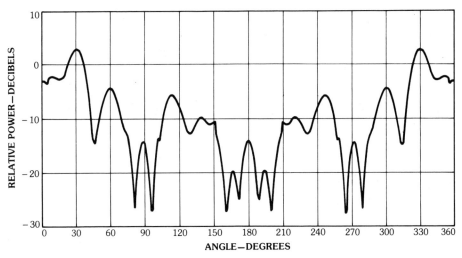

b

Fig. 49. Measured and calculated $E_{\phi p}$ near-zone radiation patterns 26.25 in (66.67 cm) from a dipole in the presence of a square plate and circular cylinder. (*a*) Patterns and test geometry. (*b*) Pattern for the coupling between a *C*-band horn at the observation point and the dipole in the presence of a square plate and a circular cylinder as in (*a*).

Plate-cylinder interaction is ignored here. The observation point is taken as 26.25 in (66.675 cm) around the dipole. The calculated pattern in Fig. 49a is the ϕ-directed **E** field with no receiver present. If a model of a C-band horn is used, the resulting calculated pattern obtained is shown in Fig. 49b. Notice that the directivity of the C-band horn causes changes in the resulting pattern. For example, the null depths around 45° are not as deep. The pattern of the horn has caused some of the rays received to be partially reduced, causing incomplete cancellation between the field terms coming from various directions.

5. Numerical Solutions for Airborne Antenna Patterns

This section is devoted to a general discussion of airborne antenna pattern calculations and the associated simulation models which are based on the following two major codes developed at the Ohio State University: (1) Airborne Radiation Pattern Code [6] and (2) NEC-Basic Scattering Code [7, 23]. The Airborne Code is used to simulate antenna problems for which the antenna is mounted on a curved surface. Thus it need not be limited to aircraft simulations although it has mainly been used to analyze such geometries. The NEC-Basic Scattering Code is used to simulate problems for which the antenna is not mounted on or extremely close to a curved surface. For airborne applications the Airborne Code is normally used to analyze fuselage mounted antenna configurations, whereas the NEC-Basic Scattering Code is applied to wing or stabilizer mounted designs. The simulations used to model a given aircraft are therefore different for the fuselage and wing mounted antenna problems. Since fuselage mounted antennas are much more prevalent, let us begin this discussion based on the Airborne Code simulations.

The Airborne Code simulates the fuselage using a composite prolate spheroid configuration as illustrated in Fig. 50. The antenna is assumed to be mounted directly on the spheroid surface such that reflections from the fuselage are not included. On the other hand one can treat a monopole, provided it is not longer than a quarter-wavelength. As discussed earlier the structure closest to the antenna plays the dominant role in dictating the pattern since the radiated field strength is largest in the antenna vicinity. With this in mind the prolate spheroid parameters should be adjusted to best represent the aircraft fuselage near the antenna location. Once the fuselage is defined, one uses flat plates to simulate the remaining appendages. The flat plates can be connected together at their corners to form a boxlike structure, or one plate can penetrate through another as shown in Fig. 51. In all cases the flat plates are defined simply by their corner locations, with the user making sure that they are flat, i.e., all corners lie in the same plane. In order to illustrate how these structures can be used to simulate an aircraft geometry let us consider several practical configurations for which measured results have been obtained. These results also allow one to evaluate the accuracy of the numerical solutions.

The Boeing 737 commercial jet is studied initially because of its simplicity. The line drawings shown in Fig. 52 are taken from *Jane's All the World's Aircraft* [24] and will be used to generate the simulation model. In the first case, a monopole antenna is mounted on the top center line of the 737 at station 220, which is just above the pilot's head as illustrated in Fig. 53. The prolate spheroid is defined to

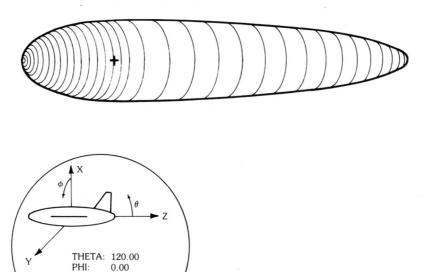

Fig. 50. Composite spheroid fuselage.

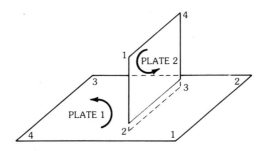

Fig. 51. A flat plate intersecting another flat plate.

represent the nose of the aircraft in a best-fit sense. The wings and horizontal stabilizers are simulated next, using finite flat plates which are attached to the fuselage. Note that one need not actually connect the plates to the fuselage, but simply indicate to the code that they should be attached. The curved junction edge is then generated internal to the code. The front view of the 737 as shown in Fig. 52 is then used to determine the x coordinate values of the simulation plates (for example, $x = 0$ for the wings). Since the horizontal stabilizers are so far away from the antenna and shadowed by the fuselage, they can be neglected. The top view of the 737 as shown in Fig. 52 is used to determine the y and z coordinate values for the wings. With this information the wing simulation plates are defined as shown in Fig. 53.

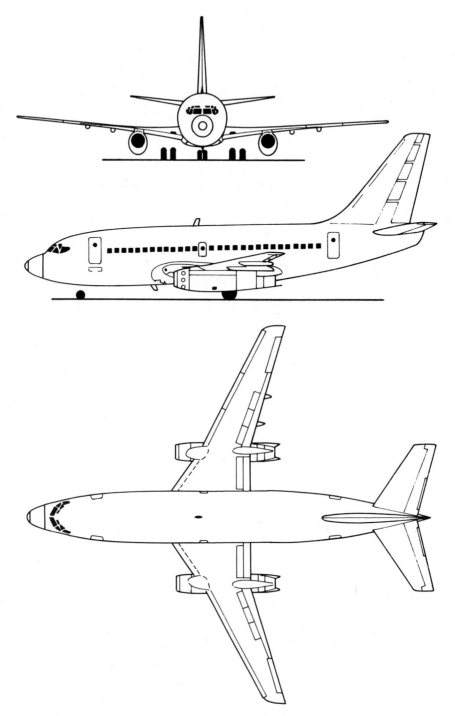

Fig. 52. Line drawing of Boeing 737 aircraft. (*After Taylor [24],* © *1974 Jane's Publishing*)

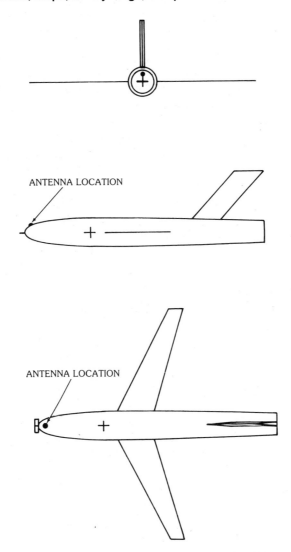

Fig. 53. Simulation model of the Boeing 737 aircraft used for calculations.

Since many airborne antennas are mounted on or near the fuselage center line the vertical stabilizer must be simulated by a structure with finite thickness. This thickness is very significant in that it tends to shadow the direct field from the antenna for aft radiation directions. In order to approximate this effect and maintain the finite flat plate representation, the vertical stabilizer is simulated by a bent plate as shown in Fig. 53. In addition the vertical stabilizer simulation must represent the location and orientation of the actual leading edge. The leading-edge geometry is important because it dictates the edge-diffracted field scattering cone.

For an antenna mounted closed to the nose of the aircraft one must simulate the nose section that extends outward from the base of the windshield. As described in [2], before the nose section can be adequately simulated, a practical representation for the radome must be found. A comprehensive study of radomes and their effect on the radiation patterns of antennas mounted in their vicinity is far too complex to be considered here. In fact the analysis of the scattering properties of radomes and the structures mounted under them is an interesting and relevant problem worthy of investigation. For simplicity it is assumed here that the radome is perfectly transparent. This is not an unreasonable assumption, as radomes are designed to be transparent at least at certain frequencies. This leaves a short blunt-looking nose section which extends out from the front of the aircraft. Various complex structures were investigated to simulate this section, all of which led to very inefficient computations. Furthermore it was found that the nose section normally has little effect on the resulting pattern. Consequently the nose section for simplicity is simulated by a bent-plate model as shown in Fig. 53. These bent plates simply model the major dimensions of the nose section as illustrated in Fig. 52.

The radiation patterns for a $\lambda/4$ monopole mounted at station 220 on top of a Boeing 737 aircraft are then calculated using the UTD solutions described in the simulation section. The three principal-plane results shown in Fig. 54 are found to be in good agreement with the measurements taken by Melvin Gilreath [2] at NASA (Langley Research Center, Virginia). The major discrepancy in these results occurs in the elevation-plane pattern and is attributed to the near-field pattern measurement versus the far-field calculation. This topic was discussed in the previous section.

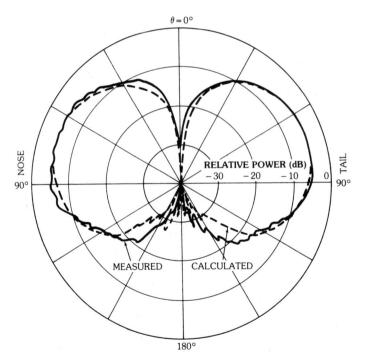

a

Fig. 54. Principal-plane radiation patterns for a $\lambda/4$ monopole at station 220 on top of a Boeing 737 aircraft. (*a*) Roll-plane radiation pattern (E_ϕ). (*b*) Elevation-plane radiation pattern (E_ϕ). (*c*) Azimuth-plane radiation pattern (E_θ).

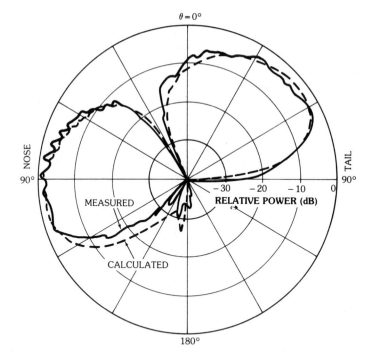

b

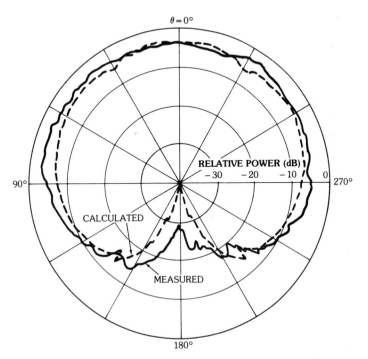

c

Fig. 54, *continued.*

The pattern coordinate system (x_p, y_p, z_p) used for the experimental results is shown in Fig. 55, in which the z_p axis is oriented vertically. In order to calculate a radiation pattern in terms of this coordinate system these coordinates must be transformed into the aircraft coordinates for analysis. The code simply allows the user to define this new coordinate system for pattern calculations and relieves the user of this complexity. The remaining patterns use this coordinate system to define the specific pattern cut. If θ is held constant, one obtains conical patterns about the vertical as shown in Fig. 55, whereas if ϕ equals a constant, one generates a great circle cut, which is also illustrated in Fig. 55. In each case the previously calculated results compare very favorably to the appropriate measurements. It is noted that the measurements have some asymmetry which might be attributed to model distortions due to the shifting of aircraft weight as it is rotated.

In order to illustrate the overall accuracy as originally presented in Reference 2, the complete volumetric patterns in terms of the directive gain are presented in Fig. 56. The various directive gain regions are indicated by the color code. For example, the red color indicates the region of space where the gain level is greater than 0 dB. In other words this is a region where the radiation intensity of the antenna of interest is greater than that of an isotropic point source. The yellow color indicates the region where the radiation intensity is greater than $-3\,$dB but less than 0 dB relative to an isotropic source. Similarly, the green, blue, purple, dark blue, and gray or black stand for -6-dB, -10-dB, -15-dB, -20-dB, and less than -20-dB gain levels, respectively. Good agreement is obtained for each of the gain levels. Note that the θ and ϕ variables used in the pattern plots are defined in Fig. 55.

The previous examples illustrate many very significant points associated with modeling complex structures using numerical solutions. Obviously the simulation

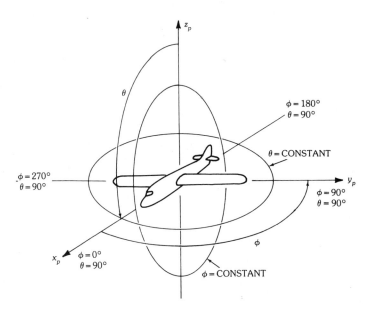

Fig. 55. Coordinate system used for experimental measurements.

a

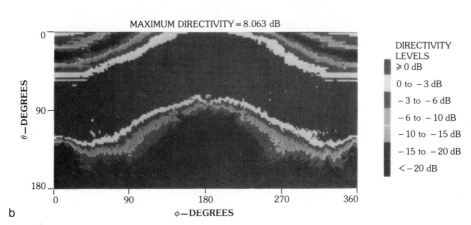

MAXIMUM DIRECTIVITY = 8.063 dB

b

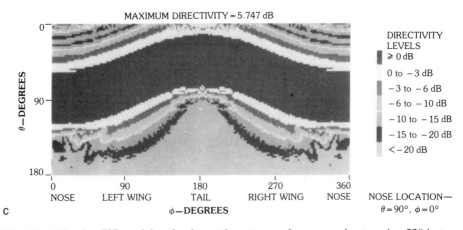

MAXIMUM DIRECTIVITY = 5.747 dB

NOSE LOCATION—
$\theta = 90°, \phi = 0°$

c

Fig. 56. A Boeing 737 model and volumetric patterns of a monopole at station 220 in terms of directive gain. (*a*) A 1:11 scale model. (*b*) Measured volumetric directive gain pattern. (*c*) Calculated volumetric directive gain pattern. (*After Yu, Burnside, and Gilreath [2]*, © *1978 IEEE*)

model represents a greatly simplified version of the actual aircraft. On the other hand the agreement achieved between the calculated and measured results indicates that this simplified version is an adequate representation. How is this possible? To understand this idealization let us examine the scattering from each structure.

The fuselage has the dominant effect on the pattern; more specifically, the fuselage nearest the antenna is most significant. The simulation structure represents this region of the aircraft very nicely, so one might expect good agreement for the scattering associated with the fuselage. When one looks at the aircraft, the wings represent the most significant appendages and must be accurately modeled. The wings affect the pattern in two dominant ways: (1) since they are large and basically flat structures, they create a large reflected field; (2) they shadow the energy flowing from the antenna as the receiver moves under the wings. Since each of these effects can be modeled using a simple flat plate, our wing simulation is not an unreasonable representation. The same conclusions can be stated for the stabilizers and nose section. Recall that the horizontal stabilizers were not included in the previous model for the 737 aircraft. One can investigate the need to model these structures by simply adding them to the model and determining their significance. If they are not needed in the model for a given antenna location, then they should be left out in order to simplify the calculations. In any event these results indicate that the simulation model shown in Fig. 53 does adequately represent the 737 aircraft even though the fuselage is not a prolate spheroid and the wings and stabilizers are not flat plates.

Whenever one attempts to simulate a new configuration one should begin the simulation using a very simplistic model to which can be added various appendages that better represent the actual structure. This is done by taking the most significant or bulky structures first. This process continues until the solution converges, which results in a great simplification of a very complex aircraft.

Since the bulk of airborne antenna designs involve military aircraft which have numerous complex appendages, it is most illuminating to discuss the modeling considerations for such an aircraft. As an example let us consider the F-4 fighter aircraft illustrated in Fig. 57. In this case a uhf blade antenna, which is simulated by a short monopole, is mounted on the belly of the aircraft. Since the antenna is mounted on the bottom of the F-4, one must turn the aircraft over in terms of the simulation model. This requirement is done to ensure that one defines the geometry associated with the belly of the aircraft differently from the top, because most aircraft, especially military ones, have a definite shape change from top to bottom. With the F-4 turned upside down, there will exist a coordinate system difference which can be corrected in terms of the pattern coordinates. For example, the $\theta = 105°$ conical pattern from the vertically upward axis is of interest, which implies that the pattern is taken below the horizon on the true aircraft. For the simulation model this corresponds to a conical pattern 75° from the vertical since the model is upside down. The fuselage of the F-4 is initially modeled by the best-fit composite prolate spheroid to represent the fuselage in the vicinity of the blade antenna. The fuselage-alone pattern is shown in Fig. 58a for a frequency of 375 MHz. Next, the flat-plate wings are added to the model, taking the data directly off the appropriate line drawings. The fuselage-plus-wings pattern, as

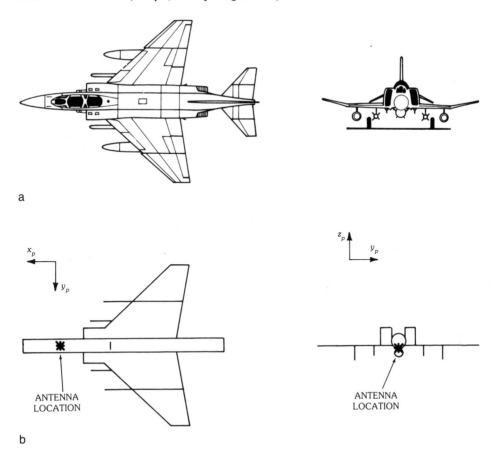

Fig. 57. The F-4 aircraft and computer-simulated model (roll plane only). (*a*) Aircraft. (*After Taylor [24], © 1974 Jane's Publishing*) (*b*) Model.

shown in Fig. 58b, indicates that the wings play very little role in this conical pattern cut. This particular F-4 had a bomb mounted along the fuselage center line, two missile racks on the inboard side of the wings, and two wing fuel tanks as shown in Fig. 57. Using our general modeling approach these structures will be simulated in a very simplistic form. Since their dominant effect on the pattern is shadowing of the incident field let us simulate these structures using single flat plates to represent each one. The plate cross-sectional shapes are chosen to represent the shadowing outline of the given structure. Considering the bomb first, its shadowing outline is basically circular, so the flat-plate simulation is represented by a hexagonal plate which is located at the midpoint of the bomb. The resulting pattern with bomb added is shown in Fig. 58c. Again note that the bomb does not play much of a role for this case. The missile racks are examined next, using a simple flat-plate representation. The composite pattern for this configuration is shown in Fig. 58d. It is apparent from these results that the missile racks play a significant role; thus one should devote additional effort to better represent them. Next, let us consider the outboard fuel tanks which are, also, initially represented

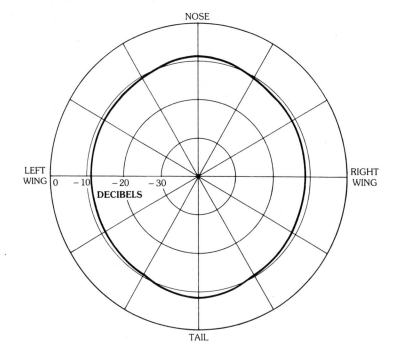

a

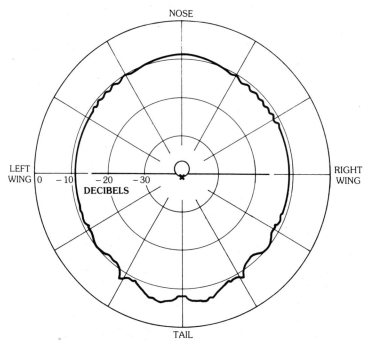

b

Fig. 58. Calculated F-4 azimuthal conical patterns at $\theta_p = 105°$ and $f = 375$ MHz. (*a*) For cylinder-only model. (*b*) For cylinder and two wings model. (*c*) For cylinder, two wings, and blockage under fuselage model. (*d*) For model with missile racks added to model (*c*). (*e*) Measured and calculated final results with outboard fuel tanks added to (*d*). (*Aircraft drawing after Taylor [24], © 1974 Jane's Publishing*)

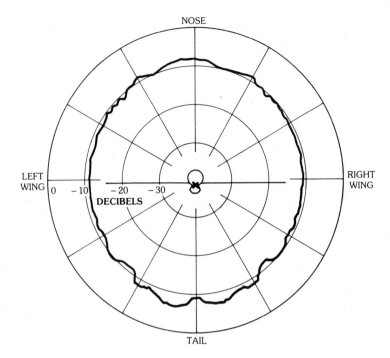

c

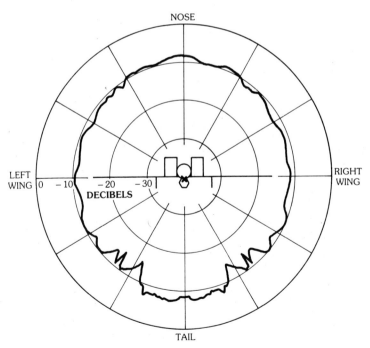

d

Fig. 58, *continued.*

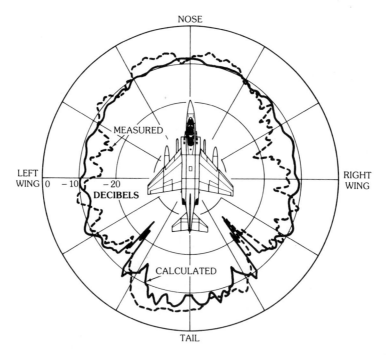

e

Fig. 58, *continued.*

by finite flat plates which simulate the shadowing effects. The complete pattern for this final structure, shown in Fig. 58e is compared with a measured result. Note that the basic pattern shape is re-created by the calculated results; however, the calculated results in the region forward of the wings could be improved. In this area, for example, a better representation of the missile racks and fuel tanks is needed so that not only the proper shadowing is accomplished but also the reflected field is better represented. Note that the earlier results indicated that both these geometries had a significant effect on the pattern such that more realistic models may be necessary. In any event one could then proceed to box out these structures using multiple flat plates until the solution again converges. Nonetheless the pattern comparison illustrated in Fig. 58e indicates the simplicity that is possible when these codes are used to represent very complex real-world structures.

A configuration which leads to some interesting pattern results involves aircraft with a high T tail, such as the C-141 shown in Fig. 59. The interest here is associated with the shadowing and reflection by the large horizontal stabilizer, which is mounted near the top of the vertical stabilizer. By using the prolate spheroid fuselage and flat-plate appendages one can simulate the C-141 as shown in Fig. 59. Fig. 60 shows various calculated elevation patterns which illustrate the effect of the vertical stabilizer as well as the horizontal stabilizer, i.e., the high T tail. Note that the horizontal stabilizer not only creates a shadow above the horizon but also a large reflected field below the horizon. For various clutter reasons this reflected lobe can be very undesirable in some applications.

As a final example of the Airborne Code the F-16 fighter illustrates the simula-

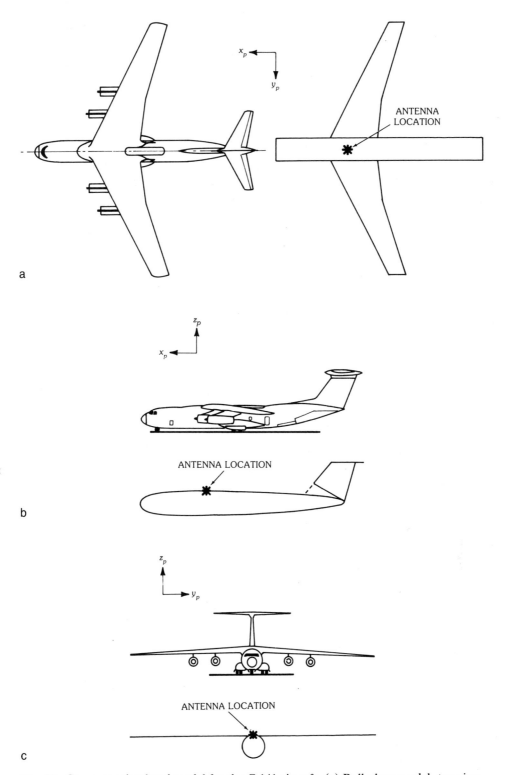

Fig. 59. Computer-simulated model for the C-141 aircraft. (*a*) Roll-plane model, top view. (*b*) Elevation-plane model, side view. (*c*) Roll-plane model, front view. (*Courtesy of USAF*)

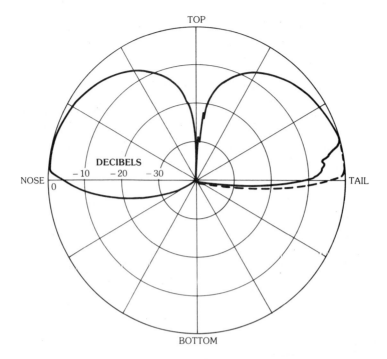

a

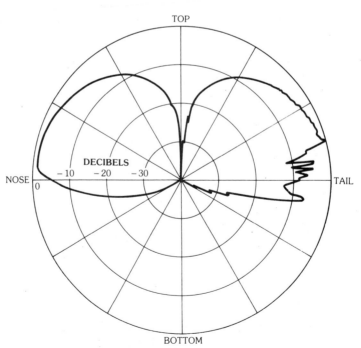

b

Fig. 60. Calculated elevation patterns. (*a*) Without T tail and vertical stabilizer (dashed line) and without T tail but with vertical stabilizer (solid line). (*b*) With T tail and vertical stabilizer.

tion capability in terms of a very complex geometry. The F-16 line drawings are illustrated in Fig. 61, and the simulation model is shown for comparison purposes in Fig. 62. Note that the flat plates are used here to complete the representation of the fuselage as well as simulating the wings, canards, and stabilizers. The three principal-plane patterns for this configuration are shown in Fig. 63 and compared with measured results obtained by General Dynamics (Dallas, Texas).

As mentioned previously, if the antennas are not mounted on the fuselage of the aircraft, the NEC-Basic Scattering Code can be used to predict the antenna patterns. For example, measurements of a slot antenna mounted on the wing of a 1:20 scale model of a Boeing 737 were taken at NASA (Langley Research Center, Virginia). An analytic model composed of a finite elliptic cylinder and flat plates is used to simulate this structure as shown in Fig. 64. In this figure the larger radius along the positive x axis is used for an antenna mounted on top of the wings, and the smaller radius along the negative x axis is used for an antenna mounted on the bottom. Note that the back of the fuselage of the aircraft is simulated by cutting the elliptic cylinder at an angle for these calculations. The curvature at the nose of the scale model has been neglected. The full scale model frequency is 1.75 GHz, which corresponds to 35 GHz for the scale model. The source is a K_a-band wave-

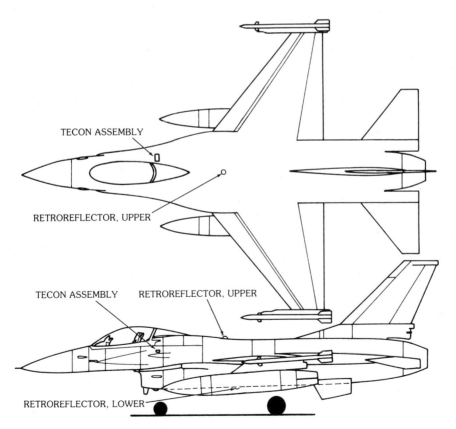

Fig. 61. Line drawing of F-16. (*Courtesy of USAF*)

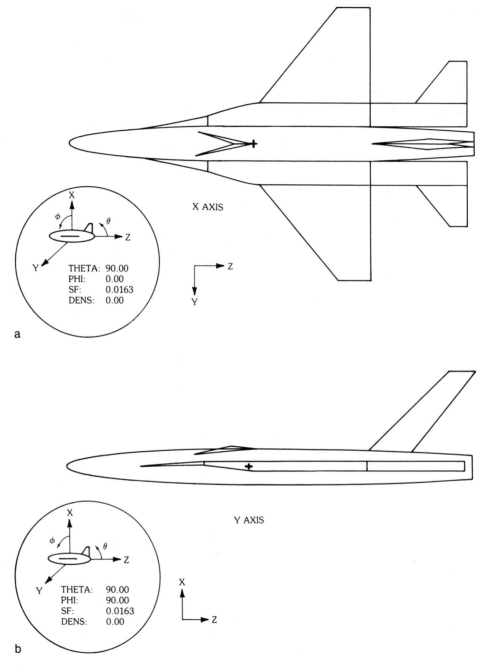

X AXIS

THETA: 90.00
PHI: 0.00
SF: 0.0163
DENS: 0.00

a

Y AXIS

THETA: 90.00
PHI: 90.00
SF: 0.0163
DENS: 0.00

b

Fig. 62. Computer-simulated model of F-16. (*a*) Top view. (*b*) Side view. (*c*) Front view.

guide mounted parallel to the y axis. This is modeled as a magnetic dipole with finite width in the calculations. The $E_{\phi p}$ pattern for an elevation cut 30° off the nose is compared to a measured result as shown in Fig. 65. The agreement is very good. The discrepancies that exist can be partially attributed to the simple elliptic cylinder representation of the fuselage in the calculations along with normal experimental

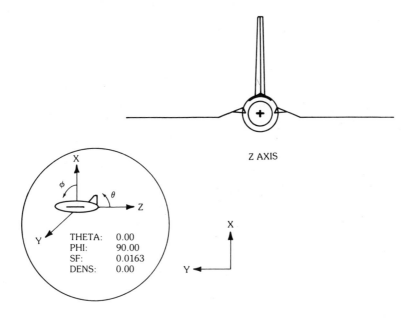

Fig. 62, *continued.*

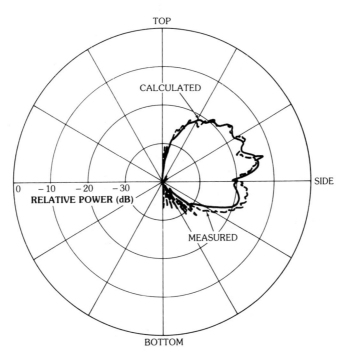

Fig. 63. Radiation patterns of a $\lambda/4$ monopole antenna mounted on top of an F-16 aircraft. (*a*) Roll-plane (E_ϕ) pattern. (*b*) Elevation-plane (E_ϕ) pattern. (*c*) Azimuth-plane (E_θ) pattern.

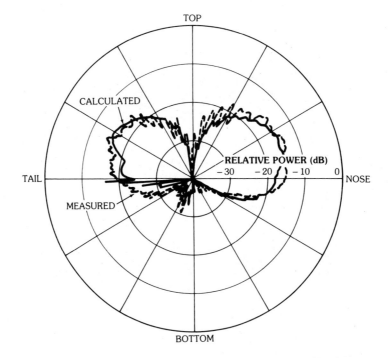

b

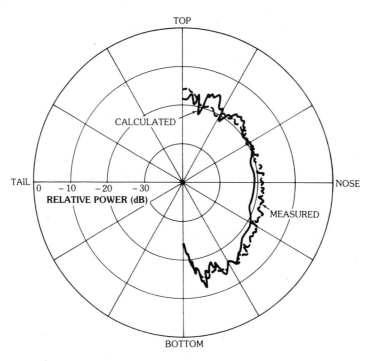

c

Fig. 63, *continued.*

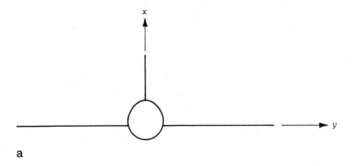

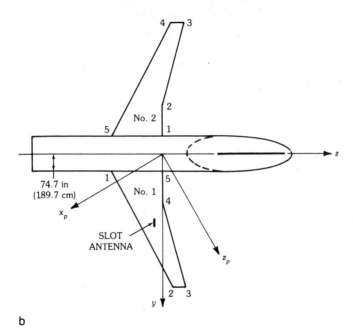

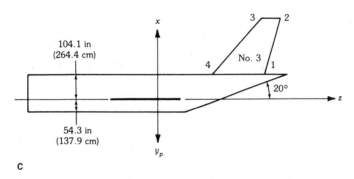

Fig. 64. Geometry of Boeing 737 aircraft model with slot antenna mounted on wing. (*a*) Front view. (*b*) Top view. (*c*) Side view.

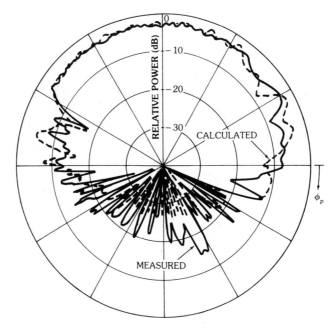

Fig. 65. Measured and calculated $E_{\phi p}$ results.

errors. These results, as well as the others, serve as a broad scope of airborne antenna configurations which illustrate the versatility and accuracy of the UTD numerical solutions.

6. Numerical Solutions for Shipboard Antenna Patterns

The previous section discussed airborne antenna patterns in particular. The UTD numerical solutions can be used to analyze many other practical type structures. For example, the plates can be combined together in such a way as to model the superstructure of a ship, the body of a truck or tank, the body of a satellite with its solar panels, or the facade of a building. The effect of the ground can be simulated by a semi-infinite half-space. Since the antennas in most of these situations are not mounted on a curved surface, the NEC-Basic Scattering Code is most suitable here.

This section intends to illustrate some examples of antennas in a shipboard environment. A typical ship presents a much more cluttered scattering environment for an antenna than an aircraft does. Such a cluttered environment is illustrated in Fig. 66. As a consequence it is more difficult to assume that a majority of the features in such an environment can be modeled. It is recommended, therefore, that a given situation be studied systematically, looking at individual pieces of the problem separately, building up experience, and developing intermediate conclusions. When the designer starts out with a simpler problem he or she can better diagnose the results to make sure that they are reasonable. The code also runs much faster using fewer components. This gives the user the feeling

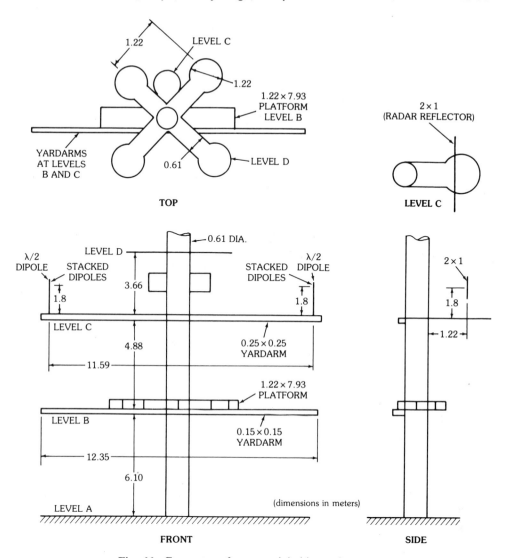

Fig. 66. Geometry of a potential ship environment.

that he or she can try many different situations cost-effectively. As experience is gained for given classes of problems, more details can be added.

The first example illustrates the modeling of an antenna of interest to a shipboard antenna designer. The antenna configuration is shown in Fig. 67. It is composed of four circularly polarized antennas that can be represented as crossed dipoles over a square ground plane. The currents on the antennas were obtained using the NEC-Moment Method Code [11] with six segments per dipole. To save time an analytic representation of the dipoles can be used in this case with equivalent accuracy. The effect of the ship's deck and ocean is modeled as an infinite ground plane here. This model is used to compare against measured results from a 1:10 scale model [25]. The results for the circularly polarized field (relative directive

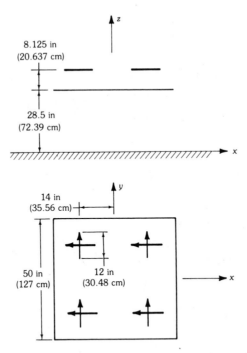

Fig. 67. Geometry for the problems of dipoles over a square plate and infinite ground plane, showing the side and top views.

gain) are compared in Figs. 68a, 68b, and 68c for the antenna pointed at $\theta = 0°$, 30°, and 60° (90°, 60°, and 30° in elevation), respectively. The calculated results are from the NEC-Basic Scattering Code. The measurements were made on a 1:10 scale model at NOSC [25]. The full scale frequency is 0.320 GHz. The basic features

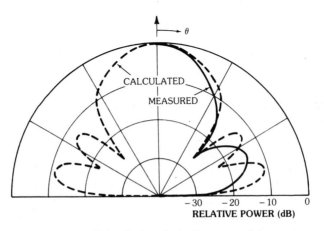

a

Fig. 68. Measured and calculated circularly polarized patterns for an antenna. (*a*) Antenna pointed at $\theta = 0°$ (90° elevation). (*b*) Antenna pointed at $\theta = 30°$ (60° elevation). (*c*) Antenna pointed at $\theta = 60°$ (30° elevation).

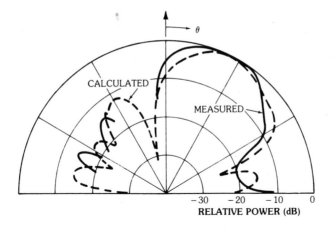

b

c

Fig. 68, *continued.*

of these patterns are reconstructed by the calculated results; however, the secondary lobes are off slightly. This is due to the simplified model used to calculate the results along with normal experimental errors. For example, it is difficult to simulate circularly polarized antennas because the low-level fields depend on the precise phasing of the elements. The model used for the computations can be greatly improved if more details are known about the experimental models and setup. For example, the polarization of the pattern could be more closely matched if more details of the receiving antenna were known. Also, the model of the antenna could be improved by adding more detail, such as using a circular cylinder to represent the antenna pedestal and using more plates to model the back plate, and so on.

The next example is for an antenna mounted on the yardarm of a mast as shown in Fig. 69a. Measurements were made for this configuration at NOSC [25] using a 1:10 scale model at a frequency of 4.0 GHz. The geometry used to model this structure for the calculations is illustrated in Fig. 69b. The antenna is simply

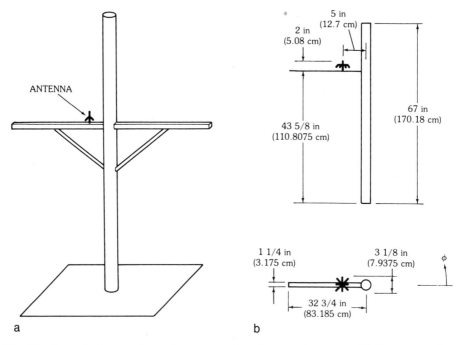

Fig. 69. Example for an antenna in a mast and yardarm environment. (*a*) Geometry of mast and yardarm. (*b*) Top and side view of computer model corresponding to (*a*). (*After Burnside, Rudduck, and Marhefka [4], © 1980 IEEE*)

modeled as a dipole of scale model length 1.925 in (4.8895 cm). The results calculated for the antenna in the presence of the circular cylinder and a flat plate representing the yardarm are compared against the measured results in Fig. 70 for the azimuth plane. The calculated result is from the NEC-Basic Scattering Code including the mast and yardarm [25]. The measurements were made on a 1:10 scale model at NOSC by L. S. Hansen [25]. The result for the cylinder alone gives the basic shape of the pattern. The plate gives the results better agreement in the null depths and in the shape of the lobe at 180°. The lack of a small feature at about 160° might be due to something present in the measurements that is not modeled by the code, or it could be higher interactions between the plate and the cylinder that have not been included. It is also possible that the feature is caused by the currents produced on the antenna due to the reflected field from the cylinder incident on the antenna structure.

The next example illustrates the blockage due to multiple masts [26] as shown in Fig. 71. The antenna system is composed of four broadband dipoles mounted around one of the masts. They are nominally supposed to produce an omnidirectional pattern. The antennas for this calculation are modeled as simple analytically defined dipoles. The frequency for this problem is 0.3125 GHz. Note that this means that the radius of the mast around which the antennas are mounted is only 0.1458 wavelength and that the dipoles are very close to the surface. This is pushing the accuracy of UTD, so the solution should be checked against a known result. This is accomplished by comparing the pattern for one of the dipoles in the

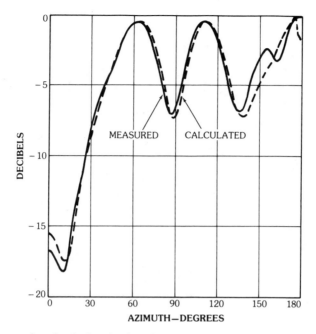

Fig. 70. Measured and calculated azimuth patterns for the mast antenna example. (*After Burnside, Rudduck, and Marhefka [4], © 1980 IEEE*)

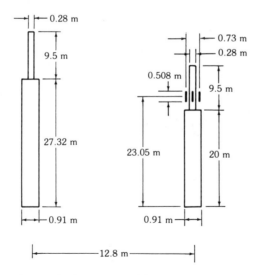

Fig. 71. Geometry used to study the pattern of four antennas in the presence of two masts.

presence of an infinite cylinder, using the exact modal solution against the UTD result for a very long cylinder. The results are compared in Fig. 72, with excellent agreement verifying the validity of the UTD solution. The azimuth-plane pattern for the vertically polarized electric field in the presence of two masts is shown in Fig. 73. This figure indicates a prediction of the amount of ripple that will

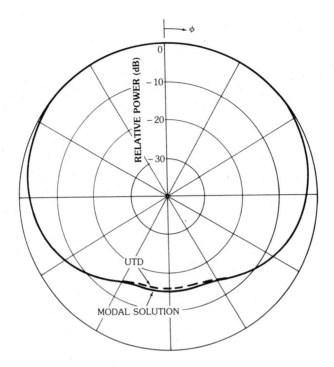

Fig. 72. The exact modal solution and the UTD solution of the E_θ pattern in the azimuth plane for one dipole on one mast.

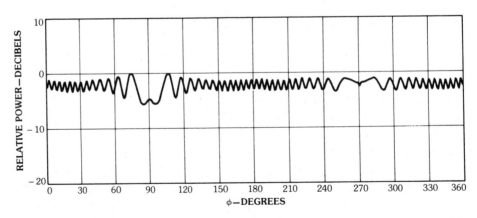

Fig. 73. Calculated azimuth plane pattern at $f = 0.3125$ GHz and a range of 250 m for four dipoles located around a mast in the presence of a second mast with higher-order cylinder interaction terms.

be produced by the presence of a second mast. The pattern for dipoles without the second mast is a straight line that goes through the average of the pattern in Fig. 73; as a result it is not shown here.

7. Summary

An antenna designer often must consider many contradictory situations when he or she is deciding where to place an antenna in a complex environment. First of all, there are usually many antennas competing for the same prime locations, such as the top of a ship's mast or the nose of an aircraft. Obviously this is not possible or even desirable from the standpoint of needed isolation between the various systems.

In many cases the environment can be used in the designer's favor. For example, if it is desired to minimize the coupling between two antennas, Friis' transmission formula says that the further the two antennas are apart, the less coupling there will be between them for the same orientation. If there is an intervening structure between them, it is possible that the minimum coupling will occur when the two antennas are closer together. This can happen when the direct path between the antennas is blocked so that the rays must travel around the structure.

In the case of radiation patterns the structure can also help shape the pattern by directing the gain away from a given direction and into a desired region; however, such terms can produce unwanted ripple. The frequency of the ripple can be minimized by placing the radiating antenna close to the reflecting or diffracting structures. The depth of the nulls in the ripple, however, can be greater because of the increased strength of the interfering fields.

In general it is not possible or even desirable from a time standpoint to try to model the complete scattering environment or even to perfectly define the antenna. It is difficult to make statements that are appropriate all the time since every situation is different; it is possible, however, to make some useful comments. First, when starting to model a new class of problems it is better to begin simple and systematically build the model up. This enables a logical evaluation of the results to make sure that they make sense, and it allows the designer to build up an intuitive understanding of the physical phenomena involved. It is also most cost-effective to start with a simple dipole test antenna or array of simple dipoles. This is usually possible since most small antennas, no matter how complicated they seem physically, radiate a pattern very similar to a dipole. This preliminary design stage often requires many trial-and-error computer runs. After the validity of the computer model is established and the most desirable location is determined, then the antenna and model can be "tweaked" up to confirm the conclusions.

When one is defining the model within the limited bounds of the computer code's capabilities it is often necessary to make compromises on the appearance of the model. The following hints can be of use in this process. The designer should emphasize the structure in the main beam of the antenna, that is, in the directions where the majority of the radiated energy strikes the structure. For low-gain antennas one should emphasize the model closest to the antenna because the power decays as $1/R^2$, where R is the range from the antenna. Always give the most consideration to the parts of the model in the direction of the pattern. It may therefore be necessary to have more than one model for different pattern regions.

As seen in the previous sections it is very useful to study the effects of the different parts of the structures separately. One of the benefits to a UTD computer model is that the major contributors to the pattern can be easily dissected and studied. This gives added insight into the problems involved as well as possible corrective measures. A rule of thumb concerning the higher-order interaction ray terms is that the higher the order of the term, the narrower its pattern effect. The magnitude of the field may be comparably large at the shadow boundaries, but on the whole these terms will be less significant than the first-order effects. This implies that these computationally time-consuming fields can be left out of the analysis until the latter stages of the design phase. Finally, whenever possible, use both numerical and experimental results to get a total picture of the problem. Either method can lead the designer to wrong conclusions if there are errors in the system. If both methods show agreement, or at least the same trends, the designer will be much more confident of his or her conclusions.

The use of high-frequency numerical techniques in solving practical engineering problems has been discussed in this chapter. The validity of these methods has been demonstrated through numerous comparisons with experimental results. The versatility of user-oriented numerical solutions has been illustrated by the wide variety of problems solved using the Airborne and NEC-BSC codes. All of this indicates that high-frequency numerical solutions are a very powerful tool for designing antenna systems located in a complex environment.

8. References

[1] W. D. Burnside, M. C. Gilreath, R. J. Marhefka, and C. L. Yu, "A study of KC-135 aircraft antenna patterns," *IEEE Trans. Antennas Propag.*, vol. AP-23, pp. 309–316, May 1975.

[2] C. L. Yu, W. D. Burnside, and M. C. Gilreath, "Volumetric pattern analysis of airborne antennas," *IEEE Trans. Antennas Propag.*, vol. AP-26, pp. 636–641, September 1978.

[3] W. D. Burnside, N. Wang, and E. L. Pelton, "Near-field pattern analysis of airborne antennas," *IEEE Trans. Antennas Propag.*, vol. AP-28, no. 3, pp. 318–327, May 1980.

[4] W. D. Burnside, R. C. Rudduck, and R. J. Marhefka, "Summary of GTD computer codes developed at the Ohio State University," *IEEE Trans. on Electromagnetic Compatibility*, vol. EMC-22, no. 4, pp. 238–243, November 1980.

[5] R. J. Marhefka, "Analysis of aircraft wing-mounted antenna patterns," *Tech. Rep. 2902-25*, June 1976, the Ohio State University ElectroScience Laboratory, Department of Electrical Engineering; prepared under Grant NGL 36-008-138 for NASA, Washington, DC. Also a doctoral dissertation, June 1976, Ohio State University.

[6] H. H. Chung and W. D. Burnside, "General 3D airborne radiation pattern code user's manual," *Tech. Rep. 711679-10*, July 1982, the Ohio State University ElectroScience Laboratory, Department of Electrical Engineering; prepared under contract F30602-79-C-0068 for Rome Air Development Center.

[7] R. J. Marhefka and W. D. Burnside, "Numerical electromagnetic code–basic scattering code, NEC-BSC (version 2), part I: user's manual," *Rep. 712242-14*, December 1982, the Ohio State University ElectroScience Laboratory, Department of Electrical Engineering; prepared under Contract No. N00123-79-C-1469 for Naval Regional Contracting Office.

[8] R. C. Rudduck and Y. C. Chang, "Numerical electromagnetic code (NEC)–reflector antenna code, part I: user's manual (version 2)," *Rep. 712242-16*, December 1982, the Ohio State University ElectroScience Laboratory, Department of Electrical Engineering; prepared under Contract No. N00123-79-C-1469 for Naval Regional

Procurement Office.

[9]   J. H. Richmond, "Radiation and scattering by thin-wire structures in a homogeneous conducting medium," *IEEE Trans. Antennas Propag.*, vol. AP-22, no. 2, p. 365, March 1974.

[10] E. H. Newman, "A user's manual for electromagnetic surface patch code (ESP)," *Rep. 713402-1*, July 1981, the Ohio State University ElectroScience Laboratory, Department of Electrical Engineering; prepared under Contract No. DAAG29-81-K-0020 for Army Research Office.

[11] G. J. Burke and A. J. Poggio, "Numerical electromagnetic code (NEC)–method of moments," *NOSC/TD 116*, Naval Ocean Systems Center, San Diego, California 92152, July 1977.

[12] E. L. Pelton, R. J. Marhefka, and W. D. Burnside, "An iterative approach for computing an antenna aperture distribution from given radiation pattern data," *Rep. 784583-6*, June 1978, the Ohio State University ElectroScience Laboratory, Department of Electrical Engineering; prepared under Contract No. N62269-76-C-0554 for Naval Air Development Center.

[13] R. Backhus, "The determination of the aperture distribution of a linear array through near-field measurements," *Rep. 713303-2*, June 1982, the Ohio State University ElectroScience Laboratory, Department of Electrical Engineering; prepared under Contract No. N62269-80-C-0384 for Naval Air Development Center.

[14] G. A. Thiele and T. M. Newhouse, "A hybrid technique for combining moment methods with the geometrical theory of diffraction," *IEEE Trans. Antennas Propag.*, vol. AP-23, no. 1, pp. 62–69, January 1975.

[15] C. H. Walter, *Traveling-Wave Antennas*, New York: Dover Publications, 1979, pp. 15–16.

[16] J. D. Kraus, *Antennas*, New York: McGraw-Hill Book Company, 1950, pp. 464–477.

[17] R. G. Kouyoumjian and P. H. Pathak, "A uniform geometrical theory of diffraction for an edge in a perfectly conducting surface," *Proc. IEEE*, vol. 62, pp. 1448–1461, November 1974.

[18] Short course notes on "The modern geometrical theory of diffraction," vols. 1, 2, and 3, the Ohio State University ElectroScience Laboratory, Department of Electrical Engineering, 1980.

[19] W. D. Burnside and K. W. Burgener, "High-frequency scattering by a thin lossless dielectric slab," *IEEE Trans. Antennas Propag.*, vol. AP-31, pp. 104–110, January 1983.

[20] C. C. Huang, N. Wang, and W. D. Burnside, "The high-frequency radiation patterns of a spheroid-mounted antenna," *Rep. 712527-2*, March 1980, the Ohio State University ElectroScience Laboratory, Department of Electrical Engineering; prepared under Contract N00019-80-C-0050 for Naval Air Systems Command.

[21] J. G. Kim, W. D. Burnside, and N. Wang, "Geodesic solution for an antenna mounted on an ellipsoid," *Rep. 713321-3*, March 1982, the Ohio State University ElectroScience Laboratory, Department of Electrical Engineering; prepared under Contract N00019-80-PR-RJ015 for Naval Air Systems Command.

[22] W. D. Burnside, N. Wang, and E. L. Pelton, "Near-field pattern computations for airborne antennas," *Rep. 784685-4*, June 1978, the Ohio State University ElectroScience Laboratory, Department of Electrical Engineering; prepared under Contract N00019-77-C-0299 for Naval Air Systems Command.

[23] R. J. Marhefka, "Numerical electromagnetic code (NEC)–basic scattering code, part II: code manual (version 2)," *Rep. 712242-15*, December 1982, the Ohio State University ElectroScience Laboratory, Department of Electrical Engineering; prepared under Contract No. N00123-79-C1469 for Naval Regional Procurement Office.

[24] J. W. R. Taylor, *Janes's All the World's Aircraft*, New York: McGraw-Hill Book Co., 1973–74 (published yearly).

[25] R. L. Mather, G. V. Vaughn, and J. C. Logan, "Computer techniques for modeling shipboard communications antennas at UHF and above," *NELC Tech. Note 3313*, February 1977, Naval Ocean Systems Center, San Diego, California 92152.

[26]   E. D. Green, "High-frequency scattering from multiple finite elliptic cylinders," *Rep. 712242-7*, June 1981, the Ohio State University ElectroScience Laboratory, Department of Electrical Engineering; prepared under Contract No. N00123-79-1469 for Naval Regional Procurement Office.

Chapter 21

Satellite Antennas

C. C. Han
Equatorial Communications Company

Y. Hwang
Ford Aerospace & Communications Corporation

CONTENTS

Ching Chun Han received the BS, MS, and PhD degrees in electrical engineering, with the doctorate being from Stanford University in 1971.

He joined Equatorial Communications Company in 1985 as Director of Microwave Systems in charge of VSET antenna and microwave converter development. He is also the project manager of Equatorial's K-220 program—the first K_u-band spread-spectrum satellite data communication network in the world.

He was employed by Ford Aerospace from 1967 to 1985, where he was the manager of the Antenna Systems Department. He worked on the INTELSAT-V frequency reuse shaped-beam antenna, the first few INTELSAT standard-A stations, the NATO-III antenna, the NASA 20/30-GHz multiple-beam antenna, low side lobe and low-cost stations, and many components, such as polarizers, OMTs, filters and diplexers, MICs, Gunn oscillators, pin and waveguide switches, VPDs, and stripline devices. He also serves as a part-time senior research associate at Stanford University's Communication Satellite Planning Center.

Yeongming Hwang received the BS in electrical engineering from the National Taiwan University, Taipei, Taiwan, in 1963, the MS in electrical engineering from the Institute of Electronics of National Chiao-tung University, Tsiuchu, Taiwan, in 1965, the masters degree in business administration from Golden Gate University, San Francisco, California, in 1982, and the PhD in electrical engineering from the Ohio State University in 1973.

He was a research associate in 1974 and Assistant Supervisor in 1975 in the Electro-Science Laboratory of the Ohio State University. He joined Ford Aerospace in 1975 and has worked in the satellite and ground antenna systems. At present he is the supervisor of the Antenna Technology Section at Ford Aerospace.

Dr. Hwang is interested in the geometrical theory of diffraction (GTD), numerical techniques, in applying the hybrid method of GTD and numerical techniques to antenna radiation problems, in antenna scattering, EMP problems, multibeam antenna systems, low side lobe ground antennas, beam optimization, frequency-selective surfaces, patched-array antennas, and microwave components for spacecraft antenna applications.

1. Introduction

The design of antennas for satellite applications differs in several respects from other applications. An antenna radiation pattern varies from omnidirectional to highly directional. It can be fixed or changed to accommodate specific needs as they arise. A satellite antenna must be designed to withstand the dynamic mechanical and thermal stresses for the satellite. The design constraints imposed by the satellite on size, shape, and weight are also important factors in design consideration. As the requirements on the side lobes and cross polarization become more stringent, the interference on antenna performance due to the presence of the satellite body, solar cell panel, and other antenna systems cannot be neglected.

The major function of a satellite antenna is communication. A communication satellite is a radio relay in space. A signal is sent from station A on earth to station B via the satellite as shown in Fig. 1. In the synchronous orbit, 22 286 mi (35 865 km) away from the earth, a satellite travels around the earth in exactly the earth's rotation time. Therefore the satellite appears to be stationary over any point on earth. Three satellites in synchronous orbit can cover over 90 percent of all inhabited regions of the earth.

The earth-subtended angle at the synchronous satellite is 17.34°. A satellite antenna can provide an earth coverage area of this size. The antenna radiation pattern can also be shaped to conform to the shape of an intended coverage area in order to improve antenna gain. Most satellite antennas are directional. Except for the tracking, telemetry, and command antenna, which must provide ranging, telemetry, and command operations throughout all mission phases, a nondirectional antenna is required to ensure the continuous reception in every mission operation.

The growing demand for increased communications satellite capacity and the crowding of synchronous orbital slots have led to more stringent demands on satellite antenna designs. The complexity and the size of satellite antennas have increased rapidly. Limited power resources in a satellite place an additional demand for designing more efficient antenna systems. Proper system design requires accurate predictions of designs for an efficient antenna system. As a result, an antenna engineer relies more on accurate theories for analysis and synthesis. Significant strides have been made in developing the analytical and synthesized techniques. Computer-aided design tools and advanced manufacturing techniques are essential not only to provide accurate antenna performance prediction, but also to reduce design cycle times.

This chapter addresses three types of satellite antenna design: communication, earth coverage, and tracking, telemetry, and command (TT&C).

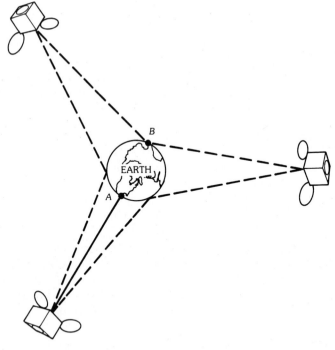

Fig. 1. Communication between two earth stations via satellite.

2. Communication Antennas

A communication satellite functions as a radio relay in space. A satellite antenna can provide the communication link not only between stations on earth but also between antennas from other satellites.

It can be shown [1] that in a communication downlink

$$A_{\text{cov}} \Delta f \cong P_T \frac{A_R}{T_S} \frac{1}{L_i} \left(\frac{C}{N}\right)^{-1} \tag{1}$$

where

$$
\begin{aligned}
A_{\text{cov}} &= \text{coverage area} \\
\Delta f &= \text{bandwidth} \\
P_T &= \text{satellite transmitter power} \\
A_R &= \text{earth station antenna effective area} \\
T_S &= \text{equivalent system temperature} \\
L_i &= \text{incidental loss} \\
C/N &= \text{carrier-to-noise power ratio}
\end{aligned}
$$

Increasing the capacity of the communication system can be directly achieved by increasing the bandwidth Δf through frequency reuse. On the other hand, for a fixed C/N and prechosen modulation system, the bandwidth can be increased by reducing the coverage area A_{cov}. The coverage area can be reduced by using a multibeam antenna. The power is divided among the beams and the bandwidth Δf remains constant for each beam. As a result, the total bandwidth available increases by the number of beams.

The demands of the frequency reuse and multiple-beam antenna systems have taken the form of requirements for more antenna pattern control for increased antenna gain, low side lobes, and cross polarizations. Low side lobes and cross polarization are required to prevent excessive interference among beams. The typical isolation requirement for a communication system is 27 dB or higher.

A typical satellite communication antenna consists of an optical aperture, a feed array, and a beam-forming network (BFN). The function of the aperture is to focus the energy from the point source to the desired direction to yield high gain. A reflector, lens, or phased array may be used as an optical aperture. A feed array is located near the focal point to direct the energy to the aperture. A BFN, which is connected to the array, distributes the energy to the proper beam port. The network includes the passive power divider, switch, or variable power divider to control the beam steering and beam shape.

The antenna beams generated can be fixed or scanned over the coverage area. A beam can be scanned mechanically or electrically. A mechanically scanning beam is achieved by gimbaling the antenna system. This approach is adopted in the application where the scanning rate is of no major concern. Certain communication systems using satellite-switched time-division multiple access require a fast scanning rate that can only be achieved by using switches electrically controlled in the BFN.

A beam is generally distorted as scanned from the antenna boresight due to the phase aberration of the antenna optics. As a result it lowers antenna gain and raises the side lobes and cross-polarization levels. In general, the antenna performance is evaluated in terms of gain, side lobes, and cross polarization achieved at the farthest scanning position.

A fixed beam can be a pencil beam or a highly shaped beam. A highly shaped beam is used to improve antenna gain over the prescribed area and to reduce interference outside the coverage area. An ideally shaped beam is one that has the highest possible antenna gain without any side lobes. The radiation pattern turns out to be flat over the coverage area. The theoretically achievable maximum value of the edge-of-coverage gain is given by

$$G = \frac{41\,253}{A_{\text{cov}}} \tag{2}$$

where A_{cov} is the coverage area in degrees square. The performance of a shaped-beam antenna is evaluated in terms of beam-shaping efficiency defined as

$$\text{beam-shaping efficiency} = \frac{G_{\min} A_{\text{cov}}}{41\,253} \tag{3}$$

or the gain-area product $G_{\min} A_{\text{cov}}$ in which $G_{\min}$ is the minimum gain over the coverage area.

The edge rolloff of an optimum elliptical beam [2] is 4.3 dB below the peak. It has a beam-shaping efficiency of 40 to 45 percent. In other words, the gain-area product (GA) is between 16 500 and 18 560. Practically, the gain-area product of a satellite antenna is ranged between 10 000 and 16 000, depending on the size of the antenna aperture and the shape of the coverage.

Antenna System Design

The major parameters involved in the design of a satellite antenna system are listed in Table 1. The requirements of a satellite antenna system are specified in terms of the minimum gain over a coverage area, the frequency band, polarization, and isolation. There is a system trade-off among the transmit power, satellite antenna gain, and the ground terminal antenna gain in a communication link. The frequency is designated from the frequency assignment. Frequency bands for satellite communications and other services have been advocated by both international and national regulatory and policy-making agencies. The range of the frequency may be from 0.1 to 150 GHz. The frequency bands in this range are listed in Table 2. Due to the spectrum crowding in the S- and C-bands, the K_u-band and

Table 1. Design Parameters of a Spacecraft Antenna System

Requirements	Optics	Feed Elements	Beam-Forming Network

Gain
Gain ripple
Gain slope
Coverage area
Frequency
Frequency band
Polarization
Scanning or
  fixed beam
Number of beams
Isolation
	Phased array
	Lens
	Reflector
	Aperture size
	Number of
	  optics
		Type of feed
		  element
		Feed spacing
		Number of feeds
		Polarizer
		Orthomode transducer
			Feed excitation coefficients
			Switch
			Variable phase shifter
			Variable power divider
			Power divider
			Diplexer
			Odd/even-mode converter
			Frequency-selective surface
			Transmission media

Table 2. Frequency Allocation for Satellite Communication

Frequency (GHz)	Frequency (GHz)
1.53–1.599	12.5–12.7
1.6265–1.6605	12.75–13.25
2.5–2.69	14.0–14.5
3.4–4.2	17.7–21.2
4.5–4.8	27.5–31.0
5.85–7.075	40.0–41.0
7.25–7.75	43.5–47.0
7.9–8.4	92.0–95.0
10.9–11.2	102.0–105.0
11.4–12.2	140.0–152.0

higher frequencies are being used for broader bandwidths. Signals may be linearly or circularly polarized. Orthogonal polarization can be reused in the same frequency band to increase satellite capacity. Isolation is specified between beam ports from the system requirements. The isolation requirements are then translated to the required side lobe or cross-polarization levels of the beams.

The design process starts with choosing the size of optical aperture. Theoretically, the larger the aperture size, the higher the antenna gain will be. An ideally shaped beam is generated by an aperture of infinite extent. Practically, the aperture size should be adequately large to meet gain and side lobe requirements. An optimal size is the minimum size that can be placed on the satellite that satisfies the system requirements. The same optical aperture may be used for multifrequency bands, dual polarization, or multibeam at the expense of increasing complexity of the antenna system. The minimum number of apertures required is determined as a result of trade-off of the achievable performance of the system.

Three basic generic types of antennas are: phased array, lens, and reflector. A phased array requires N corporate feed networks for N beams and therefore is generally heavy and complex. A lens is attractive for its good scanning performance and compactness. However, it is also heavy in the low-frequency application. A reflector is widely used for its light weight and structural simplicity.

The choice of feed elements is governed by the achieved gain and cross-polarization level over the required subtended angles, the frequency bandwidth, and the type of optical aperture. Horn is widely used in most applications. A ridged waveguide, helix, turnstile, cross yagi, small reflector, and slot array have also been used. An orthomode junction is required for dual polarization. A polarizer is used to generate circularly polarized waves for the antenna system. A polarization reflector can be used to separate two linearly polarized signals or to provide lower linear cross-polarization signals.

The design of a beam-forming network is heavily influenced by the loss, weight, and size in a satellite antenna system. Most power dividers and transmission lines are waveguides typed for frequencies above K-band because of the low insertion loss. Air stripline is widely used as a transmission medium in the C-band or lower frequency because of its compactness.

There are three main types of switches: mechanical, ferrite control, and diode switches. The choice of the switches is mostly determined by the insertion loss and switching speed. The variable phase shifters and variable power dividers are categorized to be electrical-mechanical, ferrite, or diode variable phase shifter/variable power dividers.

The odd/even converter is basically a two-to-N port network, which is used to combine the odd and even channels of a transponder output multiplexer. In order to alleviate the severe restrictions on the design of a closely separate output multiplexer, odd-numbered channels are often combined to a single output, and even-numbered channels to a second output. These two outputs are then connected to the input ports of an odd/even mode converter.

A diplexer is used to combine two frequency bands in the beam-forming network. It is a two-to-one filter microwave component. A frequency selective surface can be used as a spatial filter to replace the filter type diplexer to alleviate the congestion and complexity of beam-forming networks.

Types of Antennas

Phased array, lens, and reflectors are the three basic antennas. A phased array has a number of advantages over lens or reflector: distribution of power amplification at the elementary radiation level, higher aperture efficiency, no spillover loss, no aperture blockage, and better reliability. Disadvantages are weight, complexity, and relatively high losses in the power distribution system. The advantages of a lens antenna are excellent optics, no feed blockage, and compactness. A lens can be made rotationally symmetrical to preserve good optical characteristics. Disadvantages are heaviness, especially in the low-frequency application, and the mismatch of the lens surface. The lens itself is also a blockage to the feed array. A reflector is the most desirable candidate for spacecraft antennas because of its light weight, structural simplicity, and design maturity. The reflector, however, has to be offset to avoid feed blockage. The offset destroys the rotational symmetry of the surface and limits the range of scan before aberrations seriously degrade the scanning performance. Table 3 summarizes the characteristics of these three antennas.

The Phased Array—A phased array consists of more than one antenna element radiating in phase coherence. Horns, dipoles, helices, spiral antennas, parabolic dishes, and many other types of antennas can be used as radiating elements. The radiation pattern of a radiating element in an array environment is called the *active element pattern*. It may be substantially different from the pattern of an isolated radiating element because of the effect of mutual coupling.

A phased array for satellite applications can be used for

(*a*) a fixed beam that can be either single or multifixed
(*b*) electronically steerable scanning beams
(*c*) a feed array of a lens or reflector antenna system

Table 3. Characteristics of Three Basic Antennas

Antenna	Advantages	Disadvantages
Phased Array	Distribution of power amplification at the elementary radiation levels Reliability No spillover losses No aperture blockage	Complexity Heavy Higher beam-forming network losses
Lens	No feed blockage Better scanning performance	Heavy in low-frequency application Aperture mismatch
Reflector	Simple Lightweight Design maturity	Offset to avoid feed blockage Poor scanning performance

(*d*) an adaptive array that can automatically null a beam in the direction of oncoming jammers

If the mutual coupling among the array elements is neglected, it can be shown [3] that for a planar array of $(2M + 1)\times(2N + 1)$ apertures embedded in an infinite conducting ground plane at $z = 0$ as shown in Fig. 2, the far-field radiation pattern can be expressed as

$$E(r) = f(T_x, T_y) S_a(T_x, T_y) \tag{4}$$

where

$$S_a(T_x, T_y) = \sum_{m=-M}^{M} \sum_{n=-N}^{N} V_{mn} e^{j(\mathbf{k}_t^i \cdot \varrho_{mn})} \tag{5a}$$

and

$$f(T_x, T_y) = \frac{je^{-jk_0 r}}{2\pi r}(k_0\sqrt{1 - T_x^2 - T_y^2}) \int_{A_{00}} \mathbf{E}_{00}(x^0, y^0)\, e^{jk_0(T_x x^0 + T_y y^0)}\, dx^0\, dy^0 \tag{5b}$$

in which k_0 is the wave number, $\varrho_{mn} = mb\hat{\mathbf{x}} + nd\hat{\mathbf{y}}$, b and d are the separations between adjacent channels in the x and y directions,

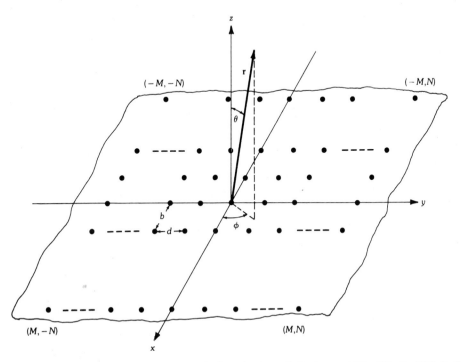

Fig. 2. Planar array in a ground plane. (*After Amitay, Galindo, and Wu [3], © 1972 Bell Telephone Laboratories, Incorporated*)

$$\mathbf{k}'_t = \hat{\mathbf{x}}k'_x + \hat{\mathbf{y}}k'_y$$
$$k'_x = k \sin\theta \cos\phi$$
$$k'_y = k \sin\theta \cos\phi$$

X^0 and Y^0 are the coordinates of the aperture A_{00} over the unit element cell, $E_{00}(x^0, y^0)$ is the field distribution in the aperture A_{00}, V_{mn} is the voltage generator of the (m, n)th element,

$$T_x = \sin\theta \cos\phi$$
$$T_y = \sin\theta \sin\phi \qquad (6)$$
$$T_z = \cos\theta$$

f is the element pattern, and S_a is the array factor. The radiation pattern is the product of the element pattern and the array factor. The element pattern determines the rolloff and the polarization of the far-zone field. It is a slowly varying function in comparison with the array factor. The array factor is a double function of U and V with periods λ/b and λ/d, respectively. Because of the periodicity in the UV plane, the peak of the array factor will be repeated every λ/b and λ/d intervals in the U and V directions, respectively. The peak of the main beam will also be repeated and its peak is tapered off following the rolloff of the element pattern. Such repetition of the main beam is called a *grating lobe*.

The grating lobe appears for any periodic planar array and is generally presented in the grating lobe diagram. It is convenient to present the grating lobe diagram in the UV plane. For a periodic planar array with parallelogram cells as depicted in Fig. 3a, the grating lobe diagram is given in Fig. 3b by invoking an affine transformation [4]. The unit cell of the grating lobe diagram is skewed by the subtended angle Ω. The small solid circle around the origin designates the broadside position of the maximum of the main beam, and the rest of the small circles at the corners of each unit cell designate the corresponding position of the maximum of each grating lobe. In all practical applications the grating lobe should not appear in the real space in the scan region.

The beam of an array is broadened while scanning. The beam is distorted by a factor of $1/\cos\theta$, which is called the *beam-broadening factor*. The beam-broadening factor is basically due to the reduction in the effective aperture area of the array. In practice the mutual coupling among elements cannot be neglected. The effects of mutual coupling in an array are the change in the radiation pattern and gain, radiation impedance, mismatch, and the cross polarization. The principle of pattern multiplication no longer holds.

The exact solution of the mutual coupling of a finite array is intractable. If the number of an array element is large, an infinite array is a good approximation model and is much easier to analyze. Several methods of analyzing an infinite array have been given in the literature [5]. They are classified as residue calculus technique, modified residue calculus, generalized scattering approach, mode matching method, complex power, integral equation, and grating lobe series.

To meet an antenna specification, trade-offs have to be made among the size of

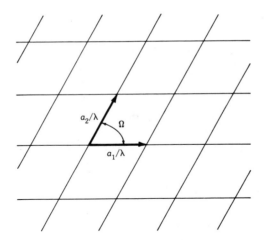

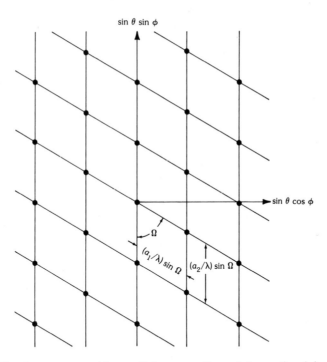

Fig. 3. Periodic planar array with parallelogram cells and its grating lobe diagram. (*a*) Periodic planar array. (*b*) Grating lobe diagram. (*After Lo and Lee [4], © 1965 IEEE*)

aperture, number of elements, element arrangement, and the type of elements. Most satellite antenna applications concern not only the peak gain of the beam but also the gain at the edge of coverage while scanning. The critical design parameters are as follows.

Size of the Array—The size of the array is mostly determined by the gain and beamwidth requirements.

Element Number—For a given array aperture the number of elements is inversely proportional to the size of element. The determinant factors are the resolution capability, grating lobe, weight, and side lobe.

To eliminate the grating lobes entirely, the element size should be kept less than a wavelength. This requires an excessive number of elements. Practical design uses the minimum number of elements to keep the grating lobes only out of the field of view. On the other hand, the number of elements cannot be too small to degrade the array sensitivity over the field of view. As the number of elements increases, the cost, weight, and complexity of the array also increase. Therefore there is a fundamental trade-off among these factors.

Element Arrangement—The elements can be arranged in a square grid, rectangular grid, triangular grid, parallelogram, or random arrangement. The choice of each arrangement is determined by the resolution, the number of elements, and the grating lobes over the field of view. It has been shown that by arranging the elements of a beam-scanning planar array in a triangular pattern [6] rather than a rectangular pattern, or in a random pattern [7] rather than a periodic pattern, the number of elements required in the array can be reduced.

Types of Element—The element types determine the achievable array gain, scan loss, cross polarization, frequency band, size, and weight. Possible candidates for satellite antenna applications are the waveguide, ridged waveguide, helix, turnstile, cross yagi, small reflector, and slot array. The relative merits of each element type are presented in Table 4.

Allowable Phase and Amplitude Quantization Errors—If the digital phase shifter and variable power divider are utilized, the effects of quantization steps on the beam-pointing accuracy, resolution, and side lobe level must be considered. The performance versus number of bits should be analyzed in the design process.

The Lens—The lens and reflector are both commonly used as collimating elements in a high-gain antenna system. A lens does not have feed blockage and its surface can be maintained rotationally symmetrical.

A lens requires two surfaces to collimate the beam. The surfaces in general have the simple form of plane, spheroid, or paraboloid. The restriction to surfaces

Table 4. Types of Radiating Elements

Type	Bandwidth	Dual Polarization	Gain	Mutual Coupling	Volume
Waveguide	10%	Yes	Medium	Medium	Medium
Ridged waveguide	Wider	More difficult	Medium	Medium	Medium
Helix	1.67 <35 GHz	One sense only	High	High	Large
Turnstile	10%	Yes	Low	Medium	Small
Crossed yagis	10%	Yes	High	High	Large
Reflector	Wideband	Yes	High	Medium	Large
Waveguide slotted array	Narrow	Difficult	High	High	Small

of simple forms imposes limitations on the ultimate performance that the systems can attain. A shaped surface or aspheric surface is proposed in some systems. The surface can be adequately represented by

$$z = \frac{\gamma \varrho^2}{1 + \sqrt{1 + (1 + K)\gamma^2 \varrho^2}} + A\varrho^4 + B\varrho^6 + C\varrho^8 + D\varrho^{10} \qquad (7)$$

where K is called the *conic coefficient*, ϱ is the distance measured from the axis of the surface of revolution, γ governs the curvature of the surface at the apex, and A, B, C, and D are the coefficients of the polynomial. The first term represents the surface of the form of hyperboloid ($K < -1$), paraboloid ($K = -1$), ellipsoid ($-1 < K < 0$), sphere ($K = 0$), or spheroid ($K > 0$).

The lowest-order aberration function of a rotationally symmetrical optical system is the primary or Seidel aberration [8]. There are five primary aberrations: spherical, coma, astigmatic, field curvature, and distortion as depicted in Fig. 4. Such aberrations, or phase errors, may arise from a displacement of the primary

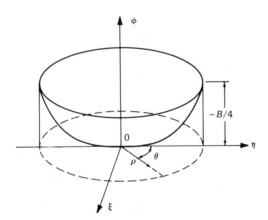

a

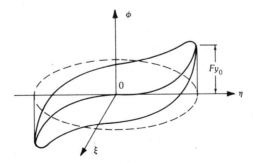

b

Fig. 4. The primary wave aberrations. (*a*) Spherical aberration: $\theta = B\varrho^4/4$. (*b*) Coma: $\phi = Fy_0\varrho^3\cos\theta$. (*c*) Astigmatism: $\phi = -Cy_0^2\varrho^2\cos^3\theta$. (*d*) Curvature of field: $\phi = Dy_0^2\varrho^2/2$. (*e*) Distortion: $\phi = Ey_0^3\varrho\cos\theta$. (*After Born and Wolf [8]; reproduced with permission of Pergamon Press, © 1980*)

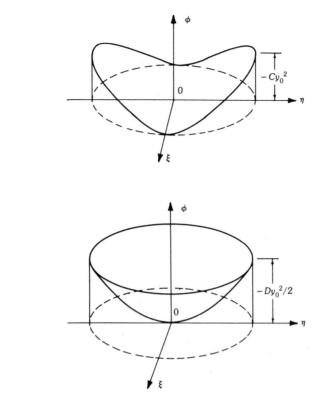

c

d

e

Fig. 4, *continued.*

feed from the focus or the phase error in the field of the primary feed. The degree of primary aberration is an excellent indication of the performance of the lens system. The effects of the two most well-known aberrations, spherical and coma, are that the spherical aberration causes null fill-in and side lobe increases, whereas the coma aberration causes imbalance of the side lobe levels. Fig. 5 presents the effects of the five primary aberrations on the radiation patterns.

The effect of aperture amplitude distribution on the radiation pattern is well studied. It has been shown that side lobes are related to the aperture field dis-

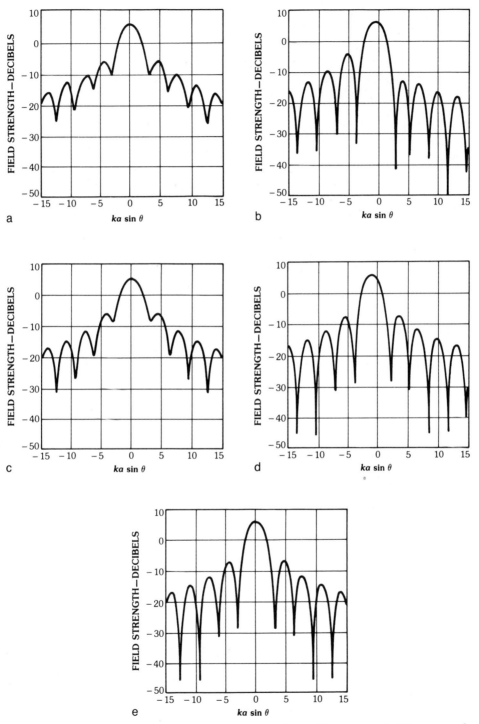

Fig. 5. Antenna pattern degradation due to aberrations. (*a*) $\phi = -\varrho^4$. (*b*) $\phi = -\varrho^3$. (*c*) $\phi = -\varrho^2$. (*d*) $\phi = -\varrho$. (*e*) $\phi = 0$.

tribution. Most of the aperture field distributions are assumed to be symmetrically tapered. An asymmetric aperture field distribution can occur when the feed is off the focus and causes asymmetric null fill-in. Fig. 6 shows the effect of the different aperture amplitude distributions and asymmetries.

Even though a lens does not have feed blockage it is basically a blockage to the feed. Reflections from lens surfaces cause feed mismatch, power loss, and gain ripple over the operating frequency band.

The key design trade-off parameters are the type of lens, size, and focal length to diameter ratio, shape of lens surfaces, surface tolerances, and lens surface mismatch.

Types of Lenses—Most widely used lenses in spacecraft applications are waveguide, TEM, and dielectric. A waveguide lens is limited to narrow frequency band operation. A TEM lens is heavier but has wider bandwidth and the dielectric lens is the heaviest.

In the high-frequency operation where the size of the lens aperture is on the order of 0.9 m or less, a dielectric lens becomes a viable candidate. For a given size of waveguide or TEM lens, there is a minimum number of elements required to adequately simulate the aperture field distribution to achieve the gain and desired side lobe level. The maximum number of elements is primarily limited by the smallest element size for which a good impedance match can be achieved. The smallest element size in a waveguide lens is also determined by its cutoff frequency. In practice, a minimum number of elements is always desired because of cost and complexity. Zoning increases the bandwidth of a waveguide lens but decreases the bandwidth of a dielectric lens. The bandwidth Δf for a waveguide zoned lens is given by [9]

$$\Delta f \cong 25 \frac{n_0}{1 + K_z n_0} \text{ percent} \tag{8}$$

where K_z is the number of zones and n_0 is the refractive index that can be expressed as

$$n_0 = \sqrt{1 - (\lambda_0/2a)^2} \tag{9}$$

in which a is the spacing and λ_0 is the wavelength.

On the other hand, the bandwidth Δf for a dielectric zoned lens is given by

$$\Delta f \cong \frac{25}{K_z - 1} \text{ percent} \tag{10}$$

In the above results it is assumed that the phase errors should not exceed 0.125λ.

Dimension—The dimension of a lens is described in terms of the diameter and the ratio of focal length over its diameter (F/D). The diameter is determined by the gain and beamwidth requirements. The longer the focal length, the better the scan performance. In general, the focal length is determined as a result of compromise between the scan performance and the spacecraft constraint.

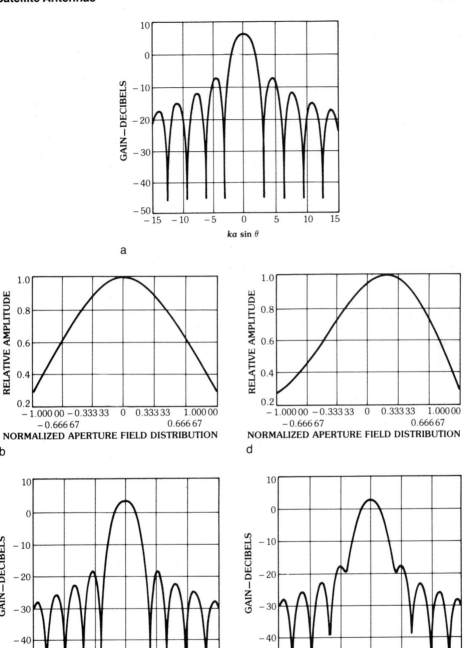

Fig. 6. Antenna pattern versus amplitude distribution. (*a*) Uniform distribution. (*b*) Tapered distribution. (*c*) Antenna pattern with amplitude distribution (*b*). (*d*) Asymmetric distribution. (*e*) Antenna pattern with amplitude distribution (*d*).

Shapes of Lens Surfaces—The surfaces of conventional lenses are planar, spherical, or paraboloidal. A shaped surface is widely used to obtain the better performance that a conventional lens design cannot attain. The principles of lens shaping can be classified into two categories. The first one is the application of the path-length constraint and Snell's law. The condition of power distribution is imposed to control the shape of the beam and the level of the side lobes [10, 11]. In some cases the Abbe sine condition is imposed instead of invoking power conservation to improve the scanning characteristics [12, 13, 14]. The second approach is to minimize the phase errors for the beams on-focus and off-focus as well. In general, it is impossible to design a system of free-of-phase error. A suitable compromise must be made to minimize all the primary as well as the higher-order aberrations of a lens system. The constraints K, A, B, C, and D in (7) are the variable parameters in the optimization process.

Surface Tolerances—The surface deformations are due to the manufacturing tolerances and thermal effects in space. The deviations of a lens surface from its ideal shape can cause loss of gain and pattern degradation. The surface deformation results in the phase front errors in the aperture field. The tolerance to be placed on a lens surface, in the first-order approximation, can then be related to the maximum allowable irregularity in the equiphase front formed by the lens aperture. If the maximum allowable phase error is taken to be $\lambda/16$ on wavefront irregularities arising from variation in either t or n, where t is the lens thickness, it can be shown [9] that

$$\Delta t \leqq \frac{\lambda}{16(1 - n)} \tag{11a}$$

and

$$\frac{\Delta a}{a} \leqq \frac{n}{16(n + 1)} \tag{11b}$$

where Δt and Δa are the tolerance on the waveguide lens thickness and plate spacing a, respectively.

For a dielectric lens,

$$\Delta t \leqq \frac{\lambda}{16(n - 1)} \tag{12a}$$

and

$$\Delta n \leqq \frac{\lambda}{16t} \tag{12b}$$

In general, the effect of the surface deformation on the gain, side lobe levels, and cross polarization should be analyzed in a more rigorous way. The radiation pattern can be obtained accurately by integrating the aperture field distribution

once the details of the phase front errors are known. The surface tolerance specifications should be examined on a case-by-case basis against system requirements.

Surface Mismatch—In order to design a good lens antenna system it is
necessary to match the surface of the lens to reduce effects of direct reflections.
Three techniques can be used for lens surface matching:

1. Quarter-wave matching layers
2. Multiple quarter-wavelength impedance-matching transformer [15, 16]
3. Artificial-dielectric quarter-wave plate

In the first technique, the permittivity of the layer should be $\epsilon^{1/2}$ to give a perfect
match at a single frequency. If a match is required over a wide bandwidth, then the
second or third technique has to be used. The problems with the first and second
techniques are the lack of materials and fabrication difficulties. The third technique requires that the dielectric surface be slotted to give the equivalent matching
effect. The slots are either parallel or perpendicular to the electric field [17, 18].

The Reflector—The reflector is the most desirable optical candidate in spacecraft
antenna systems because of its light weight, structural simplicity, and design
maturity. The disadvantage is that the reflector has to be offset to avoid the feed
blockage. The offset destroys the rotational symmetry of the optical aperture and
limits the range of scan to a few beamwidths before aberrations seriously degrade
the performance.

A reflector antenna system consists of one or more reflector surfaces. The
surface can take the form of a paraboloid, hyperboloid, spheroid, ellipsoid, or
general shape. A single offset parabolic reflector is used most often because it is the
simplest in the reflector antenna system.

Geometry—The geometry of a single-offset parabolic reflector antenna is shown in
Fig. 7. It can be described by the aperture size, offset distance, and the focal length.
The key parameters to determine the characteristics of the system are focal region,
scan loss, and cross polarization.

Focal Region—For a center-fed, circularly symmetrical reflector the secondary
beam remains symmetrical about the reflector axis when the feed is displaced along
the axis. The secondary beam scans when the feed is laterally displaced. For an
offset-fed parabolic reflector the offset axis, which is equivalent to the reflector axis
of a center-fed reflector, is in the direction of θ_0 [19]:

$$\theta_0 = \frac{1}{2}(\theta_u + \theta_L) \qquad (13)$$

where θ_u and θ_L are the angles subtended at the focal point by the upper and lower
edges of the reflector in the xz plane. The focal surface is the plane perpendicular to
the offset axis.

Polarization—Let an offset reflector be illuminated by a feed. If the feed
radiation is circularly polarized everywhere, no cross polarization will appear in the
radiation from the reflector. A small beam squint will occur because of variation in
the phase shift across the reflector [20]. If the feed radiation is linearly polarized

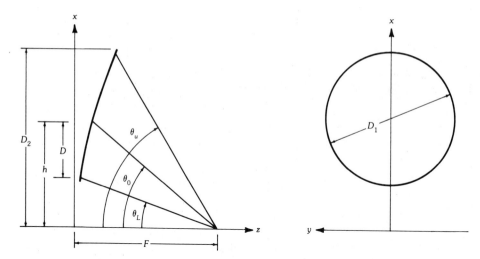

Fig. 7. An offset-fed parabolic reflector antenna system.

everywhere, a small polarization rotation will occur except when the feed is displaced in the plane of asymmetry (xz plane). However, there will be a small cross-linear polarization off the beam axis.

Lateral Feed Displacement—The beam generated by the reflector scans when the feed moves laterally from the offset axis. The beam degradation occurs when it scans. The degradation includes beam shape distortion, gain, loss, beamwidth, and higher side lobes. Typical beam shape degradation is depicted in Fig. 8. The gain decreases as the beam scans away the offset axis. The scan loss of a center-fed parabolic reflector antenna can be represented as a function of F/D and the number of half-power beamwidths scanned [21]. The scan loss of an offset-fed parabolic reflector antenna can be represented in a similar way. Fig. 9 shows the scan loss of an offset-fed parabolic reflector antenna for F/D of 1.0 and 1.3.

Design Parameters—The key design parameters are the aperture size, type of reflector, focal length, offset distance, and surface tolerance.

Aperture Size—The reflector aperture size is determined by the gain and the beamwidth required. In a multibeam antenna application the performance for generating the required shaped beam is better if the reflector is larger. The larger reflector allows the radiation energy to concentrate on a small area and provides steeper slope to reduce interference between closely spaced beams and between satellites. Practical aperture size is the smallest aperture that meets the performance requirements.

Type of Reflector—A single reflector is always the first choice in a reflector antenna system. A single offset parabolic reflector is a favorite candidate because of its better performance and design maturity. It is well known that the spherical reflector is a coma- and astigmatism-free optics when the entrance pupil is at the center of curvature. It is therefore suggested that a surface having the form between a paraboloid and a sphere may have better scan performance than a paraboloid. Such a surface can be expressed as

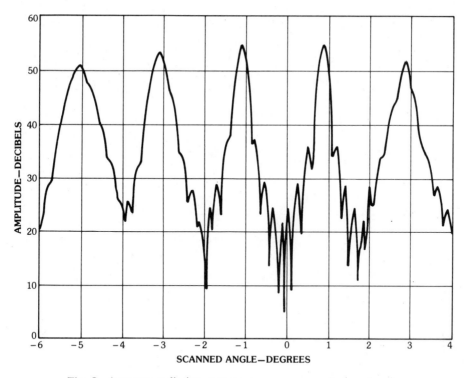

Fig. 8. Antenna radiation pattern versus scan angle (xz plane).

$$z = \frac{x^2}{4f} + (1 + b)\frac{x^4}{8(2f)^3} + \cdots \tag{14}$$

in which $b = -1$ for a paraboloid and $b = 0$ for a sphere, and f is a constant.

The reflector surface can be solid for circular polarized waves or gridded for linearly polarized waves. A gridded reflector surface can enhance the purity of linear polarization. Two orthogonal gridded surfaces can also share a common aperture area to form a compact antenna system. The front surface reflects one sense of linearly polarized waves while it allows the other orthogonally polarized wave to pass through with little attenuation. The rear reflector reflects the orthogonally polarized waves and filters the unwanted cross polarization.

Because of the limitation of the scan performance of a single reflector, a dual reflector may be used to replace the single reflector for a better optical system. A dual reflector has one more degree of freedom in the number of surfaces and should be able to reduce the phase aberration for better scan performance. The classical dual-reflector antenna system consists of Cassegrain and Gregorian antenna systems.

Cassegrain antenna reflector surfaces have the form of a paraboloidal primary reflector and a hyperboloidal secondary reflector. The system is electrically equivalent to a parabolic reflector antenna with a longer focal length mF, where m is the magnification factor of the Cassegrain antenna system and F is the focal

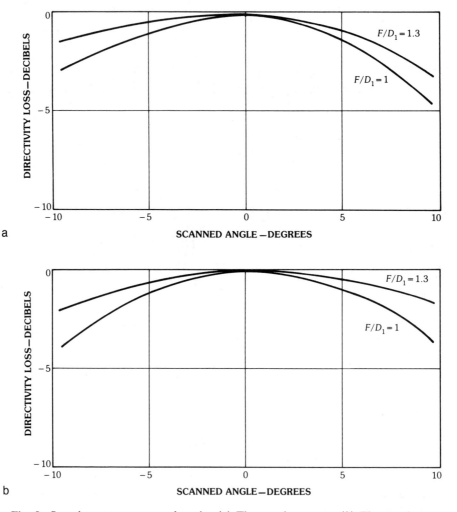

Fig. 9. Scan loss versus scanned angle. (*a*) The *xz* plane scan. (*b*) The *yz* plane scan.

length of the paraboloid. Therefore it can provide a better scan performance than a single parabolic reflector with the focal length F. Even though the equivalent principle no longer holds when the feed is not on-focus, a Cassegrain antenna is still a better optical system and compact in spacecraft integration.

A Gregorian antenna reflector surface has the paraboloid form for both the primary and secondary reflectors. Such an optical system forms an image at infinity and requires an additional optical aperture to focus the image at a finite distance. A shaped dual-reflector antenna is also used to improve the scan performance. A semishaped dual-reflector system, such as the surface, has two off-axis foci-bifocal reflectors. A generally shaped dual-reflector antenna is also under study. The idea is to give up the foci in order to reduce the most harmful aberrations for the scanning beams.

Focal Length—One of the most important parameters of a reflector antenna system is the ratio of focal length to parent aperture diameter (F/D), where the parent aperture is the largest circular aperture of the symmetrical paraboloid from which an offset reflector is derived. The larger the F/D is, the better performance for a scanning beam. However, a large F/D results in a small subtended angle for the feed element and the reflector and in turn requires a relatively large feed element. In a multibeam antenna application it implies that the total feed array size is also large. There is a trade-off of the number of feed elements versus the focal length. The aberration, which is worse for a shorter F/D reflector system, can be reduced and compensated by more feed elements. The improvement is due to the fact that more feed elements have more degrees of freedom for beam optimization. From the spacecraft viewpoint a larger F/D means bigger physical size, which is not cost-effective.

Offset Distance—The offset distance, defined as the distance from the reflector axis to the center of the reflector aperture, should be large enough to eliminate the feed blockage. The feed blockage increases the side lobe levels and the cross polarization as well.

Simple approximation formulas have been given [22] for designing an offset parabolic reflector with a circular aperture and feed array configurations. For given SL dB side lobe level of a single pattern, half-power beamwidth $2\theta_0$, maximum scan angle θ_3 with allowable GL dB scan loss, the spacing d between adjacent elements, and the offset distance h, one can determine the aperture taper and efficiency, reflector diameter, focal length, beam deviation factor, and element number in simple formulas. The formulas are given in Table 5 and in Figs. 10 and 11. They are reasonably accurate under the following constraints:

$0 < \Delta < 0.85$
$1 \leqslant GL \leqslant 5$
$F/D \leqslant 1.5$
$\theta_0 < 30°$
θ_3 can be approximated by $\tan \theta_3$ with reasonable accuracy

Surface Tolerance—The reflector surface errors due to the reflector surface distortion can be classified in two categories: random surface errors and deterministic surface errors. Random surface errors are due to manufacturing distortions, treated as a random error because of a lack of precision in constructing the reflector surface. Deterministic surface errors are due to the thermal distortion of a reflector surface and the surface deformation of an unfurlable or mesh reflector.

Classical antenna tolerance theory [23] shows that the effect of random surface distortion results in reduction in the peak gain and increase in side lobe levels. The perturbed radiation field due to the random surface distortion is given as

$$P_s = (2\pi c/\lambda)^2 \bar{\delta}^2 e^{-[(\pi c/\lambda)\sin\theta]^2} \tag{15}$$

where

$$\bar{\delta}^2 = \left(4\pi a^2 \frac{\Delta z}{\lambda}\right)^2 \tag{16a}$$

Table 5. Approximated Design Formulas for an Offset Parabolic Reflector

Design Parameters	Formulas	Type A Feed	Type B Feed
Aperture taper (Δ)	$\displaystyle\sum_{n=0}^{3} \alpha_n \left(\frac{\text{SL}^n}{10}\right)$	$\alpha_0 = -26.55$ $\alpha_1 = 35.17$ $\alpha_2 = -15.59$ $\alpha_3 = 2.37$	$\alpha_0 = -8.87$ $\alpha_1 = 9.32$ $\alpha_2 = -3.0$ $\alpha_3 = 0.32$
Aperture efficiency (η)	$\displaystyle\left(\sum_{n=0}^{3} \beta_n \Delta^n\right) \times 100\%$	$\beta_0 = 1$ $\beta_1 = -0.026$ $\beta_2 = 0.039$ $\beta_3 = -0.263$	Numerical curves
Reflector diameter (D_1/λ)	$\displaystyle\frac{1}{\pi \sin\theta_1} \sum_{n=0}^{3} \gamma_n \Delta^n$	$\gamma_0 = 1.609$ $\gamma_1 = 0.245$ $\gamma_2 = -0.259$ $\gamma_3 = 0.396$	$\gamma_0 = 1.61$ $\gamma_1 = 0.57$ $\gamma_2 = -1.43$ $\gamma_3 = 1.47$
Focal length (F)	$\displaystyle\frac{\pi(\sin\theta_3/\sin\theta_1)D}{190C\cos^{-1}[1 - (\text{GL}/5)]}$	$C = 1 - e^{-0.12\sqrt{D_1/\Lambda}}$	
Beam deviation factor	$\tau(1 - 0.72e^{-3.2(F/\tau D_1)})$	$\tau = \dfrac{\cos\theta_0 + \cos(\theta - \theta_L)}{1 + \cos(\theta_0 - \theta_L)}$ $\tan\theta_0 = \left(\dfrac{h}{F}\right)\left[1 - \dfrac{1}{4}\left(\dfrac{h}{F}\right)^2\right]^{-1}$ $\tan\theta_L = \left(\dfrac{h_1}{F}\right)\left[1 - \dfrac{1}{4}\left(\dfrac{h_1}{F}\right)^2\right]^{-1}$	
Number of feed elements (N)	Nearest integer of ($\theta_3/2\theta_2$)		

$$a = \frac{\tan^{-1}(D/4F)}{D/4F} \tag{16b}$$

and

$\quad c$ = length of correlation interval,

$\quad \lambda$ = wavelength,

$\quad D$ = antenna diameter,

$\quad \theta$ = pattern angle,

$\quad \bar{\delta}^2$ = phase error variance,

$\quad \Delta z$ = surface tolerance,

$\quad F$ = focal length

An improved and more accurate approach is to assume that the rms surface errors are given in a prescribed annular region. A closed-form expression for the

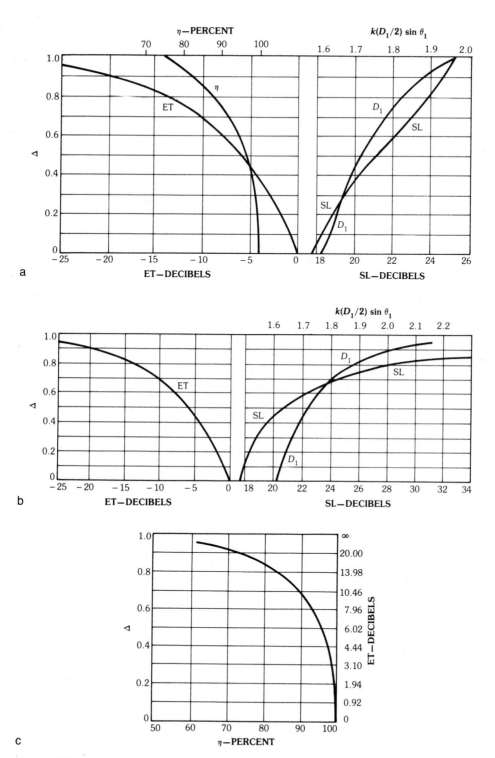

Fig. 10. Universal curves for designing an offset-fed parabolic reflector antenna. (*a*) Relation between Δ and reflector performance parameters when feed element is of type A. (*b*) Relation between Δ and reflector performance parameters when feed element is of type B. (*c*) Aperture efficiency versus edge taper when feed element is of type B. (*After Lee and Rahmat-Samii [22], © 1981 IEEE*)

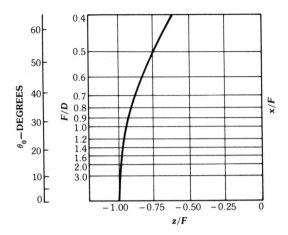

Fig. 11. Universal chart of parabola. (*After Lee and Rahmat-Samii [22], © 1981 IEEE*)

average power pattern of two-dimensional aperture distributions can then be derived [24]. The average power pattern is a function of radial surface random errors, illumination taper, and *F/D*. The results of this study indicate a strong dependence of low side lobe levels on the surface random errors (ϵ_{rms}). Fig. 12 shows the side lobe level increase versus the surface random errors.

Performance degradation caused by deterministic surface errors is evaluated via the aperture field method or induced current method. The surface tolerance requirement of a reflector antenna system should be placed after carefully analyzing the performance degradation against the system requirements in a case-by-case basis.

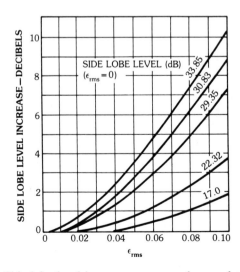

Fig. 12. Side lobe level increase versus random surface errors.

The Feed Array

The design of the feed array differs in scanning beam and fixed beam antenna systems. In a scanning beam antenna system the key design parameters are gain, gain ripple, side lobe, and cross-polarization levels. Gain ripple is due to the crossover of two discrete scanning beams generated by two adjacent feed elements. Gain and side lobe levels are controlled by the optics and the feed illumination taper. In order to improve gain, a larger feed element is preferred to reduce spillover energy. A larger feed element, however, results in a wider feed separation and hence lowers the crossover point of the two adjacent scanning beams. This in turn increases the gain ripple. In order to maintain high gain and reduce gain ripple at the same time, overlapping cluster feed is used. A single beam is generated by a number of feed elements. Some feed elements are used for both of the two adjacent scanning beams. The other approach is to use a continuous scanning system. Variable phase shifters and variable power dividers are used to control the beam scanning continuously from one position to another position. Fig. 13 presents three basic approaches of the scanning beam system.

In a fixed beam antenna system, not only the peak gain, gain ripple, and side lobe level, but also the minimum gain in the coverage area, interference and transmitter power between beams, and the complexity of BFNs have to be considered.

Design Parameters—Key design parameters are the location of feed elements, number of feed array elements, lattice spacing, feed array configuration, radiating elements, and materials.

Location of Feed Elements—From the optical viewpoint it is desirable to place the feed elements on the best focal surface. In general, the best focus surface is not necessarily planar and may be curved. A curved feed array surface introduces design difficulties, such as increased mutual coupling, depolarization, and unavailable design-analytical tools. Trade-offs among the achieved performance, feed array complexity, and design difficulty have to be conducted.

Number of Feed Elements and Lattice Spacing—For a fixed F/D, there is a minimum number of elements required for a given coverage area. The number of elements is inversely proportional to element spacing. The element spacing affects the cross polarization because of mutual coupling. From the beam optimization viewpoint, a greater number of feed elements improves beam shaping and provides higher spatial isolation but results in a more complex beam-forming network.

Grating lobes associated with lattice spacing should be considered. The grating lobes should not fall into the coverage area in a phased array antenna system or fall into the optical aperture to degrade the antenna performance.

Feed Array Configuration—The feed array can be arranged in a periodic or nonperiodic configuration. A nonuniform feed configuration is often used in a multibeam antenna system when the number of feed elements is small. Different feed spacing and different feed size are used for better beam shaping to improve the gain at the edge of coverage.

Radiating Elements—The choice of radiating elements is determined in a

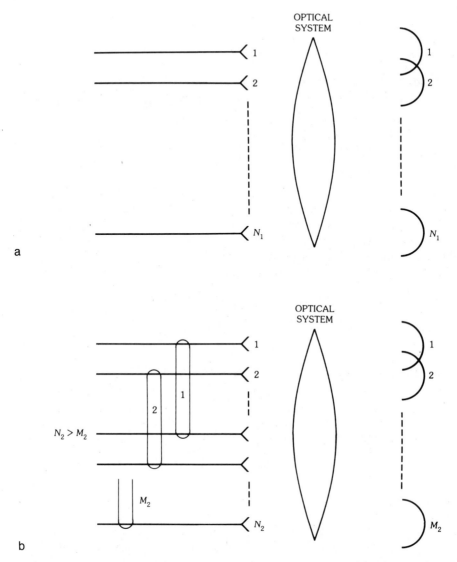

Fig. 13. Three approaches of scanning beam antenna systems. (*a*) Scanning beam feed array. (*b*) Overlapping scanning beam feed array. (*c*) Overlapping scanning beam feed array (variable power dividers).

similar way as for the phased array given under "Types of Element" in the phased array section.

Materials—Feed elements have typically been fabricated using (*a*) cast of machined aluminum alloys, (*b*) electroformed copper or nickel, or (*c*) have been layed up using composite materials such as graphite-epoxy.

Thermal distortion has to be considered in the extreme temperature cycling environment. Among three candidates, copper-clad graphite-epoxy is the most stable material over a wide temperature range. Weight is also a major design

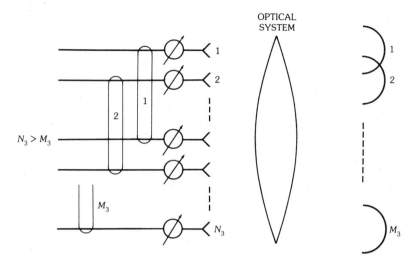

Fig. 13, *continued.*

criterion in feed design. Permanent-mold cast aluminium feed horns have the comparatively thick wall required for manufacturing. Electroformed nickel may provide the required thermal stability and can be accurately manufactured with very thin walls. However, it is heavy because of nickel's high density. The graphite-epoxy feed horn provides the lightest weight.

Polarizer—The next key component of a feed array is a polarizer that is required to provide for circular polarization. The choice of polarizer is decided by performance specifications such as frequency bandwidth, axial ratio, insertion loss, network loss, isolation between ports, and compactness in feed array integration. Fig. 14 shows the most commonly used polarizers: septum, multiprobe launcher, pin, dielectric-slab, pinched-guide, and Meanderline. A Meanderline polarizer uses a square-wave printed-circuit pattern to provide reactive loading to the orthogonal linear component of an electric field. By orienting the Meanderline print at 45° in space with respect to the incident field, one field component, parallel to the printed pattern, is inductively loaded and its orthogonal counterpart is capacitively loaded. This causes a differential electrical phase shift between two orthogonal fields. Multiple sheets are used to provide the required 90° phase differential of two orthogonal incident waves to and from circularly polarized waves. It can be designed to operate in multifrequency bands and be placed in front of a feed array to form a compact system. It has, however, a higher insertion loss and axial ratio than the other types of polarizers.

Beam-Forming Networks

The beam-forming networks (BFNs) are *n*-to-*m* port networks and their purpose is to interconnect the *n* input ports to the individual *m* ports with required amplitudes and phases. The BFNs can be divided into two categories: scanned and fixed.

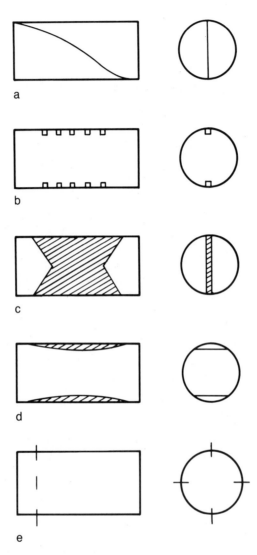

Fig. 14. Types of polarizers. (*a*) Septum polarizer. (*b*) Pin polarizer. (*c*) Dielectric-slab polarizer. (*d*) Pinched-guide polarizer. (*e*) Multiprobe launcher polarizer.

Scanned Beam-Forming Networks—Classical methods of beam steering used either variable phase shifters or interconnecting switches to select a particular individual beam. To these methods can be added scanning with variable amplitude networks, scanning with both variable amplitude and phase networks, and scanning space-fed networks (such as lenses).

Switch Beam Networks—Switch beam networks are composed of many fixed networks, each with an output port connected to a terminal, that represent a unique beam position as shown in Fig. 15.

The BFN for nonoverlapping beam clusters is simple and straightforward. A switch matrix, which allows the selection of one of N available beams formed by

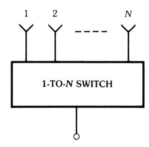

Fig. 15. Nonoverlapping switch beam-forming network.

an array, is shown in Fig. 16. For an overlapping beam the BFNs become more complex since a particular feed element may have to be shared with two or more beams as shown in Fig. 17.

Other variations of switched circuits are Butler and Blass matrices. The Butler matrix (Fig. 18) consists of identical junctions (3-dB hybrids) and fixed phase shifters. This beam-forming technique can be used for planar arrays by combining columns of antenna elements in one set of matrices, and combining the outputs of the column matrices in a group of row matrices. Beams formed by the total matrix may be selected through an auxiliary switching matrix. Such a configuration would be appropriate for feeding a near-field plane wave type of optical system. Characteristics of this configuration are as follows:

1. The number of different beams that can be formed is equal to N, and the number of antenna elements (or clusters) must generally be equal to a power of 2.

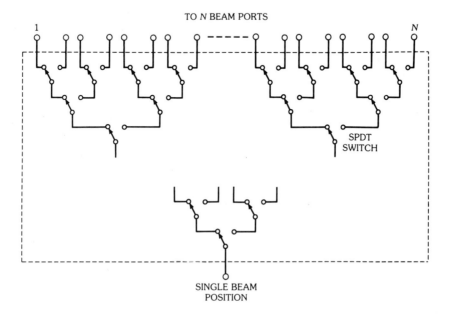

Fig. 16. Multiport switched beam-forming network.

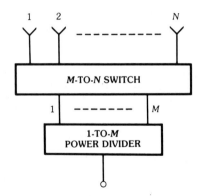

Fig. 17. Overlapping switch beam-forming network.

2. The BFN can be theoretically lossless and matched. The insertion loss of the matrix will be determined by the rf losses of the matrix components.
3. The bandwidth is limited by the components, such as phase shifters and directional couplers. However, the positions of the antenna beam peaks will vary with frequency. The amount of variation with frequency is a function of beam location.
4. The antenna array illumination is uniform. It is possible to achieve (cosine)^n aperture distribution using the weighted addition of selected beams. For example, it is possible to obtain a cosine distribution, but the beam crossover level is 9.5 dB below the peaks. Thus, for lossless beams, the aperture distributions are limited to a uniform distribution that has relatively high first side lobe or tapered distribution that has far-field patterns with undesirably low crossover levels. Another major drawback of this beam-forming technique is its complexity. For example, a 64-element matrix requires 92 directional couplers and 160 fixed phase shifters.

    The Blass network (Fig. 19) is based on a series-fed matrix scheme. Each junction is a directional coupler and the phase to each element is depicted by a different line length. Therefore, by selecting the proper phase to each element, a progressive phase front across the aperture can be obtained which produces a beam in the desired direction. The coupling factors of the directional coupler provide the appropriate aperture amplitude tapers. A serious limitation of the Blass matrix is that each coupler on any given radial line must have a different coupling value; also, the series-fed array requires more couplers than other approaches and consequently implies greater cost and weight.

Variable Phase Shifter Networks—The variable phase shifter networks are used in a phased array. Phased array networks vary the relative phases of feed elements in order to scan antenna beams. A phased array feed network can be either a parallel-fed or a series-fed type depending on the distribution between the input and the output individual array elements, each with its controlling phase shifter.

Variable Amplitude Networks—The variable power dividers can be used in the scanned beam networks for beam steering. A switch is essentially operated in one of the special states of a variable power divider in which the power is either on or

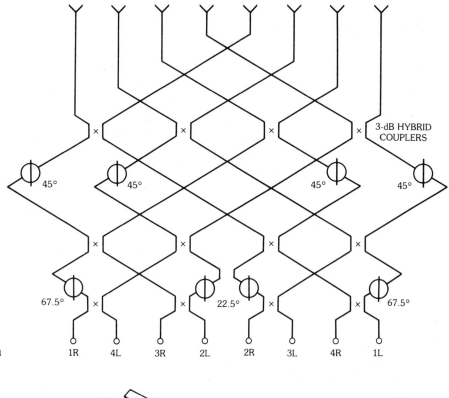

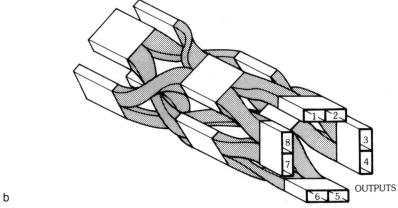

Fig. 18. Butler matrix. (*a*) Eight-element Butler matrix. (*b*) Eight-port Butler matrix in waveguide.

off. Consequently, an antenna excited by a BFN using variable power dividers can produce more flexible coverage than the BFN using switches. However, it requires more sophisticated control circuits than do switch networks.

Variable Phase and Amplitude Networks—The BFNs using both variable phase and amplitude networks provide the maximum degree of beam control. The scanned beam can be optimized at any position by controlling array amplitude and

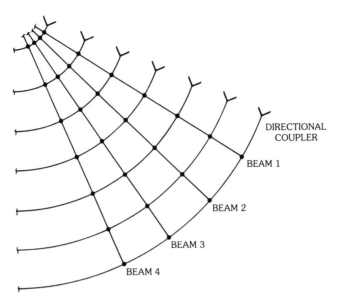

Fig. 19. Blass network.

phase. The beam can also be scanned with finer scanning resolution. This approach requires the most sophisticated control circuits and devices in the BFNs.

Space-Fed Arrays—Beam scanning can be accomplished with optical feed systems in the form of space-fed arrays. These systems avoid the need for corporate feed networks, and typically employ a horn or ensemble of horns illuminating a lens type device, which contains an array of radiating elements and associated phase shifters. In most cases the phase shifters are used to focus the beam as well as to steer it.

Space-fed arrays are divided into transmission arrays and reflection arrays as shown in Fig. 20. The transmission and reflection arrays require a combination of lens and array design techniques because they have the properties of conventional lenses, while the secondary radiator functions as a phase-controlled array.

Fixed Beam-Forming Networks—A fixed beam is formed by feeding one or more feed elements, the number required depending on the shape of the beam, gain, and/or side lobe levels. The BFNs can be divided into three categories: constrained, unconstrained, and the hybrid of constrained and unconstrained networks. The constrained BFNs use transmission lines exclusively to transfer energy from input port to output ports. The unconstrained BFNs use free space as a transmission medium. A hybrid of these two types can also be used where energy is constrained in one dimension but not in the other.

The advantages of the constrained BFNs are the accurate controls of amplitude and phase. The physical size can be a small fraction of the overall antenna system. The networks must utilize components which, even though they are separately well matched, are difficult and practically impossible to match to the entire network after assembly. The advantage of the unconstrained BFNs is simple. The physical

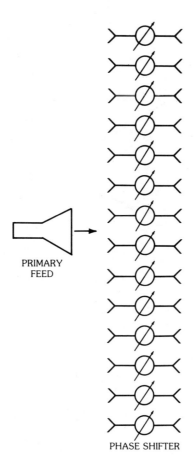

PRIMARY
FEED

PHASE SHIFTER

a

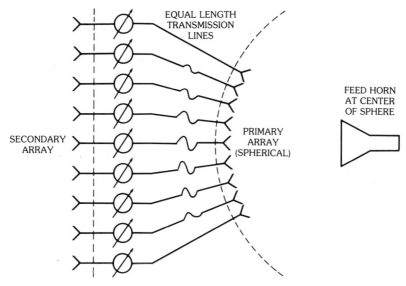

EQUAL LENGTH
TRANSMISSION
LINES

SECONDARY
ARRAY

PRIMARY
ARRAY
(SPHERICAL)

FEED HORN
AT CENTER
OF SPHERE

b

Fig. 20. Types of space-fed beam-forming networks. (*a*) Variable-phase planar-planar transmission array. (*b*) Variable-phase spherical-planar transmission array. (*c*) Variable-phase reflection array. (*d*) Optical parallel-plate phase-scanned array.

21-35

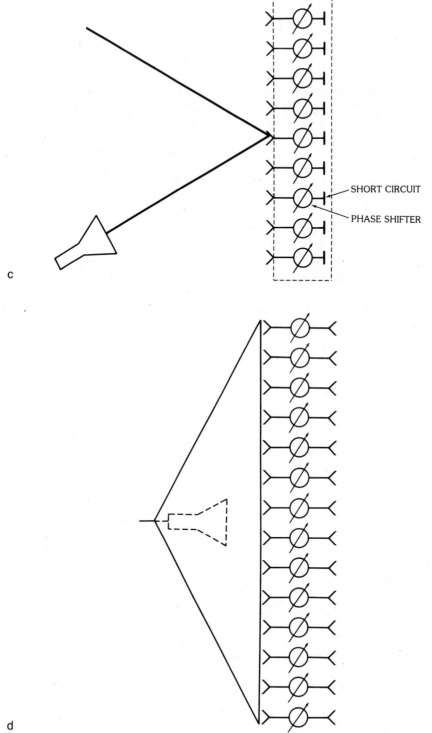

c

SHORT CIRCUIT

PHASE SHIFTER

d

Fig. 20, *continued.*

size is at least half the overall antenna system. However, the networks provide less control of amplitude and phase. The unconstrained feed system may be practical if a large number of components is required in a large array.

Both scanned and fixed BFNs consist of components and feed network media. Components that are generally used include switches, phase shifters, power dividers, hybrids, diplexers, variable phase shifters, variable power dividers, rotary joints, and frequency-selective surfaces. Low insertion loss is extremely important in the BFN for efficient antenna performance. The choice of feed network media is determined by the rf performance requirement, frequency band operation, size, weight, reliability, thermal stability, and manufacturability.

Design Considerations—Selection of an optimum BFN for a given antenna system involves the trade-offs of types of components and feed network media. It also depends on the operating frequency bands. Striplines are often used at C-band or lower frequencies and waveguide transmission lines are used at X-band or higher frequencies. Various components, such as power dividers, switches, variable phase shifters and power dividers, and amplifiers, have to be chosen based on the trade-off of the method of network reconfigurability. The selection of the most suitable feed network transmission media is determined by the insertion loss and the availability of the appropriate components for transmission media.

The following are general trade-off criteria:

Allowable insertion loss
Compactness and weight
Design maturity
Power-handling capacity
Manufacturability
Thermal stability

The first of these criteria is important since the highest achievable performance is the ultimate goal of an efficient antenna system. As a satellite is tightly packed, the BFN must occupy the smallest area as possible. The design must be mature so that essential design information is available. The BFN must also be able to handle the maximum power transferred, must be fabricated with provided tolerances, and must be stable in the thermal environment.

Scanned Beam-Forming Network—The key components are switches, variable phase shifters, and variable power dividers. The design parameters are vswr, insertion loss, isolation, switching time, and size.

1. *Switches*—There are three types of switches: electromechanical, ferrite, and diode. Electromechanical switches offer the lowest insertion loss of the various switch types available with a reasonable degree of reliability. Manufacturers often claim lifetimes of 10^6 actuations or more. Both coaxial and waveguide designs are available, but the latter are usually large and heavy. Actuating mechanisms are usually magnetically controlled, using solenoids to produce mechanical motion of one or more contact bars or levers and permanent magnets to hold them in position to form a latching device. Use of dielectrics to prevent multipacting in many critical

areas is precluded by the mechanical motion. High-power designs are therefore either hermetically sealed or vented to vacuum for space applications.

Single-pole, double-throw (spdt) ferrite switches are available based on the switching circulator principle. The direction of circulation around a circulator structure depends on the direction of the applied magnetic bias field. If this direction is reversed, output power is switched between two alternate output ports. Isolation of 20 to 25 dB is achievable over 10-percent bandwidth, and insertion loss of stripline-type 4- and 6-GHz switches may be as low as 0.15 dB. Cascaded junctions could be used if greater isolation is desired. The switch times are generally under 1 μs. It has been demonstrated that the switch time can be in the range of 200 to 400 ns. Switching energy is in the range of 50 to 100 μJ per switch event. Satisfactory operation over temperature ranges of −40°C to 90°C have been demonstrated.

Microwave diode switches are attractive for their compactness, light weight, moderate control power requirements, and high switching speeds. In addition, multiple-port designs are feasible. Their losses are considerably higher than those of the other two switch candidates, especially at high frequencies. The basic switching diode is a pin junction, which can approximate a short circuit or a small lumped capacitance in the forward or reverse bias states. Such an element can be used either in series or in shunt with a transmission line to form a basic spst switch. Tuning structures can be incorporated to match out diode reactive elements over reasonable bandwidths, and multiple diodes can be incorporated to meet enhanced isolation requirements. High peak and average powers can be handled using diodes with adequate heat dissipation construction and peak inverse voltages in the kilovolt range. Harmonic and intermodulation distortion are possible at higher power levels. Insertion losses are generally 2 dB for 35-dB isolation in the 18-GHz frequency band.

2. *Variable Phase Shifters*—There are three main types of variable phase shifters: ferrite, diode, and electromechanical. Design parameters are: frequency band, insertion loss, phase range, phase steps, phase slope, maximum peak-to-peak phase variation versus frequency band, power-handling capacity, temperature range, size, and weight. The phase shift of a ferrite phase shifter is dependent on the magnetization of the material. Two forms of phase control are analog and digital. Ferrite phase shifters can be divided into the following categories [25]:

Waveguide nonreciprocal
Reciprocal dual-mode
Helical
Reciprocal and nonreciprocal strip transmission line
Microstrip
Latching Reggia-Spencer

Of the types listed, the waveguide nonreciprocal phaser and reciprocal dual-mode phasers have been proved to be electrically superior to the others.

The basic nonreciprocal waveguide phase shifter consists of a toroidal ferrite, rectangular in cross section, located in the center of a rectangular waveguide as shown in Fig. 21. The energy is phase-shifted in the ferrite section. The ferrite

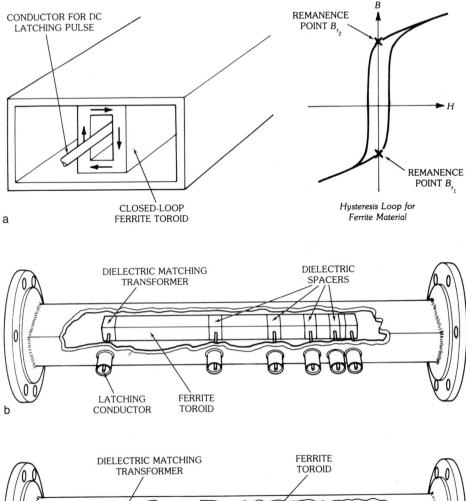

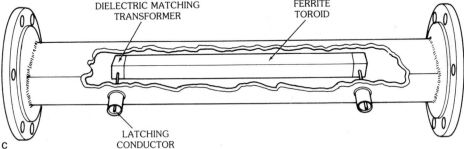

Fig. 21. Nonreciprocal toroidal-waveguide phase shifter. (*a*) Cross section and *B/H* curve. (*b*) With dielectric spacers. (*c*) Without dielectric spacers. (*After Whicker [25], © 1974; reprinted with permission of Artech House*)

section is a loaded waveguide in which a circular or square ferrimagnetic rod is metallized. Good electrical performance has been observed in the toroidal waveguide phase shifters.

The desirable features of a dual-mode phase shifter are as follows:

Parts for many phase shifters may be machines

All non-rf circuits are external to the fully loaded waveguide

The phase shift element is accessible for heat sinking

The phase may be characterized accurately and computational techniques may be utilized

The disadvantages are slower switching speeds and the unit's heavier weight as compared to a corresponding nonreciprocal phaser.

A C-band dual-toroid ferrite phase shifter, as shown in Fig. 22, was developed [26]. The design approach was to develop a transition to match the 50-Ω input/output SMA connection to the ferrite-loaded rectangular waveguide. The measured insertion loss is shown in Fig. 23. The input and output port vswr's are presented in Fig. 24. Fig. 25 illustrates the phase shifter's frequency characteristic for various phase shift settings from 0° to 360°. Achieved performance of two types of ferrite variable phase shifters is given in Table 6.

Diode phase shifters can be constructed using either varactor or pin switching diodes to form either continuously variable or digital type phase shifters. The varactor is a pin junction, usually with a relatively low breakdown voltage, which exhibits a change in reactance (capacitance) with reverse-bias control voltage. The pin diode usually has a relatively high breakdown voltage, but it is useful only in either of two states: a low-impedance (forward-biased) or high-impedance (reverse-biased) state. Either type may be used (in pairs, combined with a quadrature hybrid) to form a reflection type phase shifter. The variable reactance of the varactors, combined with a fixed tuner, appears to be a variable line length terminating two ports of the hybrid, causing equal reflections of variable phases that combine at the hybrid output port. The pin version operates in a similar manner, with the phase states alternating between two fixed values differing approximately

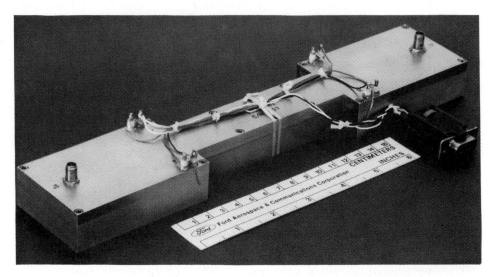

Fig. 22. C-band dual-toroid phase shifter. (*Courtesy Ford Aerospace and Communications Corp.*)

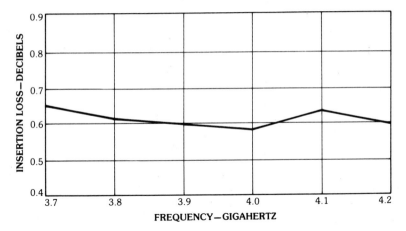

Fig. 23. C-band dual-toroid phase-shifter insertion loss versus frequency. (*After Smith, Mathews, and Boyd [26],* © *1982 AIAA; reprinted with permission*)

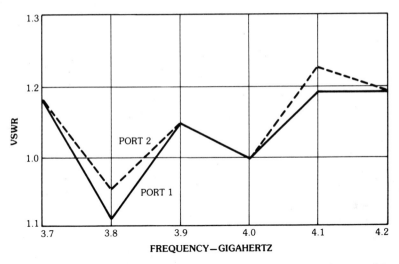

Fig. 24. C-band phase-shifter vswr versus frequency. (*After Smith, Mathews, and Boyd [26],* © *1982 AIAA; reprinted with permission*)

by twice the line length from the diodes to a pair of fixed short circuits. The pin devices are thus used in combinations of "bits" representing differential phases of 108°, 90°, 45°, etc., depending on the resolution required.

Diode designs are amenable to low-cost quantity production using rf printed circuits and simple transistor drives. The major deficiency of diode phase shifters at microwave frequencies is their relatively high loss compared to other devices. These losses generally increase with the number of bits in the circuit, thus penalizing accuracy of setting. They are attractive, however, in the active antenna system in which the loss can be compensated by the amplifiers.

The phase shift of an electromechanical phase shifter relies on the mechanical

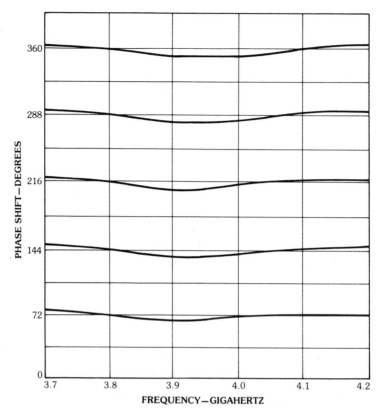

Fig. 25. *C*-band phase-shifter phase shift versus frequency. (*After Smith, Mathews, and Boyd [26], © 1982 AIAA; reprinted with permission*)

Table 6. Achieved Performances of Ferrite Variable Phase Shifters

Type	Frequency (GHz)	Insertion Loss (dB)	VSWR Maximum
Toroidal	5.4–5.9	0.8	1.2:1
	3.2–4.2	0.7	1.2:1
Dual-mode	8.0–9.0	0.9	1.2:1
	9.0–10.0	0.75	1.2:1
	2.9–3.25	1.0	1.2:1
	5.275–5.725	0.9	1.25:1

movement of a physical component in the device to change the phase. Generally, a line length as in "line stretcher" is changed or a dielectric portion of the rf circuit between regions of high and low electric field strength is moved to change the effective phase velocity of the circuit. All electromechanical phase shifters are bulky and heavy because they require electric drive motors. Many mechanically moving parts severely limit the reliability of the component. Even the low insertion loss cannot compensate for the above disadvantages. Thus, electromechanical phase shifters are not used in satellite applications.

3. *Variable Power Dividers* — There are three main types of variable power dividers: ferrite, diode, and electromechanical. Design parameters are insertion loss, setting accuracy, temperature stability, power handling capacity, isolation to off-port, reset characteristics, vswr, phase characteristics, intermodulation products, gain slope, group delay, size, weight, and reliability.

(*a*) *Ferrite variable power dividers*—The ferrite variable power dividers may be designed for either continuously driven operation or a latched operation using remnant magnetization to maintain set states. The variable power divider may be constructed in waveguide, stripline, or some other transmission medium in the integration of the BFN. The most widely used ferrite variable power dividers are

dual-hybrid,
driven variable polarizer, and
latching Faraday rotator

The basic dual-hybrid variable power divider network, as shown in Fig. 26, consists of a 3-dB 180° hybrid, two variable phase shifters, and a 3-dB 90° hybrid. The output signals at ports 1 and 2 can be shown by

$$V_1 = \cos\left(\frac{\phi_1 - \phi_2}{2} + \frac{\pi}{4}\right) e^{j[(\phi_1 + \phi_2)2 + \phi\pi/4]} \tag{17a}$$

$$V_2 = \sin\left(\frac{\phi_1 - \phi_2}{2} + \frac{\pi}{4}\right) e^{j[(\phi_1 + \phi_2)/2 + \pi/4]} \tag{17b}$$

Latching toroidal phase shifters are used to provide the insertion phase shift values ϕ_1 and ϕ_2.

A *C*-band VPD as shown in Fig. 27 was developed [26]. The insertion loss measured for the case of equal power division ratio is shown in Fig. 28. The measured vswr over the 3.7- to 4.2-GHz band is presented in Fig. 29.

The driven-ferrite variable-polarizer variable power divider as shown in Fig. 30 consists of two main sections: a magnetically variable differential phase shift section and a septum polarizer. An incoming signal, which is linearly polarized, is converted into an elliptically polarized wave by the phase shift section. It is then

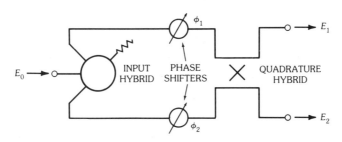

Fig. 26. Dual-hybrid variable power divider circuit.

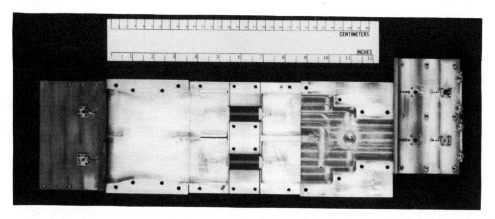

Fig. 27. *C*-band coupler/phase-shifter variable power divider. (*Courtesy Ford Aerospace and Communications Corp.*)

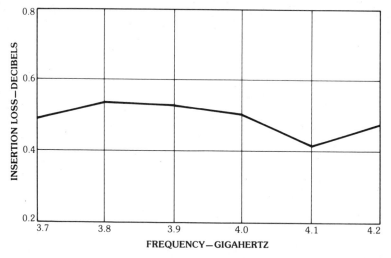

Fig. 28. Coupler/phase-shifter variable power divider insertion loss versus frequency. (*After Smith, Mathews, and Boyd [26], © 1982 AIAA; reprinted with permission*)

converted into two linearly polarized components in separate output ports. The phase shift section is filled with a ferrite rod. The phase shift is controlled by applying a transverse quadruple magnetic bias field to the ferrite rod. The principal axis of the differential phase shift section is oriented 45° with respect to the linearly polarized incoming wave. Equal power output results when no drive current is applied. The change of output power ratio is determined by the degree of ellipticity of the converted circularly polarized waves.

The latching Faraday rotator variable power divider as shown in Fig. 31 consists of an input matching transformer, rotation field section, and an output matching transformer into an orthomode junction as two output ports. The rotation field section is a magnetic yoke that includes a ferrite rod. The input signal is

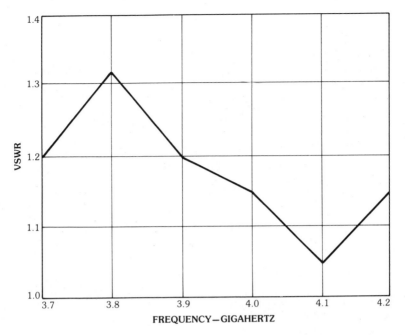

Fig. 29. Coupler/phase-shifter variable power divider vswr versus frequency. (*After Smith, Mathews, and Boyd [26], © 1982 AIAA; reprinted with permission*)

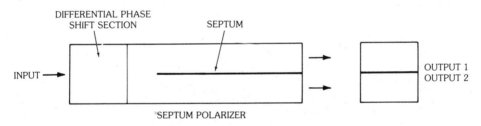

Fig. 30. Driven-ferrite variable-polarizer variable power divider.

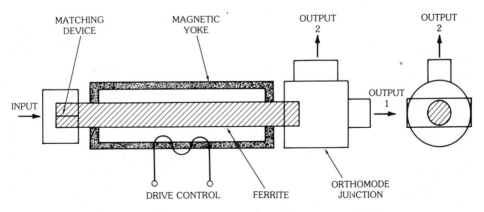

Fig. 31. Latching Faraday rotator variable power divider.

coupled into the ferrite rod and is rotated by an externally magnetic field. The output power is determined by the degree of the field rotation. The yoke is used to maintain the magnetic field after the termination of the drive pulse for latching purpose.

A *C*-band Faraday rotator VPD as shown in Fig. 32 was developed [26]. The insertion loss versus frequency for the case of equal power diversion is shown in Fig. 33a. The input vswr is presented in Fig. 33b. The amplitude variation for both output ports is plotted in Fig. 33c, and the phase variation in Fig. 33d.

Test performances of three types of variable power dividers operated over 7.25 to 7.75 GHz and over the temperature range of −10°C to 40°C are given in Table 7.

(*b*) *Diode variable power dividers*—The diode variable power dividers, which have been used in satellite applications, are a voltage-controlled type and a pin

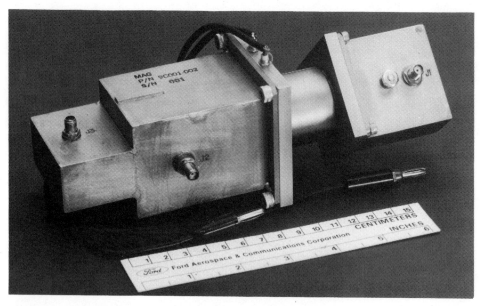

Fig. 32. *C*-band Faraday rotator variable power divider. (*Courtesy Ford Aerospace and Communications Corp.*)

Table 7. Achieved Performances of Ferrite Variable Phase Dividers

Type	Insertion Loss (dB)	VSWR	Isolation (dB)
Dual hybrid	0.34	1.32	30
Driven variable polarizer	0.21	1.56	25
Latching Faraday rotator	0.31	1.91	25

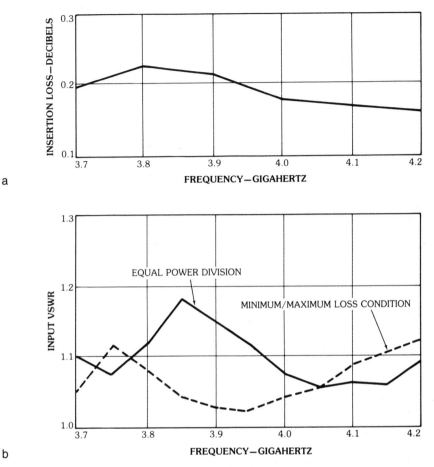

a

b

Fig. 33. Faraday rotator variable power divider characteristics. (*a*) Insertion loss versus frequency. (*b*) Input vswr versus frequency. (*c*) Amplitude variation versus power division ratio. (*d*) Phase variation versus power division ratio. (*After Smith, Mathews, and Boyd [26], © 1982 AIAA; reprinted with permission*)

type. Diode variable power dividers operate on the same principle as the dual-hybrid ferrite variable power dividers. The phase shifters are the diode phase shifters, which can be either the continuously variable voltage-controlled varactor type or the pin diode stepped-increment type.

A voltage-controlled diode variable power divider consists of a three-port split-T power divider, two reflection type diode phase shifters, and a 90° hybrid as shown in Fig. 34. A voltage-controlled diode variable power divider without phase shifter has also been designed. It consists of two 3-dB hybrids, two varactor diodes, and two 50-Ω lines one-quarter of a wavelength long as shown in Fig. 35. The input port is port 1 and the two output ports are ports 2 and 3. The varactor diode capacitances control the reflection and transmission of the incident power.

Typical insertion loss for diode variable power dividers is 0.8 dB or higher for *S*-band. The advantage is the fast switching time. As indicated in the application of

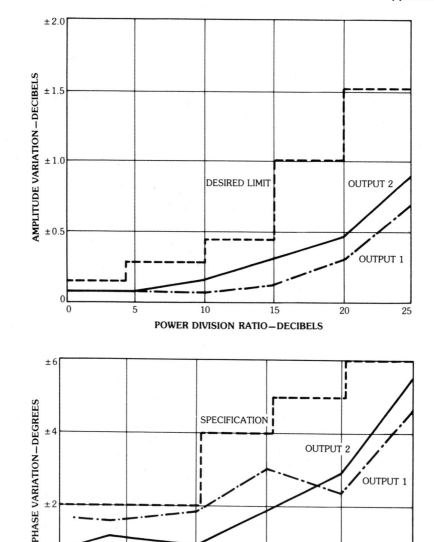

Fig. 33, *continued.*

diode variable phase shifter, the diode variable power dividers are lossy. However, they are attractive in an active antenna system. Diode devices which are operated in the higher frequency bands are being developed.

(*c*) *Electromechanical variable power dividers*—Several types of electromechanical variable power dividers have been developed. They are not suitable for satellite application for the same reasons given at the end of the variable phase shifter subsection (2).

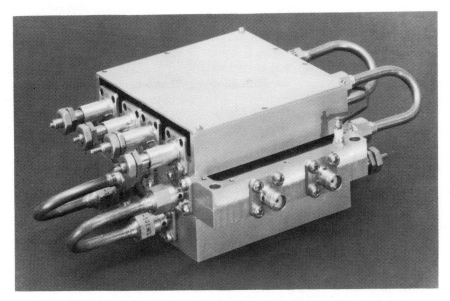

Fig. 34. Voltage-controlled diode variable power divider. (*Courtesy Ford Aerospace and Communications Corp.*)

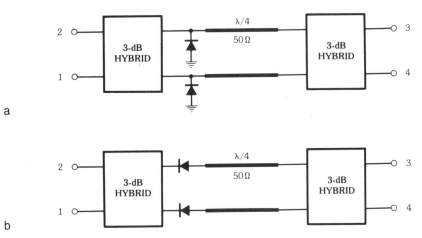

Fig. 35. Voltage-controlled variable power divider without phase shifters. (*a*) With shunt diodes. (*b*) With series diodes.

Fixed Beam-Forming Networks—Fixed beam-forming networks include power division elements, directional couplers, transitions, phase shifters, diplexers, rotary joints, and odd/even mode converters. Most of these components are also used in the scanned beam-forming networks.

1. *Power Division Elements*—The choice of the types of power division elements is based on considerations of the physical layout, manufacturing techniques, the feed transmission medium, and the component insertion losses.

Two main types of power division elements have been widely used: air stripline and waveguide. The air stripline components are used in combination with the air stripline transmission medium, where the waveguide components are used with the waveguide transmission medium. The air stripline type of BFN is compact but more lossy. Its application is generally limited at the C-band or lower-frequency bands. The waveguide BFN is bulky but with lower loss and is mostly used at the K-band or higher-frequency bands.

Several components are suitable for use as air stripline power division elements as depicted in Fig. 36:

Hybrid-ring directional coupler
Split-T power divider
Branchline directional coupler
Symmetrical directional coupler
Semicircular-rod directional coupler
Gysel hybrid

Waveguide power division elements may be the following:

Magic-T junction
Folded-T junction

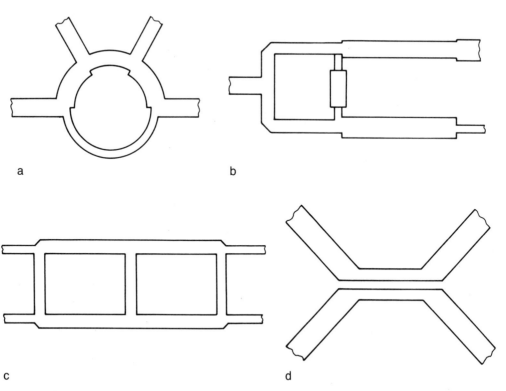

a   b

c   d

Fig. 36. Stripline power division elements. (*a*) Hybrid-ring directional coupler. (*b*) Split-T power divider. (*c*) Branchline directional coupler. (*d*) Symmetrical directional coupler. (*e*) Semicircular-rod directional coupler. (*f*) Gysel hybrid ring.

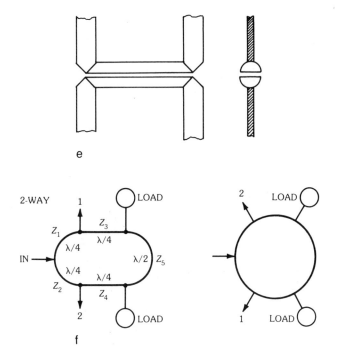

e

f

Fig. 36, *continued.*

Top-wall hybrid junction
Narrow-wall coupler
Aperture couplers (single-hole or multihole)
Branch-guide coupler
H-plane septum T
Waveguide hybrid ring

These are depicted in Fig. 37.

Major trade-offs of the types of power division elements are the insertion loss, frequency band, and power split ratios. A combination of types may be required to provide a wide range of power distributions required in a multibeam antenna system. For air stripline division elements, hybrid-ring directional couplers and split-T power dividers are suitable for moderate power split ratios. The Gysel hybrid is more complex but suitable for wider frequency band operation because of the symmetrical treatment of the hybrid. Symmetrical directional couplers are more suitable for the power split ratio that is greater than 10 dB. They can also be cascaded to improve the coupling bandwidth. For waveguide power division elements the magic-T, folded-T, top-wall hybrid junction, and narrow-wall couplers are widely used as 3-dB couplers. The others are suitable for different power split ratios. The hybrid ring's bandwidth is comparatively narrow. Its awkward geometry is not as compact as the corresponding ring-hybrid stripline coupler and does not lend itself very well to compact layouts of BFNs.

Since a wide range of power division ratios is required, a complete design curve

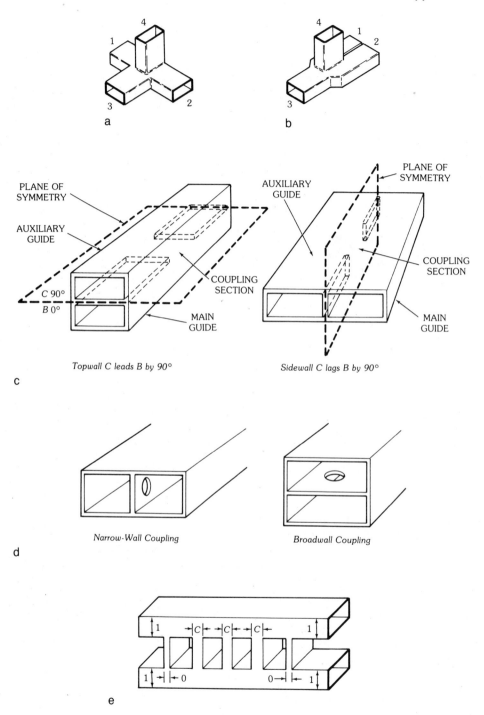

Fig. 37. Waveguide power division elements. (*a*) Conventional magic-T. (*b*) *H*-plane folded T. (*c*) Top-wall hybrid junction. (*d*) Aperture couplers. (*e*) Typical branch-guide coupler with five branches. (*f*) *H*-plane septum T. (*g*) Waveguide hybrid ring.

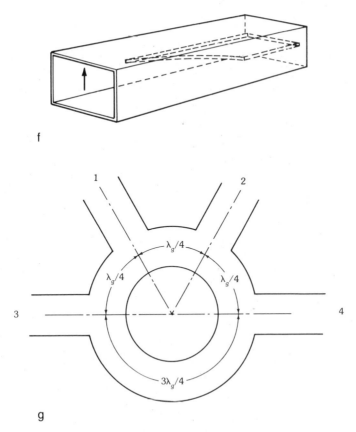

f

g

Fig. 37, *continued.*

is not available. The components are usually designed, fabricated, and tested for a wide range of power division ratios. The design for arbitrary ratios are then extrapolated from the measured data.

2. *Diplexer*—A diplexer can be a filter type or a frequency selective surface. The former approach sometimes imposes severe constraints on the design of the BFN, which is complex and congested in a frequency reuse multibeam antenna system. A frequency selective surface can be designed to be transparent for one frequency band and reflective for the other band, to properly direct and redirect the energy to the same optical aperture as shown in Fig. 38. Trade-offs of the two approaches are the difficulty in the design of the BFN, size of the frequency selective surface, and its thermal stability.

3. *Rotary Joint*—The rotary joint is used to provide a single-channel or multichannel transmission path between a spin spacecraft and its despun antennas. The design trade-off is the choice among waveguide, coaxial, and combined waveguide-coax types with an auxiliary choice of waveguide or coax as an output. Design parameters are insertion loss, performance variation with respect to rotation, size, and weight, and ease of design and construction. Fig. 39 depicts a three-channel rotary joint.

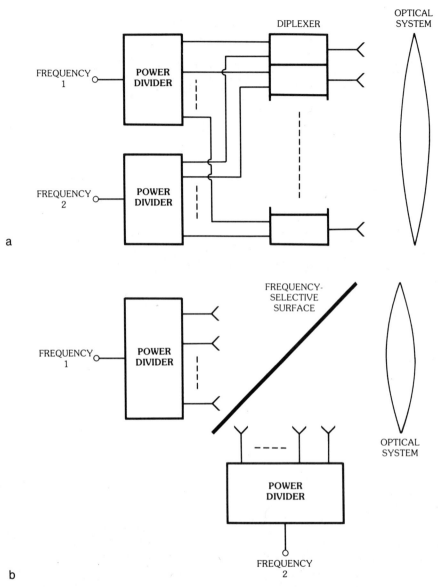

Fig. 38. Types of diplexers. (*a*) Filter-type diplexer. (*b*) Spatial diplexer.

4. *Odd/Even-Mode Converter*—An odd/even-mode converter [27] is used to alleviate the severe design constraint on an output multiplexer. The odd channels of an output multiplexer are combined into a single output, and even channels into a second output. It is generally desirable to use a common optical aperture for both ports. An odd/even-mode converter is a 2-to-*N* port network to combine the two ports to a feed array that illuminates an optical aperture as depicted in Fig. 40.

Let [*S*] be the scattering matrix of the network. It can be shown that in a lossless network,

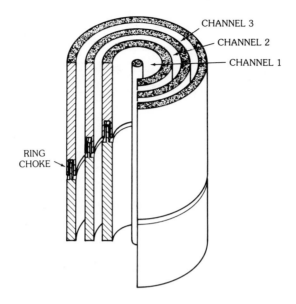

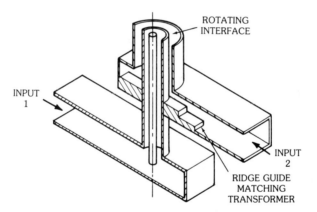

Fig. 39. Three-channel rotary joint. (*a*) Concentric section. (*b*) Transitions, center channel to waveguide.

$$S_{11'}^* S_{1'2}^* + S_{12'}^* S_{2'1} + \cdots + S_{in'}^* S_{n'i} = 0$$

and

$$P_{1'} + P_{2'} + \cdots + P_n = 1$$

where * represents the complex conjugate of the complex number, the prime refers to the input port, and nonprimed numbers the output ports; $P_{i'}$ is the power at the ith input port.

For an equal power distributed mode converter in which the output power ratios among the output ports are all equal, a two-to-two dual-mode converter is a

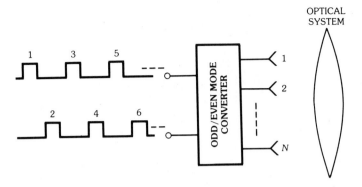

Fig. 40. Odd/even-mode multibeam antenna system.

conventional quadrature hybrid, and a two-to-three dual-mode converter has a phase progression of 60°, 0°, −60° for one mode and −60°, 0°, 60° for the other.

A dual-mode converter can be realized by a combination of orthomode transducer and circular polarizer. The odd channel output and even channel output are connected to the input ports of the orthomode transducer. The odd mode generates one sense of circularly polarized waves while the other mode generates another sense of circularly polarized waves. The signals are then coupled out of the output ports of the polarizer as shown in Fig. 41a. An alternate approach uses hybrids and differential line lengths as shown in Fig. 41b.

Feed Transmission Media—Three main types of feed transmission media are waveguide, ridge waveguide, and TEM waveguide as depicted in Fig. 42a. The

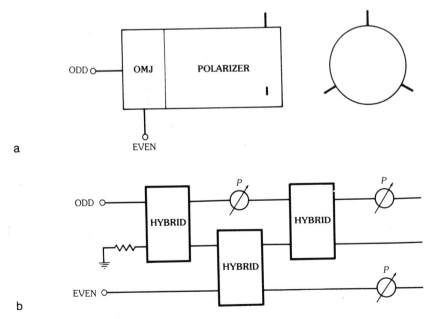

Fig. 41. Odd/even-mode converter. (*a*) Approach 1. (*b*) Approach 2.

waveguides have the advantage of high power-handling capacity. A wide range of commercial waveguide components is also available. The disadvantages are the frequency dispersion of the waveguide, rigidity, and size. The bandwidth capability can be improved by using ridge waveguides. However, ridge waveguide component design is complicated due to the complex structure. The TEM waveguide's bandwidth is much greater than is possible with the other two waveguides since it propagates only the TEM mode. A coaxial line is one of the TEM waveguides. Its modification from a circular to rectangular one yields a planar strip transmission line. The high-Q triplate type of stripline can use printed strips that are compact and easy to construct. As indicated earlier, this air stripline transmission medium is very attractive in the C-band and in lower-frequency–band applications.

The variety of network media spinned off these three main types are shown in Fig. 42b. The relative advantages and disadvantages of each type are summarized in Table 8. Comparisons among the types are given in Table 9.

Beam-Forming Network Topology—The starting point for the circuit layouts are the coordination of the feed array and power division elements.

The initial goal of the network layout is to make all line lengths equal from the common input port to each output element port in order to minimize the effects of frequency dispersion. Lengths can be adjusted slightly to provide the exact phases desired at each element, and to compensate the electrical length of the network paths, such as switches and power divider components. This design of the feasibility layout is an iterative process and is done via a computer graphics software program in the case of designing a complex multibeam antenna system.

Beam-Forming Network Distortion—The distortion of a BFN is due to the manufacturing distortion and thermal distortion. The beam-forming network distortion results in the deviation of amplitude and phase desired for the network. Such deviation yields the performance degradation in an antenna system. The gain will be lower, and side lobe and cross-polarization levels will be higher.

The amplitude and phase variations due to the network distortion can be analyzed. The variations of each component can be measured and used as design information to study the total variations of a beam-forming network. Either the antenna system will be designed with additional margin to compensate the degradation or the BFN will be designed to have smaller tolerance to reduce the distortion in the manufacturing process and thermal environment.

Multibeam Antenna System

A multibeam antenna is the most widely used in satellite antenna systems. It consists of a focusing optics illuminated by an array of feed elements. Each feed element illuminates the optical aperture and generates a constituent beam. Any shaped beam can be formed from a number of these constituent beams by the principle of superposition. Fig. 43 shows a multibeam antenna system. The key features of this antenna system are that it

- (*a*) generates a multibeam pattern from one optical aperture,
- (*b*) provides pattern shaping and pattern weighting for the radiation pattern, and
- (*c*) yields steeper pattern rolloff and results in a higher spatial isolation for the communication system.

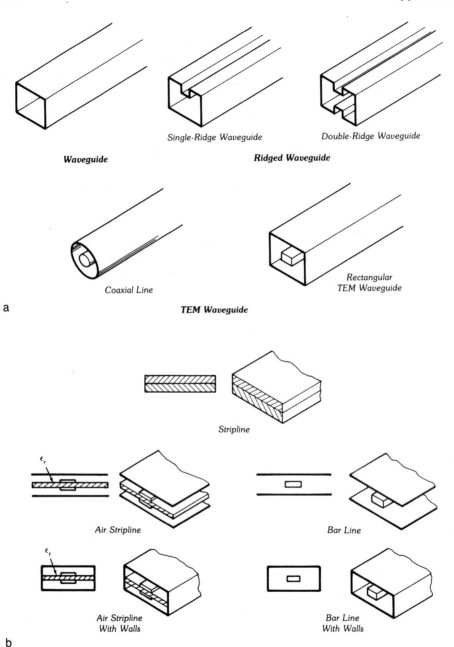

Single-Ridge Waveguide

Double-Ridge Waveguide

Waveguide

Ridged Waveguide

Coaxial Line

Rectangular
TEM Waveguide

a

TEM Waveguide

Stripline

ϵ_r

Air Stripline

Bar Line

ϵ_r

Air Stripline
With Walls

Bar Line
With Walls

b

Fig. 42. Feed transmission media. (*a*) Three main types. (*b*) Variety of media.

The beamwidth of a constituent beam is determined by the size of the optical aperture. The position and angular separation of these constituent beams are determined by the feed element separation and the beam deviation factor of an optics. Several possible optical configurations for a multibeam antenna system are

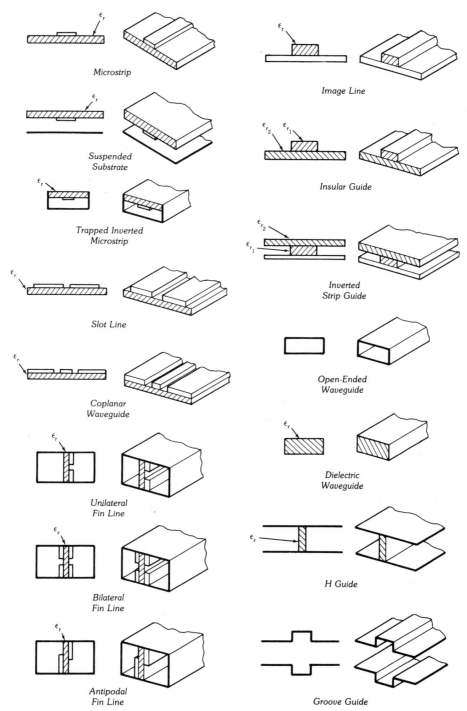

Microstrip

Suspended Substrate

Trapped Inverted Microstrip

Slot Line

Coplanar Waveguide

Unilateral Fin Line

Bilateral Fin Line

Antipodal Fin Line

Image Line

Insular Guide

Inverted Strip Guide

Open-Ended Waveguide

Dielectric Waveguide

H Guide

Groove Guide

b

Fig. 42, *continued.*

Table 8. Relative Advantages and Disadvantages of Transmission Line Types

Transmission Line Type	Advantages	Disadvantages
Rectangular waveguide	Very low insertion loss Dominant mode rectangular Conventional approach Characteristics are well known High-power capacity Does not radiate	Relatively large Relatively expensive Circuitry for beam-forming networks is complicated and difficult to fabricate
Oversized waveguide	Lower loss than conventional waveguide	Too large Can propagate higher-order modes
Circular waveguide	Lowest loss Characteristics are well known	Larger cross section Few components can be made in circular guide
Coaxial line	Small cross section Characteristics are well known Moderate loss for air-filled line	Beam-forming network components are difficult to fabricate Dielectric-filled lines are heavy, lossy, and unreliable Requires many interconnections Reliability problems
Stripline	Characteristics are well known Fabricated with printed-circuit techniques Relatively compact Nondispersive	High insertion loss Thermal instability problems
Air stripline, solid conductor	Low loss Compact and lightweight Excellent thermal properties Design information is readily available	Needs support of center conductor
Air stripline, dielectric supported	Low loss Lightweight and compact Printed-circuit fabrication techniques	Care must be taken to avoid thermal instabilities and undesirable modes
Microstrip line	Very small size Printed-circuit fabrication techniques Properties are well known High capability for component integration	High loss Tends to radiate at discontinuities Susceptible to higher-order modes Dispersive
Suspended substrate	Less loss than microstrip Printed-circuit fabrication techniques Tolerances less critical than for microstrip	Suffers same radiation problems as microstrip Dispersive

Table 8, *continued.*

Transmission Line Type	Advantages	Disadvantages
Trapped inverted microstrip	Moderate loss Tolerances not as critical as for microstrip Less radiation at discontinuities than for microstrip Moderate degree of potential for integration Printed-circuit techniques	Somewhat difficult to fabricate circuits
Slot line	Printed-circuit fabrication techniques Compact and lightweight	Very high loss Tends to radiate Not suitable to beam-forming networks
Coplanar waveguide	Printed-circuit fabrication techniques All conducting elements are on the same side of the substrate	Very high loss Tends to radiate Not suitable for beam-forming networks
Fin line	Moderate loss Partially fabricated with printed-circuit techniques	Large cross section Not suitable for beam-forming networks
Dielectric waveguide	Low attenuation loss Relatively simple to fabricate	Radiation occurs at discontinuities, such as bends Difficult to support mechanically Relatively large dimensions Limited bandwidth Not suitable for beam-forming networks
Image guide	Moderately low loss Overcomes support problem of dielectric waveguide	Radiation occurs at discontinuities, such as bends Small range of impedances around 26 Ω available Difficult to achieve zero gap between the ground plane and the dielectric waveguide, causing rf problems Relatively large dimensions Not suitable for beam-forming networks
Insular guide	Moderately low loss Overcomes gap problems of image guide	Radiation problems Relatively large dimensions Not suitable for beam-forming networks

Table 8, *continued.*

Transmission Line Type	Advantages	Disadvantages
Strip dielectric guide	Radiation loss is less than for other dielectric waveguides	Large dimensions Relatively heavy Not suitable for beam-forming networks
Inverted strip guide	Moderate loss Small radiation	Poor guidability at bends Large dimensions Heavy Not suitable for beam-forming networks
Trapped image guide	Does not radiate at bands	Large dimensions Not practical for beam-forming networks
H-guide	Low loss Sections can be joined without connectors	Large cross section dimensions Not practical for beam-forming networks
Groove guide	Low loss Broadband operation	Very large dimensions Little design information available Not suitable for beam-forming networks
Fence guide	Moderately low loss	Little information available Not suitable for beam-forming networks

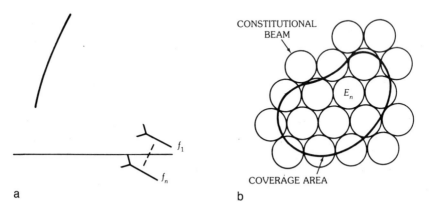

Fig. 43. Multibeam antenna system. (*a*) Feed elements and focusing optics. (*b*) Shaping of beam.

shown in Fig. 44. The lens and reflector are interchangeable in the configurations. The design of the system is the same as those presented previously. The excitation (amplitude and phase) of each beam has to be determined by antenna pattern synthesis.

Table 9. Comparisons of Transmission Lines

Type of Line	Insertion Loss	Impedance Levels (ohms)	Power Handling	Tendency to Radiate	Suitability for Beam-forming Networks	Cross-Sectional Dimensions	Availability of Design Information
Standard waveguide	Very low		Moderate	Zero	Poor	Medium	Excellent
Circular waveguide	Very low		High	Zero	Poor	Large	Excellent
Oversized waveguide	Very low		High	Zero	Poor	Large	Excellent
Coaxial line	Moderate	40–120	Low–mod	Zero	Poor	Small–medium	Excellent
Stripline	High	30–120	Low–mod	Zero	Excellent	Small	Excellent
Air stripline	Low	30–130	Low–mod	Zero	Excellent	Small	Good
Microstrip line	High	20–125	Low	Moderate	Good	Very small	Excellent
Suspended substrate	Moderate	25–130	Low	Moderate	Good	Small	Fair
Trapped inverted microstrip	Moderate	30–140	Low	Small	Poor	Small	Fair
Slot line	Very high	60–200	Low	High	None	Very small	Poor
Coplanar waveguide	Very high	40–150	Low	High	None	Small	Poor
Fin line	Moderate	10–400	Low	Zero	None–poor	Moderate	Poor
Image line	Very low	$\cong 26$	Moderate	Mod–high	None–poor	Moderate	Poor
Insular guide	Low		Moderate	Mod–high	None–poor	Moderate	Poor
Inverted strip guide	Moderate		Moderate	Low	None–poor	Moderate	Poor
Dielectric waveguide	Moderate		Moderate	High	None	Moderate	Fair
H-guide	Low		High	Small	None	Large	Fair
Groove guide	Very low		High	Small	None	Large	Poor

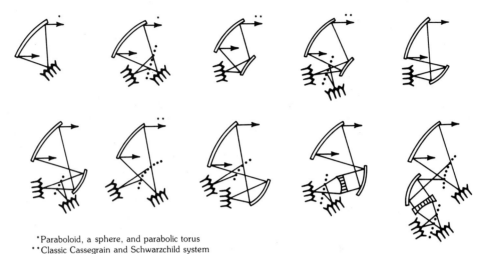

*Paraboloid, a sphere, and parabolic torus
**Classic Cassegrain and Schwarzchild system

Fig. 44. Multibeam antenna system optical configurations.

The effect of weighting the excitations of multifeed elements can be expressed in the power gain calculation. If unit excitation of the ith feed element produces the secondary field $E_n(r, \theta, \phi)$, and voltage coefficients f_n are applied, the power gain is

$$P^2(\theta, \phi) = \frac{4\pi r^2 \left| \sum\limits_{n=1}^{N} f_n E_n(r, \theta, \phi) \right|^2}{n_0 \Sigma |f_n|^2} \tag{18}$$

where $E_n(r, \theta, \phi)$ can be computed via the aperture field method, induced current method, or uniform method [9, 28], and n_0 is the free-space impedance. Equation 18 can be expressed in the following way:

$$[E][f] = [P] \tag{19}$$

in which

$$[f] = \begin{pmatrix} f_1 \\ \vdots \\ f_N \end{pmatrix} \qquad [E] = (E_1, \cdots, E_N), \qquad \text{and} \qquad P = \begin{pmatrix} P_1 \\ \vdots \\ P_M \end{pmatrix} \tag{20}$$

describes the radiation field sampled at M points. The synthesis problem can be represented by

$$[E][f] \cong [P^d]$$

where $[P^d]$ is the desired pattern vectors. The above equation represents a system of M linear equations with N unknowns. If $[E]$ is a square matrix and nonsingular, the solution is unique. If the rank of $[E]$ is less than N, then more than one solution may exist. If M is greater than N, which is generally the case in a multibeam antenna system, then no exact solution exists but there will be a unique least-squares solution.

Most multibeam antenna systems specify only the power of a required radiation pattern. Let $(P_m^d)^2$ be the desired radiation power at point m; then the deviation from the desired value, which is defined as the *pattern error*, is

$$\epsilon_m = \omega_m \left| \left| \sum_{n=1}^{N} f_n E_{mn} \right| - P_m^d \right|, \qquad m = 1, \ldots, M \tag{21}$$

where ω_m is a weight factor. The sampling points are taken in the coverage and outside the coverage.

The gradient, minimax, and regularization methods of synthesis are used most widely. The *gradient method* minimizes the total errors defined as

$$\epsilon = \sum_{m=1}^{M} |\omega_m (P_m - P_m^d)|^2 \tag{22}$$

by advancing (f_n) along the $-\Delta\epsilon$ direction. The *minimax method* [29, 30] minimizes the maximum errors, which are given in the form of

$$\max \epsilon_m = \omega_m \left| \frac{P_m^d - P_m}{P_m^d} \right| \tag{23}$$

The *regularization method* [31] minimizes the total error

$$\epsilon = \sum_{m=1}^{M} |\omega_m (P_m - P_m^d)|^2 \tag{24a}$$

subject to the constraint of

$$\| f \| \leqslant C \tag{24b}$$

where C is a positive constant. The solution obtained is then used to design the power distribution of the beam-forming network.

The design procedure can be divided into key steps as shown in Fig. 45. The first step is to set requirements. The requirements are given in terms of frequency band and bandwidth, polarization, coverage area, antenna gain, and side lobe and cross-polarization levels. The coverage area is transformed in the antenna coordinate system as viewed from the satellite. The optical aperture size is determined by the resolution of the antenna radiation pattern, gain requirement, and spacecraft constraint. The focal length, as indicated earlier, is determined by the scan loss, side lobe and cross-polarization levels, and the number of feed elements used in the antenna pattern synthesis. The design trade-off parameters of a feed array include the feed array configuration and the shape and size of feed elements. The feed array configuration directly influences the element spacing and the re-

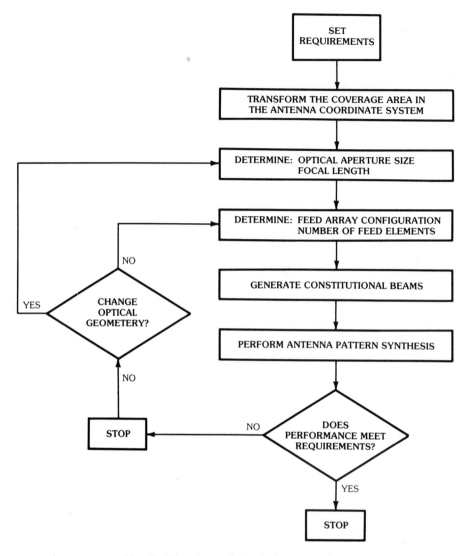

Fig. 45. Flowchart of the design procedure.

quired number of feed elements. It can be periodic or nonperiodic. A periodic configuration is used in a circularly polarized antenna system, while a nonperiodic configuration is sometimes used in a linearly polarized antenna system for pattern control. The constituent beams are then generated. They can be approximated by $(\sin x)/x$ or $J_1(x)/x$, expressed by a Fourier series, computed by rigorous antenna pattern analysis, or measured in a satellite environment. The number of the beams required can be approximately determined by overlaying the beams on the top of the coverage area. Antenna pattern synthesis is applied to obtain the required excitation of each beam. The achieved performance is evaluated against the requirements. The whole procedure is an iterative process.

Design Examples

An S-Band Phased Array—An S-band multiple-access phased array antenna [32, 33] has been designed for the Tracking and Data Relay Satellite System (TDRSS). The antenna as shown in Fig. 46 consists of a 30-element array with 10 elements commonly used for both transmit and receive frequency bands and 20 elements used only for the receive frequency band. The number of elements is chosen so that minimal numbers are used to reduce weight and simplify the complexity of the beam-forming network.

The requirements on each array element are that each array element must provide 13 dB of gain over ±13.5° field of view. The scan loss of the beam at the edge of field of view is to be 3 to 4 dB. Based on these requirements, three array element configurations have been considered: single element, large subarray of low-gain elements, and a small subarray of relatively high gain elements. The single element is chosen because of the simplicity of the beam-forming network. The reflector, horn helix, and short backfire antenna are considered for an array element. The reflector and horn are precluded because of weight and size. The short backfire antenna is eliminated because of its narrow bandwidth performance. The helix is chosen because of its narrow bandwidth performance, inherent circular polarization, acceptable axial ratio, gain, and simplicity.

The end of the helix is terminated in a cone spiral to improve the axial ratio over the field of view and over the transmit and receive frequency bands. A cup is also used to reduce mutual coupling and suppress the strong normal mode radiation near the base of the helix. Fig. 47 shows the measured active element patterns and the single element pattern. The helix in the array environment behaves almost the same, compared to its single element in free space. The design parameters of the element performance are axial ratio and gain versus array spacing. The measured performances are presented in Fig. 48. Even though the optimum minimum spacing is 3.8 cm, 34.3 cm is chosen for conservative reasons. Fig. 49 depicts the ten-element array. Fig. 50 shows the boresight beam and the scanned beam at $\theta = 15°$, $\phi = 15°$.

Lens Antenna—A waveguide lens has been built and tested [34, 35] for a variable earth coverage X-band satellite antenna system. This lens is excited by a variable beam-forming network capable of producing radiation patterns varying from a narrow high-gain beam to the earth coverage beam. The waveguide lens has a 76.2-cm aperture with a 76.2-cm focal length and contains approximately 700 titanium waveguides with a 2.54 × 2.54-cm cross section and a 0.0127-cm wall thickness as shown in Fig. 51 [34]. The lens surface facing the feed is a segment of a sphere centered at the feed and the opposite surface is a segment of a spheroid. This surface is a limiting case of a two-focus design, a design with two focal points. It is expected that the lens has a better scanning performance than those conventional lenses with other different surface shapes. The concave surface facing the feed will improve aperture illumination efficiency. The lens is stepped only on the surface opposite the feed to reduce shadowing effects of the steps.

The waveguides have a square cross section because the circular polarization is required. A 2.54-cm waveguide size is chosen as a compromise of the dominant

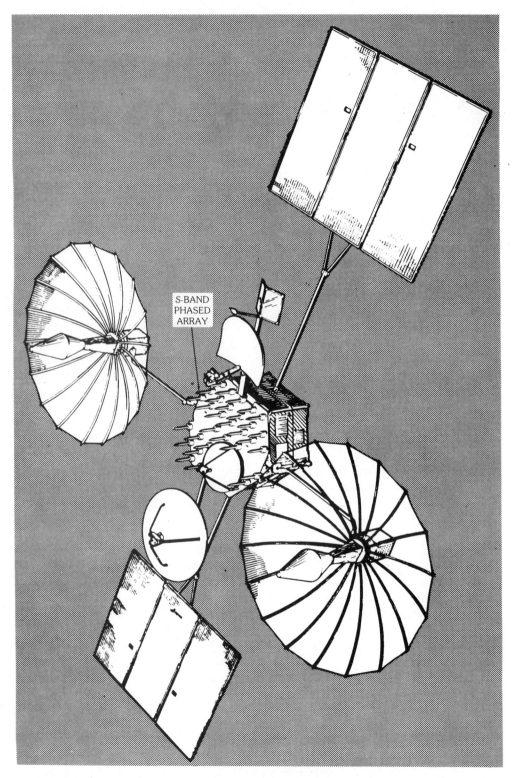

S-BAND
PHASED
ARRAY

Fig. 46. Tracking and data relay satellite system. (*Courtesy C. Donn*)

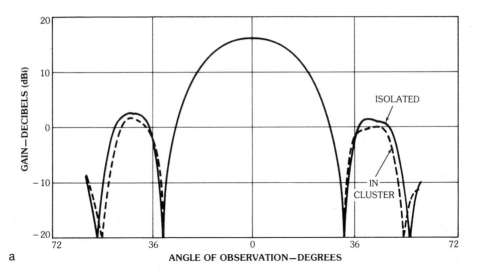

a

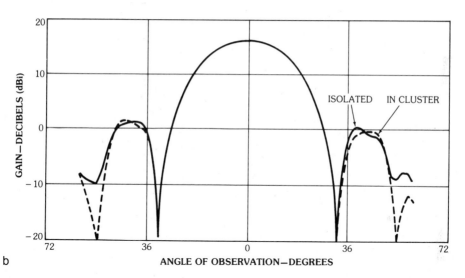

b

Fig. 47. Active element patterns of helix array measured isolated and in seven-element cluster. (*a*) Spacing equal to 13.5 in (34.29 cm). (*b*) Spacing equal to 11.5 in (29.21 cm). (*Courtesy C. Donn*)

mode propagation, higher-order mode suppression, and reflection loss of the lens surface.

Nonoverlapping and overlapping feed clusters are considered in the beam-forming network. The achievable performance of a scanned beam that is generated with either one, two, or three feeds is under study. In order to simplify the beam-forming network, the beam is assumed to scan in a discrete fashion and the feed cluster is excited with equal amplitude and phase. The diameter of the lens and the feed horn spacing are then varied to determine the minimum value of antenna gain anywhere in the field of view. The range of the lens diameter is chosen from 50.8

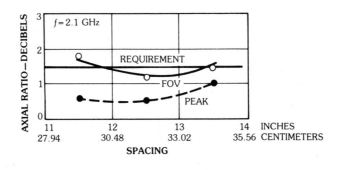

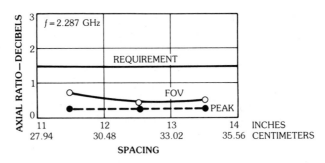

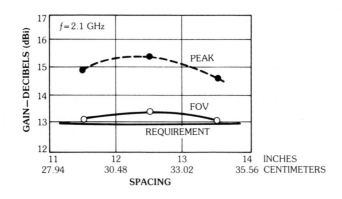

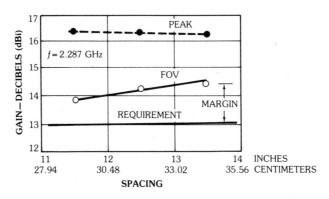

Fig. 48. Design paramenters of helix seven-element cluster. (*a*) Measured axial ratio. (*b*) Measured gain. (*Courtesy C. Donn*)

Fig. 49. Breadboard ten-element array. (*After Donn [33], © 1980 IEEE*)

to 86.4 cm. For a given diameter and 19 feed elements, there is unique feed horn spacing that yields a maximum antenna gain achievable over the field of view. The minimum gains achieved over the coverage area versus lens diameters are plotted in Fig. 52. The gain, which is increased with phase correction for the feed cluster, is also given in the figure.

Fig. 53 shows the measured and computed patterns of the on-focused beam while Fig. 54 is for an off-focused beam. The earth coverage radiation pattern obtained by exciting all 19 feed elements with equal amplitude is shown in Fig. 55.

A Reflector Antenna System—A reflector antenna system [36] is designed for Arabsat to provide transmission of *C*-band signals to operate between 3.2 and 4.2 GHz for fixed satellite service. Frequency reuse is achieved through dual circular polarizations. The antenna system consists of an offset parabolic reflector fed by an array of circular feed elements as depicted in Fig. 56. The reflector has a projected aperture of 1.52 × 1.57 m with focal length of 1.58 m. The feed elements are excited by an air-supported bar-line feed network using a direct plug-in interface as

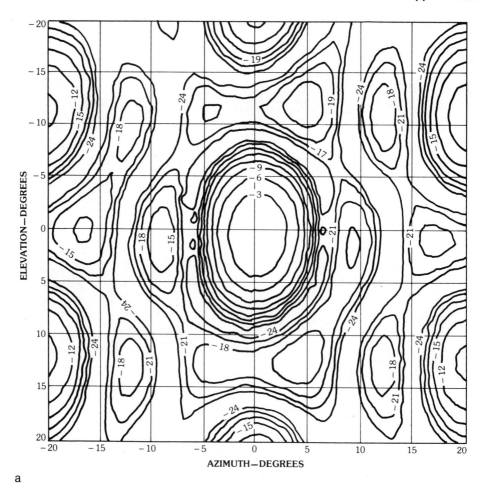

a

Fig. 50. Beam performances. (*a*) Boresight beam. (*b*) Scanned beam. (*After Donn [33],* ©
1980 IEEE)

shown in Fig. 57. The feed network consists of a two-to-four dual-mode converter
and a four-to-thirteen power divider for two circular polarizations. The circuit
diagram of the beam-forming network is presented in Fig. 58.

A circular open-ended waveguide is chosen because of good polarization
performance, wide bandwidth, simplicity of mechanical integration, simple and
mechanically strong structure, and extremely low loss.

Two different sizes of feed spacing are considered: 1.1λ and 1.6λ. The smaller
feed spacing, 9.53 cm, is selected because of better axial ratio performance and
better beam shaping. The septum polarizer is used because of its compactness
to integrate in the feed array and its good performance. The polarizer has a
square-cross-section waveguide at one end and a sloping septum that gradually
divides the square waveguide into a separate rectangular-cross-section waveguide
with a common broadwall at the other end. This septum polarizer also functions as

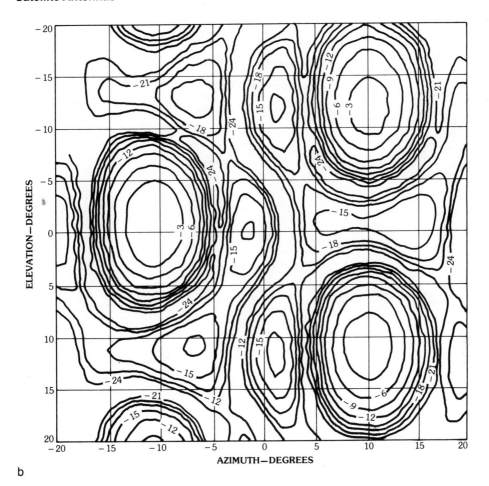

Fig. 50, *continued.*

an orthomode transducer in separating two circularly polarized signals. The achieved performance is summarized in Chart 1. Two types of transition from the septum polarizer to feed horn aperture are considered: stepped and tapered. A stepped transition is physically shorter than the tapered, but it generates both propagating and evanescent higher-order modes. A tapered transition is chosen because it is simpler and will not generate significant higher-order modes that may cause a degradation of the feed patterns.

Chart 1. Measured Performance of the Septum Polarizer

Frequency band	3.7–4.2 GHz
Maximum axial ratio	0.2 dB
Insertion loss	0.1 dB
Return loss	28 dB
Isolation	32 dB

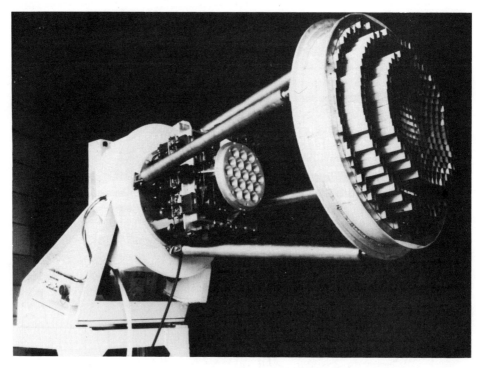

Fig. 51. LES 7 multibeam lens antenna (*After Dion and Ricardi [34], © 1971 IEEE*)

The axial ratio performance of an active feed element pattern of feed horn and polarizer is shown in Fig. 59 at ambient prethermal, hot, cold, and ambient postthermal. The performance presented is based on the figure of merit of the pattern in which the axial ratios are taken at different pattern cuts and then averaged.

Thirteen feed elements are required to cover the Arabic coverage area. They are grouped as shown in Fig. 60 with respect to the four output ports of the odd/even mode converter to provide better beam shaping and to minimize mode shift. A mode shift is due to the fact that the radiation pattern for the odd mode is slightly shifted and different from that for the even mode. The whole feed network consists of four separate circuit segments: RHCP power divider, LHCP power divider, RHCP dual-mode converter, and LHCP dual-mode converter (Fig. 57). The LHCP dual-mode converter, which consists of ring hybrids and differential line lengths, is shown in Fig. 61. The measured and computed patterns are given in Fig. 62.

3. Earth Coverage Antennas

Horns are widely used for earth coverage antennas. The design difficulty lies in the polarization purity requirement over the earth coverage, which is approximately a ±9° circle. This requirement precludes the conventional conical horn

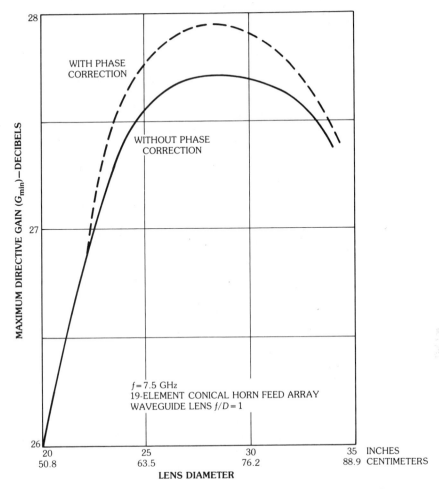

Fig. 52. Minimum directive gain over field of view versus lens diameter.

antenna even though it has the simplest structure. A conical horn has a poor axial ratio at the edge of the beam and relatively high side lobe levels.

In order to meet the polarization purity requirement an earth coverage antenna must have a good polarizer and a rotationally symmetrical beam over at least ±9° for the earth subtended angle over the whole frequency bandwidth. There are two approaches to achieve a rotationally symmetrical beam: corrugation of the surface and generation of a higher-order mode inside the horn. A corrugated horn antenna has wider operating frequency characteristics but it is difficult to fabricate and is relatively heavy. A dual-mode horn is used more frequently because of its simplicity.

A dual-mode horn with an abrupt waveguide discontinuity at the throat [37] is strongly frequency dependent and therefore is a narrow-band radiator. There are two ways to improve the bandwidth of a dual-mode horn: dielectric loading [38] and introduction of dual steps. In the former approach the horn is loaded with a

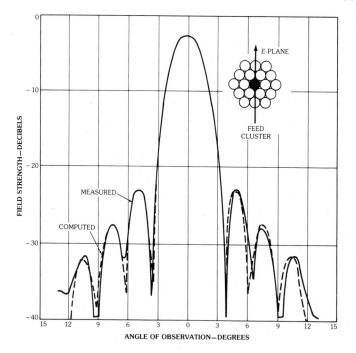

a

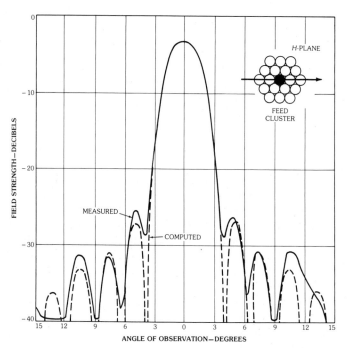

b

Fig. 53. Patterns of center beam. (*a*) *E*-plane. (*b*) *H*-plane. (*After Dion and Ricardi [34],* © *1971 IEEE*)

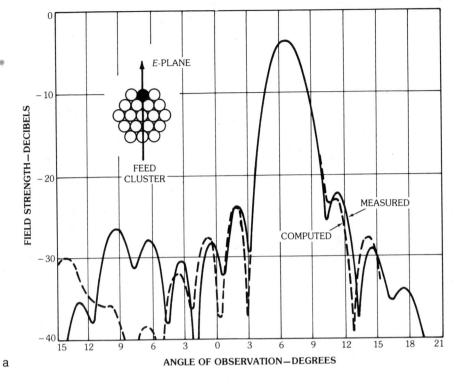

a

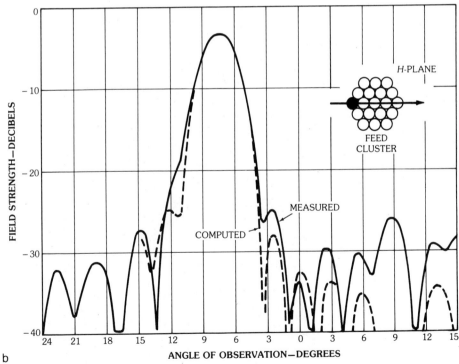

b

Fig. 54. Patterns of off-focused beam. (*a*) *E*-plane. (*b*) *H*-plane. (*After Dion and Ricardi [34], © 1971 IEEE*)

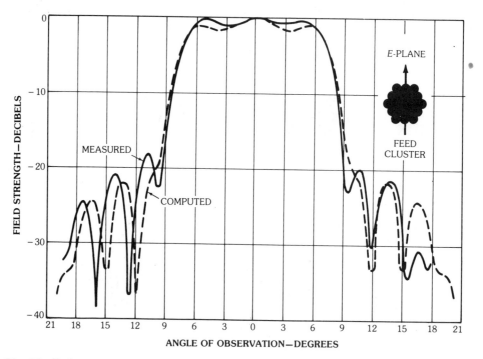

Fig. 55. *E*-plane pattern of earth coverage beam (*After Dion and Ricardi [34], © 1971 IEEE*)

Fig. 56. Arabsat *C*-band transmit antenna—breadboard model. (*Courtesy Ford Aerospace and Communications Corp.*)

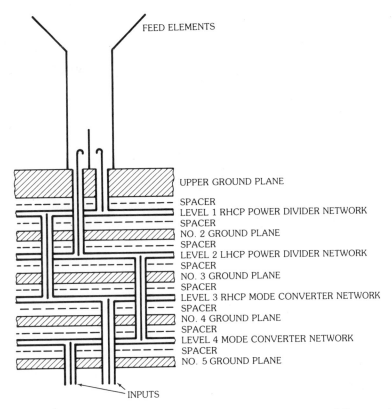

Fig. 57. A 4-GHz transmit feed assembly. (*Courtesy Ford Aerospace and Communications Corp.*)

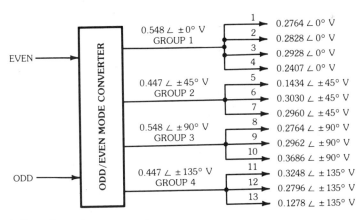

Note: Plus angles are even mode
Minus angles are odd mode

Fig. 58. Diagram of 4-GHz transmit beam-forming network. (*Courtesy Ford Aerospace and Communications Corp.*)

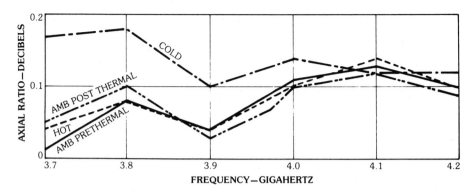

Fig. 59. Axial ratio performance of five-screw graphite feed element. (*Courtesy Ford Aerospace and Communications Corp.*)

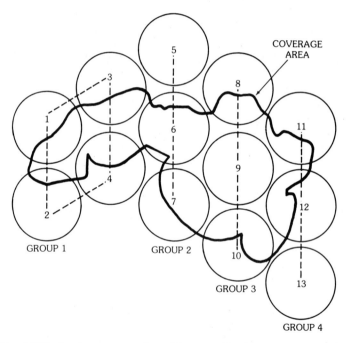

Fig. 60. The 4-GHz feed array grouping. (*Courtesy Ford Aerospace and Communications Corp.*)

dielectric sleeve inside the horn. In the latter the higher-order mode is excited with multi-irises to reduce the frequency dependence of the mode excited.

A shaped-beam earth coverage antenna is proposed, based on the fact that the paths tangential to the earth are longest. As a result the gain should be highest in this region to compensate for the atmospheric attenuation and decrease to a minimum for the path normal to the earth. A nine-horn array [34] has been proposed to achieve the required beam shape. Alternate approaches are the shaped lens and a multibeam reflector antenna system. Table 10 summarizes the advantages and disadvantages of each design.

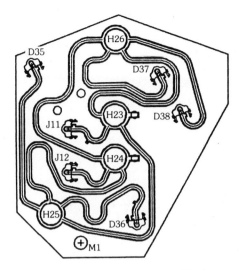

Fig. 61. Level LHCP 2–4 mode converter network. (*Courtesy Ford Aerospace and Communications Corp.*)

Table 10. Alternate Types of Earth Coverage Antennas

Type	Advantages	Disadvantages
Conical horn	Simple Easy to fabricate	Poor axial ratio at the beam edge
Potter horn	Simple Easy to fabricate	Narrow band
Dielectric-loaded horn	Wide band Good axial ratio	
Multistepped dual-mode horn	Wide band Good axial ratio over 11° circle	
Corrugated horn	Wide band Excellent axial ratio	Highest cost Difficult to fabricate Relatively heavy
Shaped-beam nine-horn	Higher gain	Complicated design Heavy

The Stepped Horn

It is well known that for a conical horn, its beamwidths in the E- and H-planes are different when operating in the dominant TE_{11} mode. The beamwidths can be equalized by introducing the higher-order TM_{11} mode. These two modes, when excited in the horn aperture with the appropriate relative amplitude and phase, can affect the beamwidth equalization. A simple step discontinuity used to excite the horn with TE_{11} and TM_{11} modes is proposed [37]. The power conversion coefficients and the launching phases of the TE_{11} and TM_{11} modes are given [39] in Fig. 63. Fig. 64 sketches a K-band earth coverage horn used as a beacon designed

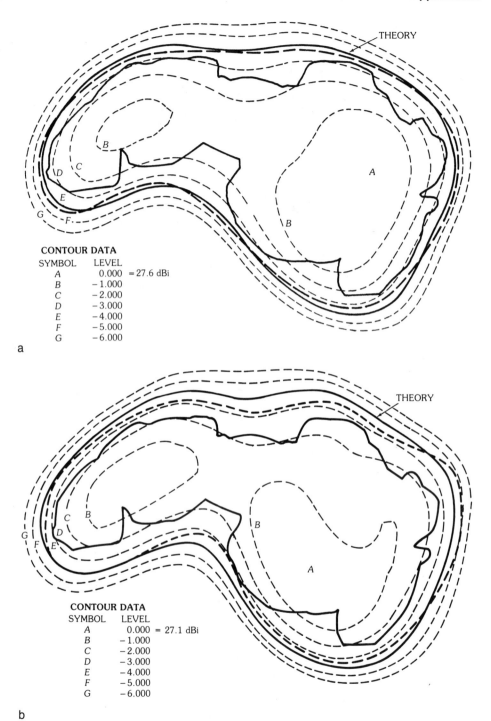

CONTOUR DATA
SYMBOL   LEVEL
   A   0.000 = 27.6 dBi
   B   − 1.000
   C   − 2.000
   D   − 3.000
   E   − 4.000
   F   − 5.000
   G   − 6.000

a

CONTOUR DATA
SYMBOL   LEVEL
   A   0.000 = 27.1 dBi
   B   − 1.000
   C   − 2.000
   D   − 3.000
   E   − 4.000
   F   − 5.000
   G   − 6.000

b

Fig. 62. Theoretical and measured patterns. (*a*) For 3.7-GHz even mode. (*b*) For 3.7-GHz odd mode. (*Courtesy Ford Aerospace and Communications Corp.*)

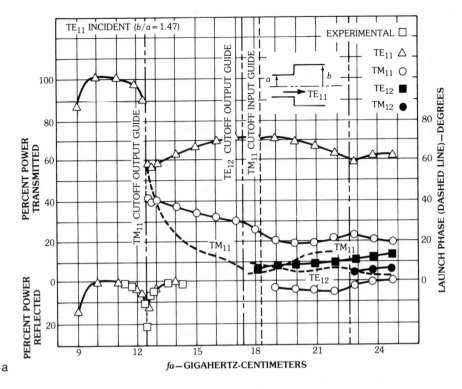

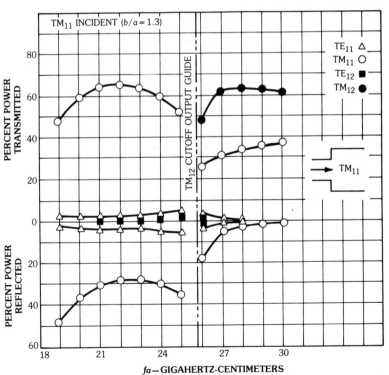

Fig. 63. Mode transducing properties of a symmetric step-discontinuity waveguide junction. (*a*) Illuminated with an incident fundamental TE$_{11}$ mode. (*b*) Illuminated with an incident TM$_{11}$ mode. (*After English [39],* © *1973 IEEE*)

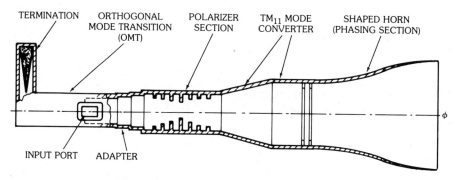

Fig. 64. An 11-GHz earth coverage antenna.

by this principle. An oversized waveguide section is used as a phasing section to ensure that both modes are radiated in proper relationship. Table 11 summarizes the achieved performance.

The Dielectric-Loaded Horn

A dielectric-loaded horn is sketched in Fig. 65a. In this approach the introduction of the dielectric sleeve excites an antiphase component of the electric field in the dielectric band and results in a field distribution similar to that of the TM_{11} mode across the cross section of the horn. Fig. 65b shows the measured pattern of a *C*-band earth coverage antenna [38]. The achieved performance is summarized in Chart 2.

Table 11. Summary of Electrical Performance of 11-GHz Beacon Antenna

Frequency	11 190 to 11 460 MHz
Coverage area	22° circular
Gain over coverage area (antenna input flange)	15.9 dBi
Waveguide loss (antenna to transponder) interface	1.1 dB
Gain over coverage (at antenna transponder) interface	14.8 dBi
Polarization	RHCP
Axial ratio (over coverage area)	0.98 dB
Voltage standing-wave ratio	1.15

Chart 2. Summary of Achieved Performance of a Dielectric-Loaded Earth Coverage Horn

Frequency	4.037–4.198 GHz
Coverage area	±9°
Antenna gain	17 dBi
Axial ratio	0.4 dB

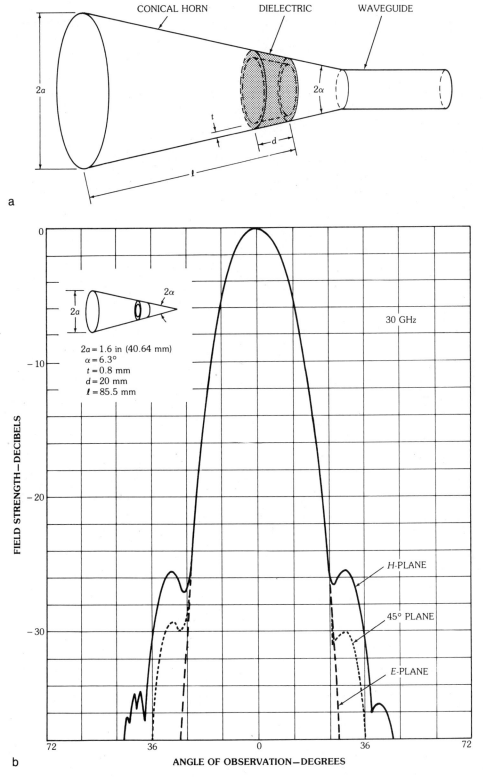

Fig. 65. Dielectric-loaded barn antenna. (*a*) Structure. (*b*) Radiation patterns. (*After Satoh [38], © 1972 IEEE*)

The Multistepped Dual-Mode Horn

In this approach the amount of energy converted into the TM_{11} mode is controlled by the flare and irises and by adjusting the phase difference between TE_{11} and TM_{11} modes in the oversized waveguide. Normally the amount of TM_{11} mode conversion increases with increasing frequency, but the phase difference between the two modes in the oversized waveguide acts to decrease the total amount of the TM_{11} mode. Therefore the total amount of energy converted into the TM_{11} mode becomes nearly constant over a broad frequency range. The difference between the wavelength of the TE_{11} and TM_{11} modes can be made smaller by using larger oversized waveguides, thus decreasing the phase difference dependence between the two modes as a function of frequency. Fig. 66 is a sketch of a *C*-band earth coverage antenna. The achieved performance is summarized in Table 12.

The Shaped Beam

An ideally shaped beam pattern for the earth coverage antenna is shown in Fig. 67. The antenna consists of a nine-horn array [40] with a large central horn surrounded by a ring of eight smaller horns. The central horn is multimoded to provide a rotationally symmetrical pattern. The power and phase distributions among the elements are to simulate aperture distributions that yield a rotationally symmetrical sector-shaped pattern. Fig. 68 shows a schematic diagram of the nine-horn array and the achieved performance of the measured patterns.

4. Tracking, Telemetry, and Command Antenna

The tracking, telemetry, and command (TT&C) antenna provides ranging, telemetry, and command operation throughout all mission phases after launch vehicle separation. The operational sequence from launch to synchronous orbit for a spin-stabilized and body-stabilized satellite is shown in Figs. 69a and 69b, respectively. A geosynchronous satellite, once in position, still needs occasional adjusting to keep it in position. In accomplishing all the missions the TT&C antenna is designed to

Table 12. Performance for a Typical *C*-Band Earth Coverage Antenna

Parameter	Performance
Gain over FOV at antenna ports (dBi)	15.6
Waveguide and isolator loss (dB)	0.6
Gain at antenna/transponder	15.0
FOV (coverage)	22°
Axial ratio (dB)	0.32
Voltage standing-wave ratio	0.4
Gain variation (dB)	3.0
Isolation, port-to-port (dB)	>40

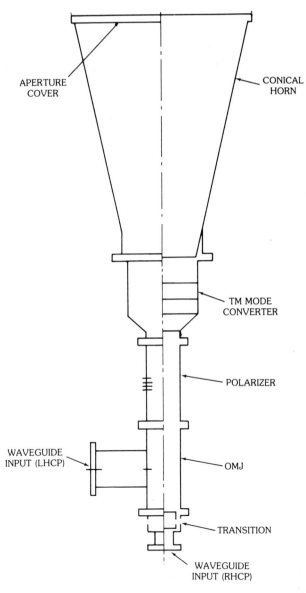

Fig. 66. A 4-GHz-band earth coverage antenna. (*Courtesy Mitsubishi Electric Corp.*)

(a) receive satellite functional commands from ground stations,
(b) transmit satellite functional data (telemetry) to ground stations,
(c) provide a beacon signal to aid ground station acquisition of the satellite, and
(d) receive and retransmit ranging signals.

The typical antenna pattern requirement for the TT&C antenna is illustrated in Fig. 70. A nondirectional and circularly polarized antenna is required to ensure

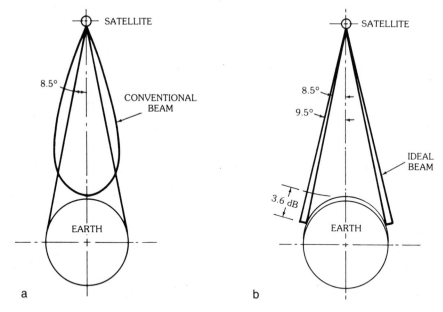

Fig. 67. Earth coverage antennas. (*a*) Conventional beam. (*b*) Shaped beam. (*After Ajioka and Harry [40], © 1970 IEEE*)

the continuous reception of command signals in every mission operation. For a spin-stabilized satellite a toroidal beam (omnidirectional in the plane perpendicular to the satellite spin axis) could provide a continuous telemetry link in almost all directions. For a body-stabilized satellite, because of the requirement of a more complex maneuvering operation from the transfer orbit to the synchronous orbit, a cardioid beam is generally used to maintain continuous coverage.

Types of Antennas

Although a number of antennas have been designed to generate a nondirectional beam, there are few suitable for satellite application. The design of the antenna is dominated by weight, complexity, and the available location in the satellite.

An ideal location for the telemetry and command antenna for a spin-stabilized satellite is at the satellite cylindrical body. A circular array is a logical choice as a result of the location restrictions and omnidirectional requirements. Because of the broad pattern of the telemetry and command antenna, interference from the other antenna systems, such as large reflectors for communication antenna systems and the satellite structures, might cause intolerable degradation on the antenna performance. A biconical antenna, which is placed in an optimized location to minimize scattering or reduce blockage of the sight of view, is also used. A cardioid beam is generated by a slotted ring antenna. The pattern shaping is achieved by using a multiring on the cylindrical waveguide or attaching a conical reflector to the waveguide structure. These three types of antennas are described below.

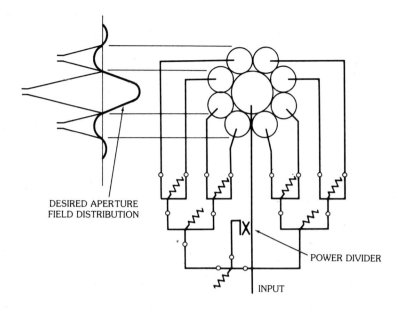

DESIRED APERTURE
FIELD DISTRIBUTION

POWER DIVIDER

INPUT

a

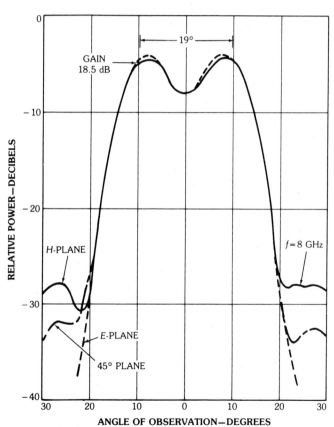

b

Fig. 68. Nine-horn array diagram and pattern. (*a*) Schematic diagram. (*b*) Measured patterns with center horn multimoded. (*After Ajioka and Harry [40], © 1970 IEEE*)

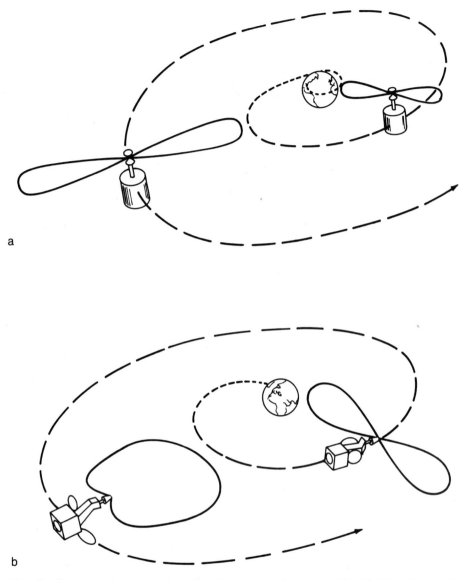

Fig. 69. Operational sequence from launch to geosynchronous orbit. (*a*) Of a spin-stabilized satellite. (*b*) Of a body-stabilized satellite.

The Circular Array—A conformal array has been designed [41, 42] to provide an omnidirectional beam in satellite application. The term "omnidirectional" refers to the azimuthal pattern in the plane of the array. The design parameters are the number of elements, radiating elements, and feed components. The number of elements S is determined to provide a nearly omnidirectional pattern. The minimum number of the elements is decided by the allowed amplitude ripple. The evaluation of amplitude ripple can be given in terms of the fluctuation, which is

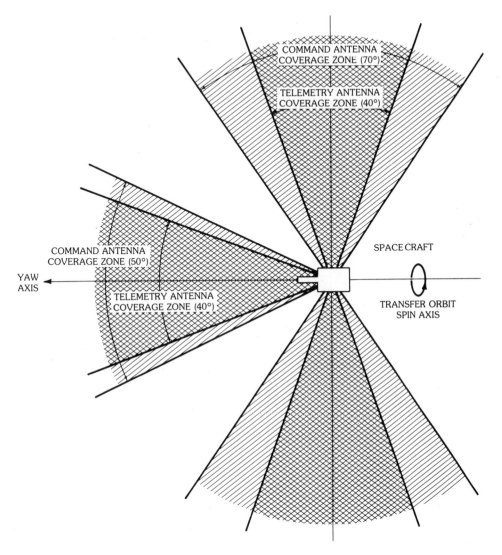

Fig. 70. Telemetry and command coverage requirements.

defined as the ratio of maximum $|\phi|$ to minimum $|\phi|$, where $|\phi|$ is the total far-field pattern of an S-element circular array and is given [41] as

$$\Phi \cong S \sum_{n=0}^{N} A_n(-j)^n \frac{d^n}{dz^n}[J_0(z) + 2(j)^S J_S(z) \cos \phi] \qquad (25)$$

where $J_S(z)$ is the Bessel function of the first kind of order S, $z = k_0 a \sin \theta$ in which k_0 is the free-space wave number, and a is the radius of the circular array and is measured from the polar axis. In this derivation it is assumed that the single-element pattern, $F(\phi')$, can be represented by the Fourier cosine series

$$F(\phi') = \sum_{n=0}^{\infty} A_n \cos^n \phi' \tag{26}$$

A practical single-element pattern can be approximated by

$$(1 + \cos\phi')/2 \qquad \text{or} \qquad (2 + 3\cos\phi' + \cos 2\phi')/6$$

as shown in Fig. 71. The pattern fluctuation as a function of size of cylinder and number of elements for the above two single-element patterns is given in Figs. 72 and 73, respectively. It can be seen that for a small fluctuation of less than 1 dB, the pattern fluctuation is less sensitive to the element pattern.

The exact solution for a single-element pattern [44] is given in the following. For an axial slot as depicted in Fig. 74a, the field in the aperture can be assumed:

$$E_\phi = \frac{V}{aa}\cos\left(\frac{\pi z}{L}\right), \qquad \begin{cases} -L/2 < z < L/2 \\ -\alpha/2 < \phi < \alpha/2 \end{cases} \tag{27}$$

The radiation pattern is

$$E_\phi \cong \frac{VL}{\pi^3 a}\frac{e^{-jk_0 r}}{r}\left\{\frac{\cos(k_0 L/2)\cos\theta}{1 - [(kL/\pi)\cos\theta]^2}\right\}\sum_{n=-\infty}^{\infty}\frac{j^n e^{jn\phi}}{H_n^{(2)\prime}(k_0 a\sin\theta)} \tag{28}$$

$$E_\theta = 0$$

where $H_n^{(2)}$ is the Hankel function of second kind and

$$\sum_{n=-\infty}^{\infty}\frac{j^n e^{jn\phi}}{H_n^{(2)\prime}(k_0 a\sin\theta)}$$

(called the cylinder space factor) can be regarded as a correction factor to the pattern of the isolated slot antenna. For a circumferential slot as shown in Fig. 74b the field in the aperture can be assumed:

$$E_z = \frac{V}{W}\cos\left(\frac{\pi\phi}{\alpha}\right), \qquad \begin{cases} -W/2 < z < W/2 \\ -\alpha/2 < \phi < \alpha/2 \end{cases} \tag{29}$$

The radiation pattern is

$$E_\phi = -\frac{Va\cos\theta}{\pi k_0 a\sin\theta}\frac{e^{-jk_0 r}}{r}\sum_{n=-\infty}^{\infty}\frac{nj^n\cos\theta\,(n\alpha/2)e^{jn\phi}}{[\pi^2 - (n\alpha)^2]H_n^{(2)\prime}(k_0 a\sin\theta)} \tag{30}$$

$$E_\theta = \frac{k_0 Va}{j\pi\sin\theta}\frac{e^{-jk_0 r}}{r}\sum_{n=-\infty}^{\infty}\frac{j^n\cos(n\alpha/2)e^{jn\phi}}{[\pi^2 - (n\alpha)^2]H_n^{(2)}(k_0 a\sin\theta)}$$

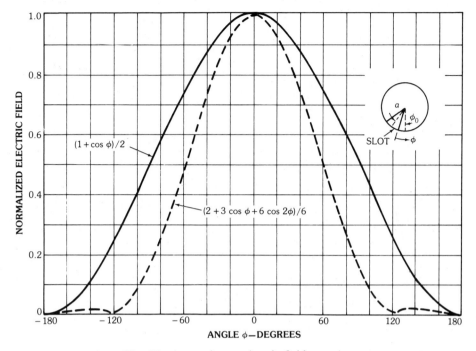

Fig. 71. Approximate electric field patterns.

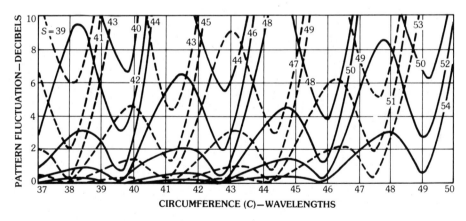

Fig. 72. Pattern fluctuation dependence on number of elements and cylinder size for feed pattern $1 + 1/2\cos\phi$.

In the azimuthal plane ($\theta = 90°$), the radiation pattern becomes

$$E_\phi = 0$$

$$E_\theta = \frac{k_0 V a}{j\pi} \frac{e^{-jk_0 r}}{r} \sum_{n=-\infty}^{\infty} \frac{j^n \cos(n\alpha/2) e^{jn\phi}}{[\pi^2 - n\alpha]^2 H_n^{(2)}(k_0 a)} \tag{31}$$

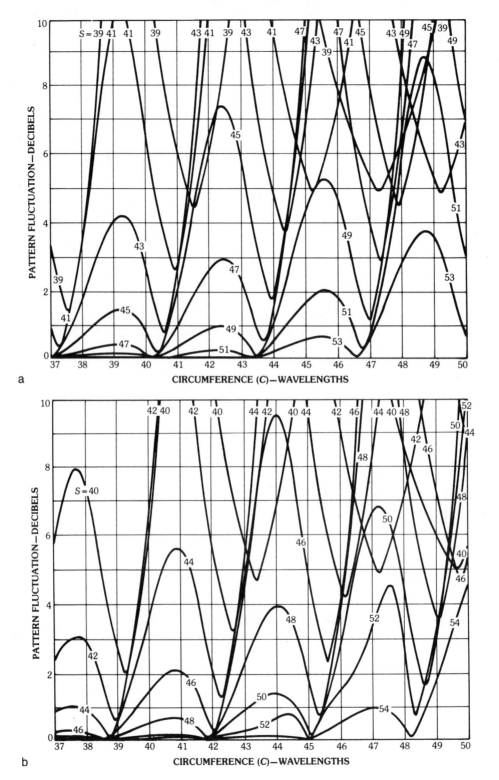

Fig. 73. Pattern fluctuation dependence on number of elements and cylinder size for feed pattern $(2 + 3\cos\phi + \cos 2\phi)/6$. (*a*) For odd numbers of elements. (*b*) For even numbers of elements.

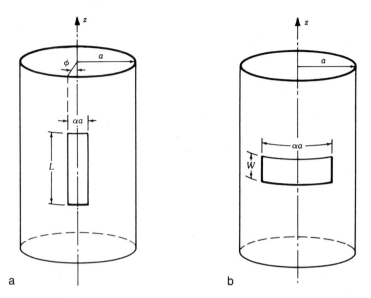

Fig. 74. Conducting cylinder. (*a*) With axial slot. (*b*) With circumferential slot.

The above series is slow convergent when k_0a (circumference of the cylinder in wavelengths) becomes large. The geometrical theory of diffraction [45] is then applied. According to the geometrical theory of diffraction, the far-field region is divided into the illuminated, transition, and shadow regions as shown in Fig. 75. For an axial slot the radiation pattern in the azimuthal plane can be given as

$$dE_\phi = -\frac{jk_0}{4\pi} dP_m F \frac{e^{-jk_0r}}{r} \qquad (32)$$

where dP_m is the infinitesimal magnetic current moment and $dP_m = E_{\phi'}^a \, da'$, where $E_{\phi'}^a$ is the electric field in the aperture, da' is an area element, and

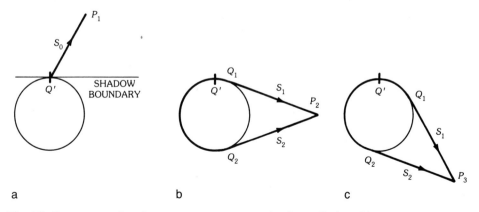

Fig. 75. Rays emanating from a source on a conducting cylinder. (*a*) In the illuminated region. (*b*) In the transition region. (*c*) In the shadow region.

$$F = \begin{cases} 2 & \text{illuminated region} \\ g(\xi)e^{-jk_0t} & \text{transition region} \\ \displaystyle\sum_{p=1}^{\infty} L_p^h(Q')A_p^h(Q)e^{-(\alpha_p^h+jk_0)t} & \text{shadow region} \end{cases} \tag{33}$$

Similarly, for a circumferential slot the radiation pattern is

$$dE_z = -\frac{jk_0}{4\pi}dP_m G \frac{e^{-jk_0r}}{r} \tag{34}$$

where $dP_m = E_{z'}^a\, da'$, with $E_{z'}^a$ the electric field in the aperture and

$$G = \begin{cases} 2\cos\theta & \text{illuminated region} \\ -j\left(\dfrac{2}{k_0a}\right)^{1/3}\hat{g}(\xi)e^{-jk_0t} & \text{transition region} \\ \displaystyle\sum_{p=1}^{\infty} L_p^s(Q')A_p^s(Q)e^{-(\alpha_p^s+jk_0)t} & \text{shadow region} \end{cases} \tag{35}$$

The $g(\xi)$ and $\hat{g}(\xi)$ are the Fock functions that can be expressed as

$$\begin{aligned} g(\xi) &= \frac{1}{\sqrt{\pi}}\int_{e^{-j2\pi/3}}^{\infty} \frac{e^{-j\tau\xi}}{W_2'(\tau)}d\tau \\ \hat{g}(\xi) &= \frac{1}{\sqrt{\pi}}\int_{e^{-j2\pi/3}}^{\infty} \frac{e^{-j\tau\xi}}{W_2(\tau)}d\tau \end{aligned} \tag{36}$$

in which $W_2(\tau)$ and $W_2'(\tau)$ are the Fock-type Airy function and its derivative, respectively. The term t is the distance parameter as shown in Fig. 76, and

$$\xi = \frac{1}{a}\left(\frac{k_0a}{2}\right)^{1/3}t$$

The terms $\alpha_p^{s,h}$, $L_p^{s,h}$, and $A_p^{s,h}$ are the attenuation constant, launching coefficient, and attachment coefficient of the p mode. Chart 3 gives the attenuation constant, launching coefficient, and attachment coefficient of the p mode in a circular cylinder.

The total field for an axial slot or circumferential slot can be obtained by

$$E_\phi = \int_{\substack{\text{over} \\ \text{slot}}} dE_\phi \quad \text{for axial slot} \tag{37}$$

or

$$E_z = \int_{\substack{\text{over} \\ \text{slot}}} dE_z \quad \text{for circumferential slot} \tag{38}$$

Chart 3. Launching Coefficient, Attachment Coefficient, and Attenuation Constant of a Conducting Cylinder

Axial slot

$$L_p^h = -j\left(jk_0\frac{\pi}{2}\right)^{1/2} H_{\nu_p}^{(2)}(k_0a)\, D_p^h$$

$$A_p^h(Q) = L_p^h(Q)$$

$$(D_p^h)^2 = \frac{\pi^{-1/2} 2^{-5/6} a^{1/3} e^{-j\pi/12}}{k_0^{1/6} \bar{q}_p [A_i(-\bar{q}_p)]^2}\left[1 + \left(\frac{2}{k_0a}\right)^{2/3}\left(\bar{q}_p/30 - \frac{1}{10\bar{q}_p^2}\right) e^{-j\pi/3}\right]$$

$$\alpha_p^h = \frac{\bar{q}_p}{a}\, e^{j\pi/6}\left(\frac{k_0a}{2}\right)^{1/3}\left[1 + \left(\frac{2}{k_0a}\right)^{2/3}\left(\frac{\bar{q}_p}{60} + \frac{1}{10\bar{q}_p^2}\right) e^{-j\pi/3}\right]$$

$A_i'(-\bar{q}_p) = 0$

$\bar{q}_1 = 1.01879$

$\bar{q}_2 = 3.2482$

$A_i'(-\bar{q}_1) = 0.53566$

$A_i(-\bar{q}_2) = -0.41902$

Circumferential slot

$$L_p^s = -\left(jk_0\frac{\pi}{2}\right)^{1/2} H_{\nu_p}^{(2)'}(k_0a)\, D_p^s$$

$$A_p^s(Q) = -L_p^s(Q)$$

$$(D_p^s)^2 = \frac{\pi^{1/2} 2^{-5/6} a^{1/3} e^{-j\pi/12}}{k^{1/6}[A_i'(-\bar{q}_p)]^2}\left[1 + \left(\frac{2}{k_0a}\right)^{2/3}\left(\frac{q_p}{30}\right) e^{-j\pi/3}\right]$$

$$\alpha_p^s = \frac{q_p}{a}\, e^{j\pi/6}\left(\frac{k_0a}{2}\right)^{1/3}\left[1 + \left(\frac{2}{k_0a}\right)^{2/3}\frac{q_p}{60}\, e^{-j\pi/3}\right]$$

$A_i(-q_p) = 0$

$q_1 = 2.33811$

$q_2 = 4.08795$

$A_i'(-q_1) = 0.70121$

$A_i' - (q_2) = -0.80311$

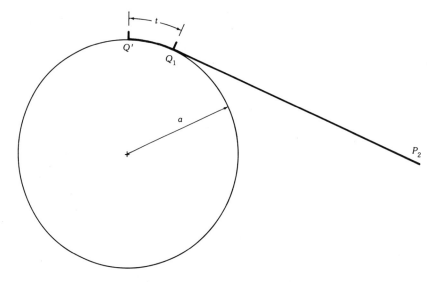

Fig. 76. Diffraction from a source Q'.

The pattern of the cylindrical space factor for a narrow axial slot for different k_0a is plotted [46] in Fig. 77a, while the pattern for a narrow circumferential slot for different k_0a is plotted in Fig. 77b.

The total radiation pattern of the array on a cylinder can be obtained by using the known field from the single slot and using superposition directly. The mutual coupling among slots is neglected in the above computation. The mutual coupling among slots on a cylinder has been studied and presented by several authors [47,48]. The contribution of the mutual coupling to the radiation pattern in this case is not significant.

A typical example of this type of antenna is a 64-element array [49] mounted on a 1.91 m diameter and 1.63 m-long cylindrical satellite design for over the 40 percent bandwidth at S-band. The radiating element is a lightweight cavity-backed turnstile as shown in Fig. 78a. Fig. 78b shows a circumferential section of the conformal array. The measured versus calculated amplitude ripple using (25) in the azimuthal plane is shown in Fig. 78c. The measured elevation radiation pattern (0.7λ interelement spacing) is given in Fig. 78d.

The Biconical Horn—A biconical horn, because of its rotationally symmetrical structure as shown in Fig. 79, could provide an omnidirectional radiation pattern in the azimuthal plane. The two lowest-order modes, the transverse electromagnetic TEM and the transverse electric TE_{01}, are excited to generate vertically and horizontally polarized waves, respectively. The field of the TEM wave is described by [50]

$$E_\theta = -\sqrt{\frac{2k}{\pi}}\frac{1}{r\sin\theta}e^{-jk_0r}$$

$$H_\phi = \sqrt{\frac{\epsilon_0}{\mu}}E_\theta$$

$$(39)$$

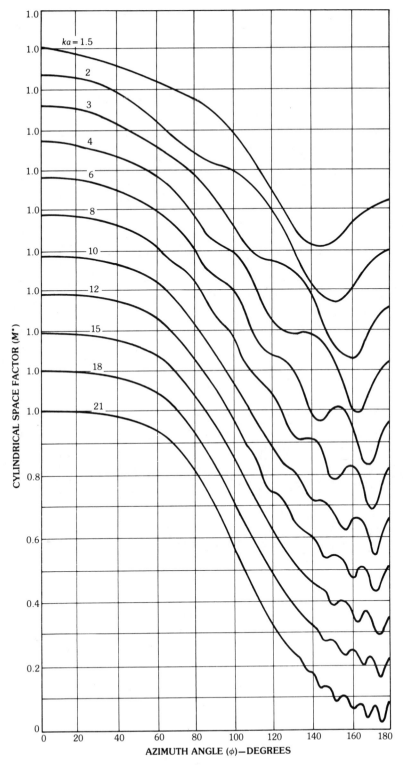

Fig. 77. Cylinder space factors, with vertical scale shifted for each curve by constant amount. (*a*) For narrow axial slot. (*After Wait [46], © 1959 Pergamon Books Ltd.*) (*b*) For narrow circumferential slot.

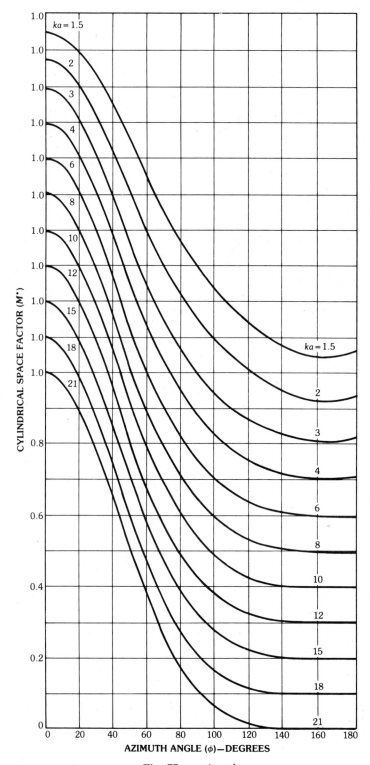

Fig. 77, *continued.*

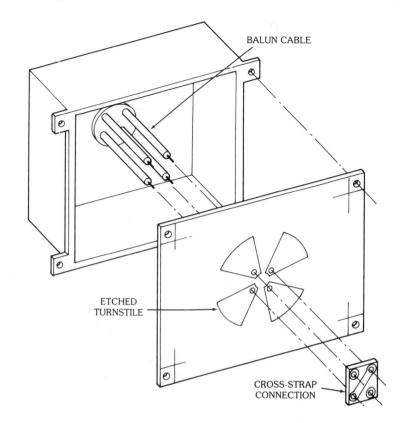

BALUN CABLE

ETCHED
TURNSTILE

CROSS-STRAP
CONNECTION

a

b

Fig. 78. Structure and characteristics of 64-element array. (*a*) Exploded view of array element. (*b*) Circumferential section of conformal array. (*c*) Pattern fluctuation in the ϕ plane. (*d*) Measured elevation (θ-plane) radiation pattern of circular array. (*After Gregorwich [49], © 1979 IEEE*)

while the field of the TE_{01} wave is given by

$$
\begin{aligned}
E_{\phi} &= -j\omega\mu L'_{\ell}(\cos\theta)\, r^{-1/2} H_p^{(2)}(k_0 r) \\
H_r &= L_{\ell}(\cos\theta)(P^2 - 1/4)\, r^{-3/2} H_p^{(2)}(k_0 r) \\
H_{\theta} &= L'_{\ell}(\cos\theta)[(P^2 - 1/2)\, r^{-3/2} H_p^{(2)}(k_0 r) - k_0 r^{-1/2} H_{p-1}^{(2)}(k_0 r)]
\end{aligned}
\tag{40}
$$

where

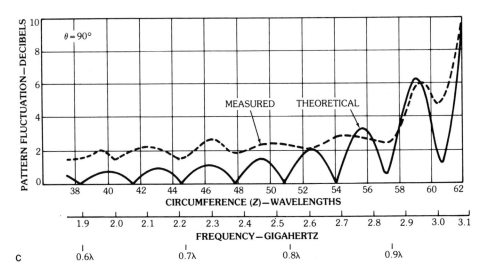

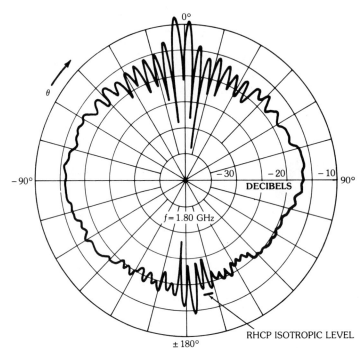

Fig. 78, *continued.*

$$L_\ell(\cos\theta) = AP_\ell(\cos\theta) + BQ_\ell(\cos\theta)$$

P_ℓ = associated Legendre's function of the first kind

Q_ℓ = associated Legendre's function of the second kind

$P = [\ell(\ell + 1) + 1/4]^{1/2}$

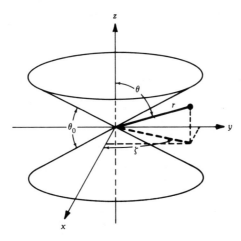

Fig. 79. Biconical horn antenna.

$H_p^{(2)}$ = Hankel function of the second kind

$k = \omega\sqrt{\mu_0\epsilon_0}$

$\mu_0 = 4\pi \times 10^{-7}$ H/m

$\epsilon_0 = 1/(36\pi \times 10^9)$ F/m

The ℓ is determined from the boundary condition that the field E_ϕ must vanish at $0 = (\pi \pm \theta)/2$, where θ_0 denotes the flat angle between the cones:

$$\frac{\partial}{\partial\theta} L_\ell(\cos\theta) = 0 \quad \text{at } \theta = (\pi \pm \theta_0)/2$$

Let the propagation constant be $\gamma = \alpha + j\beta$. Then

$$\alpha_{\text{TEM}} = \frac{1}{r}$$

$$\beta_{\text{TEM}} = \omega\sqrt{\epsilon_0\mu_0} \tag{41a}$$

and

$$\alpha_{\text{TE}_{01}} = \frac{2P + 1}{2r} - \text{Re}\,\{k_0 H_{p-1}^{(2)}(k_0 r)/H_p^{(2)}(k_0 r)\}$$

$$\beta_{\text{TE}_{01}} = -\text{Im}\,\{k_0 H_{p-1}^{(2)}(k_0 r)/H_p^{(2)}(k_0 r)\} \tag{41b}$$

Fig. 80 shows $\alpha_{\text{TE}_{01}}$ and $\beta_{\text{TE}_{01}}$ as functions of r, where the angles are the corresponding flare angles of the horn.

A circular-polarized wave is obtained by using the property of the different phase constants of the two modes so that when the TEM and TE_{01} modes are excited at the throat of the horn, the phase difference between modes will be $\pm 90°$

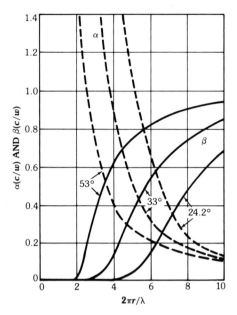

Fig. 80. Variations of attenuation constant α and phase constant β of TE_{01} wave with radial distance. (*After Barrow, Chu, and Jansen [50], © 1939 IRE*)

at the horn aperture. The two modes are generated by the skewed slots on a circular cylinder. The field excited by the slot is then decomposed into vertically polarized and horizontally polarized components. In all practical applications the radius R of the biconical horn is greater than 2 wavelengths. The gain of the horn can be calculated [51] by the following approximations:

$$G_{dB} = 10 \log(2a/\lambda) - L_e \quad \text{for TEM mode} \tag{42a}$$

and

$$G_{dB} = 10 \log(2a/\lambda) - (L_h + 0.91) \quad \text{for } TE_{01} \text{ mode} \tag{42b}$$

where L_e and L_h are the gain correction factors as shown in Fig. 81. The gain plotted against the flare angle for the TEM and TE_{01} waves is shown in Fig. 82. Values of R/λ and the flare angle exist that yield the maximum gain for a biconical horn. The radiation pattern of the biconical horn is isotropic in the azimuthal plane. In the elevation plane ($\phi = $ constant), the radiation pattern can be given by the universal curves of Fig. 83a for the TEM wave and Fig. 83b for the TE_{01} wave. In the vicinity of the axial direction ($\theta = 0°$ or $180°$), the edge of the horn gives rise to a focusing effect due to the formation of a caustic of the edge-diffracted rays. The equivalent edge currents [52] should be used to evaluate the radiation field. This focused field radiates into the satellite body and radiates into the main beam region. It is very important to reduce the edge currents so that the interference in

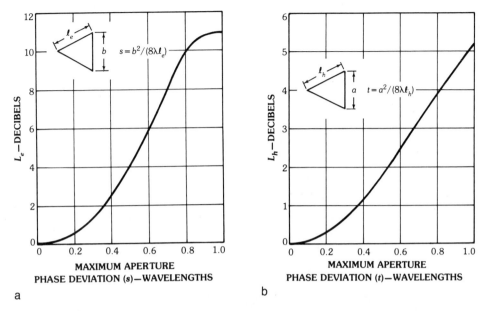

a   b

Fig. 81. Gain correction factors. (*a*) For *E*-plane flare. (*b*) For *H*-plane flare. (*After Jasik* [51], © 1961 McGraw-Hill Book Co.)

the main beam region will be minimal. The edge currents could be reduced by curving or serrating the edge. The bottom and the top of the edge can also be designed to radiate out of phase so that their contributions cancel in the axial direction.

The Slotted-Cylinder Reflector Antenna—A slotted-cylinder reflector antenna is used to provide a cardioid shaped pattern. The bottom of the cardioid shaped pattern is facing the satellite body so that the scattering of the satellite can be minimized. The antenna consists of a ring of slots placed near the end of a circular waveguide and a conical reflector as depicted in Fig. 84. The slots are excited by a rotating TE_{11} mode so that the field in the axial direction of the cylinder is circularly polarized.

The pattern shaping in the elevation plane can be achieved by using more than one ring of slot array on the circular waveguide. It is simpler, however, to use a conical reflector attached to the circular waveguide to shape the pattern [53]. The design problem is then divided into two parts: the external, to determine the excitation of slots and the inclined angle of the reflector to yield the desired pattern; and the internal, to decide how to launch the TE_{11} mode in the waveguide and to match the excited mode to the slots to reduce vswr. The radiation pattern of the finite waveguide structure and the conical reflector can be obtained via the method of moments [54]. In this particular case the solution can be simplified by invoking the rotational symmetry of the antenna. Only the fundamental mode is considered in the slot aperture. The rotating TE_{11} mode is launched by two orthogonal probes

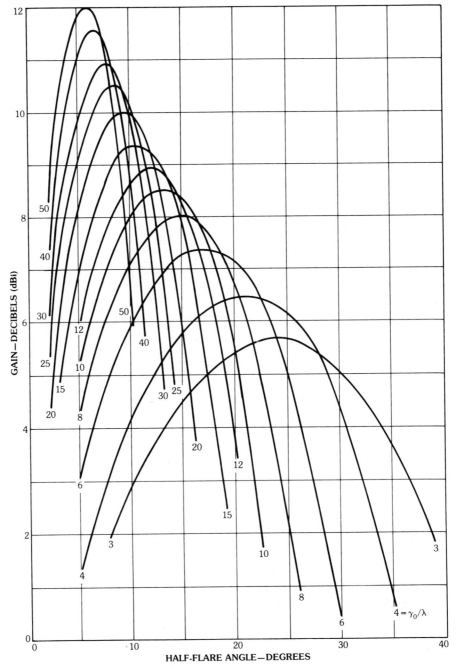

a

Fig. 82. Biconical horn gain. (*a*) For TEM mode. (*b*) For TE_{01} mode.

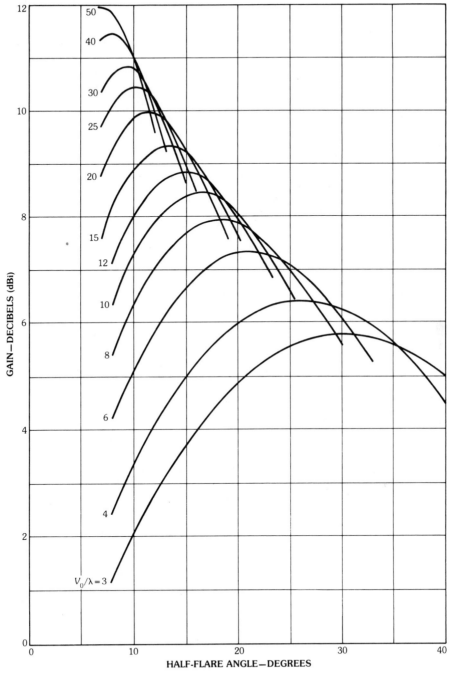

b

Fig. 82, *continued.*

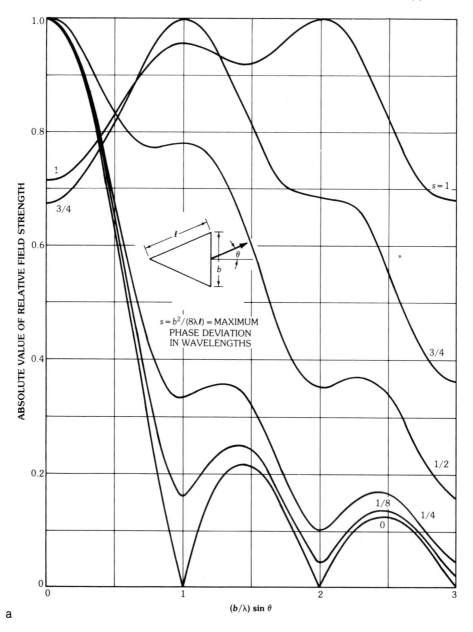

a

Fig. 83. Universal radiation patterns. (*a*) Of horns flared in the *E*-plane. (*b*) Of horns flared in the *H*-plane. (*After Jasik [51], © 1961 McGraw-Hill Book Co.*)

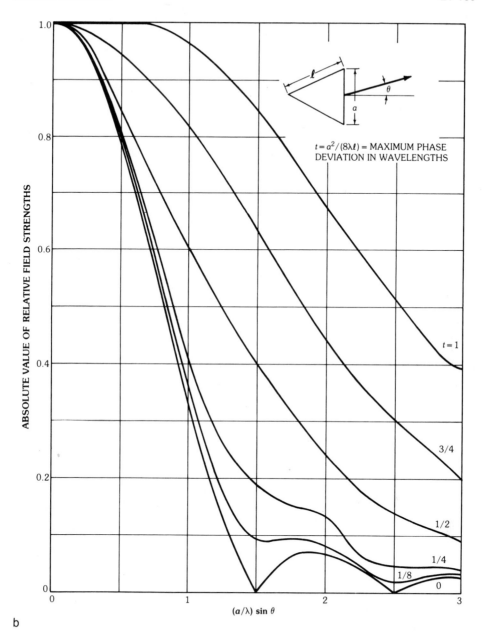

$t = a^2/(8\lambda\ell) =$ MAXIMUM PHASE DEVIATION IN WAVELENGTHS

b

Fig. 83, *continued.*

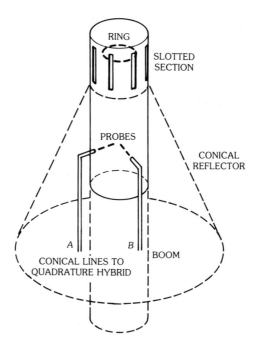

Fig. 84. Slotted-cylinder reflector antenna. (*After Albertsen, Balling, and Laursen [53]*)

excited in phase quadrature. The calculated and measured patterns are given in Fig. 85.

5. References

[1] W. L. Pritchard, "Satellite communication—an overview of the problems and programs," *Proc. IEEE*, vol. 65, no. 3, pp. 294–307, March 1977.

[2] J. W. Duncan, "Maximum off-axis gain of pencil beams," *Proc. IEEE* (letters), vol. 57, pp. 1791–1792, October 1969.

[3] N. Amitay, V. Galindo, and C. T. Wu, *Theory and Analysis of Phased Array Antennas*, New York: Wiley-Interscience, pp. 6–8, 1972.

[4] Y. T. Lo and S. W. Lee, "Affine transformation and its application to antenna arrays," *IEEE Trans. Antennas Propag.*, vol. AP-13, no. 6, pp. 890–896, November 1965.

[5] A. A. Oliner and G. H. Knittel, eds., *Phased Array Antennas*, Dedham: Artech House, pp. 68–82, 1972.

[6] E. D. Sharp, "A triangular arrangement of planar array elements that reduces the number needed," *IEEE Trans. Antennas Propag.*, vol. AP-9, pp. 126–129, March 1961.

[7] Y. T. Lo, "A mathematical theory of antenna arrays with randomly spaced elements," *IEEE Trans. Antennas Propag.*, vol. AP-15, pp. 231–235, March 1967.

[8] M. Born and E. Wolf, *Principles of Optics*, New York: Pergamon Press, 1964, pp. 203–232.

[9] S. Silver, ed., *Microwave Antenna Theory and Design*, New York: McGraw-Hill Book Company, 1949, pp. 389–412.

[10] J. J. Lee, "Numerical methods make lens antennas practical," *Microwaves*, pp. 81–84, September 1982.

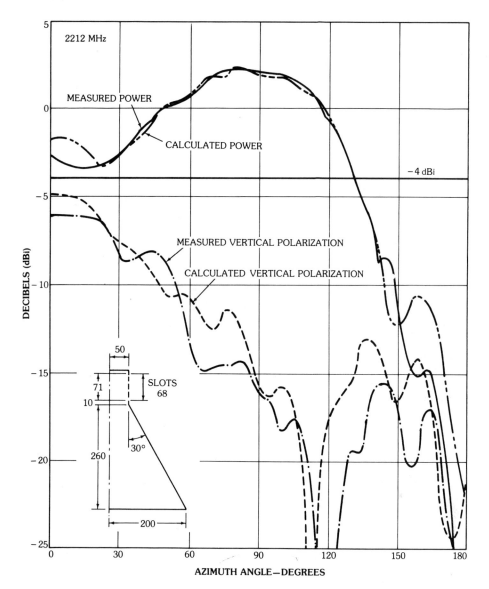

Fig. 85. Measured and computed patterns of slotted-cylinder reflector antenna. (*After Albertsen, Balling, and Laursen [53]*)

[11] D. Waineo, "Lens designed for arbitrary illumination," *Proc. of 1976 IEEE AP-S Symp.*, p. 476, 1976.
[12] F. G. Friedlander, "A dielectric lens aerial for wide-angle beam scanning," *J. Inst. Electr. Eng.*, vol. 93, pt. 3A, pp. 658–662, 1946.
[13] J. A. Jenkins and H. E. White, *Fundamentals of Optics*, chapter 9, New York: McGraw-Hill Book Company, 1957, p. 154.
[14] D. H. Shinn, "The design of a zoned dielectric lens for wide-angle scanning," *Marconi Rev.*, no. 117, p. 37, 1953.

[15] L. Young, "Tables for cascaded homogeneous quarter-wave transformers," *IRE Trans. Microwave Theory Tech.*, vol. 7, pp. 233–237, 1959; vol. 8, pp. 243–244, 1960.

[16] L. Young, "Synthesis of multiple antireflection films over a prescribed frequency band," *J. Opt. Soc. Am.*, vol. 51, pp. 967–974, 1961.

[17] A. F. Harvey, "Optical techniques at microwave frequencies," *Proc. Inst. Electr. Eng.* (London), C106, pp. 141–157, 1959.

[18] R. H. Garnham, "Optical and quasi-optical transmission techniques and component systems for millimeter wavelengths," *RRE Rep. no. 3020*, Royal Radar Establishment, Malvern, England, March 1958.

[19] P. G. Ingerson and W. C. Wong, "Focal region characteristics of offset-fed reflectors," *IEEE AP-S Intl. Symp. Dig.*, pp. 121–123, June 1974.

[20] T. Chu and R. H. Turrin, "Depolarization properties of offset reflector antenna," *IEEE Trans. Antennas Propag.*, vol. AP-21, pp. 339–345, May 1973.

[21] J. Ruze, "Lateral-feed displacement in a paraboloid," *IEEE Trans. Antennas Propag.*, vol. AP-13, pp. 660–665, September 1965.

[22] S. W. Lee and Y. Rahmat-Samii, "Simple formulas for designing an offset multibeam parabolic reflector," *IEEE Trans. Antennas Propag.*, vol. AP-29, no. 3, p. 472, May 1981.

[23] J. Ruze, "Antenna tolerance theory—a review," *Proc. IEEE*, vol. 54, pp. 633–640, April 1966.

[24] Y. Rahmat-Samii, "An efficient computational method for characterizing the effects of random surface errors on the average power pattern of reflectors," *IEEE Trans. Antennas Propag.*, vol. AP-31, pp. 92–98, January 1983.

[25] L. R. Whicker, *Ferrite Control Components, Volume 2*, Dedham: Artech House, 1974.

[26] T. M. Smith, E. W. Mathews, and C. R. Boyd, "C-band variable power divider and variable phase shifter development," *Proc. AIAA 9th Communications Satellite Systems Conf.*, pp. 693–697, March 1982.

[27] J. L. Janken, W. J. English, and D. F. DiFonzo, "Radiation from 'multimode' reflector antennas," *IEEE G-AP Intl. Symp.*, pp. 306–309, August 1973.

[28] Y. Hwang, A. Tsao, and C. C. Han, "Uniform analysis of reflector antenna for satellite application," *1983 IEEE AP-S Intl. Symp.*, vol. 1, pp. 88–90.

[29] J. E. Heller and J. B. Cruz, Jr., "An algorithm for minimax parameter optimization," *Automatica*, vol. 8, New York: Pergamon Press, 1972, pp. 325–335.

[30] K. Madsen, O. Nielson, H. S. Jacobsen, and L. Thrane, "Efficient Minimax design of networks without using derivatives," *IEEE Trans. on Microwave Theory Tech.*, vol. MTT-23, pp. 507–512, 1975.

[31] J. R. Mantz and R. F. Harrington, "Computational method for antenna pattern synthesis," *IEEE Trans. Antennas Propag.*, vol. AP-23, no. 4, pp. 507–512, July 1975.

[32] C. Donn, W. A. Imbriale, and G. G. Wong, "An S-band phased array design for satellite application," *IEEE Intl. Symp. Antennas Propag.*, pp. 60–63, 1977.

[33] C. Donn, "A new helical antenna design for better on-and-off boresight axial ratio performance," *IEEE Trans. Antennas Propag.*, vol. AP-28, no. 2, pp. 264–267, March 1980.

[34] A. R. Dion and L. J. Ricardi, "A variable-coverage satellite antenna system," *Proc. IEEE*, vol. 59, no. 2, pp. 252–262, February 1971.

[35] L. J. Ricardi, A. J. Simmons, A. R. Dion, L. K. DeSize, and B. M. Potts, "Some characteristics of a communication satellite multiple-beam antenna," *MIT Tech. Note 1975-3*, January 1975.

[36] Ford Aerospace & Communications Corporation, Palo Alto, California, "Design for Arabsat C-band communication antenna system."

[37] P. D. Potter, "A new horn antenna with suppressed side lobes and equal beamwidths," *Microwave J.*, vol. Vl, pp. 71–78, June 1963.

[38] T. Satoh, "Dielectric-loaded horn antenna," *IEEE Trans. Antennas Propag.*, vol. AP-20, pp. 199–201, March 1972.

[39] W. J. English, "The circular waveguide step-discontinuity mode transducer," *IEEE Trans. Microwave Theory Tech.*, vol. MTT-21, pp. 633–636, October 1973.

[40] J. S. Ajioka and H. E. Harry, Jr., "Shaped beam antenna for earth coverage from a stabilized satellite," *IEEE Trans. Antennas Propag.*, vol. AP-18, no. 3, pp. 323–327, May 1970.

[41] T. S. Chu, "On the use of uniform circular arrays to obtain omnidirectional patterns," *IRE Trans. Antennas Propag.*, vol. AP-7, pp. 436–438, October 1959.

[42] V. Galindo and K. Green, "A near-isotropic circular polarized antenna for space vehicles," *IEEE Trans. Antennas Propag.*, vol. AP-13, no. 6, pp. 872–877, November 1965.

[43] W. F. Croswell, C. M. Knop, and D. M. Hatcher, "A dielectric-coated circumferential slot array for omnidirectional coverage at microwave frequencies," *IEEE Trans. Antennas Propag.*, vol. AP-15, no. 6, pp. 722–727, November 1967.

[44] R. F. Harrington, *Time-Harmonic Electromagnetic Fields*, New York: McGraw-Hill Book Company, 1961, pp. 245–250.

[45] P. H. Pathak and R. G. Kouyoumjian, "An analysis of the radiation from apertures in surfaces by the geometrical theory of diffraction," *Proc. IEEE*, vol. 62, pp. 1438–1447, November 1974.

[46] J. R. Wait, *Electromagnetic Radiation From Cylindrical Structures*, New York: Pergamon Press, 1959.

[47] G. E. Stewart and K. E. Gorden, "Mutual admittance for axial rectangular slots in a large conducting cylinder," *IEEE Trans. Antennas Propag.*, vol. AP-19, pp. 120–122, January 1971.

[48] S. W. Lee, "Mutual admittance of slots on a cone: solution by ray techniques," *IEEE Trans. Antennas Propag.*, vol. AP-26, no. 6, pp. 768–773, November 1978.

[49] W. S. Gregorwich, "An electronically despun array flush-mounted on a cylindrical spacecraft," *IEEE Trans. Antennas Propag.*, vol. AP-22, no. 1, January 1974.

[50] W. L. Barrow, L. J. Chu, and J. J. Jansen, "Biconical electromagnetic horns," *Proc. IRE*, pp. 769–779, December 1939.

[51] H. Jasik, ed., *Antenna Engineering Handbook*, New York: McGraw-Hill Book Company, 1961, pp. 10-13 to 10-14.

[52] C. E. Ryan and L. Peters, Jr., "Evaluation of edge-diffracted fields including equivalent currents for the caustic region," *IEEE Trans. Antennas Propag.*, vol. AP-17, no. 3, pp. 292–299, May 1969.

[53] N. C. Albertsen, P. Balling, and F. Laursen, "New low-gain *S*-band satellite antenna with suppressed back radiation," *Sixth Eur. Microwave Conf.*, Rome, Italy, pp. 14–17, September 1976.

[54] R. F. Harrington and J. R. Mantz, "Radiation and scattering from bodies of revolution," *Appl. Sci. Res.*, vol. 20, pp. 405–435, 1969.

Chapter 22

Remote Sensing and Microwave Radiometry

J. C. Shiue
Goddard Space Flight Center

L. R. Dod
Swales and Associates

CONTENTS

James Chen-Chi Shiue holds a BSEE from Taiwan University, a MSEE from the University of Vermont, an electrical engineer's degree from Stanford University, and a PhD from New York University. He is also a registered professional engineer.

He has more than 20 years of experience in the aerospace industry and in research and academic institutions. Since joining NASA's Goddard Space Flight Center in 1971 he has been working in microwave remote sensing of the earth. He has conducted experiments ranging from testing microwave radiometric and radar techniques for measuring snow depths and of mapping soil moisture. At Goddard he is the group leader of the Microwave Sensors Group in the Laboratory for Oceans and is the instrument scientist in charge of developing the advanced microwave sounding unit, a temperature and humidity sounder for the NOAA weather satellites.

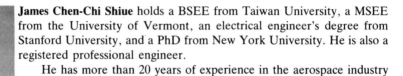 **Louis Dod** holds a BSEE from Virginia Polytechnic Institute and a MSEE from George Washington University. He has over 33 years experience in aerospace engineering while a member of the technical staff at NASA Goddard Space Flight Center. He has been involved in the design and development of antennas and microwave systems for the early Explorer Satellites and the development of antennas and RF systems for the NASA ground tracking network. He was also involved in the development of the early communications and weather satellites experiments. He was also involved in the development and flight tests of airborne and spacecraft microwave remote sensing instruments both active and passive. Currently, he is a member of the technical staff of Swales and Associates, Aerospace Consultants.

1. Basic Principles of Microwave Radiometry

In this introductory section we will consider blackbody microwaves, microwave radiative transfer, and surface emissivity and reflectivity.

Blackbody Radiation at Microwave Frequencies

According to Planck's law of radiation the spectral radiant emittance R_f of a blackbody is isotropic and is given by

$$R_f = \frac{2hf^3}{c^2} \frac{1}{e^{hf/kT} - 1} \tag{1}$$

where

R_f = spectral radiant emittance in (watts)(meter)$^{-2}$(steradian)$^{-1}$(hertz)$^{-1}$ is the power emitted per unit solid angle, per unit area of the emitter, per unit frequency

c = velocity of light, $3 \times 10^8 \, \text{m·s}^{-1}$

h = Planck's constant, $6.63 \times 10^{-34} \, \text{J·s}$

k = Boltzmann's constant, $1.38 \times 10^{-23} \, \text{J·K}^{-1}$

λ = wavelength in meters

f = frequency in hertz

T = temperature in kelvins

When

$$hf \ll kT \tag{2}$$

then

$$R_f \cong \frac{2kT}{\lambda^2} \tag{3}$$

When the condition in (2) is satisfied, then (3) is valid. This is called the *Rayleigh-Jeans approximation*. In the microwave region (1 to 200 GHz) and with the terrestrial temperatures the Rayleigh-Jeans approximation is valid. This approximation makes the radiation power P directly proportional to temperature T. The incremental power received by an antenna from $\Delta\Omega$ is

$$\Delta P = \frac{1}{2} R_f \Delta f \, \Delta\Omega \, \Delta A_s \tag{4}$$

where

$\Delta\lambda$ = incremental change in wavelength

$\Delta\Omega$ = incremental solid angle subtended by the receiving antenna

ΔA_s = incremental area of the radiation source

The factor ½ in (4) results from the fact that only a single polarization is received. The receiving antenna effective area is

$$A_R = \frac{\lambda^2}{4\pi}G \tag{5}$$

where G is the antenna gain over isotropic media. Since

$$\Delta\Omega = \frac{A_R}{r^2} = \frac{\lambda^2}{4\pi r^2}G \tag{6}$$

where r is the distance between the emitting surface element and the receiver,

$$\frac{\Delta A_s}{r^2} = \Delta\Omega_s \tag{7}$$

where $\Delta\Omega_s$ is the solid angle subtended at the receiver by ΔA_s. Therefore, by using (3), (6), and (7),

$$\Delta P = \frac{1}{4\pi}kTG\,\Delta\Omega_s\,\Delta f \tag{8}$$

From (8), the total power collected by a receiving antenna, for uniform T, is

$$P = kT\Delta f\left(\frac{1}{4\pi}\int G(\Omega)\,d\Omega\right)$$

or

$$P = kT\Delta f \tag{9}$$

Equation 9 is the Nyquist formula for noise power in bandwidth Δf. As we shall see in a later section, for a nonideal blackbody radiating surface, T should be replaced by $T_b = \epsilon T$, where T_b is the brightness temperature and ϵ the surface emissivity.

Microwave Radiative Transfer

The differential equation governing the transport of radiative energy (or power) in a medium can be expressed in the following form:

$$\frac{dW(x)}{dx} = -a(x)[W(x) - J(x)] \tag{10}$$

where

$W(x)$ = the power density in watts per unit area propagating along path x

$a(x)$ = the attenuation factor per unit distance, which is indicative of the rate at which power is being diminished (by absorption and scattering)

$J(x)$ = the source function (power per unit area) representing the thermal emission of the medium. If the source function $J(x) = 0$, then the solution for the power density $P(x)$ is an exponential decay along the direction x.

The formal solution to (10) is

$$W(x) = W(0)\, e^{-\tau(x,0)} + \int_{u=0}^{u=x} a(u)\, J(u)\, e^{-\tau(x,u)}\, du \tag{11}$$

where

$$\tau(x,x') = \int_{u=x'}^{u=x} a(u)\, du \tag{12}$$

For the special case when $a(x) = a_0$ and $J(x) = J_0$, then

$$W(x) = W(0)\, e^{-a_0 x} + J_0(1 - e^{-a_0 x}) \tag{13}$$

Since the power density W is proportional to temperature T, we can replace W by T in (13) to obtain the relationships between the temperatures.

Similarly, the source function J_0 is replaced by the physical temperature T_p of the emitting medium. Thus

$$T(x) = T(0)\, e^{-a_0 x} + T_p(1 - e^{-a_0 x}) \tag{14}$$

(In the conventional engineering unit of decibels, $L_{dB} = 10 \log_{10} L$.) Set x equal to d and let L be the loss factor of the path length d, such that

$$L = e^{a_0 x} = e^{a_0 d} \tag{15}$$

Then (14) becomes

$$T(d) = T(0)\frac{1}{L} + \left(1 - \frac{1}{L}\right) T_p \tag{16a}$$

Equation 16 represents the temperature $T(d)$ that results after the energy travels through a medium with loss factor L. This is depicted schematically in Fig. 1. The

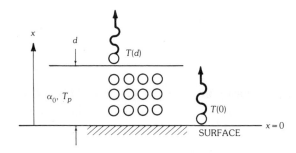

Fig. 1. Radiative transfer model.

brightness temperature $T(0)$ is attenuated by L while the medium is reemitting a component of temperature, $(1 - 1/L)\,T_p$. The radiative transfer theory, as represented by (10), governs only the amplitude, not the phase of a wave propagating in a medium. No coherent phase information of the electromagnetic wave is included in this theory. The radiative transfer theory, however, is an accurate and useful description of many phenomena in microwave radiometry for remote sensing applications.

The following example will demonstrate the use of (16) in the remote sensing of the atmosphere. The presence of high humidity in the atmosphere can be measured by a microwave radiometer looking down at the earth from space. The humid air is detected by the change in the brightness temperature T_b as compared with the adjacent dry air area. As we shall see, it turns out that the change in T_b, or contrast, is larger when detecting the humidity against the ocean background than against the land background. In (16) let $T(d) = T_b$ be the brightness temperature seen by a spaceborne radiometer, and $T(0) = T_s$ be the surface brightness temperature. Further, let $T_p = \overline{T}_A$ be the average temperature of the atmosphere and L be the transmissivity (loss factor) of the atmosphere. Then

$$T_{b_1} = \frac{T_s}{L_{A_1}} + \overline{T}_A\left(1 - \frac{1}{L_{A_1}}\right) \tag{16b}$$

where L_{A_1} is the loss factor of the atmosphere with high humidity. Similarly, the brightness temperature seen by the radiometer over an area with dry air (whose loss
factor is L_{A_2}) is given by

$$T_{b_2} = \frac{T_s}{L_{A_2}} + \overline{T}_A\left(1 - \frac{1}{L_{A_2}}\right) \tag{16c}$$

The change in T_b between the two cases is

$$\Delta T_b = T_{b_1} - T_{b_2} = (T_s - \overline{T}_A)\left(\frac{1}{L_{A_1}} - \frac{1}{L_{A_2}}\right) \tag{16d}$$

Since the humid air has a higher loss factor than dry air, $L_{A_1} > L_{A_2}$, and for land surface $T_s > \overline{T}_A$. Hence $\Delta T_b < 0$. This means that the high-humidity area will appear as a "cooler" spot over land.

To increase the contrast ΔT_b, one can choose the frequency of observation to maximize the second parenthetical term in (16d). For example, the frequency of the water vapor absorption line near 22 GHz is a good frequency for humidity observation. It turns out that because of the difference in brightness temperature between land and air is small the resultant ΔT_b is relatively small (typically 1 to 2 K), making the use of 22 GHz for humidity detection rather difficult over land. Table 1 lists the result of ΔT_b with the assumed values of T_s and $\overline{T}_A$.

High-humidity air against an ocean background produces much larger contrast compared with the land background. Because the surface brightness temperature of the ocean is much *cooler* than the air, the high-humidity area appears as a *warm* spot against an ocean background. In the case of an ocean background, because the difference between $\overline{T}_A$ and T_s is large the contrast is strong. Table 1 shows that the same humidity yields a brightness temperature change greater than 16 K.

Equation 16a can also be used to represent the remote sensing of a snowpack over land. In this case the term T_S stands for the emission from the background soil, and the L of the first term is the extinction factor due primarily to the snow scattering. The second term in (16a) is usually very small for snow. The general principles are similar in both the water vapor and snowpack cases, but the mechanisms contributing to the extinction are quite different. Water vapor absorbs and reradiates the emission from the background surface (whether ocean or land). For the case of detection of humidity over the ocean, where the radiometry technique works most effectively, water vapor absorbs cool ocean background and reemits at its own warmer temperature, thus making the humid air appear as a *warm* spot. In contrast, snow particles (assuming dry snow) absorb very little of the background soil radiation. Their extinction mechanism is primarily due to scattering. For this reason the frequencies suitable for snowpack sensing are usually in the range of 20 to 40 GHz. From (16d) it is clear that in order to infer the property of the intervening media, such as water vapor in the atmosphere or snowpack on the ground, the background radiation T_S must also be determined. This is frequently achieved by multichannel measurements.

Surface Emissivity and Reflectivity

The brightness temperature T_b of an emitting surface is the product of its thermal temperature T_p and the surface emissivity ϵ:

$$T_b = \epsilon T_p \tag{17}$$

The emissivity ϵ is a function of the radiation polarization. The emissivity may be subscripted to account for this polarization dependence:

$$\epsilon_j = 1 - R_j \tag{18}$$

where

Table 1. Brightness Temperatures Expected from a Spaceborne Microwave Radiometer

Condition	T_b (K)	ΔT_b (K)	Assumed Values
Dry air over ocean	187.71		$T_s = 170\,\text{K (ocean)}$
Humid air over ocean	171.54	+16.17	$\overline{T}_A = 280\,\text{K (air)}$
Dry air over land	289.86		$T_s = 290\,\text{K (land)}$
Humid air over land	288.39	−1.47	$L(22\,\text{GHz}) = \begin{cases} 0.06\,\text{dB (dry air)} \\ 0.76\,\text{dB (humid air)} \end{cases}$

$$
\begin{aligned}
j &= v, h \text{ for vertical or horizontal polarization} \\
R_j &= |\varrho_j|^2, \text{reflectivity} \\
\varrho_j &= \text{voltage reflection coefficient of the radiation}
\end{aligned} \tag{19}
$$

For a specular surface the voltage reflection coefficient ϱ_j is obtained from the following Fresnel equations [1]:

$$
\varrho_h = \frac{\cos\theta_i - \sqrt{(\epsilon_2/\epsilon_1)[1 - (\epsilon_1/\epsilon_2)\sin^2\theta_i]}}{\cos\theta_i + \sqrt{(\epsilon_2/\epsilon_1)[1 - (\epsilon_1/\epsilon_2)\sin^2\theta_i]}} \tag{20}
$$

$$
\varrho_v = \frac{\cos\theta_i - \sqrt{(\epsilon_1/\epsilon_2)[1 - (\epsilon_1/\epsilon_2)\sin^2\theta_i]}}{\cos\theta_i + \sqrt{(\epsilon_1/\epsilon_2)[1 - (\epsilon_1/\epsilon_2)\sin^2\theta_i]}} \tag{21}
$$

where ϵ_1 and ϵ_2 are the dielectric constants of the incident and transmitted media, respectively, and θ_i is the angle of incidence in medium 1. Horizontal polarization is the mode where the electric field vector is perpendicular to the plane of incidence (hence it is tangent to the earth's surface and horizontally oriented). Vertical polarization is the mode in which the electric field vector lies in the plane of incidence. (These definitions of horizontal and vertical polarization will be discussed further in Section 4.) The vertical polarization has total transmission at the Brewster angle in which $\theta_i = \theta_B$, and the emissivity is a maximum:

$$
\tan\theta_B = \sqrt{\epsilon_2/\epsilon_1} \tag{22}
$$

For nonspecular (rough) surfaces the reflectivity must include not only the scattering in the specular direction but also the power scattered in all other directions by the rough surface. The effective reflection coefficient ϱ is the summation of "differential scattering cross section per unit area" σ^0 over all directions

$$
\varrho_j(\theta_i) = \iint \{[\sigma_{jj}^0(\theta_i, \theta_s, \phi_s)] + [\sigma_{jk}^0(\theta_i, \theta_s, \phi_s)]\} \frac{d\Omega}{4\pi\cos\theta_i} \tag{23}
$$

where $j = v, h$ for vertical and horizontal polarizations, $d\Omega$ is the differential solid angle, and $\sigma 2_{jk}^0$ is the differential cross section per unit area, from j polarization into k polarization:

$$\sigma_{jk}^0(\theta_i, \theta_s, \phi_s) = 4\pi r_s^2 \frac{\overline{\Delta P_s}}{P_0 \Delta A_0} \tag{24}$$

where

θ_i = the angle of incidence (elevation)

θ_s = the scattering angle (elevation)

ϕ_s = scattering azimuthal angle

ΔP_s = scattered power density in the direction (θ_s, ϕ_s) in watts per square meter

r_s = distance at which P_s is measured

P_0 = incident power density in watts per square meter

ΔA_0 = differential incident area in square meters

The overbar $(\overline{})$ represents the statistical average of many measurements of power density. The subscripts j and k represent the polarizations of incident and scattered waves. The differential scattering cross section σ_{jk}^0 can be obtained by solving the wave equation and matching the boundary conditions incorporating surface roughness characteristics.

Strictly speaking, the formal definition of emissivity is valid only when the temperature of the underlying medium is uniform. If, however, the temperature is uniform to within several skin depths at the measuring frequency, the concept of emissivity can still be applied. An exact approach for the nonuniform temperature case is the solution of the radiative transfer equation for the upwelling energy.

The general emissivity characteristics derived from the Fresnel relations of (20) and (21) have been verified by observations of soil and water surfaces, although the surfaces are modified from the ideal smooth case by nonspecular rough surface effects. Fig. 2 is a plot of the emissivity versus incidence angle for a flat soil surface with two different soil moisture values [2]. Fig. 3 shows the theoretical brightness temperatures of calm sea water [3].

2. Applications of Microwave Radiometry to Remote Sensing

Microwave radiometry is the detection of thermal radiation power at microwave frequencies. A wealth of information can be derived from radiometric observations. In addition to the intensity of the radiation its dependence on frequency, angle of incidence, and polarization can also provide additional information about the source.

The development of microwave radiometry for remote sensing (also called passive microwaves, as compared with active radars) derives its background from radioastronomy. Since the advent of satellites the use of passive microwaves in

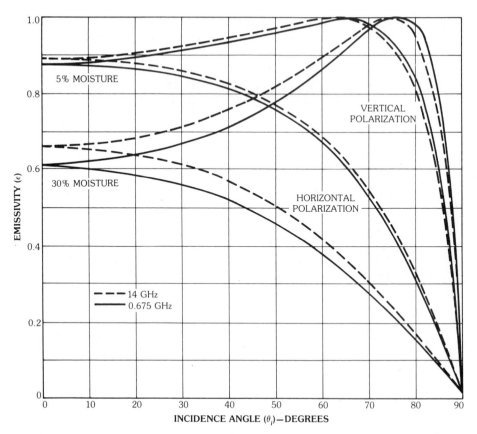

Fig. 2. Theoretical emissivity of bare soil (flat). (*After Njoku and Kong [2], © 1977 American Geophysical Union*)

remote sensing of the earth from space has gone through a rapid evolution in the last two decades, from a laboratory curiosity to daily operational systems.

As we enter the space shuttle era, which will allow one to transport larger and heavier satellites into space at cheaper costs, the use of passive microwaves for earth remote sensing will witness an even faster growth in the future. Fig. 4 is a picture of the microwave brightness temperature of the world obtained from a microwave radiometer [4] called the electrically scanning microwave radiometer (ESMR). From Fig. 4 it is evident that the microwave brightness temperature distribution is quite different from the physical temperature. (See the color insert for Fig. 4.)

Although the physical or thermodynamic temperature (based on the absolute scale, in kelvins) of the earth's surface is fairly uniform, in terms of the microwave brightness temperature scale the land masses stand out much hotter (250 to 280 K) compared with the cool oceans (200 K or less). This is because the land area, in general, has high emissivities, in the range of 0.7 to 0.9. Vegetation and forest covers increase the emissivity over bare soil; surface moisture, on the other hand, reduces the emissivity. A calm ocean surface has a low emissivity, of about 0.3 to

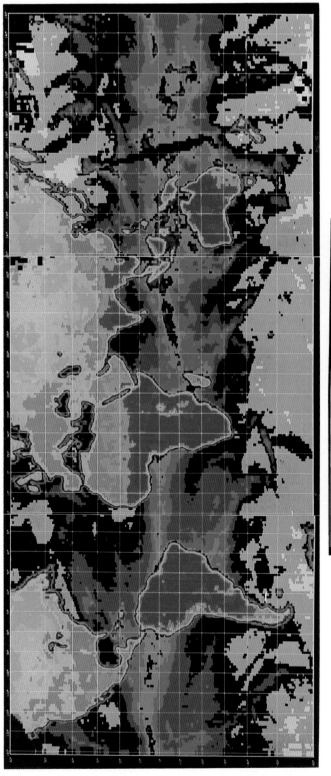

Fig. 22.4. Radio brightness of the world. (*Courtesy NASA*)

12 – 16 January 1973

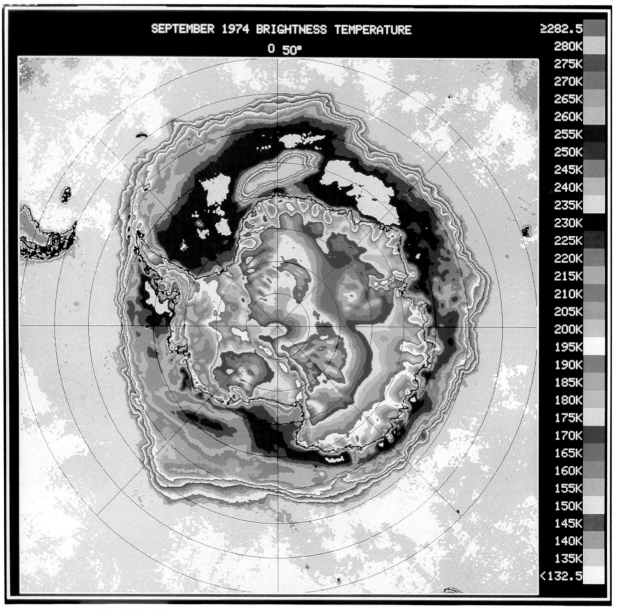

Fig. 22.6. Microwave brightness temperature of Antarctica (August 1974). *(Courtesy NASA)*

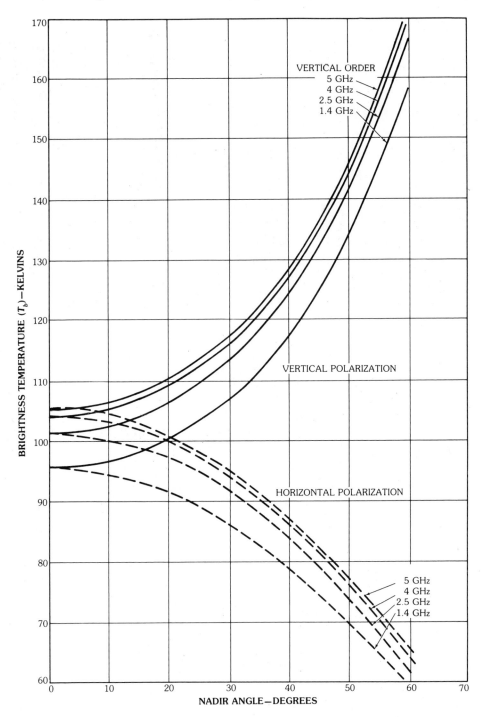

Fig. 3. Specular brightness temperature versus nadir angle of calm sea surface. (*Courtesy NOAA*)

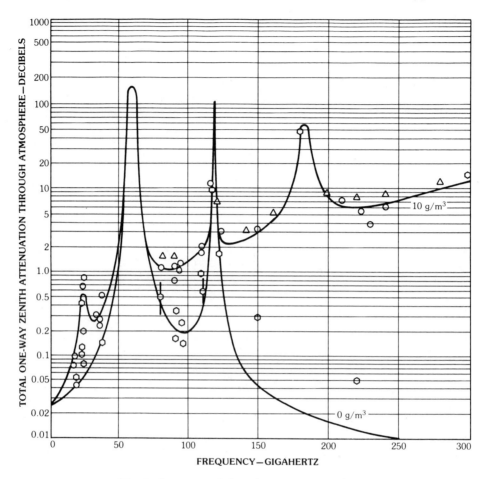

Fig. 5. Spectrum of the microwave atmosphere.

0.6 (1 to 10 GHz, nadir). Humid air and rain show up as warm areas against the cool ocean background. This is particularly evident in the microwave atmospheric spectrum of the intertropical convergence zone (ITCZ) around 10°N, S latitudes.

The particular frequencies used for a spaceborne microwave radiometer depend on the physics of the problem concerned, as well as the transmission characteristics of the atmosphere. Fig. 5 shows the microwave spectrum of the atmosphere due to oxygen and water vapor. The major opaque lines are the oxygen lines at 60 and 118 GHz, and the two water vapor lines are at 22 and 183 GHz. For radiometers whose primary purpose is sensing the earth's surface, one would use the so-called window regions between the opaque lines, such as 150, 90, and 30 GHz, or lower frequencies, below the 22-GHz line. Water vapor lines are used to sense atmospheric humidity and the oxygen lines to sense atmospheric temperature.

Like their infrared counterparts, microwave radiometers operate day or night since they rely on thermal emission rather than sunlight reflections. By virtue of their longer wavelengths, as compared with visible and infrared, microwaves have an important advantage because they can penetrate through the clouds. Therefore they work in nearly all weather conditions. In addition, there are unique features in the microwave spectrum. For example, the presence of water in soil decreases its emissivity. This fact can be used to measure soil moisture. (See Fig. 2.) The longer wavelengths can better penetrate the vegetation coverage to sense the underlying soil moisture. Because of the long wavelengths, however, microwave radiometers need relatively large antennas for good angular resolution.

Furthermore, in order to cover large areas of earth from an orbiting satellite (in a typical low earth polar orbit) the antennas must be able to scan large angular limits (such as ±50°). Also, remote sensing radiometers usually need several frequencies that are widely separated from one another. To top off these demanding requirements, remote-sensing radiometer antennas must have extremely high beam efficiency and low side lobes. The radiometer must be well calibrated and maintain good stability. In addition, there are the usual spaceflight constraints. It must be lightweight, compact in size, and produce a minimum of heat and mechanical perturbation. In short, the overall performance and engineering requirements of a satellite microwave radiometer antenna for earth remote sensing can be very stringent.

A few of the more important applications of passive microwaves in earth remote sensing are discussed next. For more detailed information, the reader is referred to a recent review article [5].

The Atmosphere

Probably the most useful application of earth satellite remote sensing is the gathering of atmospheric data for meteorological purposes. Because the atmosphere is in a continuous state of change, frequent samples (in both time and space intervals) are needed for weather forecasting. Measurements from orbiting satellites are the most economical methods used to meet this type of data requirement. Microwaves also have an advantage of being able to operate in cloudy regions, where most meteorological actions occur.

Oxygen and water vapor play dominant roles in shaping the absorption spectrum of the atmosphere. Fig. 5 shows a calculated attenuation characteristic for the atmosphere [6, 7]. The lower curve is for dry air (US Standard, 1976) and the upper curve is for humid air with $10\,g/m^3$ surface water vapor content.

Temperature Sounders—Microwave temperature sounders typically use the 60-GHz oxygen band to measure the atmospheric temperature profiles. This is based on the principle that the oxygen mixing ratio is fairly uniform in space and constant in time; hence the magnitude of the microwave brightness temperature is uniquely related to the atmospheric temperature. By having several frequencies (or channels) spreading down the wings of the 60-GHz oxygen band, each channel will sense a different layer of the atmosphere. The very opaque channels sense only the very top of the atmosphere, as the radiation from the lower layer of the atmosphere is highly attenuated and never arrives at the satellite. The more transparent a channel is, the deeper it will probe into the atmosphere. The complete temperature profile can be retrieved from the brightness temperatures and the associated "weighting functions" of all the channels.

Microwave temperature sounders are in routine use today [8]. For example, the microwave sounder units (MSUs) are aboard the National Aeronautics and Space Administration's (NASA's) TIROS-N, and the US National Oceanic and Atmospheric Administration's (NOAA-series) weather satellites. A similar sounder called the special sensor microwave/temperature (SSM/T) is on the US Air Force's defense meteorological satellites (DMSs). The hardware aspects of these sounders will be discussed in more detail in Section 3. The oxygen line at 118.75 GHz can also be used for temperature sounding [9]. A major advantage of this frequency is that it needs smaller antennas (as compared with the 60-GHz band), which can be very important for geosynchronous satellites. (See Section 6.)

Humidity Sounders—The 22-GHz water vapor line has been used for sensing total water vapor content (humidity) over the ocean [10]. The very opaque water vapor line at 183 GHz can be used to ascertain humidity profiles. In addition to the temperature and humidity profiles, window frequencies can also be used to measure precipitation distributions. The measurement is based on brightness temperature contrast between the precipitation and the "background." For example, at 18 GHz rain cells will appear as warm areas, in contrast to the rather cold ocean, because of their absorption. The rain cells will appear to be colder than the background at high frequencies and high intensities, when scattering is dominant [11]. In addition, satellite radiometers can be used to monitor other species in the atmosphere, such as O_3, CO, and so on [12].

The Ocean Surface

Sea Surface Temperature—The microwave brightness temperature of the ocean is the product of the sea-water thermal temperature and its surface emissivity. The latter is also a function of thermal temperature. The sensitivity of emissivity to thermal temperature change is maximum at 4 to 5 GHz; therefore this is the region of frequency for sea surface temperature measurement [13]. One must also account for emissivity variations caused by other reasons, such as surface roughness (due to waves) and/or atmospheric water. The theoretical sea surface brightness temperature was shown in Fig. 3 (Section 1).

Sea Surface Wind Speed—When the wind disturbs a calm sea the surface emissivity increases from that of a smooth plane surface determined by the Fresnel equations. As the sea becomes very rough, patches of foam begin to form. This also increases the surface emissivity. These relationships can be applied in determining the sea surface wind speed from radiometric observations [14].

Sea Ice—Sea ice has a microwave emissivity value of 0.8 to 0.9 (at nadir), as compared with the value of about 0.3 to 0.6 for calm sea water. Hence, radiometrically, sea ice appears as a warm island against a cold sea-water background. Moreover, microwave radiometers can differentiate between new ice, which is warmer, and the old, multiyear ice, which is comparatively colder. Sea ice concentration maps, derived from orbital microwave radiometer measurements, are very useful in guiding ship routing near the polar regions. Repeated time-series images of the polar region allow one to study and monitor annual polar ice boundary evolution [15]. Fig. 6 is an image of the microwave brightness temperature of Antarctica in the winter of 1974. The data were taken from the ESMR on the Nimbus-5 satellite at a wavelength of 1.55 cm. (See the color insert for Fig. 6.)

Land Applications

There are two major applications of remote sensing with microwave radiometry over land: soil moisture and snow cover.

The need for regional or global data of soil moisture at frequent time intervals can be more efficiently met by remote sensing with microwave radiometry. The conventional point-by-point in situ snow survey is inadequate and expensive.

The uses of soil moisture information are numerous. For example, in hydrology, area-wide soil moisture measurements are needed to assess regional drought conditions. Soil moisture information is also a basis for computing the watershed runoff coefficient, which is used for flood predictions. The evapotranspiration of soil moisture is a part of climate study. The soil moisture, at some critical period of the growth cycle of a plant, determines the yield of that crop at harvest. Timely soil moisture information can be used for irrigation control and yield forecasting.

Timely information on snow-covered areas and water equivalent (amount of water depth or water stored per unit area) in mountain watersheds, such as the western states of the United States, is difficult and expensive to gather during the winter. This snowpack information, however, is important in forecasting the amount of water runoff, which is the basis for the management of this limited and precious natural resource. The passive microwave remote-sensing technique is capable of providing such snowpack information.

Soil Moisture—Water has a much larger dielectric constant than that of dry soil, particularly at lower microwave frequencies, below 5 GHz. The presence of water in the soil increases the dielectric constant of the mixture and consequently lowers its surface emissivity. For example, at 1.4 GHz the emissivity can change from 0.9 for dry soil to 0.6 for wet soil at 30-percent water by dry weight. This range of emissivity variation is the underlying principle [16] for remote sensing of soil moisture with microwave radiometry. A change in emissivity by 0.3 corresponds to a change in microwave brightness temperature of 80 to 90 K. An orbiting microwave radiometer can achieve measurement precision on the order of 1 K. Therefore a microwave radiometer can differentiate not only dry from wet areas, but it is also capable of resolving many levels of soil moisture content.

There are, however, factors that tend to complicate the quantitative determination of moisture content, and research is currently under way to resolve them. The soil dielectric constant is also dependent on the type of soil, because the bounding force between water and the host soil depends on the type of soil. Surface roughness also affects the emissivity. The presence of a vegetation canopy increases the emissivity. The net results are that both vegetation canopy and surface roughness tend to reduce the "sensitivity" of microwave radiometry techniques to measure soil moisture. (Sensitivity is defined as change in microwave brightness temperature per unit change in soil moisture.)

The lower microwave frequencies, from 1 to 5 GHz, are better suited for soil moisture sensing. This is primarily because the difference between dielectric constants of water and dry soil is larger at lower frequencies. Also, longer wavelengths in this range can penetrate deeper into soil and are less vulnerable to the masking effects of vegetation cover and surface roughness. However, a drawback of this low frequency range is that it requires a large antenna for use from satellites. The 1.41-GHz (21-cm wavelength hydrogen line) protected radioastronomy band is a good compromise frequency for the previously mentioned factors.

Snow Hydrology—When a land area is covered by a layer of dry snow its brightness temperature decreases. This is because snow particles scatter background land emission. The snow particles also absorb and reemit the background radiation but this is relatively unimportant; the scattering is the dominant loss for dry snow. The brightness temperature of a snowpack decreases as the snow depth increases. This relationship is used for remote sensing of snow depth by microwave radiometers [17, 18]. A decrease in brightness temperature of 60 K has been measured for a snow depth of 60 cm. When a snowpack begins to melt, the presence of liquid water drastically increases the absorption. As a result the snowpack brightness temperature increases substantially. This fact can be used to monitor the onset of snowpack melting.

As in the case of soil moisture, there are also complicating factors in quantitative determination of snow depth or its water equivalence. The dominant scattering loss implies that the snowpack brightness temperature also depends on snow grain size. Snowpack with smaller grain size scatters less and is, therefore, warmer as compared with larger-grain snowpack. Ice layers embedded in the snow also modify the brightness temperature. Because of these factors multiple frequencies are needed to resolve ambiguities.

3. A Survey of Existing Spaceborne Microwave Radiometer Antennas

Table 2 is a listing of the characteristics of most existing satellite microwave radiometers. Unless mentioned otherwise the country of origin of the satellite is the United States. Most of the radiometers listed are for earth remote sensing, with the exception of Mariner-2.

4. Antenna Requirements for Remote-Sensing Microwave Radiometry

Fig. 5 shows the atmospheric spectrum up to 240 GHz. Most of the present remote-sensing microwave radiometers use frequencies to the left of the 60-GHz oxygen band. This oxygen band has been used for atmospheric temperature sounders because of the well-behaved oxygen mixing ratio in the atmosphere, as mentioned previously in Section 2. The 60-GHz oxygen band in Fig. 5 actually contains a complex band of many individual lines which manifest themselves at higher altitudes. Fig. 7 shows the fine structures of the dominant lines of this band and the frequencies used by many of the microwave temperature sounders. Fig. 7 is presented in a one-way zenith opacity in units of optical depth (OD, where 1 OD = 4.3 dB) versus frequency. These 60-GHz band lines are the rotational lines of the oxygen molecules.

In Fig. 7 the down-pointing arrows indicate the line center frequencies. The numbers with a right superscript (e.g., 9^-, 3^+, etc.) are the line designations. The

Table 2. Existing Satellite Microwave Radiometers

Year of launch	1962	1968	1972
Satellite name	Mariner-2 [19]	Cosmos-243 (USSR) [20]	Cosmos-384 (USSR) [21]
Frequency (GHz)	15.8 22.2	3.5 8.8 22 37	3.5 8.8 22 37
Bandwidth (MHz)			
Antenna type	Parabola, 42-cm diam.	Parabolas	Parabolas
Beamwidth (degree)	2.7 2.0	10 4 4 4	10 4 4 4
Scanning	±60°	Nadir viewing	Nadir viewing
Polarization			
Calibration	Space viewing horns	Spaceview horn and noise generator	Spaceview horn and noise generator
Temperature sensitivity, ΔT, (K)			
Equivalent ΔT at 1 s			
Parameter measured	Limb darkening of planetary emission, temperature of Venus	Atmospheric water vapor and liquid water, sea ice, sea surface temperature	Atmospheric water vapor and liquid water, sea ice, sea surface temperature

Year of launch	1972	1972	1973
Satellite name	Nimbus-5	Nimbus-5	Skylab
Instrument name and acronym	Nimbus-E microwave spectrometer (NEMS) [22]	Electrically scanning microwave radiometer (ESMR) [4]	S-193
Frequency (GHz)	22.235 31.40 53.65 54.90 58.80	19.3	13.9
Bandwidth (MHz)	220 220 220 220 220	240	50
Antenna type	5 lens loaded scalar horns	Phased array, 83 cm × 85 cm	Parabola
Beamwidth (degree)	10 10 10 10 10	1.4° × 1.4°, nadir, 1.4° × 2.2° at ±50° (d) × (c)	1.5
Scanning	Nadir viewing	Electrically scanning ±50°, cross track	Mechanically scanned 0 to ±48° cross- or down-track
Polarization		Horizontal	Horizontal and vertical
Calibration	Microwave load embedded in radiation cooled plate and warm loads	Spaceview horn and hot load	
Temperature sensitivity, ΔT, (K)	0.28 K to 0.56 K (at 2 s)	Better than 1.5 K at 46 ms	1.8 K at 32 ms
Equivalent ΔT at 1 s	0.4 K to 0.8 K	Better than 0.3 K	0.32 K
Parameter measured	Atmospheric temperature profile, total water vapor and liquid water, sea ice	Precipitation, sea ice, and snow cover	Sea surface wind speed
Remarks	Weight = 33-kg, power = 33 W	Weight = 31 kg, power = 41 W	Part of a combined passive/active system

	1973 Skylab S-194	1974 Meteor (USSR) / 1977 Meteor-28	1975 Nimbus-6 ESMR
Year of launch	1973	1974 / 1977	1975
Satellite name	Skylab	Meteor (USSR) / Meteor-28	Nimbus-6
Instrument name and acronym	S-194		Electrically scanning microwave radiometer (ESMR) [4]
Frequency (GHz)	1.414	≅37	37
Bandwidth (MHz)	27	1 GHz	220
Antenna type	64-element phased array (102 × 102 cm)		Phased array, 77 cm × 81 cm
Beamwidth (degree)	15	≅1°	1.17° × 0.73° down-track × cross-track
Scanning	Nadir viewing only	35° from nadir ±40°, mechanically scanned	Conical scan, half-cone angle = 45° ±34° azimuth limit
Polarization		V&H	V&H
Calibration	Radiation cooled load (200 K) and hot load (370 K)		Spaceview horn and a hot load
Temperature sensitivity, ΔT, (K)	Better than 0.5 K, at 1 s	0.5 K at 1 s	1 K at 60 ms
Equivalent ΔT at 1 s	Better than 0.5 K	0.5 K	0.25 K
Parameter measured	Soil moisture, sea water salinity	Atmospheric	Atmospheric liquid water, sea ice
Remarks	Weight = 25 kg		Weight = 44.6 kg, power = 55 W

	1975 Nimbus-6 SCAMS	1978 Block 5D SSM/T	1978 TIROS-N MSU
Year of launch	1975	1978	1978
Satellite name	Nimbus-6	Block 5D	TIROS-N
Instrument name and acronym	Scanning microwave spectrometer (SCAMS) [23]	Special sensor, microwave/temperature (SSM/T) [24]	Microwave sounder unit (MSU) [25]
Frequency (GHz)	22.235 31.60 52.80 53.80 55.40	50.5 53.2 54.35 54.9 58.4 58.825 59.4	50.30 53.74 54.96 57.9
Bandwidth (MHz)	220 220 220 220 220	400 400 400 400 115 400 250	220 220 220 220
Antenna type	3 hyperbolic reflectors	1 parabolic reflector	2 hyperbolic reflectors (for upper frequencies and lower halves)
Beamwidth (degree)	7.5°	14°	7.8×7 7.3×7 7.6×7.3 7.3×7
Scanning	±43°, cross track, rotating reflector, fixed feed	±36° cross track, scanning reflector, stationary feed	±47.4° cross track, scanning reflectors, fixed horns
Polarization	V* V* V* H**	V*/H** OMT*** separates lower 3 channels from the rest	V*/H**, OMT*** separates V from H behind each horn
Calibration	Reflector views space and an ambient temperature target	Reflector views space and a hot target	Reflector views space and an ambient temperature target
Temperature sensitivity, ΔT, (K)	0.2 K to 0.5 K at 2 s	0.4 to 0.6 K (at 2.7 s)	0.21 0.22 0.18 0.21 (at 1.8 s)
Equivalent ΔT at 1 s	0.28 to 0.71 K	0.66 to 0.99	0.28 0.30 0.24 0.28
Parameter measured	Atmospheric temperature profile liquid water and water vapor	Atmospheric temperature profile	Atmospheric temperature profile
Remarks	Weight = 31 kg, power = 41 W	Subsequent launches carry the same SSM/T	Weight = 29 kg, power = 30 W, subsequent launches of TIROS-N (renamed NOAA-1, -2, -3, etc.) carry same MSU

	1978	1979	1981
Year of launch			
Satellite name	Seasat and Nimbus-7	SEO-1 or Bhaskara-1 (Indian)	SEO-11, Bhaskara-II (Indian)
Instrument name and acronym	Scanning multichannel microwave radiometer (SMMR) [26]	Satellite microwave radiometer (SAMIR) [27]	Microwave radiometer satellite (SAMIR) [28]
Frequency (GHz)	6.60 10.69 18.00 21.00 37.00V 37.00H	19.1 19.6 22.235	19.35 22.235 31
Bandwidth (MHz)	220 220 220 220 220 220	250 250 250	250 250 250
Antenna type	80 cm diam. aperture parabolic reflector with a multifrequency feed horn	Horn Horn	Horn Horn
Beamwidth (degree)	4.0 2.5 1.5 1.2 0.7 0.7	14° 14° 23°	16.5° 26.5° 14°
Scanning	Conical scan, half-cone angle = 42° ±25° azimuthally (Nimbus-7), −3° to +47° (right of track) for Seasat	Spin scan (by spinning satellite)	Spin scan (by spinning satellite)
Polarization	Dual linear	Single linear polarization	Single linear polarization
Calibration	Spaceview horns and hot loads	Views cold space during each spin cycle	Views cold space during each spin cycle
Temperature sensitivity, ΔT, (K)	0.6 0.7 0.8 0.9 1.2 1.3 at 126 ms 62 ms 62 ms 62 ms 30 ms 30 ms	Better than 1 K (350 ms at 19 GHz, 470 ms at 22 GHz)	Better than 1 K (350 ms at 19 GHz, 470 ms at 22 GHz)
Equivalent ΔT at 1 s	0.21 0.17 0.19 0.22 0.21 0.23	Better than 0.55 K	Better than 0.55 K
Parameter measured	Sea surface temperature, sea ice concentration, sea surface wind speed		
Remarks	Weight = 40 kg, power = 55 W		

	1987	1991
Year of launch	1987	1991
Satellite name	DMSP Block 5D	DMSP Block 5D
Instrument name and acronym	Special Sensor Microwave/Imager (SSM/I), (34)	Special Sensor Microwave/Temperature-2 (SSM/T-2), (36)
Frequency (GHz)	19.35 22.235 37 85.5	91.65+/−1.25 150+/−1.25 183/−1,3,7
Band width, RF (MHz)	500 500 2000 3000	1.5 1.5 0.5, 1.0, 1.5
Beam width (deg)	1.9 1.6 1.1 0.4	6 3.6 3.3
Polarization	V,H V V,H V,H	Mixed (fixed feed, rotating reflector)
Antenna type	60 cm aperture diameter, offset parabolic reflector with one multifrequency feedhorn	7 cm dia. (rotating) offset parabolic reflector and fixed multi-band feed horn
Scanning method	Conical scan, half-cone angle 45 deg. Incidence angle: 53 deg. Reflector and feed rotates together at 30.6 RPM	Cross-track, +/− 40.5 deg. in 28 positions 3 deg. steps, once per 8 sec.
Polarization		Mixed, rotates with scanning
Calibration	Feed horn views an ambient target and cold space (through sub-reflector), once per revolution	Reflector views an ambient target and cold space, once per 8 second
Temperature sensitivity (K)	0.4 0.7 0.4 0.7	0.57 0.75 0.54 0.35 0.35
Integration time (ms)	8 8 8 4	150 150 150 150 150
Remarks	Weight = 43 kg, Power = 42 watts (Total power type radiometer) Subsequent launches carrying same SSM/Is	Weight = 14 kg, Power = 30 watts (Total power type radiometer)

d = down-track, c = cross-track.
*Vertical.
**Horizontal.
***Orthomode Transducer (OMT).

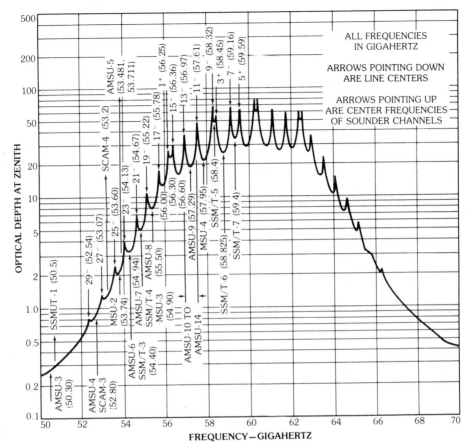

Fig. 7. Oxygen-band spectrum and various microwave sounder frequencies. (*Courtesy NASA*)

base numbers are the rotational angular momentum quantum numbers, and the superscript signs indicate the sign of the total angular momentum changes in a transition. The up-pointing arrows indicate the center frequencies of microwave temperature sounders.

The numbers following each acronym are the channel numbers of that sensor. Generally, the numerical designations of the channels are arranged in the order of ascending sensing height. This feature can also be seen from the opacity curve. The greater the opacity of a channel, the higher it senses above ground.

Note that most of the frequencies are situated at a "valley" between two lines. This is because of two conflicting requirements of a microwave radiometer to be used as a temperature sounder. As we shall see later, the radiometric measurement precision ΔT (generally referred to as the *temperature sensitivity*) is inversely proportional to the square root of the bandwidth. Hence the larger the bandwidth, the better the precision. However, a given point on the opacity curve is related to a particular height of the atmosphere. The higher the opacity, the higher is the height. In order to obtain good vertical resolution in temperature profile it is better for a channel to receive the energy from only a very narrow bandwidth (approaching a point on the spectrum) so that it senses only a very thin layer of the atmosphere (at a chosen height). A too-narrow bandwidth, however, can result in poor (large-value) temperature sensitivity. The valleys are locations where the opacity varies slowly with frequency; hence they allow the use of a maximum of bandwidth to improve the temperature sensitivity but pay little penalty in degrading the vertical resolution. There is also a single oxygen line, 1^-, at 118.75 GHz, which can also be used for temperature-sounding purposes, but this line has not been fully explored yet. Because of its shorter wavelength (by a factor of 2), as compared with 60 GHz, this 118.75-GHz line has the advantage of affording a smaller antenna size. This feature will be a substantial factor in consideration of temperature sounding from a geosynchronous orbit. (See Section 6 for a discussion of a proposed geosynchronous satellite for severe weather monitoring from a geosynchronous orbit.)

Other dominant atmospheric lines are the water vapor rotation lines at 22.2 and 183.3 GHz. Either of these lines can be used to sense the atmospheric humidity. The 22.2-GHz line is a weak line and can only yield the total (integrated) precipitable water. This line (or its vicinity) has been used by many remote sensing radiometers, e.g., Nimbus-7 and Seasat's scanning multichannel microwave radiometer (SMMR). The stronger 183.3-GHz water vapor line can be used to obtain humidity profiles. Because of the state of the art of millimeter-wave technology the use of this line is still in the experimental stage. But it is anticipated that a satellite humidity sounder, based on the 183.3-GHz water vapor line, will be developed soon. (See Section 6.) There are the so-called window regions of lesser absorption around 30, 90, and 150 GHz, with varying degrees of opacity. The 30-GHz window has been a popular one (e.g., Nimbus-6 ESMR and Nimbus-7 SMMR) because the microwave components are readily available at this frequency and reasonably high spatial resolution can be obtained. The 90-GHz window is starting to be used for its high-resolution capability as the microwave components at this frequency become more available. The region around 5 to 6 GHz is an important one, because it is the optimum frequency for sea surface temperature measurement, as mentioned in Section 2. The region between 1 and 2 GHz is sensitive to soil moisture because of the dispersion of water molecules.

Fundamentals of a Microwave Radiometer

A microwave radiometer is similar to a communication receiver except that its main signal is not a coherent carrier signal; its "signal" is the antenna (noise) temperature that a communication receiver is trying to minimize. A radiometer measures the magnitude of the noise power (brightness temperature) radiated by a target or scene.

Referring to the schematic diagram in Fig. 8, a microwave radiometer consists of an antenna for collecting the incoming radiation, whose intensity (in watts per square meter) is represented by the antenna noise temperature T_a (to be called simply "antenna temperature" for short) and a receiver for detecting and determining magnitude of the noise power. The receiver may consist of a preamplifier followed by a detector, or just a detector. It could also be a heterodyne system, in

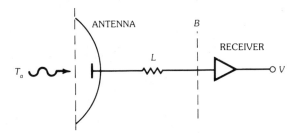

Fig. 8. Total power radiometer.

which the incoming noise power is mixed with a local oscillator and down-shifted to an intermediate frequency before it is detected by a detector.

The output of the receiver (in voltage or digital count) is linearly related to T_a. The ohmic loss of the transmission line between the antenna and receiver is represented by L. More details on radiometer fundamentals can be found in the references [29, 30, 31].

As in any instrument there are errors in a measurement. The errors in a radiometer measurement can be divided into two categories. One is the random, short-term (fast) fluctuations of the output, mainly associated with the noise of the front-end detector of the radiometer system (e.g., the mixer or first-stage amplifier). The other is the systematic errors (which may be slowly varying), resulting from calibration bias and component degradations. The former is commonly known as *temperature sensitivity* (or temperature resolution), and the latter is termed *calibration accuracy*.

Temperature Sensitivity—Temperature sensitivity, commonly represented by the symbol ΔT, is the precision of the radiometer. It is defined as the "minimum detectable change of antenna temperature at the collecting aperture." The "minimum detectable" change is taken to be the standard deviation of the radiometer output when its antenna is viewing a specified constant–brightness-temperature target.

Total Power and Modulating Radiometers—The radiometer, shown schematically in Fig. 8, is a "total power" radiometer. Its temperature sensitivity ΔT, referred to the radiometer input point B, is given by

$$\Delta T = K_1 T_{\text{sys}} \sqrt{\frac{1}{\Delta f t_I} + \left(\frac{\Delta G}{G}\right)^2} \qquad (25)$$

where

$$T_{\text{sys}} = T'_a + T_{rn} \qquad (26)$$

K_1 = radiometer modulation constant whose value depends on the type of radiometer system, with $K_1 = 1$ for the total power radiometer

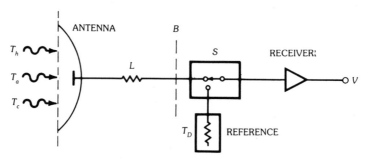

Fig. 9. Modulating radiometer.

T_{sys} = the total system noise temperature referred to the input of the receiver

T_{rn} = the receiver noise temperature referred to its input

T'_a = the effective antenna noise temperature at the input of receiver [see (28)*]

Δf = predetection bandwidth

t_I = integration time of an ideal integrator

G = radiometer gain

ΔG = gain fluctuation

In order to remove the contribution to ΔT due to gain fluctuations, a "modulating" method was introduced by R. M. Dicke and is now commonly known as the *Dicke radiometer*. Fig. 9 is a block diagram of a simplified Dicke radiometer. The receiver is switching between the antenna T'_a and a reference load kept at a known temperature T_D. Since in each half-period of a switching cycle the gain of the radiometer is measured by switching it to the known-temperature T_D load, any gain fluctuation slower than the switching rate is removed. The spectrum of gain fluctuation normally has a $1/f$ dependence on the frequency f, and as long as the switching rate is high enough, the ΔG noise can be removed or substantially reduced. The temperature sensitivity of a balanced Dicke radiometer is

$$\Delta T = \frac{2T_{sys}}{\sqrt{\Delta f t_I}} \qquad (27)$$

where the symbols are as given for (25, 26).

Most spaceborne microwave radiometers are the modulating type, and their modulating frequencies are in the 400- to 1000-Hz range. For example, the ESMR has a modulation frequency of 600 Hz; the SMMR and MSU have 1000 Hz. The

---

*Equation 28 shows that T'_a contains a term T_a, which is the antenna noise temperature received at the aperture. Strictly speaking, ΔT is defined by setting $T_a = 0$ so that it is independent of the scene temperature. This is frequently the case in radioastronomy. For remote-sensing radiometers, however, it is customary to set $T_a \cong 300$ K or some specified scene temperature value.

choice of a switching rate also depends on the dwell or integration time t_I per resolution cell. There must be at least several cycles per t_I. The switching rate must also be slow enough to avoid the transients of the switch itself.

The Dicke type microwave radiometer has an important advantage of being able to remove the receiver gain fluctuation and it has been the popular choice of most of the existing spaceborne radiometers. However, there are conditions under which the temperature sensitivity of a Dicke radiometer can be worse than a corresponding total-power type radiometer. This can be seen by comparing (27) with (25); there is an extra factor of 2 in (27). (In a Dicke radiometer the same available dwell time is split into two halves; only one half is used in viewing the scene, the other half is spent on the reference target. Hence a Dicke radiometer spends only half as long viewing the scene as compared with a total-power radiometer. Since the errors from both the scene and reference half-periods add in a root-square-sum sense, the end result is doubling the temperature sensitivity value in a Dicke radiometer.)

When the receiver gain is stable enough and/or frequent calibrations are possible to remove the gain fluctuation, then the total-power radiometer may yield a better temperature sensitivity than the Dicke type. Two examples of this condition are the special sensor microwave/imager (SSM/I) and the advanced microwave sounding unit (AMSU) (see Section 6). Both are total-power radiometers. The SSM/I is calibrated once every 2 s and the AMSU every 8 and 2.67 s.

Calibration—To remove systematic errors the complete radiometer system shown in Fig. 9 must be calibrated externally from time to time by introducing a known temperature at the antenna aperture. This is needed because during either of the two half-cycles of a modulating period the radiometer output always includes a component which is the receiver noise temperature T_{rn}, which is usually much larger than either T_a or T_D. Therefore, if one were to rely on the gain determination from the reference half-cycle alone, the precise knowledge of T_{rn} and its constancy are essential for accurate calibration. In practice it is sometimes more convenient to devise frequent onboard calibrations which will eliminate the need to know the exact T_{rn} value.

A commonly used calibration method is for the antenna to view two external targets at T_h and T_c (for hot and cold temperatures, respectively) so that the entire radiometer system, including the antenna, can be calibrated. A second, often less satisfying, method is to use two "matched" loads maintained at T_h and T_c, respectively, and connected between the receiver and the antenna as shown in Fig. 10. Switch S oscillates at a typical rate of about 1000 Hz. Switch S_1 is normally connected to the antenna. During calibration S_1 is first switched to T_h, then to T_c. The corresponding output voltages, V_h and V_c, determine the calibration equations. With the incorporation of the two-point calibrations one only needs to relate the amplitude of the demodulated voltage (which is proportional to the difference between T_a and T_D) and T_h and T_c; it is not necessary to know T_{rn} and the gain values explicitly. As long as the radiometer system transfer function is linear, any desired antenna temperature T_a can be interpolated from a calibration curve, as shown in Fig. 11.

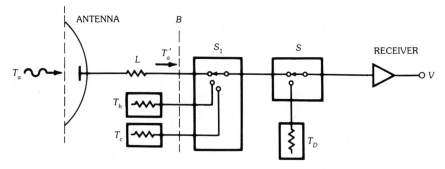

Fig. 10. Modulating radiometer with matched load calibrations.

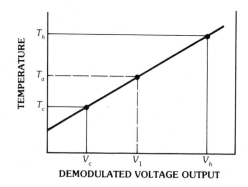

Fig. 11. Linear calibration of a radiometer with two targets: T_h and T_c.

In principle the two-point calibration procedure is needed only once in a pre-launch laboratory test. In practice, however, because of the possibility of slow drift (component degradation) it is a better idea to provide periodic in-orbit calibrations with external calibration targets or matched loads, especially for total power types.

The matched loads calibration method is less desirable because it does not include the antenna characteristics in the procedure, and it also introduces additional losses due to the switch S_1. In satellite operations, however, it is not always feasible to have onboard calibration targets due to their bulkiness. Hence, sometimes the more compact matched loads must be used. For example, in the cases of the Microwave Sounder Unit (MSU) and the scanning microwave spectrometer (SCAMS), onboard reference targets were used for through-the-antenna calibration. In these two cases the sizes of the antennas involved are relatively small, and the required calibration accuracy is high. However, because the antenna aperture sizes were too large (0.8 m in diameter) in the cases of scanning multichannel microwave radiometer (SMMR) and the Electrically Scanning Microwave Radiometer (ESMR), it was impractical to construct external targets, and matched loads were used as hot references. And space-viewing horns were used as cold references for both SMMR and ESMR.

At microwave frequencies the S and S_1 switches usually are ferrite circulators and their configurations (not necessarily those of Fig. 10) must be designed to minimize the ohmic as well as mismatch losses from the antenna to the input of the receiver. The values of these losses as well as their physical temperature must be carefully monitored to provide for accurate calibration.

When the ohmic loss L is taken into account, the theoretical temperature sensitivity ΔT, referring to the antenna aperture, is

$$\Delta T = K_1 \frac{T'_{\text{sys}}}{\sqrt{\Delta f t_I}} \tag{28a}$$

with

$$T'_{\text{sys}} = T_a + (L - 1) T_p + T_{rn} L \tag{28b}$$

where $L > 1$ is the loss factor [see (15)], and T_p is the physical temperature of the lossy element. The loss L represents the sum of total ohmic losses of the transmission line and the antenna lumped into one. It is seen that the ohmic loss L not only produces a noise temperature component $(L - 1) T_p$ due to self-emission, but also increases the effective receiver noise temperature. For radiometer systems with large values of T_{rn}, this latter factor, which causes an increase in T_{rn}, is the dominant term in degrading the system ΔT. As a consequence of the loss L, the effective antenna noise temperature is

$$T'_{\text{ant}} = \frac{T_a}{L} + \left(1 - \frac{1}{L}\right) T_p \tag{29}$$

Similarly, if there are ohmic losses associated with loads T_h and T_c, then their effective noise temperature at S_1 will be

$$T'_h = \frac{T_h}{L_h} + \left(1 - \frac{1}{L_h}\right) T_{ph} \tag{30}$$

$$T'_c = \frac{T_c}{L_c} + \left(1 - \frac{1}{L_c}\right) T_{pc} \tag{31}$$

where L_h and L_c are the losses in the T_h and T_c paths to S_1, respectively, and T_{ph} and T_{pc} are their respective physical temperatures. In practice it is rather difficult to keep track of all the losses and their temperatures. This is why the matched-load type calibration method cannot provide a high degree of accuracy.

For a multiport device, such as switch S_1, the general expression for effective noise temperature at the output of the ith port is

$$T_i = \sum_{j \neq i} \alpha_{ij} T_j + \left(1 - \sum_{j \neq i} \alpha_{ij}\right) T_p \tag{32}$$

where $\alpha_{ij} = 1/L_{ij}$ is the attenuation factor from j to i at a given switch state and T_p is the physical temperature of the switch.*

Special Requirements for Antennas for Remote-Sensing Microwave Radiometers

There are several special features of antennas that are very important to remote-sensing radiometry, although they may not be important in other fields, such as communications. These features are the following:

High beam efficiency
Low ohmic loss
Large scanning angle limits
Provisions for accurate calibration
High polarization purity

Beam Efficiency and Spatial Resolution—Power received by a radiometer antenna is the total sum of power from all directions. Since the power is proportional to temperature we can relate the antenna temperature $T_a(\Omega)$ to the brightness temperature $T_b(\Omega)$ as follows:

$$T_a(\Omega') = \frac{1}{4\pi} \int G(\Omega, \Omega') \, T_b(\Omega) \, d\Omega \qquad (33)$$

where $\Omega' = (\theta', \phi')$ is the direction the antenna main beam is pointing, $d\Omega = \sin\theta \, d\theta \, d\phi$, with the integration being over 4π steradians, and the antenna directive gain over isotropic media is such that

$$\oint G(\Omega) \, d\Omega = 4\pi \qquad (34)$$

The *main beam efficiency* (or simply *beam efficiency*) ε_{MB} is the ratio of the power in the main beam to the total power received by the antenna. And the antenna is assumed to be in an isotropic environment (i.e., the brightness temperature is not a function of angular direction). The extent of the main beam has been customarily defined as the null-to-null beamwidths (NNBWs). However, the "2.5 times half-power beamwidth" definition has also been used frequently in place of the null-to-null definition because in practical antennas, due to phase errors, there may not be distinct first nulls. In the following the 2.5 (HPBW) is defined as the main beam width for beam-efficiency computation purposes. The beam efficiency is

$$\varepsilon_{MB} = \frac{1}{4\pi} \int_m G(\Omega) \, d\Omega$$

or   (35)

---

*For a discussion on the effect due to mismatch reflections, see [31], pp. 404–412.

$$\varepsilon_{MB} = \frac{1}{4\pi} \int_{\phi=0}^{2\pi} \int_{\theta=0}^{\theta_{MB}} G(\theta, \phi) \sin \theta \, d\theta \, d\phi$$

where θ_{MB} is $1.25 \times$ half-power beamwidth.

The *stray efficiency* ε_{ST} is the fraction of power outside of the main beam [29]:

$$\varepsilon_{ST} = \frac{1}{4\pi} \int_{ST} G(\Omega) \, d\Omega \tag{36}$$

where the limit ST is $4\pi - MB$ (i.e., the angles outside the main beam), and

$$\varepsilon_{MB} + \varepsilon_{ST} = 1 \tag{37}$$

From (33), (34), and (35),

$$T_a(\Omega) = \bar{T}_{bMB}\varepsilon_{MB} + \bar{T}_{bSL}\varepsilon_{ST} \tag{38}$$

where

$$\bar{T}_{bMB} = \frac{\displaystyle\int_{MB} G(\Omega, \Omega') \, T(\Omega) \, d\Omega}{\displaystyle\int_{MB} G(\Omega, \Omega') \, d\Omega} \tag{39}$$

So $\bar{T}_{bMB}$ is the "average" brightness temperature within the main beam. Similarly, $\bar{T}_{bSL}$ is the average brightness temperature of the side lobes.

The desired quantity is $\bar{T}_{bMB}$ but the direct radiometer antenna measurement yields only T_a, which includes contributions from side lobes.

In designing the antennas for microwave radiometers, it is important to achieve high main-beam efficiency (frequently 90 to 97 percent is required). In an ideal case, $\varepsilon_{MB} = 1$ and $\varepsilon_{SL} = 0$; then, $T_a = \bar{T}_{bMB}$, the desired brightness temperature of the main-beam area, can be obtained directly from the radiometer measurement.

Usually this is not the case, and the term T_{bMB} must be solved from (38) in terms of T_{bSL} with the main beam and side lobe (stray) efficiencies, ε_{MB} and ε_{SL}, respectively, obtained from careful measurements of the antenna. The side lobe temperature term in (38) must be provided by some other means. For example, if the radiometer antenna scans the complete area cell by cell, then an interactive algorithm can be set up such that the side lobe term of one cell can be computed from the measurements of its immediate neighboring cells.

The beam efficiency as seen in (35) can be further defined for a single polarization. For example, if a vertically polarized beam is desired, then $G(\theta, \phi)$ in (35) can be changed into $G_v(\theta, \phi)$, to signify that only the directive gain of the vertically polarized wave in the main beam is to be counted.

The polarized beam efficiency is slightly lower than the nonpolarized beam efficiency, because there are always some mechanisms which tend to produce cross-polarized components of energy at the expense of the main polarization. For waveguide slotted array antennas the cross-polarized component could result from stray coupling and mechanical imperfections from the slot radiators. For reflector-type antennas the curvature of the reflector will always give rise to some cross polarization even if everything else is perfect. The curvature-related cross polarization decreases with the increase of the focal length to diameter ratio. Imperfections in feed horns and orthomode transducers may also contribute to the cross polarization.

The lowering of beam efficiency due to cross polarization is not a major source of concern in most reflector antenna designs for radiometry because the change caused by it is small. The more detrimental effect is the cross-polarization component of directive gain that will leak some of the orthogonally polarized emission from the earth's surface. This mixing of the wrong polarization can deteriorate the accuracy for some applications. For example, a radiometer antenna designed for the horizontally polarized brightness temperature of a calm sea surface, at an incident angle of 50°, is expected to see a brightness temperature of about 80 Kelvins. The brightness temperature of the vertical polarization component at this same angle will be about 150 K. Hence a 2-percent contribution from cross polarization could cause a 1.4-Kelvin error, which is appreciable if not accounted for. Of course, leakage in the orthomode transducers following the feed horn or the switches (if used) will result in the same effect.

Both accurate calibration and high beam efficiency are important features of a microwave radiometer to ensure accurate mapping of scene brightness variation. However, each affects the radiometer performance differently. The calibration accuracy affects the bias error of the brightness throughout an entire area containing many spatial resolution cells—all being affected equally. The effect due to low beam efficiency, on the other hand, is to degrade the scene brightness contrast. The effect of high beam efficiency is similar to a low integrated side lobe in a synthetic aperture radar (SAR), which prevents low-contrast targets from being "washed out." For the SAR, only the contrast is of importance in most cases; the absolute radar cross section is not always crucial. In radiometry, however, one needs both the relative contrast and the correct absolute value of the brightness temperatures of individual resolution elements.

Spatial Resolution—Spatial resolution is the "footprint" size, or the diameter of the antenna's main beam projected on the earth's surface. The term *instantaneous field of view* (IFOV) is also commonly used in satellite remote sensing to mean the spatial resolutions. If the antenna beam is a right circular cone of beamwidth θ_b, then the spatial resolution is the diameter of the intersection of the cone and the earth's surface. In general, the intersection is a pear-shaped figure, and its size can be specified by a major and minor diameter, ϱ_M and ϱ_m, respectively. For a scanning antenna the IFOV may vary with scan angle. And the minimum values of ϱ_M and ϱ_m are customarily taken as the spatial resolution values. When the antenna is pointed at nadir, the footprint at nadir is a circle with a diameter $\varrho = h\theta_b$, where h is the satellite orbital height above the earth's surface (see Fig. 12). The beamwidth θ_b is defined as the half-power beamwidth of the antenna main lobe. It is related to the Rayleigh criterion, which states that two point sources are

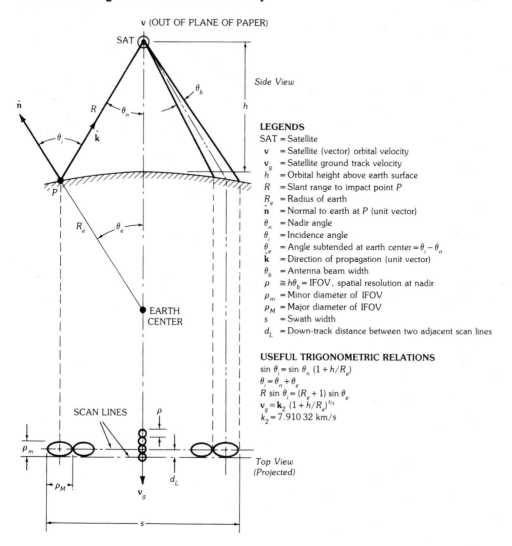

v (OUT OF PLANE OF PAPER)

SAT

Side View

n̂

R

h

θ_n

θ_b

θ_i

k̂

P

R_e

θ_e

EARTH
CENTER

SCAN LINES

ρ

ρ_m

ρ_M

d_L

v_g

Top View
(Projected)

s

LEGENDS

SAT $=$ Satellite
v $=$ Satellite (vector) orbital velocity
v_g $=$ Satellite ground track velocity
h $=$ Orbital height above earth surface
R $=$ Slant range to impact point P
R_e $=$ Radius of earth
n̂ $=$ Normal to earth at P (unit vector)
θ_n $=$ Nadir angle
θ_i $=$ Incidence angle
θ_e $=$ Angle subtended at earth center $= \theta_i - \theta_n$
k̂ $=$ Direction of propagation (unit vector)
θ_b $=$ Antenna beam width
ρ $\cong h\theta_b =$ IFOV, spatial resolution at nadir
ρ_m $=$ Minor diameter of IFOV
ρ_M $=$ Major diameter of IFOV
s $=$ Swath width
d_L $=$ Down-track distance between two adjacent scan lines

USEFUL TRIGONOMETRIC RELATIONS

$\sin \theta_i = \sin \theta_n \, (1 + h/R_e)$
$\theta_i = \theta_n + \theta_e$
$R \sin \theta_i = (R_e + 1) \sin \theta_e$
$v_g = k_2 \, (1 + h/R_e)^{2/3}$
$k_2 = 7.910\,32$ km/s

Fig. 12. Planar (cross-track) scan geometry.

resolvable if the angular separation between them is at least such that the first null of one source coincides with the maximum of the other. A uniform aperture distribution for a circular aperture has a Rayleigh resolving power of $1.22\lambda/D$ radians, where λ is the wavelength, and D the aperture diameter. The same uniform circular aperture would have a half-power beamwidth of $1.02\lambda/D$. Hence, for the circular aperture with uniform (nontapered) distribution the resolution as defined by the half-power beamwidth gives a slightly smaller value (i.e., better resolution) than the Rayleigh criterion–defined resolution. The ratio between the half-power beamwidth definition and the Rayleigh criterion is 0.84 for a circular aperture with no taper. Highly tapered antennas tend to increase this ratio as the beam broadens. Spatial resolution determines how small a scale the scene spatial variation can be resolved. But in order to faithfully reproduce scene brightness variation, a radiometer must have a high-beam-efficiency antenna. High resolution (narrow beamwidth) can be achieved with a large-aperture antenna. For a given antenna aperture size, high beam efficiency can be achieved by highly tapered aperture illumination (in addition to other design precautions, such as minimizing phase errors). High taper, however, is an inefficient way to use the aperture and leads to low aperture efficiency and consequently broadened beamwidth. In many microwave radiometer antennas, high beam efficiency is deemed to be of greater importance than narrow beamwidth, and it is often obtained at the expense of lowered aperture efficiency by using highly tapered aperture ilumination. The relationship between the two is shown in Fig. 13 ([29], pp. 219, 221).

Losses—The losses may be categorized as (1) ohmic or (2) scattering. Ohmic loss results from reflector surface resistivity, waveguide feed losses, filter losses, and so on. The scattering losses result from redistribution of energy from the main lobe into other regions of the side lobes and back lobes. The scattered energy may also occur because of undesired cross-polarized energy due to reflector curvature, feed horn cross polarization, reflector surface distortion, and the like.

The ohmic loss degrades the radiometer temperature sensitivity ΔT by increasing the effective system noise temperature as indicated in (28a). Ohmic loss also tends to deteriorate the calibration accuracy of a radiometer due to the self-emission term in (28d) because both the physical temperature of the loss element and the magnitude of the loss contain some uncertainties. Ohmic loss, however, does not affect the beam efficiency ε_{MB} [see (35)] as long as it can be considered a lumped element so that the loss does not depend on direction. Even though ohmic loss diminishes the antenna gain, it does not affect the spatial resolution either, as long as it is not direction (angle) dependent.

Generally the nonohmic losses involve redistribution of energy and may affect any or all of the three radiometer performance parameters: beamwidth, beam efficiency, and temperature sensitivity. Any scattering loss that reduces the energy received by the antenna also degrades the radiometer temperature sensitivity by the same factor. For example, any impedance mismatch causing reflection will lower the energy received, therefore increasing the ΔT value. The mismatch loss L' affects the temperature sensitivity as does the ohmic loss in (28a), except that there is no self-emission term. In other words, one can obtain ΔT from (28b) due to reflection by replacing L with L' and setting $T_p = 0$.

An example of the scattering loss is the antenna reflector surface roughness effect. The roughness produces a random scattering of energy into wide angles (as compared with the coherent main beam) and increases the side lobe envelope. The end result is to reduce the main beam efficiency. The reduction in beam efficiency can be calculated from Ruze's expression for gain reduction [32]:

$$G = G_0 e^{-(4\pi\varepsilon_{\mathrm{rms}}/\lambda)^2} \tag{40}$$

where

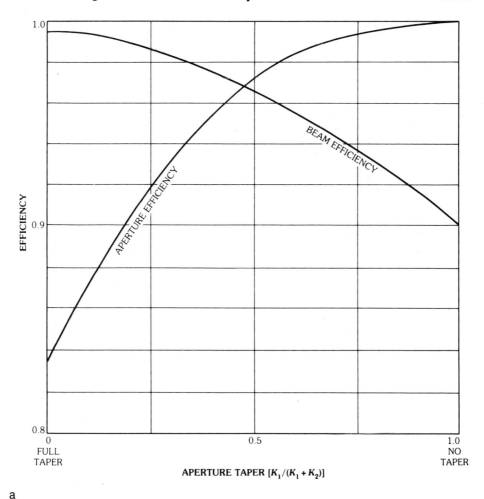

a

Fig. 13. Aperture taper and aperture and beam efficiency. (*a*) For a one-dimensional aperture as a function of taper. (*b*) For a circular aperture as a function of taper and phase error. (*After Kraus [29], from R. T. Nash, Beam efficiency limitations of large antennas, IEEE Trans. Mil. Electron., vol. MIL-8, pp. 252–257, July–October 1964; © 1964 IEEE*)

G = antenna gain of antenna with surface roughness

G_0 = antenna gain of a perfect antenna with no surface roughness

ε_{rms} = rms surface roughness

λ = wavelength

Beam Scanning—Most of the remote-sensing microwave radiometer antennas are required to perform scanning of some kind. The purpose of scanning is to produce a two-dimensional image of an area of the earth. Different types of scanning are discussed in the following paragraphs.

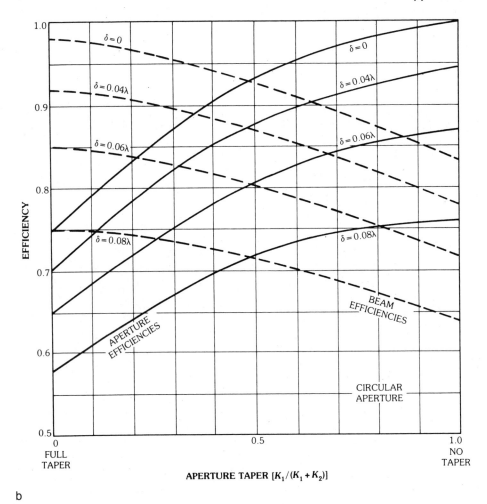

b

Fig 13, *continued.*

For a spinning satellite, such as the Geostationary Operational Environmental Satellite (GOES), which spins about an axis parallel to the earth's polar axis, the spinning action provides an east-west scan motion and a radiometer on board only has to provide a north-south stepping motion at the end of each scan line.

But the spin-scan is an inefficient scan method in the sense that most of the "available" time is not fully utilized for viewing the earth scene. For example, for each revolution the GOES satellite spins 360° but only a maximum of about 17° (which is the angle the earth subtends from the geosynchronous orbit) can be used for observation. Therefore the "spin-scan" efficiency (which is the ratio of observation time to available time per spin period) can be 4.7 percent, at best. If one only wishes to map a small portion of the earth disk, then the spin-scan efficiency e_{sp} is even less than 4.7 percent. The future GOES satellites are most likely to be of the three-axes stabilized type and the radiometers on board must be able to scan their antenna beam in both E–W and N–S directions.

Scanning Requirement for Polar Orbiting Satellites—Radiometers flying on polar orbiting satellites (typical orbit height 700 to 1000 km above the earth's surface) known as low earth orbiting (LEO) satellites only have to scan in one dimension. The orbital motion provides the scan action in the north-south or the down-track direction. Two types of scanning are commonly used in LEO satellites:

planar, or cross-track, scanning
conical scanning

The purpose of scanning is to create an image of an area by successive scan lines of a narrow beam. In principle the scan line (movement or trace of beam) can move in both directions (senses) alternately in a zigzag motion. In other words the beam can move from east to west first and then west to east in the second line, and so on. Or, the beam can be scanned only in one direction, say east to west, during which time the radiometer takes data. The beam then retraces back from west to east quickly for the beginning of a second scan line of data collection.

Since the LEO satellites continuously move in orbit, scanning in both directions (with a beam that moves only in the cross-track direction) will result in a zigzag footprint track on earth. This problem can be rectified by providing a beam motion in the down-track direction to compensate for the satellite orbital motion. While this compensation can be realized with relative ease for optical imagers (e.g., multispectral scanners [MSSs] on Landsats), it is much more complicated to provide the motion compensation in the case of a microwave antenna which usually has a much larger aperture. As a result microwave radiometers usually scan in the cross-track direction only.

Along a scan line (i.e., in the cross-track direction) the adjacent resolution cells can be spaced in a variety of ways. If the adjacent resolution cells are tangent to one another, it is called *contiguous* in the cross-track direction.

When the beam is scanned by mechanically slewing the antenna in a continuous motion the resolution cell also moves continuously, and it automatically results in a contiguous pattern in the cross-track direction. But if the beam is scanned with a stepping motion (i.e., it "dwells" at a resolution cell position for a length of time and then moves quickly to the next resolution cell), the cell spacing in the cross-track direction can be arbitrarily chosen to be either contiguous, overlapping, or undercoverage (leaving gaps between cells).

As we shall see later, the choice of resolution cell spacing is not completely arbitrary. Since the total available time per scan period is fixed, a smaller number of resolution cells per scan line (undersampling) will lead to more integration time per resolution cell, which yields better temperature sensitivity. But spatial under-sampling will lose some details of scene spatial variation (aliasing). On the other hand, oversampling (overlap between cells) will result in less integration time per cell and poor temperature sensitivity. The choice of spatial sampling frequency involves a trade-off between spatial resolution and temperature sensitivity of a mapping radiometer. Within a given time t, a scanning radiometer (assuming a single beam for the moment) must cover an area of $A = sv_g t$. (See Fig. 12.) As the sampling frequency increases, individual resolution cell integration time t_I decreases, resulting in poorer temperature sensitivity. The choice of sampling frequency depends on the antenna beamwidth and the degree of cell overlapping.

Let $p = d_L/\varrho$ be the down-track contiguity coefficients; when $p = 1$, there is (down-track) contiguity at nadir. When $p > 1$, there is a gap in the down-track direction, i.e., under coverage at nadir. When $p < 1$, there is some overlap in the down-track direction.

The scan time t_{scan} per line is

$$t_{\text{scan}} = d_L/v_g \tag{41}$$

Let $q < 1$ be the scan efficiency, which is the fraction of t_{scan} actually used for taking scene data, and let n be the number of resolution cells per line. Then

$$t_I = \frac{q t_{\text{scan}}}{n} = \frac{pq\varrho}{v_g n} \tag{42}$$

For a given total nadir angle scan limit of $2\theta_{nM}$, the number of resolution cells per line is, assuming cross-track continuity,

$$n = \frac{2\theta_{nM}}{\theta_b}, \qquad \varrho = h\theta_b \tag{43}$$

and

$$t_I = \frac{pqh\theta_b^2}{2v_g\theta_{nM}} \tag{44}$$

For earth remote-sensing applications it is usually required that the sensor completely map the earth in a short time period. The consequence is that the scan angle limit θ_{nM} must be as large as practical. For example, in a LEO polar satellite, such as TIROS-N and the NOAA-series weather satellites, it is desired that the onboard sensors map the earth's atmosphere once every 6 to 12 hours in order to update the state of the atmosphere for weather forecasts. This requires two simultaneous satellites (in two polar orbits whose orbital planes are 90° apart), each having to scan to a limit of $\theta_{nM} \cong \pm 50°$. For other earth resource applications, large swath width is also frequently needed in order to completely map the earth once in two to three days.

Equation 44 shows that the integration time t_I increases with the square of the beamwidth θ_b or, equivalently, the square of the spatial resolution ϱ. Since the temperature sensitivity ΔT is inversely proportional to the square root of the integration time t_I, it is therefore inversely proportional to θ_b. In other words, in a mapping radiometer the spatial resolution must be traded off against the temperature sensitivity (unless one can improve the radiometer system noise). Also, the large magnitude of the scan angle limit θ_{nM} makes "small-angle" scan techniques for a reflector antenna (such as mechanical feed displacement in the transverse plane about the focal point) impractical.

Planar scan has its advantages. It is easier to implement, as compared with conical scan. One can design an offset paraboloid reflector geometry in such a manner that the reflector is the only moving part while the feeds and the radiometer receivers are stationary (no rotary joints are required). This scheme has an important advantage because it drastically reduces the mechanical disturbance to the spacecraft, which can generally absorb very little mechanical disturbance. Fig. 14 is a schematic diagram of such a design. An offset paraboloid is driven by a motor to scan about an axis perpendicular to the axis of symmetry of the parent parabola. The planar scan can be easily designed to incorporate onboard calibrations, particularly when the cold space background radiation is used as a reference, simply by rotating the antenna (or reflector) about the same scan axis to view the cold space at T_c (approximately 2.7 K). One disadvantage of a planar scan is that its incidence angle and footprint size both vary with scan angle due to the geometry. Examples of planar scan instruments include MSUs, on TIROS-N, and scanning microwave spectrometers (SCAMSs) and ESMRs, both on Nimbus-5. Both the MSU and the SCAMS have antenna designs similar to that of Fig. 14, except hyperboloids instead of paraboloids were used for reflectors to make the configurations more compact in the SCAMS.

The Nimbus-5 ESMR, on the other hand, is a planar-slotted waveguide array as shown in Fig. 15. A traveling wave is fed into each waveguide "stick." As the wave travels along the guide, its energy is gradually radiated out through the slots. (It is described here as a transmitting antenna.) The beam is scanned by using ferrite phase shifters to change the phase of each waveguide stick with respect to one another.

The advantages of electronic scanning are obvious; it is motionless and agile. It hardly wastes any time for flyback to the starting angle after each scan line. However, there are drawbacks: (a) it has larger ohmic losses due to waveguides, terminating resistors, phase shifters, and power dividers as compared with a

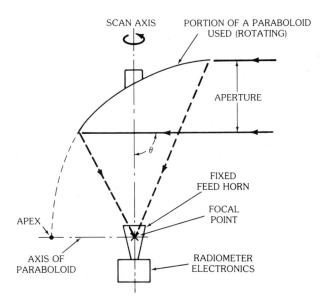

Fig. 14. Schematic of an offset reflector scanned by rotation.

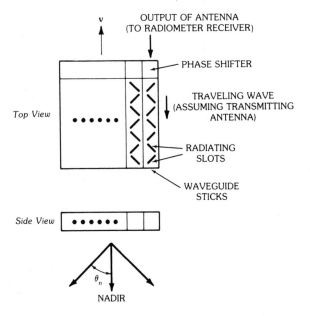

Fig. 15. Schematic of a planar scanning phased array.

reflector, (b) the beamwidth broadens as the beam is scanned away from broadside, because the effective aperture is then reduced by a factor of the cosine of the scan angle, and (c) it is difficult to share the same aperture in two or more frequencies. In many remote-sensing applications not only are multiple frequencies needed, but both linear polarizations at each frequency are needed as well. This would require many separate phased arrays which would result in heavy weight and large power and require a large earth-viewing area on the satellite. The latter is a premium quantity on a satellite. Reflectors with multiple frequency feeds tend to be the best compromise for many remote-sensing applications.

Conical Scan*—In conical scanning the beam moves on the surface of a cone. Normally, the cone axis is pointed at the earth's center. This results in constant incidence angle θ_i and constant footprint size, both features being advantageous in imaging and data interpretation. Fig. 16 depicts the conical scan geometry.

For a given orbital height h, the ground spatial resolution of a conical scan is not as good as planar scan because the slant range R in the conical system is larger than h. In the conical system, however, the slant range R is constant throughout a scan line. The cross-track contiguity definition is the same as in a planar scan, except that the direction of cross track is really along the scan line (which is not necessarily perpendicular to the ground track). If the scan time t_{scan} is chosen for the nadir resolution cells to be contiguous in the down-track direction between

*The term "conical scan" is used here in a different sense from the *IEEE Standard Definition of Terms for Antennas*, no. 145, IEEE, New York, p. 3, 1969.

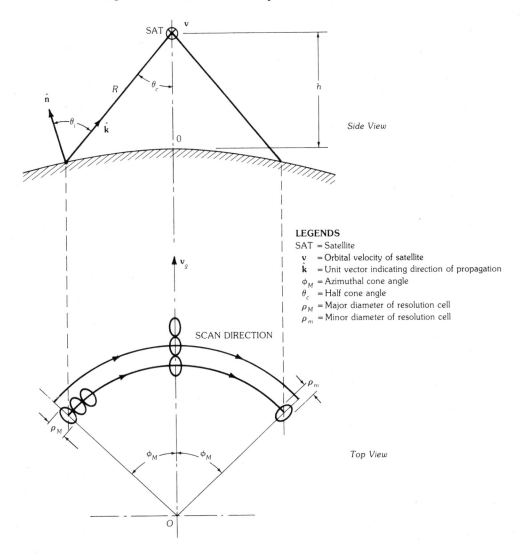

Fig. 16. Conical scanning geometry.

adjacent scan lines (i.e., $p = 1$), then, due to the curvature of the scan line, the swath edge resolution cells will have some overlap, even though the resolutions do not change. The swath width is determined by the scan limit ϕ_M in an azimuthal plane (a tangent plane to the subsatellite point O in Fig. 16 on the earth's surface).

The spatial resolution in a conical scan is represented by the "cross-track" and "down-track" diameters ϱ_m and ϱ_M, respectively.

For the conical scan system the expressions corresponding to the planar scan case are

$$p = d_L / \varrho_M \qquad (45)$$

$$t_{\text{scan}} = p\varrho_M/v_g \tag{46}$$

$$\varrho_M \cong R\theta_b/\cos\theta_i \tag{47}$$

$$n = \frac{2\phi_M}{\theta_b \sin\theta_c} \tag{48}$$

and

$$t_I \cong \frac{pqR\theta_b^2 \sin\theta_c}{2v_g\phi_M \cos\theta_i} \tag{49}$$

Equation 49 shows that the individual cell integration time for conical scanning is also proportional to θ_b^2, similar to the planar scan case.

As in the case of planar scan the conical scan can also be realized either by mechanically rotating a reflector or by an electronically scanning phased array. The scanning multichannel microwave radiometer (SMMR) is an example of conical scan by mechanical rotation of a reflector, and the Nimbus-6 ESMR is an example of conical scan by a phased array. The latter's geometry is similar to that shown in Fig. 15, except that the waveguide sticks radiate both vertical and horizontal polarizations simultaneously. Unlike Fig. 15, the planar array of the Nimbus-6 ESMR is mounted vertically on the Nimbus-6 satellite's sensory ring and its beam is pointed at θ_c from nadir. The position of the beam is selected by controlling the phasings of each vertical waveguide stick. Each beam is dual polarized. An orthomode transducer separates vertical from horizontal polarization at the output of the array.

The Nimbus-6 ESMR has a spatial resolution on earth of 42 km (down-track) by 20 km (cross-track) from its 1100-km orbit. The scan time per line is $t_{\text{scan}} = 5.33$ s, corresponding to $d_L = 33$ km, hence $p = d_L/\varrho_M = 33/42 = 0.79$, so that there is 21-percent cell overlap in the down-track direction at nadir. The Nimbus-6 ESMR takes data while scanning in only one direction and, because it is electronically scanning, the retrace time needed to swing the beam 70° azimuthally is a very small fraction (a few milliseconds) of the 5.33-s scan time.

The SMMR (on both Nimbus-7 and Seasat) is an example of mechanical implementation of conical scanning. There were many reasons and trade-offs leading to the choice of mechanical scanning of a reflector-type antenna for the SMMR. Chief among them was the fact that the SMMR required five frequencies, ranging from 6.6 GHz to 37 GHz, and each frequency had both linear polarizations. To satisfy this requirement with multiple phase arrays would be impractical for the large earth-viewing areas they would need; heavy weight and high power consumption would also result from multiple-phased array designs.

For a large mechanically scanning antenna such as the SMMR, the "retrace time" can be an appreciable fraction of the scan time and the zigzag shape of the footprint trace on earth becomes a problem. As can be seen in the following description of the SMMR, the overlapping zigzag traces were implemented only in its highest-frequency (37-GHz) channels. For the four lower frequencies, only one direction of the bidirectional scan was used for data taking, thus avoiding the zigzag pattern for these frequencies. Also, because a single reflector is shared by all five diversely different frequencies, the footprint sizes at each frequency are vastly different. In the SMMR design the scan time t_{scan} is chosen to make the higher-frequency (37-GHz) channel contiguous at nadir, hence there are varying degrees of overlapping (down-track) at the lower frequencies. The design of the SMMR is similar to that of Fig. 14, except that the angle θ between the scan axis and the symmetry axis of the parent parabola is no longer 90° but the half-cone angle $\theta_c = 42°$.

Fig. 17 is a sketch of the SMMR, and Fig. 18 is a photograph of the actual hardware.

SMMR design features are described in the following. A stationary five-frequency dual linear polarization feed horn projects the axially symmetric primary radiation patterns vertically upward at an offset paraboloid reflector (assume

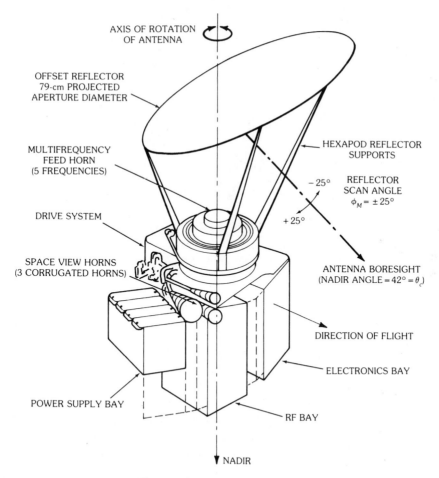

Fig. 17. SMMR configuration.

Fig. 18. Photograph of an SMMR. (*Courtesy NASA*)

transmitting). A beam is formed in the direction of the symmetrical axis of the parent parabola, whose focal point is the phase center of the feed horn. The reflector is rotated in a back-and-forth manner about the nadir axis by a drive motor which also drives a counterrotating mass to compensate the spacecraft for the angular momentum disturbance caused by the oscillating reflector. The azimuthal scan angle limit ϕ_M is $\pm 25°$.

As a result of this design (in which only the reflector moves) the total moving mass (consisting of the graphite epoxy reflector, with major diameter 108 cm by minor diameter 79 cm, ballast and thermal shield) is only 3.4 kg. The total weight of the instrument is 40 kg. The reflector first rotates in a clockwise direction for 2 s and then counterclockwise, also for 2 s. A total scan time period is $t_{\text{scan}} = 4$ s (actually 4.096 s). This 4-s scan period is chosen so that the 37-GHz footprints (ϱ_M

= 27 km, ϱ_m = 16 km) are contiguous in the down-track direction (at nadir). The subsatellite speed is v_g = 6.4 km/s, and d_L = 6.4 × 4.096 = 26.2 km, or $p = d_L/\varrho_M$ = 0.97 [see (17)]. Hence it is almost (down-track) contiguous at 37 GHz.

Step Scan Versus Continuous Scan—In a step scan the beam dwells at a given position for a length of time t_I, then moves to the next position and repeats the dwell of t_I seconds. For large antennas this step scan mode becomes impractical if an appreciable amount of time is required to accomplish the stepping motion; or it may consume too much power in order to move the antenna quickly to a new dwell position. An alternative to step scan is to slew the antenna continuously across the total scan angle limit, namely the "continuous scan." Both the SCAMS and the MSU are step scan types, because their antennas are small and their scan periods are relatively long. The SMMR, on the other hand, is a continuous scan type because its antenna is much larger than that of the MSU or the SCAMS, and the scan period at 4 s is much shorter. In a continuous scan the antenna moves (ideally) at a constant angular velocity, and the resolution cell size along the scan direction is determined by the length of integration time. (Sometimes even a constant angular speed is difficult to obtain mechanically, and some kind of velocity variation with time must be accepted. For example, the SMMR antenna actually has a sinusoidal velocity variation with time.) If, in a continuous scan, the integration time is infinitesimally short, then the IFOV along the scan direction is the antenna beamwidth θ_b. For a continuous scan with a finite integration time the effective field of view (EFOV) along the scan direction is larger than the IFOV. In other words the scan motion introduces some smearing effect along the cross-track direction. The EFOV depends on the antenna directive gain pattern $G_D(\phi)$ and the length of time a given point of the scene is viewed by the antenna. Normally the integration time is set so that the antenna moves one beamwidth during the integration time (assuming constant scan velocity).

Scanning Requirements for Geosynchronous Orbiting Microwave Radiometers— Most of the previous discussions on scanning pertain to LEO microwave radiometers. The same radiometer can certainly be used at the geosynchronous orbits. The advantage of this type of orbit is that the satellite appears stationary with respect to earth; it allows one to observe an area continually or repeatedly with high temporal frequency. This could be important for some applications, such as observing severe storms. From the radiometric viewpoint there are two types of geosynchronous earth orbiting (GEO) satellites: the spinning and the three-axis stabilized type. The present GOES satellites are of the spinning type. The other type of GEO satellite is three-axis stabilized, in which the orientation of the satellite with respect to the earth remains unchanged. In this case a sensor must scan in two orthogonal dimensions in order to obtain a map of a given area of the earth.

There is at present no microwave radiometer on a GEO satellite, because the required size of the antennas at a 36 000-km orbit is relatively large; this makes it difficult and expensive for most launch vehicles to carry. With the advent of the space shuttle, however, microwave radiometers for GEO satellites, such as GOES, will soon follow.

Because of the large orbital height, antennas used for GEO satellites will be much larger in order to achieve spatial resolution. Instead of being tens of centimeters in diameter, the antenna diameter will be in meters. There appears to be a very limited utility in having a microwave radiometer on a spinning GEO satellite because of its inherently poor spin-scan efficiency. In addition, large moving antennas may present difficult dynamic problems for the spacecraft attitude control system. For these reasons, microwave radiometers will likely be used on future three-axis stabilized GEO satellites but not on spinning GEO satellites.

Scanning requirements for a microwave radiometer, from a three-axis stabilized geostationary satellite, are quite different from those of polar LEO satellites. The angular scan limits from a GEO satellite are small, since the maximum extent of the full earth disk is only 17° but the pointing accurary must be high. The scan velocity can be much slower, although it depends on the size of a scene area to be covered and the temporal repeat frequency needed. If a 2500-km × 2500-km area near nadir has to be covered in 15 min, with a resolution cell of 42 km, then each cell has about 0.25 s of dwell time. The scan velocity is only 0.26°/s, which is much slower than the 4.3°/s MSU scan speed in low earth orbit.

Polarization—Most microwave radiometers require the reception of linearly polarized waves of either vertical or horizontal polarization, or both. The reason for this is that surface emissivity characteristics of the two modes are distinctly different from each other. Signals from vertical and horizontal channels can be used to delineate the surface from the atmospheric phenomena. For example, the absorption due to the presence of moisture in the atmosphere attenuates both vertical and horizontal polarizations equally, while wind-driven sea surface waves affect vertical and horizontal polarizations differently. Referring to Fig. 19, the vertical polarization is defined as the mode in which the electric field vector **E** lies entirely in the plane of incidence (formed by the propagation unit vector **k** and the normal unit vector **n̂**). In other words, the magnetic field **H** lies transverse to the incidence plane. Hence the term *transverse magnetic (TM) mode* is also used. The horizontal polarization is defined as the mode in which the electric field vector **E** is "horizontal" (i.e., transverse to the incidence plane, hence it is also called the TE mode).

To state it more precisely (see Fig. 19), let **ĥ** and **v̂** be the unit vectors representing the directions of horizontal and vertical polarization, respectively. Then

$$\hat{\mathbf{h}} = \frac{\hat{\mathbf{n}} \times \hat{\mathbf{k}}}{|\hat{\mathbf{n}} \times \hat{\mathbf{k}}|} \tag{50}$$

and

$$\hat{\mathbf{v}} = \frac{\hat{\mathbf{h}} \times \hat{\mathbf{k}}}{|\hat{\mathbf{h}} \times \hat{\mathbf{k}}|} \tag{51}$$

Note that while the electric field vector direction in horizontal polarization (i.e., **n̂**) is truly horizontal, the electric field of the vertical polarization may not be

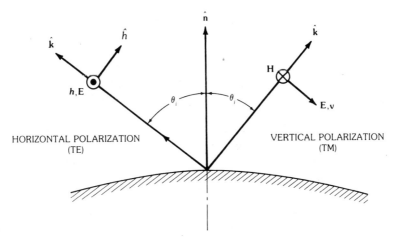

Fig. 19. Horizontal and vertical polarizations.

vertical. In fact, as the incidence angle θ_i approaches zero (approaching normal incidence) the electric field vector in vertical polarization approaches horizontal. At normal incidence there is no distinction between the two polarizations.

For applications where separate vertical and horizontal polarizations are required, the radiometer antenna must be designed for high polarization purity. That is, the amount of orthogonally (cross) polarized component leaking into the main polarization must be kept small. In other words the isolation between the two modes must be good. In general, isolation on the order of 25 dB or better is needed. This isolation is the total amount of energy leaked into vertical polarization from horizontal polarization or vice versa. It includes reflector curvature induced cross polarization (if a reflector type antenna is used), switch and/or orthomode transducer imperfections, and other leakages.

In scanning antennas such as the SMMR, where only the reflector is rotating and the feed is stationary, each of the two orthogonal feeds receives a linear combination of the vertical and horizontal polarizations, as defined in Fig. 19:

$$T_{bf_1} = T_{bv} \cos^2\phi + T_{bh} \sin^2\phi \qquad (52)$$

and

$$T_{bf_2} = T_{bv} \sin^2\phi + T_{bh} \cos^2\phi \qquad (53)$$

where ϕ is the azimuthal scan angle, T_{bv} and T_{bh} are the vertical and horizontal polarization brightness temperatures, and T_{bf_1} and T_{bf_2} are the output brightness temperatures of the two feeds, respectively. Temperatures T_{bv} and T_{bh} can be computed from the measured T_{bf_1} and T_{bf_2} and scan angle ϕ. This is inconvenient and can also introduce additional errors in retrieving in T_{bv} and T_{bh} individually.

The alternative is to scan the reflector and the feed as a unit, in which case T_{bv} and T_{bh} will be decoupled from each other. Since the front end of a radiometer is normally hardwired to the antenna by waveguides, in scanning with the whole antenna the masses of the radiometer front end must also be carried with the antenna. The penalty in scanning with the whole antenna is that the moving mass is increased and must be compensated for by the spacecraft attitude control system. An example of this type of scan is the SSM/I. (See Section 6.)

For atmospheric sounders most of the channels do not "see" the earth surface; therefore a pure linear polarization for these opaque channels is not essential. Because of this, antenna configurations such as in Fig. 14 (where only the reflector rotates) can be used. Orthomode transducers are often used as a convenient low-loss channel-diplexing technique. This is a particularly useful antenna design technique in the case of sounders where a single antenna must be shared by a large number of channels. For example, the SCAMS, SSM/T, MSU, and AMSU all have this type of design in which only the reflectors are scanning. The feed horns are stationary, with orthomode transducers at the throat of the horn as diplexers to separate signals for different channels. With this type of scan antenna design the output of each port of the orthomode transducer is, in effect, a linear combination of the pure vertical and horizontal polarizations. [See (52) and (53).] To put it in another way, the output of each port represents a rotating polarization. However, this rotation affects only those channels which see the earth's surface (or see substantial effects of the surface). As it turns out in the case of the MSU, the variation in brightness temperature (due to this) for its window channel (at 50.3 GHz) is very small throughout the entire scan angle limit. The smallness in variation is a result of the compensatory nature of the emissivity change and the change in cosine or sine functions. For example, the vertical polarization emissivity ϵ_v increases with the scan angle for angles between zero degrees and the Brewster angle. In the same range of scan angles the cosine-squared function decreases monotonically. Therefore the first term of (52), which is the product of the two, remains nearly constant. Similarly, for the second term, the horizontal emissivity ϵ_h decreases with the angle and is compensated by the increasing sine-squared function.

5. Spacecraft Constraints

Any satellite instrument is subject to the usual constraints of weight, volume (shape), and power limitations as well as thermal and dynamical interactions with the host spacecraft. These constraints vary, depending on the type of satellite and other sensors on board.

Dynamical Interactions

Any momenta produced by scanning motion of an antenna on board a satellite must be compensated. Otherwise, the antenna motion will cause a reaction by the host spacecraft, resulting in a change of its attitude. For small motion disturbances the excess momenta can usually be absorbed by the attitude control system (ACS) gyros of the host spacecraft. For larger antennas, however, the motion may be beyond the capacity of ACS gyros, and momentum compensation devices must be included in the antenna scan system. Both the SMMR and the SSM/I contain momentum compensation devices in the form of counterrotating masses driven by the same scan motor.

Thermal Considerations

The thermal environment affects satellite microwave radiometers in several ways:

1. Temperature gradients across a spacecraft antenna can cause shape distortion, which can lead to antenna performance degradation (reduced gain and beam efficiency); this is an important consideration for larger antennas and higher frequencies.
2. Ambient temperature fluctuation can affect the electronic gain stability of a radiometer (especially important for a total-power type radiometer); it can also introduce errors in the calibration by changing the self-emission part of an ohmic loss element.
3. For a radiometer mounted externally to a satellite, the instrument frequently has to be almost thermally isolated from the host spacecraft. Frequently the radiometer must dissipate its heat with its own radiators into the cold space to maintain a suitable temperature.

Weight, thermal, and mechanical considerations have led to the popular choice of graphite epoxy composite material for many spaceborne microwave radiometer antennas, especially for those with larger apertures (1 m in diameter or larger) and shorter wavelengths (1 cm or smaller). For example, both SMMR and SSM/I radiometers have graphite epoxy antennas.

6. Future Needs and Trends

As the field of remote sensing with microwave radiometry matures, and with the rapid advancement of the microwave technology itself, we are likely to see some new and more sophisticated spaceborne microwave radiometers in the near future, especially in the following two areas. We shall touch on each of them briefly.

1. Multifrequency, multibeam imaging microwave radiometers—After the success of the SSM/I (34), similar systems of this type are on the horizon. Two examples to be cited here are TMI and MIMR.
2. Millimeter-wave (100 GHz and higher in frequency) sounders—The sounders are microwave radiometers (also known as spectrometers for their many channels clustering about an atmospheric absorption line to probe the line shape) for measuring atmospheric temperature or humidity profiles. The frequencies of the temperature sounders tend to cluster around the 5-mm oxygen complex (see Figure 7) (or the 2.5-mm oxygen line); the humidity sounder uses the strongly opaque 1.65 mm water vapor line (see Figure 5).

The multifrequency, multibeam imaging microwave radiometers—The success of the SSM/I (first unit launched in 1988 marked a major milestone in the engineering of this type of spaceborne microwave radiometers. Several similar instruments, in various stages, are now under development. They include the

TRMM Microwave Imager (TMI), the Multifrequency Imaging Microwave Radiometer (MIMR), and the Advanced Multifrequency Scanning Radiometer (AMSR).

Salient features for the above type design are:

- Conical scanning—Conical scanning is typically implemented with a single rotating reflector shared by all the frequencies and polarizations. The cone axis is along the nadir direction. This feature preserves the (often quasi-elliptical) footprint shape and size, rendering the data interpretation task easier.
- Reflector and feeds scan together—The reflector and its feed-horns are rotating together as one single unit. This feature preserves the pure vertical and horizontal polarizations. This feature is to be compared to other type of designs (such as the SMMR), where only the reflector is rotating and the feed-horns are stationary. In that case, the polarization vector "rotates" with the beam position, and the output of the so-called "vertical" or "horizontal" polarization channels each contain a mixture of both components, their proportions varying with the beam position.

    The penalty of the "pure" linear polarization feature is that the feed-horns (and therefore the R.F. front-end of the microwave radiometer as well) must be on the spinning section of the satellite, thus drastically increasing the "moving mass."

    With a heavy rotating mass, the mechanical scanner is practically limited to a constant speed continuous (360 degrees) rotation, requiring a counter-rotating momentum wheel to cancel the angular momentum generated by the scanner. Since radiometers on a moving satellite need only to view the earth scene in a sector usually of 180 degrees or less, the constant speed continuously rotating scanner does not fully utilize the available earth viewing time.

- Total power microwave radiometer and external calibration—To make up the shortage in the available earth viewing time as indicated in (b) above, the "total power" type microwave radiometer is chosen to increase the signal to noise ratio. To reduce the gain-fluctuation induced noise, frequent and accurate on-orbit calibration is introduced. Typical the radiometer is calibrated once per antenna revolution, whose period is about 2 seconds. In the calibration mode, the feed-horns is "switched" from the antenna (reflector) to two known calibration temperatures. One of them is ambient temperature "target," also known as the "hot" or "warm" target. The other is a subreflector redirecting the feed-horns to view the "cold sky" (cosmic background microwave radiation at 3 K).

This type of "external calibration" indeed excludes the element of the reflector itself, and is therefore not a complete calibration. However, the reflector is usually a very low (ohmic type) loss and fairly stable in nature and its effects can be accounted for by pre-launch calibrations. This type of the external calibration has been proven highly stable in the experience of the SSM/I.

The TMI—The TMI is a microwave radiometer being built in the U.S. for NASA's Tropical Rainfall Measurement Mission (TRMM), a joint U.S.–Japanese experiment to study rainfalls in the tropics and hydrologic cycles and other global changes of the earth. This mission is scheduled for launch in the 1977–98 time frame.

TMI will have five frequencies, at 10.65, 19.35, 21.30, 37.00, and 85.50 GHz, and nine channels. Each frequency contains both vertical and horizontal polarizations, with the exception of the 21.30 GHz, which has only a horizontal channel.

It is seen that the TMI frequencies are very similar to that of the SSM/I. In fact, its designs are nearly identical to the SSM/I, having the same type of reflector antenna, scanning, and calibration methods.

The only new frequency for TMI is the 10.65 GHz, which is needed for probing the heavy precipitations in the tropics. Also, the water vapor channel's changed from the 22.235 GHz vertical polarization of the SSM/I to 21.30 GHz horizontal polarization. Again this change is designed for better total precipitable water vapor measurement in the tropics.

Although the antenna aperture diameter of the TMI is the same as the SSM/I (60 cm), but because the lower orbital height of the TRMM satellite (350 km), TMI has much better surface spatial resolutions than SSM/I. Table 3 shows TMI's IFOVs.

Table 3. TMI Characteristics

Freq (GHz)	10.65	19.35	21.30	37.00	85.50
Polarization	V, H	V, H	H	V, H	V, H
HPBW (deg)	3.95	1.90	1.60	1.00	0.45
IFOV (km × km)	63 × 38	30 × 18	23 × 16	16 × 10	7 × 5
Bd Wh (MHz)	100	500	200	2000	2000
NEdt (K)	0.5	0.5	0.8	0.7	1.0

Notes: V = Vertical, H = Horizontal, HPBW = Half-power Beam Width, IFOV = Instantaneous Field of View = Surface space resolution, Bd Wh = Band Width, NEdt = Noise equivalent delta Temperature = Temperature sensitivity.

The MIMR—The MIMR is under study by the European Space Agency (ESA) as a potential payload for one of NASA's Earth Observation System satellites, scheduled for mission around 2000.

MIMR is perhaps the most advanced second-generation multifrequency multibeam satellite-borne microwave radiometer. It will have six frequencies and a total of 20 independent channels. Table 4 shows some essential characteristics of MIMR.

Table 4. Characteristics of MIMR

Freq (GHz)	6.8	10.65	18.70	23.80	35.60	89.90
Polarization	V, H	V, H	V, H	V, H	V, H	V, H
No of feeds	1	1	1	1	2	4
HPBW (deg)	2.16	1.38	0.80	0.73	0.42	0.17
IFOV (km × km)	60 × 39	39 × 25	22 × 14	20 × 13	12 × 8	5 × 3
Bd Wh (MHz)	200	100	200	400	1000	5400
NEdt (K)	0.2	0.4	0.5	0.5	0.5	0.7

Notes: All the abbreviations are the same as in Table 3. The IFOVs are from an antenna with 143 cm aperture diameter and from a 720 km circular orbit. (Incidence angle = 50 degree)

Table 5. AMSU Channel Characteristics

Channel Number	Center Frequency	Number of Passbands	Bandwidth (MHz)	Temperature Sensitivity (K) $NE\,\Delta T$	Beam Diameter $\theta_B(°)$	Polarization Polar Angle $\theta_p(°)$
1	23 800 MHz	1	150	0.30	3.3	$90° - \theta$**
2	31 400 MHz	1	160	0.30	3.3	$90° - \theta$
3	50 300 MHz	1	160	0.40	3.3	$90° - \theta$
4	52 800 MHz	1	380	0.25	3.3	$90° - \theta$
5	53 596 MHz ± 115 MHz	2	170	0.25	3.3	*
6	54 400 MHz	1	380	0.25	3.3	*
7	54 940 MHz	1	380	0.25	3.3	*
8	55 500 MHz	1	310	0.25	3.3	*
9	57 290.344 MHz ≡ LO	1	310	0.25	3.3	*
10	LO ± 217 MHz	2	78	0.40	3.3	*
11	LO ± 322.2 ± 48 MHz	4	36	0.40	3.3	*
12	LO ± 322.2 ± 22 MHz	4	16	0.60	3.3	*
13	LO ± 322.2 ± 10 MHz	4	8	0.80	3.3	*
14	LO ± 322.2 ± 4.5 MHz	4	3	1.20	3.3	*
15	89.0 GHz	1	2000	0.50	3.3	$90° - \theta$
16	89.0 GHz	1	2000	0.60	1.1	$90° - \theta$
17	150 GHz	1	2000	0.60	1.1	$90° - \theta$
18	183.31 ± 1.00 GHz	2	500	0.80	1.1	*
19	183.31 ± 3.00 GHz	2	1000	0.80	1.1	*
20	183.31 ± 7.00 GHz	2	2000	0.80	1.1	*

Notes: θ_p = the angle between (1) the electric-field vector of the incoming radiation and (2) the line which is the intersection of a plane perpendicular to the propagation vector direction and an earth-tangent plane at the resolution cell center
*Unspecified, single polarization
**Scan angle = look angle from nadir
** θ = Scan Angle = ...

Compared to TMI, the MIMR only adds one more lower frequency, at 6.8 GHz. However, it populates the higher two frequencies, 37 and 89 GHz, with two and four feeds each, respectively. This feature allows for more integration time, thus improving the overall radiometric temperature sensitivity.

Millimeter-wave Sounders—The next generation atmospheric sounder following the current MSU (and the SSM/T) will be the Advanced Microwave Sounding Unit (AMSU), now being developed for the National Oceanic and Atmospheric Administration's (NOAA) operational polar orbiting satellites called NOAA-series satellites. The ASMU will replace the MSU starting NOAA-K satellite, estimated to launch in the mid 90s.

The AMSU, together with its IR counter part (HIRS-2 or its equivalent), form a combined microwave/IR sounding system for the U.S. civilian weather satellite systems.

The AMSU is a 20 channel radiometer system (35). It is further divided into two subsystems: Channel 1 through 15 is called AMSU-A and is primarily for temperature sounding; channels 16 through 20, called AMSU-B, are primarily for humidity profiling. Table 5 shows the characteristics of the AMSU.

The functions of the channels are as follows:

Channel 1 is set at the high side of the 22.235 GHz water vapor line. This channel is used to measure the total precipitable water vapor. Channels 2, 15, and 16 are the relatively transparent "window" channels. They are used to measure precipitation and surface emissivity and to account for their effects on temperature retrieval.

Channels 4 through 14 use the 5-mm oxygen complex for temperature sounding of the atmosphere. The lower seven channels are for the troposphere; the remaining channels are for the stratosphere. Channel 3 is a quasi-surface channel. It provides a direct surface emissivity measurement at a frequency close to the oxygen band. The weighting-functions' peak heights of the 4 through 14 channels are nearly uniformly distributed from surface to about 40 km. Channels 5 through 9 are the "valley" frequencies, i.e., they are at the valley between two oxygen line peaks. The valley channels have the favorable characteristics that permit the use of wider bandwidth and still have narrower weighting-function widths (or sharper vertical resolution) than the nonvalley type. Channels 10 through 14 are for the troposphere temperature. Because narrower bandwidths are required from a given part of a line at these channels in order to avoid too broad weighting functions, energies from two or four similar portions of a line (or two lines) are combined to form a channel. This increases the signal-to-noise ratio by increasing the total bandwidth but not broadening the weighting function. Channels 10 through 14 exploit the symmetry between lines 13^- and 11^-. Channel 10 combines two portions in the valley between the two lines. Each of the channels 11 through 14 combines energies from four passbands, two each from both sides of lines 13^- and 11^- (see Fig. 7). Channels 17 through 20 use the strongly opaque water vapor absorption line at 183.3 GHz for obtaining the humidity profile. Channels 18, 19, and 20 also combine two portions of energy from both sides of the line to enhance the signal-to-noise ratio. Channels 18, 19, and 20 are 1, 3, and 7 GHz, respectively, from the 118.3-GHz line center. Because of its increasing distance from the line center, each succeeding channel has decreasing opacity to the atmospheric water vapor and consequently each is sensing primarily a layer of the water vapor closer to the earth surface. Channel 17 contains a single passband and is located far away from the 183-GHz line. It senses water vapor down to the earth surface. Both channels 15 and 16 serve essentially the same function except that the latter has a

3:1 surface spatial resolution advantage, and hence will be better suited to delineate fine scene features, such as a weather front.

The AMSU is a total-power microwave radiometer system. The high spatial resolution required by the AMSU results in relatively short integration times for each IFOV; they are about 160 msec for channels 1 through 15 (called AMSU-A), and about 18 msec for channels 16 through 20 (AMSU-B).

The AMSU-A antenna is similar to that shown schematically in Figure 14. Because of the large spread in the frequency range in AMSU-A 23 GHz to 89 GHz), a total of the three separate reflector type antennas are used: one reflector with a 28 cm diameter aperture, and two having 15 cm diameter apertures.

Channels 1 and 2 share the 28 cm aperture. Channels 3, 4, 5, and 8 share a 15 cm aperture, while Channels 6, 7, and 9 through 15 share the second 15 cm aperture.

All of the three reflectors are scanning in synchronism at a rate of one revolution per 8 second. The antennas stop and dwell at each of the 30 beam positions (earth scene stations) for the 160 msec integration time. They also view the "cold space" (cosmic background radiation) and an orboard ambient target for "cold" and "warm" calibrations, respectively when they are outside the 30 scene stations.

The AMSU-B has only one single reflector antenna of 22 cm in diameter, shared by three group of frequencies at 89.00, 150 and 183 GHz. A quasi-optical multiplexing technique is used to separate the three frequencies to their respective feed-horns (37). The AMSU-B antenna scans three revolutions per eight second, and is synchronized with the AMSU-A. The AMSU-A is estimated to weigh 38 kg and takes 125 watts. The AMS-B is estimated to weighs 60 kg and needs 90 watts of power.

In the current design, Channels 9 through 14 of AMSU-A will sense the atmospheric temperature up to about 40 km. It is also possible to sense the temperature up to even higher regimes of the atmosphere. Higher sensing height can be reached by chosing radiometer frequencies closer and closer to the peak of one of the absorption lines shown in Figure 7. With the low-noise mixer/amplifiers gradually becoming available at the 5 mm range, the next generation microwave temperature sounder can reach up to at least 70 km above the earth surface.

However, at the low atmospheric pressure regime of the mesosphere, the Zeeman splitting of the oxygen lines (due to terrestrial magnetic field effect) becomes important and it must be properly accounted for in the retrieval models (38).

7. References

[1] W. L. Weeks, *Electromagnetic Theory for Engineering Applications*, New York: John Wiley & Sons, 1964, p. 235.

[2] E. G. Njoku and J. A. Kong, "Theory for passive remote sensing of near-surface soil moisture," *J. Geophys. Res.*, vol. 82, no. 20, pp. 3108–3118, 1977.

[3] R. A. Porter and T. J. Wentz, "Microwave radiometric study of sea surface characteristics," *NOAA NESS Report*, no. NOAA 71082701-1, July 1971.

[4] T. T. Wilheit, "The electrically scanning microwave radiometer experiment," *The*

Nimbus-5 User's Guide, NASA/Goddard Space Flight Center, Greenbelt, Maryland, pp. 59–105, 1972.

[5] E. G. Njoku, "Passive microwave remote sensing of the earth from space—a review," *Proc. IEEE*, vol. 70, no. 7, pp. 728–750, July 1982.

[6] E. K. Smith, "Centimeter and millimeter wave attenuation and brightness temperature due to atmospheric oxygen water vapor," *Radio Sci.*, vol. 17, pp. 1455–1464, 1982.

[7] International Radio Consultative Committee (CCIR), "Attenuation by atmospheric gases," *Rep. 719*, Proc. CCIR XV Plenary Assembly, Geneva, 1982.

[8] J. W. Waters et al., "Remote sensing of atmospheric temperature profiles with the Nimbus-5 microwave spectrometer," *J. Atmos. Sci.*, vol. 32, pp. 1935–1969, October 1975.

[9] A. D. Ali et al., "Atmospheric sounding near 118 GHz," *J. Appl. Meteorol.*, vol. 19, pp. 1234–1238, October 1980.

[10] D. H. Staelin et al., "Remote sensing of atmospheric water vapor with Nimbus-5 microwave spectrometer," *J. Appl. Meteorol.*, vol. 15, pp. 1204–1214, November 1976.

[11] T. T. Wilheit et al., "Monitoring of severe storms," *High-Resolution Microwave Satellites*, ed. by D. H. Staelin and P. W. Rosenkranz, Cambridge: MIT Press, 1978, p. 57.

[12] J. W. Waters and S. C. Wofsy, "Applications of high-resolution passive microwave satellite systems to the stratosphere, mesosphere, and lower thermosphere," *High-Resolution Microwave Satellites*, ed. by D. H. Staelin and P. W. Rosenkranz, Cambridge: MIT Press, 1978, pp. 724–734.

[13] R. L. Berstein et al., "Seasat scanning multichannel microwave radiometry—results of the Gulf of Alaska workshop," *Science*, vol. 204, pp. 1415–1417, June 1979.

[14] J. P. Hollinger, "Passive microwave measurements of sea surface roughness," *IEEE Trans. Geosci. Electron.*, vol. GE-9, p. 165, July 1977.

[15] H. J. Zwally and P. Gloersen, "Passive microwave images of the polar regions and research applications," *Polar Res.*, vol. 18, pp. 116, 431–450, 1977.

[16] T. J. Schmugge, "Microwave approaches in hydrology," *Photogrammetric Engineering and Remote Sensing*, vol. 46, no. 40, pp. 495–507, April 1980.

[17] J. C. Shiue et al., "Remote sensing of snowpack with microwave radiometers for hydrologic applications," *Proc. Twelfth Intl. Symp. Remote Sensing of the Earth Environment*, Environment Research Institute of Michigan, Ann Arbor, pp. 877–886, 1978.

[18] A. Rango, A. T. C. Chang, and J. L. Foster, "The utilization of spaceborne microwave radiometers for monitoring snowpack properties," *Nordic Hydrology*, vol. 10, pp. 25–40, 1979.

[19] F. T. Barath et al., "Mariner-2 microwave radiometer experiment and results," *Astronomical J.*, vol. 69, no. 1, pp. 49–58, 1964.

[20] A. E. Basharinov et al., "Some results of microwave sounding of the atmosphere and ocean from the satellite Cosmos 243," *Space Res. XI*, Berlin: Akademia-Verlag, 1971, pp. 593–600.

[21] A. E. Basharinov et al., "Satellite measurements of microwave and infrared radio brightness temperature of the earth's cover and clouds," *Proc. Eighth Intl. Symp. Remote Sensing of the Earth Environment*, Environment Research Institute of Michigan, Ann Arbor, pp. 291–296, 1971.

[22] J. W. Waters et al., "Remote sensing of atmospheric temperature profiles with the Nimbus-5 microwave spectrometer," *J. Atmos. Sci.*, vol. 32, pp. 1953–1969, October 1975.

[23] D. H. Staelin et al., "Microwave spectroscopic imagery of the earth," *Science*, vol. 197, pp. 991–993, 1977.

[24] H. E. Lauapre and K. A. Paradis, "A multichannel passive microwave atmospheric temperature sounding system," *Proc. Eleventh Intl. Symp. Remote Sensing of the*

Earth Environment, Environment Research Institute of Michigan, Ann Arbor, pp. 1212–1228, 1977.

[25] P. N. Swanson et al., "The TIROS-N microwave sounder unit," *Proc. IEEE MTT-S Intl. Microwave Symp.*, IEEE, New York, pp. 123–125, 1980.

[26] P. Gloersen and F. T. Barath, "A multichannel microwave radiometer for Nimbus-C and Seasat-A," *IEEE J. of Ocean Eng.*, vol. OE-2, no. 2, pp. 172–178, 1977.

[27] B. S. Gohil et al., "Remote sensing of atmospheric water content from Bhaskara, Samir data," *Int. J. Remote Sensing*, vol. 3, no. 3, pp. 235–241, 1982.

[28] P. D. Phavsar, G. T. Joseph, and O. P. N. Calls, "Developments of remote-sensing sensors at ISRO," *Proc. and Asian Conf. on Remote Sensing* (ACRS), Beijing, China, pp. 2-2-1 to 2-2-22, 1981.

[29] J. D. Kraus, *Radio Astronomy*, New York: McGraw-Hill Book Co., 1966.

[30] G. Evans and C. W. McLeish, *RF Radiometer Handbook*, Dedham: Artech House, 1977.

[31] N. Skou, "Microwave Radiometer Systems: Design and Analysis," Artech House, Inc., Publisher, 685 Canton St., Norwood, MA 02062, 1989.

[32] F. T. Ulaby, R. T. Moore, and A. K. Fung, "Microwave Remote Sensing, Active and Passive," vol. 1, Artech House, Inc. Publisher, 685 Canton St., Norwood MA 02062, 1981.

[33] J. Ruze, "Antenna tolerance theory," Proc IEEE, v. 54, pp. 533–540, 1966.

[34] J. P. Hollinger, J. L. Peirce, and G. A. Poe, "SSM/I Instrument evaluation," IEEE Geoscience and Remote Sensing, v. 28, no. 5, pp. 781–790, 1990.

[35] J. C. Shiue, "The next generation microwave sounder for weather satellites," Proc. National Telesystems Conference, IEEE Cat. No. 82CH1824-2, pp. C4.43.1–C4.4.7, 1982.

[36] J. D. Pickle, R. G. Isaacs, M. K. Griffin, V. J. Falcone, J. F. Morrissey, R. Kakar and J. Wang, "Comparison of collocated SSM/T-2 and MIR measurements: Results from the calibration study," SPIE Aerospace Science and Sensing '93, Microwave Instrumentation Conference, Orlando Florida, April 12–16, 1993. (Proceedings of the Conference).

[37] R. J. Martin and W. J. Hall, "Three dimensional design of quasi-optical systems at British Aerospace Space Systems," SPIE Aerospace Science and Sensing '93, Microwave Instrumentation Conference, Orlando, Florida, U.S.A., April 12–16, 1993 (Proceeding of the Conference).

[38] P. W. Rosenkranz and D. H. Staelin, "Polarized thermal microwave emission from oxygen in the mesosphere," Radio Science, v. 23, no. 5, pp. 721–729, 1988.

Chapter 23

Antennas for Geophysical Applications

D. A. Hill
National Bureau of Standards

CONTENTS

David A. Hill was born in Cleveland, Ohio, on April 21, 1942. He received the BS and MS degrees from Ohio University, Athens, in 1964 and 1966, and the PhD degree in electrical engineering from Ohio State University, Columbus, in 1970.

Since 1970 he has been a member of the scientific community in Boulder, Colorado. From 1970 to 1971 he was a visiting fellow with the Cooperative Institute for Research in Environmental Sciences, where he worked on pulse propagation. From 1971 to 1982 he was with the Institute for Telecommunication Sciences, where he worked on theoretical problems in antennas and propagation. Since 1982 he has been an electronics engineer in the Electromagnetic Fields Division of the National Bureau of Standards, where he has been working on electromagnetic compatibility and interference problems. He is also a professor adjoint in the Department of Electrical and Computer Engineering of the University of Colorado.

Dr. Hill is a member of URSI Commissions B, E, and F and a Fellow of the IEEE. He has served as a technical editor for the *IEEE Transactions on Geoscience and Remote Sensing* and is now an associate editor for the *IEEE Transactions on Antennas and Propagation*.

1. Introduction

The use of electrical methods in geophysics has expanded greatly in the past two decades, and during the same time period an interest in subsurface communication has developed. The most comprehensive book on electrical methods in geophysics was written in 1966 by Keller and Frischknecht [1], and it is still an excellent reference today. The review article by Murphy and Parkinson [2] is an excellent reference on underground communication. Both geophysical prospecting and underground communication require transmission of signals into the earth and, as a result, the same antenna types are used for both applications.

Because the methods and antennas used in geophysical probing are so varied, it is not possible to attempt a comprehensive discussion in one chapter. However, if we limit the applications to deep, subsurface probing and to through-the-earth communication, then the antennas used are primarily of two types: straight wire antennas which are grounded at the end points and wire loop antennas. Sections 2 and 3 discuss grounded wire antennas for direct current and time-varying excitations, respectively. Section 4 discusses loop antennas. In the analysis and discussion of these antennas some applications in geophysics and underground communication are described for illustrative purposes, but many other applications cannot be mentioned for lack of space. The references should be consulted for a more complete description of the applications. The primary purpose of this chapter is to describe how these antennas perform in the presence of a conducting earth.

In order to penetrate the earth to depths on the order of a hundred meters or more, it is necessary to employ extremely low frequencies (elf) below about 3 kHz. At such frequencies the free-space wavelength is greater than 100 km, and the antennas are electrically small even though they could be physically large (dimensions on the order of a kilometer in some cases). Consequently, the analyses in Sections 2 through 4 utilize the quasi-static assumption that neglects displacement currents in the air. However, no assumption is made regarding the antenna dimensions and separations in terms of the skin depth in the earth, $(2/\omega\mu_0\sigma)^{1/2}$. Here, ω is the angular frequency, σ is the earth conductivity, and the earth permeability is taken to be the free-space value μ_0, which is normally the case. In geophysics it is common to work with the earth resistivity ϱ, which is the reciprocal of the conductivity σ. The dielectric constant of the earth is unimportant at elf because conduction currents are dominant. In the transient results which are given in Section 4 the earth conductivity σ is assumed to be independent of frequency. This is not a good assumption at high frequencies but is a fairly good assumption at elf. The actual value of earth conductivity can commonly vary over a range from about 10^{-1} S/m to 10^{-4} S/m, depending on the moisture content and type of rock [3].

In Section 5, some other antenna types are discussed in much less detail. Many of these antennas are used for shorter ranges and higher frequencies where the quasi-static assumption is not valid.

2. Electrode Arrays for Resistivity Measurements

The most commonly used methods for measuring the direct-current (dc) resistivity of the earth utilize a four-electrode array. An electric current is driven through one pair of electrodes, and the potential established in the earth is measured with the second pair of electrodes. In order to study the variation of the earth resistivity with depth, the spacings between the electrodes are gradually increased. This type of measurement is called *vertical sounding*. In order to study lateral variations of earth conductivity the electrode spacings are kept fixed, and the entire array is moved as a whole along a transverse line. This type of measurement is called *horizontal profiling*.

Theory of Four-Electrode Arrays

We consider a homogeneous half-space with resistivity ϱ as our earth model. If a dc current I is injected at point 1 and removed at point 2, as shown in Fig. 1, then the scalar potential ϕ_P at any point P in the earth is given by

$$\phi_P = \frac{\varrho I}{2\pi}\left(\frac{1}{r_{1,P}} - \frac{1}{r_{2,P}}\right) \tag{1}$$

where $r_{1,P}$ is the separation between points 1 and P and $r_{2,P}$ is the separation between points 2 and P. If we now consider a pair of receiving electrodes 1 and 2, then the voltage difference V between the electrodes is

$$V = \phi_1 - \phi_2 = \frac{\varrho I}{2\pi}\left(\frac{1}{r_{1,1}} - \frac{1}{r_{1,2}} - \frac{1}{r_{2,1}} + \frac{1}{r_{2,2}}\right) \tag{2}$$

where $r_{i,j}$ is the separation between current electrode i and potential electrode j. Normally, the four electrodes are arranged in a straight-line configuration as shown in Fig. 2, but this is not necessary. It is possible to express the resistivity ϱ in (2) in the following form:

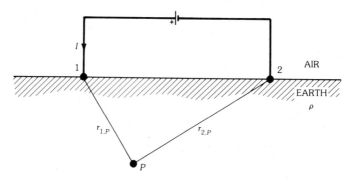

Fig. 1. Direct-current electrodes in a homogeneous earth of resistivity.

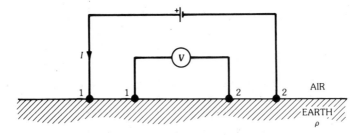

Fig. 2. Four-electrode configuration.

$$\varrho = KV/I \tag{3}$$

Here K is a strictly geometric factor which depends on the array configuration and, in general, is given by

$$K = \frac{2\pi}{1/r_{1,1} - 1/r_{1,2} - 1/r_{2,1} + 1/r_{2,2}} \tag{4}$$

The Wenner array is one of the most commonly used arrays for measuring resistivity. It employs an equal spacing, a, between any two electrodes as shown in Fig. 3. The geometric factor K for the Wenner array is

$$K = \frac{2\pi}{1/a - 1/2a - 1/2a + 1/a} = 2\pi a \tag{5}$$

The Schlumberger array shown in Fig. 4 is also commonly used in measuring earth resistivity. The closely spaced potential electrodes approximately measure the potential gradient at the center of the current array. The geometric factor K for the Schlumberger array is

$$K = \frac{2\pi}{(a - b/2)^{-1} - (a + b/2)^{-1} - (a + b/2)^{-1} + (a - b/2)^{-1}}$$
$$= \pi\left(\frac{a^2}{b} - \frac{b}{4}\right) \tag{6}$$

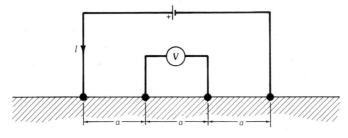

Fig. 3. Wenner array.

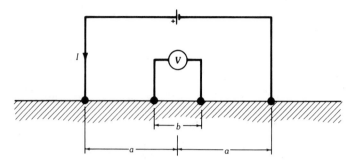

Fig. 4. Schlumberger array.

In the limit of b/a approaching zero, the geometric factor K approaches $\pi a^2/b$, and the potential gradient is measured.

In some cases dipoles (closely spaced electrodes) are used for both the current and potential arrays. In the dipole-dipole array shown in Fig. 5, both b and c are assumed to be small compared with a. The geometric factor K for the dipole-dipole array is

$$K = \frac{2\pi}{(a + b/2 - c/2)^{-1} - (a + b/2 + c/2)^{-1} - (a - b/2 - c/2)^{-1} + (a - b/2 + c/2)^{-1}} \tag{7}$$

For the special case of $b = c$, (7) simplifies to

$$K = \pi a \left(\frac{a^2}{b^2} - 1 \right) \tag{8}$$

When the earth is not homogeneous, the situation becomes more complicated and the simple expression (2) for the potential is no longer valid. However, it is still possible to derive an apparent resistivity ϱ_a in terms of the measured current I and potential difference V:

$$\varrho_a = KV/I \tag{9}$$

where K is still given by (4). When the earth is horizontally stratified, as in Fig. 6, it is possible to formulate the apparent resistivity in terms of integrals which must be

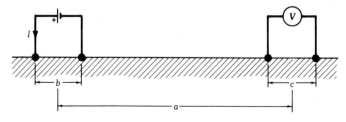

Fig. 5. Dipole-dipole array.

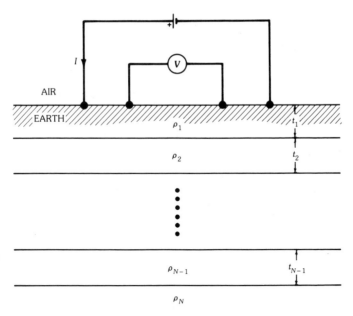

Fig. 6. Four-electrode configuration over a horizontally stratified, N-layer earth.

evaluated numerically. Apparent resistivity curves for Wenner, Schlumberger, and dipole-dipole arrays have been computed for two-, three-, and four-layer earths [4], and a computer code has been published for dipole-dipole arrays for any number of layers up to twelve [5]. The inverse problem of attempting to determine the layer parameters (vertical sounding) from apparent resistivity usually involves some type of curve matching, but that subject is beyond the scope of this chapter.

Instrumentation of Four-Electrode Arrays

The current circuit consists of an insulated wire which is grounded at each end as in Fig. 1. In practice, the wire lies on the ground, and the insulation is important to prevent leakage currents. Grounding can be achieved by driving steel or copper-clad steel stakes into the ground to a depth of several inches or deeper. copper-clad steel stakes into the ground to a depth of several inches or deeper. Spiral-blade electrodes can also be screwed into the ground, and these have the advantage of a larger contact area. The contact resistance can be as low as 10 Ω in moist soil, but can be orders of magnitude larger in dry soils. In dry ground it may be helpful to wet the area around the grounding stake to improve contact between the stake and the ground. It is also possible to employ multiple grounding stakes in parallel to decrease the grounding resistance.

In practice the current is not actually direct current but is usually a low-frequency square wave. A low-frequency sinusoidal current could also be used, but a low-frequency square wave is easier to generate. By avoiding the transient which occurs at the start of each half-cycle of the square wave, the resistivity which is measured is still the desired dc resistivity.

The contact resistance of the potential electrodes is less important because a high-resistance voltmeter is used to measure the potential difference. It is important, however, that the potential electrodes be stable and that polarization potentials between the electrodes and the ground be minimized. Nonpolarizing electrodes can be made by using a metal bar immersed in a solution of one of its salts in a ceramic vessel. Such electrodes are called *porous pots* [6]. Copper and copper sulfate are the most commonly used metal and solution.

3. Grounded Wire Antennas

In the previous section on resistivity measurements we considered only dc excitation. In this section we consider time-harmonic and transient excitation of horizontal wire antennas which are grounded at the ends. In addition to geophysical sounding, such antennas also have application for uplink and downlink communication in mines and for elf communication with submarines [7]. In the application for long-distance communication with submarines the antenna is actually used to excite the quasi-TEM mode in the earth-ionosphere waveguide. This application is beyond the scope of this book, and here we neglect the effect of the ionosphere.

Our model is a homogeneous, conducting half-space of conductivity σ as shown in Fig. 7. The antenna is of length 2ℓ, and the current is assumed to be constant over the length of the antenna. This assumption is valid for insulated wires grounded at the ends when the length of the antenna is much less than a free-space wavelength [8]. The actual input impedance of such antennas is primarily resistive, and the resistance is the sum of the two grounding resistances and the wire resistance. Normally the grounding resistance is dominant.

Time-Harmonic Excitation

In this subsection we consider time-harmonic excitation, and the $\exp(j\omega t)$ time dependence will be suppressed. The subsurface electric and magnetic fields are of interest both in mine communication and in subsurface probing of geophysical

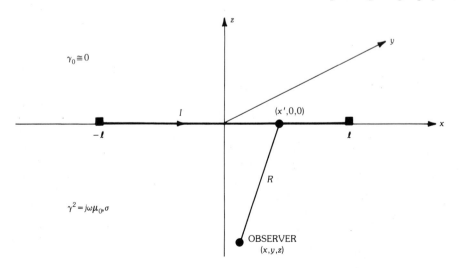

Fig. 7. Finite line source on a homogeneous half-space. (*After Hill and Wait [9]*)

features. In mine communication the subsurface magnetic field is received with a loop antenna, or the subsurface electric field is received with a grounded wire antenna.

We first consider the fields produced by an incremental source of length dx' located at x' as shown in Fig. 7. For the low frequencies of interest here, the free-space propagation constant γ_0 can be set equal to zero and the Sommerfield integral forms for an incremental source of current moment $I\,dx'$ can be greatly simplified [9]. As a result the magnetic field components are

$$dH_x = \frac{I\,dx'}{2\pi\gamma^2}\left(\frac{\partial^4 N}{\partial x\,\partial y\,\partial z^2} - \frac{\partial^3 P}{\partial x\,\partial y\,\partial z}\right) \tag{10a}$$

$$dH_y = \frac{I\,dx'}{2\pi\gamma^2}\left(\frac{\partial^3 P}{\partial z^3} + \frac{\partial^2 P}{\partial x^2\,\partial z} + \frac{\partial^4 N}{\partial z^2\,\partial y^2}\right) \tag{10b}$$

$$dH_z = \frac{I\,dx'}{2\pi\gamma^2}\left(\frac{\partial^4 N}{\partial y\,\partial z^3} - \gamma^2\frac{\partial^2 N}{\partial y\,\partial z} - \frac{\partial^3 P}{\partial y\,\partial z^2}\right) \tag{10c}$$

where

$$N = I_0[(\gamma/2)(R + z)]\,K_0[(\gamma/2)(R - z)] \tag{11a}$$

$$P = R^{-1}\exp(-\gamma R) \tag{11b}$$

$$R = \sqrt{(x - x')^2 + y^2 + z^2} \tag{11c}$$

$$\gamma = \sqrt{j\omega\mu_0\sigma} \tag{11d}$$

The terms I_0 and K_0 are modified Bessel functions of order zero. Similarly, the electric-field components are

$$dE_x = \frac{-I\,dx'}{2\pi\sigma}\left(\frac{\partial^2 P}{\partial z^2} + \frac{\partial^3 N}{\partial y^2\partial z}\right) \tag{12a}$$

$$dE_y = \frac{I\,dx'}{2\pi\sigma}\frac{\partial^3 N}{\partial y\,\partial x\,\partial z} \tag{12b}$$

$$dE_z = \frac{I\,dx'}{2\pi\sigma}\frac{\partial^2 P}{\partial x\,\partial z} \tag{12c}$$

To obtain the fields of a cable of finite length 2ℓ carrying a constant current I, we integrate (10) and (12) over the range of x' from $-\ell$ to ℓ. The dominant components of interest are H_y, H_z, and E_x because all other components vanish in the plane $x = 0$. Also, these are the only nonzero components everywhere if the line source is of infinite length. For normalization purposes it is convenient to write the fields in the following manner:

$$H_y = \frac{I}{2\pi h} A(H, Y, X, L) \tag{13a}$$

$$H_z = \frac{I}{2\pi h} B(H, Y, X, L) \tag{13b}$$

$$E_x = \frac{-j\omega\mu_0 I}{2\pi} F(H, Y, X, L) \tag{13c}$$

where $H = \sqrt{\omega\mu_0\sigma}\, h$, $Y = y/h$, $X = x/h$, $L = \ell/h$, and $h = -z$. Note that Z, B, F, H, Y, X, and L are dimensionless. The specific forms for A, B, and F are

$$A(H, Y, X, L) = \frac{h}{\gamma^2} \int_{-\ell}^{\ell} \left(\frac{\partial^3 P}{\partial z^3} + \frac{\partial^3 P}{\partial x^2 \partial z} + \frac{\partial^4 N}{\partial z^2 \partial y^2} \right) dx' \tag{14a}$$

$$B(H, Y, X, L) = \frac{h}{\gamma^2} \int_{-\ell}^{\ell} \left(\frac{\partial^4 N}{\partial y \, \partial z^3} - \gamma^2 \frac{\partial^2 N}{\partial y \, \partial z} - \frac{\partial^3 P}{\partial y \, \partial z^2} \right) dx' \tag{14b}$$

$$F(H, Y, X, L) = \frac{1}{\gamma^2} \int_{-\ell}^{\ell} \left(\frac{\partial^2 P}{\partial z^2} + \frac{\partial^3 N}{\partial y^2 \partial z} \right) dx' \tag{14c}$$

The integral forms in (14) simplify for both the low-frequency (small H) and high-frequency (large H) cases [9] but, in general, numerical integration is required. Typical numerical results are shown for $H = 2$ in Fig. 8. Although A, F, and B are complex for H greater than zero, only the magnitudes are plotted. The phases are relatively constant as a function of L. For very small values of L, the fields are essentially those of a short dipole and are proportional to L as indicated by (10) and (12). For large L, the field components must eventually reach those of an infinite line source. The results in Fig. 8 were found to agree well with earlier results computed for an infinite line source by Wait and Spies [10]. In some geophysical applications it is desirable to make the grounded wire long enough to simulate an infinitely long line source. For practical purposes this is achieved for L greater than about 2. This means that the antenna length 2ℓ should be about 4 times the depth h of interest in the particular geophysical sounding application of interest.

Straight, grounded wire antennas can also be used underground. For example, long wire antennas have been laid out in mine tunnels for uplink transmission as indicated in Fig. 9. The surface fields of such antennas have been computed for the case where the antenna is not parallel to the air-earth interface [11] in order to model cases where either the tunnel or the earth surface is not level. Since the antenna must usually be located at the surface in geophysical applications, we will not go into detail on the subsurface case. The antenna current is normally constant over the length of the antenna, and the surface fields are qualitatively similar to the subsurface fields of surface antennas. Also, the grounding resistance is again normally greater than the wire resistance.

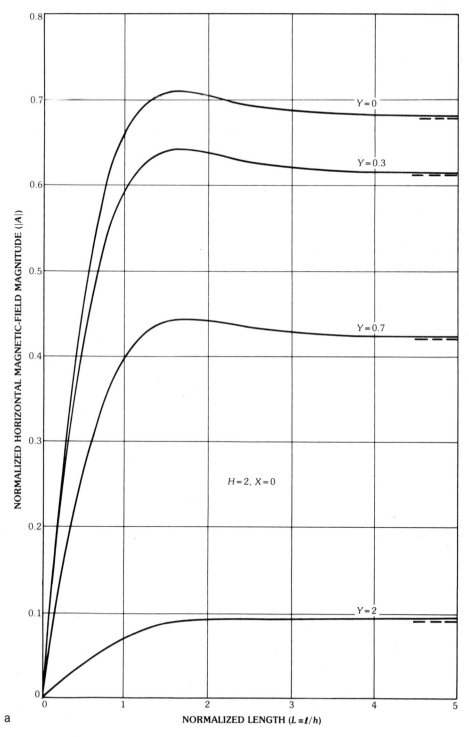

Fig. 8. Magnitudes of the normalized fields as a function of Y and L. (*a*) Horizontal magnetic field. (*b*) Vertical magnetic field. (*c*) Horizontal electric field. (*After Hill and Wait [9]*).

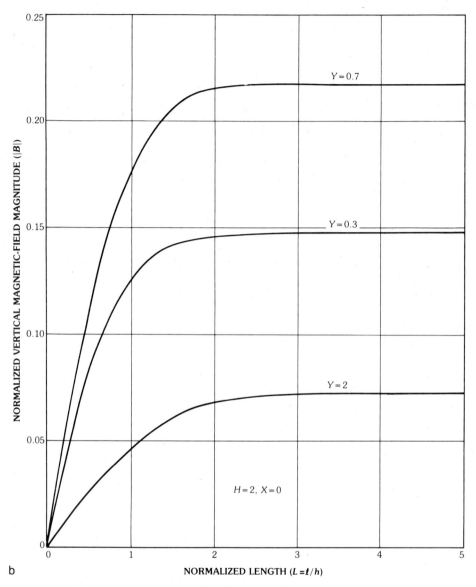

b

Fig. 8, *continued*.

Transient Excitation

The straight, grounded wire is also used to excite transient fields in the earth for geophysical applications. For example, a grounded wire several hundred meters in length can be excited with a step-function current, and the vertical magnetic field at some remote location can be received with a large horizontal loop antenna. We assume that the transients are sufficiently band limited that the current along the antenna is independent of position.

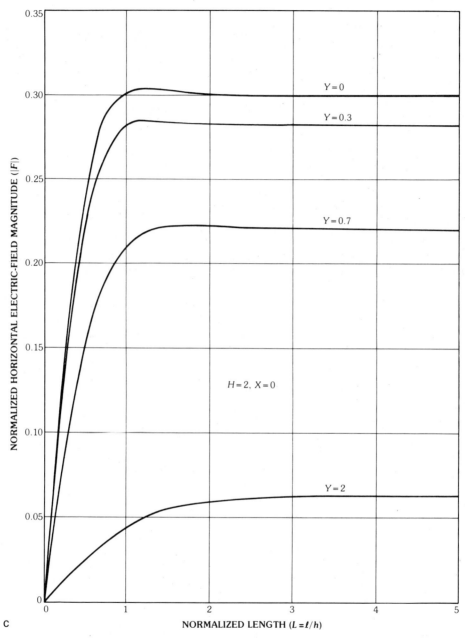

Fig. 8, *continued.*

To illustrate the dispersive nature of the earth we first examine the frequency dependence of the subsurface electric fields. The frequency dependence of all three components has been shown [12], but here we consider only the dominant component E_x:

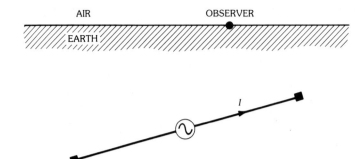

Fig. 9. Subsurface insulated antenna grounded at the ends.

$$E_x = -[I/(2\pi\sigma h^2)]\, E_{xn}(W, L, X, Y) \tag{15}$$

where

$$E_{xn} = h^2 \int_{-\ell}^{\ell} \left(\frac{\partial^2 P}{\partial z^2} + \frac{\partial^3 N}{\partial y^2\, \partial x} \right) dx' \tag{16}$$

$$W = \omega\mu_0\sigma h^2 \tag{17}$$

Equations 15 through 17 are consistent with the earlier results in (13c) and (14c), but here the frequency dependence is explicitly displayed through the dimensionless frequency parameter W. Some numerical results for $|E_{xn}|$ as a function of W are shown in Fig. 10. As W approaches zero, E_{xn} approaches the dc result, E_{xn}^{dc}, which is obtained from the gradient of a scalar potential:

$$E_{xn}^{dc} = (L - X)\, R_1^{-3} + (L + X)\, R_2^{-3} \tag{18}$$

where

$$R_1 = [(X - L)^2 + Y^2 + 1]^{1/2} \tag{19}$$
$$R_2 = [(X + L)^2 + Y^2 + 1]^{1/2}$$

Equation 18 is consistent with the result from potential theory in (1).

When a step-function voltage is applied to the antenna the antenna current is approximately a step current $I\, U(t)$ because the input impedance is approximately resistive and constant over the frequency range of the band-limited step function. The transient electric field $\tilde{E}_x^s(t)$ is given by the following inverse Fourier transform:

$$\tilde{E}_x^s(t) = \frac{1}{2\pi} \int_{-\infty}^{\infty} \frac{E_x}{j\omega} \exp(j\omega t)\, d\omega \tag{20}$$

It is more convenient to rewrite (20) in the following normalized form:

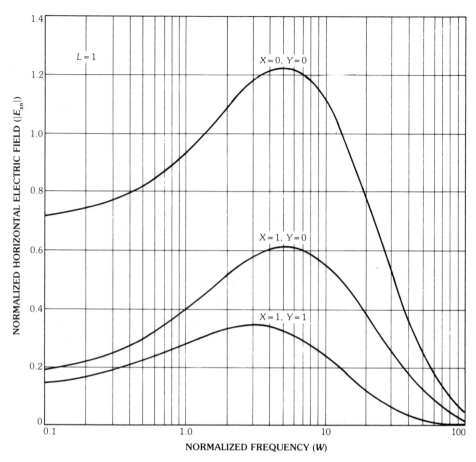

Fig. 10. Magnitude of the normalized horizontal electric field as a function of normalized frequency W. (*After Hill and Wait [12]*)

$$\tilde{E}_x^s(t) = -[I/(2\pi\sigma h^2)]\,\tilde{E}_{xn}(T, L, X, Y) \tag{21}$$

where

$$\tilde{E}_{xn}(T, L, X, Y) = \frac{1}{2\pi}\int_{-\infty}^{\infty}\frac{E_{xn}(W, L, X, Y)}{jW}\exp(jWT)\,dW \tag{22}$$

$$T = t/\sigma\mu_0 h^2 \tag{23}$$

The term T is a normalized time, and $\sigma\mu_0 h^2$ can be thought of as a depth-dependent time constant. The inverse transform in (22) has been carried out numerically, and results for $\tilde{E}_{xn}^s$ are shown in Fig. 11.

In all cases the waveforms in Fig. 11 have a fairly rapid rise which overshoots the final value, followed by a slow decay to the dashed final values. The final values are the dc values as given by (18) and (19). The impulse response is the time

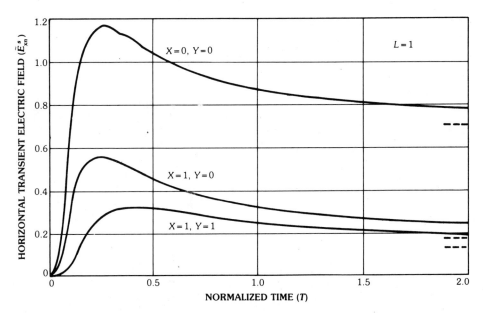

Fig. 11. Horizontal electric field for a step-function current, where the final dc values are dashed. (*After Hill and Wait [12]*)

derivative of the step response in (13), and it consists of a spike followed by a slow tail of opposite polarity. This behavior is well known, and it has been illustrated [13] for the impulse response of an infinite line source ($L = \infty$). Responses for other values of L tend to be qualitatively similar.

Receiving Application

The straight grounded wire antennas shown in Figs. 7 and 9 can also be used as receiving antennas. Typically the subsurface antenna in Fig. 9 would be used to receive a downlink signal as in mine communication or submarine communication. The surface antenna in Fig. 7 could be used to receive an uplink signal, or it could be the receiving antenna in an earth-sounding system. The receiving antennas for dc resistivity in Figs. 2 through 5 are special cases of this receiving application.

Because the current distribution on the antenna is constant over the length, the received voltage is simply the integral of the incident electric field over the length of the antenna. For example, the received voltage v for the antenna in Fig. 7 when used as a receiving antenna is

$$ v = \int_{-\ell}^{\ell} E_x(x) \bigg| dx_{z=y=0} \qquad (24) $$

The electric field E_x could be either a time-harmonic or a transient field. For example, Wait [14] has used (24) to compute the transient coupling between a pair of grounded wires. In the magneto-telluric method [15], the electric field that is measured is a transient signal produced by lightning or some other natural source.

For the special case of dc fields, (24) simply yields the potential difference between the end points of the antenna as in (2).

4. Loop Antennas

Loop antennas are commonly used in geophysical sounding and subsurface communications, and they have the advantage that no grounding is required. In induction methods, loop antennas transmit a time-varying magnetic field into the earth, and eddy currents are excited in conducting bodies. These eddy currents generate a secondary magnetic field which can be received by a second loop antenna. In mine communications [16], transmitting loops can be used either at or below the earth surface.

Horizontal transmitting loops are typically a large, single turn of insulated wire laid out on the earth. The loop dimensions can be up to a kilometer, and the current can be up to several amperes. Various shapes, such as circular or rectangular, are used depending on the application. When the loop dimensions are small compared with the skin depth in the earth and the observer distance, then the horizontal loop radiates as a vertical magnetic dipole.

Vertical transmitting loops are usually multiple turns of insulated wire on some type of frame with dimensions on the order of meters. Normally such vertical loops radiate as horizontal magnetic dipoles. The inductance of such multiturn loops can be fairly large, and a series capacitor is usually used to tune to the operating frequency.

Time-Harmonic Excitation

In this subsection we consider time-harmonic excitation of a circular loop at the earth surface as shown in Fig. 12. For small loops the results depend only on the magnetic dipole moment IA where I is the loop current and A is the loop area, and the shape is unimportant. Here we wish to consider the effect of finite loop size, and a circular loop is the simplest shape to consider. For a circular loop with constant current I, the nonzero field components are H_z, H_ϱ, and E_ϕ.

The subsurface fields are of interest in mine communications and induction sounding, and here we consider a homogeneous half-space model. The vertical magnetic field is of primary interest for downlink communication between horizontal loops, and it is given by [17]

$$H_z = \frac{-IA}{2\pi h^3} Q \tag{25a}$$

where

$$Q = \int_0^\infty \frac{x^3 e^{-(x^2+H^2)^{1/2}}}{x + (x^2 + iH^2)^{1/2}} J_0(xD) \frac{2 \, J_1(xa/h)}{xa/h} \, dx \tag{25b}$$

$$H = \sqrt{\omega\mu_0\sigma}\, h \tag{25c}$$

$$D = d/h \tag{25d}$$

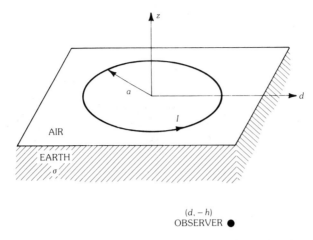

Fig. 12. Circular loop on a conducting half-space with a subsurface observer.

and J_0 and J_1 are the zeroth- and first-order Bessel functions. When a/h is sufficiently small, then the factor $2J_1(xa/h)/(xa/h)$ approaches unity over the important range of x. In that case the dependence on a enters only through the loop area $A\ (=\pi a^2)$. If the loop is buried at a depth h and the observer is located at the surface (as in uplink communications), then the result for H_z is identical except that the minus sign in (25a) becomes a plus.

For the special case of $H = a/h = 0$, the term Q reduces to the following result for a static magnetic dipole:

$$Q|_{H=a/h=0} = \frac{2 - D^2}{2(1 + D^2)^{5/2}} \tag{26}$$

If, in addition, $D = 0$, then $Q = 1$. Thus Q is the magnetic field normalized to the on-axis field of a static dipole. For $D = 0$, both H_ϱ and E_ϕ are zero.

Numerical results for a magnetic dipole ($a/h = 0$) are shown in Fig. 13. Note that for the static case ($H = 0$), there is a null at $D = \sqrt{2}$ in accordance with (26). For other values of H, the term Q is complex and the null is filled in. The peak value of $|Q|$ always occurs for $D = 0$, and this peak value decreases as H increases. Since H is proportional to the square root of frequency times the depth h, the frequency must be decreased in order to transmit to greater depths. This has been shown experimentally for mine communications. For geophysical probing, vertical sounding [18] is accomplished by varying the frequency, and low frequencies are used to obtain information on the earth conductivity at great depths. The dependence of $|Q|$ on H is shown more explicitly in Fig. 14 for the case of $D = a/h = 0$. For large values of H, the field Q decays exponentially with H just as a plane wave in a lossy medium does. This exponential decay with H will be shown to lengthen the rise time for transient fields of a loop in the following subsection because the high frequencies are severely attenuated.

The field results in Figs. 13 and 14 are actually valid for small loops of any shape which can be represented by a vertical magnetic dipole. When the loop

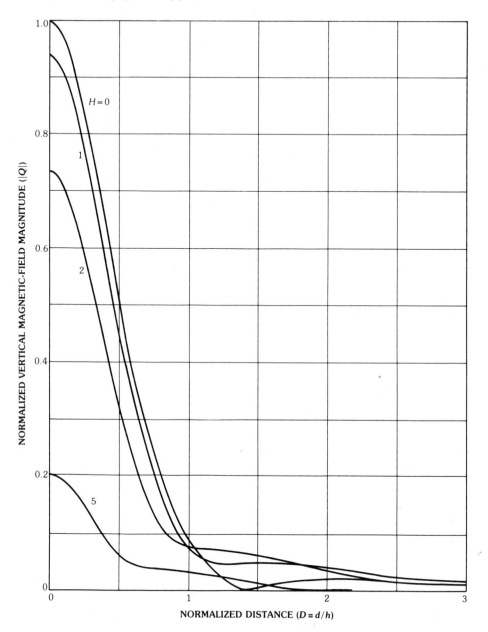

Fig. 13. Magnitude of the normalized vertical magnetic field for a vertical magnetic dipole source ($a/h = 0$). (*After Wait [17], © 1971 IEEE*)

dimensions are large, then the field depends strongly on shape, and the theory has been developed for loops of arbitrary shape in a conducting medium which is either homogeneous or layered [19]. Results for a finite circular loop are shown in Fig. 15 for the static case ($H = 0$). As the loop radius is increased, the vertical field is reduced on the axis ($D = 0$), but is increased at the larger horizontal distances. A

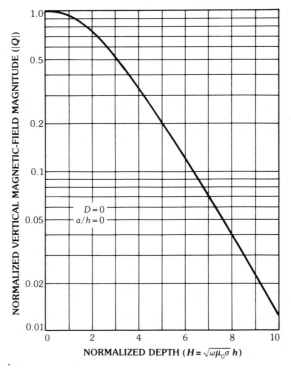

Fig. 14. Magnitude of the normalized vertical magnetic field as a function of H on the axis ($D = 0$).

similar behavior occurs for nonzero values of H. Also, similar results have been calculated for rectangular loops. Rectangular loops are of interest in mine communications where subsurface loops are laid out around rectangular coal pillars.

Transient Excitation

Loops are also used to excite transient fields in the earth for subsurface probing. Again, we assume that the frequency spectrum of the transient is band-limited and the current is independent of position along the loop. The fields have been computed for transients of various shapes, but the most physically realistic is the step response. For a step current $I U(T)$ in the loop, the transient subsurface field $\tilde{H}_z(t)$ can be expressed as [20]

$$\tilde{H}_z(t) = \frac{-IA}{2\pi h^3} Y(D, T) \tag{27a}$$

where

$$T = t/\sigma\mu_0 h^2 \tag{27b}$$

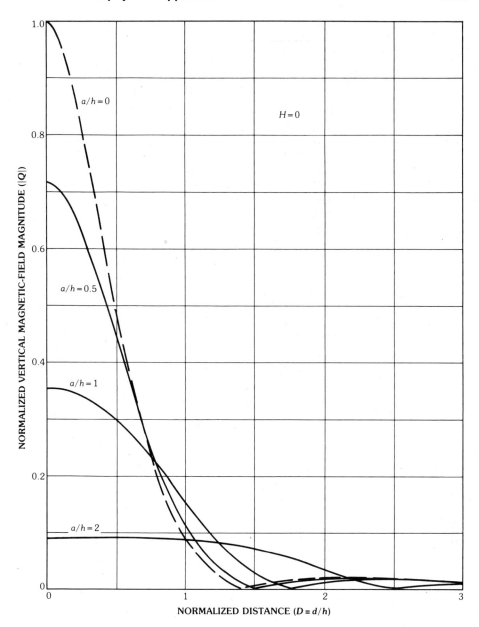

Fig. 15. Magnitude of the normalized vertical magnetic field of a finite circular loop. (*After Wait and Hill [19]*)

Again T is a normalized time, and Y is a normalized step response which can be obtained numerically from the frequency-domain solution in (25).

Numerical results for Y are given in Fig. 16 for the magnetic dipole case ($a/h = 0$). The final values of the curves are obtained from the low-frequency limit of the time-harmonic solution in (26):

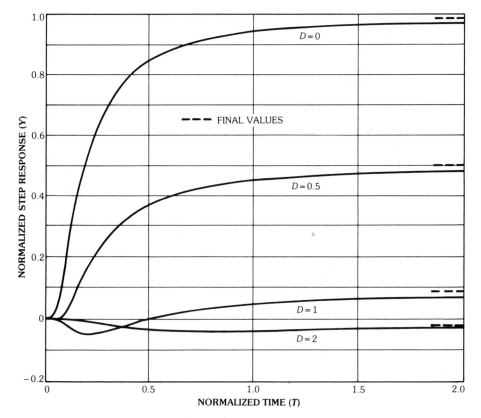

Fig. 16. Normalized vertical magnetic field for a step-function loop current. (*After Wait and Hill [20]*)

$$Y(D, \infty) = \frac{2 - D^2}{2(1 + D^2)^{5/2}} \tag{28}$$

The rise time of the step response is on the order of $\sigma\mu_0 h^2$, and it therefore increases with depth. In vertical sounding applications the information on conductivity at greater depths is obtained from the late time portion of the transient response.

Other loop geometries [21] and excitations have been studied, but we will not cover these here. The excitation by a rectangular pulse of current has been analyzed for both vertical and horizontal loops [22]. This type of pulse is of interest because many transient systems actually employ a periodic square wave.

Receiving Application

Receiving loops are normally small multiturn coils of small wire which can be hand held. Sometimes receiving loops are shielded electrostatically by wrapping them in foil, and the shielding should be grounded at some point to prevent capacitive coupling between the loop and the earth or the operator. Sometimes it is desirable to find the direction of the magnetic field by rotating the loop axis, and then a means of measuring the loop inclination is required. For example, a pair of orthogonal loops is used to measure the tilt of the natural magnetic field in audio-frequency (AFMAG) systems [23].

Since receiving loops are normally small they simply respond to the time derivative of the magnetic flux. For example, the voltage induced in a horizontal loop as in Fig. 12 is given by

$$\bar{v}(t) = A_R N_R \mu \frac{\partial \bar{H}_z(t)}{\partial t} \tag{29}$$

where A_R is the receiving loop area, N_R is the number of turns, and $\bar{H}_z(t)$ is the time-varying vertical magnetic field. If the loop has an air core, then μ is the free-space permeability μ_0; but if the loop has a ferrite core, then μ is some larger effective value. If the loop has some other orientation, then $\bar{H}_z$ is simply the magnetic field along the loop axis. For example, in the magnetotelluric method the loop is oriented in a vertical plane in order to measure the time-varying horizontal magnetic field.

For time-harmonic fields the received voltage is proportional to $j\omega$ times the axial magnetic field. For a horizontal loop the Fourier transform of (29) yields the following received voltage:

$$v(\omega) = j\omega\mu A_R N_R H_z(\omega) \tag{30}$$

An interesting application of the small receiving loop is the location of a horizontal subsurface transmitting loop in the trapped-miner problem [2]. As indicated in Fig. 17, the horizontal magnetic field has a null and the vertical magnetic field is a maximum directly above the transmitting loop. The transmitting loop is excited with a pulsed cw signal with a carrier frequency of about 1 kHz, and the search coil is equipped with earphones. The null in the horizontal magnetic field can typically be located by the search coil within an accuracy of a few meters for a transmitter depth of a couple hundred meters. The depth of the transmitter is not determined by the null location, but can, in principle, be located from more extensive surface measurements [24].

5. Miscellaneous Antennas

The preceding sections were devoted to the commonly used grounded wire and loop antennas. In this section we discuss a number of antennas and applications, but in much less detail.

Magnetometers are sometimes used in place of loop antennas for receiving magnetic fields. The problem with loop antennas is that the induced voltage is proportional to the time derivative of the magnetic field as shown in (29) or $j\omega$ times the time-harmonic magnetic field as shown in (30). As a result loop antennas are not very sensitive to slowly varying (or low-frequency) fields, and this can be a serious problem in magnetotelluric applications. Several types of magnetometers ([1], pp. 231–243) respond to the magnetic field itself rather than to the time

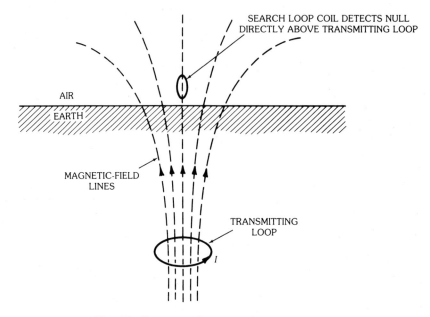

Fig. 17. Geometry for trapped-miner location.

derivative and are useful for measuring slowly varying signals. In the magnetic balance method the torque on a bar magnet due to the magnetic field is balanced against gravity. In the flux-gate magnetometer the nonlinear permeability of a ferromagnetic material is utilized to generate a second harmonic which is proportional to the exciting field. Optical pumping magnetometers utilize the energy level differences within an atom which depend on the ambient magnetic field, and these can have very good sensitivity.

The insulated antenna which is grounded at the end can be assumed to carry a uniform current when the length of the antenna is short compared with the wavelength in the insulation. This is always the case for the elf applications discussed in Section 2, but is not always true for higher frequencies where the current is essentially that of a transmission line whose electrical length is not short. Also, there are cases where the antenna has no insulation, and the current distribution is more complicated. Such cases, where the current distribution on linear antennas in lossy media has to be determined, are discussed thoroughly in the book by King and Smith [25].

Electrical well logging has been used extensively for more precise information on rock conductivity as a function of depth ([1], Chapter II). Single electrodes or various configurations of multiple electrodes are lowered into wells or drill holes on insulated wire, and the apparent resistivity is a function of the material surrounding the electrode or multiple electrodes. In order to have electrical continuity between the electrodes and the surrounding rock, the drill hole must be filled with water or drilling mud. In induction logging, an induction coil excites current in the surrounding rock via a time-varying magnetic field, and electrical contact is not required. Consequently, induction logs can be run in either dry or fluid-filled holes.

The voltage induced in a secondary coil is directly related to the conductivity of the surrounding rock, and the dielectric constant plays no role at the typical frequencies below about 20 kHz. More recently, however, dielectric logging has become quite popular at higher frequencies all the way from high frequencies to microwaves. The crossover frequency, where displacement currents are equal to conduction currents, is determined from

$$\sigma = \omega \epsilon_r \epsilon_0 \qquad (31)$$

where ϵ_r is the relative permittivity of the ground and ϵ_0 is the permittivity of free space. For example, for $\sigma = 10$ S/m and $\epsilon_r = 10$, the crossover frequency is approximately 18 MHz. Thus systems operating above high frequency are primarily
sensing the dielectric constant.

When shallow depths on the order of several meters down to a fraction of a meter are of interest, then frequencies all the way up to microwaves can be used because a relatively small skin depth can be tolerated. Frequencies in the range from 1 to 2 GHz have been used in an fm-cw system [26] to sense the thickness of coal remaining on the roof of a mined haulageway for thicknesses up to approximately a half meter. A pair of broadband rectangular-aperture horns which utilize double-ridged waveguide techniques were used for transmission and reception. The same system has been used for probing of soil to shallow depths and for probing the properties of snowpack. A pulsed system with a frequency spectrum from about 100 MHz to 500 MHz has been used for similar applications involving the detection of targets at shallow depths. The antennas used were a pair of broadband crossed dipoles [27] with transmission and reception on orthogonal polarizations. It appears that the use of broadband vhf and microwave systems for high-resolution probing to shallow depths is a promising area which will continue to develop.

6. References

[1] G. V. Keller and F. C. Frischknecht, *Electrical Methods in Geophysical Prospecting*, Oxford: Pergamon Press, 1966.

[2] J. N. Murphy and H. E. Parkinson, "Underground mine communication," *Proc. IEEE*, vol. 66, pp. 26–50, 1978.

[3] E. I. Parkhomenko, *Electrical Properties of Rock*, New York: Plenum Press, 1967.

[4] P. K. Bhattacharya and H. P. Patra, *Direct Current Geoelectric Sounding*, Amsterdam: Elsevier Publishing Co., 1968.

[5] E. F. Laine and R. J. Lytle, "A computer program for four probe resistivity measurements in a horizontally layered earth," *IEEE Trans. Geosci. Electron.*, vol. GE-14, no. 4, pp. 232–235, 1976.

[6] O. Koefoed, *Geosounding Principles, 1*, Amsterdam: Elsevier Publishing Co., 1979.

[7] M. L. Burrows, *ELF Communications Antennas*, Stevenage, Herts., UK: Peter Peregrinus, 1978.

[8] E. D. Sunde, *Earth Conduction Effects in Transmission Systems*, Chapter V, New York: D. Van Nostrand Co., 1949.

[9] D. A. Hill and J. R. Wait, "Subsurface electromagnetic fields of a grounded cable of finite length," *Can. J. Phys.*, vol. 51, pp. 1534–1540, 1973.

[10] J. R. Wait and K. P. Spies, "Subsurface electromagnetic field of a line source on a conducting half-space," *Rad. Sci.*, vol. 6, pp. 781–786, 1971.

[11] D. A. Hill, "Electromagnetic surface fields of an inclined buried cable of finite length," *J. Appl. Phys.*, vol. 44, pp. 5275–5279, 1973.

[12] D. A. Hill and J. R. Wait, "Subsurface electric fields of a ground cable of finite length for both frequency and time domain," *Pure Appl. Geophys.*, vol. 111, pp. 2324–2332, 1973.

[13] D. A. Hill and J. R. Wait, "Diffusion of electromagnetic pulses into the earth from a line source," *IEEE Trans. Antennas Prop.*, vol. AP-22, pp. 145–146, 1974.

[14] J. R. Wait, "Propagation of electromagnetic pulses in a homogeneous conducting earth," *Appl. Sci. Res.*, vol. 8, pp. 213–253, 1960.

[15] A. A. Kaufman and G. V. Keller, *The Magnetotelluric Sounding Method*, Amsterdam: Elsevier Publishing Co., 1981.

[16] D. G. Large, L. Ball, and A. J. Farstad, "Radio transmission to and from underground coal mines—theory and experiment," *IEEE Trans. Commun.*, vol. COM-21, pp. 194–202, 1973.

[17] J. R. Wait, "Electromagnetic induction technique for locating buried source," *IEEE Trans. Geosci. Electron.*, vol. GE-9, pp. 95–98, 1971.

[18] A. A. Kaufman and G. V. Keller, *Frequency and Transient Sounding*, Amsterdam: Elsevier Publishing Co., 1983.

[19] J. R. Wait and D. A. Hill, "Fields of a horizontal loop of arbitrary shape buried in two-layer earth," *Rad. Sci.*, vol. 15, pp. 903–912, 1980.

[20] J. R. Wait and D. A. Hill, "Electromagnetic surface fields produced by a pulse-excited loop buried in the earth," *J. Appl. Phys.*, pp. 3988–3991, 1972.

[21] D. A. Hill, "Transient signals from a buried horizontal magnetic dipole," *Pure and Appl. Geophys.*, vol. III, pp. 2264–2272, 1973.

[22] H. J. Tsaknakis and E. E. Kriezis, "Transient electromagnetic field due to a circular current loop perpendicular or parallel to a conducting half-space," *IEEE Trans. Geosci. Remote Sensing*, vol. GE-20, pp. 122–130, 1982.

[23] S. H. Ward, J. O'Donnell, R. Rivera, G. H. Ware, and D. C. Fraser, "AFMAG—applications and limitations," *Geophys.*, vol. XXXI, pp. 576–605, 1966.

[24] J. R. Wait, "Criteria for locating an oscillating magnetic dipole buried in the earth," *Proc. IEEE*, vol. 59, pp. 1033–1035, 1971.

[25] R. W. P. King and G. S. Smith, *Antennas in Matter*, Cambridge: MIT Press, 1981.

[26] D. A. Ellerbruch and D. R. Belsher, "Fm-cw technique of measuring coal layer thickness," *IEEE Trans. Geosci. Electron.*, vol. GE-16, pp. 126–133, 1978.

[27] L. C. Chan, L. Peters, Jr., and D. L. Moffat, "Improved performance of a subsurface radar target identification system through antenna design," *IEEE Trans. Antennas Propag.*, vol. AP-29, pp. 307–311, 1981.

Chapter 24

Antennas for Medical Applications

C. H. Durney
University of Utah

M. F. Iskander
University of Utah

CONTENTS

Carl H. Durney was born in Blackfoot, Idaho, on April 22, 1931. He received the BS degree in electrical engineering from Utah State University, Logan, in 1958, and the MS and PhD degrees in electrical engineering from the University of Utah, Salt Lake City, in 1961 and 1964, respectively.

From 1958 to 1959 he was employed as an associate research engineer with the Boeing Airplane Company, in Seattle. He has been with the University of Utah since 1963, when he was appointed assistant research professor in electrical engineering.

While on leave in 1971 he studied microwave biological effects at the University of Washington. From 1977 to 1982 he was chairman of the Electrical Engineering Department at the University of Utah, where he is presently a professor of electrical engineering and engaged in teaching and research in electromagnetics, engineering pedagogy, and microwave biological effects. Dr. Durney has received several awards in teaching and research and is a member of many technical societies, editorial boards, and committees.

Magdy F. Iskander was born in Alexandria, Egypt, on August 6, 1946. He received the BSc degree in electrical engineering, University of Alexandria, Egypt, in 1969. From the University of Manitoba, in Winnipeg, Manitoba, Canada, he received the MSc and PhD degrees, both in microwaves, in 1972 and 1976, respectively.

In 1976 Dr. Iskander was awarded a National Research Council of Canada postdoctoral fellowship at the University of Manitoba. Since March 1977 he has been with the Department of Electrical Engineering and the Department of Bioengineering at the University of Utah, in Salt Lake City, where he is currently a professor of electrical engineering.

Dr. Iskander has contributed chapters to four research books. He has also presented and published in technical journals more than 140 papers. He is a member of the editorial boards of the *IEEE Transactions on Microwave Theory and Techniques* and the *Journal of Microwave Power*. His interests include the scattering and diffraction of electromagnetic waves, antenna design, and the biological effects as well as the medical applications of electromagnetic waves.

1. Introduction

Rapid advances in technology have been accompanied by an increasing application of technology to health care. More and more interaction of scientists and engineers with medical people has resulted in significant developments in improved health care using highly sophisticated equipment. Correspondingly, the use of antennas of one kind or another in the practice of medicine has increased. Most of the medical applications of antennas involve coupling electromagnetic energy into the human body or into other biological systems, such as animals used in experimental measurements.

Primarily, these applications can be divided into two categories: therapy and diagnostics. In therapy, antennas have been used mainly for producing hyperthermia for treating cancer. This therapy consists of heating tumors, alone or in combination with either X-ray irradiation or chemotherapy. In diagnostics, electromagnetic energy is coupled into and out of the body to monitor various physiological parameters. Tomographic imaging of the body by electromagnetic measurements has been investigated, but not yet developed to the point of practical implementation.

The problems encountered in designing antennas for medical applications are somewhat different from those involved in the design of antennas for other applications. For example, in coupling electromagnetic energy into biological systems the near fields are often more important than the far fields, in contrast to other applications. Near-field problems are especially important in the design of hyperthermia applicators. Since biological systems usually have high dielectric constants the antennas for coupling into them must be designed differently from those radiating into free space. It is also very important that antennas used for coupling electromagnetic energy into the human body be designed to produce minimum leakage radiation outside the body. Leakage radiation can sometimes be hazardous to medical personnel, and it can produce artifacts in measurements.

Furthermore, since biological systems are usually quite lossy, there is a fundamental trade-off between depth of penetration and localization of the fields inside the body. In many applications, localization of the internal fields is important. In hyperthermia, for example, it is often desirable to concentrate the internal fields at the tumor, thereby heating it without appreciably affecting the surrounding normal tissue. Concentrating the fields, however, amounts to having narrow-beam radiation, which requires an electrically large aperture. To restrict apertures to physical sizes convenient for use with patients and still get narrow-beam radiation, higher-frequency operation, in the high-megahertz or low-gigahertz range, is required. At these high frequencies, however, the penetration of the electromagnetic radiation is too shallow to be useful. Some insight into this problem can be gained from the data presented in Fig. 1, which shows the plane-wave penetration depth versus frequency in a dielectric half-space with a permittivity equal to the average permittivity of the human body. The power magnitude is normalized to the value at the surface. The profile of electromagnetic field penetration into the body is, of course, a strong function of aperture size. Since the data in Fig. 1 is for an incident plane wave, it corresponds to that of a very large aperture. The curves show that even for a very large aperture the penetration at high frequencies is too shallow to be useful for any application that requires transmission of a signal through the body or heating of tissue in the center of the body. Given the severe attenuation at high frequencies, there is no alternative for some applications except to go to lower frequencies. But at lower frequencies the radiators must often be electrically small, resulting in radiation profiles similar to that of a point source, which means that concentrating the fields in a small region inside the body (such as in a tumor) would be very difficult. Thus there is a fundamental trade-off between depth of penetration and localization. At frequencies high enough that apertures can be electrically large enough to produce good localization of the internal fields, the penetration is too shallow to be useful for many applications. At lower frequencies, while the penetration will be deep enough, it is difficult to make apertures electrically large enough to localize the internal fields to small regions. Electrically small apertures at low frequencies often are described as nonradiating because

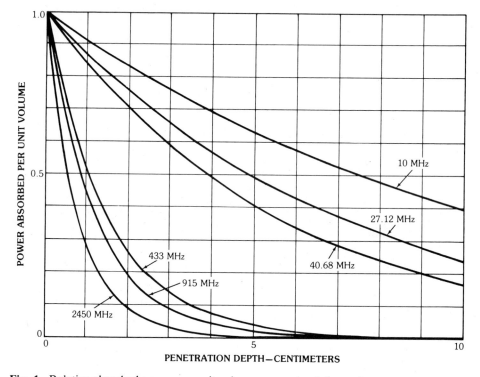

Fig. 1. Relative absorbed power per unit volume versus depth for a plane wave propagating into a dielectric half-space having the properties of tissue. (*After Iskander [1], © 1982 American Institute of Physics*)

their fields are mostly near fields that decay rapidly with distance away from the applicator. These devices are usually thought of as inducing currents, rather than radiating. As explained in the various applications described in this chapter this fundamental problem is encountered over and over again. In producing hyper-thermia, for example, the shallow penetration at higher frequencies forces the designer to use low frequencies to get deeper heating; otherwise the surface over-heats, even with surface cooling, before sufficient heating occurs deeper inside the body. At lower frequencies, though, the radiation produced by the electrically small apertures is often dominated by near fields, which, because they die away rapidly with distance from the applicator, also cause the surface to overheat. The main problem in hyperthermia applications, therefore, is how to obtain deep heating without overheating the surface.

Another factor that is important in coupling electromagnetic energy into biological systems is the reflection at the interfaces between different kinds of tissues. This reflection depends not only on the specific types of tissues at the interface, but also on the frequency, the polarization, and the angle of incidence of the propagating waves. Fig. 2 illustrates the variation of the magnitude of the reflection coefficient as a function of frequency for various tissue interfaces [1]. These calculations, although they are made for planar dielectric layers exposed to an incident plane wave, nevertheless illustrate what large reflections can occur at

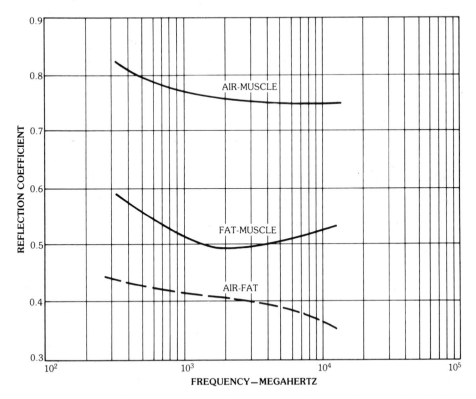

Fig. 2. Plane-wave reflection coefficient for normal incident at the interfaces between different tissue layers. (*After Iskander [1], © 1982 American Institute of Physics*)

the tissue interfaces. The interference between the incident wave and these large reflected waves results in significant standing-wave patterns that may result in excessive heating (hot spots) of certain regions in the body.

Extension calculations of absorbed power in various combinations of tissues have been made by Schwan [2]. Typical results [3] for a planar model of a fat layer in front of a large muscle mass is shown in Fig. 3. Note that the curves are normalized to the value at the fat-muscle interface. Calculations show that the relative absorption is not very sensitive to the thickness of fat, which for subcutaneous fat in people may vary from less than a centimeter to about 2.5 cm. The data in Fig. 3 clearly shows that there is a severe discontinuity in the power absorption at the interface. Furthermore, as the frequency increases, the penetration of the electromagnetic fields into the muscle decreases, and standing waves begin to occur in the fat.

Based on these results, it is clear that significant advantages can be gained by choosing frequencies below 1 GHz for many medical applications. But, once again, the near-field problems as well as the difficulty of localizing the electromagnetic radiation at these lower frequencies often limit the usefulness of devices operating at frequencies much lower than 500 MHz.

This chapter begins with a description of waveguide-type antennas, some used at frequencies as low as 27 MHz. Next, the microstrip and transmission-line kinds of antennas are discussed. Then implantable antennas used in hyperthermia applications and those used to measure the permittivity of tissue *in vivo* are treated.

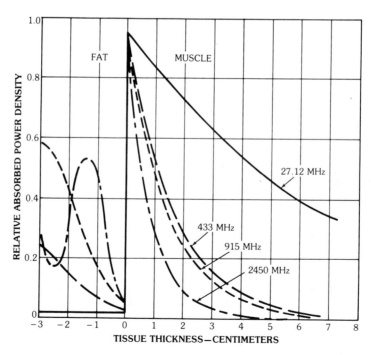

Fig. 3. Relative absorbed power density patterns in planar fat and muscle layers exposed to a plane-wave source. (*After Guy, Lehmann, and Stonebridge [3], © 1974 IEEE*)

This is followed by a description of antennas used to monitor electromagnetic radiation, primarily in relation to hazard assessment, and a discussion of other antenna applicators used for producing hyperthermia. Finally, some experimental procedures for characterizing antennas used in medical applications are discussed.

2. Waveguide- and Radiation-Type Antennas

One class of commonly used antennas for coupling electromagnetic energy into the human body consists of open-ended waveguides of one kind or another. Such radiators have been used primarily for inducing hyperthermia for cancer therapy, which requires effective heating of muscle tissue without excessive heating of the fat layer that overlies the muscle. Waveguide radiators generally provide good coupling of the electromagnetic energy into the tissue, but they tend to be bulky and awkward. Since the waveguides must be physically very large at low frequencies, they have been used mostly at frequencies of 915 MHz and above, although a ridged waveguide loaded with water has been used at frequencies as low as 27 MHz for inducing hyperthermia. Other radiation-type applicators that are used in electromagnetic hyperthermia applications include various forms of reflector antennas and the annular phased array (APA) applicator.

Direct-Contact Waveguide Applicators

Perhaps the simplest effective waveguide applicator is an open-ended dielectric-loaded waveguide. Loading the waveguide with a lossless dielectric of permittivity approximately equal to that of tissue provides a good impedance match and ensures good transmission into the tissue. The dielectric loading also reduces the size of the waveguide. However, the effective heating of the tissue depends strongly on the size of the aperture and the field distribution in it. In an extensive analysis of the internal fields produced by a TE_{10}-mode rectangular aperture in a two-layer semi-infinite tissue model, Guy [4] calculated internal fields as a function of the size of the aperture and the thickness of the fat layer. He showed that of the two frequencies, 915 MHz and 2450 MHz, that are authorized for medical applications in the higher-frequency ranges in the United States, 915 MHz produces a higher ratio of muscle-to-fat heating, which is desirable for therapeutic applications. At 915 MHz, the optimum aperture size was found to be 13 cm high and between 13 and 26 cm in width. On the basis of these results, a 13-cm by 13-cm aperture applicator that includes provisions for cooling the skin was designed and tested [5]. Details are shown in Fig. 4. Measured heating patterns correlated well with calculated values.

This applicator offers the advantage of providing optimal heating of a fat-muscle configuration at 915 MHz. The disadvantage of the applicator is that it is not suitable for heating muscle at depths of more than about four cm because it operates at 915 MHz.

TEM Waveguide Antenna

Van Koughnett and Wyslouzil [6] suggested the use of an inhomogeneously loaded dielectric waveguide to simulate TEM-mode wave propagation for biological studies. Cheung, Dao, and Robinson [7] used open-ended TEM wave-

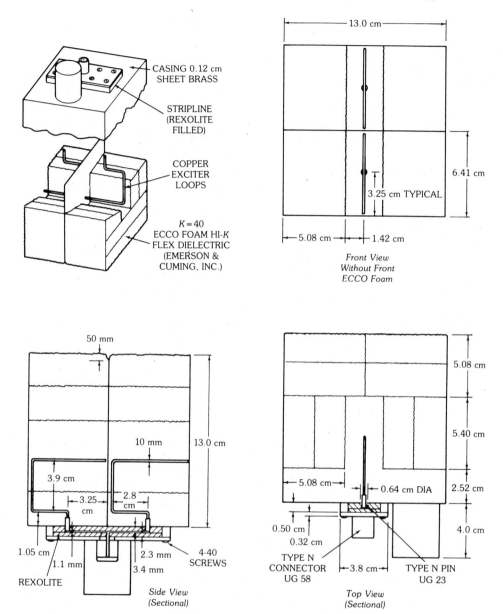

Fig. 4. Details of the air-cooled applicator designed by Guy and others. (*After Guy et al.* [5], © 1978 IEEE)

guide as an applicator for hyperthermia to take advantage of the uniformity of the fields across the aperture of the waveguide. Usually it is desirable to have a power absorption pattern that is uniform in the cross section of the region to be heated. The TEM mode would therefore be more desirable than the TE_{10} waveguide

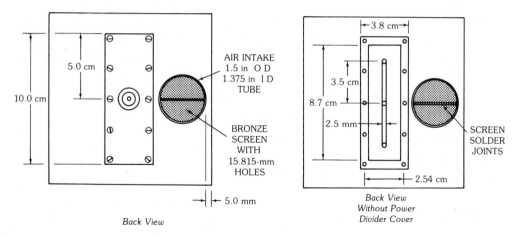

Fig. 4, *continued.*

mode, for example, which has a cosine variation in one direction perpendicular to the direction of propagation.

A diagram of the cross section of the TEM-mode waveguide is shown in Fig. 5. For the field patterns in the central portion of the waveguide to approximate those of the TEM mode, the following conditions must be satisfied [6]:

$$t\sqrt{\epsilon_1/\epsilon_2 - 1} = \lambda_1/4 \qquad (1)$$

where λ_1 is the wavelength in an unbounded region of permittivity ϵ_1, and t, ϵ_1, and ϵ_2 are as defined in Fig. 5. This relationship was obtained by writing the field equations in the regions inside the waveguide, matching the boundary conditions, and then choosing the conditions for which the velocity of propagation in the loaded waveguide is the same as the velocity of light in an unbounded region of permittivity ϵ_1, which is one principal characteristic of the TEM mode. The other principal characteristic, that **E** and **H** are uniform, was found to be true in the central region when (1) is satisfied.

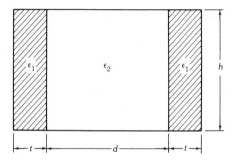

Fig. 5. Cross section of the inhomogeneously dielectric-loaded TEM-mode waveguide.

The height h of the waveguide should not exceed $\lambda_1/2$ so that higher-order modes will not propagate [6]. Also, to ensure that higher-order modes will be cut off, d should not exceed a maximum value d_{m1} that satisfies

$$\sqrt{\epsilon_1/\epsilon_2}\tan\left(\frac{\pi}{2}\sqrt{\frac{\epsilon_1/\epsilon_2}{(\epsilon_1/\epsilon_2)-1}}\right) = -\tan\left(\frac{\pi d_{m1}}{\lambda_1}\right)$$

If, however, the applicator is symmetrical about the midline at $t + d/2$, the first higher-order mode above the TEM mode will be cut off, and then the maximum value of d to ensure cutoff of the higher-order modes is given by the d_{m2} that satisfies

$$\sqrt{\epsilon_1/\epsilon_2}\tan\left(\frac{\pi}{2}\sqrt{\frac{\epsilon_1/\epsilon_2}{(\epsilon_1/\epsilon_2)-1}}\right) = \operatorname{ctn}\left(\frac{\pi d_{m2}}{\lambda_1}\right)$$

In practice, d should be about 10 percent less than d_{m1} or d_{m2} so that the higher-order modes will be sufficiently attenuated.

The impedance of the TEM-mode applicator could be matched to that of tissue by using an ϵ_2 approximately equal to that of tissue. That would cause three problems, however, all related to the rather high permittivity of tissue. First, a high ϵ_2 would probably require a heavy material that would be expensive and would make the applicator very heavy and awkward to use. Secondly, although deionized water could perhaps be used, it is difficult to find a solid material with a high relative permittivity and low enough loss to avoid excessive heating in the large volume occupied by ϵ_2. Thirdly, since ϵ_1 must be greater than ϵ_2, and since the thickness t is related directly to the ratio ϵ_1/ϵ_2, finding a low-loss material with a high enough permittivity for ϵ_1 with a reasonable thickness could be difficult. Cheung, Dao, and Robinson [7] avoided these problems by using air for the central dielectric and using a quarter-wave transformer to match into tissue. Their design for a 2450-MHz applicator is shown in Fig. 6.

The TEM-mode waveguide applicator has the advantage of providing a very uniform field distribution in the cross section. Its disadvantages are that it is relatively narrow-band and, like most other waveguide applicators, is usually practical only at higher frequencies.

Ridged-Waveguide Antennas

As pointed out previously, waveguide applicators often must be too large and bulky to be practical at frequencies low enough to get sufficiently deep penetration into the body. One method of reducing the size of waveguide that will propagate waves at a given frequency is to add ridges. In addition to lowering the cutoff frequency the ridges also reduce the wave impedance, which makes it easier to match into tissue.

Design curves for ridged waveguide are given by Cohn [8], with the dimensions for single- and double-ridged guide denoted as in Fig. 7. Curves for the cutoff frequency for two different values of b_1/a_1 are given in Fig. 8 as a function of a_2/a_1,

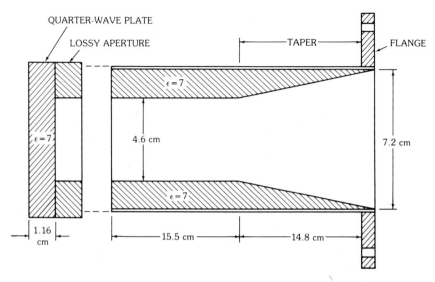

QUARTER-WAVE PLATE

LOSSY APERTURE

TAPER

FLANGE

$\epsilon = 7$

$\epsilon = 7$

$\epsilon = 7$

4.6 cm

7.2 cm

1.16 cm

15.5 cm

14.8 cm

Fig. 6. Top sectional view of the TEM applicator and quarter-wave matching transformer designed by Cheung, Dao, and Robinson. (*After Cheung, Dao, and Robinson [7], © 1977 American Geophysical Union*)

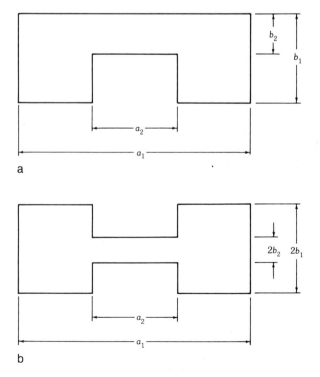

b_2

b_1

a_2

a_1

a

$2b_2$ $2b_1$

a_2

a_1

b

Fig. 7. Dimensions for ridged and double-ridged waveguide. (*a*) For single-ridged waveguide. (*b*) For double-ridged waveguide.

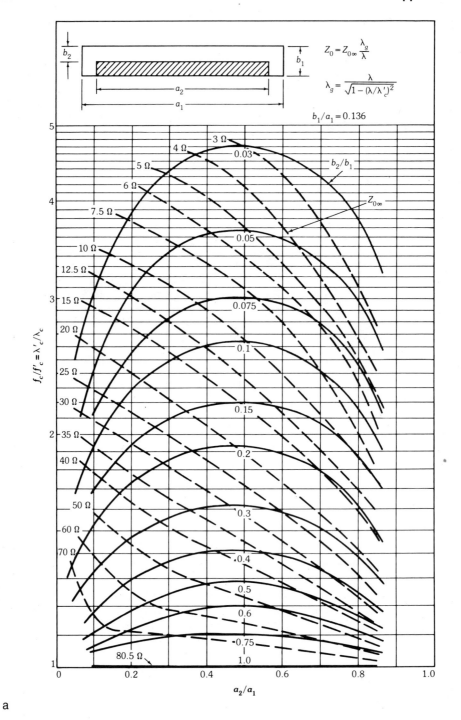

a

Fig. 8. Characteristic impedance and cutoff wavelength of ridged waveguide. (*a*) For b_1/a_1 = 0.136. (*b*) For b_1/a_1 = 0.5. (*After Cohn [8], © 1947 IRE*)

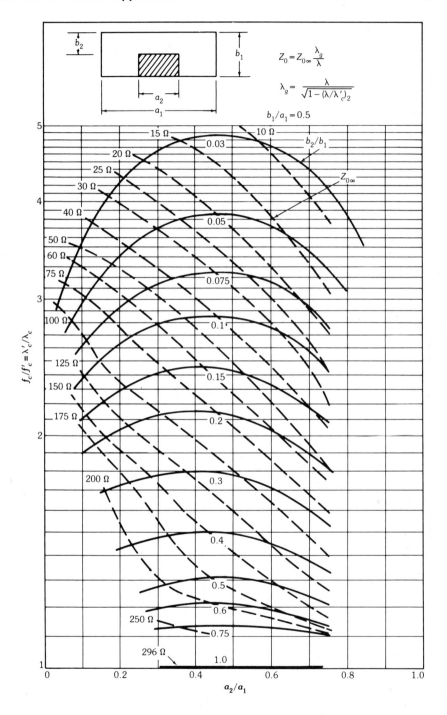

b

Fig. 8, *continued.*

with b_2/b_1 as a parameter. The definitions of the terms in the ordinate are as follows:

f'_c = cutoff frequency of the ridged waveguide

f_c = cutoff frequency of the waveguide without the ridge

$\lambda'_c = c/f'_c$

$\lambda_c = c/f_c$

c = velocity of light in free space

The characteristic impedance of the TE_{10} mode at any frequency f is given by

$$Z_0 = \frac{Z_{0\infty}}{\sqrt{1 - (f'_c/f)^2}}$$

where $Z_{0\infty}$ is the value of Z_0 at infinite frequency, as given in Fig. 8. The wavelength inside the ridged waveguide is given by

$$\lambda_g = \frac{c/f}{\sqrt{1 - (f'_c/f)^2}}$$

The curves in Fig. 8 give good accuracy when $(a_1 - a_2)/2 > b_1$. For double-ridged waveguide the curves in Fig. 8 for cutoff frequency also apply if the definitions for b_1 and b_2 as given in Fig. 7 are used. The characteristic impedance for a double-ridged waveguide may be found by doubling the value of $Z_{0\infty}$ given in Fig. 8 for single-ridged waveguide. Homogeneously dielectric-loaded ridged waveguides may be designed by replacing the free-space wavelength λ_0 by the wavelength in an infinite dielectric medium and using the same design curves.

Ridged-waveguide applicators have been used by several investigators to produce hyperthermia at frequencies as low as 200 MHz. They have the advantage of being more compact and having lower impedance than ordinary waveguides. The lower impedance makes it easier to match into biological tissue, which because of its relatively high dielectric constant has a much lower impedance than free space. Ridged waveguides have the disadvantage of concentrating the electric field in the space above the ridge, which sometimes leads to undesirable hot spots, especially when b_1 is very small.

Sterzer and colleagues [9] developed a large ridged-waveguide applicator that operates at 27.12 MHz. It is loaded with deionized water to further reduce its size, but it is still so large that the patient must sit or lie on the aperture. Because it operates at a relatively low frequency it produces good deep regional heating.

Annular Phased Array Applicator [10]

The annular phased array is an applicator which is specifically designed for hyperthermia applications. It is a large, octagonal, multidirectional applicator with a central opening of 51 cm in diameter. It is designed to treat large body areas, such as the thorax, abdomen, pelvis, and thighs. The basic radiating element in this annular phased array can be any one of many kinds of local applicators, such as those discussed previously. Fig. 9 illustrates the locations of the radiating apertures along the inside surface of the octagonal opening. Fig. 10 shows that the annular phased array applicator actually consists of sixteen applicators arranged in annular form around the patient's body.

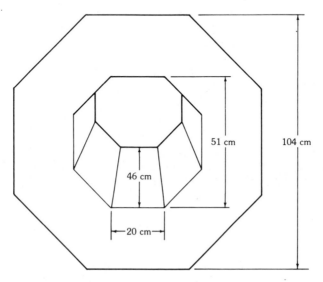

Fig. 9. Annular phased array applicator. (*After Turner [10]*)

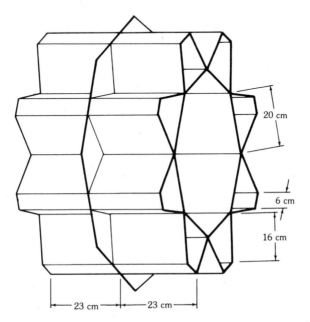

Fig. 10. Annular phased array: view of sixteen apertures. (*After Turner [10]*)

The resultant wavefronts are shown in Fig. 11. The frequencies of operation are usually chosen to be in the range between 50 and 110 MHz. All eight wavefronts operate synchronously in phase with equal amplitudes so that a region of constructive interference occurs at the axis of the array. The array is powered by four coaxial inputs, each activating a quadrant of the array. The apertures are also lined with water-filled pvc plastic bags which are used to direct the energy efficiently into the region of the body to be heated. The heating pattern of the annular phased array has been extensively tested experimentally [11] and numerically [12].

3. Microstrip Antennas and Applicators

Antennas consisting of radiating structures on microstrip transmission line are attractive applicators for coupling electromagnetic energy into biological systems because they can be very flat, compact, and mechanically convenient. These kinds of applicators are also relatively easy to design and fabricate. However, they tend to be narrow-band, produce some leakage radiation, and would be physically too large to be useful at lower frequencies where the penetration is best. They have been found to be useful for heating tumors near the body surface, typically using frequencies of 433, 915, or 2450 MHz, and they have been used for diagnostics.

Since microstrip antennas are treated extensively elsewhere in this book, the discussion in this chapter is limited to their medical applications. The primary difference between designing microstrip antennas for medical applications and other applications is that for medical applications the antenna must usually be designed to radiate into a lossy medium of relatively high permittivity, whereas in other applications the antenna often is designed to radiate into free space.

Microstrip Patch Antennas

A microstrip patch antenna consists of a conducting patch on the side opposite the ground plane of a microstrip transmission line, as shown in Fig. 12a. The patch

Fig. 11. Annular phased array radiated wavefronts. (*After Turner [10]*)

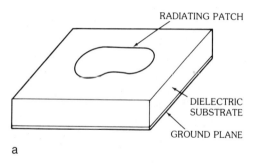

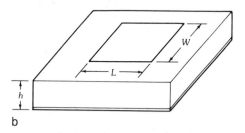

Fig. 12. Microstrip patch antennas. (*a*) Geometry of an arbitrarily shaped radiating patch antenna on microstrip transmission line. (*b*) Rectangular patch antenna geometry and parameters as defined by Bahl and others [13].

can be of arbitrary shape, but since analysis of antenna characteristics is easier for regular shapes, the rectangular, circular, loop, triangular, and elliptical shapes have been used most often. Only the rectangular patch (Fig. 12b) and the loop have been used much in medical applications. The rectangular patch radiator is discussed in this section and the loop radiator is discussed in the next section.

Bahl and his colleagues have developed design procedures for microstrip patch antennas for producing hyperthermia. Their equations for patch antennas radiating into free space are given first, and then the modifications to these equations for radiation into biological tissue are described.

The design procedure outlined by Bahl and Bhartia [13] for microstrip patch antennas to radiate into free space starts with choosing a practical value for the patch width W, which is given by

$$W = \frac{c}{2f_r}\left(\frac{\epsilon_r + 1}{2}\right)^{-1/2} \tag{2}$$

where f_r is the resonant frequency, c is the velocity of light, and ϵ_r is the relative permittivity of the substrate. For smaller values of W than given by (2), the radiation efficiency is lower, and for larger values, higher-order modes begin to appear. The effective dielectric constant ϵ_{eff} is given by

$$\epsilon_{\text{eff}} = \frac{\epsilon_r + 1}{2} + \left(\frac{\epsilon_r - 1}{2}\right)\left(1 + \frac{12h}{W}\right)^{-1/2} \tag{3}$$

and $\Delta\ell$ is given by

$$\Delta\ell = 0.412h\frac{(\epsilon_{\text{eff}} + 0.3)(W/h + 0.264)}{(\epsilon_{\text{eff}} - 0.258)(W/h + 0.8)} \tag{4}$$

where $2\Delta\ell$ is the difference between the effective length and the physical length L of the radiator. The physical length can be found from

$$L = \frac{c}{2f_r\sqrt{\epsilon_{\text{eff}}}} - 2\Delta\ell \tag{5}$$

These equations do not apply when the antenna is placed near to or in contact with biological tissue. However, the equations for W, $\Delta\ell$, and L can be used for an antenna in contact with a dielectric medium if the proper ϵ_{eff} is obtained for that case. The calculation of ϵ_{eff} is not easy, but Bahl and Stuchly [14] give results obtained from a variational method for particular cases. The geometry is shown in Fig. 13. Their results for two specific cases are reproduced in Fig. 14.

Microstrip Loop Radiators

Bahl and colleagues [15] designed and tested microstrip transmission-line loop radiators in the configuration shown in Fig. 15. The ring conductor, which is the radiating element, is fed on one side by a coaxial line protruding through the substrate. A pin shorting the ring to the ground plane through the substrate is located 180° from the feed point. In practice, it was found that best results for heating biological tissue were obtained when a water bolus was placed between the loop radiator and the biological body. Although the relationship beween the size of the ring and the resonant frequency is rather complex, Bahl and colleagues [15] developed a simple empirical formula that applies when the ring is placed on a thick, lossy, dielectric medium. For this case the mean radius R of the ring is given in centimeters by

$$R \cong \frac{6}{f\sqrt{|\epsilon_r|}}$$

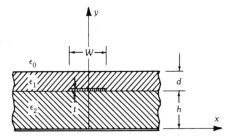

Fig. 13. Geometry of a microstrip covered with dielectric as used by Bahl and Stuchly [14] to calculate the effective dielectric constant. (*After Bahl and Stuchly [14], © 1980 IEEE*)

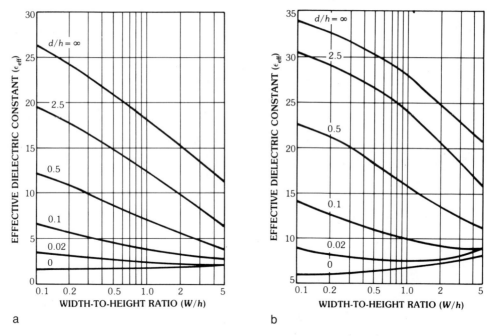

Fig. 14. Effective dielectric constant as a function of W/h for $\epsilon_{r1} = 75.0$ and various values of d/h. (*a*) For $\epsilon_{r2} = 2.32$. (*b*) For $\epsilon_{r2} = 10.0$. (*After Bahl and Stuchly [14], © 1980 IEEE*)

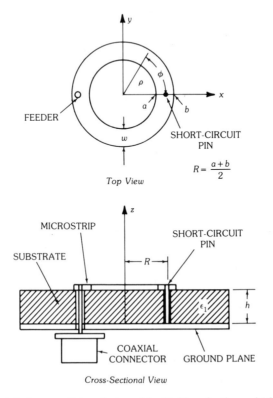

$$R = \frac{a+b}{2}$$

Top View

Cross-Sectional View

Fig. 15. The microstrip loop radiator designed by Bahl and others. (*After Bahl et al. [15], © 1982 IEEE*)

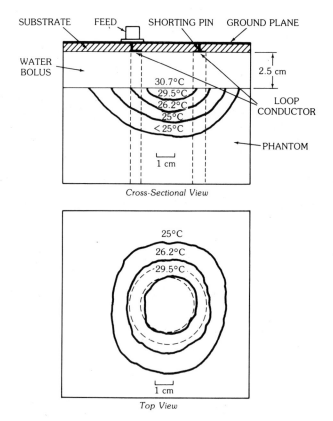

Fig. 16. Temperature distribution in the muscle phantom irradiated through a water bolus. (*a*) By the 433-MHz radiator. (*b*) By the 915-MHz radiator. (*After Bahl et al. [15]*, © *1982 IEEE*)

where f is the resonant frequency in gigahertz, and ϵ_r is the complex relative permittivity of the medium on which the ring is placed. The width of the ring conductor is chosen to match the impedance of the radiator to that of the coaxial transmission line. Typical design values for structures tested by Bahl's group [15] are $R = 1.6$ cm and $W = 0.5$ for $f = 433$ MHz, and $R = 0.9$ cm, $W = 0.3$ cm for $f = 915$ MHz. These values are for a duroid substrate with a relative permittivity of 2.32 and with the radiating loop placed on a water bolus. Measured heating patterns in muscle phantom [15] for these two cases are shown in Fig. 16. The heating was effective to a depth of approximately 1.6 cm at 433 MHz and 1.1 cm at 915 MHz.

Microstrip Slot Antennas

Microstrip slot antennas have also been designed for hyperthermia applications. Bahl, Stuchly, and Stuchly [16] used a rectangular slot in the ground plane of a microstrip transmission line, as shown in Fig. 17. A metal reflector was used on the opposite side of the ground plane and on the sides of the substrate to restrict the radiation to the half-space above the slot. The distance d was adjusted

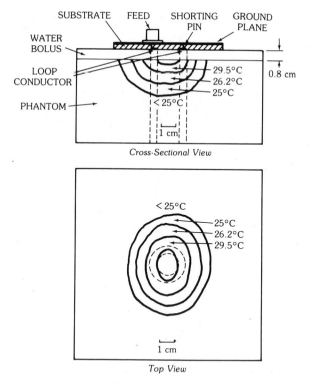

Fig. 16, continued.

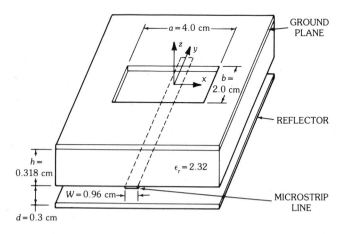

Fig. 17. Geometry of microstrip slot radiator. (*After Bahl, Stuchly, and Stuchly [16]*, © *1980 IEEE*)

to obtain a good impedance match between the transmission line and the antenna. Their dimensions for an applicator at 2.45 GHz are shown in Fig. 17. In this case the dimensions of the slot are approximately 1λ by 2λ when the applicator is in contact with muscle tissue ($\epsilon_r \cong 48 - j14$), where λ is the wavelength in muscle. The usual design procedures for a slot antenna radiating into a dielectric medium can be used to determine dimensions of the slot. The radiator in Fig. 17 produced a heating pattern in simulated muscle material similar to that of an open-ended waveguide aperture. The heating was relatively uniform over an area of about 1.5 cm by 1 cm. The leakage radiation was about 35 dB below the transmitted power.

The microstrip slot radiator has the advantages of being flat, compact, and lightweight, with good impedance matching and a relatively uniform heating pattern. Since the slot dimensions must be an appreciable fraction of a wavelength, this kind of applicator will probably be useful only at the higher frequencies and therefore only for relatively shallow heating.

Coplanar Transmission-Line Applicator

Iskander and Durney [17] have used an applicator for coupling electromagnetic energy into and out of the body that is based on a coplanar transmission line, as shown in Fig. 18. The coplanar transmission line consists of a center conductor separated from a coplanar ground plane by an air gap, with both the center conductor and ground plane located on a very low permittivity, thin substrate. A small coaxial cable is used to feed the coplanar transmission line, which is designed to be matched to the coaxial line when the applicator is placed on biological tissue. A resistor equal to the characteristic impedance of the coplanar transmission line when it is on tissue is located at the end of the coplanar line to terminate it.

The design equation for the coplanar transmission line is [18]

$$Z_0 = 60\pi \frac{K'}{K\sqrt{(\epsilon_{r1} + 1)(\epsilon_{r2} + 1)}}$$

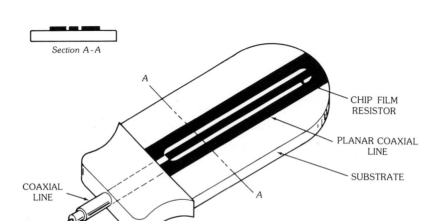

Fig. 18. Geometry of the coplanar transmission-line applicator. (*After Iskander and Durney [17], © 1979 IEEE*)

where

ϵ_{r1} = relative permittivity of region 1

ϵ_{r2} = relative permittivity of region 2

$k = OA/OC$

$k' = \sqrt{1 - k^2}$

K = complete elliptic integral of k

K' = complete elliptic integral of k'

and the dimensions are as given in Fig. 19. To design the transmission line, a convenient width OA is first chosen. Then the gap between the strips is calculated from the design equation. To obtain a good match between the coaxial line and the coplanar line, a taper in the gaps is used, as indicated in Fig. 17, and the transition is adjusted by trial and error using a time-domain reflectometer until the desired match is obtained. In practice, a good match can be obtained without much difficulty.

At 915 MHz this applicator has been found to provide good coupling into the human body, with low leakage radiation. It has the advantages of being light-weight, compact, and easy to tape on the skin. Although it has not been tested as a hyperthermia applicator it would not be expected to provide a very desirable heating pattern because of the nature of the field pattern.

Arrays of Microstrip Antennas

The compactness of microstrip antennas suggests their use in arrays for biological applications. Such arrays can be readily designed using the techniques described elsewhere in this book, but arrays have not been widely used in biological applications because they are practical only at the higher frequencies where penetration of electromagnetic energy into biological tissue is shallow. Sterzer and colleagues [9] have used an array of printed-circuit dipole arrays for heating superficial tumors at 2450 MHz. Other applications, such as receiving antennas for radiometry in biological applications, will probably occur.

4. Implantable Antennas (Radiators) for Localized Cancer Treatment

Heating deep-seated tumors noninvasively using external electromagnetic radiation is extremely difficult because, as explained in the introduction, both the

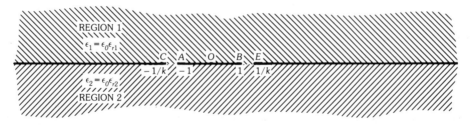

Fig. 19. Dimensions of the coplanar transmission line for use in the design equation. (*After Iskander and Hamid [18], International Microwave Power Institute, Clifton, Virginia*)

shallow penetration of the fields at higher frequencies and the near-field dominance of the radiation from electrically small applicators at lower frequencies tend to overheat the surface before the tumor is heated sufficiently. A large hyperthermia machine, such as the annular phased array system described in Section 2, produces deep regional heating but tends to be expensive and is limited to specialized centers. For these reasons the use of implantable antennas [19] that would produce localized deep heating for cancer treatment without overheating the surface is certainly attractive in situations where the implantation of the antennas is practical.

Several designs of these implantable radiators are available. Fig. 20 shows some antennas suitable for heating tumors in solid organs or in a hollow organ that can be reached through a body orifice [19]. Specifically, Figs. 20a and 20b describe antenna geometries that have been used to heat tumors implanted in solid organs. The radiator in Fig. 20a consists simply of the center conductor extended an appropriate distance from the end of a coaxial cable. The coaxial line may be made of the subminiature semirigid coax that is available in a variety of materials and sizes down to a diameter of 0.2 mm. The extended length of the center conductor may be adjusted to match the impedance of the implanted radiator to the 50-Ω impedance of the feed cable. To some extent the size of the heated area can also be controlled by adjusting the length of the center conductor. The purpose of the dielectric sheathing is to extend the propagation distance along the needle. The radiator shown in Fig. 20b has also been used for invasively heating deep tissues. Fig. 21 shows an alternative structure consisting of a hollow stainless-steel hypodermic needle that serves as a monopole antenna and permits simultaneous chemotherapeutic treatment of tissue [20]. The chemotherapy is accomplished by introducing the chemicals through an inlet port located at the end of the needle. By proper choice of the length-to-radius and the length-to-operating-wavelength ratios of the antenna, both satisfactory radiation patterns and impedance matching can be obtained. In this structure, however, the outer conductor, which is flanged to serve as a ground plane flush to the skin, limits the penetration of the device.

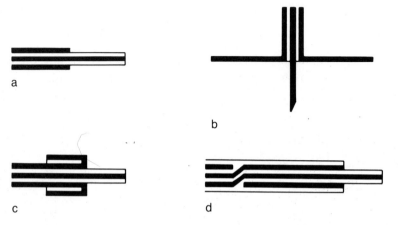

Fig. 20. Implantable radiator types (not drawn to scale). (*a*) Needle radiator. (*b*) Hypodermic monopole and ground-plane flange. (*c*) Sleeve antenna. (*d*) Cross-switch section plus needle. (*After Taylor [19], © 1980 IEEE*)

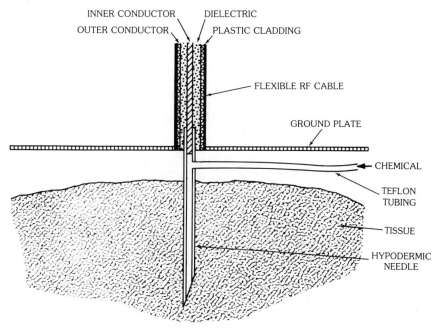

INNER CONDUCTOR DIELECTRIC
OUTER CONDUCTOR PLASTIC CLADDING

FLEXIBLE RF CABLE

GROUND PLATE

CHEMICAL

TEFLON
TUBING

TISSUE

HYPODERMIC
NEEDLE

Fig. 21. Hypodermic needle monopole radiator. (*After Bigu-del-Blanco and Romero-Sierra [20], © 1977 Pergamon Journals Ltd.*)

In applications where the tumor is in a hollow organ that can be reached through a body orifice, a coaxial radiator located at the center of a body cavity is usually used to heat the organ wall. Fig. 20c illustrates such a radiator, in which the needle radiator configuration is replaced by a coaxial sleeve radiator to terminate transmission back to the generator. The dipole antenna in this case is usually formed of a quarter-wavelength section of the coaxial center conductor with the outer conductor removed and a quarter-wavelength choke (sleeve) placed to minimize the leakage radiation along the outer conductor (see Fig. 22a). Either a metallic semicylindrical reflector (see Fig. 22b) or a polytetrafluoroethylene (ptfe) bulb covering the dipole antenna may be used to direct the energy [21]. Fig. 23 shows a cutaway view of the bulb. This specific bulb was made of ptfe because it has a low dielectric constant. Its cross section is pear shaped to produce the uneven radiation sometimes needed for directional heating [22]. For example, the eccentric position of the antenna within the bulb (see Fig. 23b) places a sector of about 90° of the external surface of the bulb's "active surface" closest to the antenna. In a typical operation the applicator is inserted into a body cavity (such as the rectum) with its active surface facing the site to be treated.

In some clinical situations the geometry of the tumor requires the radiation pattern of the antenna to be cylindrically symmetric and to extend for a length up to 10 cm. A needle radiator (Fig. 20a) is not expected to produce a heating pattern this long, even if a dielectric sheathing is used to extend the propagation distance along the needle. To get a longer pattern it is suggested [19] that the inner and outer conductors of the cable be cross-connected at a point a half-wavelength (in the dielectric) away from the needle radiator (see Fig. 20d). The presence of

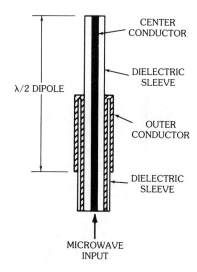

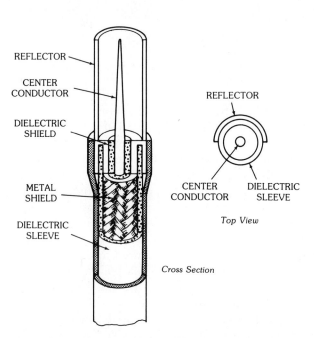

Fig. 22. Dipole antenna and applicator with reflector. (*a*) Longitudinal section of dipole antenna. (*b*) Coaxial applicator with directive reflector. (*After Mendecki et al. [21], © 1980 Pergamon Journals Ltd.*)

this excited gap extends the current longitudinally on the outer conductor, thus resulting in a longitudinal energy deposition pattern. The geometry of this radiator together with its radiation pattern is shown in Fig. 24. From Fig. 24 it is clear that the radiation pattern displaying two principal peaks with an interference maximum in between is very roughly that of a collinear array of two half-wave dipoles [19].

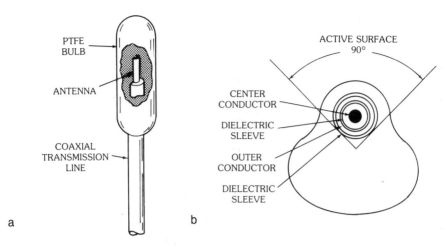

Fig. 23. The ptfe bulb. (*a*) Cutaway view showing the position of the antenna within the bulb. (*b*) Cross section showing the pear-shaped bulb and the eccentric location of the antenna. (*After Mendecki et al. [22], © 1978 Pergamon Journals Ltd.*)

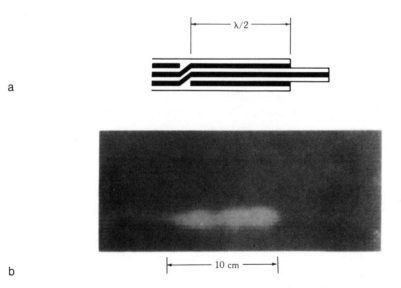

Fig. 24. A cross-switch radiator. (*a*) Diagram of radiator. (*b*) Thermogram of its energy deposition pattern. (*After Taylor [19], © 1980 IEEE*)

The radial pattern is symmetric, which might be required to fit the geometry of the tumor.

Implanted radiators provide the significant advantages of highly localized and highly controlled heating. However, they can be used only in certain cases where the implantation is practical. Detailed results that illustrate the usefulness of these radiators and their ability to produce controllable localized heating of malignant internal structures are described elsewhere.

5. Antennas for In-Vivo Measurement of the Complex Permittivity of Tissue

Measurement of the dielectric properties of tissue is important in evaluating biological effects of electromagnetic radiation, since it provides information necessary to calculate the rf power absorption by biological models and also to simulate these tissues in phantom experiments. These phantom experiments are useful not only in evaluating the biological hazards, but also in evaluating the electromagnetic heating patterns of devices used to produce hyperthermia. Also, many biophysical interaction mechanisms of electromagnetic fields with biological systems can be inferred from accurate knowledge of tissue permittivity over a broad frequency band.

In spite of the many available sophisticated techniques suitable for making dielectric measurements, there is still a significant need to make these measurements *in-vivo* and without extracting tissue samples. This is because of the extreme difficulty in preparing suitable tissue samples and also because of the possible deterioration of the electrical properties of tissue with time in the *in-vivo* measurements. This section describes two of the recently developed procedures for measuring the dielectric properties of tissue. These procedures basically utilize open-ended sections of transmission lines with the extended portion of the center conductor being embedded in the dielectric under test. Also described is a simple procedure for optimizing the dimensions of the sample holder so as to provide the least measurement uncertainties in the desired frequency range.

Theoretical Basis for the In-Vivo Probes

There are two types of probe configurations that have been used for making *in-vivo* dielectric measurements. These are the short monopole antenna [23] and the open-ended coaxial line [24]. In addition, there has been some effort to measure dielectric properties at X-band and millimeter wavelengths using open-ended waveguides. The following is a summary of the theoretical basis for the measurement procedure using coaxial sample holders.

The Short Monopole Antenna as a Dielectric Probe [23]—An open-ended coaxial line probe with an extended center conductor embedded in the dielectric under test has been successfully used for the *in-vivo* measurement of the dielectric properties of tissue. Results have been obtained in the frequency band between 10 MHz and 10 GHz. The primary theoretical basis for the *in-vivo* probe measurement concept is found in an antenna modeling theorem [25] that can be made applicable to a short (antenna length much less than 0.1 wavelength) monopole antenna. This theorem simply relates the impedance of a short antenna operating at a frequency ω, and radiating in the material under test, to the impedance value at a frequency $n\omega$ when surrounded by free space. When non-magnetic materials are considered, this theorem has the mathematical form

$$\frac{Z(\omega,\epsilon)}{\eta} = \frac{Z(n\omega,\epsilon_0)}{\eta_0} \tag{6}$$

where

Z = impedance

ϵ = complex permittivity of the material being measured

ϵ_0 = permittivity of free space

$\eta = \sqrt{\mu_0/\epsilon}$ = intrinsic impedance of the material being measured

$\eta_0 = \sqrt{\mu_0/\epsilon_0}$ = intrinsic impedance of free space

$n = \sqrt{\epsilon/\epsilon_0}$ = index of refraction of the material being measured relative to free space

In order for the theorem to be valid, the following conditions must be satisfied:

1. The short antenna or probe must be completely immersed in the material to be measured
2. The material sample size must be larger than the penetration depth $(1/e)$ of the field pattern within the material (otherwise, the reaction theorem must be solved for the difference in impedance due to measuring a sample of finite volume)
3. An analytical expression based on the current distribution along the probe must be obtained for the impedance of the probe

When a short monopole antenna is used as the probe the necessary analytical expression for probe impedance has the form

$$Z(n\omega, \epsilon_0) = A\omega^2 + \frac{1}{j\omega C} \tag{7}$$

where A and C are constants determined by the probe's dimensions. This expression is valid so long as the proble length is less than 10 percent of the wavelength in the material being measured.

With a knowledge of the antenna constants A and C as well as the magnitude and phase of the complex impedance $Z(\omega, \epsilon)$ of the short monopole antenna in the material being measured, the values of relative dielectric constant and conductivity can be obtained from (6). In order to characterize the electrical properties of the material in terms of the relative dielectric constant ϵ' and conductivity σ, the complex index of refraction is defined in terms of the loss tangent of the material as

$$n = \sqrt{\epsilon/\epsilon_0} = \sqrt{\epsilon'(1 - j\tan\delta)} \tag{8}$$

where $\tan\delta = \sigma/\omega\epsilon'$. Using the antenna impedance in (7) in the modeling theorem of (6), the following expressions [23] are obtained for the resistance and reactance of the complex impedance $Z(\omega, \epsilon) = R + jX$:

$$R = \frac{\sin 2\delta}{2\epsilon'\omega C} + A\sqrt{\epsilon'}\,\omega^2 \sqrt{\frac{\sec\delta + 1}{2}} \tag{9}$$

and

$$X = \frac{\cos^2\delta}{\epsilon'\omega C} + A\sqrt{\epsilon'}\,\omega^2 \sqrt{\frac{\sec\delta - 1}{2}} \tag{10}$$

All the parameters in the above pair of equations except ϵ' and δ are known or can be determined from experimental measurements. Because simultaneous solution of these equations would be quite difficult, an iterative method of solution is usually utilized. The second term in both (9) and (10) is small at low frequencies. When these terms are neglected the following equations result:

$$R = \frac{\sin 2\delta}{2\epsilon'\omega C} \tag{11}$$

and

$$X = \frac{\cos^2\delta}{\epsilon'\omega C} \tag{12}$$

Solutions to these equations are easily obtained by noting that $\tan\delta = R/X$; therefore, by measuring the input impedance of a short monopole antenna inserted into a material, it is possible to calculate both the relative dielectric constant ϵ' and the conductivity σ.

The Open-Ended Coaxial Line as a Dielectric Probe—The other probe that has been used in the *in-vivo* measurements of dielectric properties is in essence a special case of the monopole antenna described previously [24]. An open coaxial line, placed in contact with a test sample, is used as a sensor. The equivalent circuit of the sensor consists of two elements (Fig. 25), a lossy capacitor $C(\epsilon_r)$ and a capacitor C_f that accounts for the fringing field in the Teflon. Here $C(\epsilon_r) = C_0\epsilon_r$, where C_0 is the capacitance when the line is in the air and ϵ_r is the complex relative permittivity. This equivalent circuit is valid only at frequencies for which the dimensions of the line are small compared to a wavelength so that the open end of the line does not radiate. At higher frequencies, increased evanescent TM modes excited at the junction discontinuity cause C_0 to increase with frequency [24]. When the evanescent modes are taken into account, C_0 should be replaced by $C_0 + Af^2$ (see the previous subsection), where A is a constant dependent on the line dimensions. Furthermore, the evanescent modes may become propagating modes when a high dielectric constant material is placed at the interface. For example, the TM_{01} mode propagates when $\lambda_\epsilon < 2.03\,(a - b)$, where λ_ϵ is the wavelength in the external dielectric and a, b are the line dimensions shown in Fig. 25a [24]. To account for the radiation produced by the propagating modes it may be necessary to add a parallel conductivity term to the equivalent circuit of Fig. 25b. When the frequency is low enough that the equivalent circuit of Fig. 25b is valid, the input reflection coefficient at the plane of the discontinuity can be calculated from

$$\hat{\Gamma} = \Gamma e^{j\phi} = \frac{1 - j\omega Z_0[C(\epsilon_r) + C_f]}{1 + j\omega Z_0[C(\epsilon_r) + C_f]} \tag{13}$$

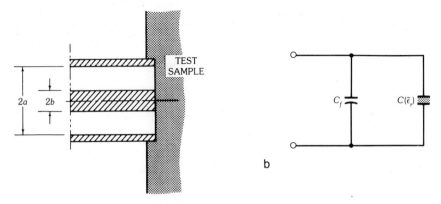

Fig. 25. Open-ended coaxial line sensor. (*a*) Coaxial line terminated with sample. (*b*) Equivalent circuit. (*After Athey, Stuchly, and Stuchly [24],* © *1982 IEEE*)

Solving for the permittivity gives

$$\epsilon_r = \frac{1 - \hat{\Gamma}}{j\omega Z_0 C_0 (1 + \hat{\Gamma})} - \frac{C_f}{C_0} \tag{14}$$

It has been shown [26] that for a given accuracy in measurement of the reflection coefficient, the greatest accuracy in determining the permittivity is obtained when

$$C_0 = \frac{1}{\omega Z_0 \sqrt{\epsilon'^2 + \epsilon''^2}} \tag{15}$$

where $\epsilon_r = \epsilon' - j\epsilon''$. Strictly speaking, this is valid only when the uncertainties in the magnitude and phase of the reflection coefficient are approximately the same, i.e., when $\Delta\phi \cong \Delta\Gamma/\Gamma$. In other cases this is still a good compromise [26].

Table 1 lists approximate values of the optimum capacitance calculated from (15) for several biological substances and reference materials. Table 2 lists the total capacitance C_T for a few commercially available coaxial lines [24].

Table 1. Optimum Capacitance (in Picofarads) for a Probe in Various Materials (*After Athey, Stuchly, and Stuchly [24],* © *1982 IEEE*)

Material	Frequency (GHz)				
	0.01	0.05	0.1	0.05	1.0
Muscle	0.25	0.25	0.25	0.08	0.035
Fat	0.8	0.8	0.8	0.45	0.45
Water, 25°C	4	0.8	0.4	0.08	0.04
0.02 N NaCL, 25°C	0.37	0.37	0.37	0.08	0.04
0.08 N NaCL, 25°C	0.19	0.19	0.19	0.08	0.04
0.25 N NaCL, 25°C	0.08	0.07	0.07	0.055	0.035

Table 2. Capacitance of Open Coaxial Lines (*After Athey, Stuchly, and Stuchly [24],* © *1982 IEEE*)

Line	a (cm)	b (cm)	C_T (pF)
14 mm, air	0.7145	0.3102	0.14
7 mm, air	0.35	0.1520	0.079
8.3 mm, Teflon	0.362	0.1124	0.055
6.4 mm, Teflon	0.2655	0.0824	0.041
3.6 mm, Teflon	0.1499	0.0455	0.027
2.2 mm, Teflon	0.0838	0.0255	0.016

An alternative and more general procedure for determining the optimum value of capacitance for measuring the permittivity of a given dielectric in a specified frequency band is described in the next section.

Detailed Construction of the In-Vivo Probe

A diagram of an *in-vivo* measurement probe [27] is shown in Fig. 26. It can be seen that the probe is essentially a section of open-ended semirigid coaxial cable with a slightly extended center conductor. The small ground plane may or may not be included to minimize fringing effects. The following procedure can be used to construct the *in-vivo* measurement probe:

Step 1
Remove the Teflon dielectric and center conductor from a short piece (approximately 3 cm length) of semirigid coaxial cable.
Step 2
Silver-solder a small circular disk to one end of the tube formed by the now-empty outer conductor. Machine the surface of this disk smooth.
Step 3
Electroplate the resulting item and the center conductor that was previously removed. Use either nickel, gold, or platinum plating.

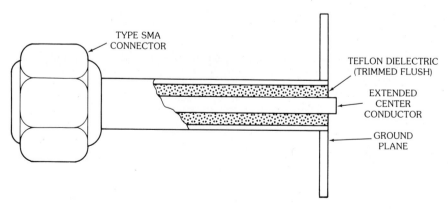

Fig. 26. Diagram of *in-vivo* measurement probe. (*After Toler and Seals [27]*)

Step 4

Solder a connector to the other end of the outer conductor tube.

Step 5

Replace the Teflon dielectric and the plated center conductor, leaving the center conductor slightly extended.

Step 6

Finally, trim any excess Teflon away from the extended center conductor.

Optimizing the Length of the Center Conductor in the Coaxial Sample Holder

In the *in-vivo* measurement procedure involving the open-ended coaxial line with or without the extended portion of the center conductor, the probe can be conveniently represented by a simple shunt capacitor terminating the transmission line at lower frequencies where the dimensions of the transmission line and, in particular, the extended portion of the center conductor are small compared to a wavelength [28]. In using this dielectric probe it is extremely important to determine the optimum value of the capacitance that can be used for measurements over a specified frequency band. There is, however, no easy procedure to determine such a value since the optimum capacitance depends on

  (*a*) the value of the complex permittivity under test, and
  (*b*) the specific frequency band of interest

One of the authors (Iskander) has devised a procedure to calculate such an optimum value of the capacitance that can be used for a given dielectric (i.e., approximate value of the complex permittivity is required) and in a given frequency range [29]. While a detailed description of the method can be obtained from the cited reference, the results generally indicated that larger values of capacitances are required at lower frequencies (which is expected since lumped-circuit elements are usually used at these frequencies), and smaller values of capacitances are preferable at higher frequencies, which is also expected because smaller capacitances in the form of distributed circuit elements are usually used at higher frequencies.

In *in-vivo* dielectric measurements using the open-ended section of the transmission line, smaller values of capacitance are achieved by decreasing the length of the extended portion of the center conductor. The open-ended coaxial transmission line sample holder, therefore, is suitable for making dielectric measurements at higher frequencies and may be used at lower frequencies (e.g., hundreds of megahertz) if dielectrics with high dielectric constants are used. For materials with lower dielectric constants, on the other hand, larger values of capacitances are required. This could be achieved by extending the length of the center conductor further to increase the value of the capacitance of the sample holder. Under this condition, however, the input impedance of the sample holder would not be approximated by a capacitance terminating the transmission line and instead the expression for the input impedance given in (9) and (10) should be used to take into account the radiation resistance of the antenna. These low-frequency extensions, particularly when dielectrics with low complex permittivities are being tested, are still in progress in our laboratories [28].

6. Antennas for Monitoring RF Radiation

In any situation that could involve exposure of people to significant amounts of electromagnetic radiation, monitoring the electromagnetic radiation levels is important. In some cases a device that measures radiation in a given area would be satisfactory. In other cases it would be desirable to have a device that could be worn by a person to monitor the radiation to which that person is exposed. Area radiation monitors are commercially available, but a satisfactory personnel dosimeter is not, probably because of the many difficulties encountered in the design and construction of such a device.

An important aspect of either kind of radiation monitor is what parameter of the electromagnetic fields should be measured. It is generally known that for a simple plane wave any one of a number of field parameters constitutes a valid index and the direction of propagation [30]. These parameters include the magnitudes of the electric field **E**, of the magnetic field **H**, and the time-average power density. Provided that the source does not show supergain, plane-wave approximations are appropriate in the practical sense when the distance from the source is large compared to $2D^2/\lambda$, where λ is the wavelength in the medium and D is the largest dimension of the source. Therefore, for quantifying this far-field plane-wave-type radiation, it is sufficient to measure the magnitude of either the electric or magnetic field. Other parameters of interest can easily be computed from these magnitudes without approximation for given properties of the propagation medium. In the near field, however, magnitudes of the electric and magnetic fields are not generally easily related; also, the other radiation parameters cannot simply be calculated from the magnitudes of these fields. For accurate quantification of the electromagnetic radiation in the near field, therefore, the instrumentation suitable for power density measurements (which is reasonably satisfactory in plane-wave radiation) is inadequate and instead independent measurement of the electric and magnetic fields should be performed.

The basic requirements for radiation monitors and some antennas to measure electric and magnetic fields are described in this section.

Basic Requirements for Field Probes

Desirable features in the design of an rf and microwave radiation monitor include the following:

1. The measuring probes should disturb the electromagnetic fields minimally.
2. The probe response should be isotropic and hence independent of the field polarization.
3. It should have a flat frequency response over a broad frequency band. This characteristic is important in making field measurements without precise knowledge of the frequency.
4. It should exhibit linear dynamic response with the field or power quantity being measured.
5. It should be suitable for operation in the near- or far-field regions. This simply requires independent measurement of the electric- and magnetic-field components in three orthogonal directions.

6.   The field probes (**E**- and **H**-field sensors) as well as the connecting leads should preferably be made of high-resistance material (carbon-loaded Teflon) to minimize their interference with the fields being measured.

In addition to these basic requirements the probe should also have fast response and be small enough to make spatially resolved measurements in small regions comparable in size to organs of the body.

Detection Methods

In designing a suitable antenna system for detecting the **E** and **H** fields of rf radiation, it is necessary to include in the probe a means for detecting the induced currents. The type of detectors used can generally be segregated into two basic types:

1.   Thermocouple type devices that measure the electromagnetic power by sensing the change in temperature between the hot and cold junctions of a thermocouple device. Examples of this thermocouple detecting device include the electromagnetic radiation monitor developed by the Narda Microwave Corporation [31]. In this device the thermocouple materials of the sensor are antimony and bismuth deposited on a substrate of plastic or mica. The sensitivity of the sensor is controlled by variation of the substrate and the dimensions of the hot junction of the thermocouple.
2.   Diode detectors, which produce a current or voltage related to the fields of the electromagnetic radiation [32].

The usefulness of thermocouple detection devices is limited by their sensitivity to changes in ambient temperature and other changes in all forms of heat. These devices also have slow response time and may even burn out if overloaded. Their main advantage is linearity of response with intensity of electromagnetic power. Diode detectors are more sensitive than thermocouple detectors, but the diode response is nonlinear with electromagnetic power. For example, if the signal level is such that the diode is operating in the square-law region, the induced current will be proportional to the square of the **E**-field intensity.

E- and H-Field Probes

As indicated in a previous section, a measurement of power density is not sufficient to characterize electromagnetic radiation in the near-field region. The question that is often raised, however, is whether or not we need to measure both the electric- and magnetic-field intensities. Since these fields are indeed related to each other by Maxwell's equations, one can generally measure either one of the fields, say, the electric field, and calculate the other, if necessary, from it. Although the above statements are certainly correct, practical problems in relating these fields make it easier to measure them both. For instance, it is known that the ratio of **E** to **H** in the near field is not the same as the intrinsic wave impedance in free space and that **H** and **E** are not generally perpendicular to each other. In other words, no simple relation exists between these near **E** and **H** fields, and full utilization of Maxwell's equations is required.

Pursuing such an option of fully utilizing Maxwell's equations is not easy, simply because phase information of the measured electric or magnetic fields (in addition to the magnitude) must be obtained. Finally, it is not obvious which fields should be measured, especially at low frequencies where the **E** and **H** fields, according to Maxwell's equations, can be very loosely coupled, and either **E** or **H** can exist almost independently of the other. Therefore, in some applications **E**-field measurements could be easier, more accurate, and more reliable, while in others the **H**-field measurements could be more sensitive and reliable. As an example of the latter case, consider an electric power transformer. It is true that both electric and magnetic fields are present, but the magnetic-field intensity, in this case, is certainly larger than the electric-field intensity and it would be impractical to measure the small **E**-field intensity in the presence of significant errors due to noise and from it find the **H**-field intensity through Maxwell's equations. This is why analyses of transformer action are traditionally based on magnetic flux linkages instead of electric-field quantities. Therefore, it is more advantageous to measure the magnetic-field intensity in this case, while in some other applications it could be advantageous to measure the **E**-field intensity.

In conclusion, it is believed that in order to avoid the prejudgment of what should and should not be measured, an electromagnetic radiation monitoring device should measure both **E**- and **H**-field intensities at the lower frequencies. At higher frequencies (say, above a few hundred megahertz) it is safe to assume that both fields are strongly coupled and the measurement of only one (say, the **E** field since it is usually the easier one to measure) should suffice.

The design of **E**- and **H**-field probes used both for area monitors and for a personnel dosimeter is described below.

Electric-Field Probe

Basic Design and Characteristics—To measure the electric-field intensity small dipole antennas with detecting devices are usually used. The commonly used short electric dipoles are often fabricated from cylindrical wires, with their length limited to a small fraction of the wavelength ($\ell < 0.1\lambda$) to allow a reasonably broad frequency response. For the short electric dipole shown in Fig. 27, with the assumed triangular current distribution I, it can be shown that the open-circuit voltage at the terminals V_{oc} is given by

$$V_{oc} = E_{\parallel} \ell / 2$$

where $E_{\parallel}$ is the component of the electric field parallel to the dipole and ℓ is the dipole length. The output voltage of the antenna can be measured by placing a detector between the terminals of the antenna. As indicated earlier, this detector may be just a thermocouple element with the heat generated in the hot junction being produced by the received rf currents in the dipole circuit [31]. Diode detectors are other possible detecting devices which, when placed between the output terminals of the antenna, would produce a rectified output voltage that can be detected by a dc voltmeter [33]. As mentioned previously, diode detectors are more sensitive than thermocouple detectors, but diodes are nonlinear. Therefore, depending on the intensity of the incident rf signal, the output voltage of the diode

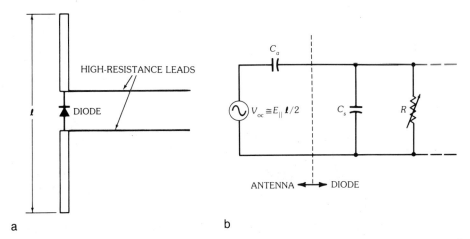

Fig. 27. Diagram of the E-field probe and its equivalent circuit. (*a*) The E-field probe. (*b*) Approximate equivalent circuit.

detectors may be in the square-law region, thus providing an output voltage that is related to the power density of the incident radiation, or in the linear region, thus providing an output voltage that is linearly proportional to the electric field associated with the incident radiation. From the equivalent circuit in Fig. 27b, it can be shown [34] that if the current in the diode can be assumed to be $I = I_{sat}(e^{aV} - 1)$, the output voltage from the probe is given by

$$V \cong -(a/2)[V_{oc}/(1 + C_s/C_a)]^2$$

for $V < 0.01$ V, and

$$V \cong -V_{oc}/(1 + C_s/C_a)$$

for $V > 1$ V.

The dipole antennas themselves may be of various kinds and shapes, including cylindrical rods, biconical antennas, triangular strips on printed-circuit board, or zigzag antennas. The zigzag dipole has a nearly flat frequency response and good sensitivity. Typical dimensions of a constructed zigzag dipole are given in Fig. 28, which also shows the high-resistance leads used to connect the output of the diode to a dc voltmeter with minimum perturbation of the rf radiation fields.

Isotropic Response—To obtain an isotropic response for the electric-field probe, three orthogonal antennas arranged as shown in Fig. 29 are usually used. A top view of the orthogonal arrangement illustrating the direction of each of the three dipoles is given in Fig. 29b. This specific angle of orientation shown in Fig. 29 coincides with the three orthogonal directions of the three orthogonal diagonals of a cube. Typically the three dipoles are arranged around a ¼-in- (6.35-mm)-diameter cylindrical tube as shown in Fig. 29b. Electrical connections can be made either to measure the sum of the responses of these antennas or to measure these responses individually.

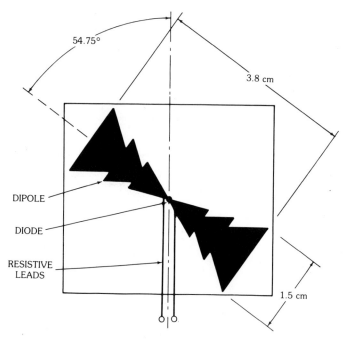

54.75°

3.8 cm

DIPOLE

DIODE

RESISTIVE
LEADS

1.5 cm

Fig. 28. Diagram illustrating the geometry of the printed-circuit zigzag antenna.

The performance of the isotropic probe can be evaluated experimentally in a Crawford TEM cell (see Section 8). By rotating the three orthogonal dipoles in the cell and measuring the total sum of the output voltages, it was found that the response is omnidirectional within a few percent. In another procedure one dipole is aligned along the direction of the electric field in the calibration (Crawford) cell and the responses of the other dipoles relative to the one in the direction of the **E** field are measured. The relative responses were found to be at least -20 dB, which indicates an adequate orthogonality.

Magnetic-Field Probe

The design of the magnetic-field probe is significantly more involved than that of the electric-field probe. The commonly used diode-loaded circular loops shown in Fig. 30 are generally not adequate for the development of a general-purpose electromagnetic radiation monitor for the following reasons [35]:

1. The frequency response of these loops is linearly proportional to the frequency, which means that the frequency of the incident radiation must be accurately known in order to measure the magnetic field. This characteristic frequency response can be shown as follows. From Faraday's law it is known that

$$\oint_C \mathbf{E} \cdot d\ell = -\frac{\partial}{\partial t} \int_S \mathbf{B} \cdot d\mathbf{S} = -\mu \frac{\partial}{\partial t} \int_S \mathbf{H} \cdot d\mathbf{S}$$

a

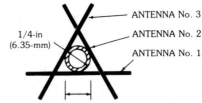

b

Fig. 29. Orthogonal arrangement of the E-field probes to achieve isotropic receiving characteristics. (*a*) Photograph of E-field probe (side view). (*b*) Diagram of E-field probe (top view).

where **E** and **H** are the electric and magnetic fields, respectively, and S is any surface bounded by the closed path C. For the sinusoidally time-varying, steady-state case, $\partial/\partial t \rightarrow j\omega$. Hence, by integrating the above equation over the area of the loop (assuming **H** to be uniform over the area of the loop), we have

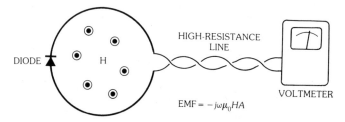

Fig. 30. Diagram illustrating the geometry of the diode-loaded circular loop. (*After Greene [35]*)

$$\oint_C \mathbf{E} \cdot d\ell = \text{emf} = -j\omega\mu HA \qquad (16)$$

where $A = \pi a^2$ is the area of the loop, $\omega = 2\pi f$, f is the frequency, and the emf is the induced voltage across the loop terminals. From (16) it is clear that the induced emf is proportional to the frequency f. Therefore, for measurements in industrial sites where the generation of higher-order harmonics has been observed, the linear frequency response of the loop would be very undesirable.

2.   These probes generally respond to the electric field as well as to the magnetic field. This, of course, is undesirable when one is trying to measure the magnetic field. Fig. 31 shows that the loop diameter should not exceed 0.05λ if

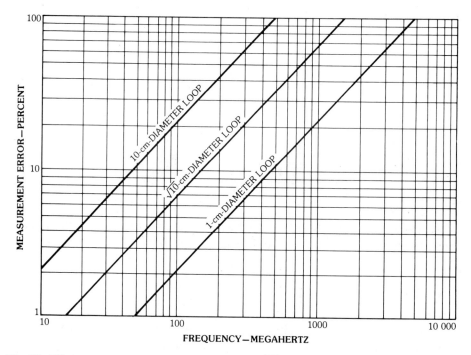

Fig. 31. Worst-case measurement error of the 1-, $\sqrt{10}$-, and 10-cm–diameter loop antennas due to their electric-dipole response. (*After Greene [35]*)

the probe's response to the electric field is to be minimized [35]. A loop of this size, however, is too small to provide the desired sensitivity. Also, at higher frequencies, where the wavelength is shorter, it becomes increasingly difficult to make such small probes sensitive enough.

The frequency response of a loop can be made flatter by making the loop the inductive part of a low-Q resonance circuit [36]. If the resonant frequency of the low-Q circuit is considerably lower than the frequency of operation, the response of the loop in the postresonance frequency range will usually be reasonably flat, but the sensitivity will also be reduced. The reduced sensitivity requires use of zero-bias diodes (HSCH-3486 or HSCH-3171) to rectify the output signal detected by the probe. A schematic diagram illustrating the probe's principle of operation [36] is shown in Fig. 32a. An isotropic **H**-field probe can be constructed by placing three magnetic-field probes on the three orthogonal surfaces of a cube.

Design of an RF Personnel Dosimeter [37]

Although there are several commercially available rf radiation area monitors, there is little available information on the development of a personnel dosimeter. The function of an area monitor is generally to detect electromagnetic radiation and sound an alarm if the level in the working area gets too high. The function of a personnel dosimeter is to measure the electromagnetic radiation to which the person wearing it is exposed, record the amount of radiation, and sound an alarm if excessive exposure occurs. The personnel dosimeter must be able to relate the fields measured on the surface of a worker's body to the free-space radiation, in terms of which the safety standard is written. This is important since this safety standard is the legally binding measurement rather than the significantly distorted fields (due to scattering) measured on the surface of an operator's body.

In summary, an rf personnal dosimeter should consist of the following items:

  (*a*) An antenna system to measure the **E**- and **H**-field intensities associated with the electromagnetic radiation. This antenna system can be basically the same as those described earlier in this section.
  (*b*) A microprocessor system to store the values of the measured fields, calculate the incident fields from the measured fields, compare the calculated values of the incident fields with the safety standard, and sound an alarm if the safe levels are exceeded [38].

It would also be desirable for the dosimeter to require minimal power, to be inexpensive, rugged, and lightweight, and to have overload protection.

7. Other Applicators Used to Produce Hyperthermia

There are some other devices that have been widely used to generate hyperthermia that should be mentioned in this chapter, even though they are not always thought of as antennas because their use is based more on near-field coupling at low frequencies than on traditional antenna radiation characteristics. These applicators are usually thought of as inducing currents in the body instead of producing

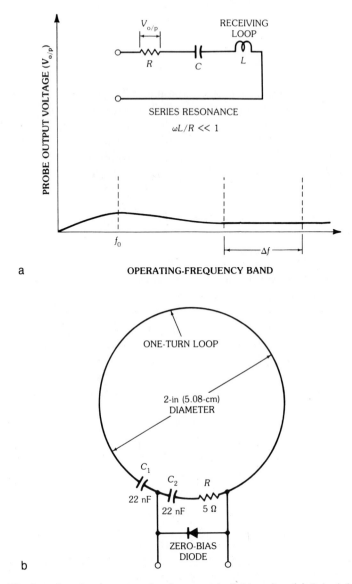

Fig. 32. The low-Q series-resonant circuit magnetic-field probe. (*a*) Principle of operation. (*b*) Design of the loop. (*After Babij and Bassen [36]*)

waves that propagate into the body. For example, one of these applicators is a coil placed coaxially around the body. It induces rf currents that are more conveniently thought of as eddy currents than as propagating waves. Another applicator consists of capacitor plates placed on the body. These plates induce rf currents that are also not conveniently characterized as propagating waves. For this class of applicators the term "depth of penetration" is often used to describe heating patterns, but it is important to note that this term does not have the same meaning as it does for radiating applicators that produce waves that propagate from the surface into the interior of the body. These nonradiating applicators can also overheat the surface, even at low frequencies, because of the spatial variation of the fields that they produce. Thus the fundamental problem in hyperthermia that is described in the introduction to this chapter, how to heat the interior without overheating the surface, is generally not solved by using nonradiating applicators, even at low frequencies. Four couplers of this kind that are described briefly in this section are coaxial current loops, capacitor-plate applicators, pancake coils, and helical coils. The following four figures illustrate the geometries of the four low-frequency couplers and the currents or the fields induced by these couplers inside the body.

Coaxial Current Loops

A simple coaxial current loop placed around a patient is attractive for producing regional hyperthermia because it requires no coupling medium, it is easy to match to a transmitter, and getting a patient into and out of it is easy. (See Fig. 33.) Also, the impedance matching is not very sensitive to patient size and position within the coil. Storm and colleagues [39] have pioneered the use of such an applicator. Their device operates at 13.56 MHz with an input power of 10 to 1000 W.

The coaxial current loop for use at low frequencies is easy to design because the dimensions are not critical if the size of the loop is small compared with a free-space wavelength. The main design task is to construct a matching network of some sort to compensate for the inductive driving-point impedance and to match the coil to the feed line. Standard rf techniques can be used to do this.

Since measurements have shown that the fields inside the coil consist mostly of an axial magnetic field that is nearly uniform in the central portion of the space inside the coil, an approximate expression for the induced **E** field inside a homogeneous body centered within the coil can be obtained from Maxwell's equation for the steady-state sinusoidal case:

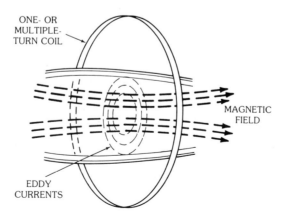

Fig. 33. Magnetic field and eddy currents produced by a coaxial coil "Magnetrode," which surrounds a portion of the patient's body. (*After Iskander [1], © 1982 American Institute of Physics*)

$$\oint_\ell \mathbf{E} \cdot d\ell = -j\omega\mu_0 \int_S \mathbf{H} \cdot d\mathbf{S}$$

where ω is the angular frequency, μ_0 is the permeability of free space, and S is any surface bounded by the closed path ℓ. Assuming circular symmetry and letting ℓ be a circular path inside the body, the equation reduces to

$$2\pi r E_\phi = -j\omega\mu_0 H_z \pi r^2$$

and

$$E_\phi = \frac{-j\omega\mu_0 H_z r}{2}$$

where r is the radius of the path. This expression shows that the induced **E** field, and hence the absorbed power, will be zero at the center and increasing toward the surface. These results thus predict that the current loop will produce primarily surface heating, and if the center is heated, it must be through thermal conduction.

Whether dielectric inhomogeneities would result in better internal heating than that predicted by the homogeneous model is a question that has been widely discussed. Hill, Christensen, and Durney [40] have calculated internal field patterns in a two-dimensional inhomogeneous model, and Paulsen and colleagues [41] combined these calculations with a finite-element calculation of temperature distributions. From their results they concluded that in most cases the current loop does not effectively heat deep-lying tumors. Their conclusions were based on both transient and steady-state temperature calculations, with various patterns of blood flow.

Capacitor-Plate Applicators

At low frequencies capacitive applicators consisting of one plate on one side of a patient and a second plate on the other side have been used for many years in diathermy. (See Fig. 34.) The plates produce a strong electric field perpendicular

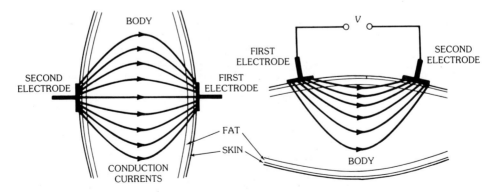

Fig. 34. Two possible arrangements for a regional heating of layered tissue using capacitor-type electrodes for short-wave diathermy. (*After Iskander [1],* © *1982 American Institute of Physics*)

to the plates and a negligible magnetic field. The only design procedure required for the applicator is some method of matching its impedance to a transmission feed line.

Although the capacitive applicator can produce good deep heating, it has the serious disadvantage that it tends to overheat the fat [42]. This occurs when the internal **E** field is perpendicular to the muscle-fat interface. At this interface the boundary conditions on the **E** field require that

$$\epsilon_f E_f = \epsilon_m E_m$$

where ϵ_f and ϵ_m are the relative permittivities in the fat and muscle, respectively, and E_f and E_m are the normal components of the **E** field in the fat and muscle, respectively. Thus, at the interface,

$$E_f = \frac{\epsilon_m}{\epsilon_f} E_m$$

and since ϵ_f is considerably smaller than ϵ_m, the field in the fat at the boundary is considerably larger than the field in the muscle. If the fat and muscle layers were approximately planar, the field in the volume of the fat would likewise be considerably stronger than the field in the muscle. Since power varies as E^2, the stronger fields in the fat cause it to overheat, even though the fat is less lossy than the muscle. For this reason capacitive applicators are useful mostly for thin patients.

Pancake Coils

Another commonly used diathermy applicator is the so-called pancake coil, consisting of a planar helical coil placed flat on the surface of the body. (See Fig. 35.) Guy, Lehmann, and Stonebridge [3] have calculated and measured heating

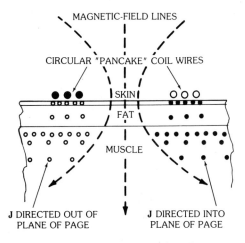

Fig. 35. Magnetically induced current in tissue exposed to short-wave diathermy "pancake" coil. (*After Guy, Lehmann, and Stonebridge [3], © 1974 IEEE*)

patterns of typical coils. Their results show that the pancake coil can heat underlying muscle tissue without overheating the intervening fat. At 27.12 MHz and with a 2-cm fat layer the coil can heat to a depth of about 4 cm (depth where power absorption is e^{-2} of the maximum). The main disadvantage of the pancake coil is the nonuniform heating pattern, which tends to be toroidal in shape.

Helical-Coil Applicators

The helical coil has recently been investigated for use as a hyperthermia applicator [43]. Although enough work has not yet been done to determine its full potential for generating hyperthermia, the helical coil does appear promising. Chute and Vermeulen [44] demonstrated that the axial electric field inside a helical coil is approximately uniform and much stronger than the other electric- and magnetic-field components. (See Fig. 36.) This makes the coil attractive for hyperthermia because the axial field would be mostly parallel to the fat-muscle boundaries in a patient placed inside the coil, and consequently overheating of the fat would not be expected. Since the field is approximately uniform it would also be expected to heat the center of the body much better than a single coaxial current loop.

Chute and Vermeulen [44] showed that when the radius of the helical coil is much smaller than its length and when both dimensions are small compared with a free-space wavelength, the fields inside the coil are given by

$$H_z \cong \frac{I_0}{d} \cos \omega t$$

$$E_\phi \cong \frac{\omega \mu_0 I_0}{2d} \varrho \sin \omega t$$

$$E_z \cong - \frac{\omega \mu_0 a I_0}{2d} \cos \psi \sin \omega t$$

where

I_0 = magnitude of the coil current

d = axial spacing of the individual turns of the coil

ω = angular frequency

μ_0 = permeability of free space

ctn $\psi = 2\pi a/d$, where ψ is the coil pitch angle

ϱ = radial distance in a cylindrical coordinate system (ϱ, ϕ, z)

The equations are valid for a coil in which the direction of twist is clockwise with increasing z when viewed in the positive z direction.

Ruggera and Kantor [43] made measurements of the temperature distribution produced by helical coils in arm-sized (8-cm diameter) and thigh-sized (12-cm diameter) fat-muscle cylindrical phantoms. In studies at the FCC-designated Instrument, Scientific, and Medical (ISM) frequencies of 13.56, 27.12, and 40.68

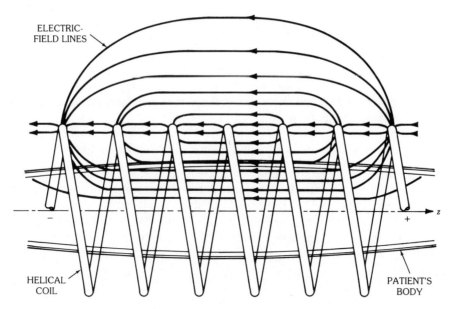

Fig. 36. Helical coil carrying a time-varying current and the electric-field lines associated with it. (*After Chute and Vermeulen [44],* © *1981 IEEE*)

MHz, they found that heating uniform within 10 percent or less over the cross section could be obtained in arm-sized phantoms under the right conditions. Heating uniform within about 20 percent over the cross section could be obtained in thigh-sized phantoms. Their results indicated that two special modes of operation produced the best results, the first when the total wire length in the helix was equal to one-half wavelength, and the second when it was equal to one wavelength. For these two conditions the driving-point impedance was purely resistive. In the full-wavelength operation a coil length-to-diameter ratio of 4 was needed to produce uniform transverse heating, and in half-wavelength operation a ratio of 2 was needed.

Although initial work indicates that the helical coil could be a very effective applicator for hyperthermia generation, much work remains to be done in this area.

8. Experimental Procedures for Characterizing Antennas Used in Medical Applications

This chapter concludes with a brief description of some commonly used techniques for measuring the properties of antenna systems used in medical applications. Since the techniques are designed to measure the most important characteristics of the antennas, it is helpful to summarize those characteristics first. In general, antennas for medical applications should have the following features:

(*a*) Impedance well matched to the human body over a broad frequency band. Good impedance matching is necessary to achieve maximum power transfer to the body, and broadband operation is sometimes useful because it provides flexibility in using the antenna under a variety of circumstances.

(*b*) Minimum external leakage radiation. This requirement is necessary first to protect patients and medical personnel from excessive electromagnetic radiation, and second to prevent leakage radiation from producing artifacts in measurements [17].

(*c*) Satisfactory internal field patterns. For hyperthermia applications, this means producing deep enough heating without overheating surfaces.

From these basic requirements it is clear that some of the experimental procedures necessary to characterize these antennas are similar to routine antenna measurements. These include measurement of the input impedance over a broad frequency band and mapping the fields around an antenna in contact with a human body to determine the leakage radiation. To evaluate the heating patterns of antennas used in hyperthermia applications, one of the procedures described below may be used.

Evaluation of Heating Patterns of Antennas Used in Hyperthermia

Nonperturbing Implantable Temperature Probes—There are several temperature probes that are especially useful for evaluating heating patterns of electromagnetic antennas because they produce almost no interference with the electromagnetic fields. Among the commercially available and most stable of these probes are those made by Vitek, Inc. [45], and Narda Corporation. Both systems use thermistor detectors and high-resistance leads (carbon-loaded Teflon) to minimize interference with the electromagnetic fields.

Another type of nonperturbing temperature probe was developed by Christensen [46]. It utilizes a small 0.25-mm gallium-arsenide semiconductor sensor at the end of a fiber-optic bundle. The probe relies on the shift in optical absorption at the semiconductor's band edge as a function of temperature. Recent versions of this probe achieved resolutions better than 0.1°C over the temperature range from 37°C to 47°C. This probe is also available commercially from Clini-Therm Corporation. Other fiber-optic probes reported in the literature include a liquid crystal probe [47], a birefringent crystal optical thermometer [48], an etalon fiber-optic probe [49], and a phosphor fiber-optic probe [50]. Another probe called a *viscometric thermometer*, which is based on measurement of the flow resistance of a fluid (a temperature-dependent property) through a small capillary at the tip of the implantable probe, has also provided promising results [51]. In this probe the probe body material and the working fluid can be chosen to match closely the electrical and thermal properties of tissue, thus minimizing perturbation of the electromagnetic and temperature fields in the tissue. This matching of properties is not generally possible in other electrical and optical probes.

In using probe temperature measurements to evaluate heating patterns it is important to keep exposure times to a minimum to minimize the error due to heat conduction and smearing of the heating patterns. With this precaution in mind, this measurement technique is otherwise straightforward and has the advantage of being the most accurate in experiments on phantoms. It also allows temperature regulation to be measured in live animals and is inexpensive. The major disadvantage is the time required to define heating patterns in large or complex geometrical bodies. With the advent of systems having multiple temperature probes [52], this problem has been largely reduced.

Thermographic Cameras—Scanning thermographic cameras can be used to provide rapid acquisition of data on heating patterns in phantoms. The experimental procedure involves using the antenna under test to heat the phantom at high power levels in the shortest possible time. The phantom models are designed to separate along planes perpendicular to tissue interfaces so that cross-sectional relative heating patterns can be measured with a thermograph. After a short exposure the model is quickly disassembled and the temperature pattern over the surface of separation is observed and recorded by means of a thermograph. Details of this technique have been developed at the University of Washington and are described in detail by Guy [53]. The procedure originally involved using a thin sheet of plastic to facilitate separating the two halves of the phantom. Thus the procedure was limited to symmetrical models exposed to a linearly polarized field (**E** field parallel to the interface) to avoid interrupting any induced currents that would normally flow perpendicular to the median plane of separation [53]. For near-field measurements and in evaluating antennas for hyperthermia applicators, however, the procedure was modified by replacing the plastic sheet with a silk screen, thus allowing easy separation without loss of electrical continuity [54]. The major disadvantage of this technique is the high cost of the required equipment.

Field Mapping Using Implantable E-Field Probes

The radiation characteristics of an antenna in tissue can be equivalently evaluated by mapping the electric-field distribution using implantable probes. There are advantages in using these probes for evaluating heating patterns instead of thermographic and thermometric techniques. For example, the latter two techniques determine internal fields only indirectly and under exposure conditions that provide moderate elevation in temperature. The thermoregulatory system of living subjects further complicates thermal dosimetry for localized determination of absorbed energy. For such cases the electric-field measurements provide much more accurate and reliable assessment of the power absorbed by the subject. A comparison of the various dosimetric techniques [55] is presented in Table 3. This comparison simply indicates that the internal-field probe is well suited for *in-vivo* measurement of the absorption characteristics of subjects.

Probes used for measuring electric fields in simulated or actual biological media often use dipole-diode combinations similar to those described in Section 6. The detailed structure of the implantable electric-field probe described by Bassen and colleagues [55] is shown in Fig. 37. The electric-field probe shown in Fig. 37b consists of a single dipole-detector configuration terminated with high-resistance leads that transmit the detected signal to the readout electronics. The use of the three orthogonal dipoles in an I-beam array (Fig. 37a) has been shown to yield a satisfactory isotropic response when the outputs of square-law detectors are summed. Interested readers can find a description of the detailed procedure for fabricating the probe elsewhere [56].

Table 3. Comparison of Internal Dosimetric Techniques (*After Bassen et al. [55],* ©
1977 American Geophysical Union)

	"Ideal" E-Field Probe	Thermometer Probe	Thermography
Directly measured parameter	Electric-field strength	Localized temperature	Spatial distribution of temperatures
Dosimetric approach	Direct, absolute E-field measurement, independent of media. Must know conductivity to compute SAR	Direct thermal measurement. Must know mass density and specific heat to compute SAR	Temperature measured after exposure. Must know mass density, specific heat, and infrared emittance to compute SAR
Sensitivity to external field	$< 10\,\mathrm{mW/cm^2}$	$> 10\,\mathrm{mW/cm^2}$	$> 100\,\mathrm{mW/cm^2}$
Suitable for *in-vivo* E-field dosimetry	Yes	Yes	No
Spatial measurement capability	Point or continuous-line scan. Arbitrary site of measurement	Single-point measurement. Arbitrary site of measurement	Rapid planar scan. Subject must be prepared in advance
Accuracy and limitations	Good absolute accuracy.* Relatively independent of boundaries and ϵ_r	Good absolute thermal accuracy ($\pm 0.05°C$). Measurement of SAR is media dependent	Good relative thermal accuracy ($\pm 0.2°C$) but limited by thermal diffusion. Measurement of SAR is media dependent

*Present prototypes are media dependent (± 2.25 to 4.25 dB).

    It is important to note that experimental measurements of internal fields within simulated or actual biological media have shown the response of some probes to be dependent on the dielectric constant of the medium in which the measurements were made, or on the position of the probe with respect to boundaries. Based on experimental and theoretical analyses, sets of principles were formulated for the design of E-field probes whose responses are ideally independent of the media [57].

    A typical result obtained by Smith [57] is shown in Fig. 38. In this figure, $V(\epsilon_r)$ is the voltage across the terminals of an electrically short probe which is proportional to the component of the local electric field E_z parallel to the axis of the probe, i.e., $V(\epsilon_r) = K_e E_z$. The proportionality constant K_e depends on the effective height of the probe, its input admittance, and the admittance of the load

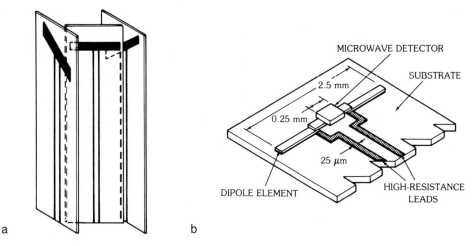

Fig. 37. Structure of implantable electric-field probe. (*a*) Orthogonal detector array. (*After Bassen, Swicord, and Abita [56]*) (*b*) Details of fabrication of probe. (*After Bassen et al. [55], © 1977 American Geophysical Union*)

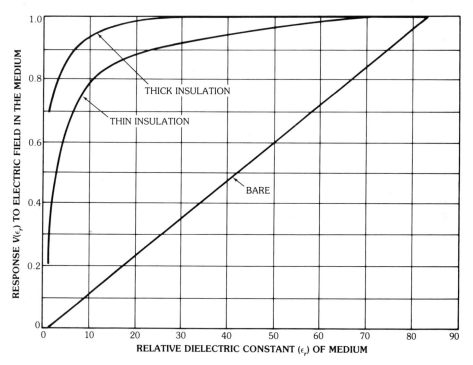

Fig. 38. Response of various probes in media as a function of dielectric constant. (*After Smith [57], © 1975 IEEE*)

connected to it. It is clear that unless a thick, lossless, insulating cylinder with dielectric constant lower than the surrounding media is used, the probe's response varies significantly with the relative dielectric constant. A very thick insulating cylinder (ten times the dipole width) around a dipole, on the other hand, can cause significant *H*-plane pattern asymmetry if the insulated cylinder is eccentric with respect to the dipole axis and if the insulation is larger than a quarter-wavelength in the media. An alternative way to reduce probe sensitivity to the media dielectric constant is to terminate the dipole with a detector having an impedance that is much larger than the dipole terminal impedance [56]. Bassen also suggested the use of a very low capacitance beam-lead diode chip rather than bulky axial-lead diodes with internal bonding of wires. It should be noted, however, that even a microminiature beam-lead diode has a finite impedance that is not significantly greater than the impedance of a small dipole, and insulation is therefore needed to minimize the effects of boundary reflections and of media dielectric properties. Also, at lower frequencies the media dependence of the probe's response increases because of the higher antenna impedance and the higher leakage current of the diode.

Use of Phantoms to Measure Power Deposition Patterns (Heating Patterns) of Antennas

Phantoms are made by first constructing a mold of radio-frequency transparent foam* material in the shape of the phantom to be investigated and then filling the mold with tissue-equivalent material.

A number of tissue-equivalent materials have been developed by various investigators. The physical properties of some of these tissue-equivalent materials are given in Table 4. Sources of phantom materials and detailed procedures for the phantom preparation are described elsewhere [58].

Calibration of E- and H-Field Probes

A number of techniques exist for establishing known, uniform levels of electromagnetic fields for testing and calibrating **E**- and **H**-field probes. For example, high-level fields can be generated accurately above a few hundred megahertz using standard-gain horns and below a few megahertz using parallel-plate lines. Both these techniques, while widely used, have the disadvantage of radiating electromagnetic energy into the surroundings, thus limiting the accuracy of the measurements and causing some concern about their possible hazardous effects. The calibration of these probes, however, can be most conveniently achieved inside a TEM chamber. This TEM cell is basically a section of a rectangular strip transmission line of 50-Ω characteristic impedance. It is therefore relatively inexpensive, versatile in size, and also broadband. A typical design [59] of the TEM cell is shown in Fig. 39. This design generally provides a suitable space where the **E** and **H** fields can be calculated and hence used to calibrate the probes. It should be noted that the dimensions of the cell given in Table 5 can be linearly

---

*Equal proportions of polyol and isocyanate, available at Utah Foam Products, Inc., Salt Lake City, Utah 84119.

Table 4. Properties of Materials for Constructing Phantom Models for Muscle, Brain, Fat, and Bone at Microwave Frequencies [58]

Physical Properties		
Material	Specific Heat	Specific Density
Muscle	0.84	0.97
Brain	0.83	0.96
Fat and bone	0.29–0.37	1.29–1.38

Electrical Properties*					
	915 MHz			2450 MHz		
Material	ϵ_r	σ	τ	ϵ	σ	τ
Muscle	50.63	1.355	0.526	49.61	2.286	0.333
Brain	34.37	0.730	0.442	33.56	1.266	0.2713
Fat and bone	5.61	0.0665	0.233	4.51	0.172	0.187

Composition			
Material	Super Stuff	Polyethylene powder	H_2O	NaCl
Muscle	8.45%	15.20%	75.44%	0.9069%
Brain	7.01%	29.80%	62.61%	0.5823%
	Laminac 4110	Aluminum powder		Acetylene black
Fat and bone**	85.20%	14.5%		0.24%

*Here ϵ_r = relative permittivity, σ = conductivity in siemens per meter, and τ = tan δ (loss tangent).

**For fat and bone add 0.375% of total weight P-102 (60% methyl ethyl ketone peroxide as a catalyst).

scaled with the desired operating frequency band. In constructing a TEM chamber special attention must be paid to the tapered transitions to ensure good impedance matches. Electric and magnetic fields should be mapped inside the chamber to determine their uniformity and ensure the absence of any significant standing waves caused by mismatches at transitions or by the presence of higher-order modes.

When high-order modes and standing waves are not present in the chamber, the electric-field strength can be found from the measurement of the total power transmitted through the cell. It has been shown that [59]

$$E = \sqrt{P/G}/d$$

in volts per meter, and

$$H = E/377$$

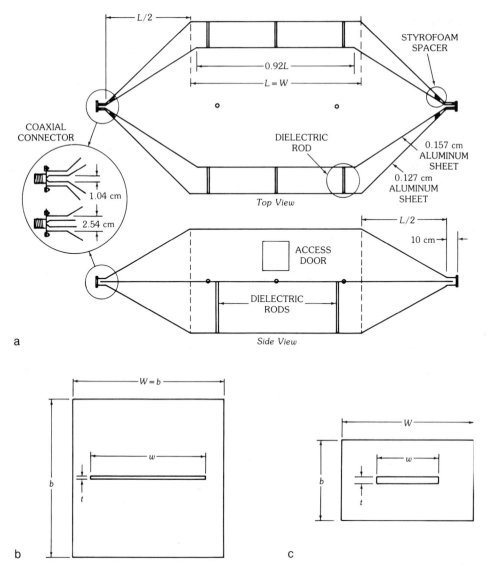

Fig. 39. Typical TEM cell design. (*a*) Design for rectangular TEM transmission cell. (*b*) Cross section of optimum geometry of rectangular transmission line for maximum test area and frequency. (*c*) Cross section of rectangular transmission line with improved **E**-field uniformity. (*After Crawford [59], © 1974 IEEE*)

Table 5. TEM Cell Dimensions

Cutoff/ Multimode Frequency (MHz)	Square Cell (Fig. 39b)				Rectangular Cell (Fig. 39c)			
	Plate Separation b (cm)	w (cm)	t (cm)	C'_f (pF/cm)	Plate Separation b (cm)	w (cm)	t (cm)	C'_f (pF/cm)
100	150	123.83	0.157	0.087	90	108.15	0.157	0.053
300	50	41.28	0.157	0.087	30	36.05	0.157	0.053
500	30	24.77	0.157	0.087	18	21.83	0.157	0.053

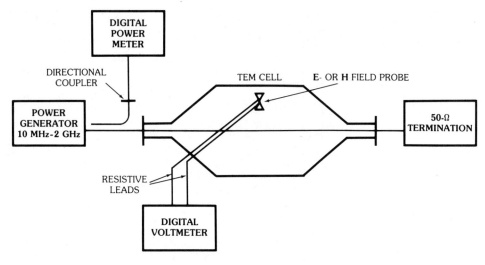

Fig. 40. Diagram of the calibration setup.

in amperes per meter, where P is the net power flowing through the cell, G is the real part (conductance) of the input admittance, and d is the separation distance between the upper and the middle conductors in the cell.

The experimental setup used to calibrate the **E**- and **H**-field probes [60] is shown in Fig. 40. It consists of a microwave power generation unit and a TEM transmission cell terminated by a 50-Ω load.

9. References

[1]   M. F. Iskander, "Physical aspects and methods of hyperthermia production by rf currents and microwaves," in *Physical Aspects of Hyperthermia*, ed. by G. H. Nussbaum, New York: American Institute of Physics, 1982.

[2]   H. P. Schwan, "Biophysics of diathermy," in *Therapeutic Heat and Cold*, ed. by S. Licht, New Haven: Licht, 1965, sec. 3, pp. 63–125.

[3]   A. W. Guy, J. F. Lehmann, and J. B. Stonebridge, "Therapeutic applications of electromagnetic power," *Proc. IEEE*, vol. 62, pp. 55–75, January 1974.

[4]   A. W. Guy, "Electromagnetic fields and relative heating patterns due to a rectangular aperture source in direct contact with bilayered biological tissue," *IEEE Trans. Microwave Theory Tech.*, vol. MTT-19, pp. 214–223, February 1971.

[5]   A. W. Guy, J. F. Lehmann, J. B. Stonebridge, and C. C. Sorenson, "Development of a 915-MHz direct-contact applicator for therapeutic heating of tissues," *IEEE Trans. Microwave Theory Tech.*, vol. MTT-26, pp. 550–556, August 1978.

[6]   A. L. Van Koughnett and W. Wyslouzil, "A waveguide TEM-mode exposure chamber," *J. Microwave Power*, vol. 7, pp. 381–383, December 1972.

[7]   A. Y. Cheung, T. Dao, and J. E. Robinson, "Dual-beam TEM applicator for direct-contact heating of dielectrically encapsulated malignant mouse tumor," *Radio Sci.*, vol. 12, pp. 81–85, November–December 1977.

[8]   S. B. Cohn, "Properties of ridge waveguide," *Proc. IRE*, vol. 35, pp. 783–788, August 1947.

[9]   F. Sterzer, R. W. Paglione, J. Mendecki, E. Friedenthal, and C. Botstein, "RF therapy for malignancy," *IEEE Spectrum*, pp. 32–37, December 1980.

[10] P. F. Turner, "An annual phased array for deep regional hyperthermia," presented at

the Second Annual Meeting of North American Hyperthermia Group, Salt Lake City, Utah, April 17–19, 1982.

[11] P. F. Turner, "Electromagnetic hyperthermia devices and methods," MSc thesis, Department of Electrical Engineering, Univ. of Utah, June 1983.

[12] M. F. Iskander, P. F. Turner, J. B. DuBow, and J. Kao, "Two-dimensional technique to calculate the EM power deposition pattern in the human body," *J. Microwave Power*, vol. 17, pp. 175–185, 1982. Also see M. F. Iskander, O. Koshdel-Milani, and P. F. Turner, "Numerical calculation of the heating patterns in realistic cross-sections of the human body," presented at the Thirty-first Meeting of the Radiation Research Society, San Antonio, Texas, February 1983.

[13] I. J. Bahl and P. Bhartia, *Microstrip Antennas*, Dedham: Artech House, 1980.

[14] I. J. Bahl and S. S. Stuchly, "Analysis of a microstrip covered with a lossy dielectric," *IEEE Trans. Microwave Theory Tech.*, vol. MTT-28, pp. 104–109, February 1980.

[15] I. J. Bahl, S. S. Stuchly, J. J. W. Lagendijk, and M. A. Stuchly, "Microstrip loop radiators for medical applications," *IEEE Trans. Microwave Theory Tech.*, vol. MTT-30, pp. 1090–1093, July 1982.

[16] I. J. Bahl, S. S. Stuchly, and M. A. Stuchly, "New microstrip slot radiator for medical applications," *Electron. Lett.*, vol. 16, no. 19, pp. 731–732, September 11, 1980.

[17] M. F. Iskander and C. H. Durney, "An electromagnetic energy coupler for medical applications," *Proc. IEEE*, vol. 67, pp. 1463–1465, October 1979.

[18] M. F. Iskander and M. A. K. Hamid, "A new strip transmission line for moisture content measurements," *J. Microwave Power*, vol. 12, pp. 16–18, 1977.

[19] L. S. Taylor, "Implantable radiators for cancer therapy by microwave hyperthermia," *Proc. IEEE*, vol. 68, pp. 142–149, January 1980.

[20] J. Bigu-del-Blanco and C. Romero-Sierra, "The design of a monopole radiator to investigate the effect of microwave radiation in biological systems," *J. Bioengineering*, vol. 1, pp. 181–184, 1977.

[21] J. Mendecki, E. Friedenthal, C. Botstein, R. Paglione, and F. Sterzer, "Microwave applicators for localized hyperthermia treatment of cancer of the prostate," *Intl. J. Radiation Oncology Biol. Phys.*, vol. 6, pp. 1583–1588, 1980.

[22] J. Mendecki, E. Friedenthal, C. Botstein, F. Sterzer, R. Paglione, M. Nowogrodzki, and E. Beck, "Microwave-induced hyperthermia in cancer treatment: apparatus and preliminary results," *Intl. J. Radiation Oncology Biol. Phys.*, vol. 4, pp. 1095–1103, 1978.

[23] E. C. Burdette, F. L. Cain, and J. Seals, "In-vivo probe measurement technique at vhf through microwave frequencies," *IEEE Trans. Microwave Theory Tech.*, vol. MTT-28, pp. 414–427, 1980.

[24] T. W. Athey, M. A. Stuchly, and S. S. Stuchly, "Measurement of radio frequency permittivity of biological tissues with an open-ended coaxial line: part I," *IEEE Trans. Microwave Theory Tech.*, vol. MTT-30, pp. 82–86, 1982. See also M. A. Stuchly, T. W. Athey, G. M. Samaras, and G. E. Taylor, "Measurement of radio frequency permittivity of biological tissues with an open-ended coaxial line: part II—experimental results," *IEEE Trans. Microwave Theory Tech.*, vol. MTT-30, pp. 87–91, 1982.

[25] G. A. Deschamps, "Impedance of an antenna in a conducting medium," *IRE Trans. Antennas Propag.*, pp. 648–650, September 1962.

[26] M. A. Stuchly and S. S. Stuchly, "Coaxial line reflection method for measuring dielectric properties of biological substances at radio and microwave frequencies—a review," *IEEE Trans. Instrum. Meas.*, vol. IM-29, 1980. See also S. S. Stuchly, M. A. Rzepecka, and M. F. Iskander, "Permittivity measurement at microwave frequencies using lumped elements," *IEEE Trans. Instrum. Meas.*, vol. IM-23, pp. 56–62, 1974, and M. A. Rzepecka and S. S. Stuchly, "A lumped capacitance method for the measurement of the permittivity and conductivity in the frequency and time domain—a further analysis," *IEEE Trans. Instrum. Meas.*, vol. IM-24, pp. 27–32, 1975.

[27] J. Toler and J. Seals, "RF dielectric properties measurement system: human and animal data," NIOSH research report, *HEW (NIOSH) Pub. No. 77-176*, July 1977.

[28] M. F. Iskander and J. B. DuBow, "Time- and frequency-domain techniques for measuring the dielectric properties of rocks: a review," *J. Microwave Power*, special issue on electromagnetics in energy applications, ed. by M. F. Iskander, March 1983. See also S. C. Olson and M. F. Iskander, "A new in-situ procedure for measuring the dielectric properties of low permittivity materials," *IEEE Trans. Instrum. Meas.*, vol. IM-35, pp. 2–7, March 1986.

[29] M. F. Iskander, "Permittivity measurements in time domain," MSc thesis, University of Manitoba, Winnipeg, Manitoba, Canada, 1972.

[30] P. F. Wacker and R. R. Bowman, "Quantifying hazardous electromagnetic fields: scientific basis and practical consideration," *IEEE Trans. Microwave Theory Tech.*, vol. MTT-19, pp. 178–187, 1971.

[31] E. Aslan, "Broadband isotropic electromagnetic radiation monitor," *IEEE Trans. Instrum. Meas.*, vol. 21, pp. 421–424, 1972.

[32] R. L. Moore, S. W. Smith, R. L. Cloke, and D. G. Brown, "Comparison of microwave power density meters," *Non. Ioniz. Radiation*, vol. 2, pp. 15–19, 1971.

[33] F. M. Greene, "A new near-zone electric field strength meter," *NBS Tech. Note 345*, November 1966.

[34] R. R. Bowman, "Some recent developments in the characterization and measurement of hazardous electromagnetic fields," *Proc. Intl. Symp. on Biol. Effects and Health Hazards*, Warsaw, October 15–18, 1973, Warsaw, Polish Medical Publishers, pp. 217–227, 1974.

[35] F. M. Greene, "Development of magnetic near-field probes," NIOSH Technical Information, *HEW (NIOSH) Pub. No. 75-127*, January 1975.

[36] T. M. Babij and H. I. Bassen, "Optimizing frequency response characteristics of an **E/H** probe," presented at the Fourth Annual Bioelectromagnetics Society Meeting, Los Angeles, June 28–July 2, 1982. Also see E. Aslan, "A low-frequency **H**-field radiation monitor," *Selected Papers on Biol. Effects of Electromagnetic Waves*, USNC/URSI Annual Meeting, Boulder, October 20–23, 1975, HEW publication (FDA) 77-8010, December 1976.

[37] M. F. Iskander, H. Massoudi, C. H. Durney, and M. Yafeh, "The development of an rf personal dosimeter," presented at the Fourth Annual Bioelectromagnetics Society Meeting, Los Angeles, June 28–July 2, 1982.

[38] M. F. Iskander, C. H. Durney, and D. L. Jaggard, "The development of a microwave personal dosimeter," presented at the Bioelectromagnetics Society Meeting, San Antonio, September 14–18, 1980.

[39] F. K. Storm, R. S. Elliott, W. H. Harrison, and D. L. Morton, "Clinical rf hyperthermia by magnetic-loop induction: a new approach to human cancer therapy," *IEEE Trans. Microwave Theory Tech.*, vol. MTT-30, pp. 1149–1158, August 1982.

[40] S. C. Hill, D. A. Christensen, and C. H. Durney, "Power deposition patterns in magnetically induced hyperthermia: a two-dimensional quasistatic numerical analysis," *Intl. J. Radiation Oncology Biol. Phys.*, in press.

[41] K. D. Paulsen, J. W. Strohbehn, S. C. Hill, D. R. Lynch, and F. E. Kennedy, "Theoretical temperature profiles for concentric coil induction heating devices in a two-dimensional axi-asymmetric, inhomogeneous patient model," presented at the North American Hyperthermia Group Meeting, San Antonio, February 27, 1983.

[42] D. A. Christensen and C. H. Durney, "Hyperthermia production for cancer therapy: a review of fundamentals and methods," *J. Microwave Power*, vol. 16, pp. 89–105. Also see A. W. Guy, J. F. Lehmann, and J. B. Stonebridge, "Therapeutic applications of electromagnetic power," *Proc. IEEE*, vol. 62, pp. 55–57, January 1974.

[43] P. S. Ruggera and G. Kantor, "Development of a family of rf helical coil applicators which produce transversely uniform, axially disturbed heating in cylindrical fat-muscle phantoms," *IEEE Trans. Biomed. Eng.*, vol. BME-31, pp. 98–106, January 1984.

[44] F. S. Chute and F. E. Vermeulen, "A visual demonstration of the electric field of a coil carrying a time-varying current," *IEEE Trans. Education*, vol. E-24, pp. 278–283, November 1981.

[45] R. R. Bowman, "A probe for measuring temperature in radio frequency heated material," *IEEE Trans. Microwave Theory Tech.*, vol. MTT-24, pp. 43–45, 1976.

[46] D. A. Christensen, "A new nonperturbing temperature probe using semiconductor band edge shift," *J. Bioengineering*, vol. 1, pp. 541–545, 1977.

[47] C. C. Johnson, O. P. Gandhi, and T. C. Rozzell, "A prototype liquid crystal fiberoptic probe for temperature and power measurements in rf fields," *Microwave J.*, vol. 18, pp. 55–59, 1975.

[48] T. C. Cetas, "A birefringent crystal optical thermometer for measurements of electromagnetically induced heating," USNC/URSI 1975 Annual Meeting, Boulder, October 20–23, 1975.

[49] D. A. Christensen, "Temperature measurement using optical etalons," 1975 Annual Meeting of the Optical Society of America, Houston, October 15–18, 1974.

[50] K. A. Wickersheim, R. V. Alves, and J. T. Christol, "Improved fluoroptic thermometry system for hyperthermia," Second Annual Meeting, North American Hyperthermia Group, Salt Lake City, April 17–19, 1982.

[51] M. M. Chen, C. A. Cain, K. L. Lam, and J. Mullin, "The viscometric thermometer: a nonperturbing instrument for measuring temperature in tissues under electromagnetic radiation," *J. Bioengineering*, vol. 1, pp. 547–554, 1977.

[52] D. A. Christensen and R. J. Volz, "A nonperturbing temperature probe system designed for hyperthermia monitoring," URSI Meeting and Bioelectromagnetics Symposium, Seattle, June 18–22, 1979.

[53] A. W. Guy, "Analyses of electromagnetic fields induced in biological tissues by thermographic studies on equivalent phantom models," *IEEE Trans. Microwave Theory Tech.*, vol. MTT-19, pp. 205–214, 1971.

[54] A. W. Guy et al., "A new technique for measuring power deposition patterns in phantoms exposed to em fields of arbitrary polarization—example, the microwave oven," *Proc. Microwave Power Symp.*, University of Waterloo, Ontario, Canada, pp. 36–40, May 1975.

[55] H. Bassen, P. Herchenroeder, A. Cheung, and S. Neuder, "Evaluation of an implantable electric-field probe within finite simulated tissues," *Radio Sci.*, vol. 12, pp. 15–25, 1977.

[56] H. Bassen, M. Swicord, and J. Abita, "A miniature broadband electric-field probe," in *Biological Effects of Nonionizing Radiation*, ed. by P. E. Tyler, Ann. N.Y. Acad. Sci., vol. 247, pp. 481–493, 1975.

[57] G. S. Smith, "A comparison of electrically short bare and insulated probes for measuring the local radio frequency field in biological systems," *IEEE Trans. Biomed. Eng.*, vol. BME-22, pp. 477–483, 1975.

[58] C. H. Durney, H. Massoudi, and M. F. Iskander, *Radiofrequency Radiation Dosimetry Handbook*, 4th ed., Report USAF SAM-TR-85-73, USAF School of Aerospace Medicine, Brooks Air Force Base, Texas 78235, October 1986.

[59] M. L. Crawford, "Generation of standard EM fields using TEM transmission cells," *IEEE Trans. Electromagnetic Compatibility*, vol. EMC-16, pp. 189–195, 1974.

[60] M. F. Iskander, H. Massoudi, and C. H. Durney, "Development of rf personnel dosimeter," final report prepared for R. S. Landauer, Jr., and Co., Department of Electrical Engineering, Univ. of Utah, May 25, 1982.

Chapter 25

Direction-Finding Antennas

R. E. Franks
ESL, a Subsidiary of TRW

CONTENTS

 **Raymond Franks** is an electrical engineer with over 30 years of experience in communication and radar systems, specializing in the antenna and propagation aspects of system analysis and design. He is currently a principal engineer at ESL, a subsidiary of TRW, in Sunnyvale, California. He has been responsible for the antenna and system design of several vhf, uhf, and microwave direction-finding systems from the conceptual stage to field evaluation. The development of antenna systems with extremely broad bandwidths, such as log-periodic arrays and ridged waveguide horns, has occupied much of his career. He has been awarded three patents involving broadband sequential lobing tracking antennas and a dual-polarized ridged waveguide horn design. He has coauthored several technical papers on broadband antennas, polarization coupling, and multiple-signal df processing (MUSIC algorithm).

Prior to joining ESL, he was an engineering specialist at GTE Sylvania Electronic Defense Laboratories, and had previous experience with the General Electric Company. He received the BS degree in electrical engineering from Rice University in 1950.

1. Introduction

The purpose of a radio direction finder (df) is to measure the direction of arrival of a radio signal. Traditionally the azimuth angle of arrival, i.e., the angle of arrival in the horizontal plane, has been considered sufficient to define the direction of arrival of the signal, but the elevation angle of arrival as measured in a vertical plane is important for some target location schemes and for other purposes, such as propagation research. In any event the elevation angle of arrival of the incoming wave will affect the performance of azimuth-only df systems, even if elevation output is not needed.

Another complication in the use of df systems arises because the signal may propagate via more than one path between transmitting and receiving antennas. This is particularly likely in the case of ionospheric propagation of hf signals. It also occurs to some degree at most ground-based df sites when signals are being received via any mode of propagation because of rf being scattered by objects near the df receiving site. Much of the complexity of df systems and the error in direction of arrival of their output results from the multipath propagation of radio signals.

A direction finder consists of three principal elements: a directional antenna or antenna system to sample the incident wave, a receiver or receivers to limit the signal bandwidth and provide system gain, and a processing and indicating subsystem to provide an output in a useful form. The antenna portion of the df system will be considered in this chapter, with some discussion of the processing and output implementation where necessary to understand the antenna function.

The fact that the signal being received is a traveling wave spreading outward from its source is the basis for all direction-finding antenna techniques. The phase progression of the signal as it travels along its path is the characteristic used to advantage by df antennas. The df antenna is made of two or more elements that are located physically displaced from each other, so that the wave may be sampled at two or more points. The phase differences between these points contain the information on the angle of arrival of the wave. A line perpendicular to the direction of travel of the wave will be a line of constant phase. If the sampling points lie along this line, all of the points will be in phase. This is the condition that is utilized by most df antenna arrays to determine the direction of travel, and thus the angle of arrival of the signal.

The phase differences between the array elements may be directly measured, as in an interferometer; or it may be more convenient to combine the signals from the elements so that the response of the array of elements varies with the direction of arrival of the signal. The directivity of the antenna system can then be used to measure angle of arrival by rotating the antenna pattern and observing how the power out of the antenna changes with angle. Several types of antenna patterns may be generated. A peak in the pattern might indicate the angle of arrival of the signal. Alternately, a null in the pattern could be used. Or the equality of power out of two ports of the array could be used to indicate signal angle of arrival.

2. Rotating Antenna Patterns

When a directive antenna pattern is formed by the df antenna, the angle of arrival of the signal is determined by rotating the pattern and noting the angle at which the distinguishing pattern feature (null or maximum) is seen in the signal output. This angle corresponds to the angle of arrival of the signal. The entire antenna system can be physically rotated about an axis, usually vertical, to rotate the directional pattern. This is usually done for antenna systems that are not large. When the antenna system is physically too large to rotate, the elements may be stationary and the antenna pattern rotated using a goniometer that rotates and couples to the elements in a sequential fashion to produce a rotating pattern.

Physically Rotating Antenna Systems

Fig. 1 shows examples of rotating df antennas. The loop and the Adcock array produce nulls when the incoming wave direction is perpendicular to the plane of

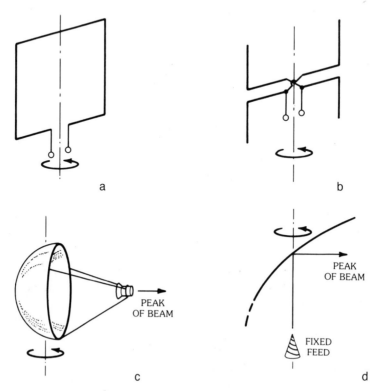

Fig. 1. Rotating df antennas. (*a*) Loop. (*b*) Adcock. (*c*) Reflector with feed. (*d*) Rotating reflector with stationary feed.

the antenna, and they are generally used at longer wavelengths. Rotating reflector antennas with feeds attached and offset-fed rotating reflectors with fixed feeds are typically used at shorter wavelengths where a high-gain pencil beam may be formed in an aperture small enough to rotate. Here the peak of the rotating pattern indicates angle of arrival.

If the antenna is continuously rotated, the instantaneous angular position of the antenna may be read at the antenna and transferred to the deflection circuits of a cathode-ray tube to drive the angular orientation of the display in synchronism with the antenna. Radial amplitude of the display is then driven by the received signal level out of the antenna and the antenna pattern is thus traced out on the display. The angle of the peak for a narrow-beam antenna or the angle of the null for a low-gain element, such as a loop, can be read off an azimuth scale around the face of the tube.

Loop Antenna DF Patterns—The open-circuit voltage of a loop antenna with its axis vertical as shown in Fig. 2 when receiving a vertically polarized signal of wavelength λ arriving in a horizontal plane at an angle θ to the plane of the loop is

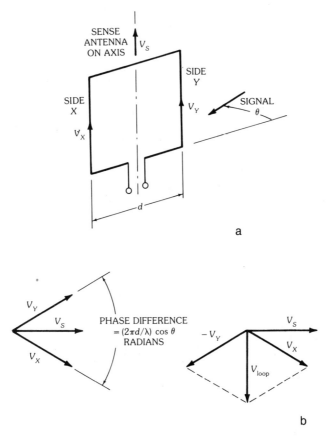

Fig. 2. Loop antenna output voltage. (*a*) Loop geometry. (*b*) Addition of voltages.

$$V_{\text{loop}} = 2V_S \sin\left(\frac{\pi d \cos \theta}{\lambda}\right) \qquad (1)$$

where d/λ is the spacing in wavelengths between the sides of the loop and V_S is the voltage induced in one vertical side. This response pattern is plotted in Fig. 3 for several values of d. The nulls at $\theta = 90°$ and $270°$ are the pattern characteristics used for direction finding.

Because there are two nulls, $180°$ apart, the loop presents an ambiguity in its df indication that must be resolved. This can be done by combining the output of the sense antenna shown in Fig. 2, which is a vertical element, with the loop to generate a cardioid pattern. The loop resultant voltage E_R is in quadrature with the voltage induced in the sense antenna as shown in Fig. 2b. If the sense antenna voltage is shifted $90°$ in phase lead and added to the loop as can be done with a $90°$ hybrid combiner, the resultant pattern will be a cardioid if the loop and sense antenna voltages are equal in magnitude. If they are unbalanced, the resulting pattern is distorted as shown in Fig. 4 but will still resolve the ambiguity. In all cases the pattern will produce a sloping response near the two nulls of the loop, and the null that is at the location where the slope increases with increasing angle of arrival of the signal will be the correct null for the signal arriving at $90°$ to the plane of the loop.

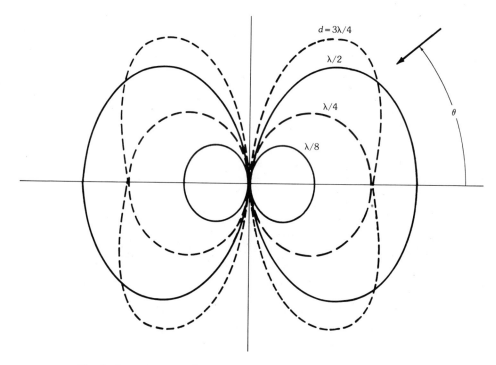

Fig. 3. Loop patterns for several values of spacing between sides.

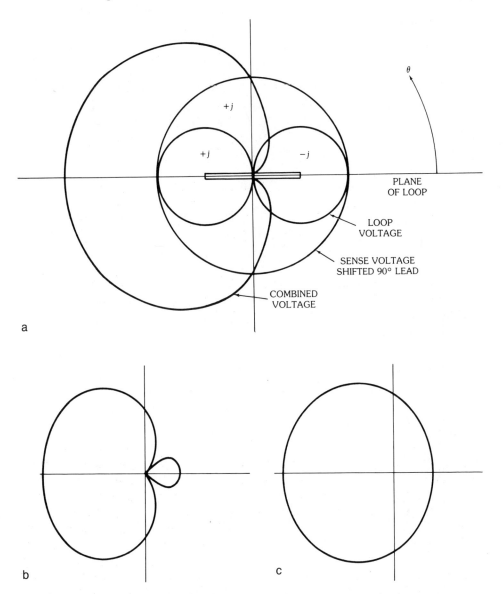

Fig. 4. Combined patterns for loop and sense antenna. (*a*) Cardioid pattern when equal loop and sense antenna voltages are combined. (*b*) Combined pattern with sense antenna voltage of half the loop voltage. (*c*) Combined pattern with sense antenna voltage of twice the loop voltage.

Loop Antenna Impedance and Efficiency—The electrical characteristics of a small, square loop antenna can be expressed as follows. The radiation resistance is

$$R_r \cong 30\,000\ N^2(d/\lambda)^4$$
$$= 30\,000\ (NA/\lambda^2)^2$$

(2)

and the inductive reactance is

$$X_L \cong 3000\ N^2(d/\lambda) \tag{3}$$

for a ratio of winding length/side = t/d = 0.2. Each side of the loop is of length d, giving an area $A = d^2$, and the number of turns in series is N. The radiation resistance and the reactance of loops for several ratios of winding length to side are shown in Fig. 5.

The open-circuit voltage induced in the loop is

$$V_{\text{loop}} = (2\pi N d^2/\lambda)\, E^i, \qquad d/\lambda \ll 1/\pi \tag{4}$$

when the loop is oriented for maximum signal. Here E^i is the incident electromagnetic-field strength. The equivalent circuit of the loop connected to a load is

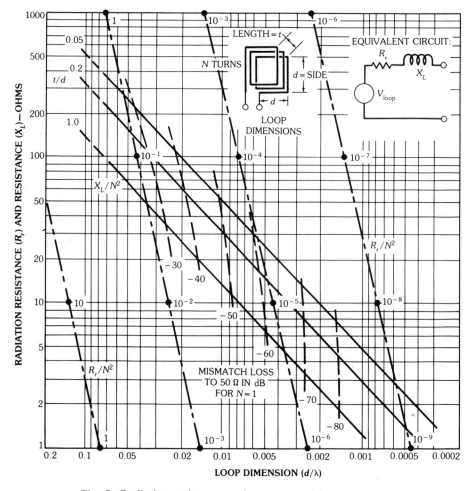

Fig. 5. Radiation resistance and reactance of loop antennas.

given in Fig. 5. The mismatch loss of a single-turn loop directly connected to a 50-Ω transmission line is plotted in the figure. For more information on the use of loops see Chapter 7.

Special Problems With Loops—The environment around the loop can upset the balance between the two vertical sides of the loop and seriously degrade the null. If the capacitance to ground is not the same for both sides, stray voltages may be coupled into the loop and unequal currents may flow into the load. To avoid this condition the loop can be enclosed in a metal shield as in Fig. 6. The shield must have a gap at the center of the loop to prevent it from shorting the loop. The transmission line feeding the loop should be balanced to ground, with a balanced-unbalanced transformer to convert the loop output to single-ended coaxial transmission line for connection to the receiver.

In the hf band the loop is unsatisfactory because of downcoming sky-wave signals. This limitation may also apply in other bands if horizontally polarized waves can fall on the loop.

Elevated H Adcock—An antenna that rejects horizontal polarization is the elevated H Adcock array shown in Fig. 7. This antenna consists of two vertical dipoles supported by a horizontal member which carries balanced shielded transmission lines. The dipoles are connected out of phase to produce nulls broadside to the array. The pattern for vertically polarized signals in the horizontal plane is similar to that given for a loop in Fig. 3. A sense antenna is needed, and this may be a third dipole located at the center of the array and connected to the output of the array in the same manner that the sense antenna is connected to the loop.

The voltage out of the array will be decreased as the cosine squared of the elevation angle because both the phase difference between the dipoles and their induced voltages decrease with elevation.

Spaced Loops—Another type of antenna useful for sky-wave signals is an array

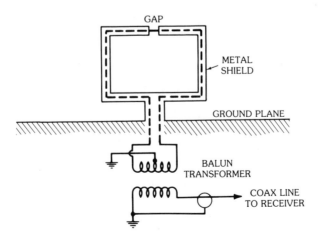

Fig. 6. Balanced shielded loop.

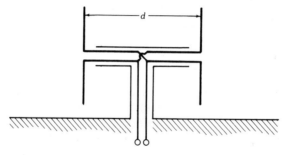

Fig. 7. Elevated H Adcock antenna.

of loops. Three configurations are shown in Fig. 8. The vertical coplanar loops of Fig. 8a generate a pattern similar to that of the single loop, except that the null is deeper because in addition to the array factor forming a null each loop of the array also has a null at 90° and 270° to the plane of the loop. The downcoming horizontally polarized wave produces identical signals in the two loops at the null, and this is canceled by the out-of-phase connection. The induced voltage of downcoming signals is not reduced as it is with Adcock antennas.

A second spaced loop array is the coaxial arrangement of vertical loops shown in Fig. 8b. This reduces the coupling of the loops to the horizontal support structure, but produces a pattern with four nulls per revolution, at 0°, 90°, 180°, and 270°, because now the loop elements generate nulls in the 0° and 180° directions.

Loops that lie in the horizontal plane can be arranged as shown in Fig. 8c. The response of the horizontal coplanar loop array is the same for horizontally polarized signals as that of the elevated H Adcock array for vertically polarized signals.

Rotating High-Gain Antennas—In the frequency range above 1 GHz, a relatively small pencil-beam antenna is feasible. Two types of high-gain aperture antennas were illustrated in Fig. 1. In each of these antennas the antenna pattern will be traced out by receiving a signal as the antenna is rotated through 360° azimuth, and the direction that the peak is pointed in when maximum energy is received will indicate the angle of arrival of the signal.

The rotating reflector with stationary feed is particularly attractive because no rotary rf joint is needed in the signal path. The feed must be circularly polarized with a symmetric pattern to produce good df results, for otherwise the polarization of the df beam will rotate as the beam scans.

Amplitude comparison between two beams that point in nearly the same direction can also be used to directionally find a target more accurately than with a single pencil beam. Any technique suitable for generating two beams can be used, such as two separate reflectors, one reflector with two feeds located on either side of the focal axis to produce two squinted beams, or an array that forms two beams simultaneously that overlap on the array boresight direction. One receiver can be alternately switched between the two beams to detect the point of equal output within the main lobe response of the antenna system [1].

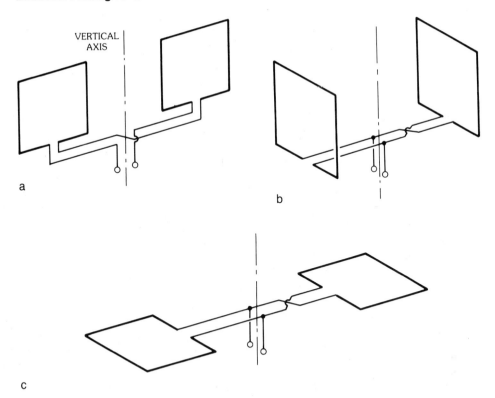

Fig. 8. Spaced loop arrays. (*a*) Vertical coplanar spaced loops. (*b*) Vertical coaxial spaced loops. (*c*) Horizontally polarized spaced loops.

Electrically Rotating Null Patterns

When the antenna array is too large to be physically rotated conveniently, the pattern from stationary elements can be rotated. An example of this using crossed loops is shown in Fig. 9, where the loops feed a goniometer having two stationary coils at right angles and enclosing a smaller rotating coil. The stationary coils simulate the external radio-frequency field within the goniometer so that the rotor behaves as if it were a loop antenna rotating in the external field. Pairs of Adcock elements may be substituted for crossed loops to produce the same performance.

Ideally the external antenna patterns must vary as $\sin \theta$ and $\cos \theta$ with the same maximum response. Coupling between stator and rotor must also vary sinusoidally with rotation. Because of symmetry no error occurs at 0°, 45°, 90°, 135°, etc., but midway between these points, at 22.5°, 67.5°, etc., error builds up as the diagonal of the array (*d* of Figs. 2, 5, or 7) exceeds 0.2λ because the element patterns depart from the ideal. At an array diagonal of 0.5λ, the peak error reaches 7°, and it is 18° at 0.7λ diagonal. If two sets of Adcock elements with an angular separation of 55° between the sets are connected to each stationary coil, the array diagonal can be increased to one wavelength before the error exceeds 3°; however, it builds up rapidly beyond this point. This error decreases with increasing elevation angle of

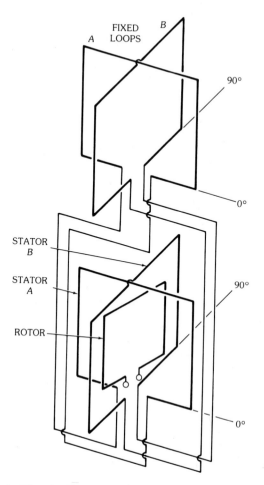

Fig. 9. Electrically rotating loop pattern using a goniometer.

arrival of the wave because the effective antenna diagonal is reduced, causing the antenna patterns to better approximate sinusoidal variation.

Instantaneous df Patterns

The Watson-Watt df system uses two crossed loops or two Adcock arrays at right angles in a manner similar to the rotating goniometer, except that the two rf signals are compared instantaneously to generate angle of arrival. The output of each array pair is amplified by a receiver, usually down-converted to a convenient intermediate frequency, and applied to one pair of deflection plates of a crt. A bi-directional line results, inclined at the signal angle of arrival. The sense antenna may be coupled into the receivers momentarily to resolve the ambiguity. This system is particularly useful for short-duration signals as no pattern scanning is required.

The two coherent receiver channels need to have good phase and amplitude balance to give accurate results. A single-channel system can be developed, but it will no longer generate instantaneous df output. The antenna signals can be combined, as shown in Fig. 10, to produce a composite signal that carries the bearing information in one rf channel on two audio frequencies. Two separate audio tones, one for the north/south (N/S) loop and one for the east/west (E/W) loop, are used to modulate each antenna output as shown in the figure [2].

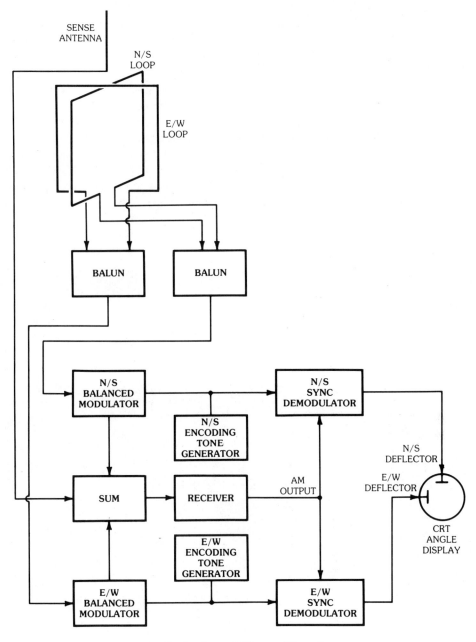

Fig. 10. Single-channel Watson-Watt df system.

The balanced modulators generate double-sideband suppressed-carrier rf outputs which are summed with the sense antenna output. Since this sense signal acts as a reinserted carrier, the composite signal applied to the receiver is an am signal with the two low-frequency audio tones carrying the amplitude and phase information from the df antennas. This modulation is imposed on top of any other modulation that may be on the signal. The receiver am output is synchronously demodulated to recover the two df information channels. Then these two voltages may be applied to a crt to indicate angle of arrival, or measured so that their ratio will give the tangent of the angle of arrival.

Electrically Rotating High-Gain Pattern

The Wullenweber array consists of a large number of vertical elements symmetrically located around a cylindrical reflecting screen. Fig. 11 shows a 120-element array with a diameter of 1000 ft (304.8 m) installed on the facilities of the University of Illinois near Urbana, Illinois. This antenna operates in the hf band. Each antenna element is connected to a segment of a capacitor rotating switch or goniometer. The rotor segments span an arc of about 100° to 120° so that 33 to 40 elements are activated at one time to form a beam. Feed delay lines of proper lengths that equalize the free-space delay differences between the elements are attached to the segment as indicated in Fig. 12. The element outputs from the two equal half-sections may be combined as shown in the figure to form sum and

Fig. 11. Wullenweber hf array at the University of Illinois. (*Courtesy Radio Location Lab., University of Illinois at Urbana*)

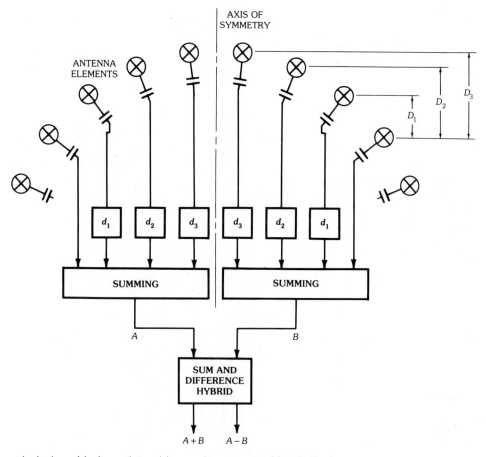

d_1, d_2, d_3 are delay lines with time delays equal to propagation delays D_1, D_2, D_3

a

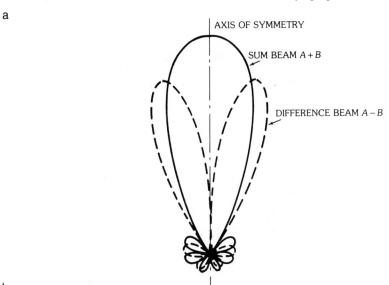

b

Fig. 12. Electrically rotating high-gain df pattern. (*a*) Goniometer circuitry. (*b*) Sum and difference patterns.

difference beams, with the sum beam giving an approximate angle of arrival and the deep null in the difference beam sharpening the estimate of the angle. The structure of a rotating goniometer is illustrated in Fig. 13.

Doppler Direction Finder

The direction of arrival of a signal can be determined by moving an antenna in a circle in the horizontal plane and noting the phase change that results. Phase change at a rate of 2π radians per second is equivalent to a Doppler frequency shift of one hertz. When the antenna is moving directly toward or away from the target the rate of phase change is maximum and will produce a maximum frequency increase (toward) or decrease (away), with zero frequency shift when the antenna is moving tangential to the wave equiphase lines, as shown in Fig. 14. Rather than physically rotating an antenna element, the outputs of each element in a circular

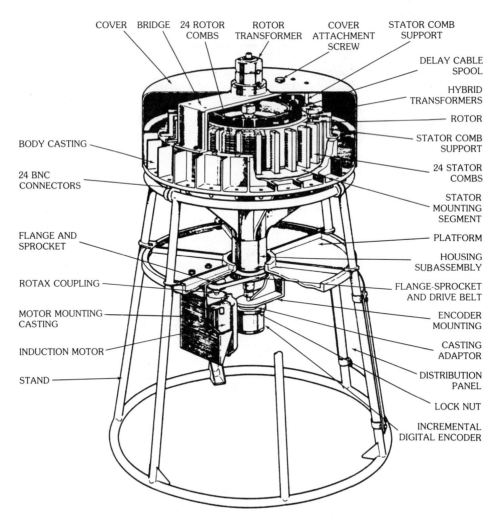

Fig. 13. Mechanical structure of a rotating goniometer. (*After Gething [11]*)

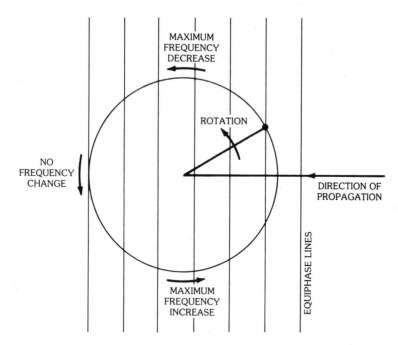

Fig. 14. Doppler shift produced by rotating antenna.

array are sampled sequentially around the circle to generate a phase shift that can be detected to extract the Doppler modulation thus introduced. This modulation will vary at the same rate as the cyclic rate of sampling of the antennas, so if the phase of the modulation is compared with a reference voltage synchronized with the antenna commutation, the angle of arrival of the signal will be indicated. Sampling may be accomplished using a mechanical capacitive goniometer or an electronic commutating switch.

The array elements must be spaced less than one-half wavelength apart around the circle or phase ambiguities will be introduced in the phase modulation. A center antenna may be used as a phase reference so that the phase difference between the rotating samples and the fixed antenna can be detected at the output of a dual-channel phase-coherent receiver.

3. Multimode Circular Arrays

The Wullenweber and the Doppler circular array antennas just described provide direction finding through a scanning process that involves rotating an antenna response through 360° azimuth and deriving angle of arrival from the timing of the antenna output within the scan cycle. Circular arrays can also be used in a nonscanning mode to give an instantaneous df response. This is done by exciting the circular array of elements with a feed network that sets up several far-field phase modes.

The nth-order phase mode pattern is defined as having constant amplitude and a phase varying linearly from 0 to $2n\pi$ as the angle of arrival varies through 360°.

Angle of arrival can then be determined by measuring the phase difference between two adjacent modes, i.e., the phase difference between the 0 and +1 modes will uniquely indicate the angle as the phase progresses through 360° for a change of 360° in azimuth. The phase difference as a function of angle of arrival will increase for higher-order modal differences.

For example, in an eight-element array the 0, ±1, ±2, and ±3 modes may be useful. The mode designated as $n = +4$ or $n = -4$ for an eight-element array represents one mode, the highest-order mode, which will not have useful characteristics. A df system could use the 0, +1, and −3 modes with a three-channel phase coherent receiver. Unambiguous df would be generated by comparing the phases of the 0 and +1 modes. Comparison of the +1 and −3 modes would also be used to give higher accuracy because of the 4° phase difference produced for each degree change in angle of arrival.

The nth-order phase mode may be generated in the far-field pattern by exciting a corresponding nth-order current phase mode on the array. A convenient way to do this is to use a Butler matrix network, such as that shown in Fig. 15 for an eight-element array. Each of the mode inputs to the network will excite a uniform phase progression around the elements of the array, with the phase differential between elements equal to n times 45°, giving n times 360° total phase progression around the eight elements. The phase progressions for the $n = 0$ and $n = +1$ modes are shown in the figure.

The relative amplitudes of the modes are given by Bessel functions for a continuous circular array of omnidirectional elements excited by a phase progression of order n:

$$\text{relative amplitude of } n\text{th mode} = J_n(\pi D/\lambda) \tag{5}$$

where

J_n = Bessel function of order n

D = diameter of array

An array of discrete omnidirectional elements will produce essentially these same amplitudes as long as the element spacings are well under a half wavelength. In general, the higher the order of the mode and the greater the element spacing, the more the mode phase or amplitude may depart from the ideal.

The Bessel functions that give the mode amplitudes predict nulls for certain combinations of mode number and array diameter, which would seriously limit the bandwidth that could be achieved for a given combination of modes. Fortunately, if directional elements are used in the array with the peak gain of each element pointed away from the center of the array, the mode amplitudes are more constant with array size. The calculated mode amplitudes for omnidirectional elements and for elements with a $1 - \cos \theta$ pattern are shown in Fig. 16 for the 0, 1, and 2 modes. Experimental results have shown that the mode amplitudes do vary smoothly without nulls as the operating frequency varies using directional elements. Dipole elements in front of a reflecting screen or slots on a conducting cylinder would yield element patterns with the desired directivity [3, 4].

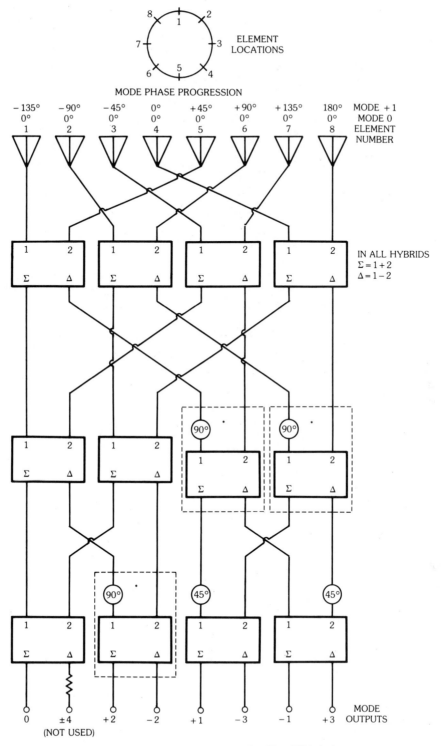

Fig. 15. Butler matrix feed network for a multimode array.

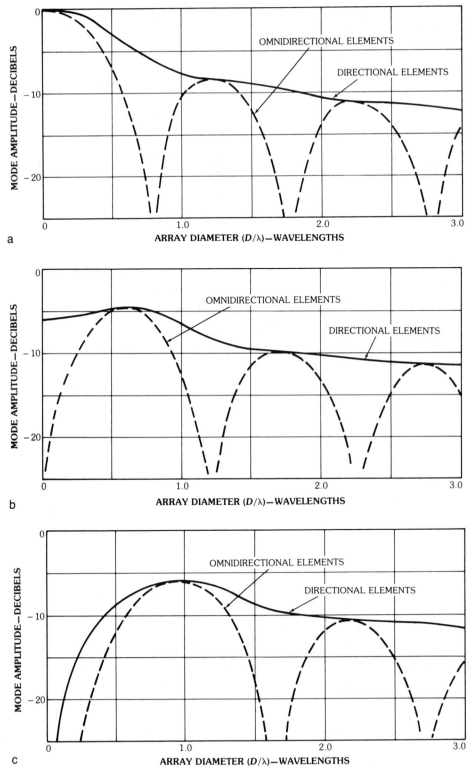

Fig. 16. Mode amplitudes versus array diameter. (*a*) Mode 0. (*b*) Mode 1. (*c*) Mode 2.

4. Interferometers

The df techniques described up to now have employed the amplitude and/or phase pattern characteristics of an antenna array to indicate angle of arrival of a signal. An interferometer df system makes a direct measurement of the phase differences between two or more points on the wavefront and converts the phase information into angle of arrival. Fig. 17 illustrates the relationship between angle of arrival, α, as measured from the normal to the base line and the phase difference, $\Delta\phi$, between two elements spaced along a base line a distance d apart. In Fig. 17a, the output of antenna 2 leads that of antenna 1 by $\Delta\phi = (2\pi d/\lambda)\sin\alpha$ and it can be seen that the phase function has mirror symmetry about the base-line direction, $\alpha = \pm 90°$

For element spacing less than one-half wavelength (short base lines) each value of phase difference corresponds to two angles of arrival. For element spacings in excess of one-half wavelength additional ambiguities are present because of "foldover" in the phase function when the phase difference exceeds 180°.

The mirror ambiguities for short base lines can be resolved by using two orthogonal base lines and combining the phase differences to get the angle α of arrival as measured from north:

$$\alpha = \tan^{-1}\left(\frac{\Delta\phi \text{ E/W base line}}{\Delta\phi \text{ N/S base line}}\right) \qquad (6)$$

Mutual coupling between antenna elements will generate an octantal error that is zero for 0°, 45°, 90°, etc., and alternately reaches positive and negative maxima at 22.5°, 67.5°, etc. The magnitude of the error generally increases with element spacing and with element radiation efficiency.

The ambiguities on long base lines can be resolved by using a combination of short and long base lines, or two long base lines whose difference in length is less than one-half wavelength. Angle of arrival measurement errors tend to be reduced in proportion to the increase in base-line length, but the probability of an ambiguity being introduced by a given phase measurement error increases as the base lines are made longer. For example, if $n(\lambda/2)$ and $(n-1)(\lambda/2)$ are the two base-line lengths, the base lines will both indicate the correct phase values at the true angle of arrival. At the angle corresponding to the nearest ambiguity on the longer base line, the other base line will differ in phase by only $360°/n$ from experiencing an ambiguity also. More antennas and more measurement base lines are needed as the overall length of the interferometer is increased [5,6].

Antenna Elements

If the interferometer has 360° angular coverage, the antenna elements ideally have omnidirectional patterns. Vertically polarized dipoles, crossed vertical loops fed in quadrature, or monopoles on a ground plane are typically used. Loop antennas in a horizontal plane or a turnstile formed by crossed horizontal dipoles fed in quadrature can give omnidirectional horizontally polarized coverage.

The relative transmission phase stability of the antenna elements must be good to avoid introducing phase errors and resulting angle of arrival errors in the inter-

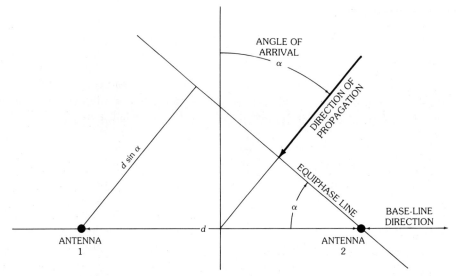

a

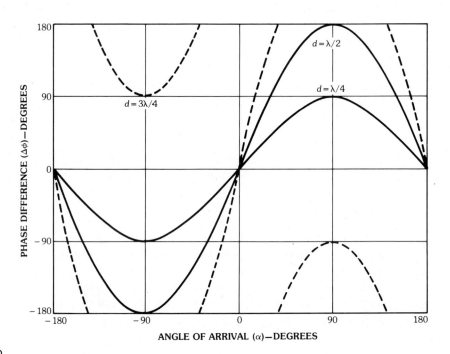

b

Fig. 17. Phase difference versus angle of arrival for a two-element interferometer. (*a*) Antenna geometry. (*b*) Phase difference.

ferometer output. If the df system is to operate over a wide frequency bandwidth, which is the usual condition, tuning of the elements to improve efficiency at the operating frequency may not be practical because of the difficulty in making the tuning circuits for all elements track in phase. Broadband impedance matching can be chosen to favor the portion of the band where antenna efficiency is lowest. As an example, electrically small dipoles or monopoles can be connected to a load impedance approximately equal to the magnitude of the antenna reactive impedance at its lowest operating frequency to give a compromise broadband match favoring the lower frequencies.

In the low vhf and the hf bands electrically small antennas may provide sufficient output in spite of their low efficiency. The external noise levels are generally high such that the noise power output from an efficient antenna would be much higher than internal system (receiver) noise. In this situation both signal and external noise will be reduced by the low antenna efficiency but the system signal-to-noise ratio will not be reduced significantly until the external noise power out of the antenna falls to a level near that of the system internal noise level.

At higher frequencies, and particularly into the microwave region, directive antennas may be needed to provide adequate system gain. In this case the angular coverage will be reduced and it may require several interferometer arrays to cover 360° in several sectors. Typically the choice between arrays is made by comparing the antenna amplitude responses to the signal being received to determine from the strongest response which angular sector contains the signal. Then the array covering that sector makes the phase measurements to determine angle of arrival.

Vehicular-Mounted Interferometers

An array of antennas mounted on a ground vehicle or an aircraft will receive energy scattered by the vehicle in addition to the direct signal wavefront. As a result the actual phase differences derived for antenna pairs will usually depart widely from the ideal sinusoidal variation with angle of arrival. Direction-finder outputs based on the antenna geometry alone may not be acceptable. The phase response will be stable, however, and can be measured to generate calibration data for accurate direction finding.

The vehicle with antennas installed may be rotated through 360° while receiving a signal from a known location. The base-line phase responses are measured and stored along with the true angle of arrival. These calibration data can be recalled when a df measurement on an unknown signal is made, and the calibration data searched to find the angle corresponding to the best fit of calibration to measured data. This angle is the best estimate of the signal angle of arrival.

Dipole antennas are preferred over monopoles for vehicle or aircraft df systems. The dipole will usually not generate as much cross-polarized radiation and will generally have a more uniform amplitude pattern. A monopole located on a long structure such as an aircraft fuselage may excite the structure in a long wire mode to radiate a multilobed, cross-polarized pattern. If two monopoles are placed diametrically opposite on the structure and driven out of phase to form a dipole, the excitations from the two monopoles tend to cancel in the structure and suppress long wire patterns.

5. Multiple-Signal Direction Finding

The df techniques described have been applied to measuring the angle of arrival of a single signal when only one wave is incident on the antenna. In general there may be several wavefronts present. These may consist of waves from one source that have traveled via different propagation paths and/or signals from more than one source.

The behavior of the df output under the condition of multiple uncorrelated (independent) signals depends heavily on the processing portions of the df system. A phase interferometer with amplitude limiting in the phase detectors may respond to the strongest signal with little effect from weaker signals. An interferometer that

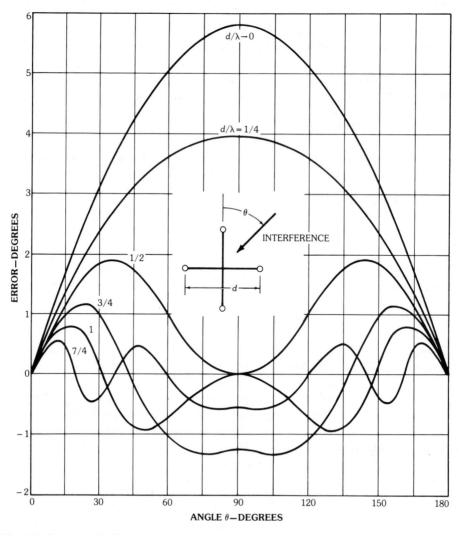

Fig. 18. Direction-finding error in the estimate of a signal at 0° in the presence of interference 10 db below the desired signal.

averages the complex cross-correlation coefficient between antenna outputs to measure phase will tend to produce an output that is the power-weighted average of the two signals; however, for long base lines the error decreases as base-line length increases as shown in Fig. 18 for interference 10 dB below a desired signal. Direction-finding systems with high-gain, narrow-beam antennas can give the angles of arrival of each of several signals if the signals are separated in angle by greater than the beamwidth.

If the signals are correlated, such as by multimode propagation from the same source, the error condition will be similar to that described above for uncorrelated signals except that the error produced in a df system will be related to the voltage and phase of the weaker signal instead of the power ratio. For example, Fig. 18 would apply to the rms error for an interferometer in the presence of multipath 20 dB below the direct path signal as the multipath phase varies over 360°.

To generate angle of arrival on each one of several signals or waves present, high-resolution processing techniques have been developed that use the phase and amplitude data from the df antennas and attempt to model the df system and the signal environment to estimate the angles of arrival in some optimal way. Generally a combination of several signal responses is selected that best coincides with the set of measured outputs from the antenna ports, and the angles of arrival of these selected responses are then an estimate of the angles of arrival of the signals [7, 8, 9, 11].

These system design techniques may impose more stringent requirements on the antenna components. Elements in an array may need to be more alike and their phase and amplitude characteristics better defined. It may be necessary to measure the overall response of the df antennas in their environment, especially where scattering from structures such as an airframe may be significant. Polarization diversity may be required in the array elements to cope with a wide range of signal polarizations [10].

6. References

References 12 through 22 are not indicated in the text of this chapter but are included here as being of general interest.
[1] R. A. Watson-Watt and J. F. Herd, "An instantaneous direct reading goniometer," *J. IEE*, vol. 64, p. 611, 1926.
[2] R. Rainer and A. J. Burwasser, "An approach to hf tactical radio direction finding and signal monitoring," *J. Electron. Defense*, vol. 6, no. 10, pp. 173–184, October 1983.
[3] D. E. N. Davies, "Circular arrays: their properties and potential applications," *Proc. IEE Conf. Antennas Propag.*, York, England, 1981.
[4] T. Rahim and D. E. N. Davies, "Effect of directional elements on the directional response of circular antenna arrays," *Proc. IEE*, vol. 129, pt. II, February 1982.
[5] W. M. Sherrill, "Bearing ambiguity and resolution in interference direction finders," *IEEE Trans.*, vol. AES-5, pp. 959–966, November 1969.
[6] R. L. Goodwin, "Ambiguity-resistant three- and four-channel interferometers," *NRL Report 8005*, Naval Research Laboratory, Washington, D.C. 20375, September 1976.
[7] W. F. Gabriel, "Spectral analysis and adaptive array superresolution techniques," *Proc. IEEE*, vol. 68, pp. 654–666, June 1980.
[8] R. Schmidt, "Multiple emitter location and signal parameter estimation," *Proc. RADC Spectrum Estimation Workshop*, RADC-TR-79-63, Rome Air Development

Center, Rome, New York, p. 243. October 1979. Also in *IEEE Trans. Antennas Propag.*, vol. AP-34, pp. 276–280, March 1986.

[9] D. H. Johnson and S. R. DeGraaf, "Improving the resolution of bearing in passive sonar arrays by eigenvalue analysis," *IEEE Trans. Acoust., Speech, Signal Processing*, vol. ASSP-30, pp. 638–647, August 1982.

[10] E. R. Ferrara, Jr., and T. M. Parks, "Direction-finding with an array of antennas having diverse polarizations," *IEEE Trans. Antennas Propag.*, vol. AP-31, pp. 231–236, March 1983.

[11] P. J. D. Gething, *Radio Direction Finding*, Stevenage, Herts., UK: Peter Peregrinus, Ltd., 1978.

[12] "IRE standards on navigation aids: direction finder measurements 1959," *Proc. IRE*, vol. 47, pp. 1349–1371, August 1959.

[13] J. A. Boyd et al., *Electronic Countermeasures*, Los Altos Hills: Peninsula Publishing, 1978, pp. 10-1 to 10-96.

[14] J. E. Browder and V. J. Young, "Design values for loop-antenna input circuits," *Proc. IRE*, vol. 35, pp. 519–526, May 1947.

[15] L. L. Libby, "Special aspects of balanced shielded loops," *Proc. IRE*, vol. 34, pp. 641–646, September 1946.

[16] H. A. Wheeler, "Fundamental limitations of small antennas," *Proc. IRE*, vol. 35, pp. 1479–1484, December 1947.

[17] J. H. Moon, "Design of electromagnetic goniometers for use in medium frequency direction finding," *J. IEE*, vol. 94, p. 69, January 1947.

[18] C. W. Earp and R. M. Godfry, "Radio df by cyclical differential measurement of phase," *J. IEE*, vol. 94, pt. IIIA, pp. 705–721, March 1947.

[19] W. Hausz, "Angular location, monopulse and resolution," *Microwave J.*, vol. 7, no. 2, p. 60, February 1964.

[20] A. D. Bailey and W. C. McClurg, "A sum-and-difference interferometer system for hf radio direction finding," *IEEE Trans. Aerospace and Navigational Electronic*, vol. ANE-10, pp. 65–72, March 1963.

[21] E. C. Hayden, "Propagation studies using direction-finding techniques," *J. Res. Natl. Bur. Stand.*, vol. 65D, pp. 197–212, May 1961.

[22] P. J. D. Gething, "Influence of ionospheric conditions on the accuracy of high-frequency direction finding," *J. Res. Natl. Bur. Stand.*, vol. 65D, pp. 225–228, May 1961.

Chapter 26

Standard AM Antennas

C. E. Smith
Carl E. Smith Electronics, Inc.

CONTENTS

Carl E. Smith received the BSEE degree from Iowa State University in 1930 and the MSEE and Professional EE degrees from the Ohio State University in 1932 and 1936, respectively. A registered professional engineer, he has gained recognition as an authority in electronics engineering for his achievements in broadcast-station antenna design and in education. In the course of his career he has been responsible for the engineering of scores of am and fm broadcasting stations here and abroad, including antennas, ground systems, and proofs of performance.

Smith Electronics was organized by Mr. Smith to provide research, development, and engineering services to government and industry. It has strong capabilities in antenna research and development, propagation studies, systems engineering, solid-state electronics, and electromagnetic compatibility testing.

The Cleveland Institute of Electronics was founded by Mr. Smith, who authored a major portion of the original advanced engineering course. It is now one of the leading electronics training institutes in this country with a large active student enrollment. Mr. Smith is still active as Chairman of the CIE Educational Committee. Mr. Smith now owns Carl E. Smith Electronics and handles his commercial work under his name.

1. Introduction

There is a world-wide medium-frequency broadcast band from 535 to 1605 kHz and a low-frequency broadcast band from 150 to 255 kHz in the Region 1 area.* At these frequencies the normal broadcasting antenna is a vertical conductor, usually a guyed or self-supporting tower. Vertical polarization is used because of its superior ground-wave propagation characteristics and the simplicity of the antenna design.

The purpose of a radio broadcasting station is to transform sound waves into radio waves that can be picked up by radio receiving sets. The utility of this service to the public depends on (1) signal strength, (2) program content, and (3) signal distortion. The broadcast antenna should efficiently radiate the energy supplied to it by the transmitter.

A simple vertical tower radiates the energy quite well equally in all directions along the ground. A secondary purpose of the antenna system may be to concentrate the amount of radiation in the directions that it is wanted and to restrict the radiation in the directions that it is not wanted. A directional antenna system is required if a nondirectional antenna causes interference to other stations or if the signal is not strong enough to adequately serve the populated areas of interest.

2. Standard Reference Antennas

The uniform, omnidirectional, or isotropic radiator in free space is taken as the *standard reference antenna* because it has no directivity. See Fig. 1. Actually such a radiator of radio waves cannot be realized, because all radio antennas have directional properties.

If a uniform radiator is placed at the surface of a perfectly conducting earth, all of the power must be radiated in the hemisphere above the earth as shown in Fig. 2.

Vertical Radiation Characteristics

A nondirectional tower has its own vertical radiation characteristic whether series or shunt fed, sectionalized or nonsectionalized, top loaded or without top loading. The vertical radiation pattern is usually computed using the assumption of a sinusoidal current distributon on the radiating portion of the tower.

The vertical radiation characteristic of a base-fed vertical tower of height H meters is given by

---

*Region 1: Europe, Africa, USSR, Turkey, and Arabia.
 Region 2: The Americas.
 Region 3: All other parts of the world.

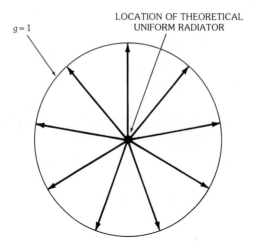

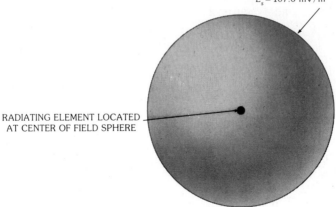

Fig. 1. Pattern of a uniform radiator which is the theoretical standard reference antenna. (*a*) Cross section of spherical pattern which has a field-strength gain of 1, a power gain of 1 (*g* = 1), and a decibel gain of 0. (*b*) Spherical radiation pattern surface of a uniform or isotropic radiator with radiated power P_r and electric-field strength E_s.

$$f(\theta) = \frac{\cos(G \sin \theta) - \cos G}{\cos \theta (1 - \cos G)} \tag{1}$$

where

$f(\theta)$ = vertical radiation characteristic

G = electrical height of tower, in degrees

θ = elevation of observation point, in degrees

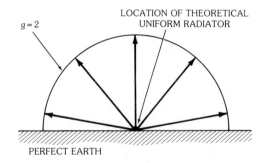

g = 2

LOCATION OF THEORETICAL
UNIFORM RADIATOR

a

PERFECT EARTH

FOR P_{ent} = 1 kW, THE ELECTRIC-FIELD
STRENGTH AT 1 MILE (1.609 km) IS
E_s = 152.1 mV/m

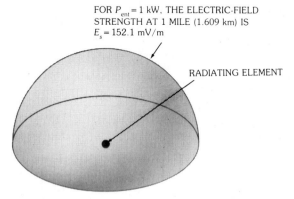

RADIATING ELEMENT

b

Fig. 2. Pattern of a uniform radiator at the surface of a perfect reflecting and conducting earth. (*a*) Cross section of a hemispherical pattern which has a field-strength gain of 1.4, a power gain of $g = 2$, and a decibel gain of 3.01. (*b*) Hemispherical radiation pattern surface of a uniform radiator.

A comparison of the vertical radiation characteristics of several standard reference antennas is shown in Figs. 3 and 4. A summary of several standard reference antennas is given in Table 1. In Fig. 5 the theoretical electric-field strength, loop radiation resistance,* and base radiation resistance are graphed as a function of antenna height.

The vertical radiation characteristic for a base-fed, top-loaded tower is given by

$$f(\theta) = \frac{\cos B \, \cos(A \sin \theta) \, - \, \sin \theta \, \sin B \, \sin(A \sin \theta) \, - \, \cos G}{\cos \theta (\cos B \, - \, \cos G)} \quad (2)$$

where

A = electrical height of the tower, in degrees

B = effective electrical length of top loading, in degrees

$G = A + B$

---

*Radiation resistance at antenna current maximum I_a as shown in Fig. 5.

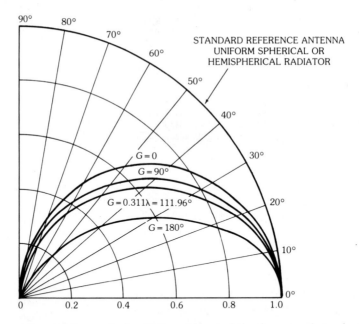

Fig. 3. Comparison of the vertical radiation characteristics for several standard reference antennas.

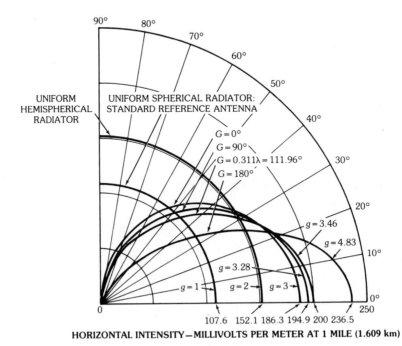

Fig. 4. Comparison of vertical radiation patterns of standard reference antennas with a radiated power of 1 kW.

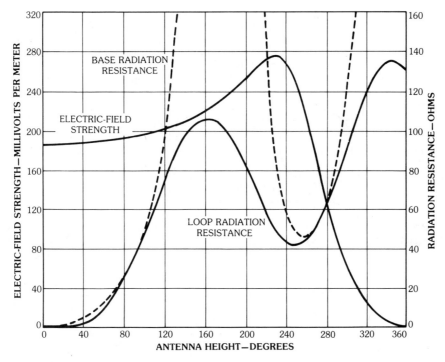

Fig. 5. Electric-field strength at 1 mi (1.609 km) for a radiated power of 1 kW, with loop and base radiation resistance as a function of tower height over a perfectly conducting earth.

The other terms are defined following (1). The theoretical current distribution is shown in Fig. 6. The loop current maximum on the tower is I_a. Equation 2 is valid only if the radiation from the top-loaded section B is negligible.

The vertical radiation characteristic for a base-fed, top-loaded sectionalized tower is given by

$$f(\theta) = \frac{\cos B \ \cos(A \sin \theta) - \cos G + \dfrac{\sin B \ \cos(H - C) \cos(C \sin \theta)}{\sin(H - A)}}{}$$

$$\frac{-\dfrac{\sin B \ \sin \theta \ \sin(H - C)\sin(C \sin \theta)}{\sin(H - A)} - \dfrac{\sin B \ \cos(H - A)\cos(A \sin \theta)}{\sin(H - A)}}{\cos \theta \{\cos B - \cos G + [\sin B / \sin(H - A)](\cos \overline{H - C} - \cos \overline{H - A})\}}$$

(3)

where

> C = electrical height of top section, in degrees
>
> D = electrical length of top loading of top section, in degrees
>
> $H = C + D$

The other terms are defined in (1) and (2). The theoretical current distribution is shown in Fig. 7. The loop current maximum of the top section is I_c.

Table 1. Summary of Standard Reference Antennas

Type of Antenna	Vertical Pattern Shape	Electric-Field Strength for 1 W at 1 mi (mV/m)* E_ϕ	Electric-Field Strength for 1 kW at 1 mi (mV/m)* E_0	Power Gain g	Decibel Gain G
Uniform spherical radiator		3.402	107.6	1	0
Current element		4.167	131.8	1.5	1.761
Half-wave antenna		4.358	137.8	1.641	2.151
0.622λ antenna		4.472	141.4	1.728	2.375
Two ends on half-wave in-phase antenna		5.283	167.1	2.411	3.822

Theoretical Self-Resistance and Field Strength

It is useful to know the theoretical loop and base resistance of a vertical radiator. This information is presented graphically in Fig. 5 along with the theoretical field strength at 1 mi (1.609 km).

Self Base Impedance Characteristics

The loop and base radiation resistances as given in Fig. 5 are not very useful because they are for a very thin conductor and the antenna is not driven at the loop very often. Their primary application is in theoretical calculations. Since most

Table 1, *continued*

Type of Antenna	Vertical Pattern Shape	Electric-Field Strength for 1 W at 1 mi (mV/m)* E_ϕ	Electric-Field Strength for 1 kW at 1 mi (mV/m)* E_0	Power Gain g	Decibel Gain G
Uniform hemi-spherical radiator		4.811	152.1	2	3.010
Vertical current element		5.893	186.3	3	4.771
Quarter-wave vertical antenna		6.163	194.9	1.282	5.161
0.311λ vertical antenna		6.324	200	3.450	5.386
Half-wave vertical antenna		7.471	236.2	4.822	6.832

*E_0 is the electric-field strength on a plane passing through the center and perpendicular to the conductor.

antennas are driven at the base, the self base resistance along with the self base reactance curves are presented in Figs. 8 and 9.

The self-resonant frequency of a tower depends on the cross-sectional size [1]. The first factor to be considered is the average characteristic impedance Z_0 as shown in Figs. 8 and 9. The value of Z_0 depends on the average cross-sectional size, which in ohms for a cylindrical antenna is

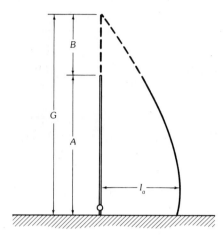

Fig. 6. Theoretical current distribution on top-loaded tower.

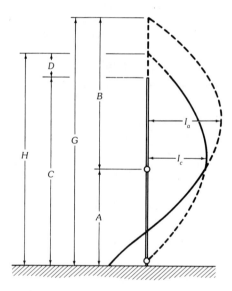

Fig. 7. Theoretical current distribution on top-loaded sectionalized tower.

$$Z_0 = 60[\ln(2G/a) - 1] \tag{4}$$

where

    G = antenna height, in degrees

    a = antenna radius, in degrees or same units as G if G is not in degrees

Since most antennas have a square or triangular cross section, a good approxima-

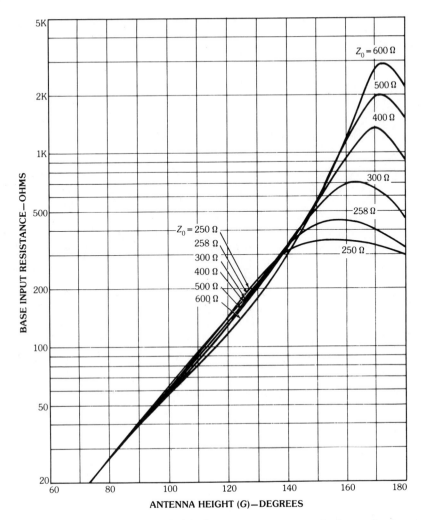

Fig. 8. Base input resistance of cylindrical antennas over a perfectly conducting ground plane.

tion is to determine the cross-sectional area and convert it to a circle with radius a to be used in (4). The equivalent radius a for the cross section of a regular polygon with n sides, each of length d, is [2]

n	2	3	4	5	6
a/d	0.25	0.4214	0.590	0.756	0.920

Example 1—Determine the average characteristic impedance of a 400-ft uniform cross-sectional tower that is 6.5 ft square. Then $a = 0.59 \times 6.5 = 3.835$ ft.
Solution—Substituting in (4),

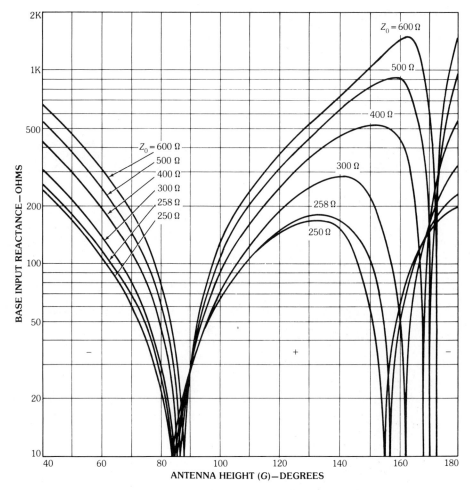

Fig. 9. Base input reactance of cylindrical antennas over a perfectly conducting ground plane.

$$Z_0 = 60 \, [\ln 2(400)/3.835 - 1]$$
$$= 60 \, (\ln 208.6 - 1)$$
$$= 60 \, (5.34 - 1)$$
$$= 260.4 \ \Omega$$

Mutual Base Impedance

The mutual impedance between vertical antennas is needed by the designer of directional antenna feeder systems. Sufficient design equations have been developed and presented in the literature [3, 4] to solve this problem; however, only a few cases of equal-height antennas are presented in Figs. 10 and 11. The loop mutual impedance can be transferred to the base by dividing by the square of the sine of the tower height, $\sin^2 G$.

The general design equation for the mutual base resistance and reactance is given by

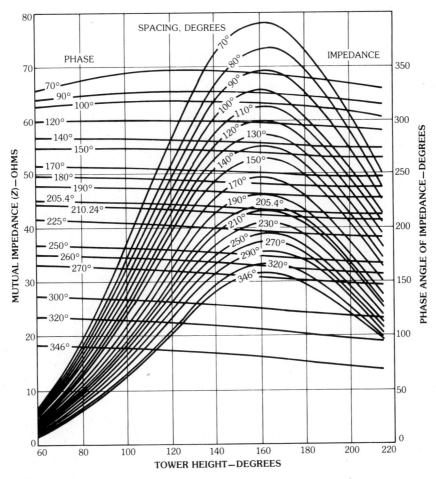

Fig. 10. Loop mutual impedance and phase angle between two towers of equal height.

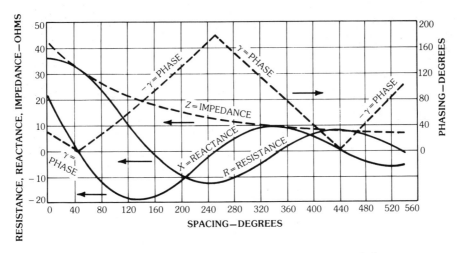

Fig. 11. Loop mutual impedance between quarter-wave vertical towers.

$$R_{12} = \frac{15}{\sin G_1 \sin G_2} \{\cos(G_2 - G_1)[Ci(u_1) - Ci(u_0) + Ci(v_1) - Ci(v_0)$$
$$+ 2Ci(y_0) - Ci(y_1) - Ci(s_1)] + \sin(G_2 - G_1)[Si(u_1) - Si(u_0)$$
$$+ Si(v_0) - Si(v_1) - Si(y_1) + Si(s_1)] + \cos(G_2 + G_1)[Ci(w_1)$$
$$- Ci(v_0) + Ci(x_1) - Ci(u_0) + 2Ci(y_0) - Ci(y_1) - Ci(s_1)]$$
$$+ \sin(G_2 + G_1)[Si(w_1) - Si(v_0) + Si(u_0) - Si(x_1) - Si(y_1)$$
$$+ Si(s_1)]\} \tag{5}$$

$$X_{12} = \frac{15}{\sin G_1 \sin G_2} \{\cos(G_2 - G_1)[Si(u_0) - Si(u_1) + Si(v_0) - Si(v_1)$$
$$+ Si(y_1) - 2Si(y_0) + Si(s_1)] + \sin(G_2 - G_1)[Ci(u_1) - Ci(u_0)$$
$$+ Ci(v_0) - Ci(v_1) - Ci(y_1) + Ci(s_1)] + \cos(G_2 + G_1)[Si(v_0)$$
$$- Si(w_1) + Si(u_0) - Si(x_1) + Si(y_1) - 2Si(y_0) + Si(S_1)]$$
$$+ \sin(G_2 + G_1)[Ci(w_1) - Ci(v_0) + Ci(u_0) - Ci(x_1) - Ci(y_1)$$
$$+ Ci(s_1)]\} \tag{6}$$

where

R_{12} = mutual base resistance, in ohms, between antennas no. 1 and no. 2
X_{12} = mutual base reactance, in ohms, between antennas no. 1 and no. 2
G_1 = height of antenna no. 1, in degrees
G_2 = height of antenna no. 2, in degrees
$Si(x)$ = sine integral function of x
$Ci(x)$ = cosine integral function of x
s = spacing between antennas, in degrees
$u_0 = \sqrt{s^2 + G_1^2} - G_1$, in degrees
$u_1 = \sqrt{s^2 + (G_2 - G_1)^2} + G_2 - G_1$, in degrees
$v_0 = \sqrt{s^2 + G_1^2} + G_1$, in degrees
$v_1 = \sqrt{s^2 + (G_2 - G_1)^2} - G_2 + G_1$, in degrees
$w_0 = v_0$, in degrees
$w_1 = \sqrt{s^2 + (G_2 + G_1)^2} + G_2 + G_1$, in degrees
$x_0 = u_0$, in degrees
$x_1 = \sqrt{s^2 + (G_2 + G_1)^2} - G_2 - G_1$, in degrees
$y_0 = s$, in degrees
$y_1 = \sqrt{s^2 + G_2^2} + G_2$, in degrees
$s_0 = y_0 = s$, in degrees
$s_1 = \sqrt{s^2 + G_2^2} - G_2$, in degrees

The desired theoretical mutual base impedance can be obtained by solving (5) and (6). If the antennas are of equal height ($G = G_1 = G_2$), then (5) and (6) reduce to

$$R_{12} = \frac{15}{\sin^2 G} \{4\,Ci(u_1) - 2\,Ci(u_0) - 2\,Ci(v_0) + \cos 2G[Ci(w_1) - 2\,Ci(v_0)$$
$$+ Ci(x_1) - 2\,Ci(u_0) + 2\,Ci(u_1)] + \sin 2G[Si(w_1) - 2\,Si(v_0)$$
$$- Si(x_1) + 2\,Si(u_0)]\} \tag{7}$$

$$X_{12} = \frac{15}{\sin^2 G} \{-4\,Si(u_1) + 2\,Si(u_0) + 2\,Si(v_0) + \cos 2G[-Si(w_1)$$
$$+ 2\,Si(v_0) - Si(x_1) + 2\,Si(u_0) - 2\,Si(u_1)] + \sin 2G[Ci(w_1)$$
$$- 2\,Ci(v_0) - Ci(x_1) + 2\,Ci(u_0)]\} \tag{8}$$

Now, if antenna no. 2 is 90° in height ($G_2 = 90°$), then (5) and (6) reduce to

$$R_{12} = 15\,\{Ci(u_1) + Ci(v_1) - Ci(w_1) - Ci(x_1) + \cot G_1[Si(u_1) - Si(v_1)$$
$$- 2\,Si(y_1) + 2\,Si(s_1) + Si(w_1) - Si(x_1)]\} \tag{9}$$

$$X_{12} = 15\,\{Si(w_1) + Si(x_1) - Si(u_1) - Si(v_1) + \cot G_1[Ci(u_1) - Ci(v_1)$$
$$- 2\,Ci(y_1) + 2\,Ci(s_1) + Ci(w_1) + Ci(x_1)]\} \tag{10}$$

Example 2—Determine the mutual base impedance between two antennas spaced at 160° with electrical heights of 120° and 90°, respectively.
Solution—This is as follows:

$$G_1 = 120°, G_2 = 90°, \text{ and } s = 160°$$
$$u_1 = \sqrt{(160)^2 + (30)^2} + 90 - 120 = 132.8°, \text{ or } 2.32 \text{ rad}$$
$$v_1 = \sqrt{(160)^2 + (30)^2} - 90 + 120 = 192.8°, \text{ or } 3.366 \text{ rad}$$
$$w_1 = \sqrt{(160)^2 + (90 + 120)^2} + 90 + 120 = 474°, \text{ or } 8.27 \text{ rad}$$
$$x_1 = \sqrt{(160)^2 + (90 + 120)^2} - 90 - 120 = 54°, \text{ or } 0.943 \text{ radi}$$
$$y_1 = \sqrt{(160)^2 + (90)^2} + 90 = 273.6°, \text{ or } 4.77 \text{ rad}$$
$$s_1 = \sqrt{(160)^2 + (90)^2} - 90 = 93.6°, \text{ or } 1.63 \text{ rad}$$

Substituting these values in (9) and (10):

$$R_{12} = 15\,\{Ci(2.32) + Ci(3.366) - Ci(8.27) - Ci(0.943) + \cot 120°$$
$$\times [Si(2.32) - Si(3.366) - 2\,Si(4.77) + 2\,Si(1.63) + Si(8.27)$$
$$- Si(0.943)]\}$$
$$= -2.935 \;\Omega$$

$$X_{12} = 15 \{Si(8.27) + Si(0.943) - Si(2.32) - Si(3.366) + \cot 120°$$
$$\times [Ci(2.32) - Ci(3.366) - 2 Ci(4.77) + 2 Ci(1.63)$$
$$+ (8.27) - Ci(0.943)]\}$$
$$= -28.85 \ \Omega$$

Therefore the mutual base impedance is

$$Z_{12} = -2.935 - j28.85 \ \Omega$$

Example 3—Determine the mutual base impedance between two towers of equal height $G = 110°$ and having a spacing of $200°$.

Solution—Since the tables of the sine integral and cosine integral functions are ordinarily tabulated with the arguments in radians, it has been found simpler to convert G and s to radians before substituting in the formulas.

$$G_1 = G_2 = 110° = 1.9199 \text{ rad}$$
$$s = 200° = 3.4907 \text{ rad}$$

Then

$$u_0 = \sqrt{3.4907^2 + 1.9199^2} - 1.9199 = 2.0639$$
$$u_1 = s = 3.4907$$
$$v_0 = \sqrt{3.4907^2 + 1.9199^2} + 1.9199 = 5.9037$$
$$w_1 = \sqrt{3.4907^2 + (2 \times 1.9199)^2} + (2 \times 1.9199)$$
$$= 9.0290$$
$$x_1 = \sqrt{3.4907^2 + (2 \times 1.9199)^2} - (2 \times 1.9199)$$
$$= 1.3496$$
$$\sin^2 G = 0.9397^2 = 0.8830$$

Substituting these values in (7) yields

$$R_{12} = \frac{15}{0.8830} \{4 Ci(3.4907) - 2 Ci(2.0639) - 2 Ci(5.9037) + \cos 220°$$
$$\times [Ci(9.0290) - 2 Ci(5.9037) + Ci(1.3496) - 2 Ci(2.0639)$$
$$+ 2 Ci(3.4907)] + \sin 220°[Si(9.0290) - 2 Si(5.9037)$$
$$- Si(1.3496) + 2 Si(2.0639)]\}$$
$$= 16.9875 \{-0.1188 - 0.8180 + 0.1666 - 0.7660[0.0524$$
$$+ 0.1666 + 0.4549 - 0.8180 - 0.0594] - 0.6428[1.6663$$
$$-2.8598 - 1.2203 + 3.2672]\}$$
$$= 16.9875 \{-0.7702 - 0.7660(-0.2035) - 0.6428(0.8534)\}$$
$$= 16.9875 \{-0.7702 + 0.1559 - 0.5486\}$$
$$= 16.9875(-1.1629) = -19.75 \ \Omega$$

For the reactance component of the mutual base impedance, substituting in (8) yields

$$
\begin{aligned}
X_{12} &= 16.9875\,\{-4\,Si(3.4907) + 2\,Si(2.0639) + 2\,Si(5.9037) + \cos 220° \\
&\quad \times [-Si(9.0290) + 2\,Si(5.9037) - Si(1.3496) + 2\,Si(2.0639) \\
&\quad - 2\,Si(3.4907)] + \sin 220°[Ci(9.0290) - 2\,Ci(5.9037) - Ci(1.3496) \\
&\quad + 2\,Ci(2.0639)]\} \\
&= 16.9875\,\{-7.3360 + 3.2672 + 2.8598 - 0.7660[-1.6663 + 2.8598 \\
&\quad - 1.2203 + 3.2672 - 3.6680] - 0.6428[0.0524 + 0.1666 \\
&\quad - 0.4549 + 0.8180]\} \\
&= -21.33\ \Omega
\end{aligned}
$$

and

$$
Z_{12} = 29.1\ \Omega\ \angle{-132.8°}
$$

It should be pointed out that when any of the towers are in excess of approximately 120° in height, and particularly in the case of self-supporting towers, the values of the base mutual impedance given by these formulas are in considerable error and therefore should not be relied upon. In such cases the use of loop values of mutual impedance and self-impedance will give a better indication of the power division in the directional antenna array.

It is of cardinal importance not to mix measured and theoretical values. If reliable measured values are available, they should be used. It is particularly pertinent not to use measured base self-impedance and theoretical base mutual impedances to determine driving-point base impedance and power division in the directional antenna array. It is usually feasible to measure the self-impedance of each tower with the other towers disabled and then measure the *magnitude* of the mutual impedance between each pair of towers. Since the mutual impedance phase angles are more difficult to measure, it is quite common practice to use the theoretical phase angles since they are usually more reliable than the measured values.

Control of Pattern Shape and Size

There are two basic problems in a directional antenna design. *First*, it is desirable to control the pattern shape to provide the necessary protections to other stations and to serve the community of interest. This is accomplished by proper selection of directional antenna parameters. *Second*, the size of the pattern for a given amount of power is very useful and often necessary in many applications. First the problem of molding the pattern shape will be treated and then methods of determining the size or gain of the directional antenna will be treated.

Space Configuration of the Array

The field strength at any given point in space is a function of the placement of the current elements, along with the magnitude and phase of the field strength produced by each current element. In other words, it is necessary to know the space configuration of the radiating elements and the field-strength magnitude and phase of each radiating element.

It is common practice to treat each tower as a radiating element over a perfectly conducting earth and select the observation point P far enough away from the radiating system to assume that lines joining the radiating elements and the observation point are parallel. This simplifies the mathematics and does not introduce appreciable error except when it is desired to deal with the nearby radiation field. For this case the more general equations must be employed.

In order to establish a system for specifying the location of each antenna in such a manner that the information can be used in the design equations, let us refer to Fig. 12, which is a plan view of the space configuration of the ith antenna. The space reference point is the space origin or point from which all distances are measured. The space reference axis is the reference line from which all azimuth angles are measured. The azimuth angle ϕ_i is measured clockwise in degrees from the space reference axis. The spacing s_i is measured in electrical degrees along the horizontal plane as shown in Fig. 12.

For an observation point P at some azimuth angle ϕ in the horizontal plane, the distance to the ith antenna s_i is shortened by the multiplication factor $\cos(\phi_i - \phi)$ as shown in Fig. 13. This space-phasing quantity is the required difference distance since the observation point P is assumed to be at a great distance; hence the lines from the reference point and the ith antenna are parallel.

When the observation point is at some elevation angle θ, the ith antenna will appear to be closer to the space reference point by the multiplying factor $\cos \theta$. Referring to Fig. 14, observe that a right triangle can be formed by dropping a perpendicular to the line connecting the space reference point and the observation point P.

The ith antenna then appears to have space phasing of $s_i \cos \theta \cos(\phi_i - \phi)$ from

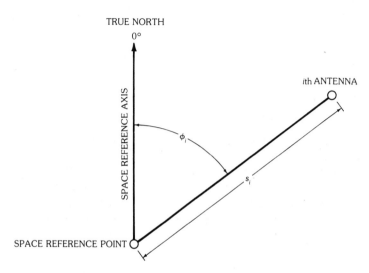

Fig. 12. Plan view of space configuration of the ith antenna.

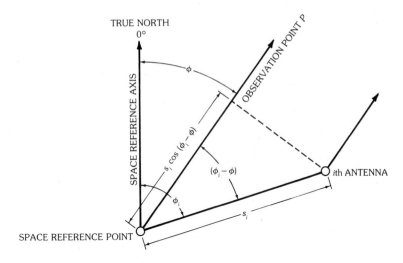

Fig. 13. Plan view of the ith antenna showing space phasing in the horizontal plane.

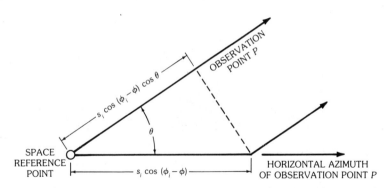

Fig. 14. Elevation angle θ shortens the spacing s_i to the value $s_i \cos \theta$.

the space reference point. This is the complete expression for the space phasing of any antenna in the system and for any observation point P in the hemisphere, as shown in Fig. 15.

Voltage Diagram

For the above generalized space configuration of the ith antenna there is a voltage phasor at the point P in space. This voltage phasor for the ith antenna depends on E_i, the field strength in the horizontal plane, the vertical radiation characteristic $f(\theta)_i$ as given by (1), (2), or (3), and the time phasing ψ_i.

The electric-field–strength vector for the ith antenna makes an angle β_i with respect to the voltage reference phasor axis, which is along the positive x axis, as shown in Fig. 16. The angle β_i is the sum of the space phasing and the time phasing. The space phasing is represented by

$$s_i \cos \theta \, \cos(\phi_i - \phi)$$

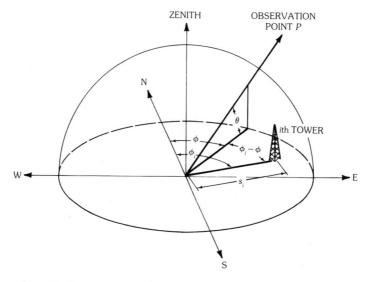

Fig. 15. Space view of observation point P and the ith tower.

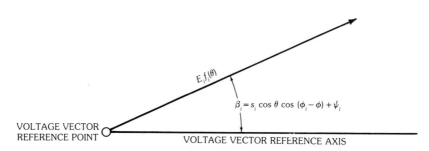

Fig. 16. Voltage vector diagram for the ith antenna.

while the time phase is simply ψ_i.

Generalized Equation

Now, if the electric-field–strength phasors of all the antennas in the directional antenna array are added in series, the resultant field strength at the observation point P can be obtained as shown in Fig. 17. If these phasors for n antennas are added together, the generalized equation can be used to express the hemispherical pattern shape of the directional antenna array with n antennas. This generalized phasor equation can be used to express the pattern shape for any directional antenna array with n antennas. The equation in condensed form is

$$E = \sum_{i=1}^{i=n} E_i f_i(\theta) \angle \beta_i \tag{11}$$

where

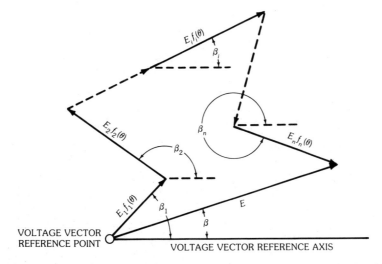

Fig. 17. Summation of electric-field–strength vectors for *n* antennas in the directional antenna array.

E = the total effective electric-field–strength phasor at unit distance (P) for the antenna array with respect to the voltage phasor reference axis. This resultant phasor makes the angle β with respect to this axis as shown in Fig. 17

i = the ith antenna in the directional antenna system

n = the total number of antennas in the directional antenna array

E_i = the magnitude of the electric-field strength at unit distance in the horizontal plane produced by the ith antenna acting alone

$f_i(\theta)$ = vertical radiation characteristic of the ith antenna as given in (1), (2), or (3)

θ = elevation angle of the observation point P measured up from the horizon in degrees

and

$$\beta_i = s_i \cos\theta \, \cos(\phi_i - \phi) + \psi_i \qquad (12)$$

which is phase relation of the electric-field strength at the observation point P for the ith antenna taken with respect to the voltage phasor reference axis, where $s_i \cos\theta \cos(\phi_i - \phi)$ is the space phasing portion of β_i due to the location of the ith antenna, for which

s_i = electrical length of spacing of the ith antenna in the horizontal plane from the space reference point

ϕ_i = true horizontal azimuth, orientation of ith antenna with respect to the space reference axis

ϕ = true horizontal aximuth angle of the direction to the observation point P (measured clockwise from true north)

ψ_i = time phasing portion of β_i due to the electrical phase angle of the voltage (or current) in the ith antenna taken with respect to the voltage phasor reference axis

3. Two-Tower Antenna Patterns

Since two-tower antenna patterns are so useful they have been systematized out to four wavelengths [5]. A few patterns are presented in Fig. 18. If only one null is desired, it must be either due north or due south since this is in line with the antenna of Fig. 18. A two-tower antenna pattern is always symmetrical with respect to the vertical plane containing the towers. Therefore those nulls which are not in line with the antennas appear even in number.

For a more general condition of producing a null in the direction ϕ_n, the azimuth angle of the null from the line of antennas is given by the equation

$$s_2 \cos \phi_n \pm \psi_2 = \pm 180° \tag{13}$$

where

ϕ_n = azimuth angle of the nulls from the line of the antenna, in degrees

    = azimuth angle, in degrees, clockwise and counterclockwise from true north to the nulls, if the antennas are on a north-south space reference axis

ψ_2 = total phasing of antenna no. 2 (north antenna) with respect to antenna no. 1 (south antenna), in degrees

s_2 = total spacing between the antennas, in degrees

This equation has been used to prepare the chart in Fig. 19.

4. Power Flow Integration Method to Determine Pattern Size

The total power radiated from a directional antenna array can be computed by integrating the energy flow outward through an imaginary spherical surface surrounding the directional antenna array. This method does not give information regarding the distribution of power to various towers of the directional antenna array. It is, however, very useful for making comparisons of pattern size.

The rate of energy flow in watts per square meter at a given point P in space can be expressed by the Poynting vector, thus

$$\mathbf{P} = \mathbf{E} \times \mathbf{H} \tag{14}$$

where

$\mathbf{P}$ = energy flow in watts per square meter

$\mathbf{E}$ = electric-field strength in volts per meter

$\mathbf{H}$ = magnetic-field intensity in ampere-turns per meter

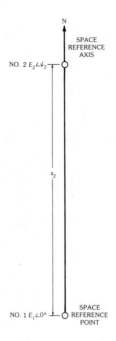

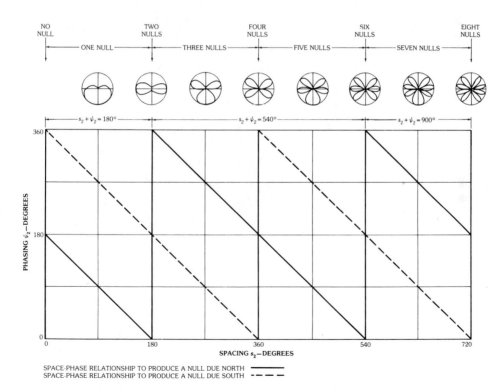

Fig. 18. Two-tower space-phase relationship to produce a null due north.

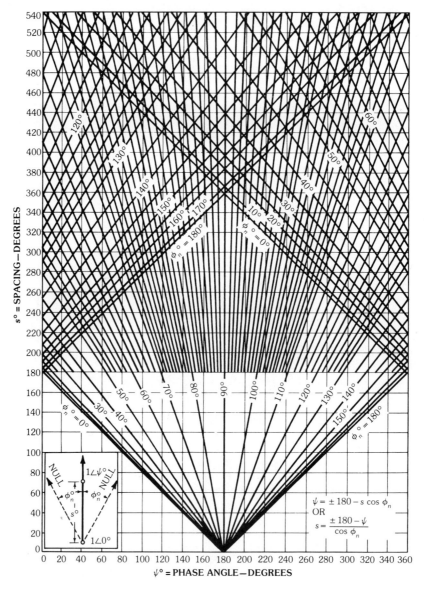

Fig. 19. Determining null directions of two-tower arrays.

Out in free space the vectors **E** and **H** are orthogonal and have the following magnitude relationship:

$$H = \sqrt{\frac{\epsilon_0}{\mu_0}} E \qquad (15)$$

where

μ_0 = the permeability of free space, $4\pi \times 10^{-7}$ henry per meter

ϵ_0 = the permittivity of free space, $1/\mu_0 c^2$ farads per meter

c = the velocity of light, 2.9979×10^8 meters per second

From these values the characteristic resistance of free space can be determined. The characteristic resistance of free space has a value of resistance such that no energy will be reflected. In other words, energy leaving the directional antenna array and flowing into free space will never return. In equation form,

$$R_c = \mu_0 c = \sqrt{\frac{\mu_0}{\epsilon_0}} \tag{16}$$

where R_c = 376.710 Ω, the characteristic resistance of free space.

Substituting (15) and (16) in (14), the power flow can be expressed as

$$P = \frac{E^2}{R_c} \ \Omega \tag{17}$$

If this power flow is integrated over an imaginary spherical surface surrounding the directional antenna array, the total power radiated is

$$P_r = \frac{1}{R_c} \int_S E^2 \, d\mathbf{S} \tag{18}$$

where

P_r = total power radiated, in watts

R_c = characteristic resistance of free space, 376.71 Ω

E = total electric-field strength at the surface of the sphere, in volts per meter

$d\mathbf{S}$ = element of area on the surface of the sphere, in square meters

If the energy flow outward through the surface of the sphere is integrated as illustrated in Fig. 20, we can write

$$dS = d^2 \cos\theta \, d\theta \, d\phi \tag{19}$$

where d is the radius in meters of the spherical surface, and the other values are defined above. Substituting this value of dS in (17) gives

$$P_r = \frac{1}{R_c} \int_0^{2\pi} \int_{-\pi/2}^{+\pi/2} E^2 \, d^2 \, \cos\theta \, d\theta \, d\phi \tag{20}$$

as the total power radiated from an antenna system in free space.

If the electric-field strength is the same in all directions, the E in (20) can be

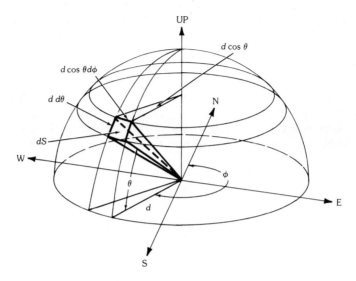

Fig. 20. Spherical surface integration.

replaced by the constant E_s. Moving the constants E_s and d^2 outside of the integral signs and performing the indicated integrations gives

$$P_r = \frac{E_s^2 d^2}{R_c} \int_0^{2\pi} \int_{-\pi/2}^{+\pi/2} \cos\theta \, d\theta \, d\phi$$

$$= \frac{2 E_s^2 d^2}{R_c} \int_0^{2\pi} d\phi = \frac{4\pi E_s^2 d^2}{R_c} \qquad (21)$$

Solving this equation for the rms electric-field strength of a uniform spherical radiator gives

$$E_s = \sqrt{\frac{P_r R_c}{4\pi d^2}} \qquad (22)$$

Example 4—Determine the electric-field strength at 1 mi (1.609 km) for an antenna power of 1 kW for an isotropic radiator.

Solution—Since there are 1609.347 m in a mile and $R_c = 376.71 \ \Omega$, the characteristic resistance of free space, (22) can be used to obtain

$$E_s = \sqrt{\frac{(1000)(376.710)}{4(3.14159)(1609.347)^2}}$$

$$= 107.584 \text{ mV/m}$$

This is the electric-field strength at 1 mi for 1 kW from the standard reference spherical radiator.

Example 5—Determine the electric-field strength at 1 mi (1.609 km) for an antenna power of 1 kW for a uniform hemispherical radiator.

Solution—In (20), the integration for θ is from 0 to $\pi/2$. Hence

$$P_r = \frac{1}{R_c}\int_0^{2\pi}\int_0^{\pi/2} E^2 d^2 \cos\theta \, d\theta \, d\phi \tag{23}$$

Substituting the constant E_s for E and solving for E_s similarly to the procedure in (21) and (22) results in

$$E_s = \sqrt{\frac{P_r R_c}{2\pi d^2}} \tag{24}$$

For 1 kW of radiated power the electric-field strength at 1 mi is

$$E_s = \sqrt{\frac{(1000)(376.71)}{2(3.1416)(1609.347)^2}}$$

$$= 152.147 \text{ mV/m}$$

This is the electric-field strength at 1 mi for 1 kW from the standard reference hemispherical radiator. This value can also be obtained by multiplying

$$107.584\sqrt{2} = 152.147$$

The electric-field strength from a center-fed conductor of length $2G$ in free space with a sinusoidal current distribution as shown in Fig. 21 has an electric-field strength of

$$E = E_0 f(\theta) = E_0\left[\frac{\cos(G\sin\theta) - \cos G}{(1 - \cos G)\cos\theta}\right] \tag{25}$$

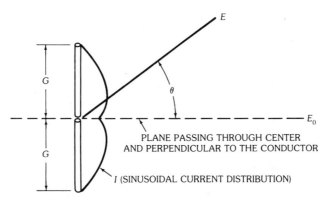

Fig. 21. Determining the electric-field strength of a conductor of length $2G$ in free space.

at any angle θ from a plane passing through the center and perpendicular to the conductor. Substituting this value of E in (20) and integrating with respect to ϕ gives

$$P_r = \frac{2\pi\, d^2\, E_0^2}{R_c(1 - \cos G)^2} \int_{-\pi/2}^{\pi/2} \frac{[\cos(G\sin\theta) - \cos G]^2}{\cos\theta}\, d\theta \tag{26}$$

Performing the indicated integration,

$$P_r = \frac{2\pi d^2 E_0^2}{R_c(1 - \cos G)^2} \{[\gamma + \ln 2G - Ci(2G)]$$
$$+ 1/2\ (\sin 2G)[Si(4G) - 2Si(2G)]$$
$$+ 1/2\ (\cos 2G)[\gamma + \ln G - 2Ci(2G) + Ci(4G)]\} \tag{27}$$

Solving this equation for E_0 results in

$$E_0 = \frac{(1 - \cos G)(P_r R_c / a\pi d^2)^{1/2}}{\{[\gamma + \ln 2G - Ci(2G)] + 1/2\ [Si(4G) - 2Si(2G)]\sin 2G}$$
$$\qquad\qquad + 1/2\ [\gamma + \ln G - 2Ci(2G) + Ci(4G)]\cos 2G\}^{1/2} \tag{28}$$

where

E_0 = the electric-field strength, in volts per meter, on a plane passing through the center and perpendicular to the conductor

P_r = total power radiated, in watts

R_c = 376.710 Ω, the characteristic resistance of free space

a = 2 for a conductor of length $2G$ in free space and $a = 1$ for an antenna of height G over a perfect ground plane

d = distance from the antenna, in meters
    = 1609.347 m in 1 mi

γ = Euler's constant, 0.577 215 66

$Ci(x)$ = cosine integral of x

$Si(x)$ = sine integral of x

$2G$ = length of the conductor, in radians or degrees

Substituting (28) into (25) gives the electric-field strength at any point in space. Thus

$$E = \{[\cos(G\sin\theta) - \cos G]/\cos\theta\}(P_r R_c / a\pi d^2)^{1/2}/\{[\gamma + \ln 2G - Ci(2G)]$$
$$+ 1/2\ [Si(4G) - 2Si(2G)]\ \sin 2G$$
$$+ 1/2\ [\gamma + \ln G - 2Ci(2G) + Ci(4G)]\ \cos 2G\}^{1/2} \tag{29}$$

where θ is the elevation angle from the plane, and the other terms are defined following (28).

Assuming a sinusoidal current distribution on a vertical antenna of height G over a perfect ground plane as shown in Fig. 22, the integration in (26) is changed to cover the hemisphere above the surface of the earth. This condition also results in $\alpha = 1$ in (28).

Along the horizon, $\theta = 0$ and the theoretical electric-field strength along the ground can be plotted as a function of antenna height G for 1 kW of antenna power at a distance of 1 mi to obtain the curve in Fig. 5.

As a matter of interest the loop radiation resistance for a thin vertical conductor is

$$R_r = 29.99776 \{[\gamma + \ln 2G - Ci(2G)] + 1/2 [Si(4G) - 2Si(2G)] \sin 2G$$
$$+ 1/2 [\gamma + \ln G - 2Ci(2G) + Ci(4G)] \cos 2G\} \tag{30}$$

where R_r is the loop radiation resistance in ohms of a thin vertical conductor over a perfectly conducting earth. This curve has also been plotted in Fig. 5.

For a vertical quarter-wave antenna the loop radiation resistance is

$$R_r = (29.9978)(1.21884) = 36.5626 \ \Omega$$

The base resistance R_b in ohms can be determined from the approximate equation

$$R_b \cong \frac{R_r}{\sin^2 G} \tag{31}$$

Thus far the power-flow integration method has been applied to the standard reference antennas and to the size of pattern produced by sinusoidal current distribution on a vertical radiator of various heights. The method is not confined to these special cases but can be applied to any directional antenna system so long as the field distribution is known from the individual elements that make up the system.

Considerable space has been devoted to determining the pattern size of a non-directional antenna, that is, an antenna which produces a constant field strength at all azimuth angles in the horizontal plane and for which at any given elevation

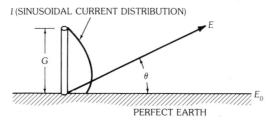

Fig. 22. Determining the electric-field strength of a vertical antenna of height G over a perfect earth.

angle the field strength is constant for all azimuth angles. The reason for this approach is that any directional antenna can be transformed to its equivalent nondirectional antenna pattern. This can be accomplished by determining the root-mean-square valve of the pattern at each elevation angle; then it is only necessary to integrate this equivalent nondirectional pattern with respect to the elevation angle since the determination of the rms values is the result of integrating the pattern with respect to the azimuth angle.

In order to develop this method consider the horizontal electric-field strength from two towers, one at the origin and one at the spacing s_2 due north. The total vector field is then

$$E = E_1 \angle 0 + E_2 \angle s_2 \cos \phi + \psi_2 \tag{32}$$

Changing to rectangular form,

$$E = E_1 + E_2 \cos(s_2 \cos \phi + \psi_2) + J E_2 \sin(s_2 \cos \phi + \psi_2)$$

The magnitude of E^2 is

$$\begin{aligned} E^2 &= E_1{}^2 + E_2{}^2 \cos^2(s_2 \cos \phi + \psi_2) + E_2{}^2 \sin^2(s_2 \cos \phi + \psi_2) \\ &\quad + 2E_1 E_2 \cos(s_2 \cos \phi + \psi_2) \\ &= E_1{}^2 + E_2{}^2 + 2E_1 E_2 \cos(s_2 \cos \phi + \psi_2) \end{aligned} \tag{33}$$

In performing the azimuth integration in (23) we substitute the value of E^2 from (34) to get

$$\begin{aligned} \int_0^{2\pi} E^2 \, d\phi &= \int_0^{2\pi} [E_1{}^2 + E_2{}^2 + 2E_1 E_2 \cos(s_2 \cos \phi + \psi_2)] \, d\phi \\ &= 2\pi \left\{ E_1{}^2 + E_2{}^2 + 2E_1 E_2 \frac{1}{2\pi} \times \int_0^{2\pi} [\cos(s_2 \cos \phi) \cos \psi_2 \right. \\ &\quad \left. - \sin(s_2 \cos \phi) \sin \psi_2] \, d\phi \right\} \end{aligned}$$

The second term in the integral is zero as can be demonstrated by plotting the function. Therefore,

$$\int_0^{2\pi} E^2 \, d\phi = 2\pi [E_1{}^2 + E_2{}^2 + 2E_1 E_2 (\cos \psi_2) J_0(s_2)]$$

where $J_0(s_2)$ is a Bessel function of the first kind and zeroth order. In this equation the phase ψ and spacing s are between elements 1 and 2; hence a more general form is to write the following:

ψ_{12} = difference in electrical phase angle of the field from the first and the second antenna

s_{12} = electrical length of spacing between the first and second antenna

With this change in nomenclature the above equation becomes

$$\int_0^{2\pi} E^2 \, d\phi = 2\pi [E_1^2 + E_2^2 + 2E_1 E_2 \cos \psi_{12} J_0(s_{12})] \tag{34}$$

In this equation the integration is performed for only two antennas in the horizontal plane. This equation can be generalized to account for any elevation angle θ if the electric-field strengths are multiplied by $f(\theta)$ and the spacing is multiplied by $\cos \theta$. By making these modifications, (34) can be written

$$E(\theta) = \frac{1}{2\pi} \int E^2 \, d\phi = E_1^2 f_1^2(\theta) + E_2^2 f_2^2(\theta)$$
$$+ \, 2E_1 f_1(\theta) \, E_2 f_2(\theta) \cos \psi_{12} J_0(s_{12} \cos \theta) \tag{35}$$

where $E(\theta)$ is the rms electric-field strength radiated at the elevation angle θ (in this case for only two antennas). If (35) is generalized for any number of elements in the directional antenna array, the rms electric-field strength at the elevation angle θ can be written

$$E(\theta) = \left[\sum_{p=1}^{p=n} \sum_{q=1}^{q=n} E_p f_p(\theta) \, E_q f_q(\theta) \cos \psi_{pq} J_0(s_{pq} \cos \theta) \right]^{1/2} \tag{36}$$

where

$E(\theta) =$ rms effective electric-field strength at the elevation angle θ

$p = p$th antenna in the system

$n =$ number of elements in the complete directional antenna array

$E_p =$ horizontal magnitude of the electric-field strength produced by the pth antenna

$f_p(\theta) =$ vertical radiation characteristic of the pth antenna

$q = q$th antenna in the system

$E_q =$ horizontal magnitude of the electric-field strength produced by the qth antenna

$f_q(\theta) =$ vertical radiation characteristic of the qth antenna

$\psi_{pq} =$ difference in electrical phase angle of the voltage (or current) between the pth and qth antennas in the directional antenna array

$s_{pq} =$ spacing in degrees or radians between the pth and qth antennas

$J_0(s_{pq} \cos \theta) =$ Bessel function of the first kind and zeroth order of the apparent spacing between the pth and the qth antennas

To clarify the application of (36), consider a three-element directional antenna system. In this case $n = 3$ and the equation can be written

$$E(\theta) = [E_1^2 f_1^2(\theta) + E_2^2 f_2^2(\theta) + E_3^2 f_3^2(\theta) + 2E_1 f_1(\theta) E_2 f_2(\theta) \cos \psi_{12}$$
$$\times J_0(s_{12} \cos \theta) + 2E_1 f_1(\theta) E_3 f_3(\theta) \cos \psi_{13} J_0(s_{13} \cos \theta) + 2E_2 f_2(\theta)$$
$$\times E_3 f_3(\theta) \cos \psi_{23} J_0(s_{23} \cos \theta)]^{1/2} \qquad (37)$$

In terms of $E^2(\theta)$ as given in (35), it is possible to rewrite (23) as follows:

$$P_r = \frac{1}{R_c} \int_0^{2\pi} \int_0^{\pi/2} E^2 d^2 \cos \theta \, d\theta \, d\phi$$
$$= \frac{2\pi d^2}{R_c} \int_0^{\pi/2} E^2(\theta) \cos \theta \, d\theta \qquad (38)$$

The standard hemispherical electric-field strength produced by the directional antenna system can be obtained by substituting (38) in (24). Thus

$$E_s = \left[\int_0^{\pi/2} E^2(\theta) \cos \theta \, d\theta \right]^{1/2} \qquad (39)$$

This is the *exact* formula for determining the size of the directional antenna pattern. To perform the integration for the general case, however, would be quite tedious.

A practical and very useful solution is to determine the value of $E(\theta)$ at a number of elevation angles and to replace the integral with the summation. Thus, for intervals of 10° of elevation the approximate equation can be written, by an application of the trapezoidal rule, as

$$E_s \cong \left\{ \frac{\pi}{18} \left[\frac{E_0^2}{2} + \sum_{n=1}^{n=8} (E_{10n})^2 \cos 10n \right] \right\}^{1/2} \qquad (40)$$

where

E_s = the standard hemispherical electric-field strength produced by the directional antenna system (152.1 mV/m for 1 kW at 1 m)

E_0 = rms effective electric-field strength in the horizontal plane

E_{10n} = rms effective electric-field strength at the specified elevation angle

n = integers from 1 to 8, which when multiplied by 10 give the elevation angles θ in degrees

Example 6—A four-tower directional antenna system has the parameters in Table 2.

Determine the values of horizontal electric-field strength from each tower for 5-kW operation, assuming the system loss is 405 W according to FCC standards. What is the value of rms field strength for each 10° elevation angle?

Solution—Initially solving for $E(\theta)$ at every 10° elevation angle by (36) in terms of field ratios F results in column 2 of Table 3.

Table 2. Parameters of Four-Tower System

No.	G (°)	ϕ (°)	s (°)	ψ (°)	F
1	90	0	0	0	1.0
2	90	274.22	176	2	0.786
3	90	302.34	211.3	275	0.841
4	90	358.38	100	260	0.786

Table 3. Values of $E(\theta)$ for Four-Tower System

θ (°)	$E(\theta)$ for $E_1 = 1$	$E(\theta)$ for $E_1 = 275$
0	1.435	395
10	1.432	394
20	1.382	380
30	1.313	361
40	1.212	333
50	1.087	299
60	0.899	247
70	0.657	181
80	0.031	86

Now, by (40) we have

$$E_s = 1.236 \qquad \text{for } E_1 = 1$$

For 5 kW radiated, the standard hemispherical electric-field strength will be

$$E_s = 152.1\,\sqrt{5} = 340 \text{ mV/m}$$

Therefore the required values of field strength are as follows:

$$E_1 = 340/1.236 = 275 \text{ mV/m}$$
$$E_2 = 275(0.786) = 216 \text{ mV/m}$$
$$E_3 = 275(0.841) = 231 \text{ mV/m}$$
$$E_4 = 275(0.786) = 216 \text{ mV/m}$$

The final rms electric-field strength for each 10° elevation angle is tabulated in column 3 of Table 3.

The input power for this problem according to FCC standards is 5405 W and the radiated power is 5 kW.

5. RMS Electric-Field Strength in the Horizontal Plane

The rms electric-field strength of a two-tower antenna array can be determined by

$$E_0 = E_1\sqrt{1 + F^2 + (2F\cos\psi)J_0(s)} \tag{41}$$

where

E_0 = rms electric-field strength at 1 mi (1.609 km), in millivolts per meter

E_1 = rms electric-field strength at 1 mi for reference tower 1 while operating in array, in millivolts per meter

F = ratio of magnitude of electric-field strength from tower 2 divided by that of tower 1

ψ = electrical phase of field from tower 2 with respect to that of tower 1, in degrees

$J_0(s)$ = Bessel function of first kind and zeroth order for tower spacing s

Now, if 90° antennas are used and the field ratio $F = 1$, then (41) can be written

$$E_0 = 195\sqrt{\frac{1 + (\cos\psi)J_0(S)}{1 + \cos(R_{12}/R_{11})}} \tag{42}$$

where

R_{12} = mutual resistance between tower 1 and tower 2, in ohms

R_{11} = self-resistance of tower 1, 36.6 Ω

The solution of this equation is shown in Fig. 23 for various values of antenna phasing ψ. It gives the theoretical electric-field strength without loss for 1-kW operation.

6. The Theoretical Pattern

The theoretical pattern, as used by the FCC, can be written

$$[E(\phi, \theta)]_{th} = \left| k \sum_{i=1}^{i=n} F_i\, f_i(\theta)\angle\beta_i \right| \tag{43}$$

where

$[E(\phi, \theta)]_{th}$ = the theoretical pattern inverse-distance electric-field strength at 1 mi (1.609 km), in millivolts per meter

k = multiplying constant which determines the basic pattern size

F_i = field strength ratio of ith element in the array

$E_i = kF_i$ in (11)

and the other terms are defined following (11) and (12).

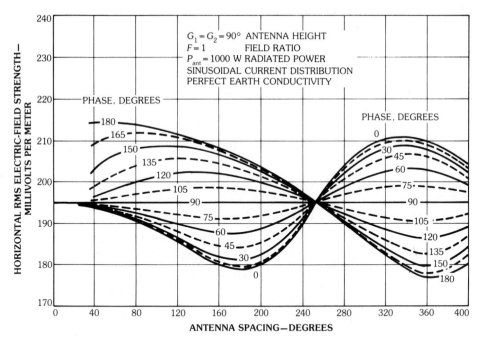

Fig. 23. Horizontal rms electric-field strength of two-antenna directional antenna.

7. The Standard Pattern

The above theoretical pattern in (43) cannot be achieved perfectly in the field, so the standard pattern was introduced by the FCC. To the theoretical value a quadrature term Q is added which primarily fills in the pattern minimums, and the resulting size is increased by 5 percent as shown in the following equation:

$$[E(\phi, \theta)]_{std} = 1.05 \sqrt{[E(\phi, \theta)]_{th}^2 + Q^2} \qquad (44)$$

where

$[E(\phi, \theta)]_{std}$ = the standard pattern inverse-distance electric-field strength at 1 mi (1.609 km), in millivolts per meter

$[E(\phi, \theta)]_{th}$ = the theoretical pattern inverse-distance electric-field strength at 1 mi (1.609 km), in millivolts per meter

Q = is the greater of the following quantities:
$0.025\, g(\theta)\, E_{rss}$ or $6.0\, g(\theta)\sqrt{P_{kW}}$

$$g(\theta) = \sqrt{\frac{f^2(\theta) + 0.0625}{1.030776}} \qquad (45)$$

$$E_{rss} = k \sqrt{\sum_{i=1}^{n} F_i^2} \qquad (46)$$

8. The Augmented Pattern

All directional antenna patterns in the United States use the standard pattern; however, if the measured pattern exceeds the standard pattern, but objectionable interference does not result, then the pattern is augmented in that discrete direction. These "patches" or augmentations may overlap. The augmented standard pattern equation is

$$[E(\phi,\theta)]_{\text{aug}} = \sqrt{[E(\phi,\theta)]_{\text{std}}^2 + A[g(\theta)\cos(180\,D_a/s)]^2} \qquad (47)$$

where

$[E(\phi,\theta)]_{\text{aug}}$ = augmented radiation value at azimuth/elevation, in millivolts per meter

$A = [E(\phi',\theta)]_{\text{aug}}^2 - [E(\phi',\theta)]_{\text{std}}^2$, augmentation constant

s = span of augmentation, in degrees

D_a = angular distance from center of span, in degrees

ϕ' = central azimuth of augmentation, in degrees

The augmentation term A specifies the magnitude of augmentation at the central angle of augmentation ϕ'. This term A is diminished to zero at the edges of the augmentation span s. An example of theoretical, standard, and augmented patterns is shown in Fig. 24.

9. Power to Provide System Losses

Because of losses in the transmission lines and matching, phasing, and power division networks, plus other losses in the system, such as resistance losses in the tower and ground system and dielectric losses in the insulators, an overfeeding of power is allowed by the FCC at the common point.

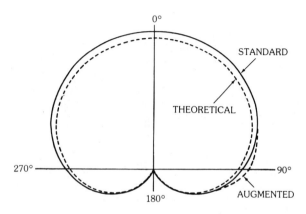

Fig. 24. Theoretical, standard, and augmented patterns.

The calculation of the amount of overfeeding of power at the common point is made as follows.

For stations with directional antennas authorized to radiate 5 kW of power or less, the measured common-point resistance is assumed to be 92.5 percent of its measured value, and for all other directional antennas it is assumed to be 95 percent. This arbitrary reduction in resistance amounts to increasing the current at the common point by $1/\sqrt{0.925} = 1.0398$, or approximately a 4-percent increase for values of transmitter power up to and including 5 kW. For transmitters with powers above 5 kW, the antenna current can be increased $1/\sqrt{0.95} = 1.026$, or approximately 2.6 percent.

Another way of saying this is that for a 1-kW station 1081 W can be fed in at the common point, while for a 50-kW station 52 632 W can be fed in at the common point. For the 1-kW station there are 81 W available for feeder-system loss, and in the 50-kW station there are 2632 W available for feeder-system loss.

10. Ground Systems

Antenna performance is standardized with reference to the ground system being a perfectly conducting flat plane. This assumption serves a useful purpose in designing directional antenna systems; however, in real life the ground-system losses must be considered and reduced to a reasonable value by using a ground system to minimize earth losses.

The **E**-field lines of force extend from the antenna out through the surrounding space to the earth. On entering the earth the current returns to the base of the antenna and produces an **H** field above the ground at right angles to the earth current. Fig. 25 illustrates both the **H**-field vector directions and earth current directions for two antennas with currents in phase. It is interesting to note that the earth currents are not radial and the **H**-field vectors do not follow a circular locus. The criterion is to design the ground system such that the earth currents will be in the copper ground system rather than in lossy earth.

The usual practice is to use no. 10 AWG bare copper wire for the ground system, consisting of 120 radials from each tower buried to a depth of 6 to 10 in (15 to 25 cm). The classic system is composed of radials 90° long, and in many cases a ground screen of expanded copper is installed at the tower base.

11. Directional Antenna Feeder Systems

In general a directional antenna feeder system, as shown in Fig. 26, will consist of a matching network from the transmitter output to the power-divider input, a power-dividing network, phase shifting networks of 0° or ±90°, transmission lines to the towers, and matching networks at the base of each tower.

The tower matching networks transform the driving-point impedance of each tower to the characteristic resistance of the transmission line. In addition, it is also desirable to provide phase shift such that 0° or 90° phase shifters can be used at the input to the transmission lines. The power divider supplies the desired amount of power to the phase shifters. Since the input impedance of the power divider may not match the output of the transmitter, an input matching network is provided.

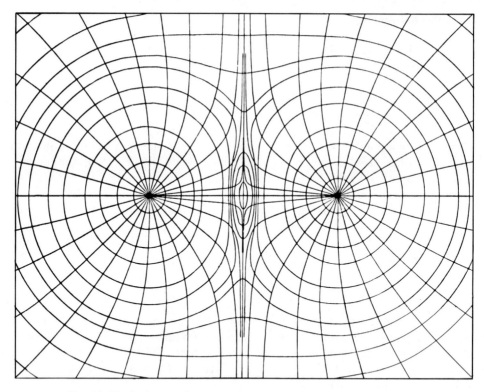

Fig. 25. Typical ground-system current and **H**-field vector direction.

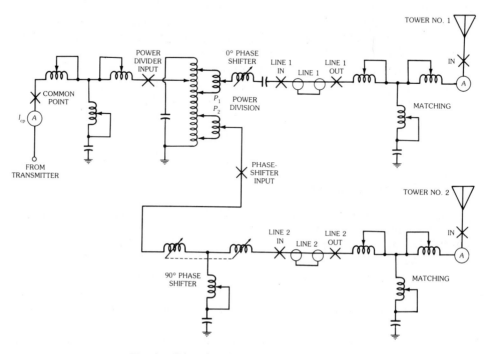

Fig. 26. Directional antenna feeder system.

The shunt coil provides the desired input resistance, while the input series coil provides zero reactance control.

In Fig. 26 there are a number of points marked X in the system. At these points an operating impedance bridge can be inserted to determine the impedance and thus maintain the desired feeder system impedances. By using an rf current meter in series with the operating bridge it is possible to make a useful inventory of the rf power in all parts of the feeder system.

Power Dividing Networks

There are several kinds of power dividing networks as shown in Fig. 27. The push-pull circuit in Fig. 27a produces outputs at terminals 1 and 2 that are 180° out

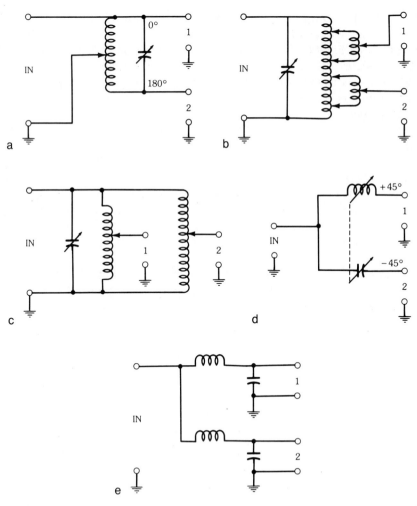

Fig. 27. Typical power-dividing networks. (*a*) Push-pull circuit, 180° feed. (*b*) Series-resonant circuit, 0° feed. (*c*) Parallel-resonant circuit, 0° feed. (*d*) Quadrature circuit, 90° feed. (*e*) Two L-sections in parallel.

of phase. Parallel resonance is achieved by tuning the capacitor while power division is varied by moving the coil tap to ground. The series resonant circuit in Fig. 27b produces output signals that are in phase. The jeep coils are connected by taps on the main coil and the adjustment of the jeep coil is often a front panel control on the phasor cabinet. The parallel variable capacitor tunes the circuit to parallel resonance and increases the input resistance. The parallel resonant circuit in Fig. 27c provides in-phase outputs and variable taps to control the power division. The parallel variable capacitor tunes the circuit to parallel resonance at the input terminal. The quadrature circuit in Fig. 27d provides output signals that are 90° in phase from each other. The loads can be matched to the input terminal in Fig. 27e by using two L-sections.

Impedance-Matching and Phase-Shifting Networks

Normally, the power dividing network input, as shown in Fig. 26, is connected through a T-network to the common point. This network matches the power dividing network input to the common point, which is usually 50 Ω pure resistance. The magnitude of the input resistance is controlled by the shunt arm of the T-network which is usually on the front control panel, and the reactance of the common point is controlled by the input arm of the T-network which is also usually on the front panel.

The T-network can be used both as a power divider and a phase shifter. In this application the shunt arm is used as the power-divider control and the input series arm is used as the phase shifter.

The T-network design equations are shown in Fig. 28. The term β is the phase shift of the T-network, the terms a, b, and c are defined in terms of the phase shift β, and r, the ratio of input resistance R_I to output resistance R_L of the load. In this figure the output load has a positive reactance, X_L; hence the output arm Z_2 of the T-network must have a negative reactance of $-X_L$ added to cancel the positive load reactance.

A design example of the T-network is illustrated in Fig. 29. The impedance transformation is from 50 Ω input to 25 Ω output and the phase shift is a delay of

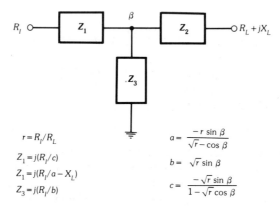

$$r = R_I / R_L$$
$$Z_1 = j(R_I / c)$$
$$Z_1 = j(R_I / a - X_L)$$
$$Z_3 = j(R_I / b)$$

$$a = \frac{-r \sin \beta}{\sqrt{r} - \cos \beta}$$

$$b = \sqrt{r} \sin \beta$$

$$c = \frac{-\sqrt{r} \sin \beta}{1 - \sqrt{r} \cos \beta}$$

Fig. 28. T-network design networks.

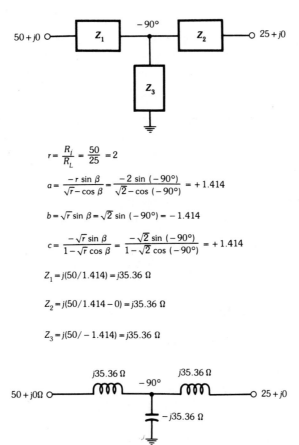

$$r = \frac{R_l}{R_L} = \frac{50}{25} = 2$$

$$a = \frac{-r \sin \beta}{\sqrt{r} - \cos \beta} = \frac{-2 \sin(-90°)}{\sqrt{2} - \cos(-90°)} = +1.414$$

$$b = \sqrt{r} \sin \beta = \sqrt{2} \sin(-90°) = -1.414$$

$$c = \frac{-\sqrt{r} \sin \beta}{1 - \sqrt{r} \cos \beta} = \frac{-\sqrt{2} \sin(-90°)}{1 - \sqrt{2} \cos(-90°)} = +1.414$$

$$Z_1 = j(50/1.414) = j35.36 \ \Omega$$

$$Z_2 = j(50/1.414 - 0) = j35.36 \ \Omega$$

$$Z_3 = j(50/-1.414) = j35.36 \ \Omega$$

Fig. 29. T-network design examples.

90° from the input to the output. This example shows how the network arms are computed.

The T-network of Fig. 30 is very useful as a 90° ± 15° phase-shifting network. It will be noted that the impedance of the shunt arm X_3 of the network is substantially constant over about 30° of phase shift. The phase shift is accomplished by varying the series arms X_1 and X_2 in unison.

The series circuit of Fig. 31 is called a 0° phase shifter. By varying the reactance X_L from resonance with X_C, the phase can be shifted ±15° without a major disturbance of the network impedance.

Antenna Sampling System

The most useful instrument in adjusting and maintaining proper operation of a directional antenna system is the antenna monitor. An approved antenna monitor system consists of single-turn unshielded loops which are rigid constructed mounted on each tower at the current loop, 90° down from the top of a non–top-loaded tower, or at least 10 ft (3 m) above the ground of short towers, and with

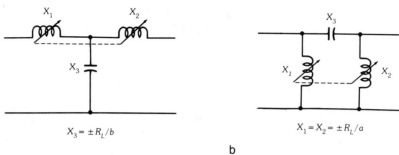

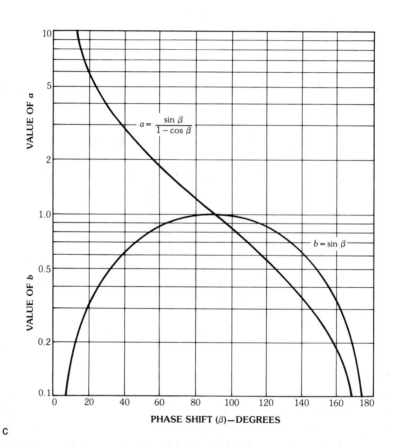

Fig. 30. Phase-shifting networks and curves (90° ± 15°). (*a*) Retarding network. (*b*) Advancing network. (*c*) Graphs of *a* and *b*.

equal electrical lengths of rigid outer conductor coaxial cables from each sampling loop to an approved antenna monitor, as shown in Fig. 32.

The antenna monitor provides magnitude and phase values of the current in each tower with respect to the reference tower. This information is very useful in making the initial adjustment and is required information to maintain proper operation of the directional antenna system.

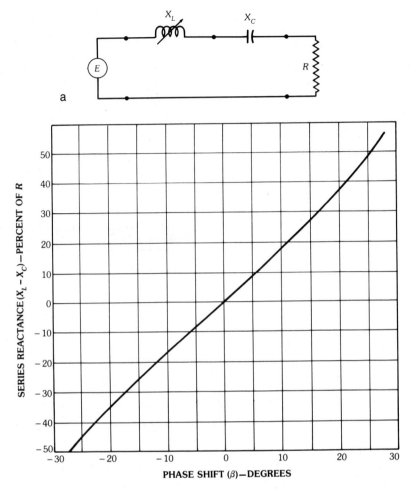

Fig. 31. Phase-shifting network and curve ($0° \pm 15°$). (*a*) Circuit containing series reactance $X_L - X_C$. (*b*) Graph of series reactance.

Driving-Point Impedance

The driving-point impedance, at the base of each tower, is needed to set up the directional antenna feeder system.

The mesh circuit equations for *n* antennas in the directional antenna system is given by

$$
\begin{aligned}
V_1 &= I_1 Z_{11} + I_2 Z_{12} + \ldots + I_i Z_{1i} + \ldots + I_n Z_{1n} \\
V_2 &= I_1 Z_{21} + I_2 Z_{22} + \ldots + I_i Z_{2i} + \ldots + I_n Z_{2n} \\
&\;\;\vdots \\
V_i &= I_1 Z_{i1} + I_2 Z_{i2} + \ldots + I_i Z_{ii} + \ldots + I_n Z_{in} \\
&\;\;\vdots \\
V_n &= I_1 Z_{n1} + I_2 Z_{n2} + \ldots + I_i Z_{ni} + \ldots + I_n Z_{nn}
\end{aligned}
\tag{48}
$$

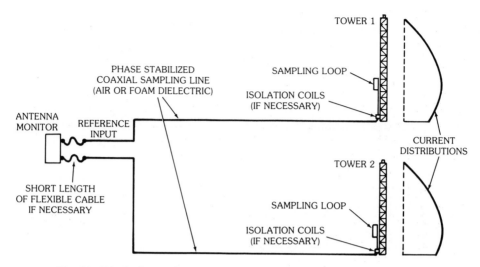

Fig. 32. Block diagram of typical directional antenna sampling system.

where

V_i = effective voltage at the input terminals of the ith antenna, in volts

I_i = effective current at the input terminals of the ith antenna, in amperes

Z_{ii} = self-impedance of the ith antenna, in ohms

Z_{in} = mutual impedance between the ith and nth antennas, in ohms

From this set of simultaneous equations it is possible to define the driving-point impedance of the ith antenna as follows:

$$Z_i = \frac{V_i}{I_i} = \frac{I_1}{I_i} Z_{i1} + \frac{I_2}{I_i} Z_{i2} + \cdots + Z_{ii} + \cdots + \frac{I_n}{I_i} Z_{in} \qquad (49)$$

The resistance component R_i of Z_i is a pure radiation resistance only if there is no loss in the system. The directional antenna system can be calculated on this theoretical basis, and then from a knowledge of the system the losses can be estimated with fair accuracy.

With the desired directional antenna characteristics, one can determine the relative required exciting currents, which in turn determine the respective driving-point impedances from (49). This information is then used to design the impedance-matching networks between the respective transmission lines and towers in the directional antenna array.

To determine the driving-point resistance of the ith antenna, the real part of (49) can be written

$$R_i = \left|\frac{I_1}{I_i}\right| \cdot \left|Z_{1i}\right| \cos(\psi_{1i} + \gamma_{1i}) + \left|\frac{I_2}{I_i}\right| \cdot \left|Z_{2i}\right|$$

$$\times \cos(\psi_{2i} + \gamma_{2i}) + \cdots + \left|Z_{ii}\right| \cos\gamma_{ii} + \cdots$$

$$+ \left|\frac{I_n}{I_i}\right| \cdot \left|Z_{ni}\right| \cos(\psi_{ni} + \gamma_{ni}) \tag{50}$$

and the reactance component is

$$X_i = \left|\frac{I_1}{I_i}\right| \cdot \left|Z_{1i}\right| \sin(\psi_{1i} + \gamma_{1i}) + \left|\frac{I_2}{I_i}\right| \cdot \left|Z_{2i}\right|$$

$$\times \sin(\psi_{2i} + \gamma_{2i}) + \cdots + \left|Z_{ii}\right| \sin\gamma_{ii} + \cdots$$

$$+ \left|\frac{I_n}{I_i}\right| \cdot \left|Z_{ni}\right| \sin(\psi_{ni} + \gamma_{ni}) \tag{51}$$

where

ψ_{ij} = phase angle of I_i minus the phase angle of I_j

γ_{ij} = phase angle of impedance Z_{ij}

Directional Antenna Feeder System Design Example

In order to design a feeder system it is first necessary to have the antenna parameters consisting of current ratios and current phases in each tower. Second, it is necessary to have the driving-point impedances of each tower. Third, the base currents are necessary. Fourth, the power distribution needs to be known. Finally, the transmission-line lengths and their characteristics of impedance and loss are needed.

The directional antenna feeder system can be described by the block diagram of Fig. 33. The various blocks have already been discussed. In this design example the driving-point impedances, powers, base currents, and phases are furnished for each tower.

The first step is to determine the phase shifts desired in each block between the power divider and each tower. Usually the first trial is to start with zero phase of the current in the reference no. 1 tower, with the vector at 0°, as shown in Fig. 33. The coupling network has a delay of 90° and the transmission line has a delay of 244°, resulting in a vector of 334° after going through the 0° phase shifter to the power divider. Assuming that a 90° phase shifter is connected to the power divider and a transmission-line delay of 165° in the transmission line to tower no. 2, the input phasor to the coupling network is 79°. Providing a delay of 59° in this network results in the correct phase of 20° for the current into the driving-point impedance of tower no. 2. This results in the best condition of phase control in the 0° and 90° phase shifters.

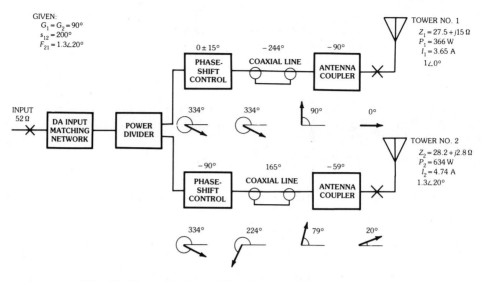

Fig. 33. Block diagram of directional antenna feeder system.

In this example let

$$G_1 = G_2 = 90°$$

$$Z_{11} = Z_{22} = 42.2 \angle 29.8° = 36.6 + j21$$

$$s_{12} = s_{21} = 200°$$

$$Z_{12} = Z_{21} = 14.9 \angle -130° = -9.7 - j11.3$$

$$F_{21} = M_{21} = I_2/I_1 = 1.3 \angle 20°$$

$$
\begin{aligned}
Z_1 &= Z_{11} + M_{12}Z_{12} \\
&= 42.2 \angle 29.8° + (0.77 \angle -20)(14.9 \angle -130°) \\
&= 27.5 + j15.8
\end{aligned}
$$

$$
\begin{aligned}
Z_2 &= M_{21}Z_{12} + Z_{22} \\
&= (1.3 \angle 20)(14.9 \angle -130°) + 42.2 \angle 29.8° \\
&= 28.2 + j2.8
\end{aligned}
$$

The power distribution as shown in Fig. 33 is then

$$P = I_1^2 R_1 + I_2^2 R_2$$

$$1000 = I_1^2(27.5) + (1.3)^2(28.2)$$

$$I_1 = 3.65 \text{ A}$$

$$I_2 = 4.75 \text{ A}$$

$$P_1 = 366 \text{ W}$$

$$P_2 = 634 \text{ W}$$

The next step is to use the design equations in Figs. 28 to 30 to determine the values of reactance in the various impedance-matching and phase-shifting T-

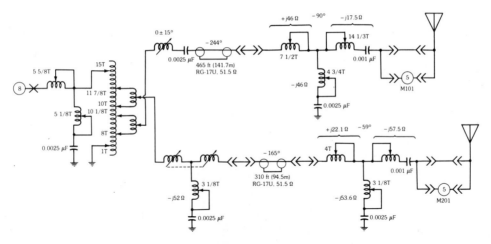

Fig. 34. Schematic diagram of feeder system.

networks. The results of the design are shown in Fig. 34. It will be noted that 0.001-μF capacitors are placed in the series T-network arms toward the antennas to isolate the feeder system from direct currents that are generated on the towers due to static charges. It will also be noted that the antenna meters are replaced with a make-before-break switch with a shorting loop having the same inductance as the meter when it is connected.

In this feeder system the power divider is a coil with jeep coils for fine tuning of the power division. The power divider is then connected to an L-network with a shunt control for the resistance magnitude and a series input arm to series-resonate the circuit, thus presenting a pure-resistance load to the transmitter.

Acknowledgments

The author wants to express his appreciation to many of his associates and seminar students for helpful suggestions. In particular he greatly appreciated the comments and suggestions of Roy E. Christen of the Cleveland Institute of Electronics, and he is indebted to his secretary, Blanche Kotalik, who typed the manuscript.

12. References

[1] S. A. Schelkunoff, "Theory of antennas of arbitrary size and shape," *Proc. IRE*, vol. 29, pp. 493–521, September 1941.
[2] Y. T. Lo, "A note on cylindrical antennas of noncircular cross section," *J. Appl. Phys.*, vol. 24, no. 10, pp. 1338–1339, October 1953.
[3] G. H. Brown, "Directional antennas," *Proc. IRE*, vol. 25, pp. 81–145, January 1937.
[4] C. R. Cox, "Mutual impedance between vertical antennas of unequal heights," *Proc. IRE*, vol. 35, pp. 1367–1370, November 1947.
[5] C. E. Smith, "Systematization of two tower patterns," in *Directional Antenna Patterns*, pt. 2, Cleveland: Smith Electronics, 1958, pp. 2.1–2.11.

Chapter 27

TV and FM Broadcast Antennas

G. W. Collins
Harris Corporation

CONTENTS

Gerald W. Collins has over 24 years of antenna and rf engineering experience. Since 1975 he has been with the Harris Broadcast Division, where he has been involved with the design and manufacture of broadcast antennas, including the Harris line of circularly polarized antennas and more recently the waveguide slot antennas. Prior to coming to the Broadcast Division he was with the Harris Government Systems Sector, where he was involved in the design of many types of antennas for military and space applications.

He received a BSEE from the University of Illinois in 1963, an MSEE from the Florida Institute of Technology in 1968, and has taken postgraduate work toward a PhD in electrical engineering through the University of Massachusetts. He is a registered professional engineer in the State of Illinois.

Mr. Collins has published several papers related to antennas and propagation, including various aspects of broadcasting. He has been issued patents for a circularly polarized zigzag antenna and a slotted cylinder antenna. He is a Senior Member of the IEEE.

1. General

Television and fm broadcasting in the United States is assigned to major portions of the vhf and uhf spectrum. Television is assigned to three bands at vhf and one at uhf, namely:

channels 2–4	54–72 MHz
channels 5, 6	76–88 MHz
channels 7–13	174–216 MHz
channels 14–83	470–890 MHz

Frequency-modulation broadcasting is at vhf from 88 to 108 MHz. The channel width for tv is 6 MHz for all channels so that the percentage bandwidth is as high as 10.5 percent at channel 2 to as low as 0.7 percent for channel 83. The fm spectrum is divided into 100 channels of 200 kHz width each, for a percentage bandwidth of about 0.2 percent.

In the design of the antennas it is important that both the pattern and impedance bandwidth meet minimum requirements over these bands. Under normal operating conditions it is generally required that the maximum vswr be less than 1.1:1 over the operating channel although values as high as 1.2:1 are acceptable for fm. For tv, more stringent requirements are placed on the vswr at the visual carrier and color subcarrier frequencies. Generally accepted values are 1.05:1 and 1.08:1, respectively. The visual frequency is 1.25 MHz above the lower band edge of each channel and the color subcarrier frequency is 3.58 MHz above the visual carrier.

The maximum effective radiated power (ERP) of fm and tv stations has been established to permit all stations, no matter the channel assignment, at a given height above average terrain (HAAT) to provide coverage to approximately equal areas. Accordingly, the maximum ERP for tv channels 2 through 6 and all fm channels is 100 kW; for channels 7 through 13 it is 316 kW, and for uhf channels it is 5000 kW. The ERP is specified for the horizontally polarized component at the rms level of an omnidirectional antenna. When circular polarization is used, equal ERP is permitted in the vertical component. The specified ERPs are readily achieved at vhf using relatively low gain antennas, with values of 2 to 6 being common for tv channels 2 through 6 and fm, and values of 6 to 12 being common for tv channels 7 through 13. (These gain values are power ratios relative to a half-wave dipole.) For uhf channels, much higher gains are usually required. This is due to limitations on available transmitter power. To achieve 5000 kW with a 220-kW transmitter usually requires an antenna gain of approximately 30. (Line efficiency is often 70 to 80 percent.)

The actual ERP of a station is dependent on its location and HAAT [1]. The United States and its possessions are divided into three zones as shown in Fig. 1 for

Fig. 1. FCC allocations and assignment zones.

tv stations. The maximum ERP for tv stations in each of these zones is a function of the station HAAT as shown in Fig. 2. Since the signal strength at a specific distance for a given ERP increases with increasing HAAT, these curves require a corresponding reduction in ERP for increasing HAAT above 1000 ft (304.8 m) for zone I or 2000 ft (609.6 m) for zones II and III.

For fm stations similar curves apply as shown in Fig. 3. The fm stations are classified as class A, B, B1, C, or C1, depending on the size of community they serve and the zone in which they are located. The zones for fm are somewhat different from those for tv. Zone I is the same as for tv allocations. Zone IA includes Puerto Rico, the Virgin Islands, and the part of California south of the 40th parallel. Zone II includes tv zone III and tv zone II except for those areas in zone IA. Class B and B1 stations may operate in zones I, IA, and II. Class C and C1 stations may operate only in zone II.

A tv station's coverage is stated in terms of distance to the city grade, grade A, and grade B contours. The signal strength at these contours for the various channel groupings is shown in Table 1 and is stated in terms of decibels over a microvolt per meter (dBu) for an ERP of 1 kW (0 dBk). The antenna should be located so that a city grade signal is provided over the principal city of service. The distances to these contours are estimated by the FCC (50/50) curves. Coverage for fm stations is determined by the distance to their 70-dBu and 60-dBu contours. It is essential to realize that signal strength values and contour distances determined from these curves are estimates only, to be expected at 50 percent of locations 50 percent of the time at a height of 30 ft (9.144 m). Actual signal strength at any time or location is dependent on local propagation conditions and receiving antenna factors [2, 3].

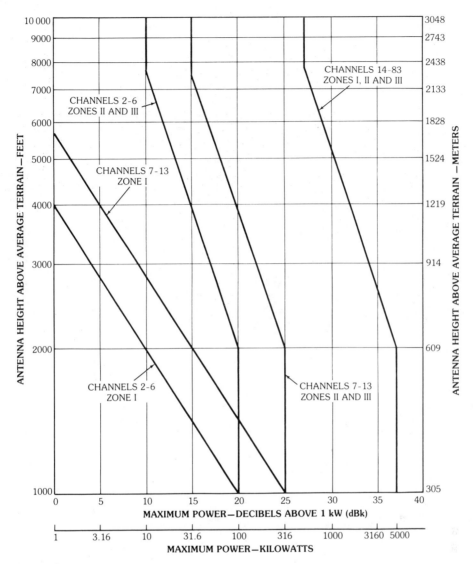

Fig. 2. Maximum effective radiated power versus antenna height for zones I, II, and III.

The usual case is for stations to utilize an omnidirectional pattern to achieve approximately uniform coverage in all directions. In practice, antennas only approximate the ideal pattern. Although there are no hard and fast generally accepted specifications, an antenna is considered to be omnidirectional if the pattern deviates from the rms level by no more than 2 or 3 dB.

It is sometimes necessary for a station to use a directional azimuth pattern. This is particularly important at uhf, where it is often desirable to concentrate the radiated power over a populated area. This permits attainment of a higher ERP over the populated areas while minimizing transmitter power and antenna aperture length. Another reason for the use of directional patterns is to prevent overlapping

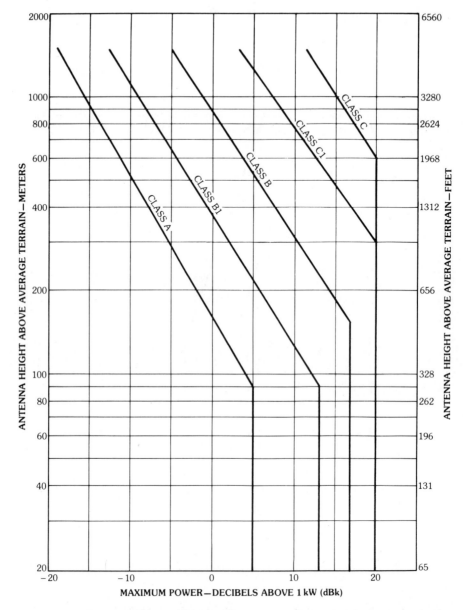

Fig. 3. Maximum effective radiated power versus antenna height for fm stations.

Table 1. Television Station Signal Strength at the Contours

Contour	Channels 2 through 6	Channels 7 through 13	Channels 14 through 83
City grade	74 dBu	77 dBu	80 dBu
Grade A	68 dBu	71 dBu	74 dBu
Grade B	47 dBu	56 dBu	64 dBu

of contours for cochannel stations operating from particular sites. If two stations on the same channel are spaced by less than the minimums specified by the FCC, the ERP along and about the radial connecting the two stations must be reduced to prevent overlapping coverage. The nulls in the pattern must be sufficiently deep to provide the protection. For fm stations the slope in the null must not exceed 2 dB/10°. When a directional antenna is used, the ERP is determined by the ERP at the azimuth of the pattern maximum. The maximum-to-minimum ratio for the azimuth pattern may not exceed 15 dB for fm and uhf tv channels 14 through 69 transmitting more than 1 kW. For tv channels 2 through 13, the maximum-to-minimum ratio is 10 dB.

The beamwidth of the elevation pattern is inversely proportional to the gain. Thus, for very high gain antennas, close attention must be paid to the stability of the beam. This is particularly true for uhf antennas, where high-gain antennas are commonly used, and to a lesser extent for vhf antennas. To ensure beam stability, both mechanical and electrical factors must be considered. The antenna must be stiff enough to minimize wind-induced deflections. In addition, the radiating elements and feed system characteristics must produce a stable beam as a function of frequency. It is also desirable to tilt the beam slightly below the radio horizon for maximum effectiveness. This is of increasing importance as the beamwidth narrows. Fig. 4 includes a plot of the depression angle to the radio horizon as a function of antenna height. A good rule of thumb is to add about 0.2° to the values read from these curves for optimum beam tilt.

Beam tilt is implemented by either electrical or mechanical means. Obviously, the beam is tilted downward in one direction when the antenna is mechanically tilted from the vertical. In the opposite direction the beam is tilted upward, and at orthogonal directions no tilt is introduced. In effect, this process creates a directional azimuth pattern. Electrical beam tilt may be used to tilt the beam downward in all directions. This is accomplished by advancing the phase of the upper portion of the antenna relative to the lower portion. Combinations of electrical and mechanical tilt may be used to increase the tilt in some direction while reducing it in others.

When the antenna is located a substantial distance from populated areas, the elevation pattern nulls need not be filled. However, when the population is close to the antenna or the antenna is unusually high, it is often necessary to fill the first null and in some cases higher-order nulls. Common amounts of first null fill range from 5 to 35 percent. Two methods are used to provide null fill. The first is to feed the upper and lower halves of the antenna with different current amplitudes. This results in incomplete field cancellation in the null regions and filling of the odd-numbered nulls. This has the advantage of minimizing radiation above the horizon and raising the antenna radiation center. The second method is to use a quadrature phase distribution over the vertical aperture. Obviously these methods may also be used together. A more general method is to determine the phase and amplitude of the currents by pattern synthesis and to design the antenna to provide these values.

When determining the power rating of tv antennas, it is necessary to recognize that tv transmitters are rated in terms of peak visual power. The average power in the visual signal with black picture is only 0.6 the peak visual power. The aural power is 10 to 20 percent of the peak visual power. Thus the worst-case average

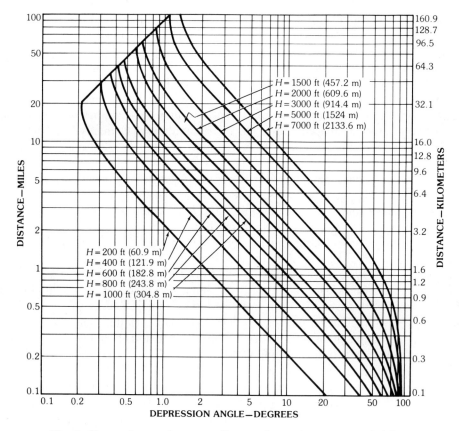

Fig. 4. Depression angle versus distance for various antenna heights.

power is 0.8 of the stated transmitter power. Peak voltage calculation must consider the sum of the voltage peaks for the combined visual and aural signals. For 20 percent aural the peak voltage is 1.477 times that due to the visual signal. Typical transmitter power ratings for tv channels 2 through 6 and fm range from 10 to 60 kW, to 10 to 100 kW for tv channels 7 through 13, and from 30 to 220 kW for uhf. For reliable service it is desirable that the antenna be rated at least as high as the maximum transmitter output.

It is essential that tv and fm antennas be designed to withstand the effects of lightning. Measures used to provide the needed protection include use of lightning rods around the beacon, designing for positive grounding of all parts, and minimizing inductance in radiating elements. To ensure that all parts are grounded, including those at high rf potential, extensive use is made of shorted quarter-wave transmission-line sections in the vicinity of radiator feed points.

2. Circularly Polarized TV Antennas

In 1977 the FCC rules were changed to permit tv stations to transmit circular polarization. These antennas were required to handle up to twice the power of conventional horizontally polarized antennas. This resulted from the FCC rules that permit maintenance of the station's ERP in the horizontally polarized component while adding an equal amount of signal in the vertical component.

In addition to providing beam shapes and impedance characteristics similar to conventional antennas, the CP antennas are required to radiate a right-hand–sense wave with a low axial ratio. It may also be important, in applications where a horizontally polarized antenna is being replaced, that the radiation center of the CP antenna be within 2 m of the antenna being replaced. Other requirements for CP antennas include means of preventing ice accumulation or deicing, lightning protection, provisions for either single or dual inputs, and provisions for either top or side mount.

3. Vee Dipole Array

The vee dipole antenna [4] is used for channels 2 through 13 and is designed for tower top-mounted application. Each bay consists of three crossed vee dipoles mounted at 120° intervals around a vertical mast as shown in Fig. 5. These dipoles are separated by three vertical grids, which isolate the vee dipoles and provide horizontal beam shaping. Each set of cross dipoles is fed in phase quadrature to produce a circularly polarized wave.

The design is rated for power up to 100 kW. A wideband flat dipole is used to handle the required power levels safely. Each dipole is mechanically supported and fed from a balun.

Radomes protect the radiating elements from exposure to moisture, ice, and

Fig. 5. Crossed vee dipoles arranged at 120° intervals.

atmospheric corrosives. Typical elevation patterns for a six-bay low-band antenna are shown in Fig. 6 for the cases of 0- and 15-percent null fill. As the null is raised, the gain is reduced. Typical values are given in Table 2.

For omnidirectional stations the shape of the azimuth pattern varies from circular by less than ±1.5 dB for channels 7 through 13; 2.0 dB for channels 2 through 6. Directional patterns may be obtained by use of nonuniform phase and amplitude around the mast or by the use of parasitic elements on the screens. A typical omnidirectional pattern is shown in Fig. 7. Two examples of directional azimuth patterns are shown in Fig. 8.

The antenna permits the use of either single or dual transmission lines. The low-band antenna is supplied with dual 3⅛-in (79.375-mm), 50-Ω input connec-

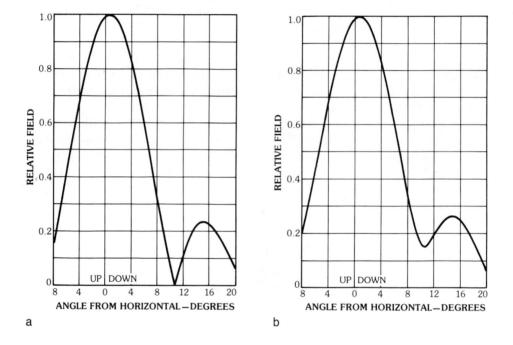

Fig. 6. Typical six-bay vee dipole array elevation patterns. (*a*) 0% null fill, −0.7° beam tilt. (*b*) 15% null fill, −0.7° beam tilt.

Table 2. Six-Bay Vee Dipole Array Gain Versus First Null Fill

Null Fill (%)	Power Gain (per Polarization)
0	2.98
5	2.96
10	2.92
15	2.82
20	2.72
25	2.66
30	2.42

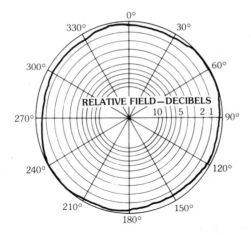

Fig. 7. Typical vee dipole array omnidirectional pattern.

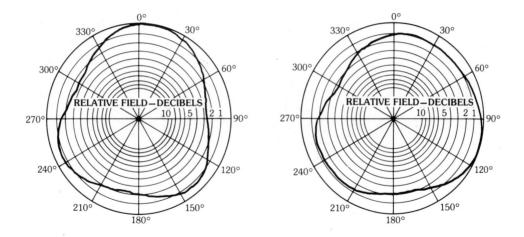

Fig. 8. Typical directional azimuth patterns for vee dipole array.

tions. The upper and lower halves of the antenna are each fed by a separate, independent transmission line to permit using one-half of the antenna in an emergency situation (with appropriate patching). It may also be supplied with a single input connection. The standard high-band antenna is supplied with a single 6⅛-in (155.575-mm), 50-Ω input connection. It may also be supplied with a dual line input which will permit feeding power to one-half of the antenna in an emergency.

4. Skewed-Dipole Antenna

This antenna makes use of circular arrays of skewed dipoles such that each dipole radiates linear polarization but in a configuration so that the complete array radiates circular polarization [5]. This is achieved by exciting N dipoles with equal currents whose phases are determined by the mode number M. That is, the current in each dipole is given by $I_m = \exp(j2\pi nM/N)$. This reduces the number of radiating elements and interbay feed lines by a factor of 2 compared with orthogonal dipole arrangements.

The approach is a generalization of the Lindenblad antenna [6, 7]. That antenna consists of four skewed dipoles arranged in a circle and fed in phase with the skew angle adjusted to give circular polarization. Excellent results were obtained since the central support mast was small in diameter and extended only to the center of the array. Larger-diameter cylinders supporting several bays of these skewed arrays degrade the axial ratio appreciably. This can be remedied by using higher-order modes of excitation and certain skew angles, or by adding short vertical elements.

The phase of the ratio E_θ/E_ϕ is shown in Fig. 9 as a function of the radius (ϱ_1) of the supporting mast. The minimum dB of the axial ratio for these phase shifts is also shown on the ordinate. In order to achieve an axial ratio of less than 3 dB it is necessary for the cylinder to have a circumference of less than 0.5λ. The axial ratio is also a function of the dipole skew angle. Fig. 10 shows the skew angle, ψ, for minimum axial ratio as a function of the dipole radius (ϱ_2). Skew angles near zero should be avoided since small alignment errors would lead to large axial

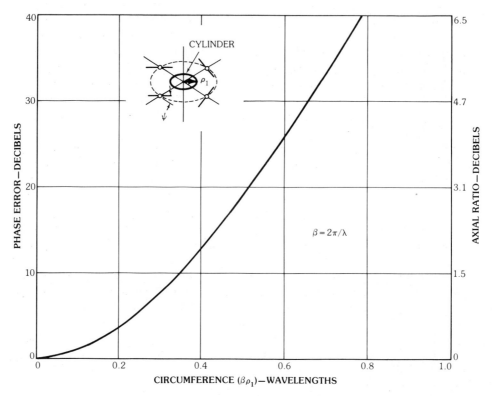

Fig. 9. Phase of E_θ/E_ϕ and axial ratio for $|E_\theta/E_\phi| = 1$ versus cylinder size. (*After DuHamel [5]*)

ratios. This places an upper limit on $\beta\varrho_2$. Fig. 10 is plotted for the case of no supporting mast present and for $\beta\varrho_1$ of 0.5. It is apparent that skew angle is not very dependent on ϱ_1 for reasonable mast diameters. Both Fig. 9 and Fig. 10 are calculated on the basis of short dipoles using the method described by Carter [9]. The results are approximately correct for half-wave dipoles. More accurate results may be obtained by using moment-method modeling of the antenna.

For broadcast applications the antenna consists of one or more bays of these circular arrays of skewed dipoles placed around a conducting cylinder as shown in Fig. 11 [8]. Each layer consists of three radiators fed in phase and mounted symmetrically around the pole. The skew angle is selected to produce equal amounts of horizontal and vertical polarization. When the radiation from all these dipoles is taken into account, an omnidirectional azimuth pattern results for both components (Fig. 12) with a low axial ratio in all azimuthal directions (Fig. 13).

Each radiator is fed by one feed line. A complete seven-layer circularly polarized antenna uses 21 feed lines fed from a branch-type feed system. One junction box feeds the upper four layers and another feeds the lower three layers. A single 3⅛-in (79.375-mm) transmission line is used to feed each junction box. The feed lines are ¾ in (19.05 mm), 50 Ω.

The tee to combine the upper and lower parts of the antenna can be located at the tower top with a single line down the tower (Fig. 14). If standby capability for operating on either the upper four or lower three layers of the antenna is desired, a transmission line from each junction box can be run down the tower and the tee can be mounted in the transmitter building (Fig. 15). The transmission lines from the two junction boxes to the combining tee must be equal in length. Since the two junction boxes are at different elevations on the pole, additional line must be inserted in the line coming from the lower junction box. This line can be inserted at the same location as the combining tee. The combining tee divides the power

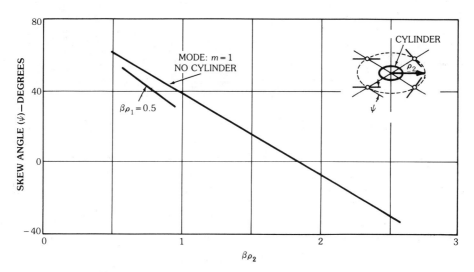

Fig. 10. Skew angle ψ versus coil circumference $\beta\varrho_2$ for mode number $m = 1$: (*After DuHamel [5]*)

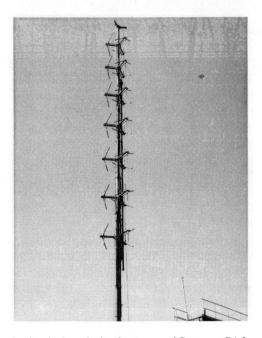

Fig. 11. Skewed-dipole circularly polarized antenna. (*Courtesy Dielectric Communications*)

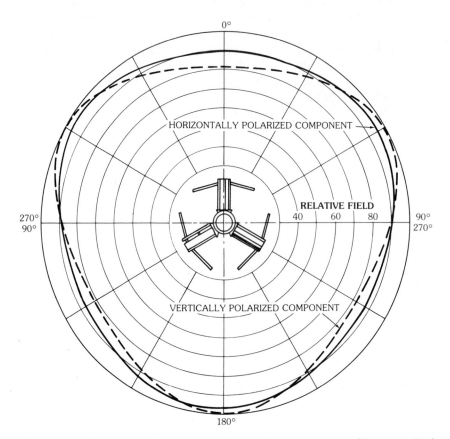

Fig. 12. Measured azimuth pattern of a skewed-dipole antenna. (*Courtesy Dielectric Communications*)

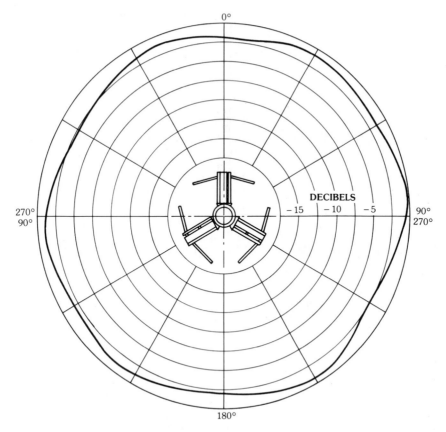

Fig. 13. Measured axial ratio of a skewed-dipole antenna. (*Courtesy Dielectric Communications*)

equally to the two output ports. Therefore each layer of the lower three layers receives more power than each layer of the top four layers. This, in addition to feed-line phasing between the layers, results in 8-percent fill in the first null and 4 percent in the second, as well as 1° beam tilt. A plot of the elevation pattern is shown in Fig. 16. The measured impedance shows a vswr of less than 1.1:1 for a bandwidth of about 12 percent. This is illustrated in Fig. 17.

5. Helical Circularly Polarized Antennas

Top-mounted circularly polarized antennas using multiple helices or spirals have been described [10, 11, 12]. Successful implementation of these using three and four conductors are known to have been made. An example using four conductors is shown in Fig. 18. Each section radiates an omnidirectional pattern broadside to a supporting mast about which four helical conductors are wound. The conductors are spaced from the mast and are equally spaced about the periphery of the mast. The signals are coupled to the conductors so that the phase of the signal coupled to one conductor is 180° out of phase with the signal coupled

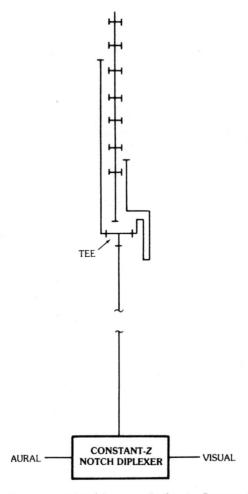

Fig. 14. Single-line feed. (*Courtesy Dielectric Communications*)

to adjacent conductors. The pitch of the helix is selected with the radius so that the radiation is circularly polarized and essentially broadside. This pitch angle ψ to achieve circular polarization and essentially broadside radiation is achieved by satisfying the condition

$$\tan \psi = \frac{J_{m-1}(\beta\varrho)}{2J_m(\beta\varrho)} \tag{1}$$

where J_{m-1} and J_m are the Bessel functions of order m and $m-1$, respectively (m is the mode number). The term $\beta\varrho$ is the circumference of the coil in wavelengths. Fig. 19 is a plot of the solution of this equation for the pitch angle versus the circumference of the helices for a mode number of 2. It is interesting to compare (1) with the corresponding curve for the skewed-dipole array (Fig. 10) for which the mode number is 1. For small values of $\beta\varrho_2$ the fit is quite good.

Radiating "end loads" are used at the end of each coil to radiate the small

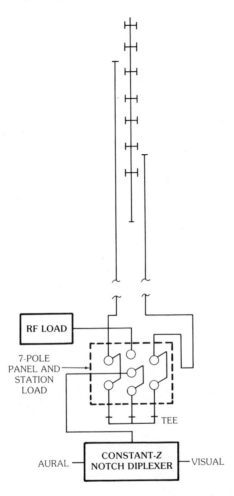

Fig. 15. Dual-line feed. (*Courtesy Dielectric Communications*)

amount of remaining energy from the main radiator. These end loads minimize the reflections of energy from the end toward the input, which otherwise would upset the traveling-wave nature of the antenna, consequently distorting the pattern. Electrical beam tilt is achieved by phasing between the vertical sections. Standard antennas are supplied with a single input of 6⅛ in (15.5 cm), 75-Ω transmission line.

The radiating elements are stainless steel tubing attached to the pole with dielectric supports. The support pole consists of either two or three sections made of galvanized steel. The antenna is designed for flange mount or bury mount at the tower top. Dielectric and steel pole steps are provided for climbing. The stainless steel helical tubes are grounded to the pole for lightning protection.

This type of antenna has also found limited application as a uhf antenna. Although each section is much smaller by virtue of the higher frequencies, more bays are required due to the higher gains usually required for uhf. The fundamental principles remain.

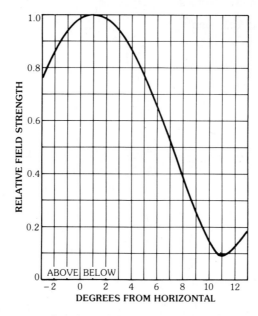

Fig. 16. Elevation pattern of a skewed-dipole antenna. (*Courtesy Dielectric Communications*)

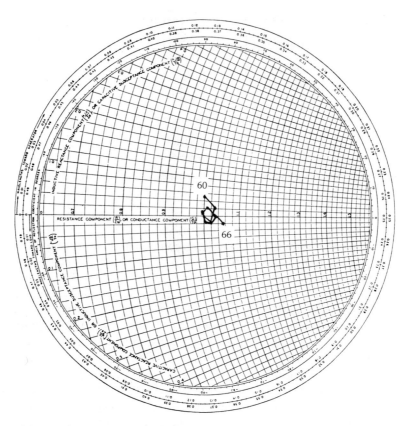

Fig. 17. Measured impedance of a skewed-dipole antenna. (*Courtesy Dielectric Communications*)

Fig. 18. Circularly polarized helical antenna. (*Courtesy Dielectric Communications*)

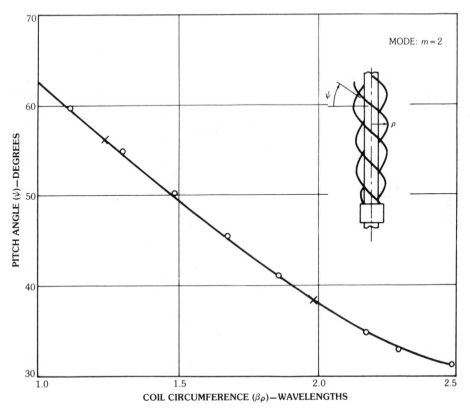

Fig. 19. Pitch angle as a function of the circumference of helices for mode number $m = 2$. (*After Ben Dov [11]*)

6. Side-Mount Circularly Polarized Antennas

When the tower top position is occupied by other antennas, it is often desirable to attach antennas to the side of the tower at lower positions. These are appropriately termed *side mount.*

A design illustrating many of the principles of the side-mount antenna design is the cavity-backed radiator [13], shown in the foreground of Fig. 20. The basic concept is the use of crossed dipoles in the aperture of a cavity. The dipoles are fed with equal currents in quadrature phase to excite the cavity in planes parallel to the dipoles. The beamwidth and radiation patterns are determined primarily by the size of the cavity. The geometry of the dipole controls the antenna impedance. The cavity performs three functions. First, it isolates the antenna from the tower, mounting structure, and adjacent cavities. Secondly, the cavity provides a sharper pattern than is achievable with dipoles alone. Thirdly, the cavity permits equalizing of the beamwidths for horizontal and vertical polarization.

The cavity is circular in shape and uses a grid construction to minimize wind load. Two designs are used, differing principally in electrical size. For a three-around array on a triangular supporting tower, a 120° half-voltage beamwidth is desired. The cavity for this is 0.6 to 0.65λ in diameter. A four-around configuration requires 90° half-voltage beamwidth. This cavity is about 0.8λ to 0.9λ in diameter. The cavities are approximately 0.25λ in depth.

CAVITY

Fig. 20. Broadcast array using cavity-backed radiator elements.

A flat, wideband dipole is used. A flat dipole may be related to an equivalent cylindrical center-fed dipole as shown in Fig. 21. The required bandwidth is achieved with a $1/w$ ratio of about 1.5. Each dipole is supported and fed from a balun. A shunt capacitance near the center of the dipole is used for impedance compensation.

The dipole has separate $1\frac{5}{8}$-in (41.275-mm), 50-Ω, coaxial inputs to the horizontal and vertical elements. The two inputs are fed with a power divider and phasing loop to provide circular polarization. The input vswr of a typical channel 7 through 13 cavity is shown in Fig. 22. The bandwidth is sufficient to permit multiplexing of two channels.

It is often desirable to provide directional azimuth patterns. A typical directional pattern, achieved by reducing power to one cavity in a three-around arrangement, is shown in Fig. 23. The serrations on the pattern are caused by a spinning source dipole to indicate the axial ratio as a function of azimuth.

Another side-mount panel design makes use of two crossed vee dipoles mounted in front of a square screen mesh as shown in Fig. 24. The pairs of dipoles are made unequal in length with the lengths chosen to produce equal quadrature currents in the dipoles. Since the dipoles are also orthogonal, circular polarization results. This design uses a single $\frac{7}{8}$-in (22.2-mm) copper feed line for each panel. The gain per polarization of each layer is slightly less than half that of a half-wave dipole. Beam tilt and null fill may be incorporated into the array pattern with a small reduction in gain.

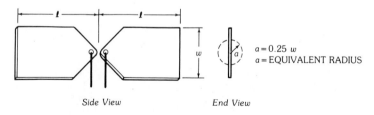

Fig. 21. Flat dipole.

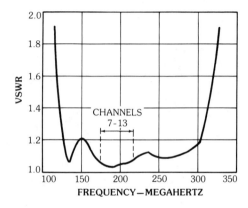

Fig. 22. Measured vswr of a single cavity.

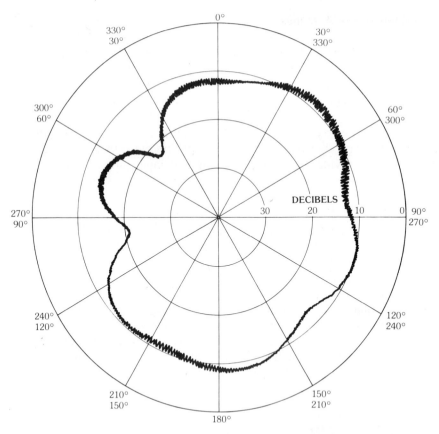

Fig. 23. Measured azimuth pattern of a directional cavity-backed radiator antenna.

Fig. 24. Circularly polarized panel antenna. (*Courtesy Dielectric Communications*)

Phase rotation is employed in the feed system to provide good vswr. This technique requires that each of the panels in a layer be phased 120° apart. The panels are then offset mechanically to restore the phase relationships and maintain good pattern circularity. The phase rotation tends to reduce changes in vswr caused by ice formation.

7. Horizontally Polarized Antennas

Even though the FCC now permits CP transmission, many stations continue to broadcast horizontal polarization (HP). Most stations were on the air prior to 1977, when the installation of HP antennas was all that was permitted. In addition, circular polarization, while improving coverage in many areas, is not a panacea for all propagation-related broadcast problems. Hence, many stations will find it desirable to continue operation in HP. As in the case of CP antennas, both top-mount and side-mount types are used.

8. Batwing Antennas

One of the most widely used designs is the batwing or superturnstile antenna [14]. The simple dipole elements of a turnstile antenna [15] are replaced with an equivalent grid of conductors forming a slot with the supporting mast [16]. (See Fig. 25.)

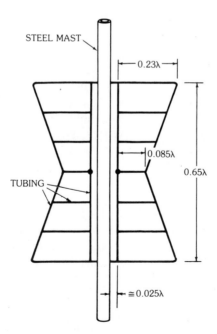

Fig. 25. Single element of batwing antenna with tubing construction, showing method of mounting on mast and typical dimensions. (*After Kraus [16],* © *1950 McGraw-Hill Book Co.*)

The radiators are bolted to the mast at about $\lambda/3$ from the feed point, thus providing structural support and a means of lightning protection. This arrangement provides an swr of less than 1.1:1 over a 35-percent bandwidth, making it convenient for tv service to as low as 47 MHz. There is very little radiation along the mast. Only one bay is required to obtain a gain equivalent to a single $\lambda/2$ dipole. To increase directivity, several bays are stacked with a separation of about λ between centers. The power gain is approximately proportional to the number of bays.

The antenna has an input resistance of about 72 Ω with low reactance over rather wide bandwidth. The bandwidth is sufficient for diplexing channels 4 and 5 or any two channels from 7 to 13. The feed-point impedance is readily matched to $\frac{7}{8}$-in (22.2-mm) coaxial lines without matching networks. Since the feed-point impedance is low, the performance is not severely affected by ordinary icing. Deicers are used where severe icing is expected.

To provide an omnidirectional azimuth pattern, orthogonal pairs of radiators are fed in phase quadrature. The pattern is generally circular and does not deviate from omnidirectional by more than ± 1.5 dB.

The antenna can be designed to provide azimuth patterns varying from the normal omnidirectional to a figure-8 shape. This is accomplished by varying the power division and/or phase relationship to the orthogonal pairs of radiators.

Usually, two transmission lines are used up the tower, although a single line feed may be used if the power division between the two halves of the antenna and 90° phasing is placed at tower top. If a single line feed is used, a notch diplexer is required. It is often feasible to feed the upper and lower bays independently. This is particularly convenient when six, eight, or twelve bays are used.

9. Sidefire Helical Antenna

A sidefire helical antenna [17, 18] utilizing a traveling-wave concept and radiating horizontal polarization is shown in Fig. 26. It is composed of a helical conductor mounted concentrically about a hollow cylindrical steel mast. The upper and lower helices are wound with opposite sense from a common feed point at the center of the section and are supported by insulators. The helices and the mast form a radiating transmission line.

Radio-frequency energy applied at the feed point travels along the transmission line, and the amplitude gradually decays (approximately 4 dB per turn) due to radiation. The helices are made long enough (usually six wavelengths) to radiate essentially all of the energy so that reflections are small. The impedance is relatively constant over the operating bandwidth and equal to the characteristic impedance of the line. The line may be shorted at the ends, thereby grounding the radiator at low frequencies and dc for lightning protection. The height of the helices above the mast determines both the characteristic impedance of the radiating transmission line and the minimum length of the helices.

In order to provide an omnidirectional azimuth pattern and broadside radiation, it is necessary that each turn be an integral number of wavelengths long. Two wavelengths per turn is generally used. If a large number of turns are used, this phase relationship is lost for small changes in frequency for turns a long way

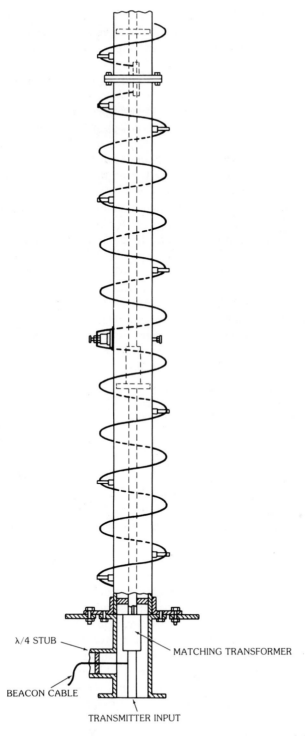

Fig. 26. Single section of sidefire helical antenna. (*After Krause and Smith [17]*)

from the feed point, thereby reducing bandwidth. Thus it is important to minimize the number of turns. On the other hand, the directivity of a section is proportional to the number of turns (for fixed turn spacing). A turn-to-turn spacing of $\lambda/2$ has been found to be optimum.

The upper and lower helices are wound with small pitch angles in opposite senses from the feed point to produce horizontal polarization. The vertical component is small and the vertical polarization from the upper helix tends to cancel that of the lower. Thus the cross-polarization loss is small.

Since the two counterwound helices are connected at the feed point, the input impedance is half the characteristic impedance of each radiating transmission line. For example, if the helices are spaced $\lambda/8$ from the mast, the parallel combination of the two transmission lines is approximately 100 Ω. This low impedance and the resulting low rf voltages result in an antenna that is relatively immune to moderate icing.

For more severe icing conditions a high current (60 Hz) is fed through the helix. The helix is constructed of a copper-clad steel core, presenting a high resistance to the deicing current and low resistance to rf. Deicing is accomplished without additional heaters while maintaining high radiation efficiency. Deicing dissipation is about 1.5 W/in^2 of helix surface. Radomes may be used to prevent icing in lieu of or in combination with electrical deicing.

Beam tilt may be introduced in a single section by shortening the upper helix fractionally and lengthening the lower helix a similar amount. For additional directivity, two to four sections may be stacked. The helices from adjacent bays are connected at their ends so that the deicing current passes through all of the radiators. The gain per bay is about 4, relative to a half-wave dipole. Beam tilt in a multibay antenna may be introduced by changing the phase between sections. This may be done by mechanically rotating about the axis of the antenna one or more sections with respect to the others, thereby causing the feed points to be in different azimuths.

The helical antenna is used at uhf channels as well [19]. At these frequencies the primary difference, aside from frequency scaling, is in the method of feeding the bays.

10. Traveling-Wave Slot Antennas

The traveling-wave slot antenna is essentially a large coaxial transmission line with a slotted outer conductor. A steel pipe comprising the outer conductor provides mechanical strength and acts as the radiating surface.

The signal is fed to successive slots through the action of capacitive probes projecting radially inward from one side of each slot. These distort the field within the line, placing voltages across the slots. An equal percentage of power is fed to each layer of slots. The result is an exponential illumination function producing a smooth elevation pattern. Two layers of "top-loading" slots are used at the upper end to radiate any remaining power. This approximates an infinitely long antenna in which no energy would be reflected.

The slots are arranged in pairs at each layer with the pairs separated by $\lambda/4$. Successive pairs occupy planes at right angles to each other. The orientation of the capacitive probes in a given plane is alternated so that all slot pairs in a given plane are excited with the correct phase.

The quarter-wave separation of layers and the space quadrature of successive layers produce a turnstile type feed which produces an omnidirectional azimuth pattern. A typical pattern is shown in Fig. 27. In addition, the quarter-wave spacing between slot layers causes reflections from adjacent slot pairs that tend to cancel each other, resulting in an inherently low vswr.

Pattern bandwidth is achieved by introducing the equivalent of a parallel resonant circuit at each slot. This produces compensation which maintains a constant electrical spacing between slot layers for frequencies within a given channel. It is implemented by shaping the slots for the required values of inductance and capacitance and by adjusting slot lengths, the shape of the ends, and the slot width at the center. The elevation beam is tilted downward by selection of the spacing between layers and the design of the capacitive probe compensation between layers.

11. UHF Antennas

Many of the principles of uhf antenna design are the same as for vhf. There are, however, differences due to a variety of factors. Because of the shorter wavelengths, propagation losses are higher, resulting in the need for higher ERP to achieve satisfactory field-strength levels. This generally requires the uhf station to transmit at a higher power level and use an antenna with higher gain. The higher gain is obtained by means of a longer aperture in terms of wavelengths. Directional azimuth patterns are also often used to increase the ERP in desired directions. The longer aperture results in decreased beamwidth so that structural stiffness becomes even more important as it affects beam stability. The stability of beam direction

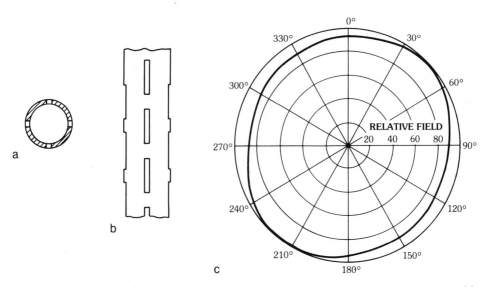

Fig. 27. Traveling-wave slot antenna. (*a*) End view of antenna. (*b*) Side view of antenna. (*c*) Pattern of antenna. (*Courtesy Dielectric Communications*)

and shape as a function of frequency also becomes more important. The increased length also makes simple feed systems more desirable. Finally, the percentage bandwidth of a uhf channel is less than that of the vhf channels, permitting the use of some antenna types (such as resonant slot arrays) that might not otherwise be suitable. Some of the same types of antennas used at vhf are used at uhf, i.e., the sidefire helical antenna. Accordingly, the antennas described most completely are those used primarily or exclusively at uhf.

12. Coaxial Slot Antennas

An antenna of this type is basically a coaxial transmission line with longitudinal radiating slots in the outer conductor fed by bar couplers bolted to the inside edge of each slot. The number of slots per layer is determined by the desired azimuth pattern, such as one slot for a cardioid pattern, two slots for a peanut-shaped pattern, three slots for a trilobe pattern, and four or more slots (depending on cylinder diameter) for an omnidirectional pattern. Typical azimuth patterns are shown in Fig. 28. Omnidirectional patterns are circular within ±1 dB. Parasitic elements have been used in conjunction with a single row of slots to produce omnidirectional patterns.

The layers are spaced at about one wavelength along the length of the antenna with the number of layers determined by the elevation directivity and pattern shape. The amplitude distribution is determined primarily by slot length and coupling bar diameter. Because of the large number of radiating elements this type of antenna lends itself readily to the use of synthesis programs to obtain a wide variety of null fills, beam tilt, and pattern smoothness. Typical elevation patterns are shown in Figs. 29a and 29b. The amount of cross polarization is quite small.

A single feed point is used. High-gain antennas are center fed by means of

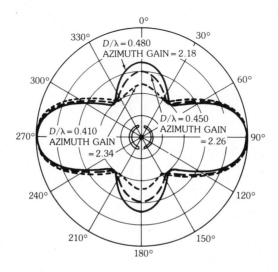

Fig. 28. Typical peanut-shaped patterns, where D is the pole outer diameter and λ is the midchannel wavelength. (*Courtesy Dielectric Communications*)

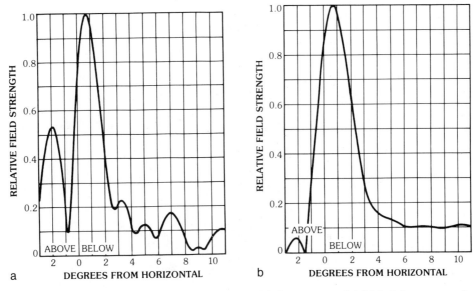

Fig. 29. Typical elevation patterns of a uhf coaxial slot antenna. (*a*) Main lobe power gain 27.5. (*b*) Main lobe power gain 28.0. (*Courtesy Dielectric Communications*)

a coaxial line terminated by a pressure seal at the center of the antenna (Fig. 30). The feed transmission line can be 3⅛ in (7.94 cm) to 9³⁄₁₆ in (23.33 cm) EIA depending on power-handling requirements and/or channel. End-fed antennas are used for moderate gains.

If the antenna is used in areas where icing is likely, it is often equipped with a deicing system. Heaters are usually clamped longitudinally to the outside of the slotted tube and used to prevent or remove ice. An ice detector is often used to detect the presence of ice and to minimize operational costs.

13. Waveguide Slot Antennas

A relatively recent approach to uhf slot antenna design has been the use of waveguide structures. The primary motivation for these designs is to utilize the inherent simplicity and high power-handling capabilities of waveguide. Two waveguide types have been used. One of these makes use of a ridged circular guide operating in the dominant TE_{11} mode. The size and shape of the ridge are selected to minimize the pipe diameter while maintaining λ/λ_g as close as possible to unity. A single row of slots is cut at small angles with respect to the ridge axis. The slot conductance and thus the aperture amplitude distribution are determined solely by this angle. Therefore, no coupling bars are needed to produce radiation from the slots. Since only one row of slots may be cut, this type of waveguide is used only for cardioid pattern requirements. A typical azimuth pattern is shown in Fig. 31. Elevation patterns may be synthesized using well-known techniques [20] to produce smooth contours with beam tilt.

Another waveguide antenna makes use of a hollow circular guide operating in the TM_{01} mode. This mode is similar to the field configuration of the TEM coaxial

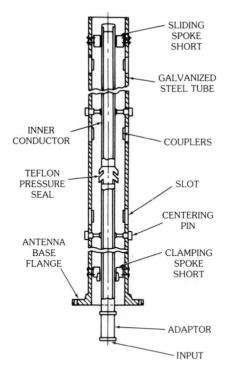

Fig. 30. Cross section of typical center-fed slot antenna. (*Courtesy Dielectric Communications*)

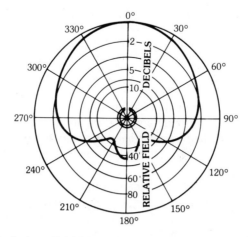

Fig. 31. Typical cardioid azimuth pattern of waveguide slot antenna.

mode. However, no center conductor is used to maintain the mode fields. Coupling devices are required at each slot to produce radiation. A variety of azimuth patterns may be generated by selecting the number and location of slots around the pipe. For an omnidirectional pattern, six equally spaced slots are used. A typical omnidirectional pattern is shown in Fig. 32.

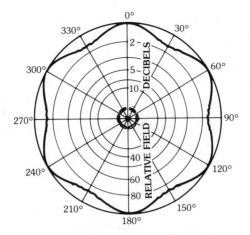

Fig. 32. Typical azimuth pattern of omnidirectional waveguide slot antenna.

The radial symmetry of the TM_{01} mode is essential to achieving a good omnidirectional pattern. The concept of this type of antenna was first described by Riblet [21] but only recently applied to broadcast use.

The construction of these antennas is such as to make approximately the bottom two-thirds a nonresonant traveling-wave array and the top one-third a resonant array. The top of the antenna acts as a matched load for the traveling-wave section while contributing to the desired aperture distribution. Use of the traveling-wave approach for the greater part of the array contributes to bandwidth.

The tendency in an end-fed array is for the beam direction θ to change as a function of frequency by an amount given by

$$\sin \theta = \lambda/\lambda_g - \lambda/2\ell \qquad (2)$$

where λ_g is the guide wavelength and ℓ is the slot spacing.

To correct for this, advantage is taken of the nonlinear and opposite phase changes that can result when a controlled standing wave is produced. Capacitive posts are located at selected positions to produce the required correction.

Either waveguide or coaxial line may be used at the input. When a coaxial input is used, the power rating of the antenna is limited to that of the transmission line.

14. Zigzag Antennas

The zigzag antenna [22, 23, 24] is a panel antenna that may be used for omnidirectional or directional applications. The antenna usually consists of four panels per bay mounted on a square mast section. Other structures of three, five, or more sides may also be used. It may be top mounted on a tower or the panels may be side mounted on a tower or other structure.

Like the sidefire helical antenna the zigzag panel uses the traveling-wave principle to excite a large aperture from a single feed point. The panel consists of a rectangular panel with two zigzag conductors mounted on insulators and spaced a fraction of a wavelength from the panel. Thus a transmission line is formed by each conductor over a ground plane. The height of the conductor is 0.1λ to 0.2λ, and there is appreciable radiation from the wire as the current decays. If the conductor is sufficiently long, it may be cut off or grounded at the far end with negligible effect on the load seen at the generator end, and a traveling wave will exist along the wire. The conductor is bent to create an in-phase horizontally polarized field component and cancellation of the vertical component on a horizontal plane normal to the axis of the panel. Two such zigzag radiators are fed at the center of the panel from a common feed point to excite an aperture of seven to eight wavelengths, as shown in Fig. 33. The radiation from the panel is similar to that of a broadside array of half-wave elements having a tapered current distribution.

The zigzag panel may be used for a large variety of directional patterns as well as an omnidirectional pattern in the aximuth plane. For directional patterns the panels facing various directions are fed with unequal amplitudes and phases. The panels may also be used in a tangential fire arrangement on large structures [25, 26, 27].

The zigzag panel has sufficient bandwidth to be used at vhf frequencies. Although the usual panel length is eight wavelengths, the panels may be shortened to about three wavelengths. This has the desirable effect of increasing bandwidth to permit diplexing of two stations [28].

15. FM Antennas

Aside from panel antennas, there are three types of antennas in general use for omnidirectional side-mount fm broadcast applications. Two of these are circularly

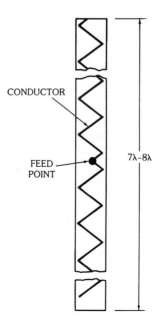

Fig. 33. A zigzag panel.

polarized, the other is horizontally polarized. The horizontally polarized antenna is primarily used for low-power educational stations and consists of one or more bays of circular loops [29] with a spacing of one wavelength between bays (see Fig. 34). They are designed for mounting on a pole having an outside diameter of 2 to 2½ in (5.08 to 6.35 cm). When mounted on such a pole the azimuth pattern is omnidirectional to within ±3 dB. The bandwidth is narrow, being sufficient to cover a single fm frequency ±1.2 MHz with a vswr of less than 1.5:1. The input connection is a uhf connector, and the power rating 800 W for two to four bays.

For circular polarization the most common types may be grouped as (a) loop/dipole combinations and (b) skewed dipoles.

A typical loop/dipole design which radiates vertical and horizontal components is shown in Fig. 35. When the phase of these components is in quadrature, circular polarization results. The rear terminal block is a balun matching the radiator impedance to the transmission line. From 1 to 16 bays may be stacked at nominal one-wavelength intervals. Elevation patterns with beam tilt and null fill are commonly used. Either radomes or electrical heaters may be used for deicing.

The azimuth pattern is omnidirectional for both the horizontal and vertical

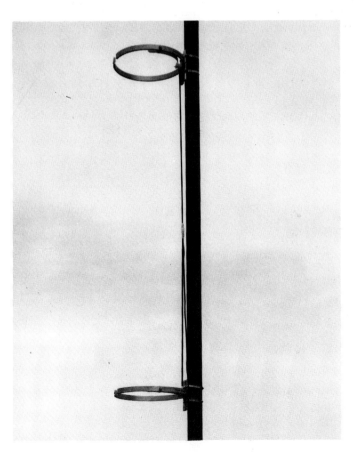

Fig. 34. Horizontally polarized loop fm antenna. (*Courtesy Electronic Research, Inc.*)

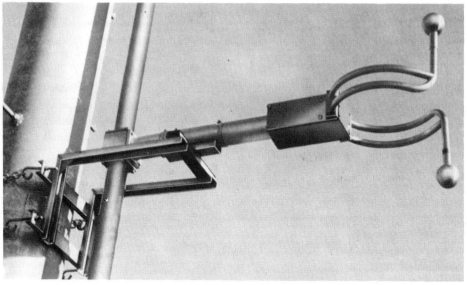

a

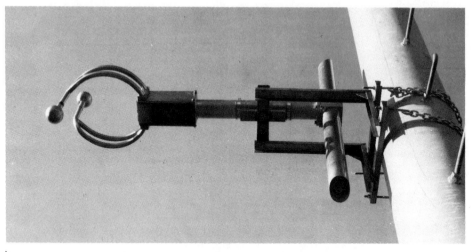

b

Fig. 35. Typical loop/dipole fm antenna. (*Courtesy Electronic Research, Inc.*)

components. The circularity is ±2 dB in free space. When the antenna is side mounted on a tower the circularity is affected. This is a common problem with most side-mount fm antennas.

The antenna is narrow band, having a vswr over a single fm channel of less than 1.1:1. It is capable of handling 10 kW per bay up to the limits of the 3⅛-in (7.94-cm) EIA feed line.

A typical skewed-dipole antenna design is shown in Fig. 36. Two skewed

FEED POINT

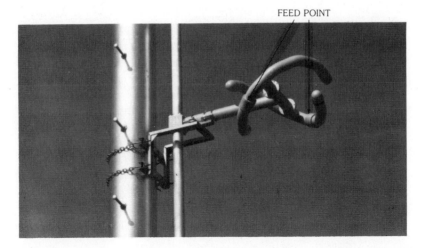

Fig. 36. A skewed-dipole fm antenna design. (*Courtesy Electronic Research, Inc.*)

dipoles are fed from a common feed line at the middle. The feed point for each dipole is offset from the center. This is a design with very high power-handling capability, radiating up to 40 kW in a single bay. The design is also broadband when compared to many fm designs. Stations having a separation of up to 5 MHz may be diplexed on a single antenna. The vswr for single-station operation is less than 1.07:1 for 200 kHz on either side of the carrier. The azimuth pattern is omnidirectional, having a circularity of less than ±2 dB when mounted on a 14-in (35.56-cm) diameter pole or a 24-in (60.96-cm) triangular tower.

16. Panel FM Antennas

The use of panel antennas has been discussed in detail for vhf tv antennas. The panel antenna is mounted on all sides of the tower, permitting a degree of pattern control not afforded by the ordinary side-mount antenna described previously. Another motivation for the use of panel antennas is the need to multiplex several stations into a common antenna. Panels such as the cavity-backed radiator are quite broadband, exhibiting an impedance bandwidth over the full fm allocation. When high power-handling capability is designed into the radiator and feed system components, it is possible to multiplex several class C stations reliably. For example, a 12-bay antenna [30] operates with 9 class C stations, each using a 25-kW transmitter. Obviously a key component in a system of this type is the multiplexer. This device must combine the outputs of the several transmitters into a common output while preserving audio quality and preventing objectionable intermodulation products.

An alternative to using a single antenna for multiplexing is to use separate antennas for each station with all antennas mounted on a common tower. This is illustrated in Fig. 37 for five stations. Each station uses a single bay of cavity-backed radiator panels to radiate an omnidirectional pattern in azimuth and a

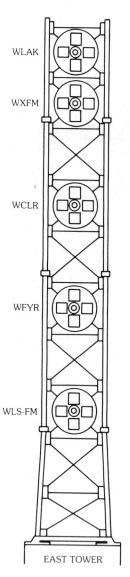

Fig. 37. Placement of fm antennas for five stations colocated at the Sears Tower in Chicago.

broad beam in elevation. Isolation is achieved by taking advantage of the natural isolation afforded by the cavities and with the addition of two- or three-section bandpass filters in the lines to each antenna [31]. The stacking plan is selected to optimize the various antenna isolations in accordance with the frequencies involved. This approach is not dependent on the use of a wideband element. Thus, narrow-band elements capable of handling the power may be used. If the pattern circularity is not extremely important, the side-mount antennas described earlier could be used. Care should be taken to provide adequate isolation by proper selection of the mounting arrangement and use of adequate filtering.

17. Multiple-Antenna Installations

The availability of wideband side-mount antennas is just one of the reasons broadcasters in a given market sometimes find it desirable to locate their antennas on a common site and tower. A desirable economic benefit is that only a single piece of favorable real estate is required and a single large tower may be used. If outdoor receiving antennas are in general use, viewers and listeners will point their antennas to the common site, thereby making it possible to receive all signals with approximately equal ease and giving all broadcasters approximately equal coverage.

Furthermore, many communities now consider tower structures to be "unsightly" and therefore prefer to place all such structures in a single locale to minimize the clutter to the landscape.

Several methods are used to accomplish the multistation installation. Two methods have already been described for use with fm stations. Yet another technique for placing multiple antennas on a common site is the use of a tee bar or candelabra at tower top. As in the case of the multiplexing, this technique places all stations at the same HAAT. All of the antennas may be essentially standard single-channel designs and each is afforded a degree of isolation due to the distance separating them. In planning a candelabra or tee-bar installation, several factors, some of which are conflicting, must be considered. These include (*a*) cost, (*b*) pattern circularity, (*c*) video response, (*d*) whether or not any of the antennas will be circularly polarized, (*e*) effects of wind sway, (*f*) isolation, (*g*) echo discernibility, and (*h*) frequency. These factors each influence the selection of the optimum separation.

Cost is a major factor for a candelabra or tee bar. While the antenna wind load and weight and transmission-line loads have some effect on the cost, the major effect is due to the physical separation between antennas. While it is difficult to give exact cost figures, it is generally recognized that, for a fixed number of support points (antennas), the cost increases approximately linearly with increasing spacing. For the antenna to be economically feasible an upper limit on allowable spacing must be set.

Pattern circularity, video response, and the effects of wind sway are all related to the azimuth pattern effects due to scattering from the other antennas. These parameters may be determined theoretically and/or experimentally, depending on the amount of time and effort available. In some cases, such as when the antennas are slotted tube designs, the scattering may be accurately determined using well-known formulas for scattering from round cylinders [32].

For more complex antennas these formulas may be used to approximate the scattered fields with the assumption that the antenna may be represented by an equivalent cylinder [33]. For example, a batwing antenna may be represented by an array of cylinders with radius equal to three-quarters the batwing radius (see Fig. 38). The maximum-to-minimum circularity values as calculated are slightly conservative when compared with measured results.

When more detailed results are required with the more complex antennas, it is necessary to model the actual structure in order to calculate or measure the scattered fields. If the antenna can be modeled as a wire antenna, the scattered

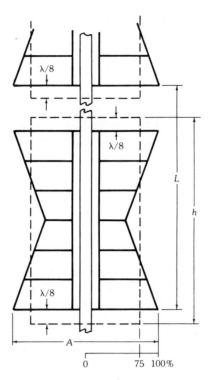

Fig. 38. Equivalent cylinder dimensions of a batwing antenna for horizontally polarized scatter pattern calculations. (*After Siukola [33]*)

fields can be calculated by means of moment methods [34]. For large antennas this approach can become complex. Furthermore, it is important to exercise great care in modeling the junctions in the wire model.

It is often simpler and more useful to build a scale model of the antennas to be used and indirectly measure the scattered fields. This is accomplished by first measuring the total field (amplitude and phase) of the radiating antenna in the presence of the scattering antenna(s) at a typical separation and the field of the radiating antenna by itself. These patterns may be subtracted vectorially to give the scatter pattern of the scattering antenna(s). This scatter pattern may then be used to compute the total fields at other separations.

To compute the circularity of the antennas it is only necessary to determine the total fields as a function of azimuth at the visual carrier frequency. To determine the video response the total fields must be determined as a function of frequency. The variation in the fields at specific azimuths are then used to determine the video response in directions of interest. For small variations the video response may be corrected for by adjustments available to the viewer. The effect of wind sway may be estimated by changing the relative spacing between antennas. The phase of the scattered fields is thereby varied, causing an effect similar to a change in frequency. However, the effect occurs as a function of time and can be observed at a specific location by a viewer.

For fixed type and number of antennas the circularity is generally improved as the separation is increased, although the variation is better described as a slowly damped oscillation as a function of distance. The effect on video response is lessened as the separation is reduced. As the separation increases, the time delay for echo signals from the scattering antennas increases so that for large delays any echos present may be discerned by the viewer. A separation of 50 ft (15.24 m) represents an echo delay of about 0.1 µs. Mitigating this effect is the fact that the magnitude of echos is reduced as separation is increased. Finally, the isolation between antennas increases directly as the spacing increases. Thus it is apparent

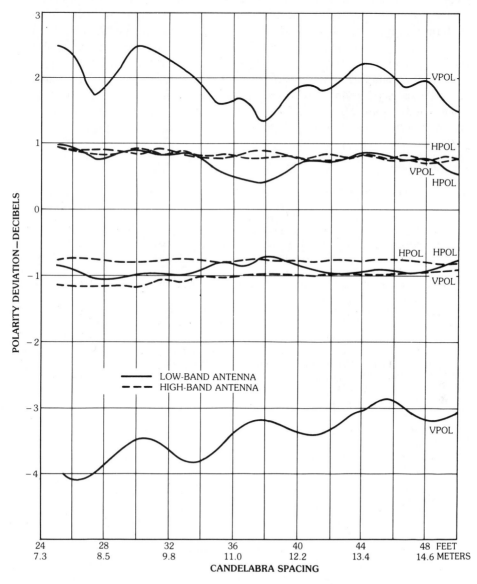

Fig. 39. Circularity deviation from rms versus spacing for three antennas. (*After Johns and Ralston [36],* © *1981 IEEE*)

the various parameters impose conflicting requirements on the configuration. In general, spacings between 10 and 100 ft (3 and 30.5 m) have been found to be desirable, depending on the number of antennas, the channels to be used, and polarization [35]. The effect of wind sway can be controlled by keeping the radiation centers as low as possible and by using very stiff antennas. In some cases, such as when stacked antennas are on a candelabra, it may be necessary to guy the antennas.

When circularly polarized antennas are used, the effect on the vertically polarized component becomes important. In addition to the factors already considered, the effect on axial ratio must be determined. In general, the effects of scattering on the vertically polarized component is greater than the effects on the horizontally polarized signal, as illustrated in Fig. 39. Thus the vertically polarized circularity and video response variations are greater for fixed separation and number of antennas. For systems using combinations of high-band, low-band, and fm antennas, separation between 30 and 40 ft (9.1 and 12.2 m) has been found to be suitable [36]. Specific determination of an optimum configuration is best determined by design studies for the specific channels to be used.

18. References

[1] *FCC Rules and Regulations*, vol. III, Part 73, March 1980. Also see E. C. Jordan (ed.), *Reference Data for Engineers: Radio, Electronics, Computer, and Communications*, 7th ed., pp. 35-9, 35-12, Indianapolis: Howard W. Sams & Co., 1985.

[2] R. A. Conners, "Understanding tv's grade A and grade B contours," *IEEE Trans. Broadcast.*, vol. BC-14, no. 4, p. 137, December 1968.

[3] J. Perini, "TV transmitting antenna selection," *IEEE Trans. Broadcast.*, vol. BC-15, no. 1, March 1969.

[4] J. A. Donovan, "Low wind load circularly polarized antenna," US Patent 4,446,465, May 1, 1984.

[5] R. H. DuHamel, "Circularly polarized antenna with circular arrays of slanted dipoles mounted around a conductive mast," US Patent 4,315,264, February 9, 1982.

[6] N. E. Lindenblad, US Patent 2,217,911.

[7] G. H. Brown and O. M. Woodward, Jr., "Circularly polarized omnidirectional antenna," *RCA Rev.*, vol. VIII, no. 2, pp. 259–269, June 1947.

[8] M. S. Suikola, "A circularly polarized antenna for channels 2–6," *Broadcast Eng.*, pp. 12–22, February 1981.

[9] P. S. Carter, "Antenna arrays around cylinders," *Proc. IRE*, vol. 31, no. 12, pp. 671–693, December 1943.

[10] R. H. DuHamel, "Circularly polarized helix and spiral antennas," US Patent 3,906,509.

[11] O. Ben Dov, "Circularly polarized broadside firing tetrahelical antenna," US Patent 3,940,772.

[12] O. Ben Dov, "Circularly polarized broadside firing, multihelical antenna," US Patent 4,011,567.

[13] R. E. Fisk and J. A. Donovan, "A new CP antenna for tv broadcast service," presented at the IEEE Symposium on Broadcasting, Washington, D.C., September 1975.

[14] R. W. Masters, "The super turnstile antenna," *Broadcast News*, no. 42, January 1946.

[15] G. H. Brown, "The turnstile antenna," *Electronics*, vol. 9, no. 15, April 1936.

[16] J. D. Kraus, *Antennas*, chapter 14, p. 428, New York: McGraw-Hill Book Co., 1950.

[17] L. O. Krause and H. G. Smith, "Wound antenna with conductive support," US Patent 2,985,878, May 23, 1961.

[18] R. E. Fisk, "The tv helical antenna adapted to structural tower shapes," *IRE Trans. Broadcast Transmission Systems*, October 1957.

[19] R. E. Fisk, "A simplified 5 megawatt antenna for the uhf broadcaster," presented at the Seventh Annual Symposium of the IRE Professional Group on Broadcast Transmission Systems, Washington, D.C., 1957.

[20] P. C. J. Hall, "Methods for shaping vertical radiation patterns," *Proc. IEEE*, vol. 116, no. 8, p. 1325, August 1969.

[21] H. H. Riblet, "Microwave omnidirectional antennas," *Proc. IRE*, vol. 35, no. 5, pp. 474–478, 1947.

[22] K. B. Hoffman and R. E. Fisk, "Directional zig-zag antenna at KERO, Bakersfield, California," *IEEE Trans. Broadcast.*, p. 12, February 1964.

[23] O. M. Woodward, "Antenna arrays," US Patent 2,759,183.

[24] L. Sin Hoi, "Theory of zig-zag antennas," Univ. of California, March 1969.

[25] J. Perini, "Improvement of pattern circularity of panel antennas mounted on large towers," *IEEE Trans. Broadcast.*, vol. BC-14, no. 1, p. 33, March 1968.

[26] J. Perini, "A method of obtaining a smooth pattern on circular arrays of large diameter," *IEEE Trans. Broadcast.*, vol. BC-14, no. 3, p. 126, September 1968.

[27] R. N. Clark, "The V-Z panels as a side mounted antenna," *IEEE Trans. Broadcast.*, vol. BC-13, no. 1, p. 31, January 1967.

[28] R. E. Fisk, "Empire State antenna installation and performance," presented at the Nineteenth Annual Broadcast Engineering Conference, NAB, Washington, D.C., 1965.

[29] C. A. Balanis, *Antenna Theory*, chapter 5, New York: Harper & Row, 1982.

[30] R. E. Fisk, "Design and application of a multiplex nine station fm antenna for the Senior Road Tower Group," presented at the 1984 NAB Convention, Las Vegas.

[31] W. G. Shulz, "Spurious emission measurements of common site transmitter installations using separate collocated antennas," presented at the 1984 NAB convention, Las Vegas.

[32] R. F. Harrington, *Time-Harmonic Electromagnetic Fields*, New York: McGraw-Hill Book Co., 1961, pp. 223–238.

[33] M. S. Siukola, "Predicting performance of candelabra antennas by mathematical analysis," *RCA Broadcast News*, vol. 97, pp. 63–68.

[34] R. F. Harrington, *Field Computations by Moment Methods*, p. 62. A book available from the author at RD2 Westlake Road, Cazenovia, N.Y. 13035.

[35] M. S. Siukola, "Size and performance trade-offs in multiple arrays of horizontally and circularly polarized tv antennas," *IEEE Trans. Broadcast.*, vol. BC-22, no. 1, pp. 5–12, March 1975.

[36] M. R. Johns and M. A. Ralston, "The first candelabra for CP broadcast antennas," *IEEE Trans. Broadcast*, pp. 77–82, December 1981.

Appendixes

CONTENTS

Appendix A

Physical Constants, International Units, Conversion of Units, and Metric Prefixes

Yi-Lin Chen
*University of Illinois**

Physical Constants

Quantity	Symbol	Value
Speed of light in vacuum	c	$2.997\,925 \times 10^{8}\,\text{ms}^{-1}$
Electron charge	e	$1.602\,192 \times 10^{-19}\,\text{C}$
Electron rest mass	m_e	$9.109\,558 \times 10^{-31}\,\text{kg}$
Boltzmann constant	k	$1.380\,622 \times 10^{-23}\,\text{JK}^{-1}$
Dielectric constant in vacuum	ϵ_0	$8.854\,185 \times 10^{-12}\,\text{Fm}^{-1}$
		$\cong (36\pi \times 10^{9})^{-1}\,\text{Fm}^{-1}$
Permeability in vacuum	μ_0	$4\pi \times 10^{-7}\,\text{Hm}^{-1}$

International System of Units (SI Units): Basic Units

Quantity	Symbol	Units
Length	ℓ	meters (m)
Mass	m	kilograms (kg)
Time	t	seconds (s)
Electric current	I	amperes (A)
Temperature	T	kelvins (K)
Luminous intensity	I	candelas (cd)

---

*On leave from the Chinese Aeronautical Laboratory, Beijing, China, during 1983.

Derived Units in Electromagnetics

Quantity	Symbol	Units
Electric-field strength	$\mathbf{E}$	volts per meter (V/m)
Magnetic-field strength	$\mathbf{H}$	amperes per meter (A/m)
Electric-flux density	$\mathbf{D}$	coulombs per meter squared (C/m^2)
Magnetic-flux density	$\mathbf{B}$	teslas (T) = Wb/m^2
Electric-current density	$\mathbf{J}$	amperes per meter squared (A/m^2)
Magnetic-current density	$\mathbf{K}$	volts per meter squared (V/m^2)
Electric-charge density	ϱ	coulombs per meter cubed (C/m^3)
Magnetic-charge density	ϱ_m	webers per meter cubed (Wb/m^3)
Voltage	V	volts (V)
Electric current	I	amperes (A)
Dielectric constant (permittivity)	ϵ	farads/meter (F/m)
Permeability	μ	henrys/meter (H/m)
Conductivity	σ	siemens per meter (S/m) = ℧/m
Resistance	R	ohms (Ω)
Inductance	L	henrys (H)
Capacitance	C	farads (F)
Impedance	Z	ohms (Ω)
Admittance	Y	siemens (S) or mhos (℧)
Power	P	watts (W)
Energy	W	joules (J)
Radiation intensity	I	watts per steradian (W/sr)
Frequency	f	hertz (Hz)
Angular frequency	ω	radians per second (rad/s)
Wavelength	λ	meters (m)
Wave number	k	1 per meter (m^{-1})
Phase shift constant	β	radians per meter (rad/m)
Attenuation factor	α	nepers per meter (Np/m)

Conversions of Units

Quantity	Symbol	SI Unit	Equivalent Number of	
			CGS Electromagnetic Unit	CGS Electrostatic Unit
Electric charge	q	coulombs	10^{-1} abcoulomb	3×10^9 statcoulombs
Current	I	amperes	10^{-1} abampere	3×10^9 statamperes
Volume current density	$\mathbf{J}$	amperes/meter^2	10^{-5} abampere/centimeter^2	3×10^5 statamperes/centimeter^2
Voltage	V	volts	10^8 abvolts	$\frac{1}{3} \times 10^{-2}$ statvolt
Electric-field intensity	$\mathbf{E}$	volts/meter	10^6 abvolts/cm	$\frac{1}{3} \times 10^{-4}$ statvolt/centimeter
Electric-flux density	$\mathbf{D}$	coulombs/meter^2	$4\pi \times 10^{-5}$ abcoulomb/centimeter^2	$12\pi \times 10^5$ statcoulombs/centimeter^2
Magnetic-field intensity	$\mathbf{H}$	amperes/meter	$4\pi \times 10^{-3}$ oersted	$12\pi \times 10^7$ oersteds
Magnetic-flux intensity	$\mathbf{B}$	webers/meter^2	10^4 gausses	$\frac{1}{3} \times 10^{-6}$ gauss
Permittivity	ϵ	farads/meter	$4\pi \times 10^{-11}$ abfarad/centimeter	$36\pi \times 10^9$ statfarads/centimeter
Permeability	μ	henrys/meter	$\frac{1}{4\pi} \times 10^7$ gauss/oersted	$\frac{1}{36\pi} \times 10^{-13}$ gauss/oersted
Magnetic flux	Φ	webers	10^8 gilberts	$\frac{1}{3} \times 10^{-2}$ gilbert
Resistance	R	ohms	10^9 abohms	$\frac{1}{9} \times 10^{-11}$ statohm
Inductance	L	henrys	10^9 abhenrys	$\frac{1}{9} \times 10^{-11}$ stathenry
Capacitance	C	farads	10^{-9} abfarad	9×10^{11} statfarads
Conductivity	σ	siemens/meter	10^{-11} absiemen/centimeter	9×10^9 statsiemens/centimeter
Work	W	joules	10^7 ergs	10^7 ergs
Power	P	watts	10^7 ergs/second	10^7 ergs/second

Conversion of Length Units

Meters	Centimeters	Inches	Feet	Miles
1	100	39.37	3.281	6.214×10^{-4}
0.01	1	0.3937	3.281×10^{-2}	
0.0254	2.540	1	8.333×10^{-2}	
0.3048	30.48	12	1	1.894×10^{-4}
1609			5279	1

Metric Prefixes and Symbols*

Multiplication Factor	Prefix	Symbol
10^{18}	exa	E
10^{15}	peta	P
10^{12}	tera	T
10^{9}	giga	G
10^{6}	mega	M
10^{3}	kilo	k
10^{2}	hecto	h
10	deka	da
10^{-1}	deci	d
10^{-2}	centi	c
10^{-3}	milli	m
10^{-6}	micro	μ
10^{-9}	nano	n
10^{-12}	pico	p
10^{-15}	femto	f
10^{-18}	atto	a

*From *IEEE Standard Dictionary of Electrical and Electronics Terms*, p. 682, The Institute of Electrical and Electronics Engineers, Inc., 1984.

Appendix B

The Frequency Spectrum

Li-Yin Chen
University of Illinois[*]

The wavelength of an electromagnetic wave in free space is $\lambda_0 = c/f$.

$$\lambda_0 = \frac{300\,000}{f(\text{kHz})}\,\text{m} = \frac{300}{f(\text{MHz})}\,\text{m} = \frac{30}{f(\text{GHz})}\,\text{cm}$$

$$= \frac{9.843 \times 10^5}{f(\text{kHz})}\,\text{ft} = \frac{9.843 \times 10^2}{f(\text{MHz})}\,\text{ft} = \frac{11.81}{f(\text{GHz})}\,\text{in}$$

The wave number of an electromagnetic wave in free space: $k_0 = \omega\sqrt{\mu_0\epsilon_0} = 2\pi f/c$.

$$k_0 = f(\text{Hz}) \times 2.0944 \times 10^{-8}\,\text{m}^{-1} = f(\text{kHz}) \times 2.0944 \times 10^{-5}\,\text{m}^{-1}$$

$$= f(\text{MHz}) \times 2.0944 \times 10^{-2}\,\text{m}^{-1} = f(\text{GHz}) \times 20.944\,\text{m}^{-1}$$

$$= f(\text{Hz}) \times 6.383 \times 10^{-9}\,\text{ft}^{-1} = f(\text{kHz}) \times 6.383 \times 10^{-6}\,\text{ft}^{-1}$$

$$= f(\text{MHz}) \times 6.383 \times 10^{-3}\,\text{ft}^{-1} = f(\text{GHz}) \times 6.383\,\text{ft}^{-1}$$

$$= f(\text{GHz}) \times 0.532\,\text{in}^{-1}$$

---

[*]On leave from the Chinese Aeronautical Laboratory, Beijing, China, during 1983.

Nomenclature of Frequency Bands

Adjectival Designation	Frequency Range	Metric Subdivision	Wavelength Range
elf: Extremely low frequency	30 to 300 Hz	Megametric waves	10 000 to 1000 km
vf: Voice frequency	300 to 3000 Hz		1000 to 100 km
vlf: Very low frequency	3 to 30 kHz	Myriametric waves	100 to 10 km
lf: Low frequency	30 to 300 kHz	Kilometric waves	10 to 1 km
mf: Medium frequency	300 to 3000 kHz	Hectrometric waves	1000 to 100 m
hf: High frequency	3 to 30 MHz	Decametric waves	100 to 10 m
vhf: Very high frequency	30 to 300 MHz	Metric waves	10 to 1 m
uhf: Ultrahigh frequency	300 to 3000 MHz	Decimetric waves	100 to 10 cm
shf: Superhigh frequency	3 to 30 GHz	Centimetric waves	10 to 1 cm
ehf: Extremely high frequency	30 to 300 GHz	Millimetric waves	10 to 1 mm
	300 to 3000 GHz	Decimillimetric waves	1 to 0.1 mm

Standard Radar-Frequency Letter Bands*

Band Designation	Nominal Frequency Range
hf	3–30 MHz
vhf	30–300 MHz
uhf	300–1000 MHz
L	1000–2000 MHz
S	2000–4000 MHz
C	4000–8000 MHz
X	8000–12 000 MHz
K_u	12.0–18 GHz
K	18–27 GHz
K_a	27–40 GHz
Millimeter	40–300 GHz

*Reprinted from ANSI/IEEE Std. 100-1984, *IEEE Standard Dictionary of Electrical and Electronics Terms*, © 1984 by The Institute of Electrical and Electronics Engineers, Inc., by permission of the IEEE Standards Department.

Television Channel Frequencies*

Channel Number[†]	Band (MHz)	Channel Number[†]	Band (MHz)	Channel Number[†]	Band (MHz)
2	54–60	29	560–566	57	728–734
3	60–66	30	566–572	58	734–740
4	66–72	31	572–578	59	740–746
5	76–82	32	578–584	60	746–752
6	82–88	33	584–590	61	752–758
7	174–180	34	590–596	62	758–764
8	180–186	35	596–602	63	764–770
9	186–192	36	602–608	64	770–776
10	192–198	37	608–614	65	776–782
11	198–204	38	614–620	66	782–788
12	204–210	39	620–626	67	788–794
13	210–216	40	626–632	68	794–800
14	470–476	41	632–638	69	800–806
15	476–482	42	638–644	70	806–812
16	482–488	43	644–650	71	812–818
17	488–494	44	650–656	72	818–824
18	494–500	45	656–662	73	824–830
19	500–506	46	662–668	74	830–836
20	506–512	47	668–674	75	836–842
21	512–518	48	674–680	76	842–848
22	518–524	49	680–686	77	848–854
23	524–530	50	686–692	78	854–860
24	530–536	51	692–698	79	860–866
25	536–542	52	698–704	80	866–872
26	542–548	53	704–710	81	872–878
27	548–554	54	710–716	82	878–884
28	554–560	55	716–722	83	884–890
		56	722–728		

*Note: The carrier frequency for the video portion is the lower frequency plus 1.25 MHz. The audio carrier frequency is the upper frequency minus 0.25 MHz. All channels have a 6-MHz bandwidth. For example, channel 2 video carrier is at 55.25 MHz and the audio carrier is at 59.75 MHz.

[†]Channels 2 through 13 are vhf; channels 14 through 83 are uhf. Channels 70 through 83 were withdrawn and reassigned to tv translator stations until licenses expire.

Appendix C

Electromagnetic Properties of Materials

Yi-Lin Chen
*University of Illinois**

Resistivities and Skin Depth of Metals and Alloys

Material	Resistivity* ($\mu\Omega$-cm)	Skin Depth[†] (μm at 1 GHz)
Aluminum	2.62	2.576
Brass (66% Cu, 34% Zn)	7.5	4.3586
Copper	1.7241	2.0898
Gold	2.44	2.4861
Iron	9.71	4.9594
Nickel	6.9	4.1807
Silver	1.62	2.0257
Steel (0.4–0.5% C, balance Fe)	13–22	5.7384–7.465
Steel, stainless (0.1% C, 18% Cr, 8% Ni, balance Fe)	90	15.0988
Tin	11.4	5.3737
Titanium	47.8	11.0036

*In solid form at 20°C; resistivity = (conductivity)$^{-1}$.
[†]Skin depth $\delta = (\pi\mu\sigma f)^{-1/2} = (20\pi)^{-1} [\sigma f(\text{GHz})]^{-1/2}$ m. The δs in the column are calculated at $f = 1$ GHz. For other frequencies, multiply them by $[f(\text{GHz})]^{-1/2}$.

---

*On leave from the Chinese Aeronautical Laboratory, Beijing, China, during 1983.

Characteristics of Insulating Materials*

Material Composition	T(°C)	Dielectric Constant† at (Frequency in Hertz)				Dissipation Factor† at (Frequency in Hertz)				Dielectric Strength in Volts/Mil at 25°C	DC Volume Resistivity in Ohm-cm at 25°C	Thermal Expansion (Linear) in Parts/°C	Softening Point in °C	Moisture Absorption in Percent
		10^4	10^6	3×10^9	2.5×10^{10}	10^4	10^6	3×10^9	2.5×10^{10}					
Ceramics:														
Aluminum oxide	25	8.80	8.80	8.79	—	0.00033	0.00030	0.0010	0.0010	—	10^{12}–10^{13}	—	1400–1430	0.1
Barium titanate†	26	1143	—	600	100	0.0105	—	0.30	0.60	75	10^{12}–10^{14}	—	—	<0.1
Calcium titanate	25	167.7	167.7	165	—	0.0002	0.0002	0.0023	—	100	—	—	1510	—
Magnesium oxide	25	9.65	9.65	—	—	<0.0003	<0.0003	—	—	—	—	—	—	—
Magnesium silicate	25	5.97	5.96	5.90	—	0.0005	0.0004	0.0012	—	—	>10^{14}	9.2×10^{-6}	1350	0.1–1
Magnesium titanate	25	13.9	13.9	13.8	13.7	0.0004	0.0005	0.0017	0.0065	—	—	—	—	—
Oxides of aluminum, silicon, magnesium, calcium, barium	24	6.04	—	5.90	—	0.0011	—	0.0024	—	—	—	7.7×10^{-6}	1325	—
Porcelain (dry process)	25	5.08	5.04	—	—	0.0075	0.0078	0.00089	—	—	—	—	—	—
Steatite 410	25	5.77	5.77	5.7	—	0.0007	0.0006	—	—	—	—	—	—	—
Strontium titanate	25	232	232	—	—	0.0002	0.0001	—	—	—	10^{12}–10^{14}	—	1510	0.1
Titanium dioxide (rutile)	26	100	100	—	—	0.0003	0.00025	—	—	100	—	—	—	—
Glasses:														
Iron-sealing glass	24	8.30	8.20	7.99	7.84	0.0005	0.0009	0.00199	0.0112	—	10^{10} at 250°	132×10^{-7}	484	poor
Soda-borosilicate	25	4.84	4.84	4.82	4.65	0.0036	0.0030	0.0054	0.0090	—	7×10^7 at 250°	50×10^{-7}	693	—
100% silicon dioxide (fused quartz)	25	3.78	3.78	3.78	3.78	0.0001	0.0002	0.00006	0.00025	410 (0.25")	>10^{18}	5.7×10^{-7}	1667	—
Plastics:														
Alkyd resin	25	4.76	4.55	4.50	—	0.0149	0.0138	0.0108	—	—	—	—	—	—
Cellulose acetate-butyrate, plasticized	26	3.30	3.08	2.91	—	0.018	0.017	0.028	—	250–400 (0.125")	—	11–17×10^{-5}	60–121	2.3
Cresylic acid–formaldehyde, 50% α-cellulose	25	4.51	3.85	3.43	3.21	0.036	0.055	0.051	0.038	1020 (0.033")	3×10^{12}	—	—	1.2
Cross-linked polystyrene	25	2.58	2.58	2.58	—	0.0016	0.0020	0.0019	—	—	—	3×10^{-5}	>125	—
Epoxy resin (Araldite CN-501)	25	3.62	3.35	3.09	—	0.019	0.034	0.027	—	405 (0.125")	>3.8×10^7	4.77×10^{-5}	109 (distortion)	0.14
Epoxy resin (Epon resin RN-48)	25	3.52	3.32	3.04	—	0.0142	0.0264	0.021	—	—	—	—	—	—
Foamed polystyrene, 0.25% filler	25	1.03	—	1.03	1.03	<0.0002	—	0.0001	—	—	—	—	85	low
Melamine—formaldehyde, α-cellulose	24	7.00	6.0	4.93	—	0.041	0.085	0.103	—	300–400	—	—	99 (stable)	0.4–0.6
Melamine—formaldehyde, 55% filler	26	5.75	5.5	—	—	0.0115	0.020	0.020	—	—	—	1.7×10^{-5}	—	0.6
Phenol—formaldehyde (Bakelite BM 120)	25	4.36	3.95	3.70	3.55	0.0280	0.0380	0.0438	0.0390	300 (0.125")	10^{11}	30–40×10^{-6}	<135 (distortion)	<0.6
Phenol—formaldehyde, 50% paper laminate	26	4.60	4.04	3.57	—	0.034	0.057	0.060	—	—	—	—	—	—
Phenol—formaldehyde, 65% mica, 4% lubricants	24	4.78	4.72	4.71	—	0.0082	0.0115	0.0126	—	—	—	—	—	—
Polycarbonate	—	2.96	—	—	—	0.010	—	—	—	364 (0.125")	2×10^{16}	7×10^{-5}	135 (deflection)	—
Polychlorotrifluoroethylene	25	2.42	2.32	2.29	2.28	0.0082	—	0.0028	0.0053	—	10^{13}	—	—	—

Material	Temp (°C)	Dielectric constant				Dissipation factor				Dielectric strength, V/mil (thickness)	Volume resistivity (Ω·cm)	Coeff. of expansion	Heat distortion (°C)	Water absorption (%)
Polyethylene	25	2.26	2.26	2.26	2.26	<0.0002	0.0002	0.00031	0.0006	1200 (0.033")	10^{17}	19×10^{-5} (varies)	95–105 (distortion)	0.03
Polyethylene-terephthalate	22	2.98	—	—	—	0.016	—	—	—	4000 (0.002")	—	—	—	—
Polyethylmethacrylate	25	2.55	2.52	2.51	2.5	0.0090	0.0075	—	0.0083	—	—	—	60 (distortion)	low
Polyhexamethylene-adipamide (nylon)	25	3.14	3.0	2.84	2.73	0.0218	0.0200	0.0117	0.0105	400 (0.125")	8×10^{14}	10.3×10^{-5}	65 (distortion)	1.5
Polyimide	—	3.4	—	—	—	0.003	—	—	—	570	—	—	—	low
Polyisobutylene	25	2.23	2.23	2.23	—	0.0001	0.0003	0.00047	—	600 (0.010")	—	—	25 (distortion)	—
Polymer of 95% vinyl-chloride, 5% vinyl-acetate	20	2.90	2.8	2.74	—	0.0150	0.0080	0.0059	—	—	—	—	—	—
Polymethyl methacrylate	27	2.76	—	2.60	—	0.0140	—	0.0057	—	990 (0.030")	—	—	70–75 (distortion)	0.3–0.6
Polyphenylene oxide	—	2.55	—	2.55	—	0.0007	—	0.0011	—	500 (0.125")	10^{17}	5.3×10^{-5}	195 (deflection)	—
Polypropylene	—	2.55	—	—	—	<0.0005	—	—	—	650 (0.125")	6×10^{16}	6–8.5×10^{-5}	99–116 (deflection)	—
Polystyrene	25	2.56	2.55	2.55	2.54	0.00007	<0.0001	0.00033	0.0012	500–700 (0.125")	10^{18}	6–8×10^{-5}	82 (distortion)	0.05
Polytetrafluoroethylene (Teflon)	22	2.1	2.1	2.1	2.08	<0.0002	<0.0002	0.00015	0.0006	1000–2000 (0.005"–0.012")	10^{17}	9.0×10^{-5}	66 (distortion) (stable to 300)	0.00
Polyvinylcyclohexane	24	2.25	2.25	2.25	—	<0.0002	<0.0002	0.00018	—	860 (0.034")	$>5 \times 10^{16}$	—	—	—
Polyvinyl formal	26	2.92	2.80	2.76	2.7	0.019	0.013	0.0113	0.0115	260 (0.125")	2×10^{14}	7.7×10^{-5}	190	1.3
Polyvinylidene fluoride	—	6.6	—	4.57	—	0.17	0.050	0.0555	—	375 (0.085")	—	12×10^{-5}	148 (deflection)	—
Urea-formaldehyde, cellulose	27	5.65	5.1	—	—	0.027	0.027	—	—	450–500 (0.125")	2×10^{11}	2.6×10^{-5}	152 (distortion)	2
Urethane elastomer	—	6.5–7.1	—	—	—	—	0.057	—	—	300 (0.125")	10^{14}–10^{16}	10–20×10^{-5}	—	—
Vinylidene–vinyl chloride copolymer	23	3.18	2.82	2.71	—	0.057	0.0180	0.0072	—	810 (0.068")	10^{16}	15.8×10^{-5}	150	<0.1
100% aniline-formaldehyde (Dilectene-100)	25	3.58	3.50	3.44	—	0.0061	0.0033	0.0026	—	277 (0.125")	—	5.4×10^{-5}	125	0.06–0.08
100% phenol-formaldehyde	24	5.4	4.4	3.64	—	0.060	0.077	0.052	—	400 (0.125")	10^{14}	8.3–13×10^{-5}	50 (distortion)	0.42
100% polyvinyl-chloride	20	2.88	2.85	2.84	—	0.0160	0.0081	0.0055	—	—	—	6.9×10^{-5}	54 (distortion)	0.05–0.15
Organic Liquids:														
Aviation gasoline (100 octane)	25	1.94	1.94	1.92	—	—	0.0001	0.0014	<0.0001	—	—	—	—	—
Benzene (pure, dried)	25	2.28	2.28	2.28	2.28	<0.0001	<0.0001	<0.0001	—	—	—	—	—	—
Carbon tetrachloride	25	2.17	2.17	2.17	—	<0.00004	<0.0002	0.0004	—	—	—	—	—	—
Ethyl alcohol (absolute)	25	24.5	23.7	6.5	—	0.090	0.062	0.250	—	—	—	—	—	—
Ethylene glycol	25	41	41	12	—	0.030	0.045	1.00	—	—	—	—	—	—
Jet fuel (JP-3)	25	2.08	2.08	2.04	—	0.0001	0.0001	0.0055	—	—	—	—	—	—
Methyl alcohol (absolute analytical grade)	25	31	31.0	23.9	—	0.20	0.038	0.64	—	—	—	—	—	—
Methyl or ethyl siloxane polymer (1000 cs)	22	2.78	—	2.74	—	<0.0003	<0.0003	0.0096	—	300 (0.100")	3×10^{12}	—	—	—
Monomeric styrene	22	2.40	2.40	2.40	—	<0.0003	<0.0003	0.0020	—	300 (0.100")	—	—	—	0.06
Transil oil	26	2.22	2.20	2.18	—	<0.0005	0.0048	0.0028	—	—	—	—	—40 (pour point)	—
Vaseline	25	2.16	2.16	2.16	—	<0.0001	<0.0001	0.00066	—	—	—	—	—	—

Characteristics of Insulating Materials* (cont'd.)

Material Composition	T(°C)	Dielectric Constant† at (Frequency in Hertz)				Dissipation Factor† at (Frequency in Hertz)				Dielectric Strength in Volts/Mil at 25°C	DC Volume Resistivity in Ohm-cm at 25°C	Thermal Expansion (Linear) in Parts/°C	Softening Point in °C	Moisture Absorption in Percent
		10^4	10^6	3×10^9	2.5×10^{10}	10^4	10^6	3×10^9	2.5×10^{10}					
Waxes:														
Beeswax, yellow	23	2.53	2.45	2.39	—	0.0092	0.0090	0.0075	—	—	—	—	45–64 (melts)	nil
Dichloronaphthalenes	23	2.98	2.93	2.89	—	0.0003	0.0017	0.0037	—	—	—	—	35–63 (melts)	—
Polybutene	25	2.34	2.30	2.27	—	0.00133	0.00133	0.0009	—	—	—	—	—	—
Vegetable and mineral waxes	25	2.3	2.3	2.25	—	0.0004	0.0004	0.00046	—	—	—	—	57	—
Rubbers:														
Butyl rubber	25	2.35	2.35	2.35	—	0.0010	0.0010	0.0009	—	—	—	—	—	—
GR-S rubber	25	2.90	2.82	2.75	—	0.0120	0.0080	0.0057	—	870 (0.040")	2×10^{15}	—	—	—
Gutta-percha	25	2.53	2.47	2.40	—	0.0042	0.0120	0.0060	—	—	10^{15}	—	—	—
Hevea rubber (pale crepe)	25	2.4	2.4	2.15	—	0.0018	0.0050	0.0030	—	—	—	—	—	—
Hevea rubber, vulcanized (100 pts pale crepe, 6 pts sulfur)	27	2.74	2.42	2.36	—	0.0446	0.0180	0.0047	0.025	300 (0.125")	8×10^{12}	—	—	—
Neoprene rubber	24	6.26	4.5	4.00	4.0	0.038	0.090	0.034	0.10	—	—	—	—	nil
Organic polysulfide, fillers	23	110	30	16	13.6	0.39	0.28	0.22	—	—	—	—	—	—
Silicone-rubber compound	25	3.20	3.16	3.13	—	0.0030	0.0032	0.0097	—	—	—	—	—	—
Woods:‡														
Balsa wood	26	1.37	1.30	1.22	—	0.0120	0.0135	0.100	—	—	—	—	—	—
Douglas fir	25	1.93	1.88	1.82	—	0.026	0.033	0.027	0.032	—	—	—	—	—
Douglas fir, plywood	25	1.90	—	—	1.78	0.0230	—	—	0.0220	—	—	—	—	—
Mahogany	25	2.25	2.07	1.88	1.6	0.025	0.032	0.025	0.020	—	—	—	—	—
Yellow birch	25	2.70	2.47	2.13	1.87	0.029	0.040	0.033	0.026	—	—	—	—	—
Yellow poplar	25	1.75	—	1.50	1.4	0.019	—	0.015	0.017	—	—	—	—	—
Miscellaneous:														
Amber (fossil resin)	25	2.65	—	2.6	—	0.0056	—	0.0090	—	2300 (0.125")	Very high	9.8×10^{-5}	200	—
DeKhotinsky cement	23	3.23	—	2.96	—	0.024	—	0.021	—	—	—	—	80–85	—
Gilsonite (99.9% natural bitumen)	26	2.58	2.56	—	—	0.0016	0.0011	—	—	—	—	—	155 (melts)	—
Shellac (natural XL)	28	3.47	3.10	2.86	—	0.031	0.030	0.0254	—	—	10^{16}	—	80	low after baking
Mica, glass-bonded	25	7.39	—	—	—	0.0013	—	—	—	—	—	—	—	—
Mica, glass, titanium dioxide	24	9.0	—	—	—	0.0026	—	0.0040	—	—	—	—	400	<0.5
Ruby mica	26	5.4	5.4	5.4	—	0.0003	0.0002	0.0003	—	3800–5600 (0.040")	5×10^{13}	—	—	—
Paper, royalgrey	25	2.99	2.77	2.70	—	0.038	0.066	0.056	0.0013	202 (0.125")	—	—	—	—
Selenium (amorphous)	25	6.00	6.00	6.00	6.00	<0.0003	<0.0002	0.00018	—	—	—	—	—	—
Asbestos fiber–chrysotic paper	25	3.1	—	—	—	0.025	—	—	—	—	—	—	—	—
Sodium chloride (fresh crystals)	25	5.90	—	—	5.90	<0.0002	—	<0.0005	<0.0005	—	—	—	—	—

Soil, sandy dry	25	2.59	2.55	2.55	—	0.017	—	0.0062	—	—	—	—	—
Soil, loamy dry	25	2.53	2.48	2.44	—	0.018	—	0.0011	—	—	—	—	—
Ice (from pure distilled water)	−12	4.15	3.45	3.20	—	0.12	0.035	0.0009	—	—	—	—	—
Freshly fallen snow	−20	1.20	1.20	1.20	—	0.0215	—	0.00029	—	—	—	—	—
Hard-packed snow followed by light rain	−6	1.55	—	1.5	—	0.29	—	0.0009	—	—	—	—	—
Water (distilled)	25	78.2	78	76.7	34	0.040	0.005	0.157	0.2650	—	10^4	—	—

†The dissipation factor is defined as the ratio of the energy dissipated to the energy stored in the dielectric, or as the tangent of the loss angle. Dielectric constant and dissipation factor depend on electrical field strength.

‡Field perpendicular to grain.

Properties of Soft Magnetic Metals*

Name	Composition (%)	Permeability Initial	Permeability Maximum	Coercivity H_c (A/m)	Retentivity B_r (T)	B_{max} (T)	Resistivity ($\mu\Omega$-cm)
Ingot iron	99.8 Fe	150	5 000	80	0.77	2.14	10
Low carbon steel	99.5 Fe	200	4 000	100	—	2.14	12
Silicon iron, unoriented	3 Si, bal Fe	270	8 000	60	—	2.01	47
Silicon iron, grain oriented	3 Si, bal Fe	1 400	50 000	7	1.20	2.01	50
4750 alloy	48 Ni, bal Fe	11 000	80 000	2	—	1.55	48
4-79 Permalloy	4 Mo, 79 Ni, bal Fe	40 000	200 000	1	—	0.80	58
Supermalloy	5 Mo, 80 Ni, bal Fe	80 000	450 000	0.4	—	0.78	65
2V-Permendur	2V, 49 Co, bal Fe	800	8 000	160	—	2.30	40
Supermendur	2V, 49 Co, bal Fe	—	100 000	16	2.00	2.30	26
Metglas[†] 2605SC	$Fe_{81}B_{13.5}Si_{3.5}C_2$	—	210 000	14	1.46	1.60	125
Metglas[†] 2605S-3	$Fe_{79}B_{16}Si_5$	—	30 000	8	0.30	1.58	125

*Reproduced with permission of the publisher, Howard W. Sams & Company, Indianapolis, *Reference Data for Engineers: Radio, Electronics, Computer, and Communications*, 7th ed., by E. C. Jordan, ed., © 1985.

[†]Metglas is Allied Corporation's registered trademark for amorphous alloys.

Appendix D

Vector Analysis

Yi-Lin Chen
*University of Illinois**

1. Change of Coordinate Systems

The transformations of the coordinate components of a vector **A** among the rectangular (x, y, z), cylindrical (θ, ϕ, z), and spherical (r, θ, ϕ) coordinates are given by the following relations (see Fig. 1):

$$A_x = A_\varrho \cos \phi - A_\phi \sin \phi = A_r \sin \theta \cos \phi + A_\theta \cos \theta \cos \phi - A_\phi \sin \phi$$

$$A_y = A_\varrho \sin \phi + A_\phi \cos \phi = A_r \sin \theta \sin \phi + A_\theta \cos \theta \sin \phi + A_\phi \cos \phi$$

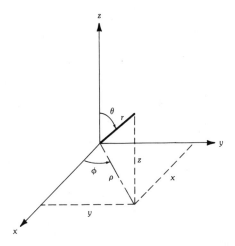

Fig. 1. Rectangular, cylindrical, and spherical coordinate systems.

---

*On leave from the Chinese Aeronautical Laboratory, Beijing, China, during 1983.

$$A_z = A_r \cos\theta - A_\theta \sin\theta$$
$$A_\varrho = A_x \cos\phi + A_y \sin\phi = A_r \sin\theta + A_\theta \cos\theta$$
$$A_\phi = -A_x \sin\phi + A_y \cos\phi$$
$$A_r = A_x \sin\theta \cos\phi + A_y \sin\theta \sin\phi + A_z \cos\theta = A_\varrho \sin\theta + A_z \cos\theta$$
$$A_\theta = A_x \cos\theta \cos\phi + A_y \cos\theta \sin\phi - A_z \sin\theta = A_\varrho \cos\theta - A_z \sin\theta$$

Differential element of volume:

$$dV = dx\,dy\,dz = \varrho\,d\varrho\,d\phi\,dz = r^2 \sin\theta\,dr\,d\theta\,d\phi$$

Differential element of vector area:

$$\mathbf{dS} = \hat{\mathbf{x}}\,dy\,dz + \hat{\mathbf{y}}\,dx\,dz + \hat{\mathbf{z}}\,dx\,dy$$
$$= \hat{\boldsymbol{\varrho}}\varrho\,d\phi\,dz + \hat{\boldsymbol{\phi}}\,d\varrho\,dz + \hat{\mathbf{z}}\varrho\,d\varrho\,d\phi$$
$$= \hat{\mathbf{r}}r^2 \sin\theta\,d\theta\,d\phi + \hat{\boldsymbol{\theta}}r \sin\theta\,dr\,d\phi + \hat{\boldsymbol{\phi}}r\,dr\,d\theta$$

Differential element of vector length:

$$\mathbf{d\boldsymbol{\ell}} = \hat{\mathbf{x}}\,dx + \hat{\mathbf{y}}\,dy + \hat{\mathbf{z}}\,dz$$
$$= \hat{\boldsymbol{\varrho}}\,d\varrho + \hat{\boldsymbol{\phi}}\varrho\,d\phi + \hat{\mathbf{z}}\,dz$$
$$= \hat{\mathbf{r}}\,dr + \hat{\boldsymbol{\theta}}r\,d\theta + \hat{\boldsymbol{\phi}}r \sin\theta\,d\phi$$

2. ∇ Operator

In rectangular coordinates (x, y, z):

$$\nabla\Phi = \left(\hat{\mathbf{x}}\frac{\partial}{\partial x} + \hat{\mathbf{y}}\frac{\partial}{\partial y} + \hat{\mathbf{z}}\frac{\partial}{\partial z} \right)\Phi$$

$$\nabla\cdot\mathbf{A} = \frac{\partial A_x}{\partial x} + \frac{\partial A_y}{\partial y} + \frac{\partial A_z}{\partial z}$$

$$\nabla \times \mathbf{A} = \begin{vmatrix} \hat{\mathbf{x}} & \hat{\mathbf{y}} & \hat{\mathbf{z}} \\ \dfrac{\partial}{\partial x} & \dfrac{\partial}{\partial y} & \dfrac{\partial}{\partial z} \\ A_x & A_y & A_z \end{vmatrix}$$

$$\nabla^2\Phi = \nabla\cdot\nabla\Phi = \left(\frac{\partial^2}{\partial x^2} + \frac{\partial^2}{\partial y^2} + \frac{\partial^2}{\partial z^2} \right)\Phi$$

$$\nabla^2\mathbf{A} = \hat{\mathbf{x}}\nabla^2 A_x + \hat{\mathbf{y}}\nabla^2 A_y + \hat{\mathbf{z}}\nabla^2 A_z$$

In cylindrical coordinates (ϱ, ϕ, z):

$$\nabla\Phi = \left(\hat{\boldsymbol{\varrho}}\frac{\partial}{\partial \varrho} + \hat{\boldsymbol{\phi}}\frac{\partial}{\varrho\partial\phi} + \hat{\mathbf{z}}\frac{\partial}{\partial z} \right)\Phi$$

Appendixes

$$\nabla \cdot \mathbf{A} = \frac{1}{\varrho}\frac{\partial}{\partial\varrho}(\varrho A_\varrho) + \frac{1}{\varrho}\frac{\partial A_\phi}{\partial\phi} + \frac{\partial A_z}{\partial z}$$

$$\nabla \times \mathbf{A} = \frac{1}{\varrho}\begin{vmatrix} \hat{\varrho} & \varrho\hat{\phi} & \hat{\mathbf{z}} \\ \frac{\partial}{\partial\varrho} & \frac{\partial}{\partial\phi} & \frac{\partial}{\partial z} \\ A_\varrho & \varrho A_\phi & A_z \end{vmatrix}$$

$$= \hat{\varrho}\left(\frac{1}{\varrho}\frac{\partial A_z}{\partial\phi} - \frac{\partial A_\phi}{\partial z}\right) + \hat{\phi}\left(\frac{\partial A_\varrho}{\partial z} - \frac{\partial A_z}{\partial\varrho}\right) + \hat{\mathbf{z}}\left[\frac{1}{\varrho}\frac{\partial}{\partial\phi}(\varrho A_\phi) - \frac{1}{\varrho}\frac{\partial A_\varrho}{\partial\phi}\right]$$

$$\nabla^2\Phi = \frac{1}{\varrho}\frac{\partial}{\partial\varrho}\left(\varrho\frac{\partial\Phi}{\partial\varrho}\right) + \frac{1}{\varrho^2}\frac{\partial^2\Phi}{\partial\phi^2} + \frac{\partial^2\Phi}{\partial z^2}$$

$$\nabla^2\mathbf{A} = \nabla\nabla\cdot\mathbf{A} - \nabla\times\nabla\times\mathbf{A} \neq \hat{\varrho}\nabla^2 A_\varrho + \hat{\phi}\nabla^2 A_\phi + \hat{\mathbf{z}}\nabla^2 A_z$$

In spherical coordinates (r, θ, ϕ):

$$\nabla\Phi = \left(\hat{\mathbf{r}}\frac{\partial}{\partial r} + \hat{\theta}\frac{1}{r}\frac{\partial}{\partial\theta} + \hat{\phi}\frac{1}{r\sin\theta}\frac{\partial}{\partial\phi}\right)\Phi$$

$$\nabla\cdot\mathbf{A} = \frac{1}{r}\frac{\partial}{\partial r}(r^2 A_r) + \frac{1}{r\sin\theta}\frac{\partial}{\partial\theta}(A_\theta\sin\theta) + \frac{1}{r\sin\theta}\frac{\partial A_\phi}{\partial\phi}$$

$$\nabla\times\mathbf{A} = \frac{1}{r^2\sin\theta}\begin{vmatrix} \hat{\mathbf{r}} & r\hat{\theta} & (r\sin\theta)\hat{\phi} \\ \frac{\partial}{\partial r} & \frac{\partial}{\partial\theta} & \frac{\partial}{\partial\phi} \\ A_r & rA_\theta & (r\sin\theta)A_\phi \end{vmatrix}$$

$$= \hat{\mathbf{r}}\frac{1}{r\sin\theta}\left[\frac{\partial}{\partial\theta}(A_\phi\sin\theta) - \frac{\partial A_\theta}{\partial\phi}\right] + \hat{\theta}\frac{1}{r}\left[\frac{1}{\sin\theta}\frac{\partial A_r}{\partial\phi} - \frac{\partial}{\partial r}(rA_\phi)\right]$$

$$+ \hat{\phi}\frac{1}{r}\left[\frac{\partial}{\partial r}(rA_\phi) - \frac{\partial A_r}{\partial\theta}\right]$$

$$\nabla^2\Phi = \frac{1}{r^2}\frac{\partial}{\partial r}\left(r^2\frac{\partial\Phi}{\partial r}\right) + \frac{1}{r^2\sin\theta}\frac{\partial}{\partial\theta}\left(\sin\theta\frac{\partial\Phi}{\partial\theta}\right) + \frac{1}{r^2\sin^2\theta}\frac{\partial^2\Phi}{\partial\phi^2}$$

$$\nabla^2\mathbf{A} = \nabla\nabla\cdot A - \nabla\times\nabla\times A \neq \hat{\mathbf{r}}\nabla^2 A_r + \hat{\theta}\nabla^2 A_\theta + \hat{\phi}\nabla^2 A_\phi$$

3. Identities

$$\mathbf{a}\cdot\mathbf{b}\times\mathbf{c} = \mathbf{a}\times\mathbf{b}\cdot\mathbf{c} = \mathbf{b}\cdot\mathbf{c}\times\mathbf{a}$$

$$\mathbf{a}\times(\mathbf{b}\times\mathbf{c}) = (\mathbf{a}\cdot\mathbf{c})\mathbf{b} - (\mathbf{a}\cdot\mathbf{b})\mathbf{c}$$

$$(\mathbf{a}\times\mathbf{b})\cdot(\mathbf{c}\times\mathbf{d}) = \mathbf{a}\cdot\mathbf{b}\times(\mathbf{c}\times\mathbf{d}) = \mathbf{a}\cdot[(\mathbf{b}\cdot\mathbf{d})\mathbf{c} - (\mathbf{b}\cdot\mathbf{c})\mathbf{d}] = (\mathbf{a}\cdot\mathbf{c})(\mathbf{b}\cdot\mathbf{d}) - (\mathbf{a}\cdot\mathbf{d})(\mathbf{b}\cdot\mathbf{c})$$

$$(\mathbf{a}\times\mathbf{b})\times(\mathbf{c}\times\mathbf{d}) = (\mathbf{a}\times\mathbf{b}\cdot\mathbf{d})\mathbf{c} - (\mathbf{a}\times\mathbf{b}\cdot\mathbf{c})\mathbf{d}$$

$$\nabla(\Phi + \psi) = \nabla\Phi + \nabla\psi$$

$$\nabla(\Phi\psi) = \Phi\nabla\psi + \psi\nabla\Phi$$

$$\nabla\cdot(\mathbf{a} + \mathbf{b}) = \nabla\cdot\mathbf{a} + \nabla\cdot\mathbf{b}$$

$$\nabla \times (\mathbf{a} + \mathbf{b}) = \nabla \times \mathbf{a} + \nabla \times \mathbf{b}$$

$$\nabla \cdot (\Phi \mathbf{a}) = \mathbf{a} \cdot \nabla \Phi + \Phi \nabla \cdot \mathbf{a}$$

$$\nabla \times (\Phi \mathbf{a}) = \nabla \Phi \times \mathbf{a} + \Phi \nabla \times \mathbf{a}$$

$$\nabla (\mathbf{a} \cdot \mathbf{b}) = (\mathbf{a} \cdot \nabla) \mathbf{b} + (\mathbf{b} \cdot \nabla) \mathbf{a} + \mathbf{a} \times (\nabla \times \mathbf{b}) + \mathbf{b} \times (\nabla \times \mathbf{a})$$

$$\nabla \times (\mathbf{a} \times \mathbf{b}) = \mathbf{a} \nabla \cdot \mathbf{b} - \mathbf{b} \nabla \cdot \mathbf{a} + (\mathbf{b} \cdot \nabla) \mathbf{a} - (\mathbf{a} \cdot \nabla) \mathbf{b}$$

$$\nabla \cdot (\mathbf{a} \times \mathbf{b}) = \mathbf{b} \cdot \nabla \times \mathbf{a} - \mathbf{a} \cdot \nabla \times \mathbf{b}$$

$$\nabla \times \nabla \times \mathbf{a} = \nabla \nabla \cdot \mathbf{a} - \nabla^2 \mathbf{a}$$

$$\nabla \times \nabla \Phi \equiv 0$$

$$\nabla \cdot \nabla \times \mathbf{a} \equiv 0$$

$$\iiint_V \nabla \cdot \mathbf{a} \, dV = \oiint_S \mathbf{a} \cdot d\mathbf{S} \qquad \text{(Gauss's theorem)}$$

$$\iint_S \nabla \times \mathbf{a} \cdot d\mathbf{S} = \oint_c \mathbf{a} \cdot d\boldsymbol{\ell} \qquad \text{(Stokes's theorem)}$$

Green's first and second identities:

$$\iiint_V (\nabla \psi \cdot \nabla \Phi + \Phi \nabla^2 \psi) \, dV = \oint_S \Phi \nabla \psi \cdot d\mathbf{S}$$

$$\iiint_V (\Phi \nabla^2 \psi - \psi \nabla^2 \Phi) \, dV = \oint_S (\Phi \nabla \psi - \psi \nabla \Phi) \cdot d\mathbf{S}$$

$$\iiint_V (\nabla \times \mathbf{A} \cdot \nabla \times \mathbf{B} - \mathbf{A} \cdot \nabla \times \nabla \times \mathbf{B}) \, dV = \iiint_V (\nabla \cdot \mathbf{A} \times \nabla \times \mathbf{B}) \, dV$$

$$= \oint_S \mathbf{A} \times \nabla \times \mathbf{B} \cdot d\mathbf{S}$$

$$\iiint_V (\mathbf{B} \cdot \nabla \times \nabla \times \mathbf{A} - \mathbf{A} \cdot \nabla \times \nabla \times \mathbf{B}) \, dV = \oint_S (\mathbf{A} \times \nabla \times \mathbf{B} - \mathbf{B} \times \nabla \times \mathbf{A}) \cdot d\mathbf{S}$$

Appendix E

VSWR Versus Reflection Coefficient and Mismatch Loss

Yi-Lin Chen
*University of Illinois**

The following relations are used in the construction of the vswr table below.

$$\text{vswr} = \frac{1 + |\Gamma|}{1 - |\Gamma|}, \qquad |\Gamma| = \frac{\text{vswr} - 1}{\text{vswr} + 1}$$

$$\text{mismatch loss (dB)} = -10 \log_{10}(1 - |\Gamma|^2)$$

VSWR Versus Reflection Coefficient (Γ) and Mismatch Loss

VSWR	$|\Gamma|$	Mismatch Loss (dB)	VSWR	$|\Gamma|$	Mismatch Loss (dB)
1.01	.0050	.0001	1.12	.0566	.0139
1.02	.0099	.0004	1.13	.0610	.0162
1.03	.0148	.0009	1.14	.0654	.0186
1.04	.0196	.0017	1.15	.0698	.0212
1.05	.0244	.0026	1.16	.0741	.0239
1.06	.0291	.0037	1.17	.0783	.0267
1.07	.0338	.0050	1.18	.0826	.0297
1.08	.0385	.0064	1.19	.0868	.0328
1.09	.0431	.0081	1.20	.0909	.0360
1.10	.0476	.0099	1.21	.0950	.0394
1.11	.0521	.0118	1.22	.0991	.0429

*On leave from the Chinese Aeronautical Laboratory, Beijing, China, during 1983.

VSWR Versus Reflection Coefficient (Γ) and Mismatch Loss (cont'd.)

VSWR	$\lvert\Gamma\rvert$	Mismatch Loss (dB)	VSWR	$\lvert\Gamma\rvert$	Mismatch Loss (dB)
1.23	.1031	.0464	1.73	.2674	.3222
1.24	.1071	.0501	1.74	.2701	.3289
1.25	.1111	.0540	1.75	.2727	.3357
1.26	.1150	.0579	1.76	.2754	.3425
1.27	.1189	.0619	1.77	.2780	.3493
1.28	.1228	.0660	1.78	.2806	.3561
1.29	.1266	.0702	1.79	.2832	.3630
1.30	.1304	.0745	1.80	.2857	.3698
1.31	.1342	.0789	1.81	.2883	.3767
1.32	.1379	.0834	1.82	.2908	.3837
1.33	.1416	.0880	1.83	.2933	.3906
1.34	.1453	.0927	1.84	.2958	.3976
1.35	.1489	.0974	1.85	.2982	.4046
1.36	.1525	.1023	1.86	.3007	.4116
1.37	.1561	.1072	1.87	.3031	.4186
1.38	.1597	.1121	1.88	.3056	.4257
1.39	.1632	.1172	1.89	.3080	.4327
1.40	.1667	.1223	1.90	.3103	.4398
1.41	.1701	.1275	1.91	.3127	.4469
1.42	.1736	.1328	1.92	.3151	.4540
1.43	.1770	.1382	1.93	.3174	.4612
1.44	.1803	.1436	1.94	.3197	.4683
1.45	.1837	.1490	1.95	.3220	.4755
1.46	.1870	.1546	1.96	.3243	.4827
1.47	.1903	.1602	1.97	.3266	.4899
1.48	.1935	.1658	1.98	.3289	.4971
1.49	.1968	.1715	1.99	.3311	.5043
1.50	.2000	.1773	2.00	.3333	.5115
1.51	.2032	.1831	2.05	.3443	.5479
1.52	.2063	.1890	2.10	.3548	.5844
1.53	.2095	.1949	2.15	.3651	.6212
1.54	.2126	.2009	2.20	.3750	.6582
1.55	.2157	.2069	2.25	.3846	.6952
1.56	.2188	.2130	2.30	.3939	.7324
1.57	.2218	.2191	2.35	.4030	.7696
1.58	.2248	.2252	2.40	.4118	.8069
1.59	.2278	.2314	2.45	.4203	.8441
1.60	.2308	.2377	2.50	.4286	.8814
1.61	.2337	.2440	2.55	.4366	.9186
1.62	.2366	.2503	2.60	.4444	.9557
1.63	.2395	.2566	2.65	.4521	.9928
1.64	.2424	.2630	2.70	.4595	1.0298
1.65	.2453	.2695	2.75	.4667	1.0667
1.66	.2481	.2760	2.80	.4737	1.1035
1.67	.2509	.2825	2.85	.4805	1.1402
1.68	.2537	.2890	2.90	.4872	1.1767
1.69	.2565	.2956	2.95	.4937	1.2131
1.70	.2593	.3022	3.00	.5000	1.2494
1.71	.2620	.3088	3.10	.5122	1.3215
1.72	.2647	.3155	3.20	.5238	1.3929

VSWR Versus Reflection Coefficient (Γ) and Mismatch Loss

VSWR	$|\Gamma|$	Mismatch Loss (dB)	VSWR	$|\Gamma|$	Mismatch Loss (dB)
3.30	.5349	1.4636	8.10	.7802	4.0754
3.40	.5455	1.5337	8.20	.7826	4.1170
3.50	.5556	1.6030	8.30	.7849	4.1583
3.60	.5652	1.6715	8.40	.7872	4.1992
3.70	.5745	1.7393	8.50	.7895	4.2397
3.80	.5833	1.8064	8.60	.7917	4.2798
3.90	.5918	1.8727	8.70	.7938	4.3196
4.00	.6000	1.9382	8.80	.7959	4.3591
4.10	.6078	2.0030	8.90	.7980	4.3982
4.20	.6154	2.0670	9.00	.8000	4.4370
4.30	.6226	2.1302	9.10	.8020	4.4754
4.40	.6296	2.1927	9.20	.8039	4.5135
4.50	.6364	2.2545	9.30	.8058	4.5513
4.60	.6429	2.3156	9.40	.8077	4.5888
4.70	.6491	2.3759	9.50	.8095	4.6260
4.80	.6552	2.4355	9.60	.8113	4.6628
4.90	.6610	2.4945	9.70	.8131	4.6994
5.00	.6667	2.5527	9.80	.8148	4.7356
5.10	.6721	2.6103	9.90	.8165	4.7716
5.20	.6774	2.6672	10.00	.8182	4.8073
5.30	.6825	2.7235	11.00	.8333	5.1491
5.40	.6875	2.7791	12.00	.8462	5.4665
5.50	.6923	2.8340	13.00	.8571	4.7625
5.60	.6970	2.8884	14.00	.8667	6.0399
5.70	.7015	2.9421	15.00	.8750	6.3009
5.80	.7059	2.9953	16.00	.8824	6.5472
5.90	.7101	3.0479	17.00	.8889	6.7804
6.00	.7143	3.0998	18.00	.8947	7.0017
6.10	.7183	3.1513	19.00	.9000	7.2125
6.20	.7222	3.2021	20.00	.9048	7.4135
6.30	.7260	3.2525	30.00	.9355	9.0354
6.40	.7297	3.3022	40.00	.9512	10.2145
6.50	.7333	3.3515	50.00	.9608	11.1411
6.60	.7368	3.4002	60.00	.9672	11.9045
6.70	.7403	3.4485	70.00	.9718	12.5536
6.80	.7436	3.4962	80.00	.9753	13.1182
6.90	.7468	3.5435	90.00	.9780	13.6178
7.00	.7500	3.5902	100.00	.9802	14.0658
7.10	.7531	3.6365	200.00	.9900	17.0330
7.20	.7561	3.6824	300.00	.9934	18.7795
7.30	.7590	3.7277	400.00	.9950	20.0217
7.40	.7619	3.7727	500.00	.9960	20.9865
7.50	.7647	3.8172	600.00	.9967	21.7754
7.60	.7674	3.8612	700.00	.9971	22.4428
7.70	.7701	3.9049	800.00	.9975	23.0212
7.80	.7727	3.9481	900.00	.9978	23.5315
7.90	.7753	3.9909	1000.00	.9980	23.9881
8.00	.7778	4.0334			

Appendix F

Decibels Versus Voltage and Power Ratios*

Yi-Lin Chen
University of Illinois[†]

The decibel chart below indicates decibels for any ratio of voltage or power up to 100 dB. For voltage ratios greater than 10 (or power ratios greater than 100) the ratio can be broken down into two products, the decibels found for each separately, the two results then added. For example, to convert a voltage ratio of 200:1 to dB, a 200:1 voltage ratio equals the product of 100:1 and 2:1. Now, 100:1 equals 40 dB; 21:1 equals 6 dB. Therefore a 200:1 voltage ratio equals 40 dB + 6 dB, or 46 dB.

$$dB = 20\log_{10}(\text{voltage ratio}) = 10\log_{10}(\text{power ratio})$$

---

*Reprinted with permission of *Microwave Journal*, from *The Microwave Engineer's Handbook and Buyer's Guide*, 1966 issue, © 1966 Horizon House–Microwave, Inc.
[†]On leave from the Chinese Aeronautical Laboratory, Beijing, China, during 1983.

Decibels Versus Voltage and Power Ratios

Voltage Ratio	Power Ratio	-dB +	Voltage Ratio	Power Ratio	Voltage Ratio	Power Ratio	-dB +	Voltage Ratio	Power Ratio	Voltage Ratio	Power Ratio	-dB +	Voltage Ratio	Power Ratio
1.0000	1.0000	0	1.000	1.000	.5309	.2818	5.5	1.884	3.548	.2818	.07943	11.0	3.548	12.59
.9886	.9772	.1	1.012	1.023	.5248	.2754	5.6	1.905	3.631	.2786	.07762	11.1	3.589	12.88
.9772	.9550	.2	1.023	1.047	.5188	.2692	5.7	1.928	3.715	.2754	.07586	11.2	3.631	13.18
.9661	.9333	.3	1.035	1.072	.5129	.2630	5.8	1.950	3.802	.2723	.07413	11.3	3.673	13.49
.9550	.9120	.4	1.047	1.096	.5070	.2570	5.9	1.972	3.890	.2692	.07244	11.4	3.715	13.80
.9441	.8913	.5	1.059	1.122	.5012	.2512	6.0	1.995	3.981	.2661	.07079	11.5	3.758	14.13
.9333	.8710	.6	1.072	1.148	.4955	.2455	6.1	2.018	4.074	.2630	.06918	11.6	3.802	14.45
.9226	.8511	.7	1.084	1.175	.4898	.2399	6.2	2.042	4.169	.2600	.06761	11.7	3.846	14.79
.9120	.8318	.8	1.096	1.202	.4842	.2344	6.3	2.065	4.266	.2570	.06607	11.8	3.890	15.14
.9016	.8128	.9	1.109	1.230	.4786	.2291	6.4	2.089	4.365	.2541	.06457	11.9	3.936	15.49
.8913	.7943	1.0	1.122	1.259	.4732	.2239	6.5	2.113	4.467	.2512	.06310	12.0	3.981	15.85
.8810	.7762	1.1	1.135	1.288	.4677	.2188	6.6	2.138	4.571	.2483	.06166	12.1	4.027	16.22
.8710	.7586	1.2	1.148	1.318	.4624	.2138	6.7	2.163	4.677	.2455	.06026	12.2	4.074	16.60
.8610	.7413	1.3	1.161	1.349	.4571	.2089	6.8	2.188	4.786	.2427	.05888	12.3	4.121	16.98
.8511	.7244	1.4	1.175	1.380	.4519	.2042	6.9	2.213	4.898	.2399	.05754	12.4	4.169	17.38
.8414	.7079	1.5	1.189	1.413	.4467	.1995	7.0	2.239	5.012	.2371	.05623	12.5	4.217	17.78
.8318	.6918	1.6	1.202	1.445	.4416	.1950	7.1	2.265	5.129	.2344	.05495	12.6	4.266	18.20
.8222	.6761	1.7	1.216	1.479	.4365	.1905	7.2	2.291	5.248	.2317	.05370	12.7	4.315	18.62
.8128	.6607	1.8	1.230	1.514	.4315	.1862	7.3	2.317	5.370	.2291	.05248	12.8	4.365	19.05
.8035	.6457	1.9	1.245	1.549	.4266	.1820	7.4	2.344	5.495	.2265	.05129	12.9	4.416	19.50
.7943	.6310	2.0	1.259	1.585	.4217	.1778	7.5	2.371	5.623	.2239	.05012	13.0	4.467	19.95
.7852	.6166	2.1	1.274	1.622	.4169	.1738	7.6	2.399	5.754	.2213	.04898	13.1	4.519	20.42
.7762	.6026	2.2	1.288	1.660	.4121	.1698	7.7	2.427	5.888	.2188	.04786	13.2	4.571	20.89
.7674	.5888	2.3	1.303	1.698	.4074	.1660	7.8	2.455	6.026	.2163	.04677	13.3	4.624	21.38
.7586	.5754	2.4	1.318	1.738	.4027	.1622	7.9	2.483	6.166	.2138	.04571	13.4	4.677	21.88

.7499	.5623	2.5	1.334	1.778
.7413	.5495	2.6	1.349	1.820
.7328	.5370	2.7	1.365	1.862
.7244	.5248	2.8	1.380	1.905
.7161	.5129	2.9	1.396	1.950
.7079	.5012	3.0	1.413	1.995
.6998	.4898	3.1	1.429	2.042
.6918	.4786	3.2	1.445	2.089
.6839	.4677	3.3	1.462	2.138
.6761	.4571	3.4	1.479	2.188
.6683	.4467	3.5	1.496	2.239
.6607	.4365	3.6	1.514	2.291
.6531	.4266	3.7	1.531	2.344
.6457	.4169	3.8	1.549	2.399
.6383	.4074	3.9	1.567	2.455
.6310	.3981	4.0	1.585	2.512
.6237	.3890	4.1	1.603	2.570
.6166	.3802	4.2	1.622	2.630
.6095	.3715	4.3	1.641	2.692
.6026	.3631	4.4	1.660	2.754
.5957	.3548	4.5	1.679	2.818
.5888	.3467	4.6	1.698	2.884
.5821	.3388	4.7	1.718	2.951
.5754	.3311	4.8	1.738	3.020
.5689	.3236	4.9	1.758	3.090
.5623	.3162	5.0	1.778	3.162
.5559	.3090	5.1	1.799	3.236
.5495	.3020	5.2	1.820	3.311
.5433	.2951	5.3	1.841	3.388
.5370	.2884	5.4	1.862	3.467
.3981	.1585	8.0	2.512	6.310
.3936	.1549	8.1	2.541	6.457
.3890	.1514	8.2	2.570	6.607
.3846	.1479	8.3	2.600	6.761
.3802	.1445	8.4	2.630	6.918
.3758	.1413	8.5	2.661	7.079
.3715	.1380	8.6	2.692	7.244
.3673	.1349	8.7	2.723	7.413
.3631	.1318	8.8	2.754	7.586
.3589	.1288	8.9	2.786	7.762
.3548	.1259	9.0	2.818	7.943
.3508	.1230	9.1	2.851	8.128
.3467	.1202	9.2	2.884	8.318
.3428	.1175	9.3	2.917	8.511
.3388	.1148	9.4	2.951	8.710
.3350	.1122	9.5	2.985	8.913
.3311	.1096	9.6	3.020	9.120
.3273	.1072	9.7	3.055	9.333
.3236	.1047	9.8	3.090	9.550
.3199	.1023	9.9	3.126	9.772
.3162	.1000	10.0	3.162	10.000
.3126	.09772	10.1	3.199	10.23
.3090	.09550	10.2	3.236	10.47
.3055	.09333	10.3	3.273	10.72
.3020	.09120	10.4	3.311	10.96
.2985	.08913	10.5	3.350	11.22
.2951	.08710	10.6	3.388	11.48
.2917	.08511	10.7	3.428	11.75
.2884	.08318	10.8	3.467	12.02
.2851	.08128	10.9	3.508	12.30
.2113	.04467	13.5	4.732	22.39
.2089	.04365	13.6	4.786	22.91
.2065	.04266	13.7	4.842	23.44
.2042	.04169	13.8	4.898	23.99
.2018	.04074	13.9	4.955	24.55
.1995	.03981	14.0	5.012	25.12
.1972	.03890	14.1	5.070	25.70
.1950	.03802	14.2	5.129	26.30
.1928	.03715	14.3	5.188	26.92
.1905	.03631	14.4	5.248	27.54
.1884	.03548	14.5	5.309	28.18
.1862	.03467	14.6	5.370	28.84
.1841	.03388	14.7	5.433	29.51
.1820	.03311	14.8	5.495	30.20
.1799	.03236	14.9	5.559	30.90
.1778	.03162	15.0	5.623	31.62
.1758	.03090	15.1	5.689	32.36
.1738	.03020	15.2	5.754	33.11
.1718	.02951	15.3	5.821	33.88
.1698	.02884	15.4	5.888	34.67
.1679	.02818	15.5	5.957	35.48
.1660	.02754	15.6	6.026	36.31
.1641	.02692	15.7	6.095	37.15
.1622	.02630	15.8	6.166	38.02
.1603	.02570	15.9	6.237	38.90
.1585	.02512	16.0	6.310	39.81
.1567	.02455	16.1	6.383	40.74
.1549	.02399	16.2	6.457	41.69
.1531	.02344	16.3	6.531	42.66
.1514	.02291	16.4	6.607	43.65

Decibels Versus Voltage and Power Ratios, (cont'd.)

Voltage Ratio	Power Ratio	-dB +	Voltage Ratio	Power Ratio
.1496	.02239	16.5	6.683	44.67
.1479	.02188	16.6	6.761	45.71
.1462	.02138	16.7	6.839	46.77
.1445	.02089	16.8	6.918	47.86
.1429	.02042	16.9	6.998	48.98
.1413	.01995	17.0	7.079	50.12
.1396	.01950	17.1	7.161	51.29
.1380	.01905	17.2	7.244	52.48
.1365	.01862	17.3	7.328	53.70
.1349	.01820	17.4	7.413	54.95
.1334	.01778	17.5	7.499	56.23
.1318	.01738	17.6	7.586	57.54
.1303	.01698	17.7	7.674	58.88
.1288	.01660	17.8	7.762	60.26
.1274	.01622	17.9	7.852	61.66
.1259	.01585	18.0	7.943	63.10
.1245	.01549	18.1	8.035	64.57
.1230	.01514	18.2	8.128	66.07
.1216	.01479	18.3	8.222	67.61
.1202	.01445	18.4	8.318	69.18
.1189	.01413	18.5	8.414	70.79
.1175	.01380	18.6	8.511	72.44
.1161	.01349	18.7	8.610	74.13
.1148	.01318	18.8	8.710	75.86
.1135	.01288	18.9	8.811	77.62
.1122	.01259	19.0	8.913	79.43
.1109	.01230	19.1	9.016	81.28
.1096	.01202	19.2	9.120	83.18
.1084	.01175	19.3	9.226	85.11
.1072	.01148	19.4	9.333	87.10
.1059	.01122	19.5	9.441	89.13
.1047	.01096	19.6	9.550	91.20
.1035	.01072	19.7	9.661	93.33
.1023	.01047	19.8	9.772	95.50
.1012	.01023	19.9	9.886	97.72
.1000	.01000	20.0	10.000	100.00
	10^{-3}	30		10^{3}
10^{-2}	10^{-4}	40	10^{2}	10^{4}
	10^{-5}	50		10^{5}
	10^{-6}	60		10^{6}
10^{-3}	10^{-7}	70	10^{3}	10^{7}
10^{-4}	10^{-8}	80	10^{4}	10^{8}
10^{-5}	10^{-9}	90		10^{9}
	10^{-10}	100	10^{5}	10^{10}

Index

Index

with Butler matrix, 19-8
on curved surfaces, 13-50 to 13-51
efficiency of. *See* Aperture efficiency
electric and magnetic fields in, 5-8
with feed systems
   radial transmission line, 19-28, 19-31,
      19-33
   semiconstrained, 19-17
   unconstrained, 19-51
and gain, 5-26 to 5-27, 17-7, 17-8
illumination of, 13-30 to 13-32, 13-36 to
   13-37
with lens antennas, 21-14, 21-16, 21-17
   phase distributions of, 16-7, 16-36
   power distributions of, 16-33, 16-36
of longitudinal array slots, 12-24
and parallel feed networks, 19-6
in perfectly conducting ground, 3-22
radiation from, 1-28
   circular, 5-20 to 5-25
   and equivalent currents, 5-8 to 5-10
   near-field, 5-24 to 5-26
   planar aperture distributions, 5-10 to
      5-11
   and plane-wave spectra, 5-5 to 5-7
   rectangular aperture, 5-11 to 5-20
reflections of, 8-67
with reflector antennas, 15-15 to 15-23,
   21-20
phase error of, 15-107 to 15-108
for satellite antennas, 21-5, 21-7, 21-14,
   21-16, 21-17
size of
   and conical horn beamwidth, 8-61
   of E-plane horn antenna directivity,
      8-16
   of lenses, 16-6
square, 5-32 to 5-33
with Taylor line source synthesis, 13-26
Aperture taper
and radiometer beam efficiency, 22-32,
   22-33
with reflector antennas, 21-24
Apparent phase center, 8-75
Appleton-Hartee formula for ionospheric
   refractive index, 29-21
Applied excitation of periodic arrays, 13-6
   to 13-7
Arabsat satellite antenna, 21-77
Arbitrary ray optical field, 4-8
Archimedean-spiral curves, 9-107 to 9-108,
   9-110

Array blindness, 13-45 to 13-49
   with flared-notch antennas, 13-59
Array collimations, 13-7 to 13-10
Array geometry, linear transformations in,
   11-23 to 11-25
Array pattern functions, 3-35, 11-8
   aperiodic, 14-4 to 14-5, 14-8 to 14-9,
      14-15, 14-22, 14-35
   Dolph-Chebyshev, 11-17 to 11-18
   linear, 11-8, 11-9, 11-11 to 11-14
   with longitudinal slots, 12-17
   phased, 21-10
   transformation of, 11-23 to 11-48
   with UTD solutions, 20-6
Arrays, 3-33, 3-35 to 3-36, 3-38 to 3-43.
   See also Aperiodic arrays; Array the-
      ory; Broadside arrays; Circular ar-
      rays; Log-periodic arrays; Log-spiral
      antennas; Periodic arrays; Planar ar-
      rays; Slot arrays
   element patterns with, 14-25
   errors in, from phase quantization, 13-52
      to 13-57
   horn antennas as elements in, 8-4
   large, design of, 12-28 to 12-34
   of microstrip antennas, for medical ap-
      plications, 24-25
   millimeter-wave, 17-23 to 17-27
   phase control of, 13-62 to 13-64
   scan characteristics of, 14-24 to 14-25
   of tapered dielectric-rod antennas, 17-47
   thinning of, with aperiodic arrays, 14-3
Array theory
   directivity in, 11-63 to 11-76
   general formulation of, 11-5 to 11-8
   for linear arrays, 11-8 to 11-23
   linear transformations in, 11-23 to 11-48
   pattern synthesis in, 11-76 to 11-86
   for planar arrays, 11-48 to 11-58
   SNR in, 11-58 to 11-63, 11-70 to 11-76
Artificial-dielectric plates, 21-19
A-sandwich panels, 31-17 to 31-19
Aspect ratio
   of dielectric grating antennas, 17-77 to
      17-78
      surface corrugations for, 17-64
   and leakage constants, 17-62, 17-63
   with log-periodic antennas, 9-12, 9-14
   and directivity, 9-21 to 9-24
Assistance with computer models, 3-68
Astigmatic lens aberrations, 16-15, 16-20,
   21-13

Wire-antennas, 7-5 to 7-11
  folded dipole, 7-37 to 7-40
  linear dipole, 7-11 to 7-23
  loop, 7-40 to 7-48
  sleeve, 7-23 to 7-37
Wire diameter and radiation patterns,
    32-76, 32-79
Wire objects, computer storage and time
    for, analysis of, 3-57
Wire-outline log-periodic antennas, 9-13
Wire problems
  subdomain procedures for, 3-60
  time-domain, 3-63
Wullenweber arrays, 25-14 to 25-16

X-band horn antennas, 8-41 to 8-42, 8-44
    to 8-45, 8-67
X-polarized antennas, 1-28
XY-scanners, 33-19

Yagi-Uda antennas, 32-76
Yardarm, shipboard, simulation of, 20-93
    to 20-94

Yield of crops, forecasting of, 22-15
Y-polarized antennas, 1-28

Zernike cylindrical polynomials, 16-13
Zero-bias diodes, 24-43
Zeroes of the Airy function, 4-59
Zeroes of Bessel functions, 1-45
Zeroes of W, 4-89
Zigzag antennas. *See also* Log-periodic
    zigzag antennas
  dipoles, as E-field probes, 24-39 to 24-40
  tv, 27-31 to 27-32
  wire, 9-33 to 9-38
Zone-plate lens antennas, 17-20, 17-22
Zones, FCC allocation and assignment,
    27-4
Zoning of lenses, 16-6, 16-28 to 16-30
  and bandwidth, 16-54 to 16-55, 21-16
  with constrained lenses, 16-43 to 16-45
  with dielectric lenses, 16-38 to 16-41
  with equal group delay lenses, 16-46
  with millimeter-wave antennas, 17-15,
    17-19 to 17-21